FUNDAMENTALS OF TELEVISION AND RADAR ENGINEERING

For
Engineering Students

By
K.K. SHARMA
Assistant Professor
Department of Electronics & Communication Engineering
Applied College of Management & Engineering (ACME)
(An Institution of PRS Education & Research Foundation)
Faridabad

S.K. KATARIA & SONS®
Publisher of Engineering Books
4885/109, Prakash Mahal, Dr. Subhash Bhargav Lane,
Opposite Delhi Medical Association, Daryaganj, New Delhi-110002 (India)
Phone: +91-11-43551243; Mobile: +91-9871775858, 7678425209
e-mail: katariabook@gmail.com, katariabooks@yahoo.com
Website : www.skkatariaandsons.com
(e-books, subscriptions and bulk buying also available on our website)

Published by :

S.K. Kataria & Sons®

4885/109, Prakash Mahal, Dr. Subhash Bhargav Lane,
Opposite Delhi Medical Association, Daryaganj, New Delhi-110002 (India)
Phone: +91-11-43551243; Mobile: +91-9871775858, 7678425209
e-mail: katariabook@gmail.com, katariabooks@yahoo.com

First Edition: 2009; Reprint: 2025

Laser Typesetting by: U.A. Graphics *Printed at : New A.S. Offset Printers*

ACKNOWLEDGEMENT

I take this opportunity to express my gratitude to the Managing Trustees of Applied College of Management & Engineering (ACME) Palwal —Mr Sunil Jain Chairman (Academics), Mr Ravinder Singh, Chairman (Resources) and Mr Pradeep Agarwal Chairman (Finance) for providing innovative and encouraging environment to work on this project.

I would like to express my sincere thanks to Dr R.P. Singh, Principal, ACME, for constant inspiration and encouragement.

I would also like to thank to the teachers, students and supporting staff of ACME.

I am grateful to my wife Sunita, my daughters Aparna and Vaishali, and son Aman for supporting me to complete this project.

I would like to express my gratitude to all the authors whose work has been freely consulted in the preparation of this text.

I deeply express my thanks to the publishers M/s S. K. Kataria & Sons and Mr Sachin Kumar, for publishing this book in such a beautiful get-up and well on time.

Lastly, I am grateful to my father and mother without their blessing this book would have not come into existence.

K. K. SHARMA

PREFACE

This book **"Fundamentals of Television and Radar Engineering"** is intended to be used as a textbook that relatively covers all topics to the subject at undergraduate and graduate engineering levels. This book attempts to present an integrated approach to the system aspect of Television and radar systems. Although the subject of radar system is of particular interest to the specialists in the radar and navigation field, it is also of interest to persons, especially the military and civilian users of radar. The author has 20 yrs of vast exposure on different types of radar system used in Indian Air Force as well as at the civil Airports. Author has experience of installation & commissioning of GCA Navigational radars. The author also has six years of teaching experience at Air Force Technical College, Bangalore and six years of teaching in various engineering colleges.

Although few books are already available on television engineering but some of the universities includes Television engineering along with Radar engineering. Author has taken a step to bring both the fields together in one book so that students and teachers would be benefited.

The book has been written in a very simple language keeping in focus the syllabi prescribed for the subject in different institutions and universities. All the topics of Television and Radar systems has been illustrated with the photograph and solved problems .The students preparing for the examinations can understand the concept in a better way by doing the solved and unsolved problems at the end of each chapter. This book is divided in two major parts, first part includes the Television and second part contains the Radar system.

I do hope that the text in the present form will meet the requirement of the students doing graduation in electronics and communication engineering. I will appreciate any suggestions and comments from students and teaching faculty, so that textbook will be more useful in the edition to come.

Author
K. K. Sharma

This book is Dedicated to

My wife

Sunita Sharma

CONTENTS

PART 1 : TELEVISION ENGINEERING .... 1—292

PART 1

TELEVISION ENGINEERING

1

CHAPTER

INTRODUCTION TO TELEVISION

INSIDE THIS CHAPTER

1.1. INTRODUCTION TO TELEVISION

The **Television** may be defined as the following ways:

1. The transmission of visual images of moving and stationary objects, generally with accompanying sound, as electromagnetic waves and the reconversion of received waves into visual images.
2. The Television is an electronic apparatus that receives electromagnetic waves and displays the reconverted images on a screen.
3. The integrated audible and visible content of the electromagnetic waves received and converted by such an apparatus.

TELE - FAR (GREAK)

VISIONIS - TO SEE

"SEE FROM A DISTANCE"

Development of TV

The development of television technology can be partitioned along two lines: those developments that depended upon both mechanical and electronic principles and those which are purely electronic. From the latter descended all modern televisions, but these would not have been possible without discoveries and insights from the mechanical systems.

Electromechanical Television

Paul Gottlieb Nipkow proposed and patented the first electromechanical television system in 1884. Nipkow's spinning disk design is credited with being the first television image rasterizer, but it is believed that he never built a prototype to prove the design (it wasn't until 1907 that developments in amplification tube technology made the design practical). Meanwhile, Constantin Perskyi had coined the word *television* in a paper read to the International Electricity Congress at the International World Fair in Paris on August 25, 1900. Perskeyi's paper reviewed the existing electromechanical technologies, mentioning the work of Nipkow and others.

A semi-mechanical analogue television system was first demonstrated in London in February 1924 by John Logie Baird with an image of Felix the Cat and a moving picture by Baird on October 30, 1925. In 1928 Baird's company (Baird Television Development Company / Cinema Television) broadcast the first transatlantic Television signal, between London and New York, and the first shore to ship transmission. He also demonstrated a Electromechanical colour, infrared, and stereoscopic television, using adding additional lenses, disks and filters. In 1931 he made the first live transmission, of the Epsom Derby. In 1932 he demonstrates ultra-short wave television. Baird's system was eventually adopted by the BBC, who later discontinued its use in 1937 in favor of purely electronic television.

Electronic Television

Although the discoveries of Nipkov, Rosing, Baird and others were extraordinary, little of their technology is used in modern television. By 1934, all electromechanical television systems were outmoded.

A.A. Campbell-Swinton wrote a letter to *Nature* on the 18th June 1908 describing his concept of electronic television using the cathode ray tube invented by Karl Ferdinand Braun. He proposed using an electron beam in both the camera and the receiver, which could be steered electronically to produce moving pictures. He lectured on the subject in 1911 and displayed circuit diagrams, but no one, including Swinton, knew how to realize the design. His system was never built.

A fully electronic system was first demonstrated by Philo Taylor Farnsworth in the autumn of 1927. At age 21, he demonstrated a working system at his own laboratory in San Francisco. His breakthrough freed television from reliance on spinning discs and other mechanical parts. All modern picture tube televisions descend directly from his design.

The controversy over whether it was first Farnsworth or Zworykin who invented modern television is still hotly debated today. Some of this debate stems from the fact that while Farnsworth appears to have gotten there first, it was RCA that first marketed working television sets, and it was RCA employees who first wrote the history of television. Even though Farnsworth eventually won the legal battle over this issue, he was never able to fully capitalize financially on his invention.

Colour Television : In 1950, a colour sequential system was developed by the Columbia Broadcasting System (CBS) for colour transmission and it was adopted by United State of America. The colour TV system used a rapidly rotating disc containing filters of the Primary colors (Red, Green and Blue) in sequence, interposed between the lens and the camera tube at the transmitter and an identical disc between the picture tube and the viewers at the receiver. Both the discs are synchronized and rotating with the same speed. There were lots of problems such as loss of luminosity due to filters, the size of the disc was large compared to the actual display and mechanical problems related to the discs.

The use of transistors and integrated circuits along with the small size TV camera tubes such as vidicon and its variance plumbicon and silicon having improved characteristics and ease of operation, had made the video camera and video recorders compact and portable. The solid state charge coupled device (CCD) sensors and the liquid crystal display (LCD) brought in the solid state technology in the flat panel LCD TV display and in the video camera.

The next great advance in television will be the adoption of a high-definition television (HDTV) system. Non-experimental analog HDTV broadcasting began in Japan in 1991. In 1994 the FCC approved a U.S. standard for an all-digital system, to be used by all commercial broadcast stations by mid-2002.

1. First demonstration of TV—1927

   By JL BAIRD in UK & CF JENKINS in USA
2. EM Scanning of both camera and picture tubes were developed in 1930.
3. Broadcasting of TV started in 1935.

Types of Monochrome Systems :

1. 525 – Lines (American)
2. 625 – Lines (European)
3. 819 – Lines (French)

In India TV transmission started in 1959 and 625-B monochrome system has been adapted.

Types of Colour Systems :

1. National Television System Committee (NTSC) used in America in 1953.
2. Phase Alteration by Lines (PAL) used in German in 1967.
3. Sequential Couleruses A Memoirs (SECAM)

In India PAL system CCIR (B Standard) is used.

Frequency Used :

VHF Band - 41 to 68 MHz & 174 to 230 MHz

Later additions-

UHF Band - 470 to 890 MHz

1.2. ELEMENTS OF A TELEVISION SYSTEM

The principle of TV transmission with a view to reproducing black-and-white pictures can be summarized as follows: the optical image of the scene to be transmitted is divided into small picture elements (pixels). An opto-electrical converter, usually a camera tube, consecutively translates the individual elements into electrical information depending on their brightness.

This signal is then transmitted at its actual frequency or after modulation onto an RF carrier. After appropriate processing at the receiving end, the information is applied to an electro-optical converter and reproduced in accordance with the brightness distribution of the pattern.

Continuous transmission is ensured by producing a defined number of frames as in cinema films.

1.2.1. Block Diagram of Monochrome TV Transmitter

A simple block diagram of Monochrome TV Transmitter is given in Fig. 1.1.

Elements of Monochrome TV Transmitter

1. Video Camera
2. Video Amplifier
3. AM Modulator (for picture)
4. Scanning & Synchronizing
5. Audio Amplifier
6. FM Modulator (for sound)
7. Combining Networks
8. Crystal Oscillator
9. Power amplifier

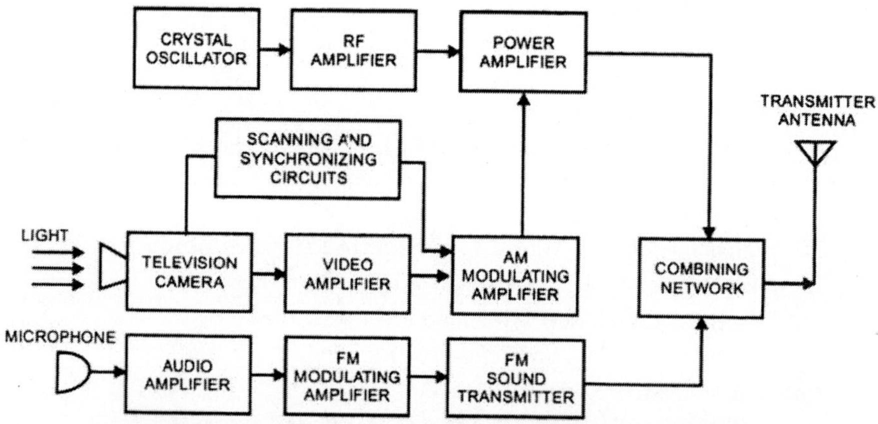

Fig. 1.1. A simple block diagram of Monochrome TV Tx.

The Picture Transmission

The picture information is optical in character and may be thought of as collection of a large number of bright and dark areas representing picture details. These elementary area into which the picture details may be broken up are known as **Picture Elements or Pixels**, which when viewed together, represent the visual information of the scene. So at a time to transmit picture details, it needs to pick up an infinite number of pixels simultaneously which is not possible. This problem is removed by a method known as scanning.

Scanning is used to convert the optical information in to electric form and its transmissions are carried out element by element, one at a time and in a sequential manner to cover the entire scene which is to be televised.

Scanning of the element is done at a very fast rate and this process is repeated at large number of times per second to create an illusion of simultaneously pickup and transmission of picture details.

A TV camera is used to convert the optical information in to electrical signal and its amplitude varies according to the variation of the brightness.

The electrical information obtained from the T V camera tube is generally referred to as video signal. This signal is amplified and modulated (Amplitude modulation vestigial side band) with the channel picture carrier frequency. The modulated output is fed to the Tx antenna for radiation along with the sound signal.

The Sound Transmission

To transmit the sound alongwith the picture information, it is must to convert the audio or voice into an electrical signal. A **microphone** is used to convert the voice or audio in to corresponding electrical signals which is normally a voltage. This electrical signal is single valued function of time so it needs a single channel for its transmission. The audio signal which we have got from the microphone is being amplified up to the adequate level. The amplified output is now Frequency-modulated over the assigned carrier frequency. Finally, it is combined with the AM picture transmitter output and fed to common antenna to transmit the signals in the form of EM wave in the space.

1.2.2. Block Diagram of Monochrome TV Receiver

A simple block diagram of Monochrome TV Receiver is given in Fig. 1.2.

Elements of Monochrome TV Receiver

1. RF Tuner
2. Common IF Amplifier
3. Video Detector
4. Video Amplifier
5. CRT
6. Sound Amplifier
7. FM Demodulator
8. Scanning and Synchronizing

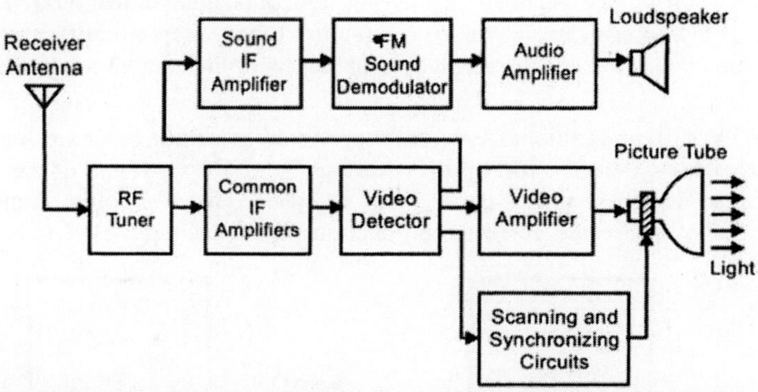

Fig. 1.2. A simple block diagram of Monochrome TV Receiver.

Reception of Picture

The receiving antenna intercepts the radiated picture and sound carrier signals and fed them to RF tuner which helps to select the desired channel.

The receiver is of the super-heterodyne type to achieve high gain and selectivity, it employs two or three stages of IF amplification. The signal is amplified by a RF (radio frequency) amplifier and then with the oscillations of high frequency with the constant amplitude, produced by local oscillator. The output of the mixer is the signal having frequency equal to the difference of incoming RF signal and output signal from local oscillator. The difference frequency signal is known as Intermediate Frequency signal (IF). The IF signal is further amplified by two or three stages of IF amplifier.

The output from the last IF stage fed to the video detector that demodulate and removes video signal from the modulated wave. This signal that carries the picture information is amplified and coupled to the picture which converts the electrical signal back into picture elements of the same degree of black and white.

Sound Reception

Sound section consists of sound IF amplifier, FM demodulator, audio amplifier and finally a loudspeaker. The path of the sound signal is common with the picture signal from antenna to the video detector section of the receiver.

In video detector, the two signals are separated and fed to their respective channels. The FM audio signal is demodulated after at least one stage of amplification.

The audio output from the FM detector is given due amplification before feeding it to the loudspeaker.

Synchronization

It is must that the same coordinates to be scanned at any instant both at the camera tube target plate at transmitter and at the raster of the picture tube at the receiver; otherwise the picture details would split and get distorted. A concept of synchronization is shown in Fig. 1.3.

Synchronization pulses are transmitted during the retrace period, i.e. the fly back intervals of horizontal and vertical motion of the camera scanning beam to make sure that perfect synchronization between the scenes be televised and the picture produced on the raster to be achieved.

Synchronization pulses which are distinct for horizontal and vertical motion controls are processed at the receiver and given to the picture tube sweep circuitry system thus ensuring that the picture tube at the receiver is in step with the camera tube beam at the transmitter.

In colour TV system additional sync pulses referred as colour burst are transmitted alongwith horizontal synchronization pulses. Colour burst are separated at the input of chroma section so that these pulses can be used to synchronize the colour demodulator carrier generator to ensure the correct reproduction of the colour.

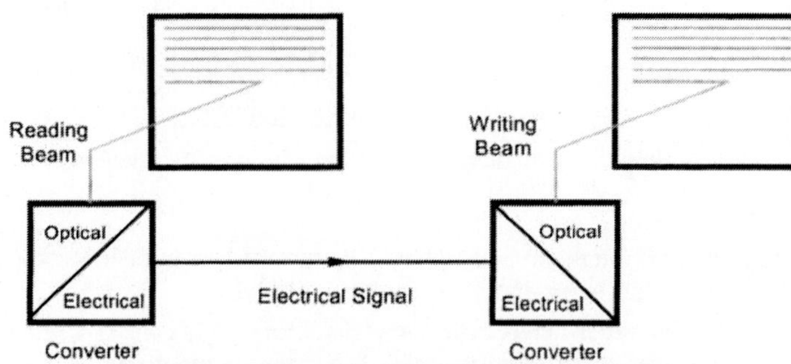

Fig. 1.3. Picture transmission and reception.

SUMMARY

- The **Television** may be defined as the following ways:

  The transmission of visual images of moving and stationary objects, generally with accompanying sound, as electromagnetic waves and the reconversion of received waves into visual images.
- **Paul Gottlieb Nipkow** proposed and patented the first electromechanical television system in 1884.
- A semi-mechanical analogue television system was first demonstrated in London in February 1924 by John Logie Baird with an image of Felix the Cat and a moving picture by Baird on October 30, 1925.
- A.A. Campbell-Swinton wrote a letter to *Nature* on the 18th June 1908 describing his concept of electronic television using the cathode ray tube invented by Karl Ferdinand Braun.
- A fully electronic system was first demonstrated by Philo Taylor Farnsworth in the autumn of 1927.
- In 1950, a colour sequential system was developed by the Columbia Broadcasting System (CBS) for colour transmission and it was adopted by United State of America.
- (*i*) First demonstration of TV—1927

  By JL BAIRD in UK & CF JENKINS in USA

  (*ii*) EM Scanning of both camera and picture tubes were developed in 1930.

  (*iii*) Broadcasting of TV started in 1935.
- **Types of Monochrome Systems :**

  (*i*) 525 – Lines (American)

  (*ii*) 625 – Lines (European)

  (*iii*) 819 – Lines (French)

  In India TV transmission started in 1959 and 625-B monochrome system has been adapted.
- **Types of Colour Systems :**

  1. National Television System Committee (NTSC) used in America in 1953.
  2. Phase Alteration by Lines (PAL) used in German in 1967.
  3. Sequential Couleruses A Memoirs (SECAM)

  In India PAL system CCIR (B Standard) is used.

  Frequency Used :

  VHF Band - 41 to 68 MHz & 174 to 230 MHz

  Later additions-

  UHF Band - 470 to 890 MHz
- **Elements of Monochrome TV Transmitter :**

  (*i*) Video Camera

  (*ii*) Video Amplifier

  (*iii*) AM Modulator (for picture)

  (*iv*) Scanning & Synchronizing

  (*v*) Audio Amplifier

(*vi*) FM Modulator (for sound)
(*vii*) Combining Networks
(*viii*) Crystal Oscillator
(*ix*) Power amplifier

- The picture information is optical in character and may be thought of as collection of a large number of bright and dark areas representing picture details. These elementary area into which the picture details may be broken up are known as **Picture Elements or Pixels**, which when viewed together, represent the visual information of the scene.
- **Scanning** is used to convert the optical information in to electric form and its transmissions are carried out element by element, one at a time and in a sequential manner to cover the entire scene which is to be televised.
- A **TV camera** is used to convert the optical information in to electrical signal and its amplitude varies according to the variation of the brightness.
- A **microphone** is used to convert the voice or audio in to corresponding electrical signals which is normally a voltage.
- Elements of Monochrome TV Receiver
  (*i*) RF Tuner
  (*ii*) Common IF Amplifier
  (*iii*) Video Detector
  (*iv*) Video Amplifier
  (*v*) CRT
  (*vi*) Sound Amplifier
  (*vii*) FM Demodulator
  (*viii*) Scanning and Synchronizing
- It is must that the same coordinates to be scanned at any instant both at the camera tube target plate at transmitter and at the raster of the picture tube at the receiver; otherwise the picture details would split and get distorted.
- Synchronization pulses are transmitted during the retrace period.

REVIEW QUESTIONS

1. Draw the basic block diagram of monochrome television receiver and explain its working.
2. Draw the basic block diagram of monochrome televisions transmitter and explain its working.
3. Define the terms Television in brief.
4. Write the development of television in brief.
5. Write the types of monochrome television systems.
6. Write the types of colour television systems.
7. Explain the steps used in picture transmission in television.
8. Explain the transmission of sound in television system in brief.
9. Discuss the reception of picture in television system.
10. Discuss the reception of sound in television system.
11. What do you understand by synchronization in television system.

12. Why amplitude modulation is preferred over FM for video signals ?

13. Why frequency modulation is preferred over AM for sound signal ?

MULTIPLE CHOICE QUESTION

1. In India, which monochrome TV system is used
(*a*) 525 line system (*b*) 625 line system
(*c*) 819 lines system (*d*) None of these

2. The colour TV system used in India is
(*a*) NTSC (*b*) PAL-B
(*c*) SECAM (*d*) None of these

3. Aspect ratio for width to height for a TV picture frame is
(*a*) 1 : 1 (*b*) 2 : 3
(*c*) 4 : 3 (*d*) 16 : 9

4. The number of interruption in projecting movie pictures on a cinema screen are
(*a*) 50 (*b*) 60
(*c*) 24 (*d*) 48

ANSWERS

1. (*b*) **2.** (*b*) **3.** (*c*) **4.** (*d*)

2
CHAPTER

BASIC PRINCIPLES OF TELEVISION ENGINEERING

INSIDE THIS CHAPTER

2.1. INTRODUCTION

The **principle of TV transmission** with a view to reproducing pictures can be summarized as follows: the optical image of the scene to be transmitted is divided into small picture elements (pixels). An opto-electrical converter, usually a camera tube, consecutively translates the individual elements into electrical information depending on their brightness. This signal is then transmitted at its actual frequency or after modulation onto an RF carrier. After appropriate processing at the receiving end, the information is applied to an electro-optical converter and reproduced in accordance with the brightness distribution of the pattern.

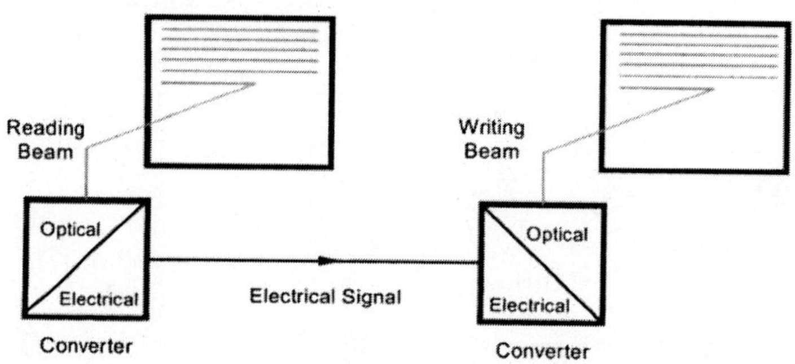

Fig. 2.1.

The primary factors which are required for successful transmission and reception of the picture are as follows:

1. Gross structure
2. Image continuity
3. Scanning lines
4. Interlaced scanning
5. Fine structure
6. Tonal gradation

2.2. GROSS STRUCTURE

In Television system the frame adopted is rectangular in shape.Width to Height ratio of a picture frame is called **Aspect Ratio** and it is 4/3. There are several reasons to adopt this aspect ratio. Width is kept longer than height because of the following facts:

1 Eyes can move with more ease and comfort in the horizontal plane than in the vertical.

2 The fovea, the surface of maximum sensitivity and resolution at the center of the retina in the eye has greater width than height. Hence the longer width of the image ensures more efficient use of the fovea.

3 As a result of intensive subjective tests by the cinema people, aspect ratio of 4:3 was found to most pleasing and less fatiguing to the eyes. Same ratio was accepted by the TV engineers.

Geometry of the field of view occupied by a TV image is shown in Fig. 2.2.

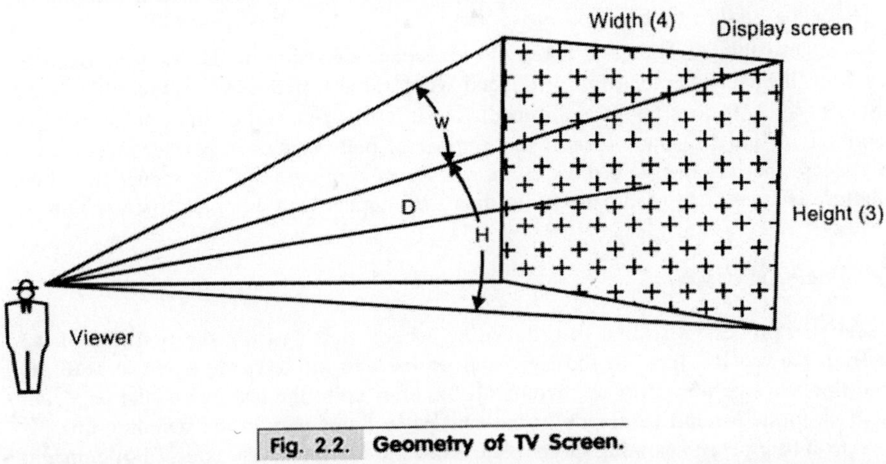

Fig. 2.2. **Geometry of TV Screen.**

The aspect ratio of the picture produced on TV screen at receiver should be the same as that is televised. The size of the TV screen and camera tube may be different but the aspect ratio should be same *i.e.* 4:3.

The aspect ratios for Television systems and motion pictures are given as follows:

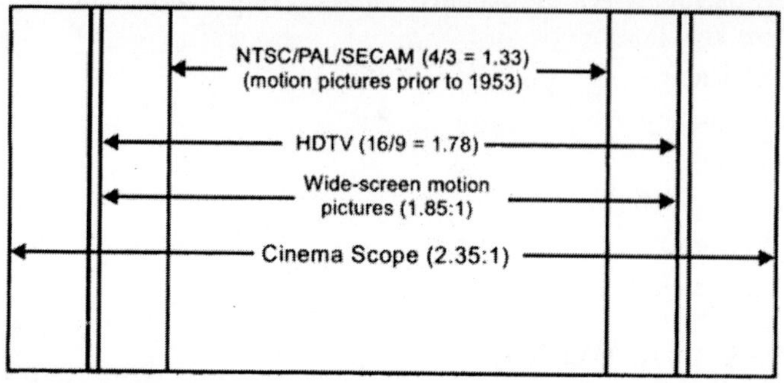

Fig. 2.3. Aspect ratio's of various TV systems and motion picture.

Dimensions of TV Receiver are specified by the diagonal length of the screen. When width is $4x$ and height, $3x$, diagonal would be $5x$ (Pythagoras theorem). Thus the diagonal will give the value of x. Hence for the aspect ratio of 4:3, the height and width of the screen can be determined, as shown in example :

Determine height and width of a TV screen of 100 cm size.

Solution. $5x = 100$ cm therefore $x = 20$ cm

Hence, Height $= 20 \times 3 = 60$ cm

and Width $= 20 \times 4 = 80$ cm.

2.3. IMAGE CONTINUITY

When we televising picture elements of the frame by means of scanning process, it is necessary to present the picture to the eye in such a way that an illusion of continuity is created and any motion in the scene appears as a smooth and continuous change on the picture tube screen.

To achieve this advantage is taken of persistence of vision of the human eye. It is arises from the fact that sensation produced when nerves of the eyes retina stimulated by incident light. It does not cease immediately after the removal of light but it persists for about $1/16^{th}$ of a second. Thus if the number of pictures shown per second is more than sixteen, the can integrate the changing level of brightness of the scene. So when the picture elements are scanned rapidly then they appear as a complete picture unit to the eyes.

2.4. SCANNING

Scanning process is similar to the reading a book starting from the first line of the page from the top left hand to the right and go back to the left side again to read the second line and like this carry on reading all the lines up to the last line of the page and come back to the top left hand side of the next page. Same way in the scanning process the electron beam start scanning target plate from the top left to the end of bottom right of the target plate. The scanning is done line by line horizontally from left to right at a fast rate and vertically from top to bottom at a slower rate and it is known as Trace as shown in Fig. 2.3. The retrace of the beam is very fast as compared to the forward scan. Scanning during retrace or flyback is blanked by making use of horizontal and vertical blanking pulses.

In a basic black and white TV, a single electron beam is used to scan a phosphor screen. The information is always displayed from left to right. After each line is written, when the beam returns back to the left, the signal is blanked. When the signal reached the bottom it is blanked until it returns to the top to write the next line as shown in Fig. 2.4.

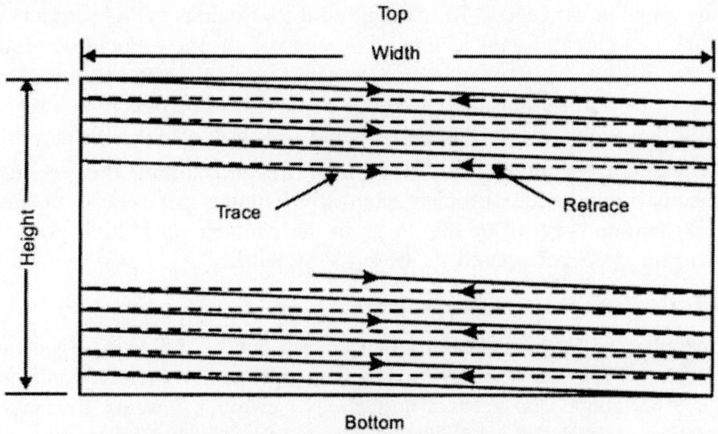

Fig. 2.4. Path of scanning beam in a frame.

It is the process of extracting signals pertaining to the individual elements of a picture in quick succession one after the other. **Scanning** may be defined as the process which converts optical signals in to electrical signals, referred as **video signals.**

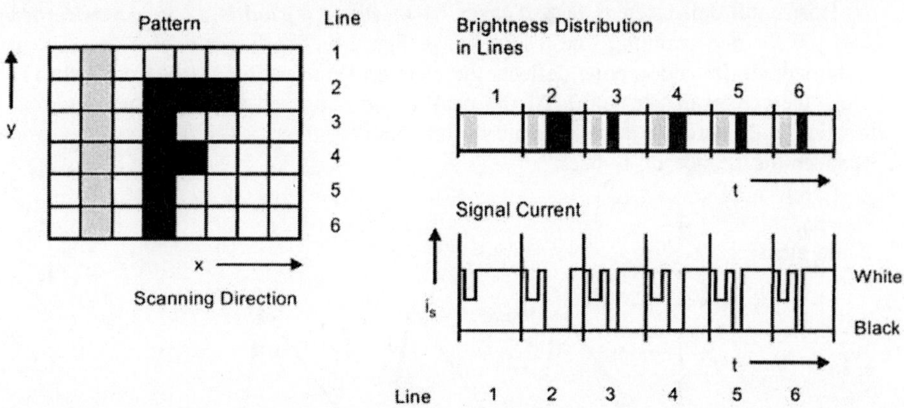

Fig. 2.5. Waveform of signal current in case of line-by-line scanning of pattern.

The pattern is divided into a number of lines which are scanned from left to right and from top to bottom

Scanning converts the individual picture elements from the geometrical into the time domain. Figure 2.5 gives a simplified representation assuming that the scanning beam returns to the left hand picture margin within a negligible period of time. In general, the signal current obtained is a train of multi shape pulses of varying mean value, corresponding to the mean brightness of the pattern. This signal current, which may contain components of very high frequency due to fine picture details, must be applied to the receiver without distortion. This requirement determines the essential characteristics of the transmission system.

2.4.1. Need of Scanning

Why scanning is needed? To transmit all the variations of light present in the space of a picture at a single instant of time, thousands of channels would be needed. It is not feasible to use so many channels to transmit a single programme. Thus the first requirement for transmitting the variations of light in a picture is to convert them into signals varying with time so that there is only a single variation at any instant.

The scene is scanned rapidly both in the horizontal and vertical direction simultaneously to provide sufficient number of frames per second to the illusion of continuous motion. Instead of the 24 as in the commercial motion picture, the frame repetition rate is 25 per second in most TV systems.

Saw Tooth Current for Scanning

Horizontal and vertical scanning is accomplished by deflecting a well focused spot of the electron beam horizontally from left to right and vertically from top to bottom. To achieve horizontal and vertical deflections, sawtooth currents are made to flow in the deflection coils in the respective camera tube at the transmitter and in the display tube at the receiver. Sawtooth currents are produced by charging a capacitor linearly through resistor of the appropriate value for trace signal and to discharge it rapidly through a low resistance transistor switch for retrace.

Horizontal Scanning

Horizontal deflection is fast. It takes 64 μs *(trace period is 52 μs +retrace period is 12 μs)* for one scanning line, as shown in Fig. 2.6. The linear rise of the current in the horizontal deflection coils deflects the electron beam continuously from left to right across the screen (Trace path). At the peak of its rise, the sawtooth wave reverses its direction and decreases rapidly to its initial value (Retrace path). The reversal process produces the retrace or flyback.

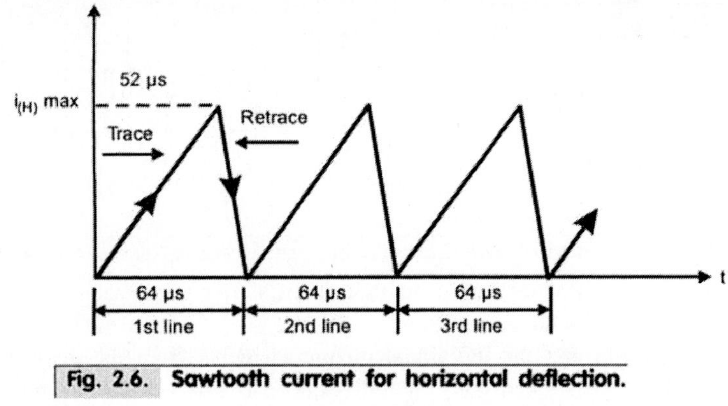

Fig. 2.6. Sawtooth current for horizontal deflection.

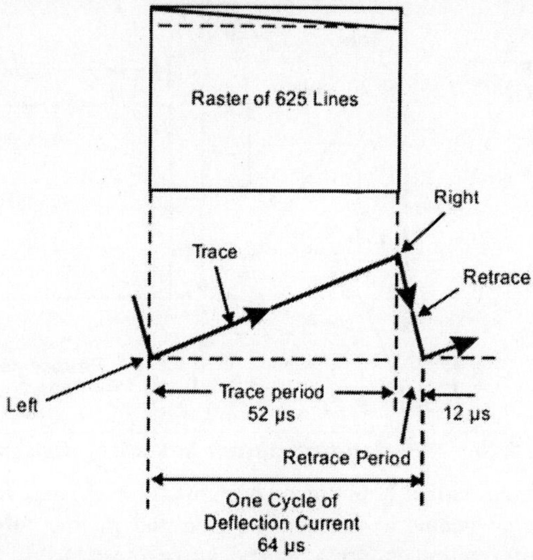

Fig. 2.7. Waveform of current in horizontal deflection coil producing linear scanning

Vertical Scanning

The sawtooth wave in vertical deflection coils moves the electron beam from top to bottom of the raster at waveform speed. Vertical trace and retrace path of electron beam is shown in Fig. 2.8.

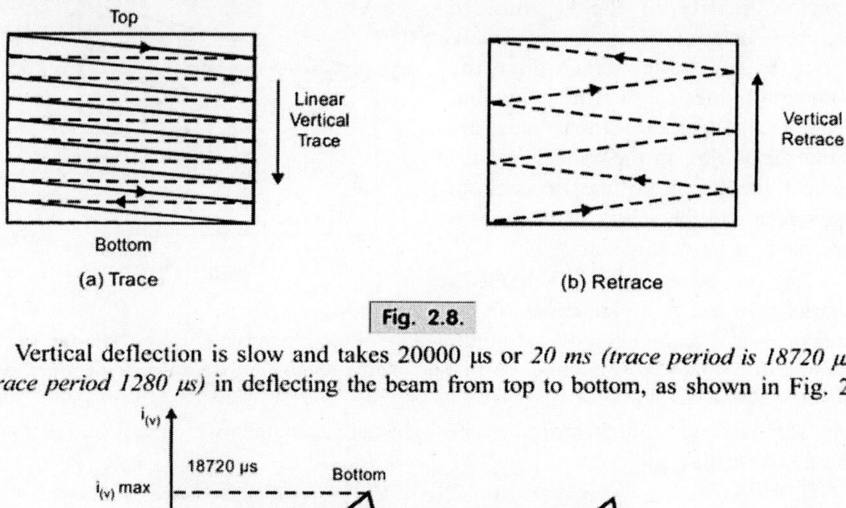

Fig. 2.8.

Vertical deflection is slow and takes 20000 μs or *20 ms (trace period is 18720 μs + retrace period 1280 μs)* in deflecting the beam from top to bottom, as shown in Fig. 2.9.

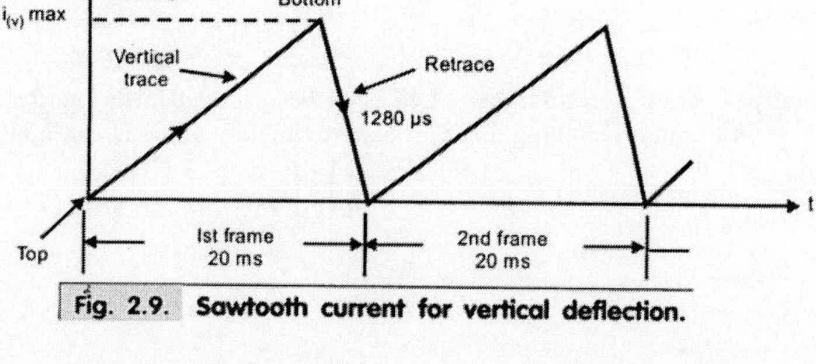

Fig. 2.9. Sawtooth current for vertical deflection.

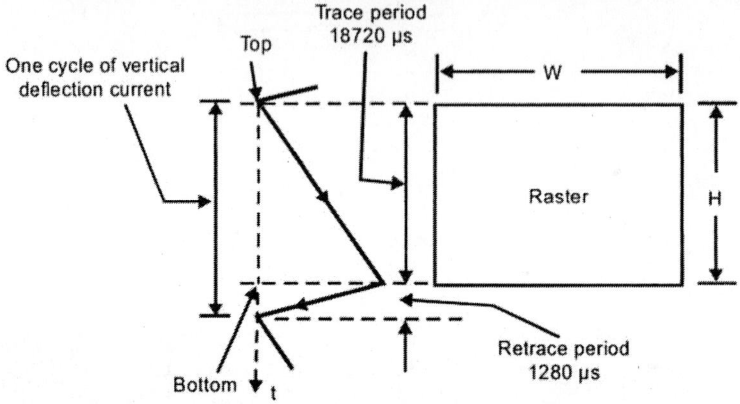

Fig. 2.10. Waveform of current in vertical deflection coil.

- It must be noted that both during horizontal retrace and vertical retrace intervals the scanning beams at the camera tube and picture tube are blanked and no picture information is either picked up or reproduced.
- On a time division basis, these short retrace intervals are utilized for transmitting distinct narrow pulses to keep the sweep oscillators of the picture tube deflection circuits of the receiver in synchronization with those of the camera at the transmitter

2.4.2. Number of Scanning Lines

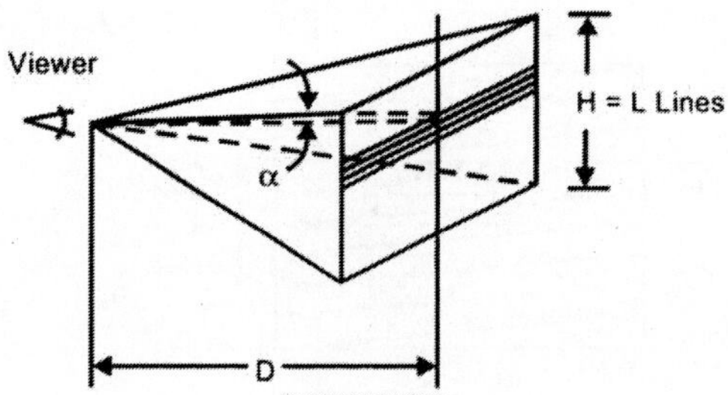

Fig. 2.11.

The quality of the reproduced picture is determined by the resolution, which is the better the higher the number of lines, a minimum number being required to ensure that the raster is not disturbing to the viewer. In this context, the distance of the viewer from the screen and the acuity of the human eye have to be considered.

The number of lines in a frame is restricted by the resolving capability of human eye. The number of scanning lines for one complete frame should be large enough to include more number of picture elements so that it can provide finer details of any picture.

The maximum number of alternate light and dark lines which can be resolved by the human eye is given by

$$N_v = \frac{1}{\alpha}\rho \qquad ...(2.1)$$

Where, N_v is the total number of lines to be resolved in the vertical direction,

α = minimum resolving angle, its approximately value is 1 minute or 60 degree and in radian it is given as $\left(\frac{1}{60}\right)\times\left(\frac{\pi}{180}\right)$,

$$\rho = \frac{D}{H}$$

The optimum viewing distance is found to be about four times the picture height, *i.e.* $D/H = 4$. At this distance, the line structure should just be no longer visible, *i.e.* the limit of the resolving power of the eye should be reached.

By substituting the values of α and ρ in Eqn. (2.1), we get

$$N_v = \frac{1}{\left[\left(\frac{1}{60} \times \frac{\pi}{180}\right) \times 4\right]} \approx 860 \text{ lines}$$

2.4.3. Kell Factor

Practically, the picture elements have random distribution of black, white and grey depending of the nature of the picture details. It is observed that 70% of the total lines get separately scanned in the vertical direction and balance 30% are merged with other elements due to the beam spot falling equally on two consecutives lines.

Elements or Pixels are not arranged in an orderly manner in a picture. These are distributed rather randomly. Hence a spot of the scanning beam may cover partly two pixels. This situation is shown in Fig. 2.12. Beam spot-I partially covers black and partially white, so the resultant luminance of this spot would be grey (white + black). Similarly spot-II would give grey. Thus instead of giving two signal outputs (black & white), the spot will give only one signal output (grey). The distinction between two pixels will not be reproduced. It is called **Kell Factor (k)** and it's approximately value is 0.7. Thus the kell effect produces degradation in the resolution of the individual pixels.

About 25 % to 35% pixels are lost due to Kell effect, only 65% to 75% of the active lines are responsible for distinguishing the individual pixels.

Hence the effective lines are worked out by multiplying the active lines by a factor of 0.65 to 0.75 (mean value = 0.7). This factor is called Kell factor and is used in estimating the actual resolution achieved in TV pictures.

Smaller the size of the beam spot, less will be kell factor and higher the effective numbers of lines.

Thus the effective numbers of lines (N_E) which can be distinctly resolved are given by,

$$N_E = N_v \times k$$
$$N_E = 860 \times 0.7 = 602$$

Fig. 2.12. Scanning beams.

In accordance with CCIR, the complete raster area has been divided into 625 lines, 575 of which are in the visible picture area due to the vertical fly back of the beam (525 lines in North America and Japan with about 475 active picture lines).

More number of lines in a frame needs large bandwidth and thickness of spot beam is small. **Channel bandwidth increases with increase in number of lines,** that add the cost of the system and reduces the number of television channel in VHF and UHF transmission band.

In India, the total numbers of lines have been chosen is 625 on the basis of quality and cost. In the America 525 line system is used in which the total numbers of the lines has been selected is 525

2.4.4. Picture Repetition Frequency

When determining the picture repetition frequency the physiological characteristics of the eye have to be considered. To reproduce a continuous rapid motion, a certain minimum frame frequency is required so that no annoying discontinuities occur. 16 to 18 frames per second, as are used for instance in amateur films, are the lower limit for this frequency. 24 frames per second are used for the cinema. This number could also be adopted for television; however, considering the linkage to the AC supply frequency, a picture repetition frequency (f) of 25 Hz for an AC supply of 50 Hz has been selected (30 Hz for a 60 Hz AC supply in North America and Japan).

However, the picture repetition frequency of 25 Hz is not sufficient for flicker free reproduction of the picture. The same problem had to be solved for the cinema where the projection of each individual picture is interrupted once by a flicker shutter, thus producing the impression that the repetition frequency had been doubled.

2.4.5. Types of Scanning

There are two types of scanning:

1. Progressive Scanning
2. Interlaced Scanning

(1) Progressive Scanning

When one cycle of vertical saw tooth current completes scanning of one frame and the next cycle starts scanning of a new frame, it is called **progressive scanning**.

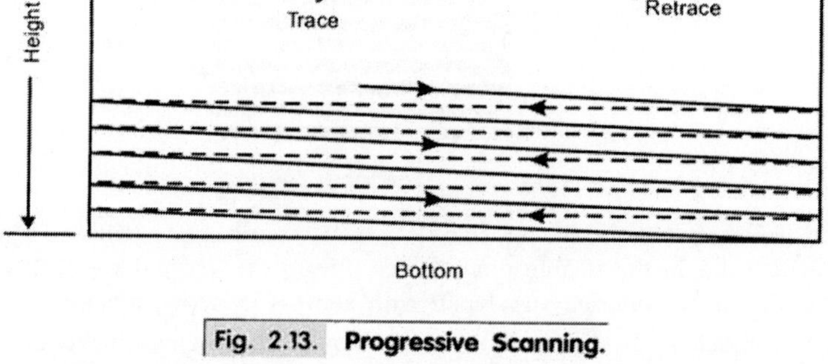

Fig. 2.13. Progressive Scanning.

During the process of scanning the horizontal and the vertical beam deflection circuits operate simultaneously, so as to cross the image in downward slant lines as shown in Fig. 2.13. On reaching the end of each line, searching spot retraces horizontally to the starting point of succeeding line. To make this retrace invisible a blanking pulse is applied to the control grid of the electron gun of picture and camera tube. For picture tube, these blanking pulses are transmitted along with video signal. To keep scanning at picture tube in step with that at camera tube, horizontal and vertical synchronising pulses are also sent along with blanking pulse and are superimposed on latter. The number of horizontal lines scanned has been standardised at a total of 625 per image of frame and 25 such frames are scanned per second.

Disadvantages of Progressive Scanning

- In progressive scanning, 25 frames per second are scanned in European system. When the scanning beam flips from bottom to the top of the picture frame, it takes about 1280μs. This retrace of 1280 μs duration is blanked on the screen of the picture tube and so the screen appears black during this time.
- When program is being displayed, dark interruptions of blanking between blank pictures occurs at 25 per second. As persistence of vision is more for dark light, these interruptions become visible as Flicker on the screen.
- If speed of progressive scanning is doubled (50 frames per second) flicker would be removed, but bandwidth will increase.

2.4.6. Persistence of Vision

- When the eye sees light, it continues to see it for about 60ms after the light source is removed. This property of eye is called persistence of vision.
- It has been possible to see movie picture because o. this property of the eyes. In TV the picture frame displayed on a TV screen changes one after the other faster than the persistence of the vision.
- Display of each complete frame takes only 40ms in Indian system of 25 frames per second and about 33ms in American system of 30 frames per second, while persistence of vision is 60ms.

2.5. FLICKER

Time for persistence of vision is more for darkness than for bright light. This result in a phenomenon called **flicker.** It means dark intervals between bright pictures become visible for a very short time and appears as flicker. The relationship between flicker and brightness is given by Ferry Poster law,

$$F_c = F + 12.6 \log_{10} B$$

F_c = Critical frequency, at which flicker will not be seen.

F = Constant related to viewing conditions (*i.e.* 37 for normal conditions)

B = Luminance of picture high lights in foot-lamberts.

For B = 10 and F= 37, F_c workouts to be 49.6.

2.5.1. Flicker Effects

The process of interlaced scanning is designed to eliminate flicker in the displayed image. Perception of flicker is primarily dependent upon two conditions:

- The brightness level of an image
- The relative area of an image in a picture

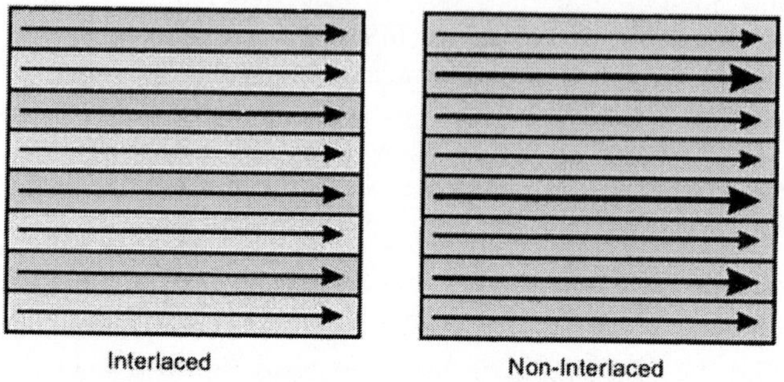

Fig. 2.14. Interlaced Vs Progressive Scanning.

2.6. INTERLACED SCANNING

- In India, 25 frames are scanned per second, *i.e.*, the frame repetition rate has been standardised at 25 per second. Despite high repetition rate, a certain amount of flicker is still present. To this residual flicker each frame is shown twice on the face of the picture tube. It is called interlaced scanning. In TV picture an effective rate of 50 vertical scans per second is utilized to reduce flicker.
- In Interlaced scanning, there are two sequences of scanning for each frame. Scanning line of one sequence occur in between the lines of the other sequence.
- This is accomplished by increasing the downward rate of travel of the scanning electron beam, so that every alternate line gets scanned instead of every successive line.
- Then, when the beam reaches the bottom of the picture frame, it quickly returns to the top to scan those lines that were missed in the previous scanning.

2.6.1. Principle of Interlaced Scanning

The total numbers of lines are divided in two groups called **Fields**. Each field is scanned alternatively. In 625 line system each frame are divided in two sets of 312.5 lines.

The lines of the complete raster are divided into two fields, which are interlaced and transmitted consecutively. Each field contains L/2 lines and is swept within the interval *T*/2. This means that lines 1, 3, 5 etc are in the first field and lines 2, 4, 6 etc. in the second field (geometrical line counting) (Fig. 2.15).

When reproducing the two fields it is essential that they be accurately interlaced since otherwise pairing of lines may cause the field raster to appear in a very annoying way. In a system using an odd line number, for instance 625, the transition from the first to the second field takes place after the first half of the last line in the first field. Thus no special auxiliary signal is required to ensure periodic offset of the two fields.

Thus 50 fields of 312½ lines each are transmitted instead of 25 pictures of 625 lines, the field repetition or vertical frequency being f_v = 50 Hz.

The resulting line or horizontal frequency is

$$f_h = 25 \times 625 = 50 \times 312\frac{1}{2} = 15625 \text{ Hz.}$$

The period of the horizontal deflection is $T_h = 64$ μs, that of the vertical deflection $T_v = 20$ ms. The horizontal and vertical frequencies must be synchronous and phase-locked.

Thus two fields are shown in each frame with repetition rate of 50 per second for each field. The picture element always bears a ratio of width to the height, called aspect ratio, of 4:3. That is width is 1.33 times the height of the picture frame.

Frame repetition rate is made equal to power line frequency (50 Hz) to reduce undesired effect of hum due to pickup from the main.

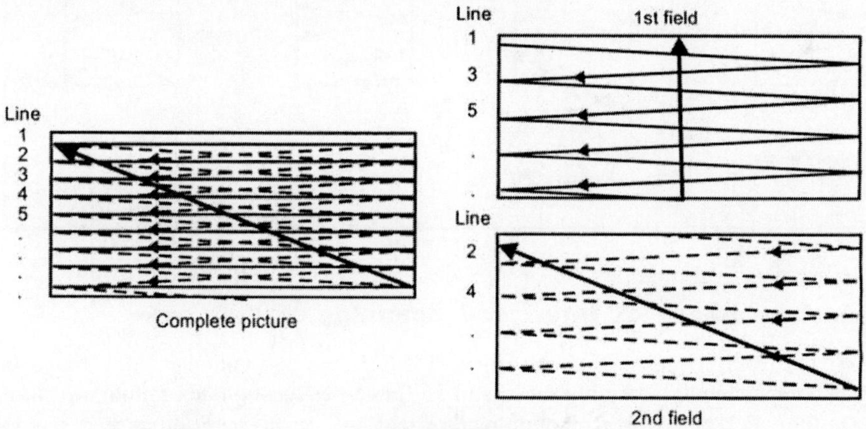

Fig. 2.15.

2.6.2. Scanning Period

- **Horizontal line:**
  Nominal duration- 64 μs (10^6 /15625)
  Active line (Trace) period - 52 μs
  Line blanking (Retrace) period- 12 μs
- **Vertical period** (field frequency 50Hz):
  Nominal duration- 20 ms (1/50)
  Active Line (Trace) period- 18.720 ms
  Line blanking (Retrace) period- 1.280 ms
- Since the horizontal and vertical sweep oscillator operates continuously to achieve the fast sequence of scanning, 20 horizontal lines (1280 μs/64 μs) get traced during each vertical retrace period.
- Thus 40 scanning lines are lost per frame, as blanked lines during the retrace interval of two fields.
- Active number of lines (Na) = 625 – 40 = 585 lines.

2.6.3. Scanning Sequence

The complete geometry of the standard interlaced scanning pattern is shown in Fig. 2.16.

First Field (312.5 lines)

First v- trace (292.5) - 1 to 292-5 lines

First v- retraces (20) – 292.5 to 312.5 lines

Second Field (312.5 lines)

Second v trace (292.5) - 312.5 to 605 lines

Second v- retraces (20) – 605 to 625 lines

One Frame (312.5 +312.5) =625 lines

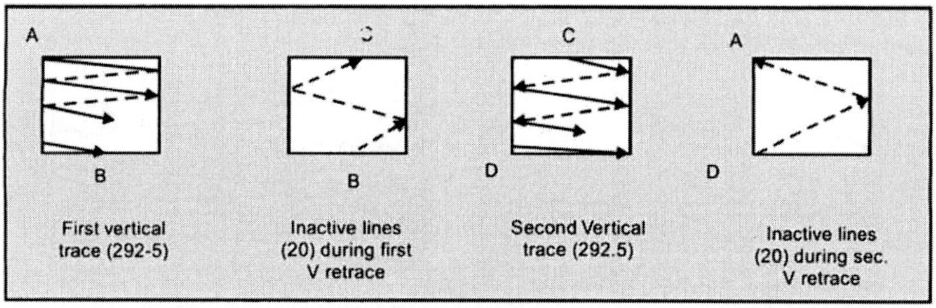

Fig. 2.16.

2.6.4. Advantages of Interlaced Scanning

- The 25 Hz transmission rate for a full 625 line conventional video frame is comparable to the highly successful 25-frame-per-second rate of motion-picture film. For comparison, motion-picture theaters, on average, produce a screen brightness of 10 to 25 ft-L (foot-lambert), whereas a direct-view CRT may have a highlight brightness of 50 to 80 ft-L.

  Through the use of interlaced scanning, single field images with one-half the vertical resolution capability of the 625 line system are provided at the high flicker-perception threshold rate of 50 Hz. Higher resolution of the full 490 lines (525 less vertical blanking) of vertical detail is provided at the lower flicker-perception threshold rate of 30 Hz. The result is a relatively flicker less picture display at a screen brightness of well over 50 to 75 ft-L, more than double that of motion picture film projection. Both 60 Hz fields and 30 Hz frames have the same horizontal resolution capability.

- The second advantage of interlaced scanning, compared to progressive scanning, is a reduction in video bandwidth for an equivalent flicker threshold level. Progressive scanning of 625 lines would have to be completed in 1/50 s to achieve an equivalent level of flicker perception. This would require a line scan to be completed in half the time of an interlaced scan. The bandwidth would then double for an equivalent number of pixels per line.

2.6.5. Degradations of Image Quality due to Interlace Scanning

- Interlace scanning, however, is the source of several. While the total area of the image flashes at the rate of the field scan, twice that of the frame scan, the individual lines repeat at the slower frame rate. This gives rise to several degradations associated with the effect known as **interline flicker**. This causes small areas of the image, particularly when they are aligned horizontally, to display a shimmering or blinking that is visible at the usual viewing distance. A related effect is unsteadiness in extended horizontal edges of objects, as the edge is portrayed by a particular line in one field and by another line in the next. These effects become more pronounced as the vertical resolution provided by the camera and its image enhancement circuits is increased.

- Interlacing also produces aberrations in the vertical and diagonal outlines of moving objects. This occurs because vertically-adjacent picture elements appear at different times in successive fields. An element on one line appears 1/50 s (actually 1/49.95 s) later than the vertically-adjacent element on the preceding field. If the objects in the scene are stationary, no adverse effects arise from this time difference.

2.7. RESOLUTION

The ability of the image reproducing system to represent the fine structure of an object is known as its resolving power or **resolution**.

Resolution means ability to distinguish between closely spaced pixels.

2.7.1. Vertical Resolution

The extent of which the scanning system is capable of resolving picture details in the vertical direction is referred to as **vertical resolution.** The no. of pixels on a vertical line on the screen will be equal to the no. of horizontal lines.

$$\boxed{V_r = Na \times k}$$

Where, V_r = vertical resolution, Na = active no. of lines,

k = Resolution factor (also called Kell factor)

Assuming reasonable value of $k = 0.69$,

$$V_r = 585 \times 0.69 = 400 \text{ lines}$$

2.7.2. Horizontal Resolution

The capability of the system to resolve maximum no. of picture elements along the scanning lines determined **horizontal resolution.**

If the aspect ratio is A, the no. of pixels on a horizontal line would be equal to

$$\boxed{H_R = V_r \times A}$$

For non-digital system $A = \frac{4}{3}$

For digital system $A = \frac{16}{9}$

In case 625 lines, $Na = 585$ lines, $k = 0.69$ and $A = \frac{4}{3}$

$$H_R = V_r \times A = 585 \times 0.69 \times \frac{4}{3} = 533 \text{ lines}$$

2.8. BANDWIDTH CALCULATIONS FOR VIDEO SIGNAL

Bandwidth is the highest video frequency related to the time taken in scanning two adjacent pixels. Fig. 2.17 shows 4 successive pixels and the corresponding 2 cycles of brightness. Time period T is time taken in scanning one cycle or 2 adjacent pixels.

A-Four Adjacent pixels of black and white shades.

B-Two cycles of brightness corresponding to pixels of –A

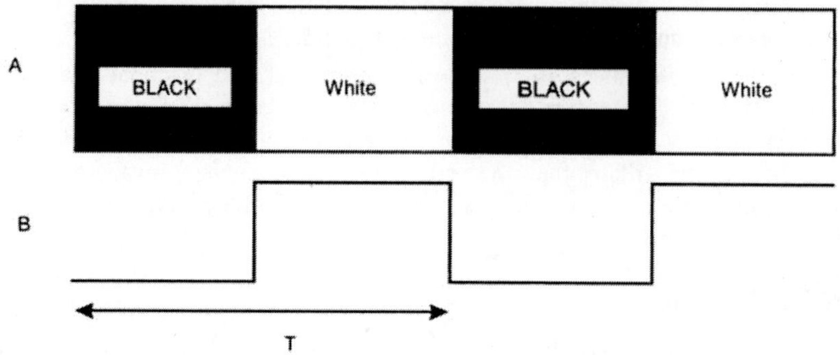

Fig. 2.17.

If t is the time in seconds for scanning one lines (excluding retrace time), then H_R pixels are scanned in t second. Therefore 2 pixels shall be scanned in $2t/H_R$.

Thus time period T is given by,

$$T = 2t/H_R$$

$$\text{Video bandwidth} = \frac{1}{T} = \frac{H_R}{2t}$$

For example (in case of 625 lines), if $H_R = 533$ and $t = 52$ μs, calculate video bandwidth for the system.

$$\text{Time period} = \text{Time required to scan 2 pixels}$$

$$= \frac{52 \times 2}{533} \ \mu s$$

Therefore,

$$\text{Bandwidth} = \frac{533}{2} \times 2 \times 10^{-6} \text{ Hz}$$

Video bandwidth = 5.125 MHz or ~ 5MHz (Highest modulating frequency)

In case 525 lines system it is 4 MHz.

If the size of pixel is small then resolution would be better but it requires more bandwidth. Whereas if the size of the pixel is bigger than the resolution became poor, however it needed less bandwidth.

The highest fundamental video frequency in a scanning system is given by

$$F_{max} = \text{Aspect ratio (AR)} \times \text{Resolution factor (k)} \times \text{Number of active lines (Na)/2}.$$

2.8.1. Some Important Points Related with the Bandwidth

We know the bandwidth depends upon the following factors:

- Number of active lines
- Aspect ratio

- Resolution factor or kell factor and
- Horizontal line trace period

Number of Active Lines

The bandwidth depends upon the number of active lines; more number of active lines means higher resolution but needs more bandwidth.

Aspect Ratio

Bandwidth and resolution depends upon the aspect ratio, higher the aspect ratio will have higher resolution but needed more bandwidth. As in High Definition Television (HDTV) the aspect ratio is 16/9 or 1.78 and in normal TV broadcasting it is 4/3 or 1.33, which shows that HDTV needs more bandwidth but have better resolution.

Resolution Factor

The maximum number of horizontal scanning lines in a frame depends upon the size of the focused spot of the scanning beam. If the scanning beam is sharp then more number of scanning lines can be accommodated on a picture frame without overlapping. Sharp beam will have larger kell factor and hence will increase the bandwidth.

Horizontal Line Trace Period

The time taken by scanning line to cover forward scan or trace is depending upon the scanning speed. Smaller the time taken in horizontal trace or forward scan, higher is the bandwidth.

2.9. TONAL GRADATION

Variation in brightness from pixels to pixels is called **tonal gradation**. If the intensity changes from one pixel to the next pixel, the rate of variations of intensity is maximum. This maximum rate of change variation is the bandwidth of variation of light, and is about 5 MHz. Tonal gradation of a picture varies from 0 Hz to 5 MHz.

Various factors that influenced the tonal quality of the reproduced picture are as follows:

- **Contrast :** This is the difference in intensity between black and white parts of the picture over and above the brightness level. This difference will depend on the strength of video signal responsible for glowing an individual pixel. This control is provided in the video signal amplifier.
- **Contrast Ratio :** The ratio of maximum to minimum relative to the original picture is called contrast ratio. In broad daylight it is as high as 10000:1, whereas as the picture tube cannot produce a contrast variation more than 50:1 or at most 1000:1.
- Human eye cannot accommodate more than 10:1 variation of light intensity.

2.10. COMPOSITE VIDEO SIGNAL

The picture information is not transmitted alone. They carries the various signals alongwith such as the blanking pulses to make the retrace invisible and synchronizing pulses to synchronize the scanning at the transmitter and at the receiver. All the components all together are known as **composite video signal** (CVS).

Composite video signal consists of

1. A camera signal corresponding to the desired picture information,
2. Blanking pulses to make the retrace invisible and
3. Sync pulses to synchronize the Tx & Rx scanning.

Composite video signal can be represented either with positive polarity or with negative polarity.

2.10.1. Positive Polarity

In the case of **Positive polarity**, whiter the scene, higher is the amplitude of the video signal. Blanking level is kept at the zero level. Below zero level is the sync pulse. Sync top is at the most negative point as shown in Fig. 2.18.

In case when video signal is produced by photo conduction type camera tubes, bright white light gives high amplitude of video signal. At the receiver; for reproducing white on the fluorescent screen, stronger signal is needed and for reproducing black, zero signals is needed.

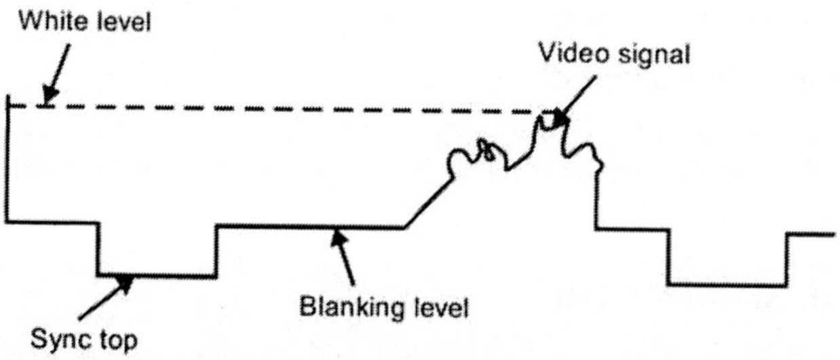

Fig. 2.18. Positive polarity.

2.10.2. Negative Polarity

In case of **negative polarity**, brighter the scene, smaller is the amplitude. Here sync pulse is positive, that is above the blanking level. Black is just below the blanking level. Brighter the scene, lower is its level below the blanking level. White is near the bottom.

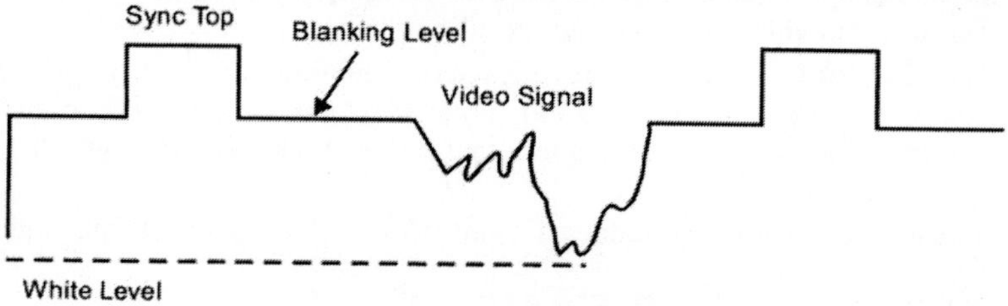

Fig. 2.19. Negative polarity.

The modulation of the RF carrier by the CVS is in the form of negative AM, where bright picture points correspond to low carrier amplitude and the sync pulse to maximum carrier amplitude (Fig 2.19).Such modulation is called **Negative Modulation.**

In most of the TV systems, negative polarity is used to modulate the video carrier.

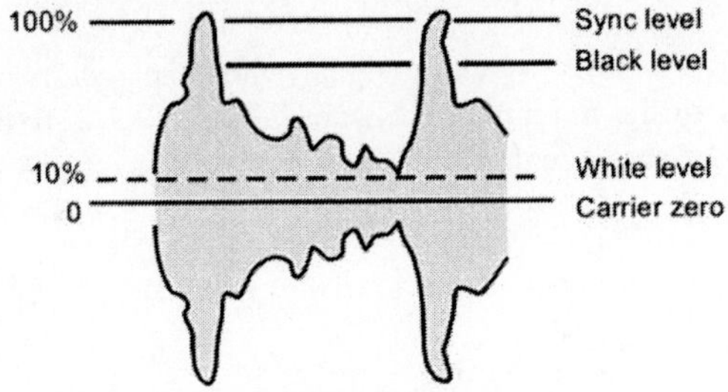

Fig. 2.20.

Advantages of Negative Modulation

1. In the case of negative modulation, the noise pulses in the transmitted signal shall increase the amplitude of the carrier which will move towards the black. Thus noise falling in dark grey or black region will not cause irritation. But in positive modulation, white blobs will be caused in the picture due to random noise signals.
2. In negative modulation the voltage of the blanking pulse is of fixed level which depends on the carrier strength. Thus this level forms the reference level for deriving AGC (automatic gain control) voltage dependent on the carrier strength.
3. The amplitude of the Carrier will remain low for most of the time as the signal content is more in white than in black. This will save the transmission power.

Disadvantages of Negative Modulation

1. Picture tube *i.e.* cathode ray tube needs positive polarity between control grid and cathode. So the negative polarity of the video signal is to be changed to positive polarity before feeding to the control grid.
2. High impulse pulse may disturb the H-sync and hence horizontal-synchronization may be disturbed for a short duration. This effect is reduced by using automatic frequency control in H- sweep circuits in the receiver.

Inspite of these disadvantages we use negative modulation in all TV systems because it has more advantages.

2.10.3. Video Signal Dimensions

The Voltage levels for sync pulse, blanking pulse and video are different, so that these pulses can be separated easily in the receiver. Voltage division multiplexing is used to keep these pulses and signal at different levels in CVS. There are two ways of indicating the levels:

(*a*) Percentage of carrier (for modulated signals).

(*b*) Institute of Radio Engineers (IRE) scale (for video base band signal)

Percentage of Carrier

For modulated signal, the various pulse levels are indicated in terms of percentage of carrier level. For positive polarity, white level is at 90%, whiter than white goes up to 100%, blanking level is at 75% and sync pulse top is at 0%.

Figure 2.21 shows the three brightness level of the composite video signal.

Peak White Level

Peak white level is the level of video signal when the picture information being transmitted corresponds to the maximum whiteness. This level is fixed at 10 to 12.5% of the maximum value of the signal.

Whiter than White

The signal level below the peak white level *i.e.* from 10% to 0% level is referred as **whiter than white level.**

Black Level

In negative modulation, the black or dark level is shown at the 72% of the maximum level.

Blacker than Black

Signal beyond the black level *i.e.* from 72% to 100% of the maximum signal level is known as **Blacker than Black level.**

Blanking Level

The blanking level in composite video signal is represented by 75% of the carrier level.

This level is also known as **pedestal level.**

Sync Pulses

Sync pulses are added at 75 % and sync top level is at 100%.

DC Components of the Video Signal

Since the amplitude of various picture elements varies continuously, the video signal has an average brightness of the scene. Rx cannot follow the changes in brightness, if the dc component is not present. Average brightness can change only from frame to frame but not from line to line.

Pedestal Height

Video signal corresponding to picture information varies in accordance with the variation of intensity light at various picture elements. Pedestal height is the difference between pedestal level and the average value of the video signal. Pedestal level may be used as a reference level to indicate average brightness of the signal.

Ratio of Picture Information to Sync Pulse Amplitude

Sync pulses are required at the horizontal and vertical traces to trigger the respective scanning oscillators. During the trace, the video signal changes between white and black levels.

Upper 25% is occupied by sync pulses and 65% of the carrier amplitude is occupied by the video signal. The transmitted signal has a picture to sync (P/S) ratio is equal to 10/4.

The overall arrangement of combining the picture signal and sync pulses may be considered as a kind of voltage division multiplexing.

If sync pulse amplitude is increased to reduce the ratio say to 10/7, the contrast of the picture will be degraded.

If the picture information is increased to increase the above ratio, say 10/3, the sync pulse amplitude would become inadequate to keep the picture stable.

Thus the ratio of 10/4 is optimum.

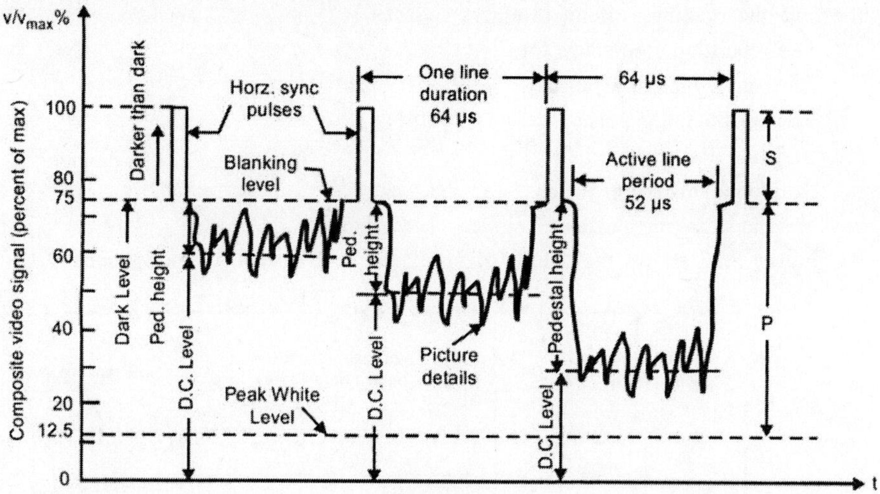

Fig. 2.21.

For negative polarity sync top is at 100%, blanking level is at 75%, white level is at 10% and whiter than white is below 10%.

- Peak white level: 10 to 12.5 %
- Blanking level: 75 % (also known as Pedestal level)
- Sync pulses: 75 to 100%
- Black level: 72 %
- Pedestal: The difference between blanking level and black level.
- Blacker than black: 72 to 100 %
- Whiter than white: signal below 10 % to 0 %

NOTE : ***To minimize the noise effect the lowest 10% of the voltage range is not used. At the Rx picture tube is biased to ensured that a received video voltage corresponding to about 10% modulation yields complete whiteness at the particular point on the screen***

IRE (Institute of Radio Engineers) Scale

- In IRE scale, blanking level is at 0 IRE ,sync top is -40 IRE, standard white is at about +86 IRE and peak white is at +100 IRE.
- The voltage signal swing from sync top to peak white level for positive polarity is indicated by IRE scale. Although IRE scale is used for positive polarity CVS, it can be used for negative polarity CVS also.
- Total swing of the signal is 140 IRE.

2.10.4. Blanking Pulses

The composite video signal contains blanking pulses to make the retrace lines invisible by raising the signal amplitude slightly above the black level (75 %) during the time the scanning circuit produces retraces.

The repetition frequency for;

Horizontal blanking pulse : 15625 Hz

Vertical blanking pulse : 50 Hz

Blanking Pulse period for

Horizontal blanking pulse : 12 μs

Vertical blanking pulse : 1280μs μs

Figure 2.22 shows the details of horizontal and vertical blanking pulses.

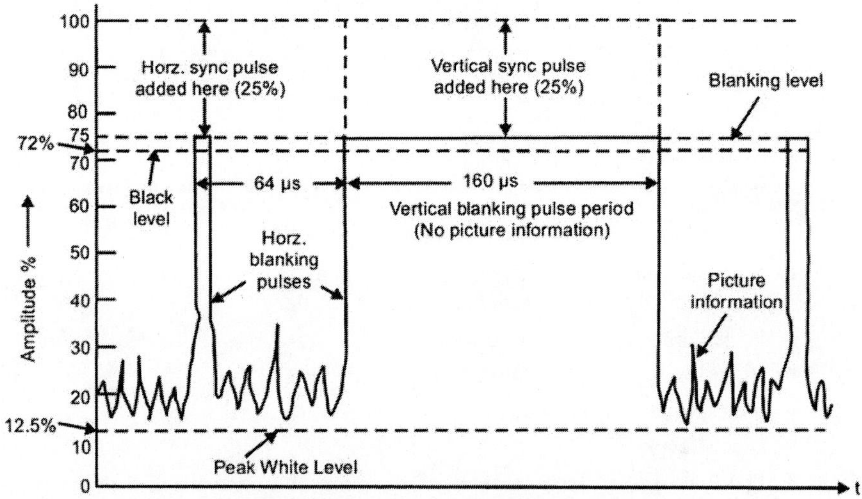

Fig. 2.22. Horizontal and Vertical blanking details.

2.10.5. Horizontal Blanking Pulse Details

The Horizontal blanking period and sync details are shown in Fig. 2.23.

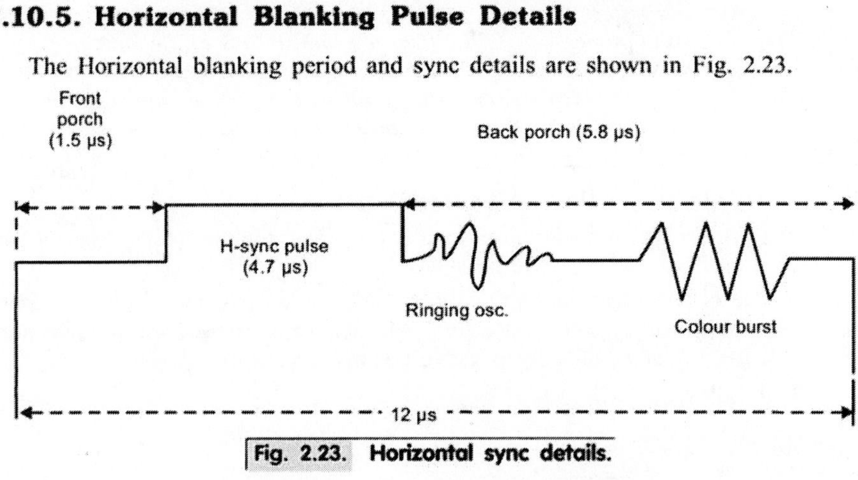

Fig. 2.23. Horizontal sync details.

The total width of H- blanking pulse is 12 μs during which H-retrace is complete and the video signals remain cut-off. It consists of three parts as shown in Fig. 2.23.

1. Front porch (1.5 μs)
2. H-sync pulse (4.7 μs)
3. Back porch (5.8 μs)

Front Porch

The width of the front porch in terms of time interval is 1.5 μs or 0.025 H.

The function of the front porch is to ensure that the H-sync pulse starts from a fixed level. In the absence of front porch, the starts of H-sync pulse would have depended on the end of the video signal.

During this period, before the sync pulse occurs, the receiver video circuit set the picture voltage level that exists at the end of picture line to the blanking level.

Sometimes it happens that when the line ends in an extreme white details, and the signal amplifier touches almost zero level, the video voltage level fails to decay to the blanking level before the leading edge of the line sync pulse occur. This results in late triggering of the time base circuit thus upsetting the horizontal line sync circuit. As a result the spot beam is late is late in arriving at the left of the screen and picture information on the next line is displaced to the left. This effect is referred as **Pulling on whites**.

Horizontal or Line Sync Pulse

Line sync pulse is used to trigger the scanning circuits to ensure synchronization for scanning at the transmitter and at the receiver. It must starts at the same instant of time in the transmitter and in receiver for correct synchronization.

Horizontal or Line sync pulse is 4.7 μs or 0.075 H wide. When the sync pulse starts after the front porch of blanking, the horizontal retrace is produced. Since the sync level is blacker than black, the fly back is blanked out.

Back Porch

Back porch is 5.8 μs wide.

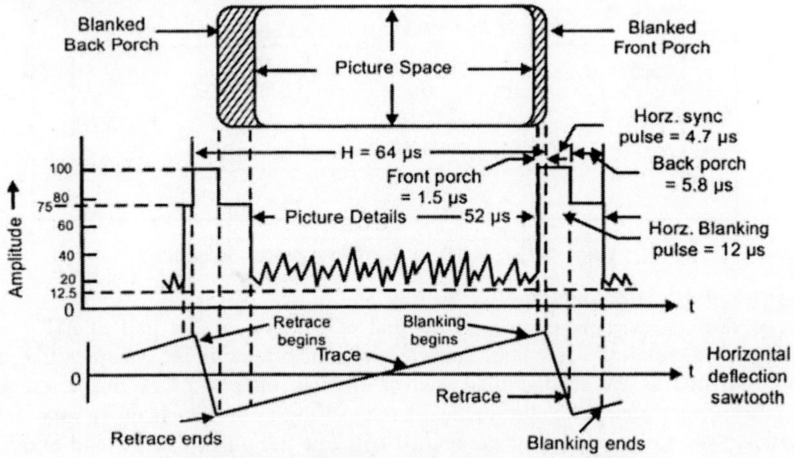

Fig. 2.24. Details of Horizontal line & sync.

Its function is to absorb the ringing oscillations which are caused during sudden fall of the voltage from the sync top during fly back.

In colour receiver, the back porch accommodates about 8 to 10 cycles (called colour burst signal) of 4.43 MHz colour sub-carrier.

The trace starts a little before the end of back porch & hence there would be black edge on the left hand side of the screen.

The retrace starts after the front porch there would be a black edge in the right hand side too for the time-width of the front porch.

Figure 2.24 shows the horizontal line and sync details.

2.10.6. Vertical Blanking Pulse Details

When the vertical scanning beam of a field reaches from the bottom line to the top, then the next field starts. The retrace of the beam from bottom to top is covered by V-blanking pulse.

During blanking period the picture signals remains cutoff. Vertical blanking pulse carries Vertical sync pulse which triggers the vertical sweep oscillator for synchronization.

The width of the vertical sync pulse must be larger than the period of horizontal sync pulse and it is because to derive a suitable field or vertical sync pulse at the receiver to trigger the vertical sweep oscillator. The period of *vertical sync pulse should be 2.5 to 3 horizontal line period.* If the length of the vertical sync pulse is less than it is difficult to distinguish between vertical and horizontal sync pulses at the receiver. And in case the length of the vertical sync is greater than the transmitter must operate at peak power for unnecessarily long interval of time.

The detail of vertical blanking pulse is as below:

- Front portion ➞ 160 μs (2.5 H)
- V-sync pulse = 160 μs (2.5 H)
- Back portion = 960 μs (15 H)
- Total vertical blanking pulse period = 1280 μs

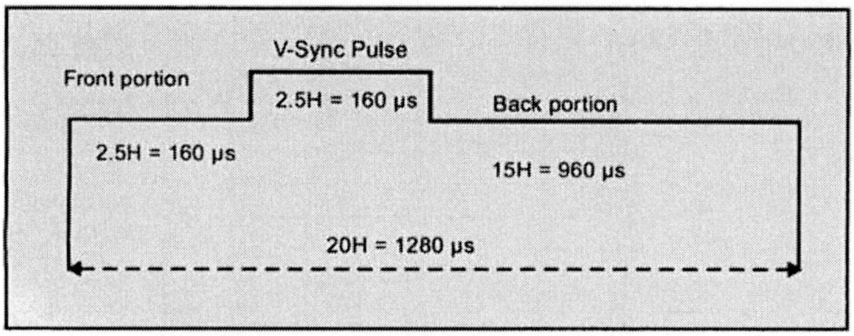

Fig. 2.25. Vertical Blanking pulse details.

The vertical sync pulse is of 160 μs or 2.5 H.

The vertical sync pulse starts at the end of 312.5th or first half of 313th line and terminate at the end of 315th line. Same way the next sync pulse occupies line number 1st to 2.5th line as the second field is over after an interval of 20 ms. These vertical sync pulses must occur after the vertical stroke of the beam that is every after 1/50 sec. A vertical sync pulse needed at the end of half line period *i.e.* 32 μs, and other after a full line period *i.e.* 64 μs. because of this arrangement the horizontal sync pulses do not appear one above the other but appears at half line interval with respect to each other.

The horizontal sync pulses continue to appear exactly at 64 μs intervals throughout the scanning period from frame to frame whereas the apparent shift of 32 μs is due to the alignments of vertical sync pulses.

Figure 2.26 shows composite waveforms with the horizontal and vertical sync pulses at the end of even field and at the end of odd field.

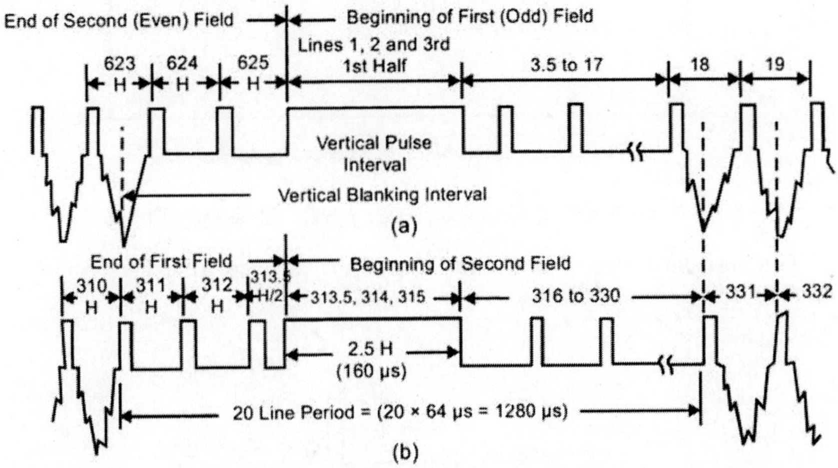

Fig. 2.26. Composite waveforms with the horizontal and vertical sync pulses (a) at the end of second or even field and (b) at the end of first or odd field.

The vertical sync pulse is of much longer duration in order to make it distinguishable from the horizontal sync pulse. We know that vertical deflection is very slow as compared to horizontal deflection, and therefore the time interval of the vertical blanking pulse is quite wide. Due to the wide interval, serrations or slots are used in the vertical sync pulse.

2.10.7. Serrations

Because of long duration of vertical sync pulse (160 μs), horizontal synchronization cannot occur for this long period as the non availability of rising edges. The rising edge is used to triggers the horizontal sweep generators. The horizontal sweep than likely to get upset and causes splitting of the picture diagonally. Now to avoid this condition it is split by serrations, each 4.7 μs wide, into five half line narrower pulses, each 27.3 μs (32- 4.7 μs) wide as shown in Fig. 2.27. Slots are made in the vertical sync pulse every 32 μs interval so that rising edges become available. The front edges of these correspond exactly with the line sync pulses they replace. The extra half line sync pulse edges that occur between each pair of pulse edges actually needed for line synchronization are unnecessary. They do not affect the line time base it is insensitive to pulses in these intermediate times. This half line pattern is necessary because, in interlaced scanning, the vertical sync pulses must occur at the middle of the lowest 313th line of the odd field or first field, and at the end of the lowest 625th line of the even or the second field.

NOTE :	***Serration allows synchronization of H-sweep periodically, so that when receiver comes out of the v-blanking pulse, it finds itself in correct synchronization.***

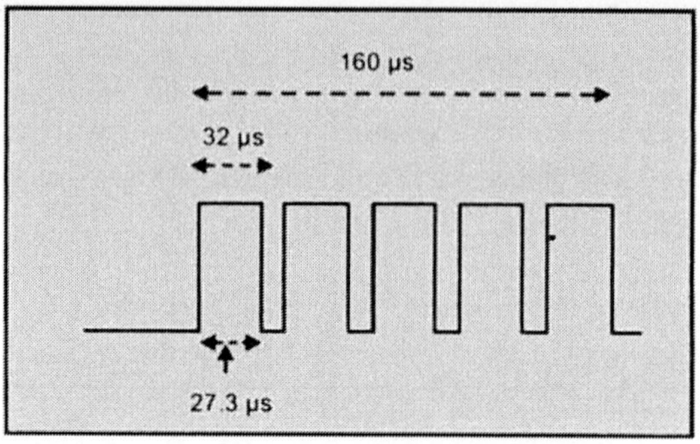

Fig. 2.27. Serrations in vertical sync pulses.

2.11. SYNC PULSE SEPARATION AND GENERATION

The horizontal and vertical sync pulses are extracted from the composite video signal by using the circuit given in the Fig. 2.28.

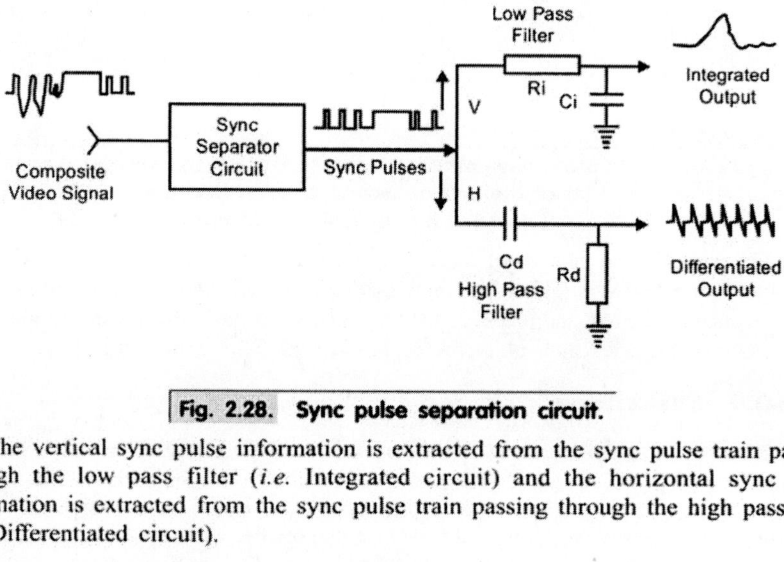

Fig. 2.28. Sync pulse separation circuit.

The vertical sync pulse information is extracted from the sync pulse train passing through the low pass filter (*i.e.* Integrated circuit) and the horizontal sync pulse information is extracted from the sync pulse train passing through the high pass filter (*i.e.* Differentiated circuit).

2.12. HORIZONTAL SYNC PULSE

The differentiated output is used to synchronize the horizontal scanning, pulse corresponding to the differentiated leading edges of sync pulses are used to synchronize the horizontal sweep oscillator as shown in Fig. 2.29.

Generally the receiver's uses a monostable multivibrator to produce horizontal scan, thus a pulse is required to trigger each cycle of the horizontal oscillator in the receiver. During the active and blanked line period, the horizontal sync pulses are generated but for the 2.5 H line period (vertical sync period) no sync pulses are produced. The horizontal sweep oscillator frequency is 15625 Hz and which tends to go out of step during each vertical sync period. This condition is difficult after an odd (first) field, since the vertical blanking period at the end of an odd field starts at a halfway through the horizontal line. The leading edge of vertical sync pulse arrived at wrong time to provide synchronisation for the horizontal oscillator. So to provide horizontal sync pulses at correct instances both after even and odd fields, it is necessary to cut slots in the vertical sync pulse at half line interval.

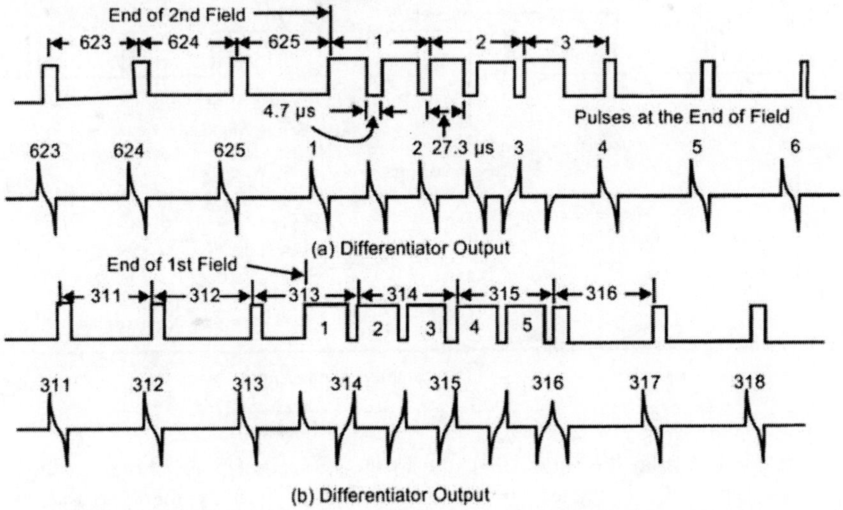

Fig. 2.29. Differentiated horizontal sync pulses for both the fields.

The insertion of short pulses of 4.7 μs is known as **serration** of the wide field pulse. Figure 2.29 shows that each horizontal sync pulse gives a positive spiked output from its leading edge and a negative spiked pulse from its trailing edge. The pulses are present at an interval of 64 μs and are marked with line numbers for both odd and even fields. Alternate vertical positive spikes are used during the interval of serrated vertical pulse train, some are used for odd field and remaining are being used for even field. It is because of the half line difference at the start of each field. The vertical sync pulse occurs at an interval of 32 μs but not at 64 μs as needed by horizontal oscillator.

2.13. VERTICAL SYNC SEPARATION

The vertical sync pulse information is extracted from the sync pulse train passing through the low pass filter (*i.e.* Integrated circuit) formed by resistance, Ri and capacitance, Ci. These integrated pulses are used to synchronize vertical sweep oscillator. The integrating circuit is a low pass filter with a cut-off frequency such that the horizontal sync pulses produce very low output and the frequency of vertical pulses should fall in the pass band of the filter. The voltage developed across the integrating circuit corresponding to the sync pulses of both even and odd fields as shown in Fig. 2.30.

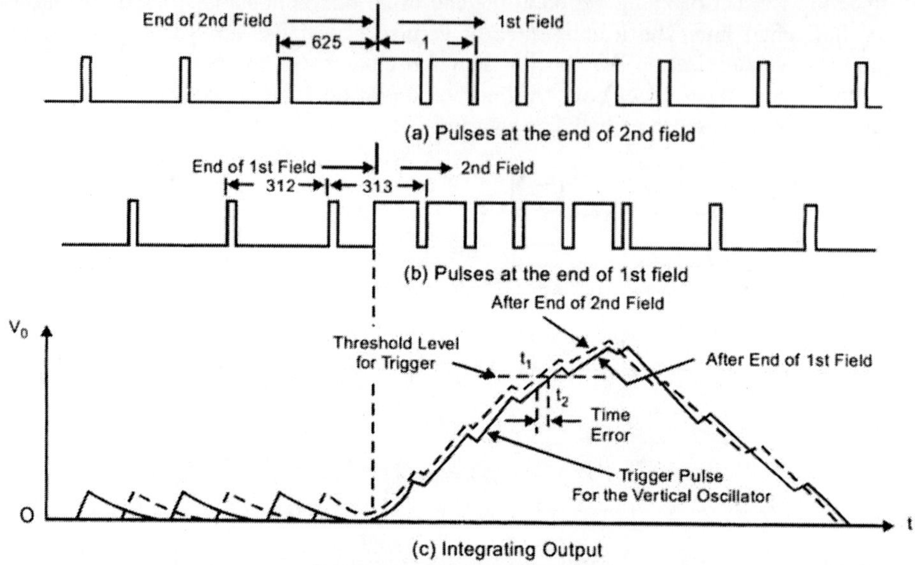

Fig. 2.30. Integrated output.

It is clear from the figure that due to the horizontal pulse there is a little rise in voltage across the capacitor but this is reduced to zero by the time the next pulse arrives. As the vertical pulse arrives the voltage across the output of filter starts increasing. The voltage developed across the filter is different for each field.

At the starting of the first field the last horizontal pulse that corresponds to the beginning of the 625th line is separated from the 2nd vertical pulse by one complete line and any voltage developed across the filter will have sufficient time to return to zero before the arrival of the first vertical pulse, and thus the filter output voltage builds up from zero in response to the five successive wide vertical sync pulses. The voltage builds up because the capacitor has only 4.7 μs to discharge whereas has more time to charge.

In the case of starting of the even field *i.e.* 2nd field, the last horizontal pulse corresponds to the beginning of 313th line is separated from the first vertical pulse by only half a line.

In this field the voltage developed across the low pass filter does not have enough time so that it can reach zero before the arrival of the first vertical pulse. The voltage build up does not start from zero as in the case of first field. The residual voltage gets added up to the voltage developed because of the vertical pulses. This output voltage across the filter at any instant is higher as compared to the voltage developed at the starting of first field as shown in figure- with the dotted line. Threshold level to trigger the vertical oscillator is also shown in the figure. For odd field (at the end of even field) reach the triggering level at time t_1 and for even field (at the end of odd field) reach the triggering level at time t_2.

Thus the timing instants are not the same for odd and even which upsets the interlacing of lines and causes lack of synchronizations and hence distortion.

2.14. EQUALISING PULSES

We have seen that the receiver deflection circuit separates the vertical field sync pulses from the line sync pulses and process integrate the vertical sync pulses to produce a composite single field pulse which is used to trigger and synchronization of vertical oscillators.

To have perfect interlaced these produced pulses should be in similar condition for both odd and even fields. Due to the unequal line period *i.e.* half line period before the odd field pulse and full line period before the even field pulse, the electrical conditions in the processing circuits are different.

2.14.1. Pre-Equalizing Pulses

This effect can be reduced by interval between the preceding line pulse and field sync pulse. This interval has been fixed at five half line period. The above problem is overcome by adding 5 narrow pulses, 2.3 μs each are inserted in this five half line period before the field sync pulses are known as **pre-equalising pulses** as shown in Fig. 2.31. Pre-equalising pulses being of 2.3 μs duration result in the discharge of the capacitors to essentially zero voltage in both the fields, besides the half line discrepancy before the voltage build up starts with the arrival of vertical sync pulses. This will equalize the discharging, as time available for discharge even after the end of odd field would be 32 μs only and not 64 μs.

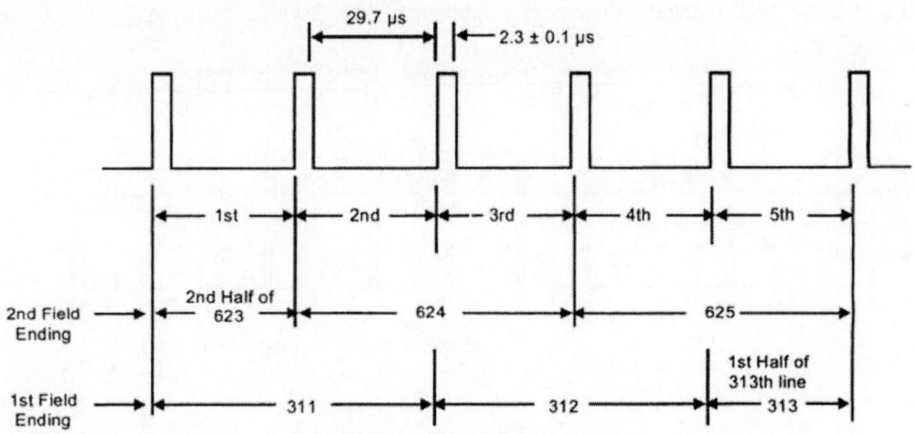

Fig. 2.31. Pre-equalizing pulses.

2.14.2. Post-Equalizing Pulses

These pulses are inserted in the back portion of the blanking pulse just after the vertical sync pulse to offset the half line effect of interlaced scanning as shown in Fig. 2.32. At the end of odd field, there is a gap of half line (32 μs) between the end of the vertical sync pulse and the next horizontal sync pulse.

Post-equalising pulses are necessary for a fast discharge of the capacitor to ensure triggering of the vertical oscillator at proper time. In the absence of post-equalising pulses, decay of voltage is slow across the capacitor and the oscillator may trigger at the trailing edge which may be far from the leading edge. This could cause an error in triggering.

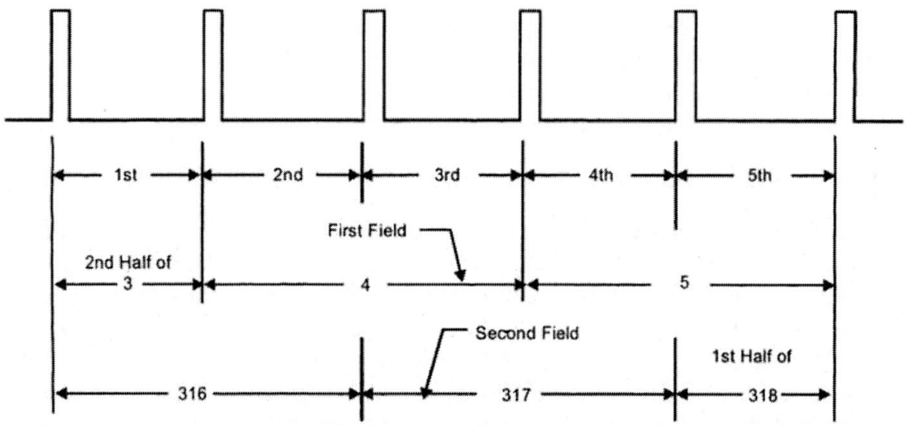

Fig. 2.32. Post-Equalising pulses.

With the insertion of narrow pulses as pre and post-equalising pulses the voltage rise or fall for both the field sequences as shown in Fig. 2.33.

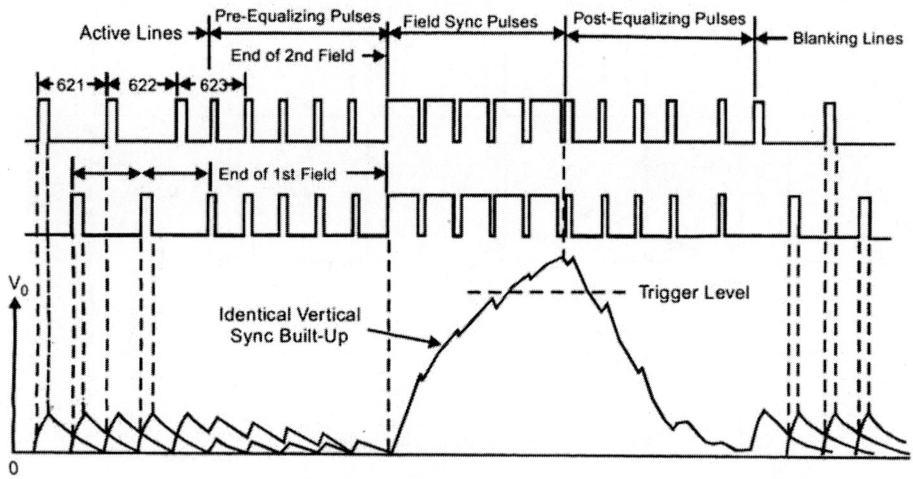

Fig. 2.33.

The complete pulse train for both the fields with the pre and post-equalizing pulses are shown in Fig. 2.34.

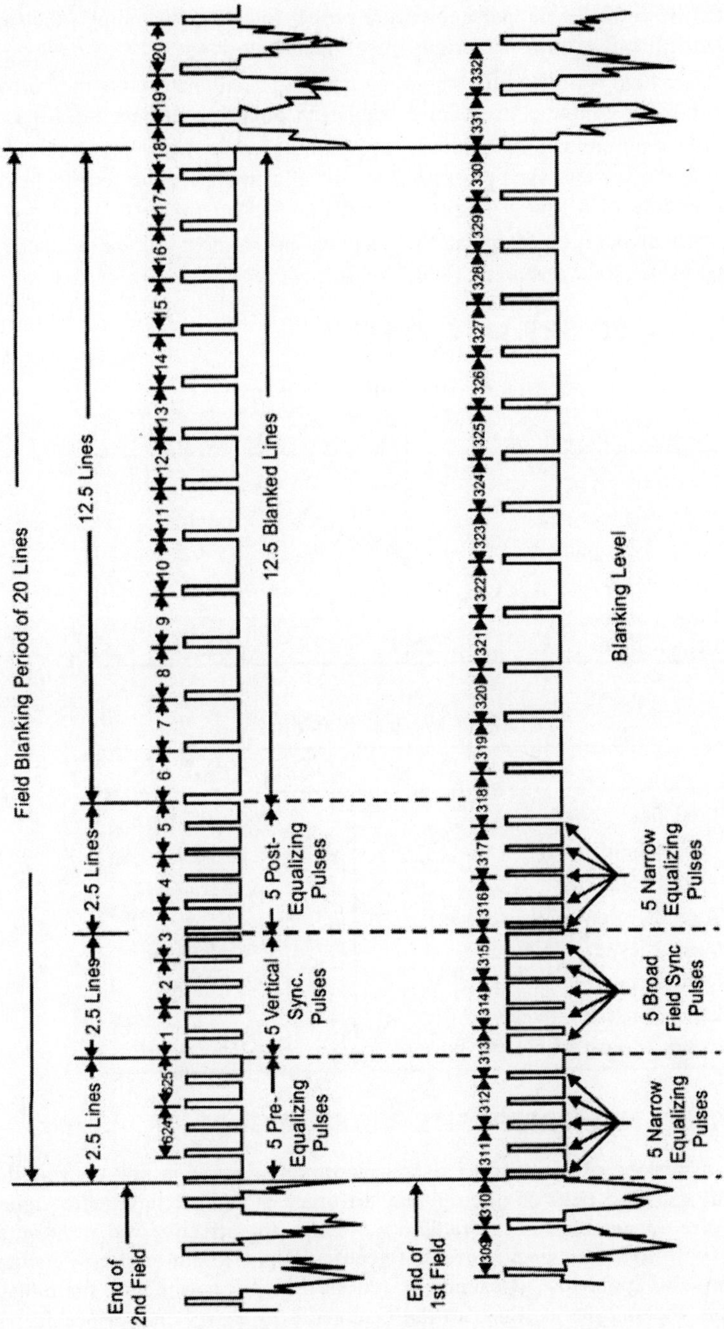

Fig. 2.34. The complete pulse train for both the fields with the equalising pulses.

2.14.3. Functions of Vertical Pulse Train

The vertical pulse has been serrated and pre and post equalilising pulse are inserted before and after the vertical sync pulse to ensure proper interlaced scanning. The various functions of vertical pulse train are briefly mentioned below:

1. A suitable field is derived for triggering the field oscillator so that the horizontal line oscillator continues to receive triggering pulses at correct intervals.
2. It becomes possible to have interlaced scanning without any interlace error by inserting the vertical sync pulses at the end of a line after the second field and at the middle of a line at the end of the first field.
3. The vertical sync build up at the receiver has almost the same shape and timing for the odd and even field.

2.15. DETAILS OF 525 LINE SYSTEM

Details of Horizontal blanking

Period	Time(us)
Field line (H)	63.5
Horz. Blanking	9.5 to 11.5
Horz. Sync pulse	4.75 $\pm$ 0.5
Front Porch	1.26
Back Porch	3.81
Visible line	52 to 54

Details of Vertical blanking

Period	Time
Total field period	1/60=16.7ms
Visible field line	15 to 16 ms
Vertical blanking	0.8 to 1.3ms
Total duration of six serrated Vertical sync pulses	3H= 190.5 us
Each serrated pulse	H/2=31.75us
Each equalizing pulse (6 Pre & Post equalizing pulses)	0.04H=2.54us

2.16. MODULATED COMPOSITE VIDEO SIGNAL

For the transmission of message to distant place, sound wave is first converted into electrical signal with the help of microphone. In order to radiate this audio signal in space, we require an antenna as the radiating system, for effective radiation antenna should be fed with high frequency currents (electrical signal). But as the frequency of audio signal remains quite low, it cannot be fed as such to the antenna for radiation. Therefore before feeding audio signal to the antenna, a process is performed in which the audio signal is superimposed on a high frequency wave, called **carrier wave**. The amplitude, or frequency or phase of the high frequency wave (carrier wave) is modified in accordance with the audio signal (called **modulating signal**), so that the resultant wave (called **modulated wave**) inherits the signal.

The process of superimposition of modulating signal (low frequency) over carrier wave (high frequency) is called **modulation.**

In telecommunications, *modulation* is the process of varying a periodic waveform, *i.e.* a tone, in order to use that signal to convey a message.

Modulation is the act of translating some low frequency or baseband signal (voice, music, data) to a higher frequency.

The aim of **analog modulation** is to transfer an analog lowpass signal, for example an audio signal or TV signal, over an analog bandpass channel, for example a limited radio frequency band or a cable TV network channel.

2.16.1. Types of Analog Modulation

(*i*) Amplitude modulation
(*ii*) Frequency modulation
(*iii*) Phase modulation.

In the expression (1) above there are two properties of the carrier that can be changed, the amplitude (A) and the angular position (argument of the cosine function).

Three cases may arise:

(*i*) When amplitude (A) is varied in accordance with the modulating wave, the process is known as **Amplitude modulation;**

(*ii*) When ω_c is varied in accordance with the modulating wave, while E_c and θ remain constant, the process is known as **Frequency modulation.**

(*iii*) When q is varied in accordance with modulating wave, while ω_c and E_c remain constant, the process is known as **Phase modulation.**

Thus we have amplitude modulation and angular modulation. Angular modulation can be further characterized as either frequency modulation or phase modulation.

A device that performs modulation is known as a **modulator** and a device that performs the inverse operation of modulation is known as a **demodulator** (sometimes detector or demod). A device that can do both operations is a modem (short for "MOdulate-DEModulate").

2.16.2. Amplitude Modulation

In amplitude modulation, the magnitude of carrier wave is varied in accordance with the amplitude and frequency of the modulating voltage.

Let the modulating signal voltage be represented as

$$e_m = E_m \cos \omega_m t, \qquad ...(2.2)$$

and unmodulated carrier signal may be written as

$$e_c = E_c \cos (\omega_c t + \theta) \qquad ...(2.3)$$

The carrier frequency ω_c is much greater than frequency of modulating signal or wave, ω_m.

The resulting modulated wave has the form

$$e = (E_c + K_a E_m \cos \omega_m t) \cos \omega_c t \qquad ...(2.4)$$

The amplitude factor $(E_c + K_a E_m \cos \omega_m t)$ respected the sinusoidal variation of amplitude of the wave. K_a is proportionality factor which determines the maximum variation in amplitude for a given single voltage E_m. Phase θ is taken zero because it plays no part in this process.

The above expression can be written as

$$e = E_c\left\{1+\left(\frac{K_a E_m}{E_c}\right)\cos\omega_m t\right\}\cos\omega_c t$$

$$= E_c\{1+m_a\cos\omega_m t\}\cos\omega_c t$$

$$e = E_c + \cos\omega_c t + \left(\frac{m_a E_c}{2}\right)\cos(\omega_c+\omega_m)t + \left(\frac{m_a E_c}{2}\right)\cos(\omega_c-\omega_m)t \quad ...(2.5)$$

where,

$$\boxed{m_a = \frac{K_a E_m}{E_c}} \quad ...(2.6)$$

and is called **modulation index**. It is the depth of the modulation or in other words extent to which carrier is modulated. The modulation index when multiplied by 100 gives percentage of modulation.

1. Sides bands : In amplitude modulation, we have three components shown in equation (2.6):

(*i*) $E_c \cos\omega_c t$; represent original carrier voltage.

(*ii*) $(m_a E_c/2)\cos(\omega_c+\omega_m)t$; represent upper sideband.

(*iii*) $(m_a E_c/2)\cos(\omega_c-\omega_m)t$; represent lower sideband

The lower side band term with angular frequency $(\omega_c - \omega_m)$ and upper side band term with angular frequency $(\omega_c + \omega_m)$ are located on either side of the carrier at a frequency interval of ω_m in there frequency spectrum as shown in Fig. 2.35.

The magnitude of the amplitude of both the upper and lower side band is $(m_a E_c/2)$. If m_a be unity, then each side band term is half the carrier voltage in amplitude.

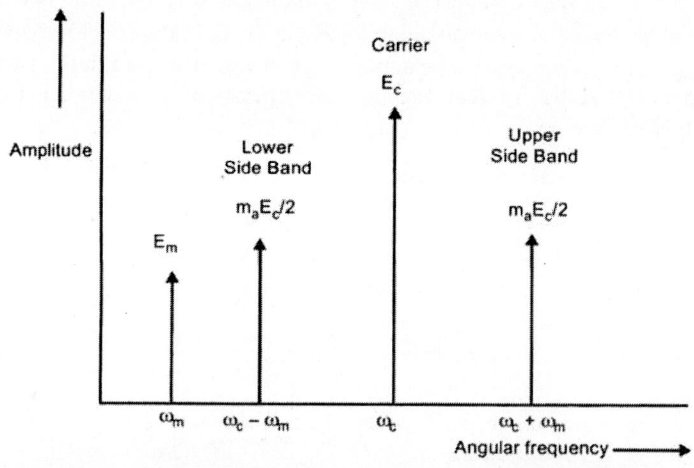

Fig. 2.35. Frequency spectrum of Amplitude Modulated wave.

The carrier voltage component transmits no information, while each of the two side bands carries complete information.

(2) Waveforms : Figure 2.36 shows the waveforms of modulating signal and that of a modulated wave called the modulation envelope. It is clear from Fig. 2.36(*b*) that

$$m_a E_c = E_{max} - E_c$$

$$\text{or } m_a = E_{max} - \frac{E_c}{E_c} \quad \text{...(2.7)}$$

Also $\quad m_a E_c = E_c - E_{min}$

or $\quad m_a = E_c - E_{min}/E_c \quad$ ...(2.8)

Equating both Eqns. (2.7) and (2.8),

$$E_{max} - E_c = E_c - E_{min}$$

or $\quad 2 E_c = E_{max} + E_{min}$

Putting this value of E_c in Eqn. (2.7), we get

$$\boxed{m_a = \left\{E_{max} - \frac{E_{min}}{E_{max}} + E_{min}\right\}} \quad \text{...(2.9)}$$

And therefore percentage of modulation is given by

$$\boxed{m_a = \left\{E_{max} - \frac{E_{min}}{E_{max}} + E_{min}\right\} \times 100} \quad \text{...(2.10)}$$

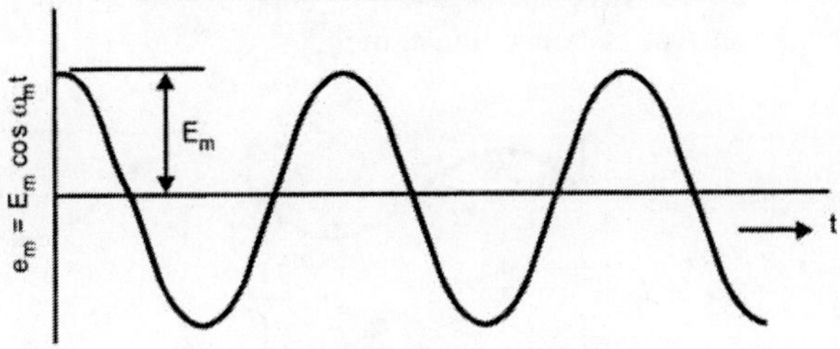

(*a*) Modulating voltage.

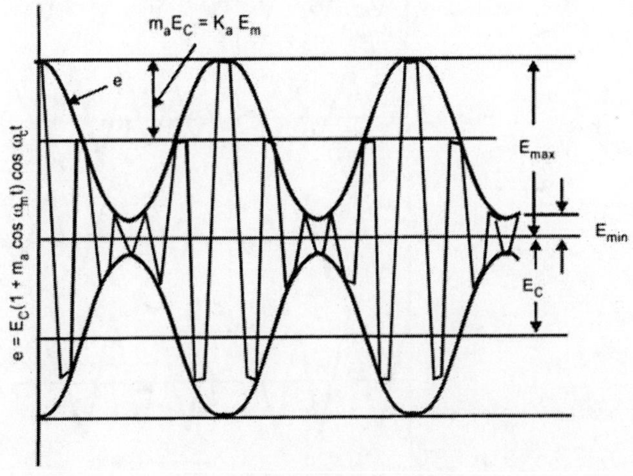

(*b*) Amplitude modulated wave

Fig. 2.36.

(3) Modulated waves with various degrees of modulation or modulation index: Modulated waves expressed as

$$e = E_c \{1 + m_a \cos \omega_m t\} \cos \omega_c t \qquad ...(2.11)$$

where, $m_a = \dfrac{K_a E_m}{E_c}$

is modulation factor or modulation index, and its value range from zero to one. We shall now take the following different cases:

(i) **When $m_a = 0$:** which mean no amplitude modulation is present and the expression for e reduces to $e = E_c \cos \omega_c t$ representing **pure carrier.**

(ii) **When $m_a < 1$, for example $m_a = 0.5$:** which means percentage of modulation is 50%. In other words, the extent to which carrier is modulated is equal to its amplitude. It is the case of **under modulation.**

(iii) **When $m_a = 1$:** which implies 100% modulation. In other words, the extent to which the carrier is modulated is equal to its amplitude. This means that at the negative peaks of the modulating signal, carrier amplitude is reduced to zero. It is also known as **Perfect modulation** or **unity modulation.**

(iv) **When $m_a > 1$:** represents the case of over modulation. For such a case expressing for e is no longer valid. At the negative peaks of modulating signals, no modulated wave exists and thus such a case results in a distortion of the intelligence that is being transmitted.

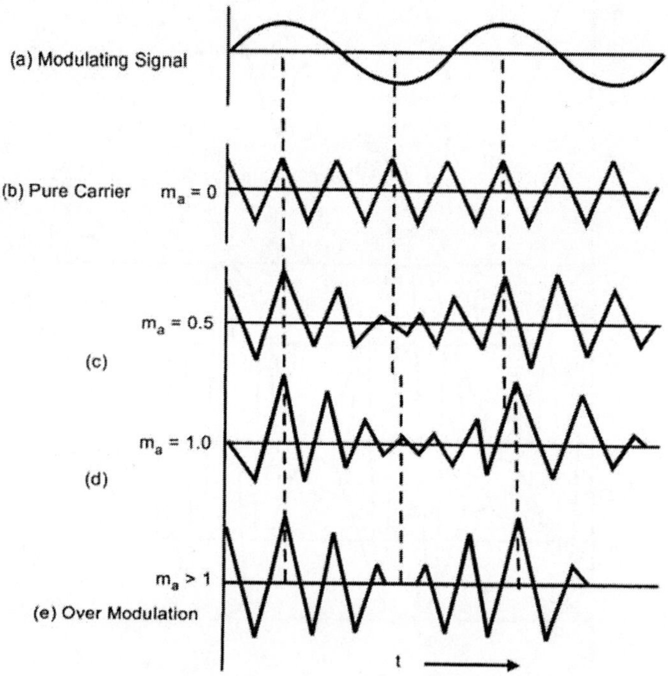

Fig. 2.37. Amplitude Modulated waves with the different modulation index.

(4) Forms of AM :

(*i*) In its basic form, amplitude modulation produces a signal with power concentrated at the carrier frequency and in two adjacent sidebands. Each sideband is equal in bandwidth to that of the modulating signal and is a mirror image of the other. Thus, most of the power output by an AM transmitter is effectively wasted: half the power is concentrated at the carrier frequency, which carries no useful information (beyond the fact that a signal is present); the remaining power is split between two identical sidebands, only one of which is needed.

(*ii*) To increase transmitter efficiency, the carrier can be removed (suppressed) from the AM signal. This produces a double-sideband suppressed carrier (DSSC) signal. If the carrier is only partially suppressed, a double-sideband reduced carrier (DSRC) signal results. DSSC and DSRC signals need their carrier to be regenerated (by a beat frequency oscillator, for instance) to be demodulated using conventional techniques.

(*iii*) Even greater efficiency is achieved—at the expense of increased transmitter and receiver complexity—by completely suppressing both the carrier and one of the sidebands. This is single-sideband modulation, widely used in amateur radio due to its efficient use of both power and bandwidth.

(5) Energy distribution among the three components of amplitude modulated wave : The voltage components contained in the wave, represents by Eqn. (2.5) have r.m.s. values equal to

$$\frac{E_c}{\sqrt{2}}, \frac{m_a E_c}{2\sqrt{2}} \text{ and } \frac{m_a E_c}{2\sqrt{2}} \text{ respectively.}$$

Therefore when such a modulated wave, having these three components, is impressed upon some resistance, R, then the total average power dissipated would be equal to

$$P_{avg} = \frac{E_c^2}{2R} + \frac{m_a^2 E_c^2}{8R} + \frac{m_a^2 E_c^2}{8R}$$

$$= \frac{E_c^2}{2R}\left\{1 + \frac{m_a^2}{2}\right\} \quad ...(2.12)$$

Where $\frac{E_c^2}{2R}$ represents the power dissipated by the carrier components alone.

Obviously,

$$P_{avg} > \frac{E_c^2}{2R} > P_c$$

That is, power content of modulated wave exceeds that of the corresponding unmodulated carrier. This additional power, supplied by modulating source, in modulated wave furnishes the energy contained in side bands of the modulated wave. It is the useful power that carries intelligence and its relative magnitude is proportional to m_a^2. The ratio of useful power contained in side bands to the total power contained in the modulated wave is

$$\frac{P_{\text{side bands}}}{P_{\text{avg}}} = \frac{\left\{\left(\frac{E_c^2}{2R}\right).\left(\frac{m_a^2}{2}\right)\right\}}{\left\{\left(\frac{E_c^2}{2R}\right).\left(1+\frac{m_a^2}{2}\right)\right\}}$$

$$\frac{P_{\text{side bands}}}{P_{\text{avg}}} = \frac{\left(\frac{m_a^2}{2}\right)}{\left(1+\frac{m_a^2}{2}\right)} \quad \text{...(2.13)}$$

Suppose modulation is 100% then $m_a = 1$. For such a case,

$$\frac{P_{\text{side bands}}}{P_{\text{avg}}} = \frac{\left(\frac{1}{2}\right)}{\left(1+\frac{1}{2}\right)} = \frac{1}{3}$$

That is, only one-third of the total power of modulated wave is contained in the two side bands. The rest two-thirds power resides in carrier component which is of no use. If in the modulated wave, carrier be suppressed then modulator would draw less power from supply source and the process of modulation will be quite economical.

2.16.3. RF Transmission of CVS by Modulation of Picture Carrier

- For radio transmission of the television signal and for some special applications, an RF carrier is modulated with the composite video signal.
- The advantage of amplitude modulation is the narrower bandwidth of the modulation product. With conventional AM the modulating CVS of BW = 5 MHz requires an RF transmission bandwidth of BWRF =10 MHz (Fig, 2.38(*a*)).
- In principle, one sideband could be suppressed since the two sidebands have the same signal content. This would lead to single sideband amplitude modulation (SSB AM) (Fig, 2.38(*b*)).

2.16.4. Vestigial Sideband Modulation

Under some circumstances it is desirable to obtain a wave consisting of a carrier plus one sideband, with the other sideband suppressed. Such a result can be approximated by transmitting an ordinary amplitude modulated wave through a filter that is so designed that the edge of the transmission band of the filter accurately coincides with the carrier frequency as shown in Fig. 2.38(*c*)

The resulting wave frequency spectrum shown in Fig. 2.38(*c*), which is characterized by a residual second sideband. This is termed a **vestigial or asymmetric sideband system**.

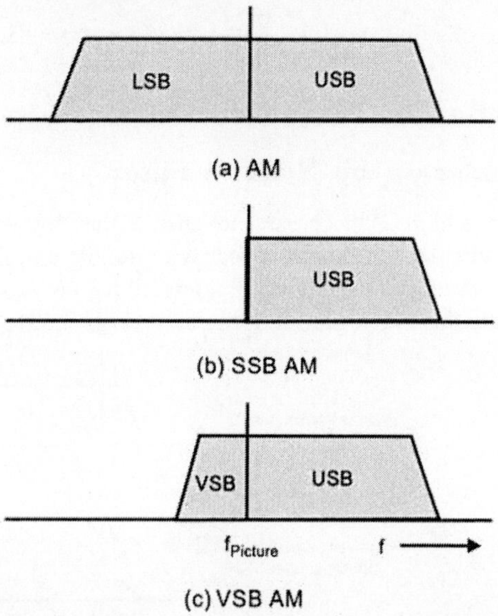

Fig. 2.38. (a) amplitude modulation, carrier with two sidebands; (b) single sideband amplitude modulation; (c) vestigial sideband amplitude modulation.

This type of modulation has the advantage of requiring less total bandwidth than does the amplitude-modulated wave with its two full sidebands. It is the standard type of signal used in television transmitters.

Due to the fact that the modulation signals reach very low frequencies, sharp cutoff filters are required; however, the group-delay distortion introduced by these filters at the limits of the pass band causes certain difficulties.

The problem is eluded by using vestigial sideband amplitude modulation (VSB AM) instead of SSB AM. In this case, one complete sideband (USB) and part (0.75 MHz of LSB) of the other are transmitted.

2.16.5. Transmission Efficiency

$$P_t = P_c + P_{usb} + P_{lsb}$$

$$= \frac{E_c^2}{2R} + \frac{m^2 E_c^2}{8R} + \frac{m^2 E_c^2}{8R}$$

$$= P_c + \frac{m^2}{4} . P_c + \frac{m^2}{4} . P_c$$

$$= P_c \left(\frac{m^2}{2} \right)$$

$$P_{t\,max} = 1.5\ P_c \quad (m = 1)$$

In SSB when $m = 1$

$$P_{SB} = \frac{P_c m^2}{4} = 0.25 \cdot P_c$$

$$\text{Saving} = 1.5 - \frac{0.25}{1.5} = 0.833 \times 100 = 83.3\%$$

2.16.6. Correction of Frequency Response in VSB Transmission by Nyquist Filter

At the receiver end it is necessary to ensure that the signal frequencies in the region of the vestigial sideband do not appear with double amplitude after demodulation. This is obtained by the Nyquist slope, the selectivity curve of the receiver rising or falling linearly about the video carrier frequency (Fig. 2.39)

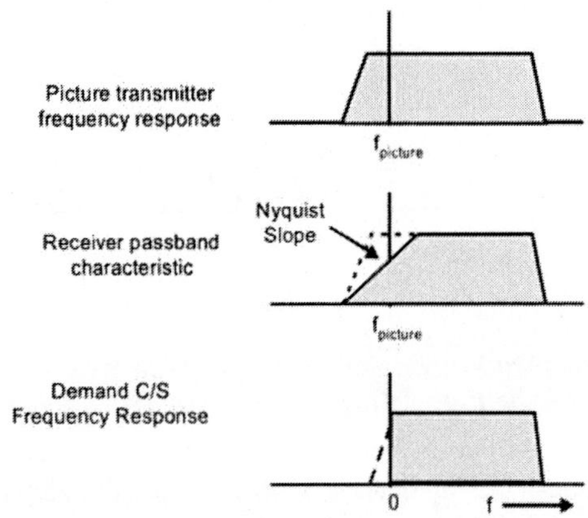

Fig. 2.39.

2.16.7. Channel BW in AM

In the 625 line system, where the frequency components present in the video signal extend from 0 Hz to 5 MHz, a double sideband AM transmission would occupy a total BW of 10 MHz.

An attenuation slope of 0.5 MHz is provided at each edge of the two sidebands, this adds 1 MHz to the required band space.

Guard frequency of 0.25MHz is also to be catered for calculating the total BW requirement.

$$\text{Total BW} = 10\text{MHz} + (0.5 \text{ MHz} \times 2) + 0.25\text{MHz} = 11.25 \text{ MHz}$$

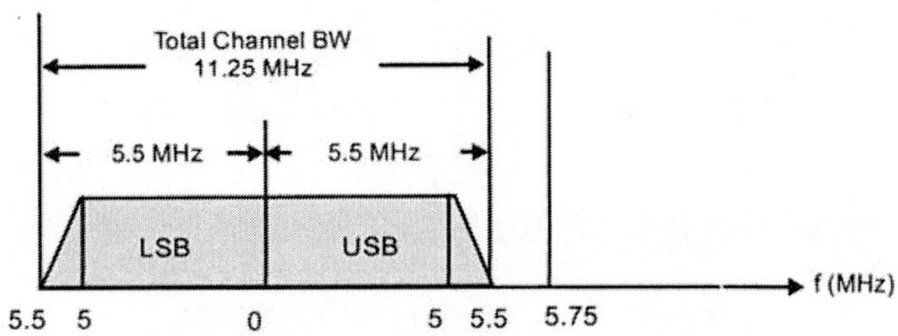

Fig. 2.40. Double Sideband AM transmission.

In accordance with CCIR, 7 MHz bands are available in the VHF range and 8 MHz bands in the UHF range for TV broadcasting. The picture transmitter frequency response and the receiver passband characteristic are also determined by CCIR standards (Fig. 2.41(*a*) and (*b*)). In most cases, both modulation and demodulation take place at the IF, the vision IF being 38.9 MHz and the sound IF 33.4 MHz.

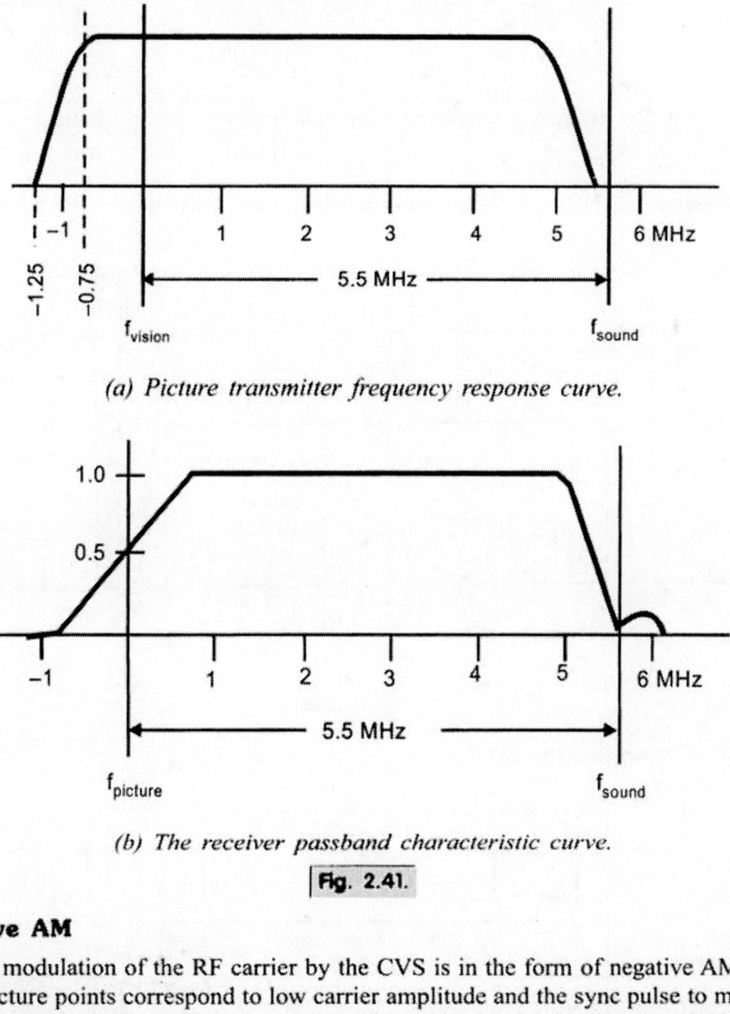

(a) Picture transmitter frequency response curve.

(b) The receiver passband characteristic curve.

Fig. 2.41.

Negative AM

The modulation of the RF carrier by the CVS is in the form of negative AM, where bright picture points correspond to low carrier amplitude and the sync pulse to maximum carrier amplitude (Fig. 2.42). A residual carrier (white level) of 10% is required because of the intercarrier sound method used in the receiver. One advantage of negative modulation is optimum utilization of the transmitter, since maximum power is necessary only briefly for the duration of the sync peaks and at the maximum amplitude occurring periodically during the sync pulses to serve as a reference for automatic gain control in the receiver.

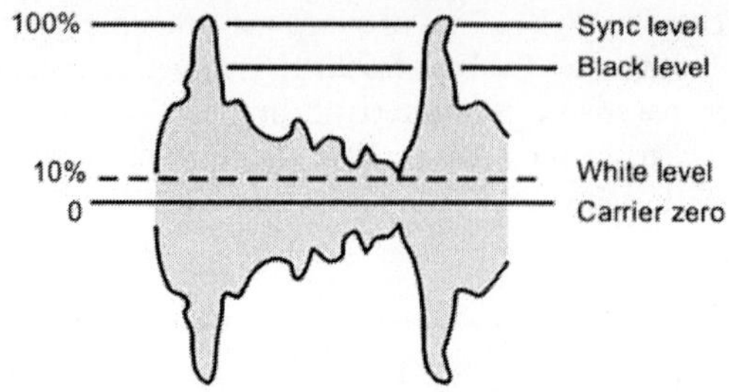

Fig. 2.42. **Negative Amplitude Modulation of RF Picture Carrier by CVS.**

TV Channel Sideband Spectrum Indian and European System

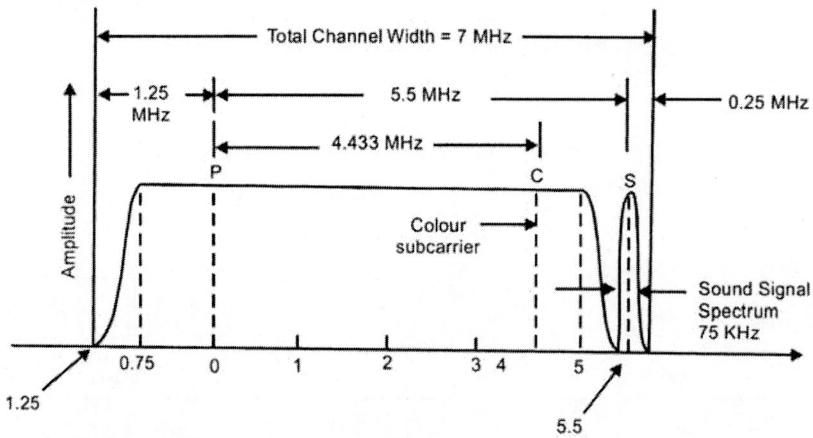

Fig. 2.43.

UK TV Channel Standard

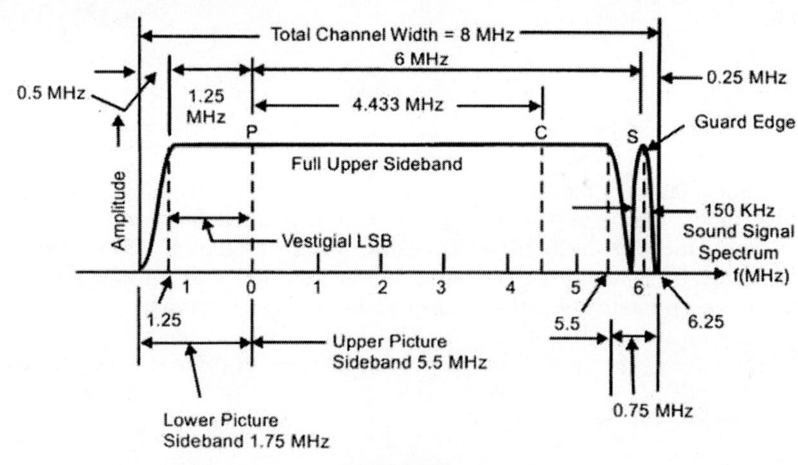

Fig. 2.44.

American TV Channel Standard

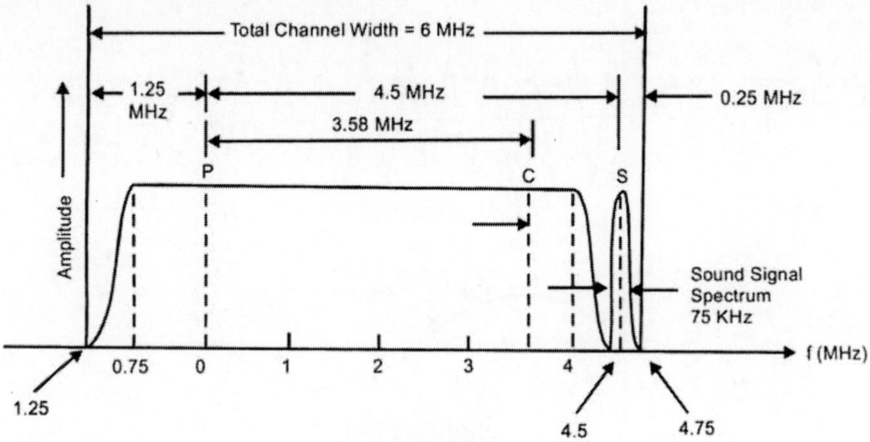

Fig. 2.45.

Sideband Spectrum of Two Adjacent Channels

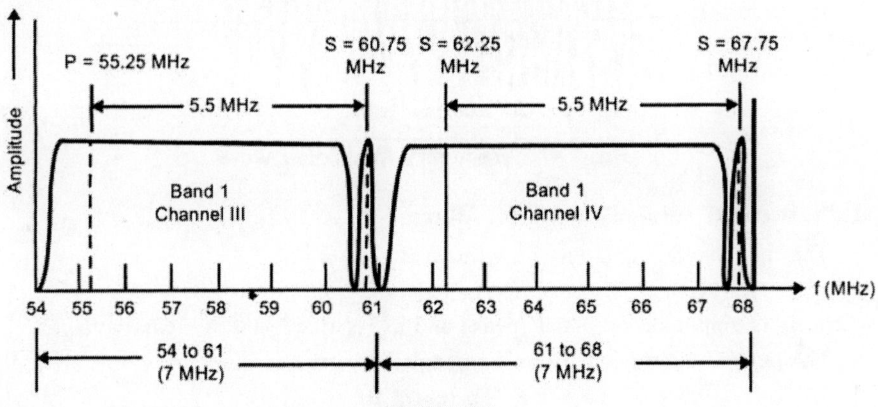

Fig. 2.46.

2.16.8. Frequency Modulation

Definition : Frequency modulation is produced when the instantaneous frequency of carrier is varied in accordance with the modulating signal, while the amplitude of the carrier remains constant.

Characteristics of FM Wave

The amount of deviation in frequency is proportional to the amplitude of the signal, while the rate is determined by the frequency of modulation source.

For example, suppose 100 MHz, carrier is being modulated by a sinusoidal signal and for given modulation system there is 2 kHz, frequency shift per volt of signal magnitude. Then, if the modulating signal has an amplitude of 20 volt and frequency of 20 kHz, the instantaneous frequency of modulated wave will vary between 99.96 MHz, and 100-04 MHz at a rate of 2000 times per second.

Figure 2.47 illustrates the form of a frequency modulated wave when a sinusoidal modulating voltage is employed.

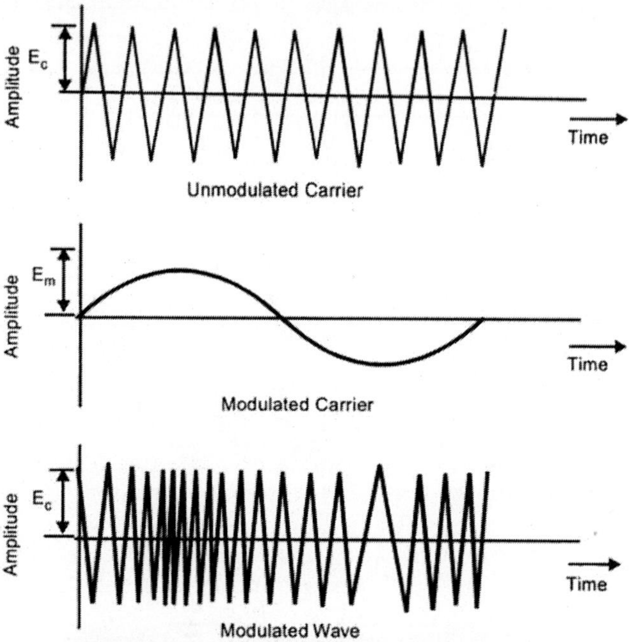

Fig. 2.47. Frequency Modulated wave.

Mathematical Analysis of F.M. Wave

The modulating signal can be written as

$$e_m = E_m \cos \omega_m t \qquad \text{...(2.14)}$$

where E_m is amplitude of signal (peak) and ω_m is frequency of signal voltage.

The general expression for an unmodulated carrier is given by

$$e_c = E_c \sin (\omega_c t + \theta),$$

where quantity $(\omega_c t + \theta) = f(t)$ ...(2.15)

is total instantaneous 'phase' of function.

So, $$e_c = E_c \sin f(t) \qquad \text{...(2.16)}$$

The instantaneous angular frequency is

$$\omega_i = \frac{df(t)}{dt} = \frac{d}{dt}(\omega_c t + \theta) = \omega_c,$$

By definition, the instantaneous angular frequency of frequency modulated wave can be written as

$$\omega_i = \omega_c + k_f E_m \cos \omega_m t, \qquad \text{...(2.17)}$$

Where proportionality factor K_f determines the maximum variation in frequency for a given signal strength E_m. Since

$$w_i = \frac{df(t)}{dt},$$

We have $$\phi(t) = \int_0^t \omega_i \, dt = \int_0^t (\omega_c + k_f E_m \cos \omega_m t) \, dt$$

$$= \left(\omega_c t + k_f \left(\frac{E_m}{\omega_m} \right) \right) \sin \omega_m t + \theta_0 . \qquad ...(2.18)$$

The initial phase θ_0 is neglected as it plays no part in modulation process. Thus FM wave can be written as

$$e = E_c \sin \left\{ \omega_c t + k_f \left(\frac{E_m}{\omega_m} \right) \sin \omega_m t \right\} \qquad ...(2.19)$$

The instantaneous frequency of FM wave is

$$f = \frac{\omega}{2\pi} = f_c + k_f \left(\frac{E_m}{2\pi} \right) \cos \omega_m t$$

Therefore, $$f_{max} = f_c + k_f \left(\frac{E_m}{2\pi} \right) \text{ and } f_{min} = f_c - k_f \left(\frac{E_m}{2\pi} \right)$$

So, **frequency deviation** (It is the maximum shift of the frequency from the carrier frequency both the side.)

$$\boxed{\Delta f = f_{max} - f_c = f_c - f_{min} = k_f \left(\frac{E_m}{2\pi} \right)} \qquad ...(2.20)$$

The instantaneous frequency can then be expressed as

$$f = f_c + \Delta f \cos \omega_m t$$

The modulation index

$$m_f = \text{Frequency deviation / modulating frequency}$$

$$\boxed{m_f = \frac{\Delta f}{f_m} = k_f \left(\frac{E_m}{\omega_m} \right)} \qquad ...(2.21)$$

Therefore the expression for FM wave becomes

$$e = E_c \sin \{ \omega_c t + m_f \sin \omega_m t \} \qquad ...(2.22)$$

The difference between maximum modulated frequency and minimum modulated frequency is known as **carrier swing (CS).**

$$CS = f_{max} - f_{min} \qquad ...(2.23)$$

or $$CS = 2 \, \Delta f$$

$$\boxed{\Delta f = \frac{CS}{2}} \qquad ...(2.24)$$

In FM broadcasting system, Δf has a maximum permissible value of ±75 kHz and f_m may vary from about 33 Hz to 15 kHz. Consequently, m_f can have values both greater than unity or less than unity whereas in amplitude modulation m_a should not exceed unity for distortion less transmission. Variation in transmitted frequency is shown in Fig. 2.48.

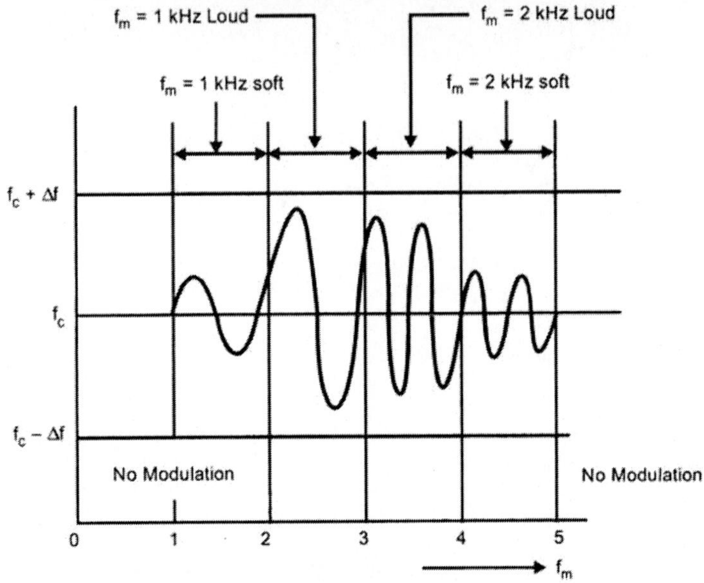

Fig. 2.48.

2.16.9. Why AM is preferred Over FM for Video Signals ?

Due to the following reasons, amplitude modulation is more suitable for video signals :

(i) AM detector detects the changes in amplitude only. It does not detect random changes in phase. So the output of the detector is free of phase noise. Human eyes are sensitive to phase noise, its absence in the output makes AM suitable for picture. Whereas, FM can detect changes in phase and hence not suitable for video signal.

(ii) Detection of AM signal is very easy in AM, whereas FM detection is more complicated.

(iii) Since in AM, baseband signal is accomodated in two sidebands only and hence bandwidth requirement is low. Maximum bandwidth for video signal could be ±5 MHz in AM and further reduced by using AM vestigial sideband. FM system requires about ten times the width of baseband for good quality video output. The high bandwidth would have consumed most of the band in a single channel only. Due to the above advantages of AM over FM, we prefer AM. The type of AM for video signal used in AMVSB (Amplitude Modulation Vestigial sideband).

2.16.10. Why FM is Preferred Over AM for Audio Signals ?

FM is used for transmission of audio signals due to following reasons :

(*i*) Signal to noise ratio is very high in FM systems. Random variations in amplitude are suppressed in FM.

(*ii*) Random variation in phase can be detected, but human ear, are not sensitive to phase changes and therefore this type of noise present in the audio output of the receiving system is harmless.

(*iii*) RF carrier can be modulated at low level when does not require high audio power.

(*iv*) In FM, the mainly information are being carried by sidebands. Power in the carrier is negligibly small. Hence the system, is more efficient and we use only 10 to 20% of power for audio transmission as compared to video transmission.

(*v*) Due to the capture effect in FM system weak interfering stations are fully suppressed.

SUMMARY

- The primary factors which are required for successful transmission and reception of the picture are as follows:
  (*i*) Gross structure
  (*ii*) Image continuity
  (*iii*) Scanning lines
  (*iv*) Interlaced scanning
  (*v*) Fine structure
  (*vi*) Tonal gradation
- In Television system the frame adopted is rectangular in shape.Width to Height ratio of a picture frame is called **Aspect Ratio** and it is 4/3.
- In the scanning process the electron beam start scanning target plate from the top left to the end of bottom right of the target plate. The scanning is done line by line horizontally from left to right at a fast rate and vertically from top to bottom at a slower rate and it is known as Trace. The retrace of the beam is very fast as compared to the forward scan. Scanning during retrace or flyback is blanked by making use of horizontal and vertical blanking pulses.
- Scanning may be defined as the process which converts optical signals in to electrical signals, referred as video signals.
- Horizontal deflection is fast. It takes 64 µs *(trace period is 52 µs +retrace period is 12 µs)* for one scanning line.
- The sawtooth wave in vertical deflection coils moves the electron beam from top to bottom of the raster at waveform speed.
- Vertical deflection is slow and takes 20000 µs or *20 ms (trace period is 18720 µs + retrace period 1280 µs)* in deflecting the beam from top to bottom.
- The quality of the reproduced picture is determined by the resolution, which is the better the higher the number of lines. The number of lines in a frame is restricted by the resolving capability of human eye.
- $$N_v = \frac{1}{\alpha}\rho$$
- $$N_v = \frac{1}{\left[\left(\frac{1}{60}\times\frac{\pi}{180}\right)\times 4\right]} \approx 860 \text{ lines}$$
- The distinction between two pixels will not be reproduced. It is called **Kell Factor (*k*)** and it's approximately value is 0.7. Thus the kell effect produces

degradation in the resolution of the individual pixels. About 25 % to 35% pixels are lost due to Kell effect.

- Smaller the size of the beam spot, less will be kell factor and higher the effective numbers of lines.
- $$N_E = N_v \times k$$
  $$N_E = 860 \times 0.7 = 602$$
- Channel bandwidth increases with increase in number of lines.
- In India, the total numbers of lines have been chosen is 625 on the basis of quality and cost. In the America 525 line system is used in which the total numbers of the lines has been selected is 525.
- Considering the linkage to the AC supply frequency, a picture repetition frequency (f) of 25 Hz for an AC supply of 50 Hz has been selected (30 Hz for a 60 Hz AC supply in North America and Japan).
- **Types of Scanning :**
  (*i*) Progressive Scanning
  (*ii*) Interlaced Scanning
- When one cycle of vertical saw tooth current completes scanning of one frame and the next cycle starts scanning of a new frame, it is called **progressive scanning**.
- Time for persistence of vision is more for darkness than for bright light. This result in a phenomenon called **flicker.** The relationship between flicker and brightness is given by Ferry Poster law,
  $$F_c = F + 12.6 \log_{10} B$$
- The process of interlaced scanning is designed to eliminate flicker in the displayed image.
- In Interlaced scanning, there are two sequences of scanning for each frame. Scanning line of one sequence occur in between the lines of the other sequence.
- The total numbers of lines are divided in two groups called Fields. Each field is scanned alternatively. In 625 line system each frame are divided in two sets of 312.5 lines.
- Thus 50 fields of 312½ lines each are transmitted instead of 25 pictures of 625 lines, the field repetition or vertical frequency being $f_v = 50$ Hz. The resulting line or horizontal frequency is
  $$f_h = 25 \times 625 = 50 \times 312\tfrac{1}{2} = 15625 \text{ Hz.}$$
- **Scanning Period :**
  - **Horizontal line:**
    Nominal duration- 64 μs (10^6 /15625)
    Active line (Trace) period - 52 μs
    Line blanking (Retrace) period- 12 μs
  - **Vertical period** (field frequency 50Hz):
    Nominal duration- 20 ms (1/50)
    Active Line (Trace) period- 18.720 ms
    Line blanking (Retrace) period- 1.280 ms
  - Active number of lines (Na) = 625 – 40 = 585 lines.

- Resolution means ability to distinguish between closely spaced pixels.
- The extent of which the scanning system is capable of resolving picture details in the vertical direction is referred to as **vertical resolution.**

$$V_r = Na \times k$$

$$V_r = 585 \times 0.69 = 400 \text{ lines}$$

- The capability of the system to resolve maximum no. of picture elements along the scanning lines determined **horizontal resolution.**

$$H_R = V_r \times A$$

In case 625 lines, $Na = 585$ lines, $k = 0.69$ and $A = \frac{4}{3}$

$$H_R = V_r \times A = 585 \times 0.69 \times \frac{4}{3} = 533 \text{ lines}$$

- **Bandwidth** is the highest video frequency related to the time taken in scanning two adjacent pixels.
- Bandwidth $= \frac{533}{2} \times 2 \times 10^{-6}$ Hz

$= 5.125$ MHz or ~ 5MHz (Highest modulating frequency)

In case 525 lines system it is 4 MHz.

F_{max} = Aspect ratio (AR) × Resolution factor (k) × Number of active lines (Na)/2.

- The bandwidth depends upon the following factors:
  - Number of active lines
  - Aspect ratio
  - Resolution factor or kell factor and
  - Horizontal line trace period
- Variation in brightness from pixels to pixels is called **tonal gradation**.
- **Contrast :** This is the difference in intensity between black and white parts of the picture over and above the brightness level.
- **Contrast Ratio :** The ratio of maximum to minimum relative to the original picture is called contrast ratio.
- Composite video signal consists of
  (*i*) A camera signal corresponding to the desired picture information,
  (*ii*) Blanking pulses to make the retrace invisible and
  (*iii*) Sync pulses to synchronize the Tx & Rx scanning.
- In the case of Positive polarity, whiter the scene, higher is the amplitude of the video signal.
- In case of negative polarity, brighter the scene, smaller is the amplitude.
- In most of the TV systems, negative polarity is used to modulate the video carrier.
- **Video Signal Dimensions :** The Voltage levels for sync pulse, blanking pulse and video are different, so that these pulses can be separated easily in the receiver. Voltage division multiplexing is used to keep these pulses and signal at different levels in CVS. There are two ways of indicating the levels:

(*a*) Percentage of carrier (for modulated signals).

(*b*) Institute of Radio Engineers (IRE) scale (for video base band signal).

- The overall arrangement of combining the picture signal and sync pulses may be considered as a kind of voltage division multiplexing.
- For negative polarity sync top is at 100%, blanking level is at 75%, white level is at 10% and whiter than white is below 10%.
  - Peak white level: 10 to 12.5 %
  - Blanking level: 75 % (also known as Pedestal level)
  - Sync pulses: 75 to 100%
  - Black level: 72 %
  - Pedestal: The difference between blanking level and black level.
  - Blacker than black: 72 to 100 %
  - Whiter than white: signal below 10 % to 0 %
- **IRE (Institute of Radio Engineers) Scale**
  - In IRE scale, blanking level is at 0 IRE ,sync top is -40 IRE, standard white is at about +86 IRE and peak white is at +100 IRE.
  - Total swing of the signal is 140 IRE.
- The composite video signal contains blanking pulses to make the retrace lines invisible by raising the signal amplitude slightly above the black level (75 %) during the time the scanning circuit produces retraces.
- The total width of H- blanking pulse is 12 μs during which H-retrace is complete and the video signals remain cut-off. It consists of three parts.

  (*i*) Front·porch (1.5 μs)

  (*ii*) H-sync pulse (4.7 μs)

  (*iii*) Back porch (5.8 μs)
- *The width of the front porch in terms of time interval is 1.5 μs or 0.025 H.*
- Line sync pulse is used to trigger the scanning circuits to ensure synchronization for scanning at the transmitter and at the receiver. It must starts at the same instant of time in the transmitter and in receiver for correct synchronization. *Horizontal or Line sync pulse is 4.7 μs or 0.075 H wide.*
- Back porch is 5.8 μs wide.
- Vertical blanking pulse carries Vertical sync pulse which triggers the vertical sweep oscillator for synchronization.
- The period of *vertical sync pulse should be 2.5 to 3 horizontal line period.*
- The detail of vertical blanking pulse is as below:
  - Front portion = 160 μs (2.5 H)
  - V-sync pulse = 160 μs (2.5 H)
  - Back portion = 960 μs (15 H)
  - Total vertical blanking pulse period = 1280 μs
- The vertical sync pulse is of 160 μs or 2.5 H.
- Because of long duration of vertical sync pulse (160 μs), horizontal synchronization cannot occur for this long period as the non availability of rising edges.

- Serrations, each 4.7 μs wide, into five half line narrower pulses, each 27.3 μs (32- 4.7 μs) wide.
- Serration allows synchronization of H-sweep periodically, so that when receiver comes out of the v-blanking pulse, it finds itself in correct synchronization.
- The horizontal and vertical sync pulses are extracted from the composite video signal.
- The vertical sync pulse information is extracted from the sync pulse train passing through the low pass filter (*i.e.* Integrated circuit) and the horizontal sync pulse information is extracted from the sync pulse train passing through the high pass filter (*i.e.* Differentiated circuit).
- 5 narrow pulses, 2.3 μs each are inserted in this five half line period before the field sync pulses are known as pre-equalising pulses.
- Post-Equalizing Pulses are inserted in the back portion of the blanking pulse just after the vertical sync pulse to offset the half line effect of interlaced scanning.
- The vertical pulse has been serrated and pre and post equalilising pulse are inserted before and after the vertical sync pulse to ensure proper interlaced scanning.
- The process of superimposition of modulating signal (low frequency) over carrier wave (high frequency) is called **modulation.**
- The aim of analog modulation is to transfer an analog lowpass signal, for example an audio signal or TV signal, over an analog bandpass channel, for example a limited radio frequency band or a cable TV network channel.
- **Types of Analog Modulation :**
  (*i*) When amplitude (A) is varied in accordance with the modulating wave, the process is known as **Amplitude modulation;**
  (*ii*) When ω_c is varied in accordance with the modulating wave, while E_c and θ remain constant, the process is known as **Frequency modulation.**
  (*iii*) When q is varied in accordance with modulating wave, while ω_c and E_c remain constant, the process is known as **Phase modulation.**
- A device that performs modulation is known as a modulator and a device that performs the inverse operation of modulation is known as a demodulator (sometimes detector or demod).
- $$e = E_c + \cos\omega_c t + \left(\frac{m_a E_c}{2}\right)\cos(\omega_c + \omega_m)t + \left(\frac{m_a E_c}{2}\right)\cos(\omega_c - \omega_m)t$$

  where, $$m_a = \frac{K_a E_m}{E_c}$$
- **Modulation index.** It is the depth of the modulation or in other words extent to which carrier is modulated. The modulation index when multiplied by 100 gives percentage of modulation.

  In amplitude modulation, we have three components :

  (*i*) $E_c \cos\omega_c t$; represent original carrier voltage.
  (*ii*) $(m_a E_c/2)\cos(\omega_c + \omega_m)t$; represent upper sideband.
  (*iii*) $(m_a E_c/2)\cos(\omega_c - \omega_m)t$; represent lower sideband

- When $m_a = 0$ representing **pure carrier.**
- When $m_a < 1$, for example $m_a = 0.5$ case of under modulation.
- When $m_a = 1$ Perfect modulation.
- When $m_a > 1$ represents over modulation.
- $$P_{avg} > \frac{E_c^2}{2R} > P_c$$
- **Vestigial Sideband Modulation :** Under some circumstances it is desirable to obtain a wave consisting of a carrier plus one sideband, with the other sideband suppressed. Such a result can be approximated by transmitting an ordinary amplitude modulated wave through a filter that is so designed that the edge of the transmission band of the filter accurately coincides with the carrier frequency.
- In accordance with CCIR, 7 MHz bands are available in the VHF range and 8 MHz bands in the UHF range for TV broadcasting.
- **Negative AM :** The modulation of the RF carrier by the CVS is in the form of negative AM, where bright picture points correspond to low carrier amplitude and the sync pulse to maximum carrier amplitude.
- **Frequency deviation** (It is the maximum shift of the frequency from the carrier frequency both the side.)

$$\Delta f = f_{max} - f_c = f_c - f_{min} = k_f\left(\frac{E_m}{2\pi}\right)$$

$$m_f = \frac{\Delta f}{f_m} = k_f\left(\frac{E_m}{\omega_m}\right)$$

- The difference between maximum modulated frequency and minimum modulated frequency is known as **Carrier Swing (CS).**

$$CS = f_{max} - f_{min}$$

$$CS = 2\,\Delta f$$

$$\Delta f = \frac{CS}{2}$$

REVIEW QUESTIONS

1. What do you understand by aspect ratio in television system and why it is 4 : 3 ?
2. What is scanning and why it is required ?
3. What do you understand by Kell factor ? How does it affect the vertical resolution of a television picture ?
4. How the vertical resolution is increased with increase of scanning lines ?
5. What is meant by interlated scanning ? Explain how it is reducing the flicker effect.
6. Explain flicker effect and how it being reduced by using interlaced scanning ?
7. What do you understand by interlace error and how does it affect the quality of the picture. If the interlace error is 50%, calculate the delay in the start of second field.
8. Why picture frames frequency is 25 per second in India and 30 per second in USA ?
9. Calculate the viewing distance for 50 cm size TV screen if the screen subtends an angle of 12° at the eye.

10. If a 50 cm size TV screen is to be viewed from a distance of 2 m, calculate the maximum number of pixels which can be resolved.

11. Define resolutions. Derive general formula for horizontal resolutions and bandwidth for TV systems.

12. What are the factors on which resolution and bandwidth depend ?

13. Discuss the factors which determine the number of scanning lines in a television system.

14. Discuss the importance of synchronization in TV broadcast. How are the sawtooth signal generators triggered for starting H-sweep and V-sweep ?

15. What do you understand by composite video signal ? Draw a neat CVS with negative polarity for one-horizontal scan, showing voltage levels and time interval width of various pulses and signals.

16. Define IRE scale.

17. What do you understand by picture to sync ratio ? Explain its effect on stability of picture.

18. What is blanking ? Draw H-blanking pulse, showing width of various sections of this pulse.

19. Explain the importance of front porch and back porch.

20. Draw V-blanking pulse, showing the width of various sections of this pulse.

21. Explain the need of serration in vertical blanking pulse.

22. Explain the need of pre-equalizing pulses and post-equalizing pulses.

23. Why flicker is not removed by progressive scanning ?

24. What would happen if the circuit bandwidth is much higher than required by the side bands ?

25. Why is the number of scanning lines per frame always odd ?

26. Draw and explain the composite video signal at the end of odd fields and at the end of even fields.

27. Why the number of scanning lines in horizontal is choosen 625 ? Why not 623 or 627 ? Explain.

28. Justify the choice of a rectangular frame with width to height ratio equal to 4/3 for television transmission and reception.

29. How is the illusion of continuity created in televisions pictures ? Why has the frame repetition rate been chosen to be 25 and not 24 as in motion pictures ?

30. Compare the progressive and interlaced scanning and show that interlaced scanning reduces the flicker and conserves bandwidth.

31. What do you understand by vertical and horizontal resolution ? Show that the highest modulating frequency that needs to handled in the 625 TV system is 5 MHz.

32. List the basic requirements that must be satisfied by the pulse train added after each frame. Why is it necessary to serrate the broad vertical sync pulse ?

33. Write the short notes on :

(*i*) Pedestal height
(*ii*) Blanking pulses
(*iii*) Aspect ratio
(*iv*) Kell factor
(*v*) Sync pulses
(*vi*) Flicker
(*vii*) Interlaced scanning
(*viii*) Field blanking period
(*ix*) Peak white level
(*x*) Pulling on white
(*xi*) Negative modulation

MULTIPLE CHOICE QUESTION

1. Field frequency in India is
   (*a*) 60 Hz (*b*) 50 Hz
   (*c*) 15625 Hz (*d*) 625 Hz
2. Horizontal scanning frequency, according to CCIR standards is
   (*a*) 15750 Hz (*b*) 15625 Hz
   (*c*) 60 Hz (*d*) 15725 KHz
3. Trace period of a horizontal line is
   (*a*) 64 μs (*b*) 12 μs
   (*c*) 52 μs (*d*) 1280 μs
4. The horizontal resolutions of a picture depends on
   (*a*) The number of horizontal lines
   (*b*) Horizontal scanning rate
   (*c*) The vertical scanning rate
   (*d*) Both (*a*) and (*b*)
5. Retrace period of a horizontal line is
   (*a*) 64 μs (*b*) 12 μs
   (*c*) 52 μs (*d*) 20 μs
6. Trace period of a vertical sweep is
   (*a*) 18720 μs (*b*) 1280 μs
   (*c*) 20 ms (*d*) None of the above
7. Each sequence of scanning in the interlaced scanning method is known as
   (*a*) Field (*b*) Scanning lines
   (*c*) Trace path (*d*) Retrace path
8. Vertical blanking period is
   (*a*) Equal to the vertical retrace
   (*b*) Greater than the vertical retrace
   (*c*) Less than vertical retrace
   (*d*) None of these.
9. According to CCIR standards the field sync occurs after lines
   (*a*) 312½ and 625 (*b*) 262.5 and 625
   (*c*) 310 and 662.5 (*d*) None of above
10. Aspect ratio of a televisions is
    (*a*) 3 : 4 (*b*) 4 : 3
    (*c*) 9 : 16 (*d*) None of these
11. In India, horizontal blanking pulse is
    (*a*) 12 μs (*b*) 4.7 μsec
    (*c*) 1280 μs (*d*) 1870 μs
12. The width of H-sync pulse in India is
    (*a*) 12 μs (*b*) 4.7 μs
    (*c*) 1870 μs (*d*) 52 μsec
13. In interlaced scanning, there is one half line spacing between the start positions for scanning odd and even fields. This is done to produce
    (*a*) Horizontal scanning (*b*) Linear scanning
    (*c*) Exact interlacing (*d*) Line pairing

14. According to CCIR standards, the televisions screen is blanked out
(a) 25 times in one sec (b) 60 times in one second
(c) 50 times in one sec (d) 15625 times in one second.

15. The value of Kell factor is about ______.
(a) 0.5 (b) 0.3
(c) 0.7 (d) 0.9

16. Bandwidth of AMVSB for video carrier is
(a) 4 MHz (b) 5.75 MHz
(c) 7.85 MHz (d) 11 MHz

17. The term pedestal is used for
(a) Sync top level (b) Blanking level
(c) Black level (d) White level

18. Video signals below 10% of the carrier are known as :
(a) Pedestal height (b) Blacker than black
(c) Whiter than white (d) Blacker than black

ANSWERS

1. (b)	**2.** (b)	**3.** (c)	**4.** (d)	**5.** (b)
6. (a)	**7.** (a)	**8.** (b)	**9.** (a)	**10.** (b)
11. (a)	**12.** (b)	**13.** (c)	**14.** (c)	**15.** (c)
16. (b)	**17.** (b)	**18.** (c)		

3

CHAPTER

TV CAMERA TUBES

INSIDE THIS CHAPTER

3.1. INTRODUCTION TO TV CAMERA TUBES

A TV camera may be called the eye of a system. A TV camera tube is a transducer which converts the variations of light intensity in to variations of electrical current or voltage, known as **video signals.** Few factors by which the most types of camera tubes are suffering given below:

1. Poor sensitivity
2. Poor resolution
3. Undesired spectral response
4. Poor contrast
5. Noise level is high
6. Instability
7. Problems in processing

Modern camera tubes have good resolution, good sensitivity and wide spectral response. These characteristics of camera tubes are fulfilled by using two types of photo electric effects such as, Photo emission and Photo conductivity.

There are some characteristics of Camera tube as follows:

1. **Sensitivity to visible light :** Sensitivity of a camera tube is represented in micro ampere per lumen. It is measured with a standard source of light, a tungsten lamp at 2870 k.

Types of camera tubes	Sensitivity	SNR
I. Image orthicon	1000 μA	40 dB
II. Vidicon	125 μA	45 dB to 50 dB
III. Plumbicon	400 μA	45 dB to 50 dB

2. **Spectral response :** The spectral response means the output of a camera tube for the range of wavelengths of light. In ideal condition, the spectral response of a camera tube must be the same as the response of the human eye. However, practically no camera tube can give ideal result. A good camera tube gives the spectral response quite close to that of the human eye in the visible spectrum.
3. **Resolution :** Ability to resolve details while viewing a multielement of a scene. Resolution of a camera tube depends upon the size of the spot formed by the scanning beam on the target plate. The diameter of the spot is called aperture.
4. **Noise level :** All electronic devices produces noise due to randomness in the number and velocity of electrons. This randomness is also referred as self noise. It is measured in signal to noise ratio.
5. **Halo Effect :** A ring of brightness or darkness is formed around a picture element in Image orthicon and Plumbicon camera tubes, this effect is referred as Halo effect.
6. **Contrast :** The ratio of the output signal of white to the output signal for is known as contrast.
7. **Lag Characteristics :** After the removal of light, the photoelectrons and holes needs a finite time to disappear and leaves a tail or smear of the previous scene. When the light intensity is too high, all the electrons released by the electron beam are not able to leave the target, and few of these electrons is left behind, resulting in slow fading of the image. It is measured by measuring residual signal current after 60ms.
8. **Dark Current :** Dark current is the current which is always available as output from the camera tube even there is no light falls on the face plate. Its value is in nanoamperes. Dark current is due to the availability of some free electrons at room temperature.
9. **Ghost Effect :** Ghost effect is an image of the same polarity displaced from the high light. It happens because of the sharp contrast between very bright and dark background. Image orthicon suffers with the ghost effect.

3.2. BASIC PRINCIPLE OF CAMERA TUBE

An optoelectrical converter is used to translate brightness variations into an electrical picture signal. Different converter systems are available, but only the pickup tubes with a photosensitive semiconductor layer are really important for TV technology.

The purpose of TV camera tube is to sense each element independently and produce a signal in electrical form, proportional to the brightness of each element. Light from the scene is focused on a photo sensitive surface known as **image plate,** and the optical image is formed with a lens system represent light intensity variation of the scene.

By making use of photoelectric properties, the image plate than convert different light intensities into corresponding electrical variation.

Now, it is required to pick up this information in the form of electrical charges as fast as possible. It is not possible to have simultaneous pick up so scanning by electron beam is employed. In the scanning process, the electron beam moves across the image plate line by line, and field by field to provide signal variation in a successive order. The image is divided in to its basic picture elements by scanning process. Fig. 3.1, illustrate the basic constructions part of the camera tube.

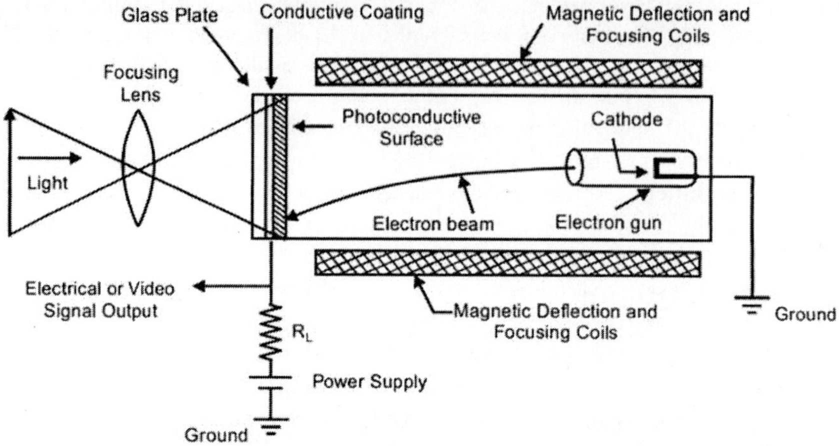

Fig. 3.1. A simple diagram of camera tube.

3.2.1. Production of Video Signal Using Photo-Electric Effect

There are two types of **photo electric** effects used for converting variations of light intensity into electrical variations are namely:

1. Photo emission
2. Photo conductivity

1. Photo Emission

There are certain metals which emits electrons when light falls on their surface, these emitted electrons are called Photo electrons and the emitting surface is known as Photo cathode. Small bundles of energy which of consists light are called photons.

When light is falls on a photo cathode, they give away their energy to the outer valence electrons to allow them to overcome the potential energy barrier at the surface. The number of electrons which can overcome the potential energy barrier and get emitted depends upon the light intensity. The energy of the electrons emitted from any point on the photocathode is proportional to the intensity of the light falling on that point. This process is known as **photo emission**. The principle of photo emission is illustrated in the Fig. 3.2.

The motion of an electron beam is controlled by the electric and magnetic fields. To avoid the secondary emission, the electric beam is decelerated to almost zero velocity before it reaches the target plate. Electron beam is moved back from the target plate due to the negative acceleration and being controlled by the focusing and deflection coils. The returned electron beam strikes the collecting electrode which is positioned close to the cathode. The amount of electrons in a returning electron beam varies according to the charge depositing on the target plate. The amplitude of the current flowing through the collecting electrode varies as per the brightness variation of the picture. Finally this current flows through a load resistance and the variation of the voltage developed across this resistance is known as the video signal.

As photo emissive surface cesium silver or bismuth silver cesium oxides are preferred because of :

1. They are sensitive to incandescent light
2. Their spectral response is very close to the human eye

3. Alkali metals have very low work function that's why they are used as photo cathode.

Image orthicon camera tube is based upon photo emission principle.

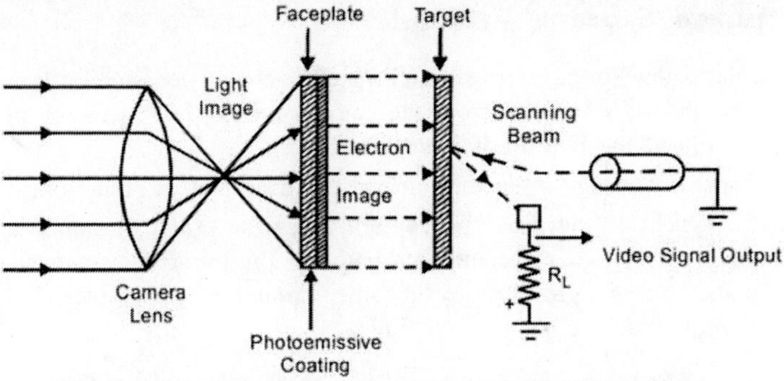

Fig. 3.2. **Production of video signal by photo emission.**

2. Photo Conductivity

In this method of producing an electrical image is by photoconduction, where the conductivity or resistivity of the photo sensitive surface varies in proportion to the intensity of the light focused on it. Semiconductor metals and their oxides have a property known as photoconductivity. The incident light creates free electrons inside a semiconductor material. This process is known as **photoconductivity**.

The principle of photo conduction is illustrated in the Fig. 3.3.

The variations in resistance at each point across the surface of the material are utilized to develop a varying signal by scanning it uniformly with an electron beam.

The number of free electrons created at any point is proportional to the intensity of the light at that point. According to the resistance offered by the surface at any point, the amplitude of current varies. This varying current follow the path through a load resistance and the instantaneous voltage developed across the load resistance is referred as video signal. Vidicon camera tube and its modified version, plumbicon camera tube is based upon photo conduction principle.

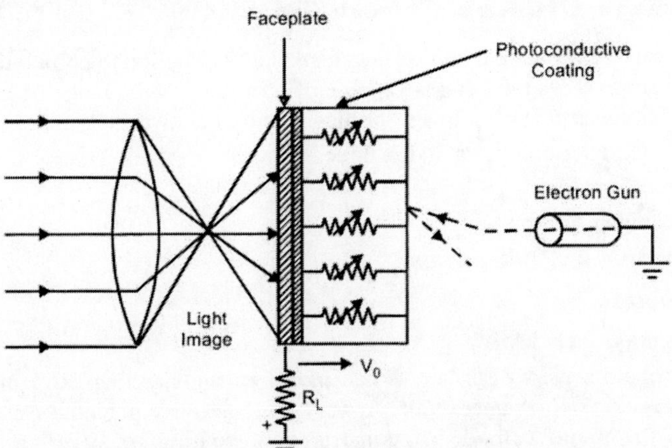

Fig. 3.3. **Production of video signal by photo conduction.**

Materials used are;

Selenium, Tellurium and Lead with their oxides, having the property of photo conductivity.

3.2.2. Image Storage Principle

TV cameras developed during initial stages were of non-storage type. In this type of TV camera the signal output from the camera for the light on each picture element is produced only at the **instant it is scanned.**

Flying spot and Image dissector cameras are examples of Non-storage type of tube.

Storage Type : In storage type camera tubes the effect of illumination on every picture element is allowed to accumulate between the times it is scanned in successive frames. By the use of light storage tube the amount of photo electric signal can be enhanced approximately ten thousand times.

3.2.3. The Electron Scanning Beam

As in the case of picture tube an electron gun produces a narrow beam of electrons for scanning. In camera tube magnetic focusing is normally employed. The electron must be focused to very narrow and thin beam because the resolving capability of the camera tube is depend upon that. The scanning of the electron beam is achieved by the deflecting the beam with the magnetic fields produced by the horizontal and vertical coils in the deflection yoke.

3.3. TYPES OF TV CAMERA TUBES

Many types of television pick-up tubes are developed, few of them are as follows:

1. Image Orthicon Camera Tube
2. Vidicon camera tube
3. Plumbicon
4. Solid state scanner
5. Charge Coupled Device (CCD) camera

3.3.1. Image Orthicon Camera Tube

Image orthicon camera tube was developed in 1945 by Radio corporation of America (RCA). It is a sensitive tube and is capable of handling a wide range of light values and contrast. The construction of image orthicon tube is shown in Fig. 3.4.

In a single envelope, it includes three sections:

1. Image section
2. Scanning section
3. Electron multiplier section

Image Section

This section includes :

(*i*) A photo sensitive surface, called **photo cathode**, acting as main transducer in image orthicon camera tube. A very large negative potential *i.e.* -600v is given to the photo cathode. It consists of a coating of silver-antimony-sesium compound on the inside of the glass face plate of the tube. When the light from a object focused on the photo cathode surface by a lens system, the optical image is formed, photoelectrons, in proportion to the amount of light impinging, are emitted. The energy of the electrons emitted from different points of the cathode depends on the incident light energy. The number of electrons emitted at any point in the photocathode has a distribution according to the optical image. Since the photocathode is a conductor so it cannot store the charge. Thus the image formed at photocathode should move towards the target plate which is positioned at a short distance from it.

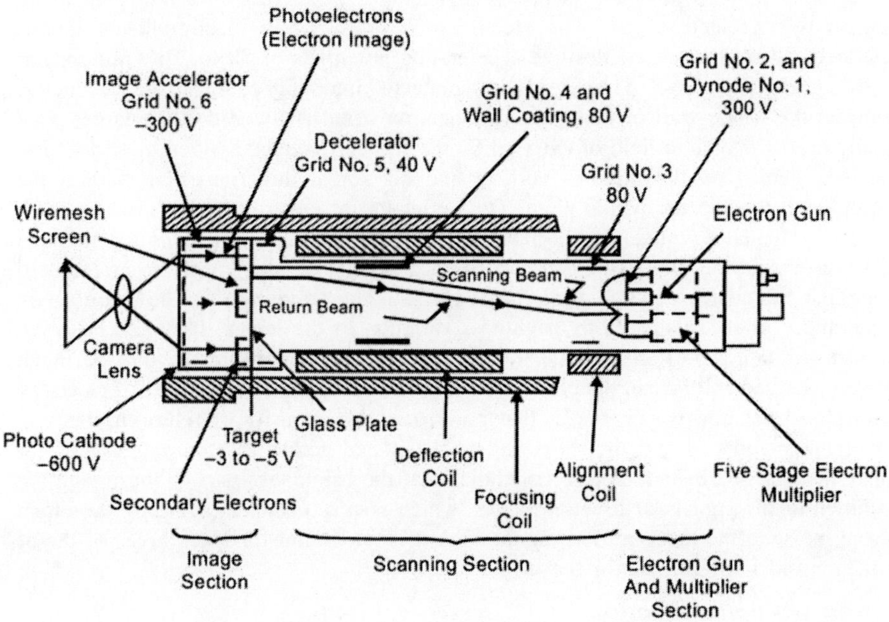

Fig. 3.4. Image Orthicon.

(*ii*) A **target plate** which is a thin plate of glass of low resistivity, having thickness about 4 microns, which can store the charge. Most of the photoelectrons pass through the screen and hit the target plate. Target plate is kept at about 400 v more positive potential with respect to the cathode. Since the electrons have tendency to repel each other and this can cause the distortion in the information which are available as charge image. This affect can be minimised by using a long focus coil which generates an axial magnetic field. Thus the emitted electrons are focused on the target plate into a well defined electron image of the original optical image.

(*iii*) As the photoelectrons are accelerated to several hundred electron volts, they liberate several secondary electrons from the target plate surface. A **wire mesh screen** is provided to collect these secondary electrons and it is kept at higher potential with respect to the target plate. The wire mesh is located very near to the target plate and has about 500000 opening per square inch. So the screen wires do not interfere with the electron image.

The emission of the secondary electrons from target plate leaves a distribution of positive charge on its surface. The low resistivity of target plate resists the lateral charge flow on its surface and thus the image charge pattern, formed on the plate, is truly restored as such. Since the target plate is thin, this charge pattern also appears on the other side (away from the screen) of the plate. Because of high secondary emission ratio, the intensity of the positive charge distribution is four to five times as compared to the charge formed at photo cathode. This increase in charge density as compared to photo cathode is known as **image multiplication.**

Scanning Section

The other side of the plate is now scanned by a beam of low velocity electrons generated by an electron gun. The electron gun produces an electron beam that is accelerated toward the target. Positive accelerating potentials of 80 to 330v are applied to grid-2, grid-3 and grid-4. These grids are connected internally to metalized conductive coating on the inside wall of the tube. The electron beam is focused at the target plate by applying the magnetic field of external forces coil and voltage applied to grid-4. The beam is deflected on the plate is vertical and horizontal directions and enables the electron beam to scan the whole plate. The deflection of electron beam is achieved by the magnetic fields of vertical and horizontal deflecting coils and these coils are mounted with yoke outside the tube. This electron beam gives the number of electrons required to neutralise the positive charge at the point and thus the returning electron beam varies in magnitude in accordance with brightness variation of the image. It should be noted that since the target portion affected by the white portion of the image will be much positively charged and hence, the electron beam has to give up large number of electrons to neutralised the positive charge at that point, *i.e.*, the intensity of returning electron beam is much reduced and the video signal developed across that output resistor for this part will be small. It therefore concludes that the brightness part of the image are transmitted as the signals of low amplitude which is very useful in avoiding the effect of strong noise at the receiver. The signal current is maximum for black areas of the of the picture and is minimum for the bright areas.

Electron Multiplier Section

An electron multiplier is located within the pickup tube for amplifying the electron density variation in the returning beam. If we use any conventional amplifier instead of electron multiplier, it may cause the problems of signal to noise ratio because the signal amplitude is very low. In electron multiplier, since there is no active device or resistors used, we obtained noise free current amplification. Cold-cathode electrodes are used to amplify the current. Since the incident electrons strike with high velocity to the metal surface, resulting emission of secondary electrons.

The image orthicon camera tube includes an electron multiplier structure which uses the secondary emitted electrons to amplify the small amount of photoelectric current.

Figure 3.5. illustrate the working of electron multiplier. It consist of number of cold anode-cathode electrodes which are known as Dynodes. These dynodes are at a progressively higher positive potential. The photocathode emits the electrons, which are accelerated towards a more positive dynodes. Secondary emission are possible due to high velocity of electrons and depending upon the surface and potential at dynodes, the secondary emission ratio is three to four. Every times the available number of electrons is multiplied by the secondary emitted electrons that strikes the next emitting surface of more positive dynode. Since each dynodes provides a multiplication of four then total gain by five dynodes is equal to $4^5 = 1024$. Finally the secondary electrons are collected by the anode which is at higher potential (1500v), resulting in anode current through a load resistor (R_L). If we keep the value of load resistor is 20Kohm then the amplitude of the camera tube output signal will be 500mV and dark current is 30μA and high light current is 5μA.

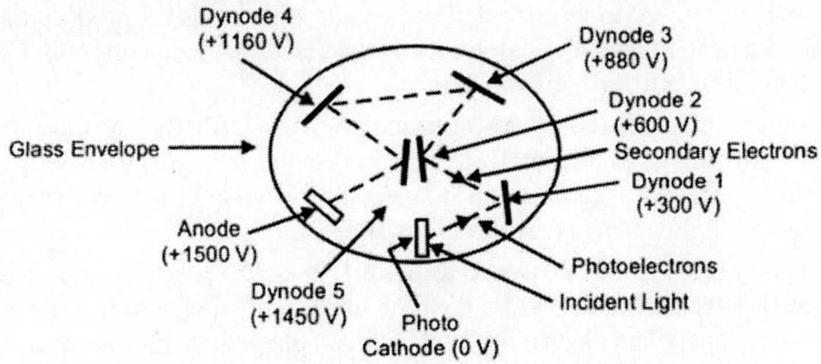

Fig. 3.5. Electron Multiplier in Image orthicon camera tube.

Applications : This camera tube is used for outdoor and indoor pickups.

3.4. VIDICON CAMERA TUBE

This camera tube is based on the photo conductive properties of semiconductors. When light falls on it, the number of free electrons created at any point is directly proportional to the intensity of light falling on that point. Photo-conductive properties of semiconductors means decrease in resistance with the amount of incident light. Brighter the light, greater is the number of free electrons. These electrons are removed from the material by using a positive voltage and hence that becomes positively charged. The value of charge on the target at any point is proportional to the intensity of light at the corresponding point in the original scene. So, the charge image of the picture is formed on the surface of the target.

This camera tube was the first camera tube based on the photoconductive principle and is developed by Radio comporation of America (RCA) in early 1950s.

Construction of Vidicon Camera Tube

The sketch of camera tube is illustrated in Fig. 3.6. It consists of :

(*i*) Target plate or signal plate

(*ii*) Scanning system.

(*i*) Target Plate : The target consist of a conducting metallic film very thin so as to be transparent. The side of this films facing cathode is coated with a very thin layer of photo-conductive material *i.e.* amorphous selenium. This side is scanned by electron beam. The optical image is focussed on the other side of this film.

The photoconductive target material is an intrinsic semi-conductor which has a very high resistivity in darkness and is decreasing it with increase in illumination. The photolayer has a thickness of about 0.0001 km and in the dark, it behaves like an insulator with a resistance of about 20 MΩ. Bright light reduces the resistance to 2 MΩ. Very less charge leaks between the successive frames and this charge is stored by the beam and the resulting current to the electrode is called **dark current.**

The target signal plate is kept approximately 40 V positive potential with respect to the cathode in series with 50 kΩ load resistance (R_L).

(*ii*) **Scanning System :** The electron beam for scanning is formed by the combination of cathode, control grid (Grid no. 1), accelerating (grid-2) and anode (grid 3). The focussing coil produces an axial field which focusses the beam on the film. Vertical and horizontal deflection of the beam, so as to scan the whole film, is accomplished by passing sawtooth current waves through deflecting coils which thus the produce transverse horizontal and vertical magnetic field respectively the alignment coils are for initial adjustment of the direction of electron beam.

An electron beam produced by an electron gun of a cathode ray tube scans the charge image horizontally as vertically and neutralise the charge fully. No discharge current flow through a load resistor and video signal voltage is produced corresponding to the variations of light intensity in the original scene.

A wire mesh at G-4 (400 V) works as a muzzle of the electron gun. The target plate is at a much lower potential (40 V) than the mesh, and so slowsdown the electrons to such an extent that when they land at the target plate with almost zero velocity, perpendicular to the target.

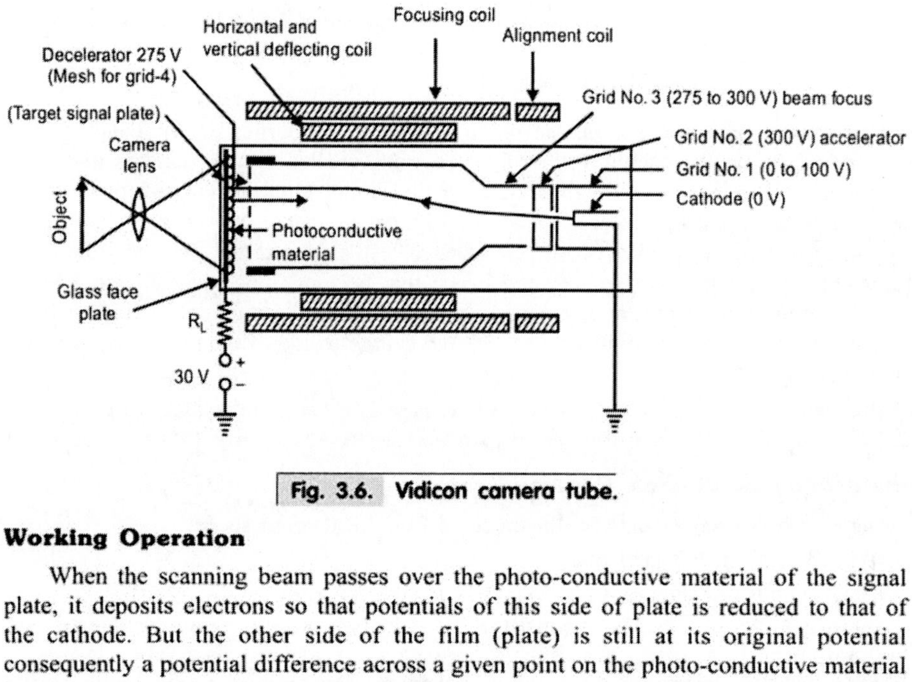

Fig. 3.6. **Vidicon camera tube.**

Working Operation

When the scanning beam passes over the photo-conductive material of the signal plate, it deposits electrons so that potentials of this side of plate is reduced to that of the cathode. But the other side of the film (plate) is still at its original potential consequently a potential difference across a given point on the photo-conductive material is created. It is approximately 40 V. Before the next scanning (which may be done after an interval of 1/50 or 1/25 sec.), the charge leaks through photo-conductive material at a rate determined by the conductivity of the material which in turn, depends upon the amount of incident light. White portions of the object will project more light on the film and make it more conductive. This change leaked to photoconductive side of the film will vary according to illumination of the object. As a result potential at every point on the photo conductive side will vary. Now the electron beam again starts scanning the photo conductive side of the film but this time the charge deposited by the beam in order to reduce its potential towards zero (cathode potential) will vary with time. Therefore current through load resistance (R_L) will follow the changes in potential difference between two surfaces of the film and hence follow the variation of light intensity of successive points in the optical image.

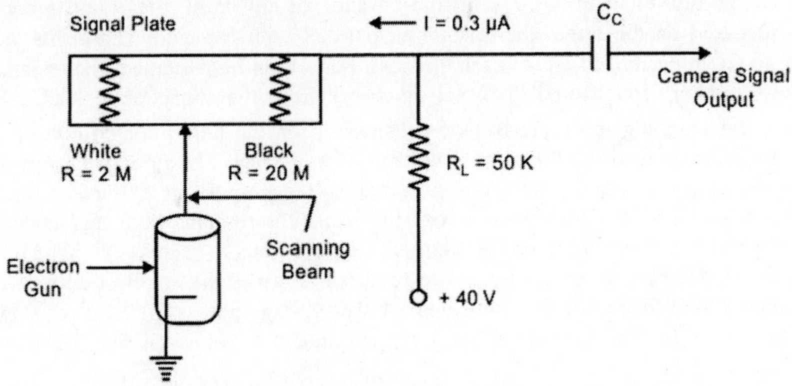

Fig. 3.7. Video Signal Output from Vidicon camera tube.

Storage Action

Since the light from the picture is continuously falling on the target, each element of photocoating is scanned at an interval to the frame time *i.e.* 1/50 sec. It gives the storage action and the net change in resistance at any point in the photoconductive layer. It depends on the time and the light information at each element or pixel is thus integrated over the time elapsed between the samples. Then it is stored in the form of light dependent positive charge image pattern on the cathode side of the plate. The storage time is same for all the points on the target plate. The net change in resistance for all the elementary areas is proportional to the light variations of the scene that televised.

Leaky Capacitor Concept

There is an another way of explaining the development of charge image on the photolayer is to consider that it is an array of individual target elements. Each elements consist of a capacitor in parallel with a light dependent resistor as shown in Fig. 3.8.

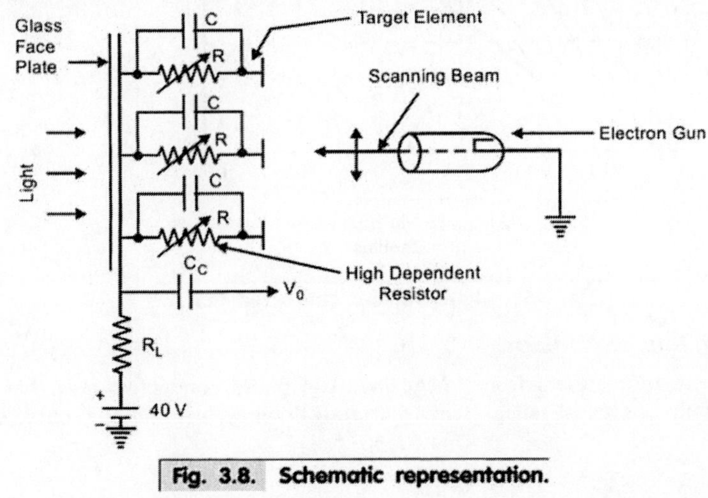

Fig. 3.8. Schematic representation.

As shown in the Fig. 3.8, the one end of these target elements is connected to the signal electrode and other with the unterminated facing the beam. These capacitors attains a charge almost to the +40 V, in the absence of any light image and when an image is focussed on the target the resistor in parallel with capacitor charge its value according to the intensity of light of each element. For a high light element the resistance across the capacitor reduces and gives lot of charge from the capacitor to leak away.

During the scanning more electrons are deposited on the unterminated end of this capacitor to recharge it to the full supply voltage of +40 volts. The flow of current that completes its path through R_L develops a signal voltage across it. Sameway for the black area of pictures, the resistance across the capacitor remains high and does not permit charge to leak from the corresponding capacitors. Thus to recharge the capacitor, less number of electrons from the beam are required. It gives the smallest current that flow, through a load resistance and develops a lower voltage across the load resistance. The video signal is fed through a blocking oscillator to an amplifier for further amplification.

Light Transfer Characteristics

Figure 3.9 shows the output characteristics called dark current of vidicon camera tube. Even in the absence of illumination on the face plate, there is some discharge of the pixels capacitors having some finite value of the resistance of the photoconductive layer. It produces some recharging signal current in the dark condition also, known as dark current. The dark curmet is set by adjusting the target voltage. Sensitivity and dark current both increases as the target voltage is increased. Typical output for the vidicon is a 0.4 μA for bright light and with a dark current of 0.02 μA.

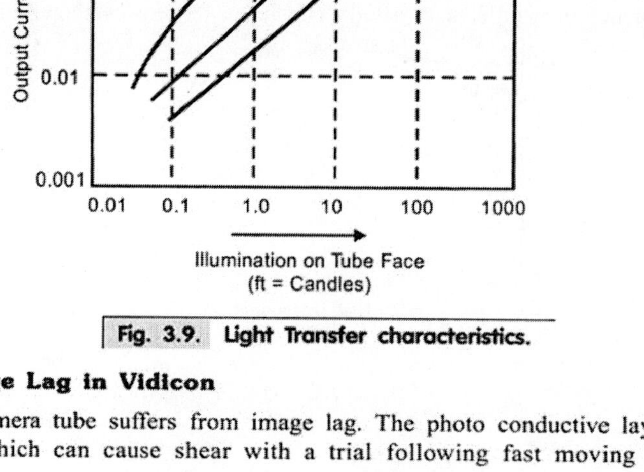

Fig. 3.9. Light Transfer characteristics.

High Image Lag in Vidicon

This camera tube suffers from image lag. The photo conductive layer has finite time lag, which can cause shear with a trial following fast moving objects. The photoconductive lag increases at high target voltages, where the vidicon has its highest sensitivity.

When bright light of a picture is incident on the target plate, large number of electrons are created and a small fraction of these electrons are not able to migrate to the conducting signal plate. Few electrons are left behind which arrive at the signal plate later. It means that the charge image pertaining to that part of the scene is retained as a faded iamge for few seconds. And hence, a faded picture of the scene would be visible on the receiver screen even after that scene has moved out. This effect is known as **image lag** and its a disadvantage for vidicon tubes.

The other variations of vidicon camera tubes are the plumbicon newvicon and chalnicon, the basic arrangement of the electron gun, electrodes and coil is similar to that of the vidicon with some improvements in the gun design.

Applications of Vidicon

This tube is quite popular for CCTV (Closed Circuit Television) applications because of low cost, small size, simplicity and ease of operations. Some more applications of vidicon camera tubes are as follows :

(*i*) Widely used for outdoor recording, domestic and industrial recording.

(*ii*) Aerospace and oceanography.

(*iii*) It is also used in slides, pictures, medicine, education etc.

Advantages of Vidicon Camera Tube

(*i*) Resolution is better than orthicon. Resolution in order of 350 line can be achieved under practical condition.

(*ii*) No ghost image.

(*iii*) There is no halo effect.

(*iv*) Signal to noise ratio is better than image orthicon.

(*v*) Spectral response is close to the human eyes.

(*vi*) No gamma corrections needed.

(*vii*) By varying the target voltage as per illumination of the scene, sensitivity can be adjusted.

(*viii*) It is compact, having a length of 12-20 cm and diameter between 1.5-4.0 cm.

(*ix*) It has long life about 5000 and 20,000 hrs.

(*x*) It is low cost and simple.

Disadvantages of Vidicon Camera Tube

(*i*) Image lag is severe.

(*ii*) Dark current is high approximately 20 μA.

(*iii*) Sensitivity is less than image orthicon.

(*iv*) The response characterstics is non-linear.

3.5. PLUMBICON CAMERA TUBE

The plumbicon camera tube is developed by Philips in 1963 to remove the problem of image lag in vidicon. It has a photo conductive target plate of lead mono-oxide (PbO) sandwitched between a n-type and P-type layer. The name of this tube is derived from the lead *i.e.* plumbicon or Pb, which is used as target material in plumbicon camera tube.

Construction

The construction of plumbicon is very similar to the standard vidicon except the target shown in Fig. 3.10. In this also focus and deflections are achieved by magnetically. This tube consists of a glass envelope with an optically flat face plate. The face plate operates effectively as P-I-N semiconductor diode. The inner surface of the face plate is coated with a thin layer of tin-oxide (SnO_2) which is transparent to light and a good conductor of electricity. This forms a N-type layer and acts as the signal plate. It is connected to load through a metallic ring around the tube.

On the scanning side of this layer is deposited a photo-conductive layer of pure lead monooxide (PbO) which is intrinsic or I-type and is a good conductor and has large number of free electrons. Finally, the pure PbO is doped to form a P-type semiconductor on which the scanning beam bands. Thus the target in plumbicon becomes a PIN diode. The thickness of target plate is about 15 mm. Figure 3.10 is illustrate the detail of the target plate.

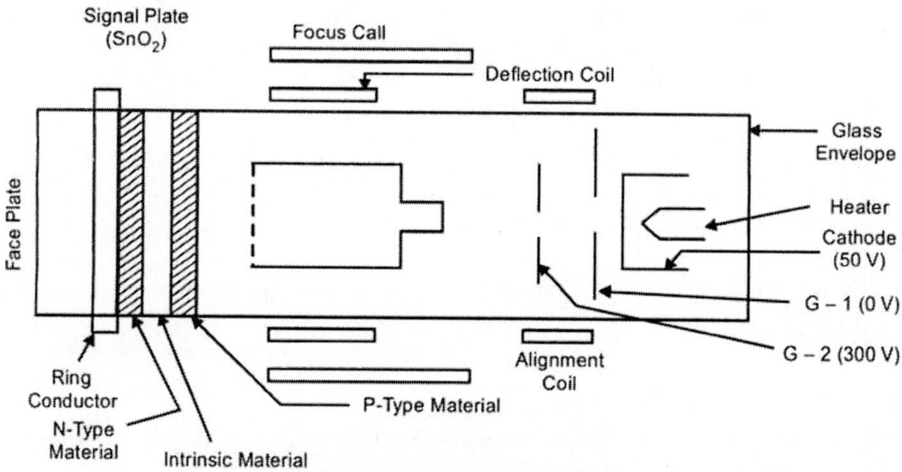

Fig. 3.10. Plumbicon camera tube.

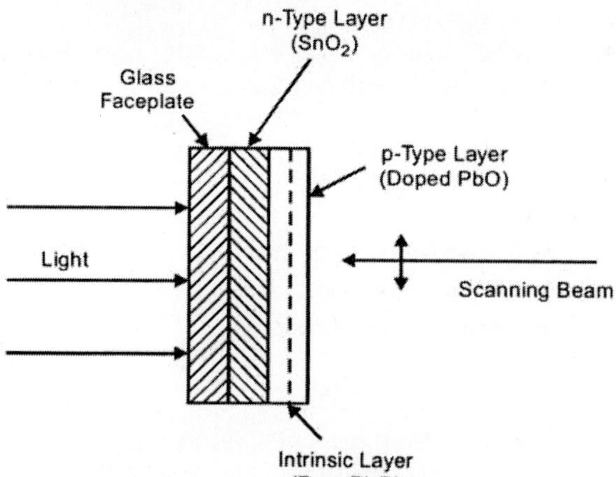

Fig. 3.11. Target plate of plumbicon camera tube.

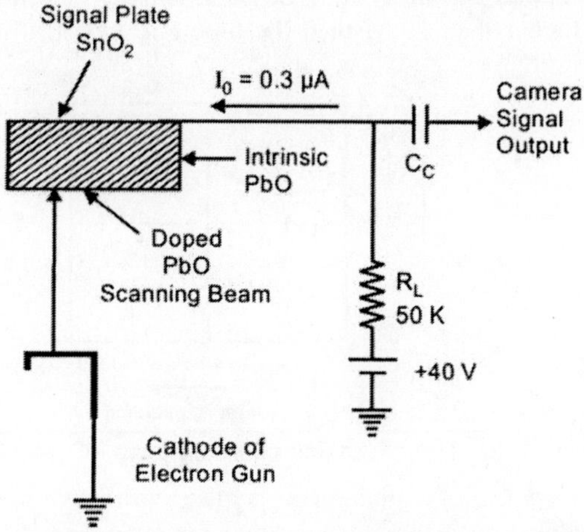

Fig. 3.12. Camera signal output of plumbicon.

The scanning of target is being done by a sharp focussed electron beam produced by an electron gun. The electron gun consists of an indirectly heated cathode, control grid (G1) and accelerating grid (G2). Focussing is achieved by applying both electrostatically by (G3) and magnetically by an external focus coil. Grid (G-3) at 260 V is at lower potential with respect to accelerating grid (G-2 at 300 V) and hence slows down the electrons. The magnetic focussing converges the electrons to the centre of the beam and resulting a sharp spot on the target. A fine wire mesh grid (G4) is connected to 400 V and acts as a muzzle for the electron gun. The target is kept at 40 V, that allows the electrons to land at the target with almost zero velocity and perpendicular to the target.

When the light incidents on the target, electrons are released. In the I-layer the electric field gradient setup is very high and due to this high gradient all released electrons sweep the target quicly so that there is no image lag.

The PIN-diode acts as a capacitor and the free electron released by light cause a resistance to be formed across the capacitor.

The conductive film of tin oxide (SnO_3) is connected to the target suppy of 40 V through a load resistor R_L to develop the camera output signal voltage. The light is focussed through the transparent layer of tin oxide on to the photoconductive lead mono oxide. The targets prevent, conductions in the absence of light because of the absence of any charge carriers. Hence there is almost nil dark current *i.e.* approximately 4×10^{-9} Amp. Because of the incident light, photo exitations of semiconductor junction occurs. Depending on the intensity of light the value of resistance varies and hence the flow of signal current. As the light intensity increases the resistance decreases and flow of current signal increases. The current is proportional to the incident light on each photo element. The video signal is developed across the load resistance.

Light Transfer Characteristics

The response of the output current and target illumination is shown in Fig. 3.13. This curve is a straight line with higher slope as compared to the slope of a vidicon camera tube. Due to the less recombination of photogenerated electrons and holes in the I-layer, hence output current is high. So its sensitivity is also high. In case when the target voltage is higher than 20 V, then the tube operates in a photo saturated mode.

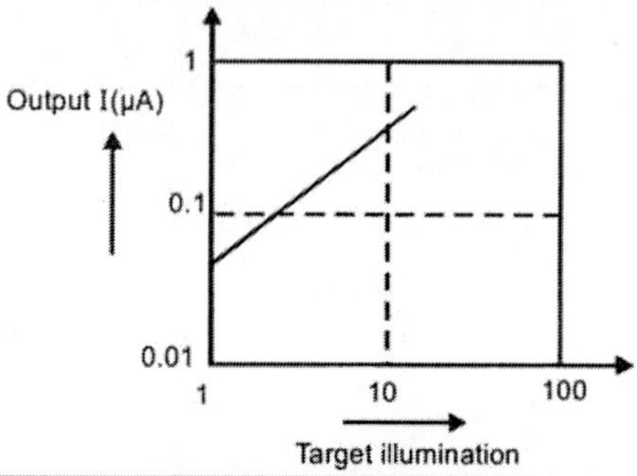

Fig. 3.13. **Light transfer characteristics of plumbicon.**

The spectral response of a plumbicon is closer to that of the human eye except in the red colour region.

Lead oxide target reflects red light towards the face light and rest all colours are absorbed. This red light is again reflected by the face plate towards the target. Due to these reflections a red halo around a white spot the scene. Figure 3.14 shows the formation of red halo around a white spot.

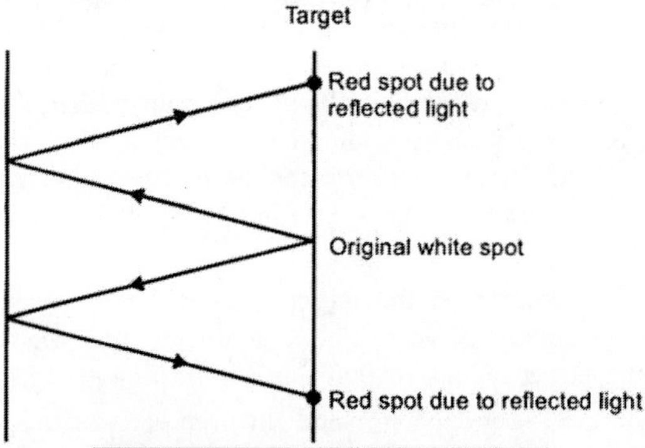

Fig. 3.14. **Formation of red halo.**

This red halo can be eliminated by an anti-halo disc whose circumference is coated black to absorb the reflected light as shown in Fig. 3.15.

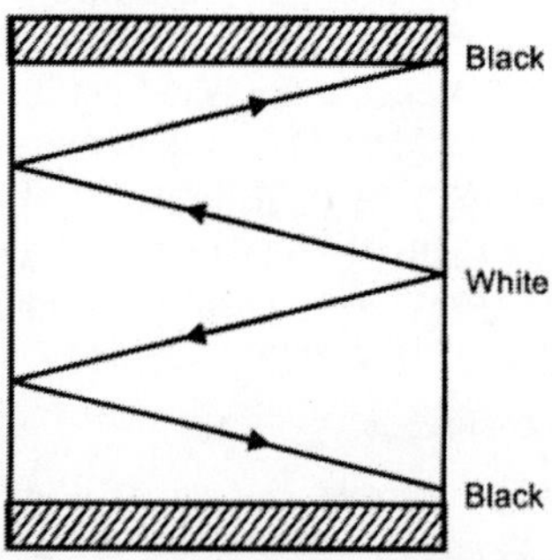

Fig. 3.15. **Elimination of red halo.**

Advantages of Plumbicon Camera Tube

(*i*) Signal to noise ratio is 47 db whereas in vidicon it is 45 dB.

(*ii*) Sensitivity is 400 μA/lumen which is more than that of vidicon (125 μA/lumen).

(*iii*) Image lag is negligible.

(*iv*) Absence of ghost image.

(*v*) Dark current is very less (mA).

(*vi*) Characteristics curve is linear.

(*vii*) Very small in size as compared to image orthicon.

(*viii*) Resolution is 50% at 400 lines.

Disadvantages

(*i*) Halo effect occurs and therefore an antihalo disc is used.

(*ii*) It sensitivity is less than that of image orthicon.

(*iii*) Spectral response is poor in red.

(*iv*) Size is bigger than vidicon.

(*v*) Needs Gamma correction.

(*vi*) Resolution is poor than vidicon.

Applications

(*i*) Indoor and outdoor shooting.

(*ii*) Suited to colour TV due to absence of image lag.

Typical Characteristics of Plumbicon Camera Tube

SNR = 47 dB
Sensitivity = 400 μA/lumen
Image lag = Negligible
Ghost = No ghost effect
Halo = Halo effect occurs
Dark current = Less than 1 nA
Resolution = 50% at 400 lines
Gamma = Unity
Size = 25 mm
IR sensitivity = Moderate.

3.6. CHARGE COUPLED DEVICE (CCD) CAMERA

A **Charge Coupled Device (CCD)** camera is an apparatus which is designed to convert optical brightness into electrical amplitude signals using a plurality of CCDs, and then reproduce the image of a subject using the electric signals without time restriction. Charge coupled devices or CCDs are arrays of semiconductor gates formed on a substrate of an integrated circuit or chip. The gates of the CCD are operative to individually collect, store and transfer charge. When used in image applications, the charge collected and stored in each gate of the array represents a picture element or *pixel* of an image. The CCD image sensor includes an image sensing section, which performs photoelectric conversion, and a storage section, which is arranged separate from the image sensing section and temporarily stores charges acquired by the photoelectric conversion.

A CCD camera includes a CCD imager 1C (integrated circuit) and control circuits typically mounted on a printed wiring assembly. In CCD cameras, either a CCD *matrix* or one or more CCD lines is/are located in the focal plane behind the optical system. A CCD is normally used for the image pickup device of a digital camera. Image data captured by it is stored in a storage medium such as a compact flash memory or an 1C memory card. To monitor the captured image, a display means such as an LCD (liquid crystal display) is generally equipped in the digital still camera. Charge coupled device (CCD) camera systems are configured to capture signals according to many different CCD output formats and *pixel* configurations.

A charge-coupled device (CCD) is an analog shift register, that enables the transportation of analog signals (electric charges) through successive stages (capacitors) controlled by a clock signal. Charge-coupled devices can be used as a form of memory or for delaying samples of analog signals. Today, they are most widely used in arrays of photoelectric light sensors, to serialize parallel analog signals. Not all image sensors use CCD technology; for example, CMOS chips are also commercially available.

"CCD" refers to the way that the image signal is read out from the chip. Under the control of an external circuit, each capacitor can transfer its electric charge to one or other of its neighbors.

3.6.1. Basics of Operation

In a CCD for capturing images, there is a photoactive region (an epitaxial layer of silicon), and a transmission region made out of a shift register (the CCD, properly speaking).

An image is projected by a lens on the capacitor array (the photoactive region), causing each capacitor to accumulate an electric charge proportional to the light intensity at that location. A one-dimensional array, used in line-scan cameras, captures a single slice of the image, while a two-dimensional array, used in video and still cameras, captures a two-dimensional picture corresponding to the scene projected onto the focal plane of the sensor. Once the array has been exposed to the image, a control circuit causes each capacitor to transfer its contents to its neighbor. The last capacitor in the array dumps its charge into a charge amplifier, which converts the charge into a voltage. By repeating this process, the controlling circuit converts the entire semiconductor contents of the array to a sequence of voltages, which it samples, digitizes and stores in some form of memory.

3.6.2. Operation of CCD Camera

The photoactive region of the CCD is, generally, an epitaxial layer of silicon. It has a doping of p+ (Boron) and is grown upon the substrate material, often p++. In buried channel devices, the type of design utilized in most modern CCDs, certain areas of the surface of the silicon are ion implanted with phosphorus, giving them an n-doped designation. This region defines the channel in which the photo generated charge packets will travel. The gate oxide, *i.e.* the capacitor dielectric, is grown on top of the epitaxial layer and substrate. Later on in the process polysilicon gates are deposited by chemical vapor deposition, patterned with photolithography, and etched in such a way that the separately phased gates lie perpendicular to the channels. The channels are further defined by utilization of the LOCOS process to produce the channel stop region. Channel stops are thermally grown oxides that serve to isolate the charge packets in one column from those in another. These channel stops are produced before the polysilicon gates are, as the LOCOS process utilizes a high temperature step that would destroy the gate material. The channels stops are parallel to, and exclusive of, the channel, or "charge carrying", regions. Channel stops often have a p+ doped region underlying them, providing a further barrier to the electrons in the charge packets (this discussion of the physics of CCD devices assumes an electron transfer device, though hole transfer is possible).

One should note that the clocking of the gates, alternately high and low, will forward and reverse bias the diode that is provided by the buried channel (n-doped) and the epitaxial layer (p-doped). This will cause the CCD to deplete, near the p-n junction and will collect and move the charge packets beneath the gates - and within the channels - of the device.

3.6.3. Architectures of CCD Camera

The CCD image sensors can be implemented in several different architectures. The distinguishing characteristic of each of these architectures is their approach to the problem of shuttering.

The most common are as follows :

1. Full-frame,
2. Frame-transfer
3. Interline transfer.

In a **full-frame device**, all of the image area is active and there is no electronic shutter. A mechanical shutter must be added to this type of sensor or the image will smear as the device is clocked or readout.

With a **frame transfer** CCD, half of the silicon area is covered by an opaque mask (typically aluminum). The image can be quickly transferred from the image area to the opaque area or storage region with acceptable smear of a few percent. That image can then be read out slowly from the storage region while a new image is integrating or exposing in the active area. Frame-transfer devices typically do not require a mechanical shutter and were a common architecture for early solid-state broadcast cameras. The downside to the frame-transfer architecture is that it requires twice the silicon real estate of an equivalent full-frame device; hence, it costs roughly twice as much.

The **interline architecture** extends this concept one step further and masks every other column of the image sensor for storage. In this device, only one pixel shift has to occur to transfer from image area to storage area; thus, shutter times can be less than a microsecond and smear is essentially eliminated. The advantage is not free; however, as the imaging area is now covered by opaque strips dropping the fill factor to approximately 50% and the effective quantum efficiency by an equivalent amount. Modern designs have removed this characteristic by adding micro lenses on the surface of the device to direct light away from the opaque regions and on the active area. Micro lenses can bring the fill factor back up to 90% or more depending on pixel size and the overall system's optical design.

The choice of architecture comes down to one of utility. If the application cannot tolerate an expensive, failure prone, power hungry mechanical shutter, then an interline device is the right choice. Consumer snap-shot cameras have used interline devices. On the other hand, for those applications that require the best possible light collection and issues of money, power and time are less important, the full-frame device will be the right choice. Astronomers tend to prefer full-frame devices. The frame-transfer falls in between and was a common choice before the fill-factor issue of interline devices was addressed. Today, the choice of frame-transfer is usually made when an interline architecture is not available, such as in a back-illuminated device.

Most common types of CCDs are sensitive to near-infrared light, which allows infrared photography night-vision devices, and video-recording/photography. For normal silicon based detector the sensitivity is limited to 1.1 μm.

CCDs containing grids of pixels are used in digital cameras, optical scanners and video cameras as light sensing devices. They commonly respond to 70% of the incident light (meaning a quantum efficiency of about 70%) making them far more efficient than photographic film, which captures only about 2% of the incident light.

Applications of CCD

CCDs are used in

1. Digital photography and astronomy (particularly in photometry),
2. Sensors, electron microscopy, medical fluoroscopy,
3. Optical and UV spectroscopy and
4. High speed techniques such as lucky imaging.

3.7. DEVELOPMENT OF THE TELEVISION CAMERA

V. K. Zworykin's iconoscope (1923) was the first successful camera tube in wide use. Its functioning involved many fundamental principles common to all television image pickup devices. The face of the iconoscope consisted of a thin sheet of mica upon which thousands of microscopic globules of a photosensitive silver-cesium compound had been deposited. Backed with a metallic conductor, this expanse of mica became a mosaic of tiny photoelectric cells and capacitors. The differing light intensities of various points of a scene caused the cells of the mosaic to emit varying quantities of electrons, leaving the cells with positive charges proportionate to the number of electrons lost. An electron gun, or "scanner," passed its beam across the cells. As it did so, the charge was released, causing an electrical signal to appear on the back of the mosaic, which was connected externally to an amplifier. The strength of the signal was proportional to the amount of charge released. The iconoscope provided good resolution, but required very high light levels and needed constant manual correction.

The orthicon and image-orthicon camera tubes improved on the iconoscope. They used light-sensitive granules deposited on an insulator and low-velocity scanning. These could be used with lower light levels than required by the iconoscope, and did not require the constant manual manipulation. The vidicon was the first successful television camera tube to use a photoconductive surface to derive a video signal.

Solid state imaging devices were first demonstrated in the 1960s. Today's solid-state television cameras use semiconductor charge-coupled devices or CCDs. Each element in a CCD stores a charge that is determined by the illumination incident on it. At the end of the exposure interval, the charge is transferred to a storage register and the CCD is freed up for the next exposure. The charges in the storage register are transferred to the output stage serially during that time. Although almost all consumer video cameras and camcorders use CCD imagers, camera tubes are still common in professional applications.

SUMMARY

- A TV camera tube is a transducer which converts the variations of light intensity in to variations of electrical current or voltage, known as **video signals.**
- Characteristics of Camera tube as follows:
  - (*i*) Sensitivity to visible light.
  - (*ii*) Spectral response.
  - (*iii*) Resolution
  - (*iv*) Noise level
  - (*v*) Halo effect
  - (*vi*) Contrast
  - (*vii*) Lag characteristics
  - (*viii*) Dark current.
  - (*ix*) Ghost effect.
- The purpose of TV camera tube is to sense each element independently and produce a signal in electrical form, proportional to the brightness of each element. Light from the scene is focused on a photo sensitive surface known as **image plate,** and the optical image is formed with a lens system represent light intensity variation of the scene.
- By making use of photoelectric properties, the image plate than convert different light intensities into corresponding electrical variation.
- There are two types of photo electric effects used for converting variations of light intensity into electrical variations are namely:
  - (*i*) Photo emission
  - (*ii*) Photo conductivity
- **Types of TV Camera Tubes**
  - (*i*) Image Orthicon Camera Tube
  - (*ii*) Vidicon camera tube
  - (*iii*) Plumbicon
  - (*iv*) Solid state scanner
  - (*v*) CCD camera

REVIEW QUESTIONS

1. Define various characteristics of a camera tube.
2. Explain the basic principle of a camera pick-up tube.
3. Describe briefly the types of effects used to converting the variations of light intensity into electrical signals.
4. Describe the construction and working principle of an image orthicon camera tube. Write their advantages and disadvantages.
5. Explain the concept of charge image and charge storage in camera tubes.
6. Describe the construction and explain the working principle of a vidicon camera tube. Discuss how it suffers from image lag.
7. What do you meant by image multiplication and signal multiplication in an image orthicon camera tube. Why is an electron multiplier preferred ?

8. What do you understand by dark current in the vidicon camera tube ?
9. Explain the working of plumbicon camera tube. How halo effect is formed and how it is eliminated ?
10. What do you understand by the resolutions of camera tube ?
11. Compare the characteristics of image orthicon, vidicon and plumbicon.
12. What is charge image ?
13. Why is the sensitivity of image orthicon the highest ?
14. Explain the working of a solid state camera tube, using charge-coupled device.
15. What are the short coming in the earlier plumbicon ? How are these removed in the modern plumbicon ?
16. Discuss the process of photo conduction and photo emission.
17. How a charge image is formed on the photolayer of vidicon camera tube.
18. What do you understand by ghost effect ? Where is it occur and how it is being eliminated ?

MULTIPLE CHOICE QUESTION

1. Electrons from the electron gun strike the target plate of a camera tube with
   (*a*) Varying velocity (*b*) Low velocity
   (*c*) High velocity (*d*) Medium velocity
2. In photo emission the number of electron emitted from the target surface
   (*a*) Varies directly as the wavelength of incident.
   (*b*) Varies directly as the frequency of light
   (*c*) is independent of wavelength
   (*d*) None of these
3. Image lag in vidicon is
   (*a*) Severe (*b*) Negligible
   (*c*) Absent (*d*) Equal to the plumbicon
4. Image orthicon is based on
   (*a*) Photo conduction principle (*b*) Photo emission principle
   (*c*) Any one of two (*d*) None of these.
5. Target plate of plumbicon consists of
   (*a*) Antimony (*b*) Trisulphide
   (*c*) Selenium tin oxide (*d*) Lead oxide
6. Spectral response of plumbicon is poor in
   (*a*) Blue (*b*) Green
   (*c*) Cyan (*d*) Red
7. Vidicon is suitable for outdoor use because
   (*a*) Its variable gamma characteristics
   (*b*) Its high contrast capability
   (*c*) Its self adjustable sensitivity
   (*d*) Its good spectral response
8. Plumbicon has gamma of
   (*a*) 0.7 (*b*) Near unity
   (*c*) 0.5 (*d*) 0.6 to 0.9

9. Halo is formed in

(*a*) Vidicon (*b*) Plumbicon

(*c*) Image orthicon (*d*) None of these

10. IR response is maximum in

(*a*) Vidicon (*b*) Plumbicon

(*c*) Image orthicon (*d*) Newvicon

11. Operation of CCD image sensors is based on

(*a*) Photoconductivity (*b*) Photovoltic principle

(*c*) Photoemission (*d*) None of these

12. The most suitable camera tube for industrial application is

(*a*) Iconoscope (*b*) Image orthicon

(*c*) Standard vidicon (*d*) Plumbicon

13. The approximate value of target voltage in a standard vidicon is

(*a*) 300 V (*b*) 30 V to 40 V

(*c*) Equal to anode voltage (*d*) None of these

14. Scanning in solid state camera is not done by deflecting on electron beam but is done by

(*a*) Principle of shift register (*b*) Principle of picture tube

(*c*) Principle of camera tube (*d*) None of these

15. The image lag in a camera tube

(*a*) Is a problem of beam aperture

(*b*) Is a problem of target plate

(*c*) Leads to appearance of ghosts when the picture is reproduced

(*d*) None of these

16. The most suitable camera for broadcast purpose is

(*a*) Plumbicon (*b*) Image orthicon

(*c*) Standard vidicon (*d*) None of these

17. Charge on pixel point on the surface of the target of a camera tube does not vary from instant to instant, but remains stored for the period of one frame True/False

18. An image orthicon camera target is

(*a*) Always anode potential stabilised

(*b*) Always cathode potential stabilised

(*c*) Either of (*a*) and (*b*)

(*d*) None of these

ANSWERS

1. (*b*)	**2.** (*c*)	**3.** (*a*)	**4.** (*b*)	**5.** (*d*)
6. (*d*)	**7.** (*d*)	**8.** (*b*)	**9.** (*b*)	**10.** (*d*)
11. (*b*)	**12.** (*c*)	**13.** (*b*)	**14.** (*a*)	**15.** (*b*)
16. (*b*)	**17.** True	**18.** (*b*)		

4

CHAPTER

MONOCHROME TELEVISION TRANSMITTERS AND RECEIVERS

INSIDE THIS CHAPTER

4.1. INTRODUCTION

In television, the term monochrome refers for black and white television. Monochrome television system does not process variations of colour in a picture but processes only the intensity of brightness of each pixel for transmission in space. The picture is reproduced on the florescent screen of the monochrome picture in a receiver, in the shades of black and white.

A simple block diagram of television transmitter and receiver is already explained in the chapter 1. We shall be discussing these topics in details in this chapter.

4.2. TELEVISION TRANSMITTERS

The RF television signal can be transmitted by two different methods as follows:

1. Direct modulation (using high level modulation) and
2. IF modulation (using low level modulation)

4.2.1. Direct Modulation

If the modulation takes place in the output stage of the picture transmitter known as **high level modulation**, the RF video carrier is first brought to the required driving power and then, with simultaneous amplitude modulation, amplified in the output stage to the nominal video carrier output power of the transmitter. The modulation amplifier boosts the wideband CVS to the level required for amplitude modulation in the output stage. The sound carrier is frequency-modulated with a small deviation at a relatively low frequency. The final frequency and the actual frequency deviation are produced via multiplier stages. The picture and sound transmitter output stages are fed to the common antenna via the video/sound diplexer.

4.2.2. IF Modulation

Second method of modulation is known as low level modulation using **IF modulation.**

When using IF modulation, first the IF vision carrier of 38.9 MHz is amplitude Modulated. The subsequent filter produces vestigial sideband AM. One or two sound carriers are also frequency modulated at the IF. Next, mixing with a common carrier takes place both in the video and in the sound channel so that the video /sound carrier spacing of 5.5 MHz is maintained at the RF. Linear amplifier stages boost the vision and sound carrier powers to the required level.

The advantage of the second method (IF modulation) is that the actual processing of the RF television signal is carried out at the IF, thus at a lower frequency, and band- and channel-independent. However, for further amplification, stages of high linearity are required, at least in the picture transmitter.

The IF modulated transmitter has many advantages such as simplicity, ease of operation and superior performance.

4.2.3. Monochrome TV Transmitter

Picture Signal Transmission

The principle of TV transmission with a view to reproducing black-and-white pictures can be summarized as follows: the optical image of the scene to be transmitted is divided into small picture elements (pixels). An opto-electrical converter, usually a camera tube, consecutively translates the individual elements into electrical information depending on their brightness.

This signal is then transmitted at its actual frequency or after modulation onto an RF carrier. After appropriate processing at the receiving end, the information is applied to an electro-optical converter and reproduced in accordance with the brightness distribution of the pattern.

Sound Signal Transmission

In TV broadcasting the sound signal is transmitted by frequency-modulating the RF sound carrier. In accordance with the relevant CCIR standard, the sound carrier is 5.5 MHz above the associated video carrier.

The maximum frequency deviation is 50 kHz. Due to certain disturbances in colour transmission, the original sound/vision carrier power ratio of 1:5 was reduced to 1:10 or 1:20. Even in the latter case no deterioration of the sound quality was apparent if the signal was sufficient for a satisfactory picture.

As mentioned above, the intercarrier sound method is used in most TV receivers. The difference frequency of 5.5 MHz is obtained from the sound and vision carrier frequencies. This signal is frequency modulated with the sound information.

The frequency of the intercarrier sound is constant and is not influenced by tuning errors or variations of the local oscillator. More recent studies have shown further possibilities of TV sound transmission, in particular as to transmitting several sound signals at the same time. A second sound channel permits, for instance, multilingual transmission or stereo operation. With the dual-sound carrier method, an additional sound carrier 250 kHz above the actual sound carrier is frequency modulated, its power level being 6 dB lower than that of the first sound carrier. A multiplex method offers further possibilities by modulating an auxiliary carrier at twice the line frequency or using the horizontal or vertical blanking intervals for pulse code modulation.

A simplified block diagram of a TV transmitter is shown in Fig. 4.1.

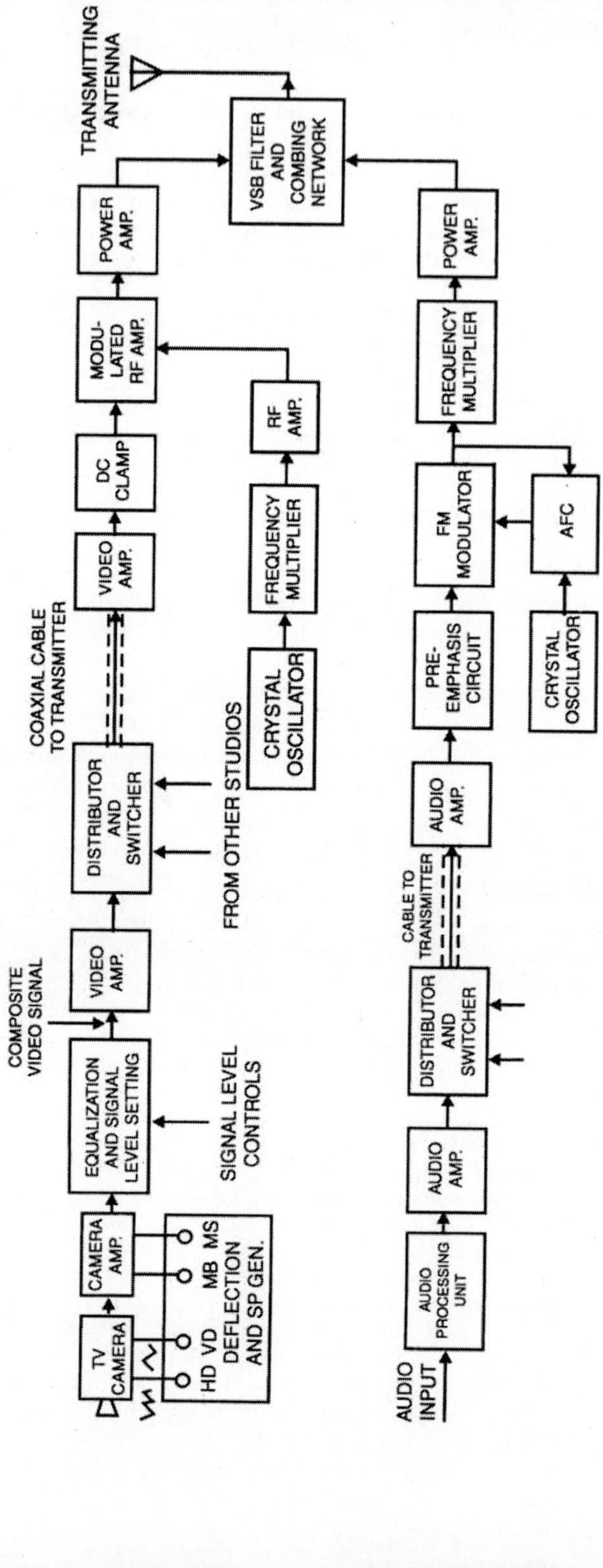

Fig. 4.1. Block diagram of TV transmitter.

Elements of Monochrome TV Transmitter

Picture Sections

1. Video Camera
2. Scanning & Synchronizing
3. AM Modulator (for picture)
4. Video pre-amplifier and Power amplifier
5. Crystal Oscillator
6. Frequency Multiplier
7. Video Modulator
8. VSB Filter

Sound and Common Sections

1. Audio Amplifier
2. FM Modulator (for sound)
3. Combining Networks or Diplexer
4. Transmitting Antenna

Picture Section

A **TV camera** is used to the optical information in to electrical signal and its amplitude varies according to the variation of the brightness.

Scanning is used to convert the optical information in to electric form and its transmissions are carried out element by element, one at a time and in a sequential manner to cover the entire scene which is to be televised. Scanning of the element is done at a very fast rate and this process is repeated at large number of times per second to create an illusion of simultaneously pickup and transmission of picture details. This is made possible by deflecting a focussed electron beam of electrons on the charge image horizontally and vertically. The deflecting circuits are linear sawtooth signals. Scanning of one horizontal line takes 64 μs including 12 μs time of retrace. Vertical scanning is relatively slow and its scanning period is 20 ms including 1.280 ms time taken in retrace from bottom to top.

Blanking and Sync Pulses

The scanning is done line by line horizontally from left to right at a fast rate and vertically from top to bottom at a slower rate and it is known as **Trace.** The **retrace** of the beam is very fast as compared to the forward scan. Scanning during retrace or flyback is blanked by making use of horizontal and vertical blanking pulses. The composite video signal contains blanking pulses to make the retrace lines invisible by raising the signal amplitude slightly above the black level (75 %) during the time the scanning circuit produces retraces.

Sync pulses are required at the horizontal and vertical traces to trigger the respective scanning oscillators.

Line sync pulse is used to trigger the scanning circuits to ensure synchronization for scanning at the transmitter and at the receiver. It must starts at the same instant of time in the transmitter and in receiver for correct synchronization. Vertical blanking pulse carries Vertical sync pulse which triggers the vertical sweep oscillator for synchronization.

The width of the vertical sync pulse must be larger than the period of horizontal sync pulse and it is because to derive a suitable field or vertical sync pulse at the receiver to trigger the vertical sweep oscillator.

Video Pre-amplifier and Power Amplifier

These amplifiers are wide band amplifiers connected in cascaded to amplify the composite video signal from a few Hz to 5 MHz. Finally by the power amplifier to give sufficient power required for amplitude modulation.

DC restorer circuit is used to provide dc clamping before the modulation.

Crystal Oscillator and Frequency Multiplier

A crystal oscillator is used to produces highly stable frequency which is a sub-multiple of the actual allotted frequency. Broadcast transmitters transmit at one allotted channel frequency. The transmitted frequency should not varies or drift and hence crystal oscillator is used which produces highly stable frequency. Frequency multipliers are employed to get the desired VHF or UHF channel frequencies known as video carrier.

Amplitude Modulator (Video)

A TV camera is used to convert the optical information in to electrical signal and its amplitude varies according to the variation of the brightness. The electrical information obtained from the T V camera tube is generally referred to as video signal. This signal is amplified and modulated (Amplitude) with the channel picture carrier frequency.

The amplitude modulator is a non-linear amplifier which changes the amplitude of the RF carrier in accordance with amplitude of the video signal. Negative modulation, having negative polarity of the video signal is used in television transmission, since it causes saving in the transmitted power because of the fact the white forms a dominant portion in a scene is of low level. It also reduces the noise in the reproduced picture. The modulated output is fed to the Tx antenna for radiation along with the sound signal. (Amplitude modulation is already explained in chapter-2)

VSB Filter

Due to the fact that the modulation signals reach very low frequencies, sharp cut-off filters are required; however, the group-delay distortion introduced by these filters at the limits of the pass band causes certain difficulties.

The problem is eluded by using vestigial sideband amplitude modulation (VSB AM) instead of SSB AM. In this case, one complete sideband (carrier frequency plus 5 MHz as Upper Sideband) and part (0.75 MHz of Lower Sideband) of the other are transmitted. A sharp filter is used to a covert the modulator's output which has double sidebands, referred as amplitude modulated vestigial sideband (AMVSB) type. The purpose of AMVSB is to accommodate more channels for transmission by reducing the overall bandwidth 5.75 MHz.

Sound Section

Microphone is used to converts the voice in to the electrical variations, known as audio signals. In the television studios, a dynamic microphone is used widely which uses the principle of electro-magnetic induction for converting the sound pressure variations in to electrical signals.

Audio Amplifier

The output of the microphone is producing very weak electrical signals, known as audio signals, are to be amplified using multistage of RC coupled, class-A amplifier.

Pre-emphasis circuit emphasises the low energy of the high audio frequencies. It enhances the frequency deviation, which results in high signal to noise ratio in FM system.

FM Modulator (Sound)

The audio signal which we have got from the microphone is being amplified up to the adequate level. The amplified output is now Frequency-modulated over the assigned carrier frequency. Before the power amplifier, low level modulation is used to save on the audio requirements. Since a crystal is used to generate the carrier frequency, direct frequency modulation is not suitable because it is difficult to change the frequency of a crystal. An indirect method is used, known as Armstrong method, in which phase modulation of the crystal is achieved. To convert the phase modulation in to frequency modulation, the change is phase is made dependent on the modulating frequency by an equalising circuit.

Frequency multipliers are used to obtain the final allotted frequency in VHF or UHF band. In accordance with the relevant CCIR standard, the sound carrier is 5.5 MHz above the associated video carrier.

Diplexer

In the combining network both the audio and video signals are combined, using a Wheatstone bridge known as **diplexer.** It provides isolation of the audio signal to the video signal and vice-versa.

Video carrier and audio carrier are kept so close to each other that they both can be accomodated in the same channel bandwidth. By this arrangement, both the signals are transmitted by a common antenna. However, there is one problem in this, that video modulated signal would enter the audio section, and audio modulated signal would enter the video sections in the transmitter. This problem is overcome by passing both the signals through a wheatstone type bridge, such bridge is known as **Diplexer**, as shown in Fig. 4.2.

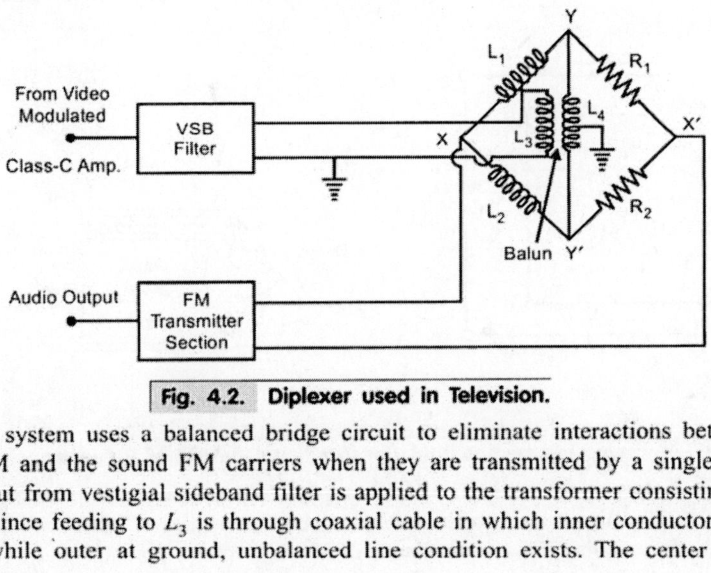

Fig. 4.2. Diplexer used in Television.

This system uses a balanced bridge circuit to eliminate interactions between the video AM and the sound FM carriers when they are transmitted by a single antenna. The output from vestigial sideband filter is applied to the transformer consisting coil L_3 and L_4. Since feeding to L_3 is through coaxial cable in which inner conductor is above ground while outer at ground, unbalanced line condition exists. The center taped L_4

converts this input to a balanced line section and is called Balun. (Balance to unbalance). Bridge circuit consists of two inductive reactances L_1 and L_2 which are coaxial cable sections. Other two areas consist R_1 and R_2 which represet radiators of the transmitting antenna.

FM sound carrier is impressed across point X and X', while picture signal from the L_4 balun section appears at Y and Y'. With balanced bridge, equal voltages exist on both the reactive as well as the resistive arms. Hence, sound modulated carrier is present in both antenna radiators R_1 and R_2. But across points Y and Y' there is zero voltage for some modulated carrier because any potential at these points will have the same polarity due to which no current flows through L_4 (from sound modulated carrier). Thus interaction between sound and video modulated carriers does not occur. On the other hand, video modulated carrier applies at Y and Y' also setup equal voltage across the arms of the bridge but because of the balanced bridge systems no voltage corresponding to this video modulated carrier will appear at X and X'. Therefore again interaction between sound and video modulated carrier is prevented.

In short, both modulated carriers develop equal voltages across arms of the bridge and at the same time no video modulated carrier is present at points at which sound modulated carrier appears and vice versa.

The turnstile antenna, usually used for transmissions consists of two dipoles perpendicular to each other. The bridge is designed to remain in balanced conditions, and hence the interference between the video and audio carriers is completely prevented. At the same time, two signal can be fed to the common turnstile antenna for further radiation in the space.

4.3. BALUN

Balun (Balance to unbalanced) is a matching transformer specially designed for impedance matching. It is used to connecting a balanced circuit to an unbalanced circuit and vice versa.

Two λ/4 lines are combined to make a balancing and impedance transforming unit shown in Fig. 4.3.

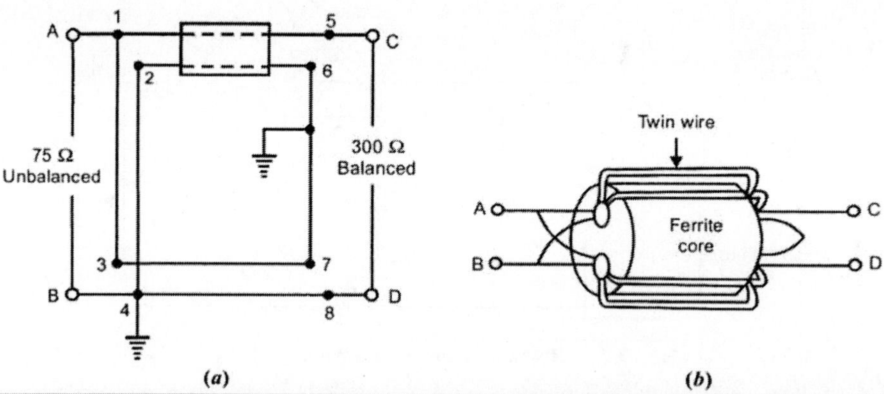

Fig. 4.3. (a) λ/4 matrching sections, (b) Constructional details of Balun transformer.

Two λ/4 lines having an impedance equal to 150 Ω are connected in parallel at one end, resulting in 75 Ω impedance across points A and B, whereas at other end they are connected in series, resulting in an impedance of 300 Ω between points C and D. It can match balanced and unbalnaced impedances. Points A or B can be grounded in order to provide an unbalanced impedance at the ungrounded point with respect to ground. On the other side, $\lambda/4$ length of the line isolates the ground point from C or D, allowing a balanced impedance with respect to ground. Thus it can match balanced 300 Ω twin feeder line to a 72 Ω unbalanced input etc.

The lines in balun are coiled to make the unit compact. Bifilar windings are used to stimualte the 150 Ω transmissions line sections and are wound on a ferrite core to increases the inductance. Generally it consist of a two-window ferrite core with two windings of four turns each, coupled to provide a balanced to unbalanced match and an impedance transformation from 75-300 Ω.

4.4. SERVICE AREA OF TRANSMISSION

Depending upon the transmitter power and other factors the service area may extend upto 120 km for the channels in VHF band and approximately 60 km for the UHF channels.

As a result of some observations, show that for a transmitting antenna height of 225 m above ground level the radio horizon is 60 km. If the height of receiving antenna is 16 m above ground level the total distance is increased to 76 km. Sometimes, a greater distance between antennas may be obtained by locating them on top of tall buildings or hills. However, links more than 120 km are rarely used for TV transmission because of limitation of radiated power, high channel frequencies and antenna heights.

There are some areas which are either shadowed by mountains or are too far away from the transmitter for satisfactory televisions reception. In these booster station can be used. A booster station must be located at such a place where it can receive and rebroadcast the programme to receivers in adjoining area.

4.5. TRANSMITTING ANTENNA

The transmitting antenna, normally a turnstile antenna, is used to transmit the combined signals connected through the coaxial cable. The antenna is mounted on a high tower situated close to the transmitter. The turnstile antenna consists of two dipoles perpendicular to each other, so that the radiation pattern is Omni directional in the horizontal plane. The antenna is mounted horizontally on the high tower, so that better signal to noise ratio is achieved.

As TV signals are transmitted by space wave, the transmitting antenna is installed at a sufficient height in order to increase the line of sight distance. Further, horizontal polarisation is preferred because it does not pick-up man made interference to a great extent *i.e.*, signal to noise ratio is favourable for horizontally polarised wave, especially when the antenna is located for remote from the ground.

4.5.1. Turnstile Antenna

A turnstile antenna is used as a transmitting antenna, consisting of two $\lambda/2$ dipoles placed at right angles to each other and fed 90° out of phase with each other as shown in Fig. 4.4. This arrangement of the $\lambda/2$ dipoles is called as **turnstile array.**

The array has two features :

(*i*) The two dipole patterns combining in the manner shown in Fig. 4.4(*b*), produce an almost circular pattern in the plane of turnstile. That is, it has an omnidirectional radiation pattern.

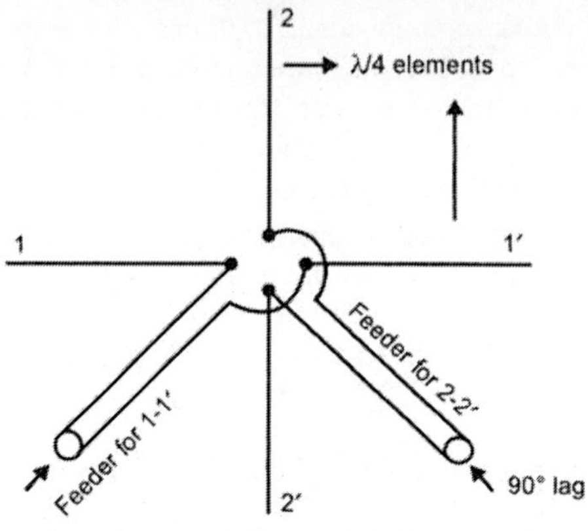

(*a*) **Turnstile array**

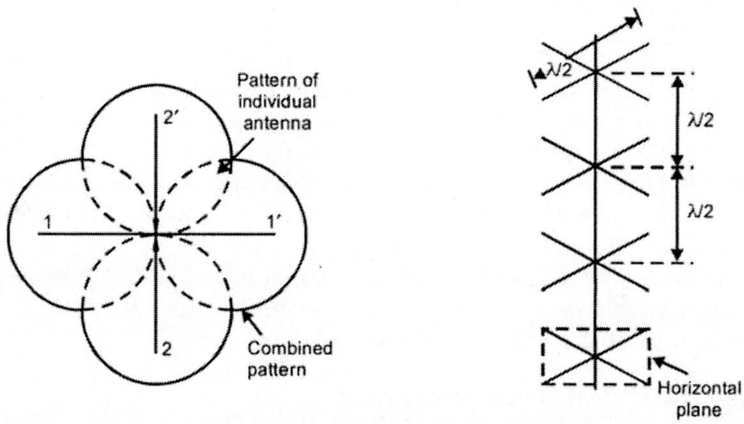

(*b*) **Radiation pattern of each dipole and resultant pattern.** (*c*) **Stacked turnstile array.**

Fig. 4.4.

(*ii*) Second features is that the radiative pattern is polarized in the same plane. That is, if array is mounted in the horizontal plane, the antenna would radiate horizontally polarised waves about equally well in all directions along the ground.

Severals of these turnstiles may be stacked along a vertical axis shown in Fig. 4.4(*c*) and phased so as to improve the radiation directivity along the ground.

4.5.2. Dipole Panel Antenna System

In this system, four panels are mounted on the four sides of the antenna tower. Each panel consists of an array of full wave dipoles mounted in front of reflection, as shown in Fig. 4.5 (*a*) and (*b*)

The combined radiation pattern of four tower mounted dipole panels is shown in Fig. 4.4(*b*). This antenna system is often used for band I (47 to 68 MHz) and band III (174-263 MHz).

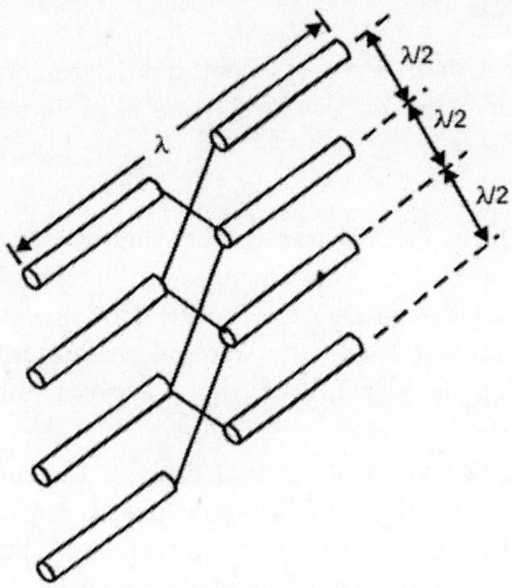

(*a*) Panel of dipoles

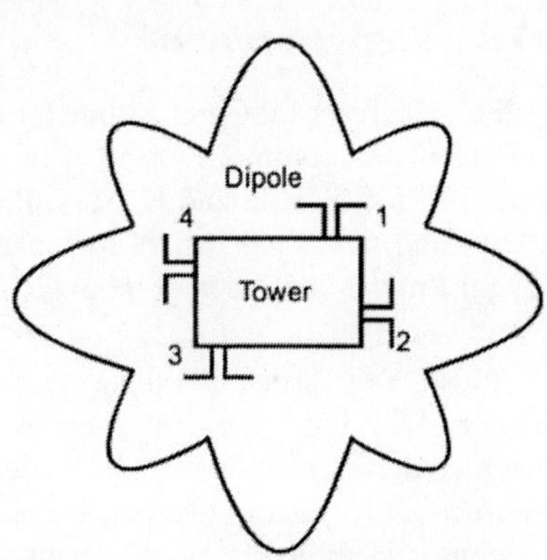

(*b*) Radiation pattern of four tower mounted dipole antenna panels.

Fig. 4.5.

4.6. COMBINING NETWORK

We know that through diplexer, AM picture signal and FM sound signal are fed to the same antenna. Equivalent bridge circuit of a diplexer for feeding picture and sound transmitters to a common turnstile array is shown in Fig. 4.6. It is a bridge configuration in which first two arms are formed by the two radiators of the turnstile antenna and the other two arms consists of two capacitive reactances. Under balanced conditions, picture and sound signals through radiated by the same antenna do not interfere with the functioning of the transmitter.

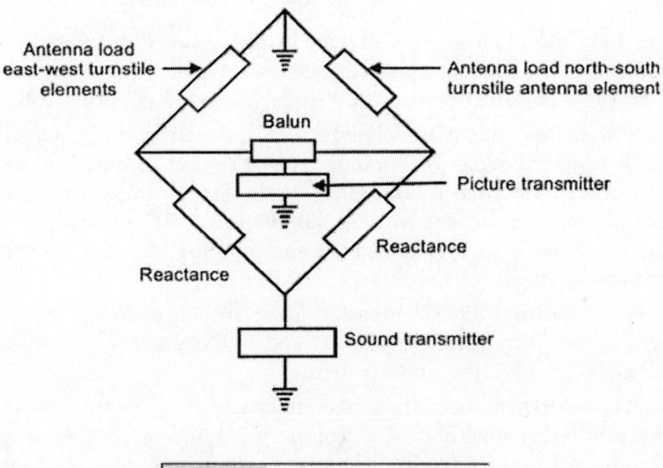

Fig. 4.6. Combining network.

4.7. TV RECEIVING ANTENNAS

For both VHF and UHF TV channels, half wave resonant dipole, remote from ground, is the basic receiving antenna. For better gain yagi antenna is preferred.

4.7.1. Yagi Antenna

For receivers that are within 60 km from the TV transmitter, yagi-uda or simply yagi antenna is commonly used. Yagi-uda antenna or yagi antennas are known after the names of Prof. S. Uda and H. Yagi. Prof. S. Uda and H. Yagi. Prof. S. Uda invented this antenna and published papers in Japanese around 1928. Afterwards it was described by H. Yagi English and became popular all over the world, Prof. Yagi has given full credit to Prof. S. Uda.

As in Yagi array, it consists of a driven element, a reflector and one or more directors. The basic element used in yagi antenna is a λ/2 dipole placed horizontally known as driven element or active element. In order to convert bidirectional dipole into unidirectional system, the passive elements or parasitic elements are used, namely reflectors and directors as shown in Fig. 4.7.

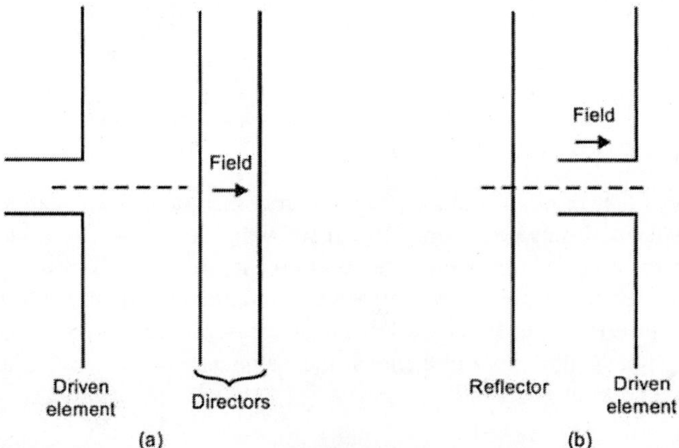

Fig. 4.7.

For the field calculations at a certain distant point developed by such parasitic element end fire array, the phases of currents in parasitic dipole antenna are of utmost concern. Phases of currents in parasitic elements depend upon the following factors :

***(i)* Length of the parasitic elements :** A parasitic dipole cut exactly a half wavelength or slightly longer will be inductive (λ/2 wave antenna can be approximated with λ/4 wave open circuited transmission line, refer impedance curve of such a line) and thus the phase of its current will lag the induced 'emf' while a parasitic dipole cut shorter than a λ/2 wave will be capacitive and thus the phase of the current in it will lead the induced emf.

***(ii)* Spacing Between the Elements :** Close spacing between the elements is used is parasitic array to obtain good excitation and in the process of excitation, induction field of the elements play the important role.

The exact analysis of such array is complicated. It is known, however that when the elements are shorter than λ/2 wave and are properly spaced, they reinforce the field of the driven element in a direction away from the driven element. Such elements are called as **directors.** Thus if a director is placed infront of a driven element, it will direct the field in a direction away from the driven element to the right in Fig. 4.8. Thus if there is line of directors, each one will excite the next one.

On the other hand, when the elements are slightly longer than λ/2 or exactly λ/2 and are correctly spaced, they reinforce the field of the driven element in a direction towards the driven element or more precisely. They reflect the field towards the driven element. Such elements are called **reflectors.**

Thus if a reflector is placed adjacent to the driven element, any element placed beyond the reflector will receive negligible field and will not be appreciably excited (to the left side of reflector in Fig. 4.8(*c*).

A typical 3 element Yagi antenna suitable for TV reception of moderate field strength is shown in Fig. 4.8.

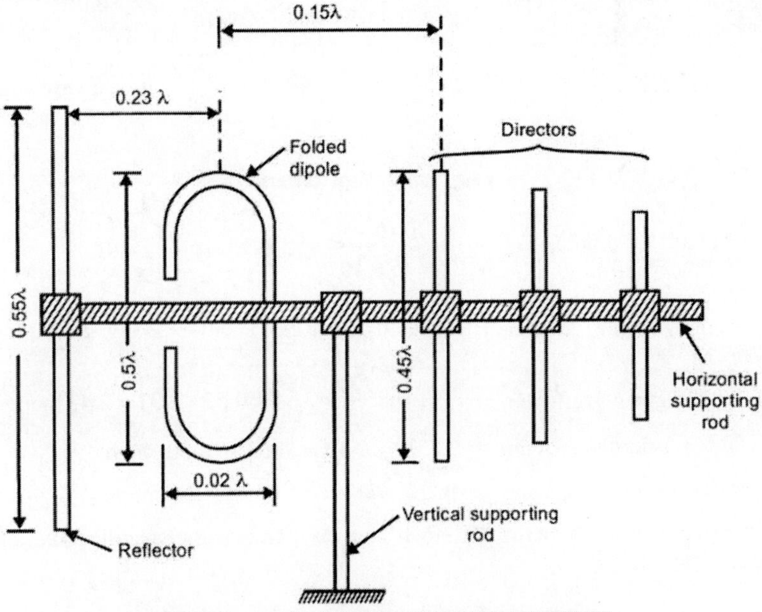

Fig. 4.8. Yagi antenna with folded dipole.

A yagi end fire array usually consists of one driven element, one reflector on one side of it and a number of directors on the other side of it as shown in Fig. 4.8. Directive gain of around 10 dB is readily achieved with a moderate number of elements (five or six).

A Yagi array has the following advantages

(*a*) They produce a uniform directional beam of moderate directivity,

(*b*) They employ an easy way of feeding the system simplicity of feed system design and low cost. The input impedance of Yagi array tends to be low and the beam width is limited to around 20% typically. To obtain higher gains, broadside array, consisting many Yagi arrays as constituent element is used.

In fringe areas where the signal level is very low, high gain antenna arrays are needed. The gain of the array increases with number of elements employed. A Yagi antenna arrays serves well The reflectors are usually 5% longer than the dipole and may be spaced from at 0.15 ‰ to 0.25 ‰ depending on the design requirment. Similarly directors may be 4% shorter than the dipole but where broad band characteristics are needed, successive directors are usually made shorter. Suppose a channel has a centre frequency 65 MHz (channel from 62.25 MHz to 67.75 MHz). Then

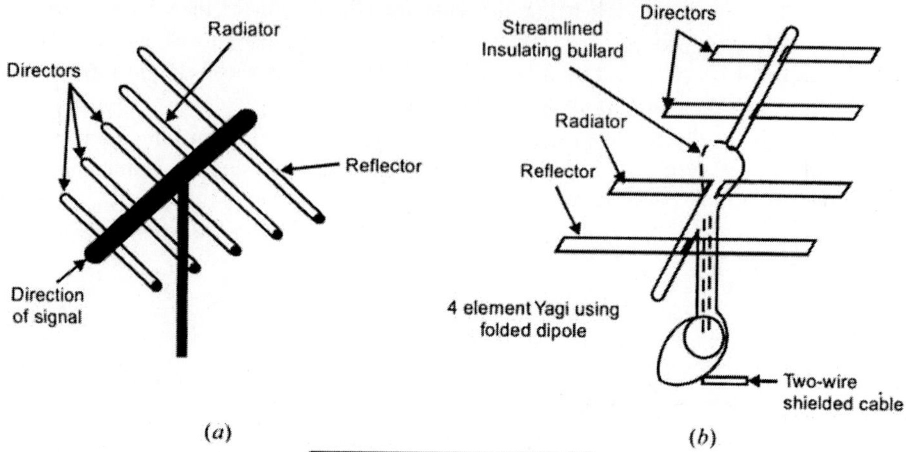

Fig. 4.9. Yagi antennas.

$$\text{Length of dipole, } \lambda/2 = \frac{3.10^8}{2 \times 65 \times 10^6} = 2.31 \text{ metre}$$

$$\text{Length of the director} = \frac{\lambda}{2} - 0.04 . \frac{\lambda}{2} = 2.31 - 0.04 \times 2.31 = 2.22 \text{ metre}$$

$$\text{Length of reflector} = \frac{\lambda}{2} + 0.05\frac{\lambda}{2} = 2.31 + 0.05 \times 2.31 = 2.43 \text{ metre}$$

$$\text{Director-dipole spacing} = 0.13\lambda = 0.13 \times 4.62 = 0.60 \text{ metre}$$

$$\text{Reflector-dipole spacing} = 0.25\lambda = 0.25 \times 4.62 = 1.15 \text{ metre}$$

In general, for conductor of 1 to 1.2 cm in diameter the specifications of any Yagi antenna array are :

$$\text{Length of dipole} = \frac{143}{f} \text{ metre } (f \text{ in MHz})$$

$$\text{Length of reflector} = \frac{152}{f} \text{ metre } (f \text{ in MHz})$$

$$\text{Length of first director} = \frac{137}{f} \text{ metre } (f \text{ in MHz})$$

Length of successive directors reduces by 2.5%

Spacing between reflector and dipole = 0.25λ

Spacing between director and dipole = 0.13λ

Spacing between director and director = 0.13λ

4.7.2. Multiband Antennas

It is not possible to receive all channels of lower and higher VHF band with one antenna because (*i*) The directional pattern of a low-band dipole splits into side lobes (hence loss in directive gain) at the third and fourth harmonics frequencies in the 174 MHz to 233 MHz band (VHF band-III). Thus it is suitable for this band.

(*ii*) On the other hand, a high-band dipole cut for a half wavelength ($\lambda/2$ long) in the 174-233 MHz band is not suitable for the 47 to 68 MHz band (VHF, band-1) because of in sufficient signal pick-up at the lower frequencies.

Thus separate dipoles are used for both VHF bands because a broad side response cannot be maintained if one dipole is used for both the bands.

4.7.3. Colour TV Antenna

In black and white (monochrome) TV receiver emphasis is on higher gain which varies from channel to channel because of wide frequency range. Gain not only varies from channel to channel but also one end of the channel to the other end.

(Figure 4.10 for channel-4, from 61 to 68 MHz, showing a variation of 4 dB upto the end). It means if this antenna is used for colour TV reception, colour signal which lies around 66-68 MHz will receive almost 2 dB less gain. This will result in poor colour picture quality. Further, in colour TV, relative phase angle of the combined colour difference signal phasor determines the colour in the picture. Therefore a change in gain, which is accompanied by a phase shift of the incoming signal, will change the hub in the picture.

For example, a slight phase shift of the colour signal phasor can turn red colour to orange, etc. Even no colours are produced if phase shift is large. Therefore, for faithful reproduction of picture, a colour TV antenna should have a flat response over the entire channel. Log periodic antenna is such an antenna.

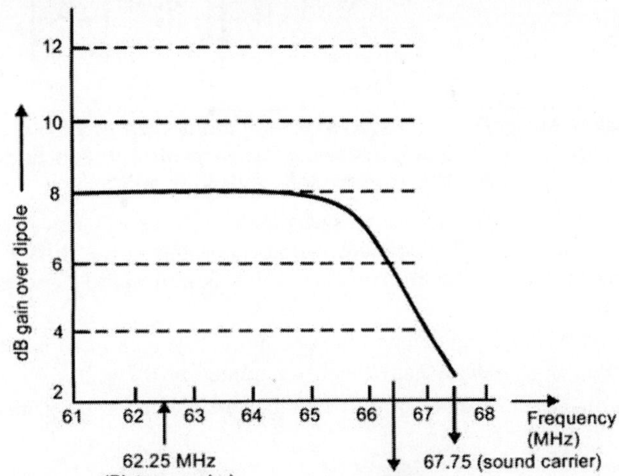

Fig. 4.10. Response curve of a wide range antenna for channel 4.

4.7.4. Log Periodic Antennas

A frequency independent antenna may be defined as the antenna for which the impedance and pattern remain constant as a function of the frequency. Log periodic antennas are broad band antennas. The bandwidth of 10 : 1 to 100 : 1 is feasible if the theoretical design is closely approximated. The antenna impedance and pattern depends upon the system shape and dimensions. For broadband characteristics, the dimensions are varied and to obtain high gain an array of different log antennas as elements can be used.

The spacing and dimensions are varied according to log and hence the system is known to be **log periodic antenna system.** The geometry of log periodic antenna structure is so selected that electrical properties must repeat periodically with the logarithm of the frequency.

The design of log periodic antenna involves a basic geometrical structure which is repeated but with a changing size of the structure. The size of the structure varies with each repetition by a constant scale factor so that it can expand or contract. The log periodic principle can be understood by the "Log period dipole array of Fig. 4.11.

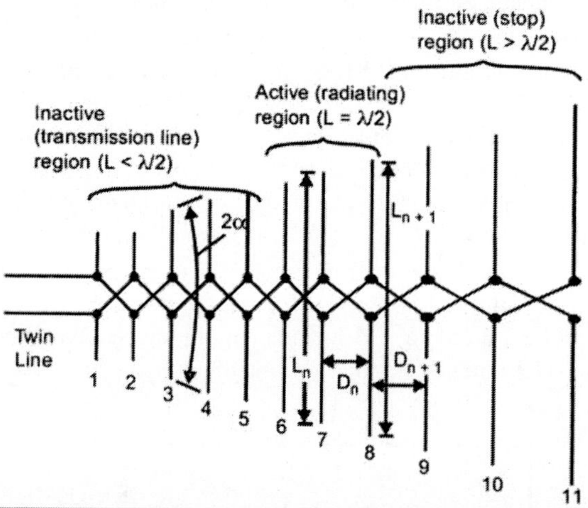

Fig. 4.11. Isbell log-periodic frequency-independent type of dipole array of 7 dB gain with 11 dipoles showing active central region and inactive regions (left and right ends).

The system consists of more than one dipole in which distance between adjacent elements and length of the elements are varied according to log. It is fed at the narrow end by a balanced 2-wire transmission line which is transposed between each adjacent pairs of dipoles.

The scale factor (τ) depends upon the operating bandwidth and number of elements. Usually the value of τ is less than 1 and maintained at 0.7.

Thus lengths (L) and spacings (D) of adjacent elements are scaled such that are related as

$$\frac{D_1}{D_2} = \frac{D_2}{D_3} = \frac{D_3}{D_4} \dots = \frac{D_n}{D_{n+1}} = \tau$$

$$\frac{L_1}{L_2} = \frac{L_2}{L_3} = \frac{L_3}{L_4} \dots = \frac{L_n}{L_{n+1}} = \tau$$

$$\frac{D_n}{D_{n+1}} = \frac{L_n}{L_{n+1}} = \tau \qquad \text{...(4.1)}$$

where τ is a constant and is also known as **design ratio or periodicity factor.**

The resonance frequency f_1 of an element of length L_1 is obtained from

$$L_1 = \frac{\lambda}{2} = \frac{c}{2f_1}$$

or

$$f_1 = \frac{c}{2L_1}$$

Similarly, the resonance frequency of an element of length L_2 is

$$f_2 = \frac{c}{2L_2}$$

For 'n' number of elements the operating bandwidth is approximately.

$$BW = f_1 - f_n$$

The frequencies should satisfy the following condition.

$$\frac{f_1}{f_2} = \frac{L_2}{L_1} = \frac{1}{\tau}$$

There are three regions in the antenna :

(*i*) **Transmission line region or inactive region :** Where $L < \lambda/2$. It is the directive region of array which acts like the director for the resonated elements ($L < \lambda/2$).

(*ii*) **Active region :** This is the region of log periodic array which resonates within the operating frequency band in which the length of each element is $L \simeq \lambda/2$.

(*iii*) **Reflective region :** Reflective region acts like reflector for the resonated elements within which L is more than $\lambda/2$ ($L > \lambda/2$).

Including the inactive regions the system acts like a parasitic array which is a perfect unidirectional. The gain for properly designed antenna is between 7.5 dB to 12 dB with respect to isotropic antenna. It's input impedance lies between 50 Ω to 200 Ω.

Consider two elements of a log periodic array which are separated by a distance D. Their respective lengths are L_n and L_{n+1} so that their half lengths are given by $L_n/2$ and $L_{n+1}/2$. The geometry of log periodic array for determing the relation of parameters is shown in Fig. 4.12. The general configuration of a log periodic antenna is defined by three parameters such as τ, D/λ and α which can be determined from

$$\tan(\alpha/2) \quad \frac{(L_{n+1} - L_n)/2}{D}$$

$$\tan(\alpha/2) = \frac{\left[1-\left(\frac{1}{\tau}\right)\right]\left[\frac{L_{n+1}}{2}\right]}{D}$$

Taking

$$L_{n+1} = \frac{\lambda}{2} \text{ (active)}$$

$$\tan\left(\frac{\alpha}{2}\right) = \frac{1-\left(\frac{1}{\tau}\right)}{\frac{4D}{\lambda}}$$

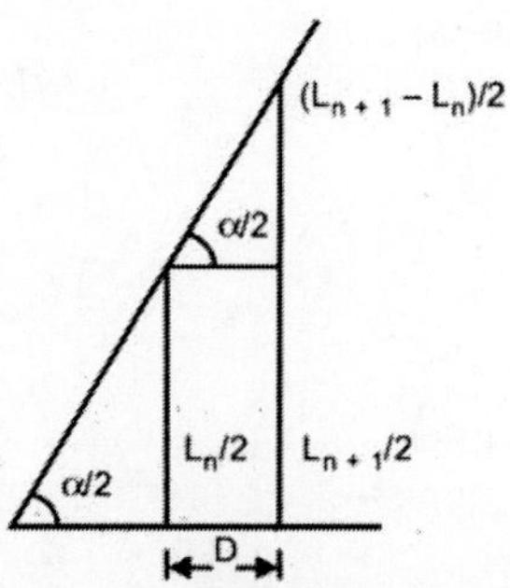

Fig. 4.12.

where α = apex angle, τ = scale factor

D/λ = spacing in wavelength shortward of $\lambda/2$ elements. Hence we see that by specifying two parameters, the third one can be determined.

General Characteristics of Log-Periodic Antenna

1. This antenna is excited from the shortest length side or high frequency side to obtain optimum directivity.
2. In active region, the magnitude and phasing of the current should be proper so that maximum radiation occurs along the backward direction, *i.e.*, towards shorter element and minimum (or zero) radiation occurs along forward direction, *i.e.*, towards longer element.
3. Typical values for unidirectional radiation are 90° phase and $\lambda/4$ spacing.
4. There should be a rapid decay of current in the inactive region.

Log-Periodic Slots

A slot in a metallic plate is a self complementary slot and can be used an antenna. The resonance frequency of log-periodic slots or complementary slots depends upon the length L provided the width 'W' is very small as comapred to λ. For wide band operation we require more than one slot and their dimensions and distances are varied according to log.

Figure 4.13 shows a log periodic array of complementary slots. This arrangement is bidirectional feeding maximum power in a plane normal to the surface of the plane.

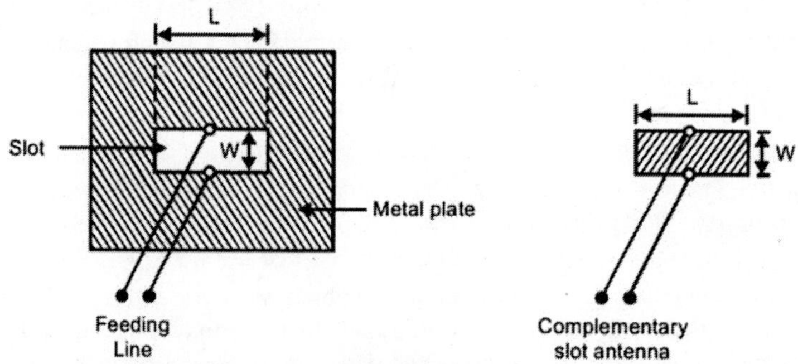

Fig. 4.13.

$$\frac{L_1}{L_2} = \frac{L_2}{L_3} = \cdots \frac{L_n}{L_{n+1}}$$

$$\frac{R_1}{R_2} = \frac{R_2}{R_3} = \cdots \frac{R_n}{R_{n+1}}$$

$$\frac{W_1}{W_2} = \frac{W_2}{W_3} = \cdots \frac{W_n}{W_{n+1}}$$

$$\frac{L}{L_{n+1}} = \frac{R_n}{R_{n+1}} = \frac{W_n}{W_{n+1}} = \tau$$

This type of log periodic slots can be converted into unidirectional by bending the system at the centre as shown in Fig. 4.14. The directivity depends upon the angle of θ between the plates.

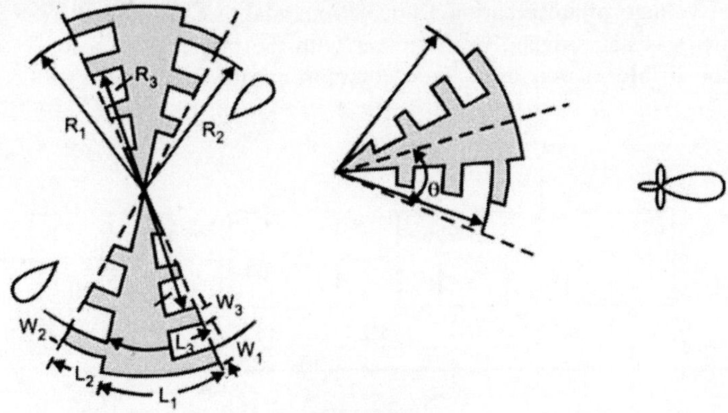

Fig. 4.14.

Uses of Log Periodic Antennas

(1) Log periodic antenna is used in television reception where a single antenna can receive all the channel in VHF and UHF band.

(2) These antennas are widely used in the field of HF communication where multiband steerable and fixed antennas are generally used.

(3) This type of antennas is best suited for all round monitoring because a single antenna can cover all high frequency bands.

4.8. TELEVISION RECEIVER

A **TV receiver** is a combination of an AM Rx for the picture signal and an FM receiver for the associated sound. In addition, the receiver also provides suitable scanning and synchronizing circuitry for reproduction of image on the screen of picture tube.

Reproduction of the image on the receiver screen is based on proper amplification of the RF signal arriving at the antenna (Fig. 4.12). To this effect, the incoming signal is converted into the IF in the VHF/UHF tuner, where also standard selection by the Nyquist filter and the required amplification are provided. The subsequent demodulator generates the CVS and the 5.5 MHz sound IF. The latter is limited in amplitude, amplified and frequency-demodulated. The CVS is applied to the video amplifier; after separation of the sync component, the signal is taken to the control section of the CRT via the video output stage. The sync signal brought out from the video amplifier by way of a sync separator is fed to the horizontal deflection system via a differentiating network and to the vertical deflection system via an integrating network.

The line deflection frequency is produced in the horizontal oscillator and compared to the incoming horizontal sync pulses in a phase discriminator. A control circuit ensures that the correct frequency and phase relation to the transmitter sync signal is maintained. In the horizontal output stage, the required deflection power is produced and the high voltage for the CRT is obtained from the line retrace pulses. The vertical oscillator is synchronized directly by the vertical sync signal.

The blanking pulses required for the beam retrace are derived from the horizontal and vertical output stages.

"The television receiver uses **superhetrodyne** principle as used in radio receivers."

The signal voltage or information from various stations modulated over different carrier frequencies is heterodyned in the mixer with the output from a local oscillator to transfer original information on a fixed common carrier frequencies called IF.

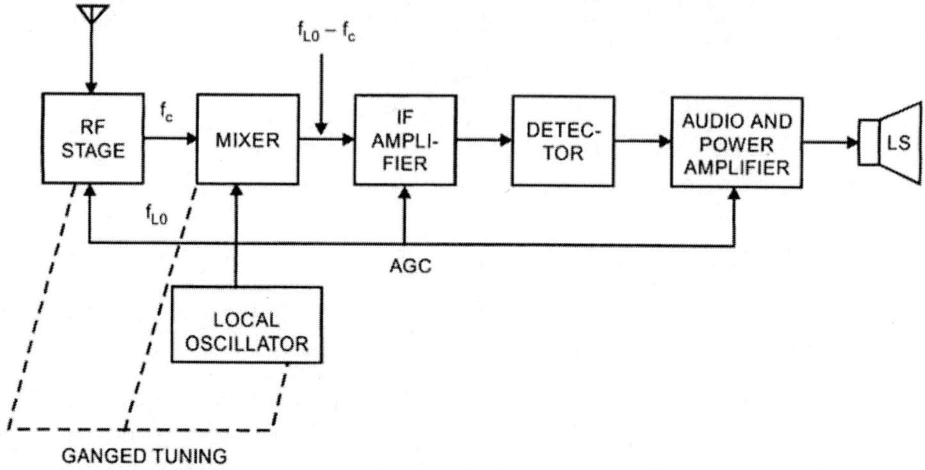

Fig. 4.15. Block diagram of simple superhetrodyne receiver.

The setting of the LO frequency enables selection of desired station.

The standard IF for the 625-B system are

Picture IF -38.9 MHz

Sound IF – 33.4 MHz

The tuning of RF and LO tuned circuits is preset for switching in different channels.

4.8.1. Factors Affecting the Choice of IF

The sound IF is always get fixed at a frequency 5.5 MHz less than picture IF frequency. Therefore we shall consider mostly to picture IF frequency.

There are some following factors which influence the choice of IF in TV receivers:

1. Image Rejection ratio
2. Pick-up due to LO radiation from TV receiver
3. Ease of separation of modulating signal from IF carrier at the Demodulator
4. Image frequency should not lie in the FM range
5. Interference or direct pickup from band assigned for other services

1. Image Rejection Ratio

The **image rejection ratio** is defined as the output due to desired station divided by output due to image signal. The signal which gets received is spaced at a gap of twice of the IF frequency, is known as **Image Signal**.

In the first example (*a*), the IF frequency is kept 10 MHz, if the desired frequency is 100 MHz than the local oscillator frequency must be 110 MHz and the undesired signal frequency is 120 MHz, which is more likely to enter through the RF amplifier. However, in the example (*b*), for the same input signal frequency (100 MHz), if the IF frequency is chosen to be 40 MHz, the LO frequency must be 140 MHz and the undesired signal frequency is 180 MHz.

The first example of IF frequency is 10 MHz is more likely to enter through pass filter in RF amp with little attenuation. But if the IF frequency is kept high, image signal frequency is 80 MHz away from the desired signal frequency and has no chance of passing through the RF amplifier. RF amplifier is used prior to mixer, to attenuate undesired image signals to enter in to further stages. Thus the use of RF amplifier help in reducing interference due to image signals and a higher IF frequency result in a very high image rejection ratio.

For example:

	I/P frequency	LO frequency	IF frequency
(*a*)	Desired -100 MHz	110MHz	10 MHz
	Undesired -120 MHz	110MHz	10 MHz (more effective)
(*b*)	Desired -100 MHz	140MHz	40 MHz
	Undesired -180 MHz	140MHz	40 MHz (less effective)

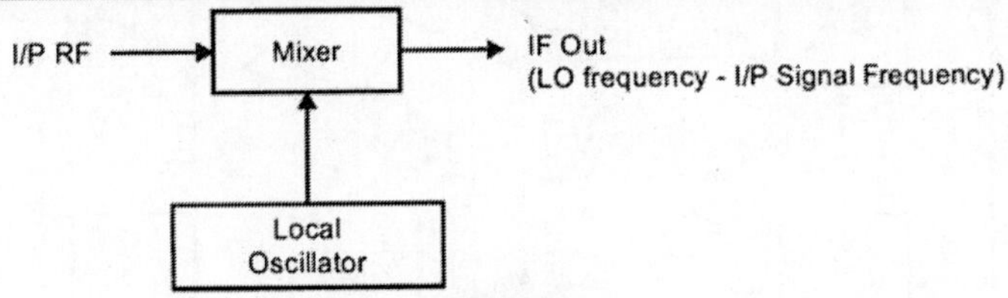

Fig. 4.16.

2. Ease of Separation of Modulating Signal from IF Carrier at the Demodulator

It is desirable to have much higher IF frequency as compared to highest modulating frequency to have ease of filtering out the IF carrier frequency.

"In TV Rx : Highest modulating frequency is 5 MHz & IF frequency is 38.9 MHz"

2. Pick-up Due to LO Radiation from TV Receiver

If the output from the LO of TV Rx gets coupled to the receiving antenna, it will get radiated and may cause interference in another nearby receiver.

So here again advantage lies with higher IF frequency, because with higher IF there is a greater separation between the resonant circuit of LO and RF amplifier circuits.

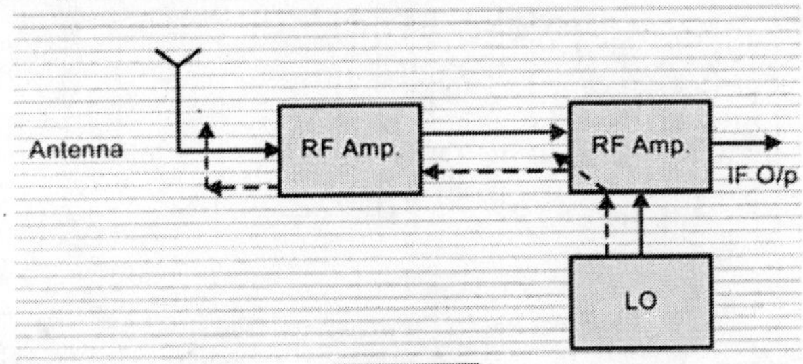

Fig. 4.17.

3. Image Frequency Should not Lie in the FM Range

The FM band is from 88 to 108 MHz with IF frequency chosen close to 40 MHz, the image frequencies of lower VHF band falls between 121 to 148 MHz and thus cannot cause interference in the FM band.

4.9. BLOCK DIAGRAM OF TV RECEIVER

A simplified block diagram of monochrome television receiver is shown in Fig. 4.18.

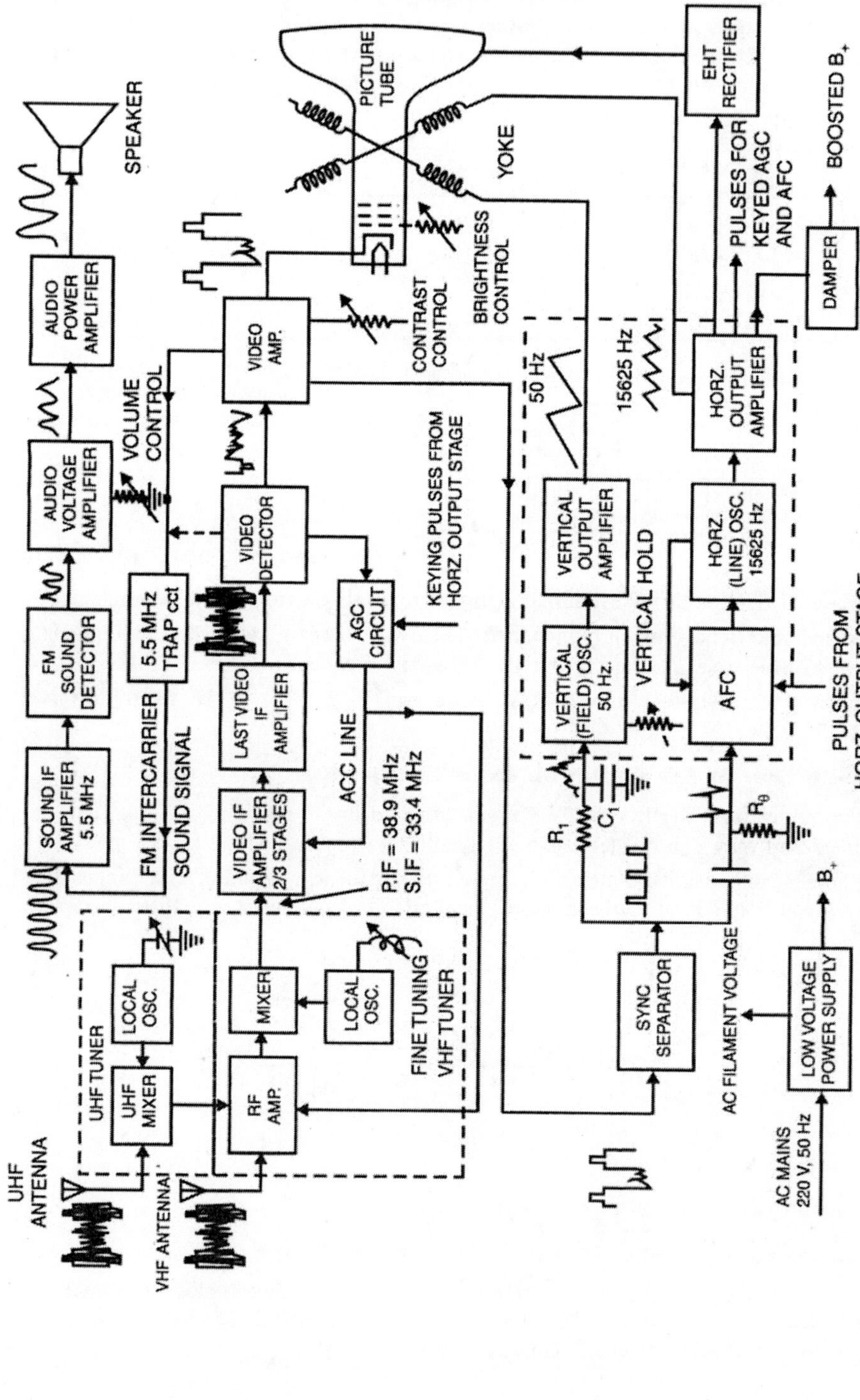

Fig. 4.18. Block Diagram of Monochrome Television Receiver.

4.9.1. Receiving Antenna

A folded dipole or yagi antenna is used as receiving antenna which offers a terminating impedance, to the connected transmission line, equal to the characteristics impedance. This provides a better matching for maximum transfer of the received energy. A broadcast receiver gets from the antenna, a weak VHF or UHF signal containing the audio and video information in the form of amplitude modulation of the picture carrier and frequency modulation of the sound carrier.

4.9.2. RF Section

RF section includes of RF amplifier, mixer and local oscillator. Normally it is mounted on a separate self-chassis, known as RF tuner. The purpose of RF tuner unit is to amplifity both sound and picture signals picked up by the antenna and to convert the carrier frequencies and their associated bands into intermediate frequencies and their sidebands. The RF tuner consists of following components :

(*i*) RF amplifier
(*ii*) Mixer
(*iii*) Local oscillator.

Figure 4.19 shows the various blocks of RF section.

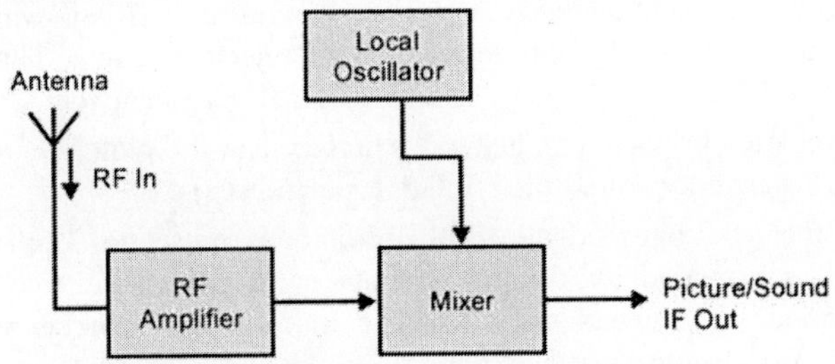

Fig. 4.19. Block diagram of RF section.

4.9.3. RF Amplifier

The RF amplifier has a sufficiently large bandwidth to amplify the full channel frequency range *i.e.* 7 MHz. This stage helps in achieving a given noise figure and providing image rejection. It has low input impedance which can match with the impedance of the receiving antenna. It selects the desired channel with the help of tuned circuits. The RF amplifier provides gain of about 20 dB to the selected RF signal has rejects the image frequency interference by 60 dB or more. The RF amplifier also decouples the local oscillator from the antenna and prevents the local oscillators frequency from radiating out in the space.

4.9.4. Local Oscillator

Local oscillator (LO) provides RF signal of constant amplitude and constant frequency for the desired channel for the reception. To have stabilized output from the local oscillator, we use synthesized tuning based on the technique of phase locked loop (PLL). LO uses high stability of a crystal and variable frequency oscillator type capability to change channels to receive different channels.

The local oscillator frequency is kept higher because of ease of oscillator design and several other merits. The ratio of highest to lower RF that the local oscillator must generate, when the local oscillator frequency is selected to be higher than the incoming carrier frequency is much less than the local oscillator frequency is kept below the incoming channel frequency. It is more convenient to design an oscillator that maintain almost a constant output amplitude and a sinusoidal wave shape when its overall frequency range is less.

The RF carrier signal and local oscillator RF sinusoidal wave, given to the mixer to provide the IF signal.

4.9.5. Mixer

The RF signal then heterodynes in the mixer with a locally generated frequency to produce a lower intermediate frequency (IF) equal to the difference between the two signals. Due to operation on the non-linear portion of the characteristics curve of the transistor, the mixer produces several intermodulation products. One of these products is the signal pertaining to the difference of the two frequencies, called **intermediate frequency** signal. The output from the mixer is passed through a resonate circuits, tuned to IF. The other signals output from the mixer such as sum of the two frequencies or their harmonics, are far off from the IF, hence are rejected by the IF tuned circuit. Only the IF signal passes to the two sets of tuned intermediate frequency amplifier sections; one of these is tuned to sound IF and other to picture IF. The unwanted signals from each channel are removed by the application of rejection filters. The picture IF and sound IF are standardised at 38.9 MHz and 33.4 MHz respectively.

As the local oscillator frequency is generally higher than the picture carrier and the sound carrier, the IF carriers reverse their relative positions.

The tuning of the RF stage and the local oscillator is preset for each channel. A fine tuning control is provided by the adjustments of the oscillator frequency done manually. All the modern receivers use electronic tuner, which can be tuned to the chosen channel by push button switches.

The above described tuner is for VHF channels, to receive UHF channels, a separate UHF tuner is used. The UHF tuner consist of a UHF selective circuit, a diode mixer and a UHF local oscillator.

4.9.6. Video IF Amplifier

The main function of video IF amplifier is to amplify modulated IF signal over its entire bandwidth with an output of about 0.5 mV from the mixer to deliver 4 volt into video detector.

The video IF amplifier provides a high gain of the order of 60 to 80 dB. It provides the requried bandwidth to amplify video signals picture IF and sound IF to the required level. The mixer output is feed to the first IF amplifier through the coaxial cable. Usually all the gain and selectivity of the receiver is provided by the IF sections. When tubes are used then it requires 2 or 3 IF stages whereas in case of transistors it requires 3 to 4 IF stages.

The picture IF is at 50% of the peak amplification for frequencies between 35 to 38 MHz containing the video band. The sound carrier is set at 5 to 10% of the picture IF so that at the video detector they produce a beat which carries the sound modulation. This beat frequency is the intercarrier sound IF that remains constant even if any drift in the oscillator frequency.

FM sound IF and AM picture IF signals need wide bandwidth for reproducing a the important sidebands. Several IF tuned circuits are used to get broad bandwidth by using staggering tuning. In the staggering tuning usually 3 IF tuned circuits are used, one IF tuning circuit is tuned to slightly lower and another one to slightly higher frequency as shown in Fig. 4.20.

Third one is broadly tuned to the correct. If the resultant response is quite wide as shown in Fig. 4.20.

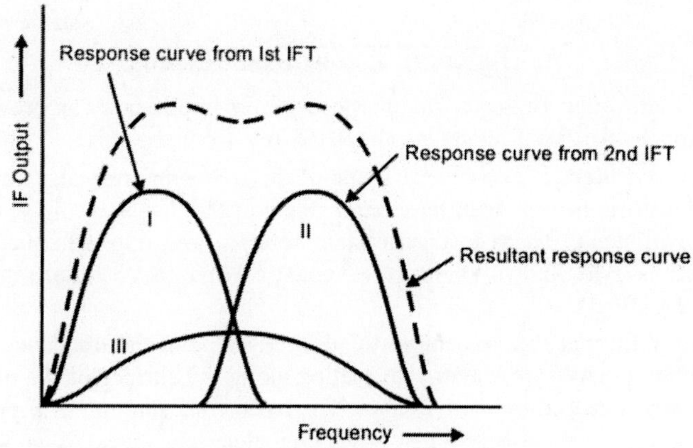

Fig. 4.20. Staggered tuning of IFT.

Figure 4.21 shows the block diagram of the IF amplifier section of a TV receiver. Blockwise brief description in given below.

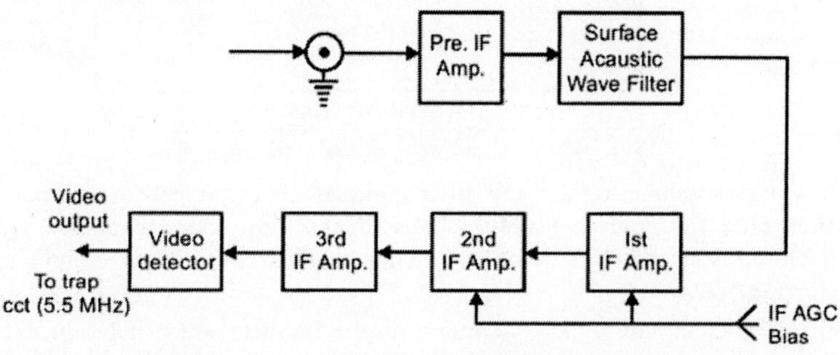

Fig. 4.21. Block diagram of the IF amplifier section.

(*i*) **Coaxial Cable :** The tuner output is coupled to the preamplifier circuit using **coaxial link coupling**. This coaxial coupling is used to couple two stages rather apart as in the case of the mixer stage in the tuner and the first IF input stage of IF amplifier as shown in Fig. 4.22. A coaxial cable at low impedance level prevents radiation from the lead as well as pick up from external interference signal. The cable in normally a mini size 75 Ω type with plug and socket at one end.

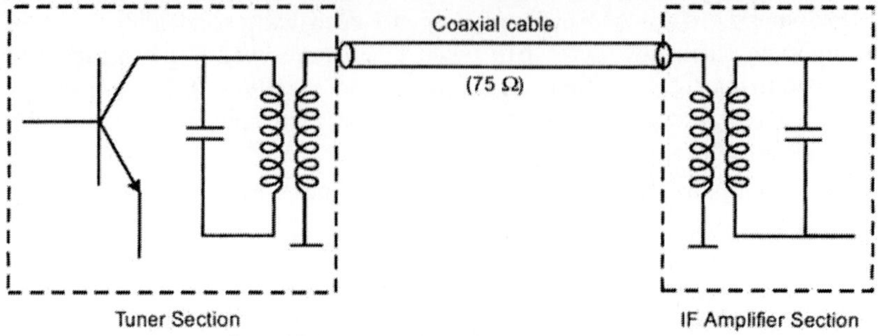

Fig. 4.22. Coaxial cable coupling.

(*ii*) **Preamplifier Stage :** The purpose of preamplifier is to provide the overall amplification to the input signal of about 0.5 mv from the mixer.

(*iii*) **SAW Filters :** The special shape of the frequency response curve of IF signal for getting distortion-free video base band signal at the detector output, the filter circuit should be sufficiently sharp to discriminate between very close frequencies a specaial type of filter is used, known as **Surface Acoustic Wave (SAW) filter,** designed in 1976 to replace LC filters.

The SAW filter is able to achieve the desired shape of the frequency response curve before detection. SAW are waves propagating along the surface of the media, similar to the elastic waves caused by earth quakes. The basic configuration of SAW filter consists of two piezoelectric crystals. When a varying electrical signal is applied across two surfaces of such a crystal, mechanical vibrations are produced across the other opposite set of faces and vice-versa. SAW filter makes use of this characteristics. Figure 4.23 shows the principle of SAW filter.

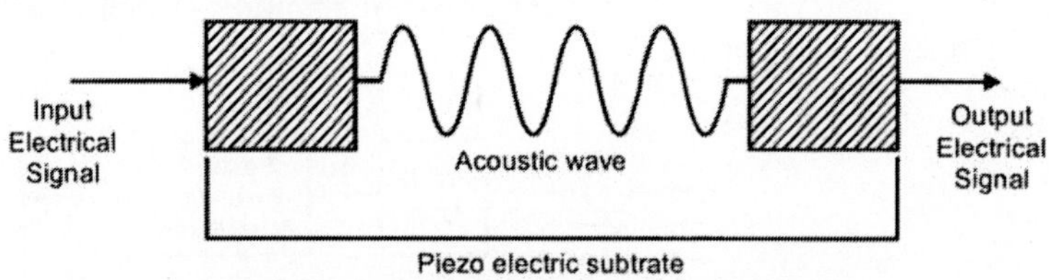

Fig. 4.23. Working principle of SAW filter.

The basic structure of a SAW filter comprises piezo-electric substrate measuring 4 to 8 mm on a side and have the thickness of 0.4 mm, upon which are deposited two interdigital transducers (IDT's) of comb-shaped electrodes, 10 to 20 μm in width, as shown in Fig. 4.24.

When the electrical signal is applied on the input crystals, it produces mechanical vibrations of the frequency equal to that of the electrical signal. These vibration generate surface acoustic wave on the surface of a substrate on which crystals are mounted. Those wave travelling to the right towards the output IDT are converted back into electrical signal by the output transducer, while those propagating to the left are absorbed by the absorber. When the pitch of the comb-shaped electrodes of the output IDT is constant than the output is maximum for SAW having $\lambda/2$ equal to the pitch. And cancellation occurs between the signal induced on each comb shaped electroue at all other wavelengths.

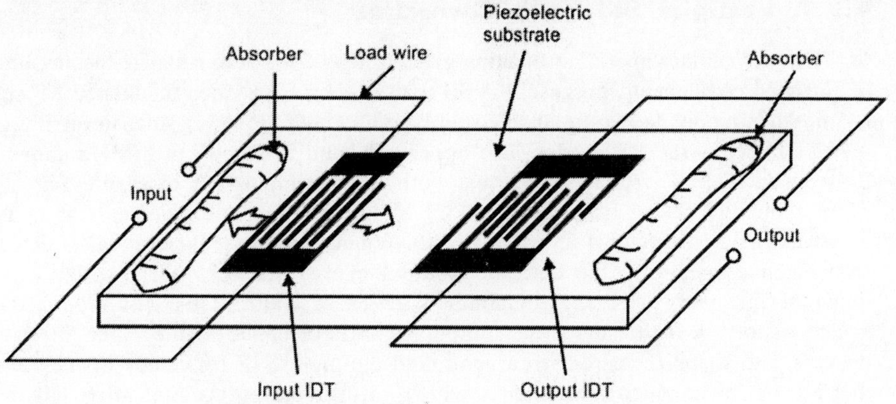

Fig. 4.24. Basic structure of the SAW filter.

The frequency characteristics of SAW filter is expressed by the (sin x)/x function. The f_0 (synchronous frequency) at which the maximum energy is produced is given as

$$f_0 = \text{Velocity of surface wave}/\lambda_0$$

where f_0 = Frequency where maximum energy is produced

λ_0 = Uniform pitch distance.

The mechanical wave travel on the surface with the velocity of sound and hence the name surface acoustic wave filter.

(*iv*) **Ist, 2nd and 3rd stage of IF amplifier using transistors :** The amplifier provides a high gain with a good stability by use of a cascade configuration interstage that provides better isolation between the stages. The frequency response and the gains of the stages are shown in Fig. 4.25, as the detector output w.r.t. signal at the base of the first, second and the third IF stages in sequence. The overall gain of the three stages is approximately 90 dB.

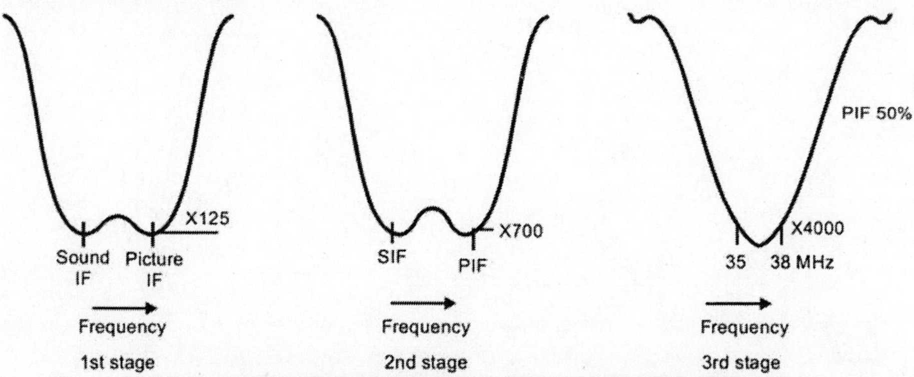

Fig. 4.25. Frequency response curve for statewise.

IF AGC bias is applied to the Ist and 2nd stage of amplifiers to control the gain of IF amplifiers. The automatic gain control bias is a dc voltage proportional to the input signal strength and its obtained by rectifying the video signal as available after the video detector cct.

4.9.7. Vestigial Sideband Correction

There is another important function given to IF section is to equalize the amplitudes of sideband components, because of VSB transmission. The video modulated RF signal is amplitude modulated vestigial sideband type in which the lower sideband is flat upto 0.75 MHz below the RF carrier. The upper sideband is flat up to 5 MHz above the RF carrier. The IF signal shows that both the sidebands are available for ±0.75 MHz about the centre frequency of 38.9 MHz, but only one sideband exists from 33.90 to 38.15 MHz (for 4.25 MHz width). When video baseband signal is detected from such a spectrum, the detector's output energy for 0.75 MHz width will be twice of the energy for the remaining 4.25 MHz width. This will produced the severe distortion in the detector's output signal and hence in the video. It is now necessary to shape IF response curve around the picture IF frequency in such a way that all the lower video frequencies, which got a boost on account of partial lower side band transmission are duly attenuated and get restored to their actual level. This is obtained by suitably tuning and shaping the response of the IF stages. The requirements of IF response curve are shown in Fig. 4.26.

(*i*) IF signal is attenuated after 38.15 MHz so that of 38.9 MHz it is half of the flat responses or 6 dB below (50%) of the maximum response.

(*ii*) And further attenuated so that it becomes zero at 39.65 MHz and onwards till the end of the channel.

(*iii*) The bandwidth should be 5 MHz between the points at which the response is 6 dB down.

(*iv*) The sound IF of 33.4 MHz should be down by 20 dB (at 10%) the level of picture IF.

(*v*) The adjacent channel picture IF at 31.9 MHz on the other side should be at least 46 dB between the maximum response.

(*vi*) The sound carrier at 40.4 MHz adjacent channel, should be suppressed below 40 dB of the maximum response.

(*vii*) The response must be flat within 2 dB between the range from 35 to 38 MHz.

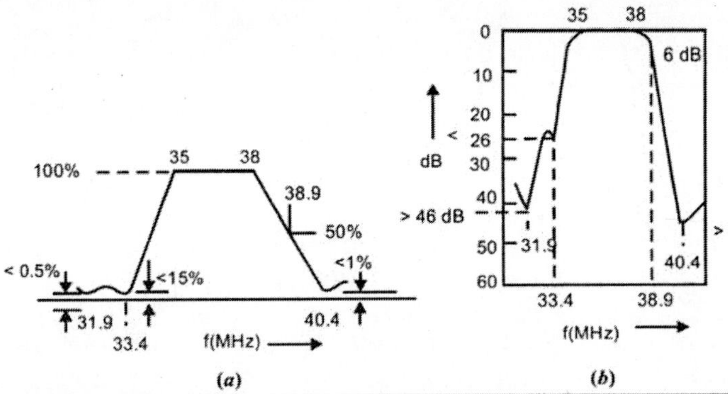

Fig. 4.26. Video IF response requirement : (a) Linear scale (b) Log scale.

4.9.8. Video Detector

Modulated IF signals after attaining required amplification in the IF sections are fed to the next stage *i.e.* video detector. It is an envelope detector using a diode, which by process of rectification and filtering, detects the base band signal of composite video's and FM inter-carrier frequency *i.e.* 5.5 MHz. To minimise undue attenuation of the video signal during filtering, it uses LC filter instead of RC filter. A simple circuit diagram of video detector is shown in Fig. 4.27(*a*).

The video detector is designed to recover composite video signal and to transform the sound signal to another lower carrier frequency. A diode is used, which is suitably connected to rectify either positive and negative peaks of input modulated composite video. The video signal shown in Fig. 4.27(*b*) is of correct polarity for feeding to the cathode of picture tube after single stage amplification.

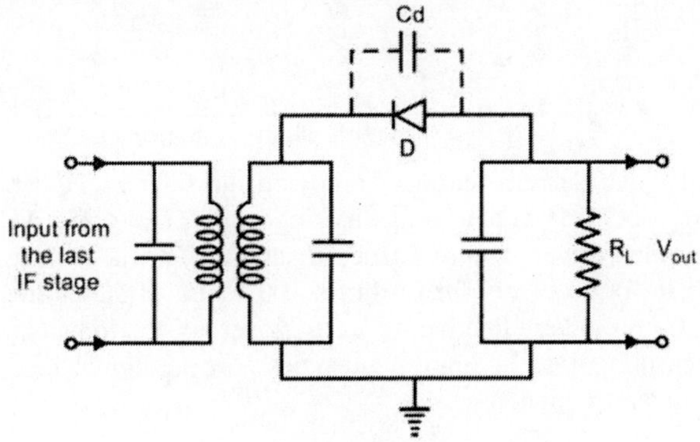

(*a*) A simple diode detector circuit.

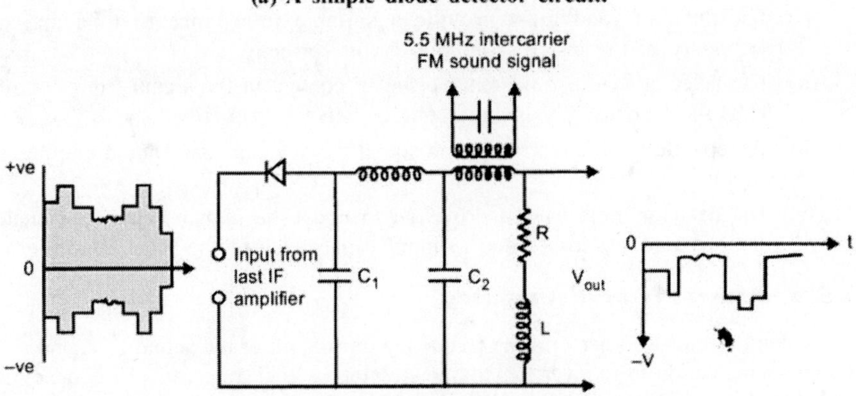

(*b*) Video detector and sound signal separation circuit.

Fig. 4.27.

The video detector output voltage consists of three components as follows :

(*i*) The required video signal

(*ii*) A dc component of the amplitude

(*iii*) IF ripple voltage superimposed on the video waveforms.

Figure 4.28 shows a practical detector circuit.

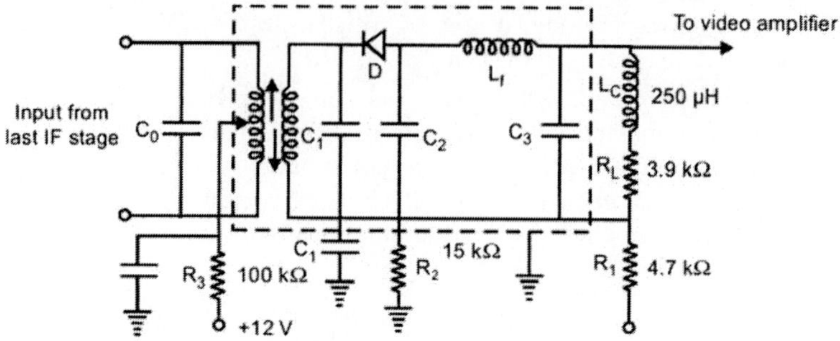

Fig. 4.28. A practical video detector circuit.

A practical video detector circuit employing the filter as *LC* filter. It consists of a load resistance = 3.9 kΩ and a compensating coil (*LC*) = 250 μH. Its output is dc coupled to the video driver. The resistance R_1 and R_2 form a voltage divider across the 12 V dc supply to fix necessary forward bias at the base of the emitter follower (driver). The capacitor C_1 provides effective ac by pass across R_2. Inductance acts as peaking coil for compensating loss at high frequencies. An additional coil may be added to suppress the higher frequencies.

Requirements of Video Detector

(*i*) The detector load must provide a suitable impedance to tune and damp secondary of the last IF coupling circuit correctly.

(*ii*) The detector load should remove the IF content in the signal from the output. Thus the load usually includes one or two low pass filter stages.

(*iii*) To separate the intercarrier sound signal the detector load should employ a trap circuit.

(*iv*) The detector must have a provision to boost the higher video frequencies to compensated the losses due to input capacitance of the video amplifier.

4.9.9. Intercarrier Frequency

Video detector also perform the frequency translation of the sound IF signal besides the rectifying aciton to recover the composite video signal. Intercarrier frequency is the difference between strong picture IF carrier (38.9 MHz) and audio IF carrier (33.4 MHz). It is 5.5 MHz in CCIRB system. Even after mixing in the tuner, the two IF's maintain the same difference. If the local oscillator frequency drifts in the tuner stage, the two IF's will drift equally, so that the difference of 5.5 MHz will be maintained. Thus the intercarrier frequency remains constant irrespective of any change in frequency of local oscillator. The video detector circuit is modified to extract the sound signal, as shown in Fig. 4.29, a parallel tuned circuit, commonly known as **sound IF trap** is inserted in the signal path. It is tuned about a center frequency of 5.5 MHz. This trap circuit offers a high impedance to the sound component of the detected signal to remove it effectively from the video signal path. A tuned secondary circuit feed the intercarrier IF to sound section of the receiver. In some of the TV receiver the intercarrier sound signal is separated, through the trap circuit, after providing the one stage of the amplification in the video amplifier.

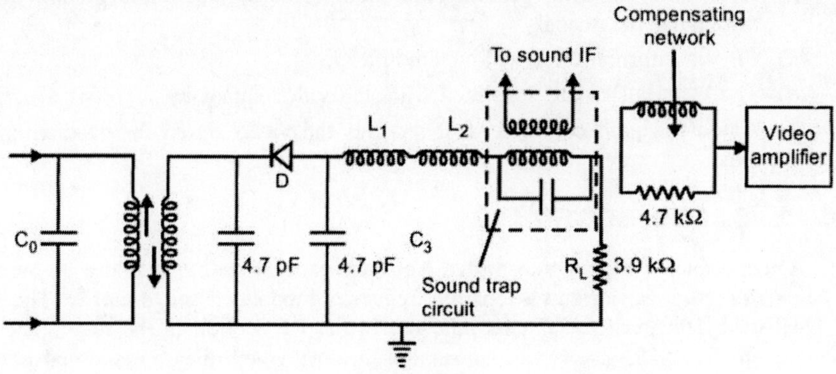

Fig. 4.29.

Video detector is AM detector and so it cannot separate out audio from the frequency modulated audio carrier. Further video carrier is at the level of 10% at the location where audio carrier is placed in the IF spectrum. Thus, there is no effect of video on the audio carrier. Finally, we get inter-carrier frequency signal containing frequency deviation of the FM signal. Figure 4.30 shows the path of video from video detector to picture tube.

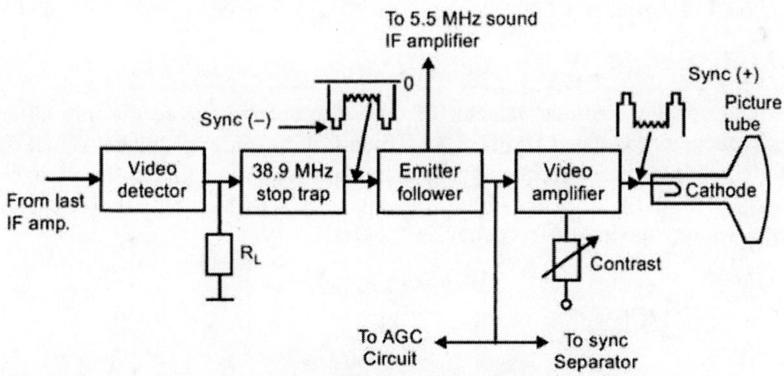

Fig. 4.30. Path of Video signal from video detector to picture tube.

4.10. FUNCTIONS OF COMPOSITE VIDEO SIGNAL

Figure 4.31 shows the various paths for composite video signal obtained from the video detector.

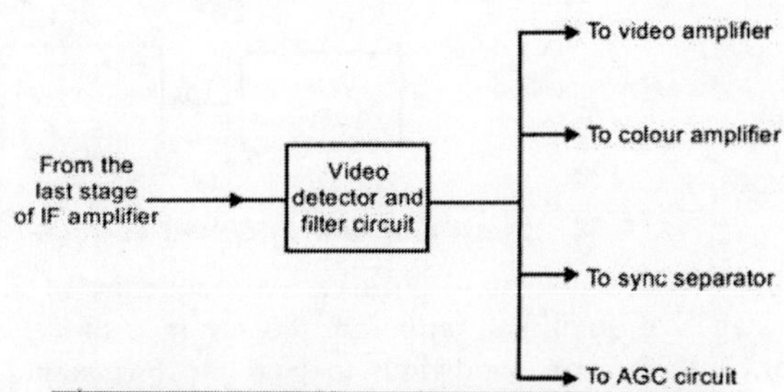

Fig. 4.31. Composite video signal function.

(*i*) The video amplifier provides complete video signal to the picture tube for displaying the scene.

(*ii*) Colour information to colour amplifier.

(*iii*) The sync pulses are separated from the video signal by the sync separator.

(*iv*) Automatic gain control circuit rectifies the video signal for developing AGC bias.

4.11. SOUND SECTION

A television RF signal transmitted from a broadcast station includes an amplitude modulated picture carrier and a separate frequency modulated sound carrier. This signal is received by the receiver antenna and fed to the RF section in the tuner. The block diagram in Fig. 4.32, shows various section through which the sound signal passes in a television receiver before it is ultimately delivered to the loudspeaker.

The sound section of a TV receiver employing the following stages:

1. Sound IF amplifier,
2. An amplitude limiter,
3. FM demodulator,
4. AF voltage amplifier
5. AF power amplifier
6. Loudspeaker.

4.11.1. Sound IF

The non linear diode detector of the video section before the trap circuit converts the frequency modulated IF of 33.4 MHz in to frequency modulated signal of 5.5 MHz, referred as intercarrier frequency signal, by the process of mixing in a non linear active device. With the help of trap filter, this signal is separated out from the detected video signal whose maximum frequency is 5 MHz.

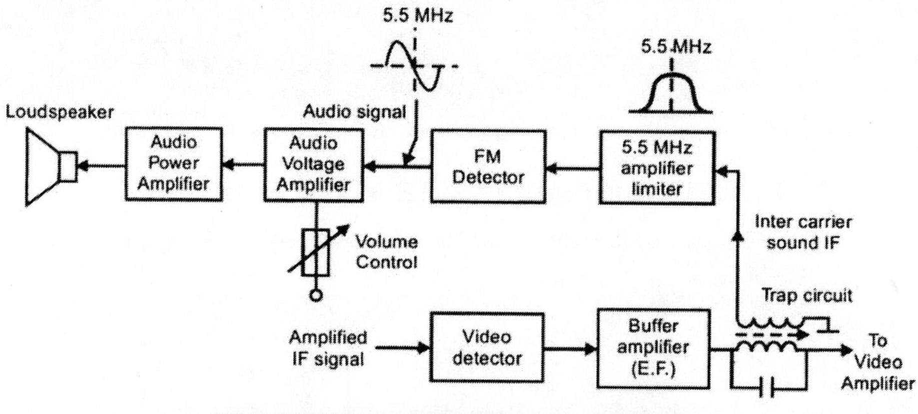

Fig. 4.32. Block diagram of Sound section.

The weak FM sound signal, now on a carrier frequency of 5.5 MHz is given at least one or two stages of amplification before feeding it to FM Detector. This stage is a tuned amplifier, with enough bandwidth to pass the FM sound signal. This tuned amplifier is known as **sound IF**. The second IF stage is also used as a limiter to eliminate any amplitude variations in the FM signal, if necessary. The IF amplifier is designed for sufficient bandwidth to pass ± 50kHz at 5.5 MHz, using single tuned or double tuned circuits to have the wide bandwidth.

4.11.2. FM Detector

- The FM detector is generally a ratio detector or a discriminator preceded by a limiter.
- The characteristics curve of a typical FM detector as shown in Fig. 4.33, the audio output is proportional to deviation from the carrier frequency.
- A de-emphasis circuit is provided at the output of FM detector that has the same time constant (50ms) as that of pr-emphasis circuit employed at the sound Tx. This restores the amplitude of higher audio frequencies to their correct level.

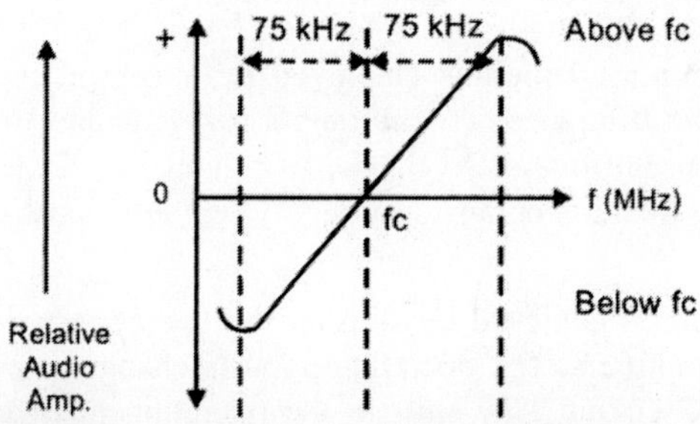

Fig. 4.33. Response curve of FM detector.

4.11.3. Video Amplifier

The video amplifier stage provides large voltage swing needed to drive display tube from cut-off for blanking to practically zero voltage for peak white. To preserve the video signal information as received by the antenna, video amplifier stage should meet the following requirements:

- The picture tube needs video signal with pp amplitude of 80 to 100v for producing picture with good contrast.
- The composite video signal at the output of the video detector requires amplification of the order of 60 to 90. With an input of about 2v from the video detector, the video amplifier is designed to have a gain that varies from 60 to 90. To obtain this much amplification two to three stages of video amplifications are needed.
- A contrast which essentially is gain control of the video amplifier is provided on the front panel of the receiver.
- The video amplifier is dc coupled from the video detector to the picture, in order to preserve the dc components for correct brightness.
- Besides gain, the response of the amplifier should ideally be flat from dc (zero) to 5 MHz to include all essential video components. To reproduce a very minute's detail; a very high video frequency would be required.

- To have a faithful reproduction of the picture, the video waveforms should be preserved during amplification. The shape of the waveform depends not only on the frequencies in the signal but also upon the relative phases. So it is needed that the relative phases of all the frequency components in the output must be the same as that at the input.

4.11.4. Coupling Methods (Video Amplifier to Picture Tube)

There are various methods of coupling the video amplifier to the picture tube as follows :

1. DC coupling (mostly used)
2. Partial DC coupling
3. AC coupling with dc restoration
4. AC coupling

4.11.5. Automatic Gain Control (AGC)

Automatic gain control circuit is employed is to control the gain of RF and IF amplifiers to reduce the difference between output for weak and strong incoming signals to maintain almost constant output level. The AGC bias is a dc voltage proportional to the input signal strength and is obtained by rectifying the video signal as available after the video detector.

The change in gain is achieved by shifting the operating points of the amplifying devices used in the amplifiers. The operating point is changed by a bias voltage that is developed in the AGC circuit. Any shift in the operating point changes in the power gain of the transistor circuit, in turn results in change in stage gain.

Basic AGC Circuit

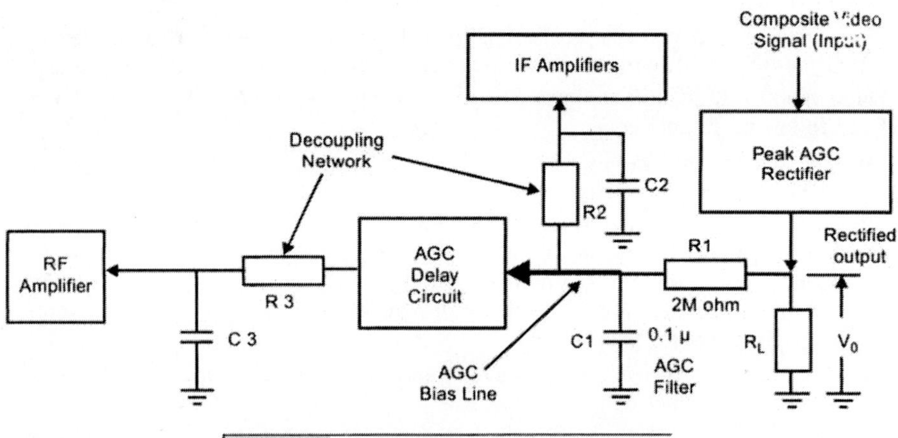

Fig. 4.34. Block diagram of AGC system.

Figure 4.34 shows the basic arrangement of AGC circuit and illustrates that how AGC bias voltage is developed and fed to IF and RF amplifier. The video signal is rectified by the AGC rectifier which develops a unidirectional voltage across load resistance R_L. To have steady dc voltage, the rectified output is passed through the RC filter formed by R_1 and C_1, known as AGC filter having the RC time constant is about

0.2μs. In case when the time constant is small than it will fail to remove low frequency variations in the rectified signal, whereas in case of too large time constant will not allow the AGC bias to change fast enough when receiver is tuned to stations having different signal strength level. A large time constant will not be able to suppress the flutter in the picture. The filtered output voltage across capacitor c_1 is the source of AGC bias to be distributed by the AGC line.

4.11.6. Delayed AGC

Since AGC voltage is proportional to the signal strength, even weak RF signals will also produce some control voltage. This is when applied to the RF amplifier will tend to reduce its gain, through the stage should provide maximum possible gain for weak signals to maintain high signal to noise ratio.

Therefore, the RF amplifier is not fed any AGC voltage till the signal strength attains a certain predetermined level *i.e.* 150 micro volts. This is achieved by providing a voltage delay circuit in the AGC line. Such a provision is referred as **Delayed AGC**.

Advantages of AGC

- Fading effects is reduced.
- Separation of sync pulses becomes easier because of constant amplitude.
- Amplitude and cross modulation distortion is avoided due to reduction in gain.
- The gain of weak signals can be increased by AGC. It is achieved by delaying the application of AGC to the RF amplifier until the signal strength attains 150 micro volts.
- When the receiver is tuned from one station to another, contrast in the reproduced picture does not change much.
- Sound signal is also controlled by AGC and stay constant at a set level

Types of AGC System

1. Peak AGC
2. Keyed AGC

4.11.7. Peak AGC

Figure 4.35 illustrates the circuit diagram of peak AGC system.

In **peak AGC system** (also known as **non-keyed AGC**), we employed a separate diode to provide the rectification to the signal, which is fed through capacitor C_1 from the output of last stage of IF amplifier, as shown in Fig. 4.35. During the positive half cycle of video signal, diode, D_1 conducts and in turn capacitor C_1, charges up to the peak value of the input signal. During the negative half cycle of the video signal, the diode will not conducts, so there is no current flow. Capacitor C_1 tends to discharge through the secondary of the IF transformer and R_1. The RC time constant of the discharge path is approximately 270 μs which is quite greater than the line period of 64 μs. The capacitor C_1 discharges partially and tends to retain charge corresponding to the each successive sync pulse. Thus the current passing through the resistor R_1 is proportional to the peak value of the video signal. The voltage drop across the resistor R_1 becomes the source of AGC bias.

The negative voltage drop across resistor R_1 is filtered by RC filter formed by R_2C_2, to remove the ripple of the horizontal sync pulses (15625Hz).

The available AGC voltage is fed to IF and RF amplifier stages as an AGC control voltage.

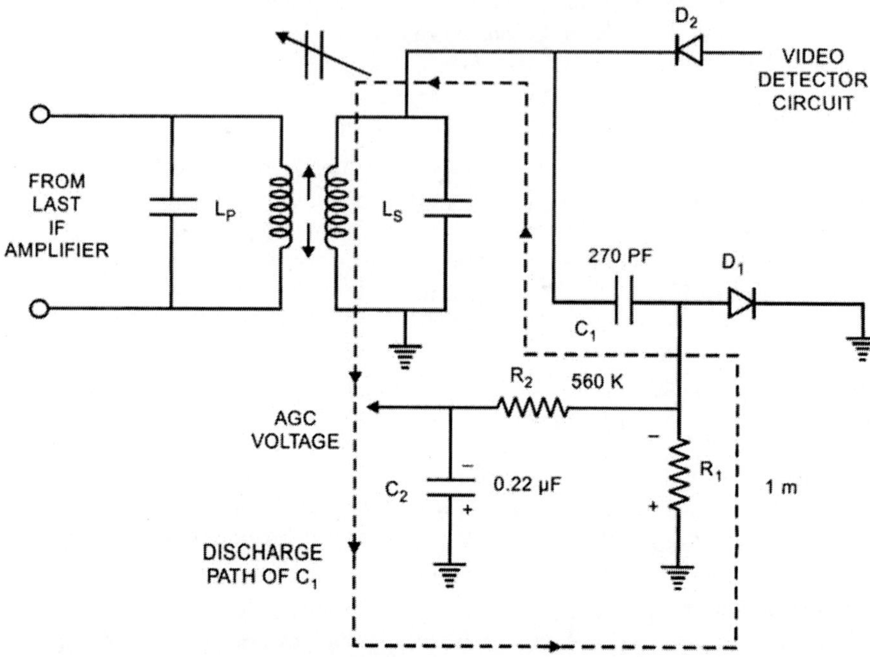

Fig. 4.35. **Peak AGC system.**

Drawbacks of Peak AGC circuit

1. Even for the weak signal strength, the AGC bias voltage is produced and resulting reduction in the gain of a receiver.
2. In fringe areas some noise pulses produces an additional AGC voltage, that will tends to reduce the overall gain of the receiver. This phenomenon is more prominent for the dark scene.

4.11.8. Keyed AGC System

The drawbacks of peak AGC system are overcome by keyed AGC system. In the keyed AGC system, the AGC rectifier is allowed to conduct only during the horizontal sync pulse periods, with the help of fly back pulses which are produced by the horizontal deflection circuit of the receiver.

Figure 4.36 shows the diagram of the keyed AGC system, which receives gating pulses from the horizontal line output transformer. These gating pulses make a transistor, receiving the video signal, to conduct during the period of horizontal sync pulses. The amplitude of the sync pulses which is in proportional to the strength of the video carrier decides the flow of current in the transistor. The AGC voltage is produced across a capacitor in proportional with the signal strength which is independent on the dc content in the signal.

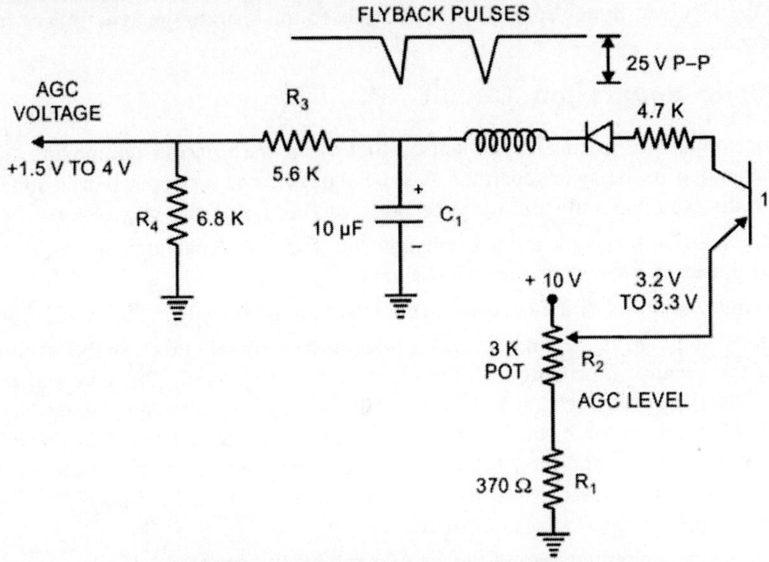

Fig. 4.36. **Transistor based keyed AGC circuit.**

Advantages of Keyed AGC system

1. AGC voltage produced in the true representation of the peak of fixed sync level. Thus it does not depend on the video variations but depends upon the actually received signal strength.
2. AGC voltage is independent on vertical sync pulse but depends upon the horizontal sync pulse.
3. The ripple to be filtered out is of 15625 Hz and hence the time constant of the AGC filter is short.
4. Fading effect is eliminated.
5. Noise effects are minimised as the conduction for AGC occurs only for short period of the horizontal pulse.
6. Sound intensity also remains constant as audio is a part of composite video signal.
7. Strong signal produce amplitude distortion and hence undesired intermodulation products. AGC reduces gain to strong signals and therefore reduces distortion.
8. Since the sync pulse remain at constant amplitude, this facilities functioning of sync separator.

4.12. DEFLECTION CIRCUITS

The synchronising pulses are also known as sync pulses which are the parts of the composite video signal as the top 25% of the signal amplitude. The sync pulses consist of horizontal, vertical and equalising pulses. Deflection circuits generate synchronised sweep currents for deflecting a beam of electrons horizontally and vertically on the screen of a picture tube to reproduce the original picture. Synchronisation is obtained by extracting the sync pulses from the composite video signal detected by the video detector in the receiver. Sync separator circuit are used to separate the sync pulses from the video signal.

4.12.1. Sync Separation Circuit

The horizontal and vertical sync pulses that form part of the composite video signal are separated in the sync separator. A **sync separator** is a clipper that is suitably biased to produce output, only during sync pulse amplitude of the video signal.

In some television receivers, a noise gate precedes the sync separator. This suppresses strong noise pulses if present in the video signal.

A sync pulse train as obtained from sync separator is shown in Fig. 4.37.

Since the sync pulses lie on the side of the video signal, this can be achieved clipping out the picture information on the other side to leave only the sync pulses in the output. The separated vertical and horizontal sync pulses are passed through the integrator and differentiator circuits respectively. The vertical sync pulses produced must be identical in shape and their timing should be such that the interval between the start of the transmitter sync pulse and the triggering of the vertical field oscillator is the same. So that a proper interlaced is ensured.

The sync separator output should not be affected by the variations in the amplitude of the video signal.

The output of the sync separator should not contain any interferences and noise pulses as they disturb the horizontal and vertical synchronisation.

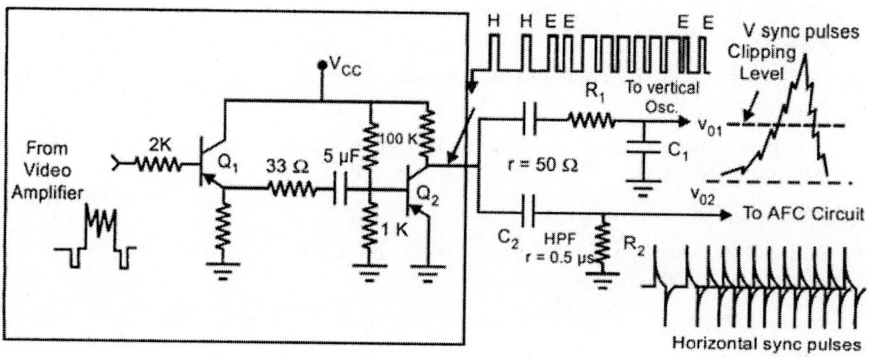

Fig. 4.37. **Sync separator circuit.**

Separation of Horizontal and Vertical Sync Pulses

As we have already discussed that in television receiver there are two types of deflections of electrons beam are required to scan the complete picture tube, one to deflect the electro n beam of the picture tube in vertical direction and other in horizontal direction. Each scanning circuit consists of an oscillator or waveform generator.

The pulse train as obtained from the sync separator is fed simultaneously to a differentiating and integrating circuits. The differentiated output provides sharp pulse for triggering the horizontal oscillator, while output from the integrator controls the frequency of the vertical oscillator. It is possible to separate the horizontal and vertical sync pulses because of the difference in the pulse time period. The width of the horizontal sync pulse is 4.7μs with the repeating frequency of 15625Hz whereas the vertical sync pulse width is 160 μs with the repeating frequency of 50Hz. By using RC filters, horizontal and vertical sync pulses are separated from each other.

As shown in the Fig. 4.37, the output of the sync pulse separator is feed parallel to A high pass filter, connected across the incoming sync pulses train will develop differentiated trigger pulses for synchronizing the horizontal oscillator. Similarly low pass filter, gives the integrated pulses for the vertical oscillator for the same pulse train. The resistor R_1 and capacitor C_1 act as low pass filter and known as integrating circuit whereas resistor R_2 and capacitor C_2 act as high pass filter and known as differentiated circuit.

The differentiated output developed across R_2 is fed to automatic frequency control (AFC). circuit which is used for correct synchronization in spite of noise pulses. The integrated output across C_1 develops upto 50 times in one second that used to trigger the vertical oscillator. An emitter follower circuit is used before the sync separator to provide isolation from video amplifier.

The RC time constant of high pass filter is kept much smaller *i.e.* $1/10^{th}$ than the width of the horizontal pulse, usually it is 0.5 μs to 1 μs. The time constant of low pass filter circuit is much larger (about 10 times) than the width of each serrated pulse.

- Integrating circuit—low pass filter (LPF)—resulting no sharp noise pulses
- Differentiating circuit - high pass filter (HPF)— develop noise pulses

4.12.2. Vertical Deflection

The vertical deflection section of a television receiver is used to provide a linear vertical deflection of the picture tube electron beam. The vertical deflection section includes a sweep oscillator with the frequency of 50Hz and a power amplifier to produce sufficient sawtooth current for the vertical deflection. The block diagram of vertical deflection section is shown in Fig. 4.38.

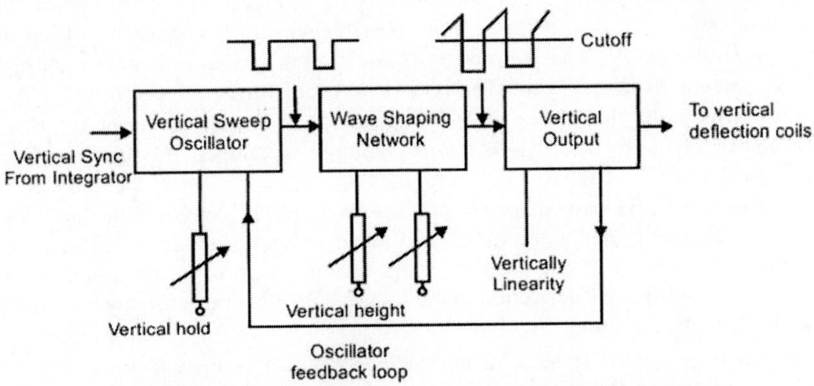

Fig. 4.38. Block diagram of vertical deflection circuit.

A Blocking oscillator or cathode coupled multivibrators are normally used as vertical deflection oscillator. The controlling time constants are suitably chosen to develop output corresponding to trace and retrace periods. The necessary saw tooth voltage is developed by charging and discharging a capacitor with different time constant. This capacitor forms parts of the wave shaping circuit which is connected across the oscillator.

The frequency of the oscillator is controlled by varying the resistance of the RC coupling network and is locked in synchronizer by vertical sync pulses.

A part of the coupling network resistance is a potentiometer that is positioned on the front panel of the receiver. This is known as **vertical hold control** and providing resetting of the vertical oscillator frequency, whenever it drifts away from 50 Hz.

The output from the oscillator-cum-wave shaping circuit is fed to a power amplifier, the output of which is coupled to the vertical deflection coils, produce vertical deflection of the beam on picture tube screen.

Some Requirements of Vertical Deflection Circuit

1. Because of incorrect vertically scanning frequency the rolling of the picture occurs. Thus the vertical oscillator is locked by the sync pulses and each frame is reproduced over the previous one, therefore the picture holds still.
2. To provide the raster on the entire screen from the top to the bottom, definite amplitude of sawtooth current must pass through the deflection coils. The vertical output amplifier is designed to fulfil this requirement. For this the height of the ramp is adjusted by a control known as **height control**.
3. The high frequency oscillation produced by the energy stored in the magnetic field when current through deflecting coils suddenly falls to zero, during the retrace period.

To suppress the undesired oscillation a shunt damping diode is connected across the primary of the transformer.

4.12.3. Horizontal Deflection

The horizontal sync pulse period is 4.7μs which is very short duration comparing to the vertical pulse period 160 μs and hence any noise or disturbance may cause loss of horizontal synchronization that result in diagonal splitting of picture. Since the reproduction of horizontal sync pulses are not possible so efforts are made to keep the horizontal sync pulse stable. **The frequency of horizontal oscillator is also very high** *i.e.* **15625 Hz, needs to keep stable.** The horizontal oscillator is similar to the vertical oscillator and is set to develop sweep drive voltage at 15625 Hz. However, the frequency of this oscillator is controlled by dc control voltage developed by the Automatic Frequency Control (AFC) circuit.

The oscillator output is wave shaped to produce linear rise of current in the horizontal deflection coils. Since the deflection coils need about one ampere of current to sweep the entire raster, the output of the oscillator is given one stage of power amplification and then fed to the horizontal deflection coils. Figure 4.39 illustrates the block diagram of horizontal deflection section.

Due to the high rate of flyback in horizontal sweep, it is used to generate EHT (extreme high tension) for the final anode of the picture tube. High tension (HT) voltages are also derived from the line output transformer of the horizontal output amplifier, for the others anodes of the picture tube.

Normally, horizontal frequency control (hold) not provided because the noise pulses in the control voltage are completely suppressed.

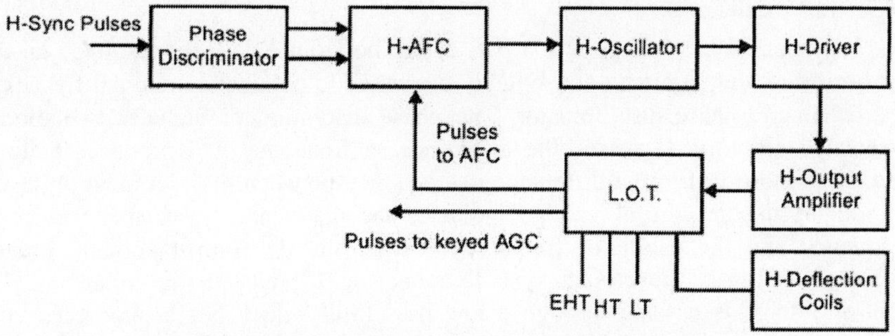

Fig. 4.39. Block diagram of H-Deflection Section.

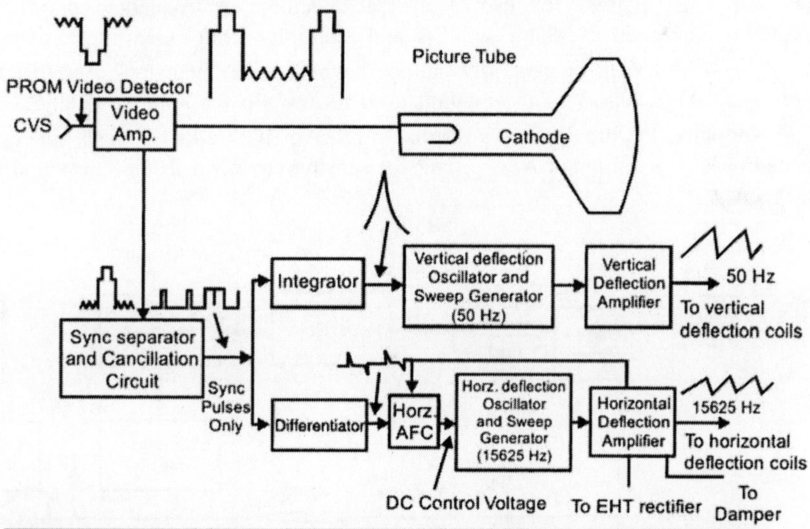

Fig. 4.40. Block Diagram of Horizontal and Vertical Deflection Stage.

4.13. AFC CIRCUIT

The horizontal oscillator driving the H-output stage must be synchronized by the narrow H-sync pulses obtained after differentiation of the sync pulses from the sync separator.

The noise pulses developed in the differentiating circuit may cause occasional wrong triggering of the horizontal oscillator, which results in diagonal tearing of the reproduced picture. It is, therefore not desirable to feed these pulses directly to the horizontal oscillator for direct triggering and synchronization as in the case of vertical synchronization. To overcome this difficulty, a special circuit known as **Automatic Frequency Control (AFC)** is employed. The AFC circuit controls the frequency of the horizontal oscillator indirectly by a dc error voltage obtained from a phase discriminator circuit that compares the phase of the horizontal sync pulses with that of the horizontal output waveform from the H-output stage. This dc error voltage is filtered by a low pass filter and is fed to voltage controlled oscillator, the frequency and phase of which is then corrected to make coincide with that of the sync pulses.

Figure 4.41 shows the block diagram of the AFC circuit used for the horizontal deflection oscillator.

Horizontal sync voltage and a part of the horizontal deflection voltage taken from the horizontal output circuit and suitably processed to obtained horizontal flyback pulses are fed in the phase discriminator. The phase discriminator includes two diodes and associated circuitry. It senses the difference in frequency and produces a dc output voltage proportional to the difference in frequency between the two input voltages. The amount of the dc control voltage indicates the difference between the correct sync frequency and the oscillator frequency. Larger the dc control voltage, greater the difference between the correct sync frequency and the oscillator frequency. This dc control voltage is passing through a low pass filter, which has a large time constant. The shunt by pass capacitor of this low pass filter eliminates the effect of noise pulses.

Thus the use of AFC circuit and a low pass filter at its output eliminates the effect of sharp noise pulses. The use of such arrangement of frequency control results in excellent horizontal oscillator stability and immunity from noise interference.

The AFC circuit is generally called flywheel sync, sync lock, stabilised sync or horizontal AFC, based on the method used to develop the control voltage.

Normally, the present day's television receiver uses either a push pull or a single ended phase discriminator AFC circuit for sensing any error in the horizontal oscillator frequency.

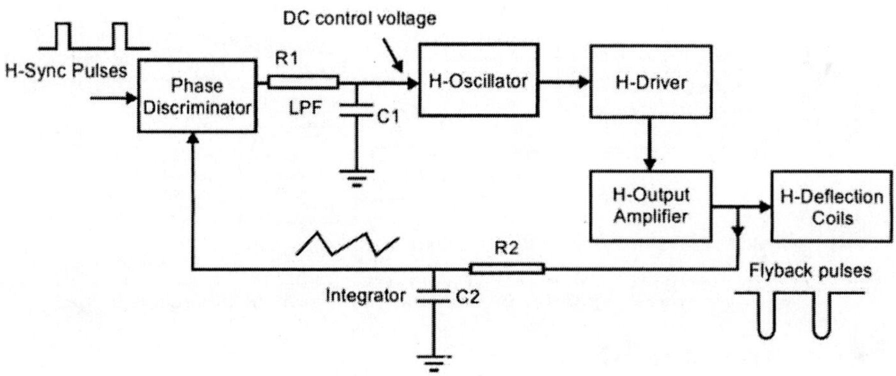

Fig. 4.41. Block diagram of AFC circuit.

4.14. HORIZONTAL OUTPUT STAGE

The horizontal output stage is a power amplifier used to feeds horizontal deflection winding to deflect the beam across the width of the picture screen. The horizontal output load is acting as inductive; the deflection current induces very high voltage pulses in the output transformer while diminishing sharply during the retrace period. These high voltages pulses are fed to the final anodes of the picture tube as **EHT** voltage after the rectification. The energy associated with the high voltage pulses produces high frequency oscillations in the output circuit. These high frequency oscillations are to be suppressed by using a **damper diode**. When the damper diode is conducting, it charges a capacitor and the voltage developed across is added to the existing B+ supply to obtained higher dc source voltage referred as boosted B+ supply. Like this part of the energy from the deflection coils is recycled to develop a voltage source for the horizontal stage circuits and other circuits.

A block diagram of the horizontal output stage is shown in the Fig. 4.42.

Following are the main sections of the horizontal output stage:

1. Output amplifier,
2. Output flyback transformer,
3. Damper diode,
4. High voltage rectifier and
5. Deflection winding

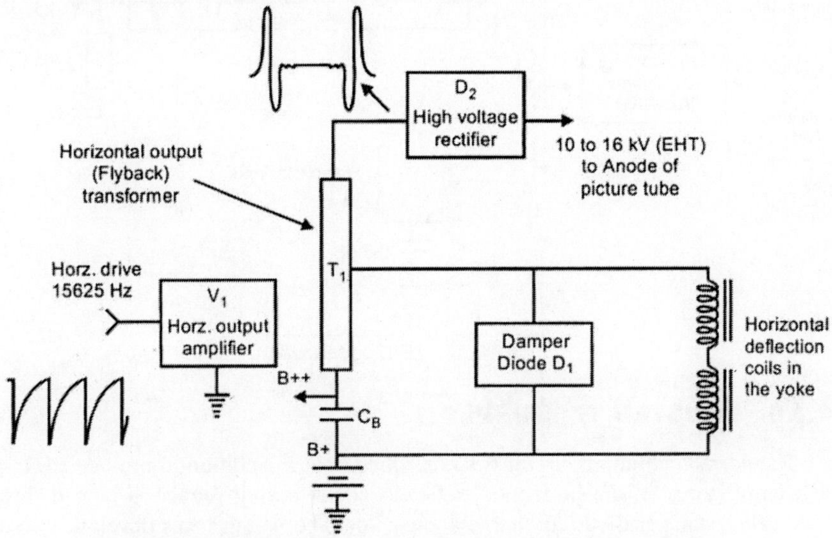

Fig. 4.42. Block diagram of the horizontal output stage.

In this stage the horizontal output transformer is designed to act as a switch which closes for a fixed period during each horizontal deflection cycle to provide power to the output transformer. The output transformer performs many functions such as to provide impedance matching between the output stage and the deflection windings and act as a step up high voltage transformer for the diode D_2 **which is functioning as EHT rectifier. The damper diode D_1 conducts soon after the flyback stroke to suppress high frequency oscillations.**

4.15. DC POWER SUPPLIES

EHT (Extra high tension) supply

- An anode voltage of the order of 15kv is needed for sufficient brightness in black and white picture tubes. This is known as **HV or EHT (Extra high tension) supply.**
- This much amplification is difficult with the transformers, so a different method is used to obtain the HV.
- During the retrace period of horizontal scanning, high voltage pulses of amplitude between 6 to 9 kv are developed across the primary winding of the horizontal output transformer.

- These are stepped up by an auto transformer winding to about 10 to 15kv and then fed to a high voltage rectifier.
- The output of the rectifier is filtered to provide required dc voltage.

Such an arrangement does not load very much the horizontal output stage because the current demands from this high voltage source are less than1mA.

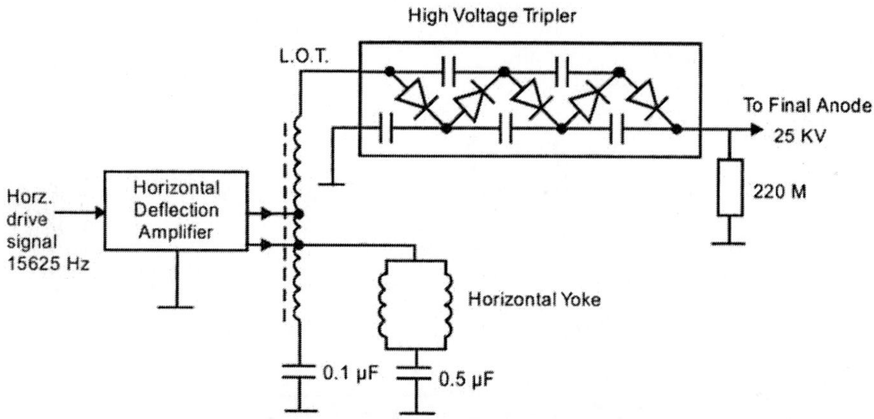

Fig. 4.43. EHT generation.

4.16. BOOSTED B^+ SUPPLY

The horizontal output circuit is so designed, that in addition to provide EHT source, the energy stored in the horizontal deflection coils during retrace is tapped through a diode called damper diode to charge a capacitor. The voltage thus developed across the capacitor, actually adds 200 to 300v to normal B+ voltage to give a boosted B+ supply of 400 to 700 v. This voltage is also suitable for first & second anode of the picture tube.

Low voltage power supply (LVPS):

About 12 to 35 volts for ICs and small signal amplifiers.

Medium voltages:

- Nearly 300 to 400v for the screen and focus grids of the picture tube.
- About 175v for the video amplifier.
- About 150v for the horizontal output stage.

4.17. LVPS (LOW VOLTAGE POWER SUPPLY)

- Usually B+ or low voltage power supply is obtained by rectifying and filtering the ac main power supply.
- If necessary the main voltage is stepped up or down before rectification.
- In some power supply design, where transformers are not used, voltage doublers are used to raise the dc voltage.

4.18. TRANSMITTING AND RECEIVING ANTENNAS

Antenna, in electronics, system of wires or other conductors used to transmit or receive radio or other electromagnetic waves; sometimes called an aerial. The idea of using an antenna was developed by Guglielmo Marconi (1897). In a transmitting antenna, the signal from an electronic circuit causes electrons in the antenna to oscillate; these moving electric charges generate electromagnetic radiation, which is transmitted through the air and space. Distribution of the waves depends on the design of the antenna; the transmitting antennas of a radio station might be designed to emit waves in all directions, while an antenna used for radar or space communications would be designed to focus the waves in a single direction.

4.19. TV STANDARDS

Various other standards are in use; for the differences in the standard specifications see Tables 1 and 2. Detailed description of transmitting and receiving antenna are already discussed in this chapter.

Table 4.1. Details of Composite Video Signal.

Standard	CCIR B, G	OIRT D	French VHF E	FCC (USA) M
Number of lines	625	625	819	525
Field repetition frequency	50 Hz	50 Hz	50 Hz	60 Hz
Line frequency	15625 Hz	15625 Hz	20475 Hz	15750 Hz
Video bandwidth	5 MHz	6 MHz	10.6 MHz	4.2 MHz
Line duration H	64 μs	64 μs	48.84 μs	63.5 μs
Field duration	20 ms	20 ms	20 ms	16.667 ms

Table 4.2. Frequency ranges, vision/sound carrier spacing, channel width, sound/modulation.

Standard	CCIR B, G	OIRT D	French VHF E	FCC (USA) M
VHF, band I/MHz	47 to 68	48.5 to 100	50 to 70	54 to 88
VHF, band III/MHz	174 to 230	174 to 230	160 to 215	174 to 216
UHF, band IV/V/MHz	470 to 853			470 to 890
Vision/sound carrier spacing	5.5 MHz	6.5 MHz	11.15 MHz	4.5 MHz
Channel width	7 MHz (B) 8 MHz (G)	8 MHz	13.15 MHz	6 MHz
Sound modulation, FM deviation	FM, 50 kHz	FM, 50 kHz	AM	FM, 25 kHz

4.20. TELEVISION RECEIVER CONTROLS

Receiver controls are needed to enhance the quality of the picture and sound. Mainly there are two types of controls available in the Television receiver.

Some controls are provided at the front panel of the television receiver for the viewer to adjust the vision and sound according to his wish. And few controls are provided on the chassis which are factory set and to be adjusted by the skilled engineer while checking the performance. These controls are normally not disturbed.

Following types of controls are used in the monochrome television receiver which is illustrated in the Fig. 4.44.

Users controls provided on front panel of the receiver

1. Power on/off control
2. Brightness control
3. Contrast control
4. Fine tuning control
5. Channel selection control
6. Volume control

Power on/off control : Power on/off control turns the receiver on and off which is located on the front panel. In modern television receiver this control is provided as **push button** switch whereas in older receiver this control is also controlling the volume of the sound.

Brightness control : This control is used to adjust the brightness of the picture on the screen by adjusting the dc voltage of the control grid or in turn controlling the number of electrons passed through the control grid. High positive voltage means more brightness.

Contrast control : This control is used to adjust the contrast of the picture on the screen. Internally this control is varying the gain of the video amplifier. Higher gain will provide brighter picture and hence increases the contrast.

Fine tuning control : The purpose of fine tuning control is to optimise the local oscillator frequency so that correct IF signal is produced for video and audio carrier.

Channel selection control : This control will facilitate the viewer to select the desired channel by varying the knob provided on the front panel. In the modern receiver the desired channels are selected by adjusting programmable dividers through push button switches.

Volume control : This control is used to vary the volume of the audio by changing the gain of the audio amplifier.

In some of the modern receiver bass and treble controls are also provided.

Factory Set Controls

Hold control : This control is provided on horizontal oscillator to prevent the diagonally tearing of the picture by adjusting the frequency of the horizontal oscillator in case AFC is not able to control the frequency.

Horizontal width control : This control is provided to adjust the width of the picture by controlling the rate of charging of the capacitor in the sweep generator in the horizontal oscillator section.

Horizontal linearity control : This control is located on the horizontal deflection amplifier, to provide the uniform display of the picture from left to right.

Vertical Hold control : This control is provided on the vertical oscillator to prevent rolling of the picture by changing its frequency.

Vertical linearity control : This control is located on the vertical deflection power amplifier, to provide the uniform spacing between the horizontal scan.

Vertical high control : This control is located on the vertical oscillator and determines the peak value of ramp voltage by charging the capacitor through a variable resistance.

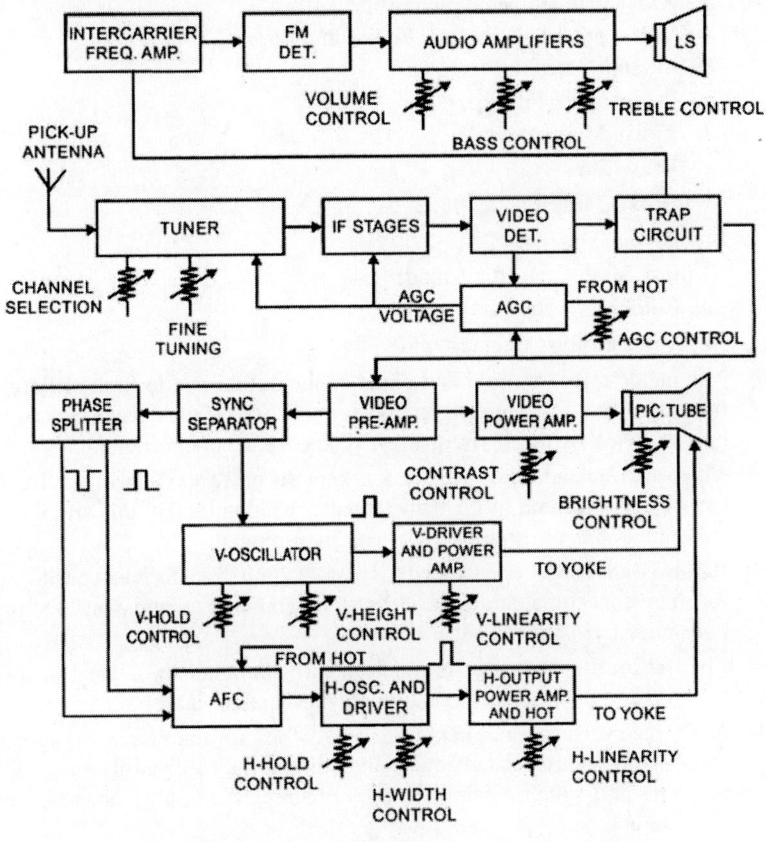

Fig. 4.44. Controls in a monochrome TV receiver.

SUMMARY

- The RF television signal can be transmitted by two different methods as follows:
  (*i*) Direct modulation (using high level modulation) and
  (*ii*) IF modulation (using low level modulation)
- **Direct Modulation :** If the modulation takes place in the output stage of the picture transmitter known as high level modulation, the RF video carrier is first brought to the required driving power and then, with simultaneous amplitude modulation, amplified in the output stage to the nominal video carrier output power of the transmitter.
- Second method of modulation is known as low level modulation using **IF modulation.** When using IF modulation, first the IF vision carrier of 38.9 MHz is amplitude Modulated.

- **Elements of Monochrome TV Transmitter**

  Picture Sections

  1. Video Camera
  2. Scanning & Synchronizing
  3. AM Modulator (for picture)
  4. Video pre-amplifier and Power amplifier
  5. Crystal Oscillator
  6. Frequency Multiplier
  7. Video Modulator
  8. VSB Filter

  Sound & Common Sections

  1. Audio Amplifier
  2. FM Modulator (for sound)
  3. Combining Networks or Diplexer
  4. Transmitting Antenna

- The image rejection ratio is defined as the output due to desired station divided by output due to image signal. The signal which gets received is spaced at a gap of twice of the IF frequency, is known as Image Signal.
- Video carrier and audio carrier are kept so close to each other that they both can be accomodated in the same channel bandwidth. By this arrangement, both the signals are transmitted by a common antenna.
- Balun (Balance to unbalanced) is a matching transformer specially designed for impedance matching. It is used to connecting a balanced circuit to an unbalanced circuit and vice versa.
- The transmitting antenna, normally a turnstile antenna, is used to transmit the combined signals connected through the coaxial cable.
- A TV receiver is a combination of an AM Rx for the picture signal and an FM receiver for the associated sound. In addition, the receiver also provides suitable scanning and synchronizing circuitry for reproduction of image on the screen of picture tube.
- The standard IF for the 625-B system are

  Picture IF -38.9 MHz

  Sound IF – 33.4 MHz
- There are some following factors which influence the choice of IF in TV receivers:

  (*i*) Image Rejection ratio

  (*ii*) Pick-up due to LO radiation from TV receiver

  (*iii*) Ease of separation of modulating signal from IF carrier at the Demodulator

  (*iv*) Image frequency should not lie in the FM range

  (*v*) Interference or direct pickup from band assigned for other services
- The image rejection ratio is defined as the output due to desired station divided by output due to image signal. The signal which gets received is spaced at a gap of twice of the IF frequency, is known as Image Signal.
- Local oscillator (LO) provides RF signal of constant amplitude and constant frequency for the desired channel for the reception.

- The local oscillator frequency is kept higher because of ease of oscillator design and several other merits.
- The mixer RF signal then heterodynes in the mixer with a locally generated frequency to produce a lower intermediate frequency (IF) equal to the difference between the two signals.
- The main function of video IF amplifier is to amplify modulated IF signal over its entire bandwidth with an output of about 0.5 MV from the mixer to deliver 4 volt into video detector. The video IF amplifier provides a high gain of the order of 60 to 80 dB.
- The SAW filter is able to achieve the desired shape of the frequency response curve before detection. SAW are waves propagating along the surface of the media, similar to the elastic waves caused by earth quakes.
- There is another important function given to IF section is to equalize the amplitudes of sideband components, because of VSB transmission.
- Therefore, the RF amplifier is not fed any AGC voltage till the signal strength attains a certain predetermined level *i.e.* 150 micro volts. This is achieved by providing a voltage delay circuit in the AGC line. Such a provision is referred as Delayed AGC.
- **Advantages of AGC**
  - Fading effects is reduced.
  - Separation of sync pulses becomes easier because of constant amplitude.
  - Amplitude and cross modulation distortion is avoided due to reduction in gain.
  - The gain of weak signals can be increased by AGC. It is achieved by delaying the application of AGC to the RF amplifier until the signal strength attains 150 micro volts.
  - When the receiver is tuned from one station to another, contrast in the reproduced picture does not change much.
  - Sound signal is also controlled by AGC and stay constant at a set level
- **Types of AGC System**
  1. Peak AGC
  2. Keyed AGC
- The horizontal and vertical sync pulses that form part of the composite video signal are separated in the sync separator.
- Integrating circuit—low pass filter (LPF)—resulting no sharp noise pulses. Differentiating circuit - high pass filter (HPF)— develop noise pulses
- The vertical deflection section of a television receiver is used to provide a linear vertical deflection of the picture tube electron beam.
- The frequency of horizontal oscillator is also very high *i.e.* 15625 Hz, needs to keep stable.
- Normally, horizontal frequency control (hold) not provided because the noise pulses in the control voltage are completely suppressed.
- The AFC circuit controls the frequency of the horizontal oscillator indirectly by a dc error voltage obtained from a phase discriminator circuit that compares the phase of the horizontal sync pulses with that of the horizontal output waveform from the H-output stage.

- An anode voltage of the order of 15kv is needed for sufficient brightness in black and white picture tubes. This is known as HV or EHT (Extra high tension) supply.
- **Low voltage power supply (LVPS):** About 12 to 35 volts for ICs and small signal amplifiers.
- **Medium voltages :** Nearly 300 to 400v for the screen and focus grids of the picture tube. About 175v for the video amplifier. About 150v for the horizontal output stage.
- **LVPS (Low Voltage Power Supply) :** Usually B+ or low voltage power supply is obtained by rectifying and filtering the ac main power supply.
- Users controls provided on front panel of the receiver
  1. Power on/off control
  2. Brightness control
  3. Contrast control
  4. Fine tuning control
  5. Channel selection control
  6. Volume control

REVIEW QUESTIONS

1. Draw a block diagram of monochrome TV transmitter and explain the function of each block.
2. Compare the television transmitters employing high level modulation and low level modulation.
3. What are the benefits of IF modulation over-direct modulation in a television transmitter ?
4. Why is a video amplifier a wideband amplifier ?
5. Discuss the advantages of vestigial sideband in transmission of video signals.
6. Draw the block diagram of a TV receiver. Explain the functions of each stage in brief.
7. Discuss the operation of super hetrodyne receiver and write its advantages.
8. What do you understand by image frequency and how it is rejected ?
9. Explain with the help of a circuit diagram how video carrier is modulated to give vestigial AM modulation.
10. Why AM is preferred for video signals and FM is preferred for audio signals ? Explain it.
11. Explain the working of displexer.
12. What do you understand by keyed AGC and write its advantages over the other forms.
13. Explain the need of AFC circuit and how it is working.
14. Explain with the help of a circuit diagram how audio carrier is directly modulated to give FM signal.
15. Describe the various controls used in a TV receiver with the help of the diagram.
16. What are the functions of video detector ?
17. Explain the working of an electronic tunner. Discuss its merits.
18. What are the factors affecting the choice of IF ? Why do IF tuned circuits use staggered tuning in TV receivers ?
19. Draw the block diagram of peak AGC and explain its working.

20. Discuss the need of attenuating IF video signal within ±0.75 MHz of the video IF carrier. Draw the shape of the frequency response curve before detection and after detection.
21. What is the need of noise cancellation in monochrome TV receiver ? Explain a method to reduce noise by noise gated diode circuits.
22. What is a trap circuit and where it is used in TV receiver ? Draw a typical trap circuit diagram to trap 5.5 MHz intercarrier frequency and prevent it from going to video amplifier.
23. Enumerate the importance of intercarrier frequency.
24. What do you understand by dc component restoration ?
25. Describe the operation of separation of vertical and horizontal sync pulses ?
26. Describe a SAW filter and explain its working.
27. Explain the importance of phase splitter and describe its working with the help of circuit diagram.
28. Draw circuit diagram of a V-oscillator, buffer amplifier and power amplifier and explain their working.
29. Draw block diagram of the H-deflections stages, starting from syn separator. Describe their functioning.
30. Draw the block diagram of horizontal deflection stages and explain the function of each block.
31. Draw circuit diagram of a horizontal output amplifier and explain how line scanning is obtained.
32. How do the stages upto video detector work as common stages for processing video and audio modulated signals.
33. What are the two most important functions of an RF amplifier in a TV receiver.
34. What are the merits of an IF stage on a TV receiver ?
35. What purpose is served by giving a special shape (frequency response) to a signal before detection.
36. Why is AFC needed for an H-sweep generator while it is not needed for V-sweep generation.
37. What is flutter fading ?

MULTIPLE CHOICE QUESTION

1. Wideband amplifiers are required for
   (*a*) Video signal (*b*) Audio modulated signal
   (*c*) Hum noise (*d*) Ultrasonic waves
2. Vestigial type modulation gives
   (*a*) High SNR (*b*) Low bandwidth
   (*c*) FM (*d*) Wide bandwidth
3. Preemphasis is used to
   (*a*) Increase RF power (*b*) Correct AM into AM VSB
   (*c*) Improve SNR (*d*) Convert FM into PM
4. In television receiver, electron beam is deflected by
   (*a*) Electromagnetic (*b*) Electrostatic
   (*c*) Any of the two (*d*) None of these.
5. The purpose of video detector is
   (*a*) To recover audio signal from modulated RF signal
   (*b*) To recover video signal from modulated RF signal

(*c*) To separate audio and video signals
(*d*) None of these

6. The gain of IF section in TV receiver is
(*a*) Is of the order of 1000 or so
(*b*) Depends upon the design and cannot changed
(*c*) Can be varied with AGC bias control
(*d*) None of these

7. Loss at low frequencies is caused by
(*a*) Shunt capacitor (*b*) Coupling capacitor
(*c*) Decompling capacitor (*d*) Miller capacitor

8. Loss at high frequencies is caused by
(*a*) Compiling capacitor (*b*) Emitter by pass
(*c*) Decoupling capacitor (*d*) Shunt capacitor

9. For negative polarity transmission, the polarity of video detector at the output of video detector is
(*a*) Positive (*b*) Negative
(*c*) Neither of two (*d*) Intermediate

10. The purpose of sync separator in television receiver is
(*a*) To separate horizontal and vertical sync pulses
(*b*) To separate sync pulses from associated noise
(*c*) To separate sync pulses from the CVS
(*d*) None of these

11. AGC bias is applied to
(*a*) RF, mixer and IF stages (*b*) IF stage
(*c*) RF and IF stages (*d*) None of these

12. Receiving antenna for television is normally a
(*a*) A folded dipole (*b*) Parabolic reflector
(*c*) Yagi antenna (*d*) Loop antenna

13. AFC is used for
(*a*) AFC of vertical and horizontal oscillator
(*b*) Horizontal synchronization
(*c*) Stabilising picture and sound IF
(*d*) Vertical synchronisation.

14. H-sync pulse is separated from V-sync pulse by employing the circuit
(*a*) Multiplier (*b*) Differentiator
(*c*) Subtractor (*d*) Integrator

15. A trap circuit is a
(*a*) Resistor (*b*) Capacitor
(*c*) Tuned circuit (*d*) Inductor

16. FM detector employed in the sound section of a television receiver is generally
(*a*) A ratio detector (*b*) A frequency discriminator
(*c*) Either of two (*d*) None of these

ANSWERS

1. (*a*)	**2.** (*b*)	**3.** (*c*)	**4.** (*a*)	**5.** (*b*)
6. (*c*)	**7.** (*b*)	**8.** (*d*)	**9.** (*b*)	**10.** (*c*)
11. (*c*)	**12.** (*c*)	**13.** (*b*)	**14.** (*b*)	**15.** (*c*)
16. (*a*)				

5

CHAPTER

ESSENTIAL OF COLOUR TELEVISION

INSIDE THIS CHAPTER

5.1. INTRODUCTION

It is desirable that a TV system should produce a picture with realistic colour, adequate brightness and good definition that can be easily perceived by our eyes.

To have a colour picture it is thus necessary to add colour to the picture produced on a white raster. Indeed the transition from black and white television simply involved transforming the black and white dots to colour, but it is a little more complicated than that. In fact, this process involves the taking of the colour, and breaking it down in to different shades of red, yellow and blue. Because these are the primary colours, any colour know derives from these. Thus, the television can portray different colours by shading the dots with red, yellow or blue.

5.2. COMPATIBILITY AND REVERSE COMPATIBILITY OF COLOUR TRANSMISSION

One of the great electrical engineering triumphs was the development of colour television in such a way that it remained compatible with black and white television. A major driving force behind the majority of current colour TV standards was to allow black-and-white TVs to continue to be able to receive a valid TV signal after colour service was in place. The compatibility constraints are:

- *Monochrome compatibility:* A monochrome receiver must reproduce the brightness content of a colour signal correctly in black and white, with no visible interference from the colour information.

- *Reverse compatibility* : A colour television receiver must reproduce a monochrome signal correctly, with no spurious colour components.
- *Scanning compatibility* : The scanning system used for colour transmissions must be identical to the one used by the existing monochrome service.
- *Transmission channel compatibility* : The colour system must fit into the existing monochrome television channel and use the same spacing between the vision and sound carriers.

The problem of colour transmission consists in maintaining the transmission method of black-and-white television and broadcasting the additional colour information as far as possible within the available frequency band of the CVS. This means for any colour TV system that a colour broadcast is reproduced as a perfect black-and-white picture on a monochrome receiver (**compatibility**) and that a colour receiver can pick up a monochrome broadcast to reproduce a perfect black-and-white picture (**reverse compatibility**).

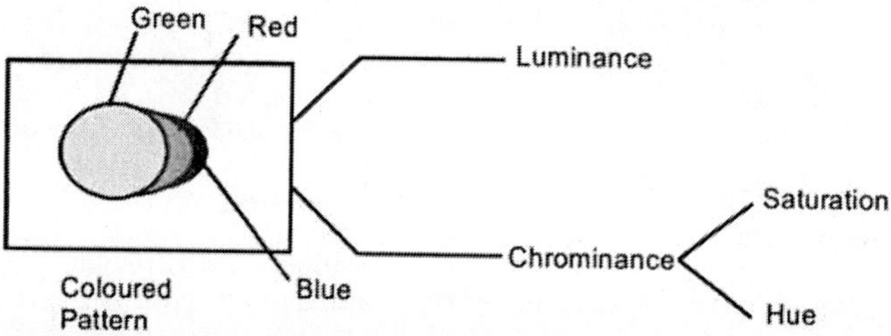

Fig. 5.1. Representation of coloured pattern by luminance and chrominance components.

These requirements can be met only if,

1. Information on the luminance distribution and
2. Information on the colour content are obtained from the coloured pattern and then transmitted.

The chromaticity is characterized by the hue – determined by the dominant wavelength of light, for instance for distinct colours such as blue, green, yellow, red – and by the saturation as a measure of spectral purity, *i.e.* of colour intensity with respect to the colourless (white). The chrominance signal cannot be obtained directly from the pattern. Instead the three primaries (red, green, blue) are used in accordance with the three-colour theory (Helmholtz). The red, green and blue signals are also required for reproducing the colour picture. Thus the scheme of compatible colour transmission is established by the luminance signal Y and the chrominance signal F (Fig. 5.2).

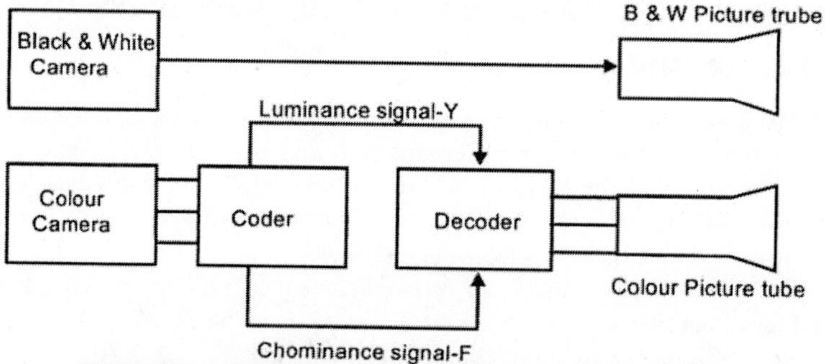

Fig. 5.2. Principle of compatible colour transmission.

Condition for Compatibility

1. Monochrome receiver should be able to receive colour transmission as black and white picture without requiring any change in the circuit.
2. The main video carrier frequency for colour transmission should remain the same as for monochrome transmission.
3. Black and white picture should not be affected by colour subcarrier produced locally in the colour receiver,
4. The baseband of the colour signals (about 1.5 MHz) should be accommodated in the baseband of the monochrome system without increasing the overall bandwidth.
5. The position of video and audio carrier frequencies and overall channel width should remain the same in colour transmission as in the monochrome transmission.
6. Scanning and sync pulses requirements of colour system should be the same for monochrome system.

5.3. BLOCK DIAGRAM OF COLOUR TELEVISION SYSTEM

A block diagram of a simplified colour television system is shown in Fig. 5.3. Light reflected (or transmitted) by an object is split into three elements so that a portion strikes each camera tube, producing outputs proportional to the tristimulus values *R*, *G* and *B*. Gain control are provided so that the three signals can be made equal when the camera is viewing a standard white object. For transmission, the signals are encoded into three different waveforms and then decoded back to *R*, *G* and *B* at the receiver. The primary purpose of encoding is to enable the signal to be transmitted within a limited bandwidth and to maintain compatibility with monochrome reception systems. The decoded *R*, *G* and *B* signals are applied to the picture tube to excite the three phosphors. Gain control are provided so equal inputs of *R*, *G* and *B* will produce a standard white on the display.

To reproduce a colour image of the pattern, additional information on the colour content, *i.e.* the "chromaticity" of the individual picture elements, must be transmitted together with the brightness or luminance distribution. This requires first the extraction of the colour information and then a possibility of reproducing the colour image.

In line with the basic principle of colour transmission, three pickup tubes are required; the image to be televised is projected, using the primaries red, blue and green, onto three photosensitive semiconductor targets via an optical beam splitter, called colour splitter, and via correcting filters for matching the signals to the spectral sensitivity of the semiconductor layers. To make the three partial images coincide accurately with their raster's, high mechanical and electro-optical precision is required. The three video signals produced by the camera represent three primaries of the colour diagram. By selective use of three signals, all colours in the visible spectrum can be reproduced on the screen of a colour picture tube. The luminance signal which represents the brightness of the elements being scanned must also be generated and transmitted along with the colour signals.

Instead of transmitting all the colours signals (red, green, blue) along with the brightness signal or monochrome signal, it is decided to send the two colour difference signals (R-Y & B-Y) and the brightness signal (Y). At the receiving end all the three colours are obtained by using the decoder system. In colour picture tubes, voltages pertaining to three primary colours give rise to three electron beams. These three beams activate three phosphor elements (one each), closely placed in a group, either in the form of dots or in the form of stripes. Thus there are lights of three primary colours coming out of these three phosphor elements about a point. The eye integrates the three colours and perceives the single resultant colour.

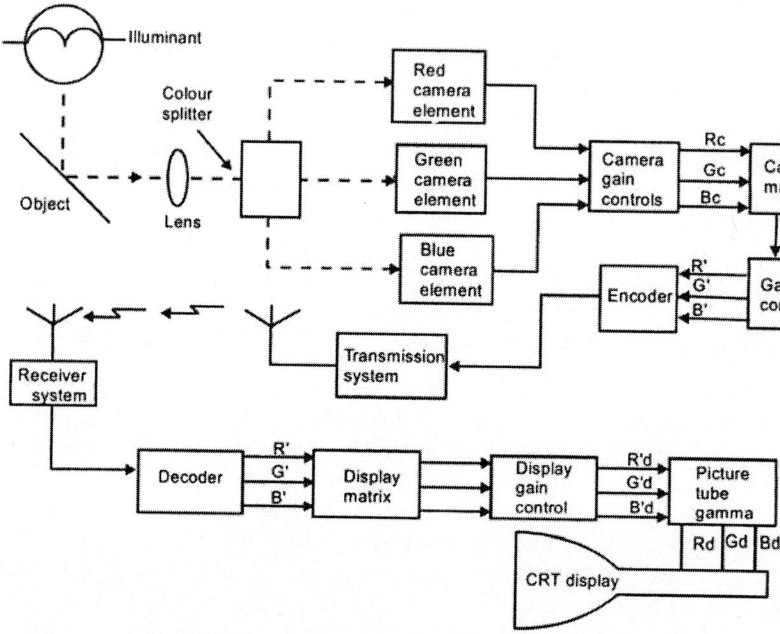

Fig. 5.3. Block diagram of colour television system.

5.4. NATURAL LIGHT

A sun light consists of a band of frequencies. A large spectrum of EM waves from low to beyond 10^{25} Hz. A light is a visible form of EM energy and it has two visible characteristics; intensity and colour. Light consists of a spectrum of frequencies of wavelength ranging from 700 nanometer (for red light) to 380 nanometer (for violet light) visible to human eye.

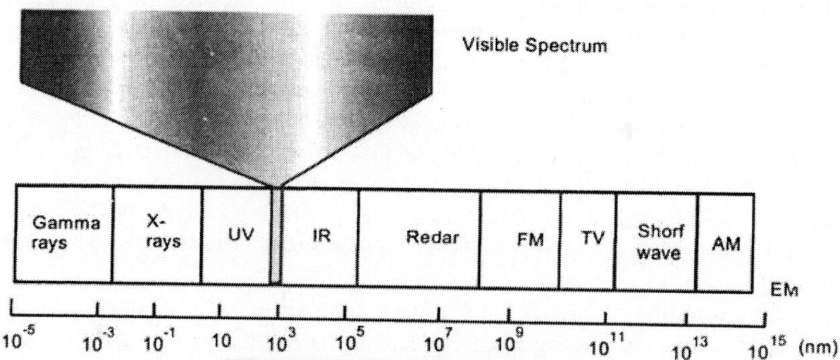

Fig. 5.4. Spectrum of sun light.

The visible spectrum is center around a frequency of the order of 5×10^{14} Hz.

Visible Spectrum

- Red –700 nm
- Orange—600 nm
- Yellow—550 nm
- Green—500 nm
- Blue—400 nm
- Violet—380 nm

5.5. COLOUR PERCEPTION

All objects are focused sharply by the lens system of the eye on its retina.

The retina which is located at the backside of the eye has light sensitive organs which measure the visual sensations. The retina is connected with the optic nerves which conduct the light stimuli as sensed by the organ to the optical center of the brain.

According to the Helmholtz, there are two light sensitive organs:

1. Rods — Brightness sensation (Gray from Black)
2. Cones – Sensitive to colour (RGB) 3 groups.

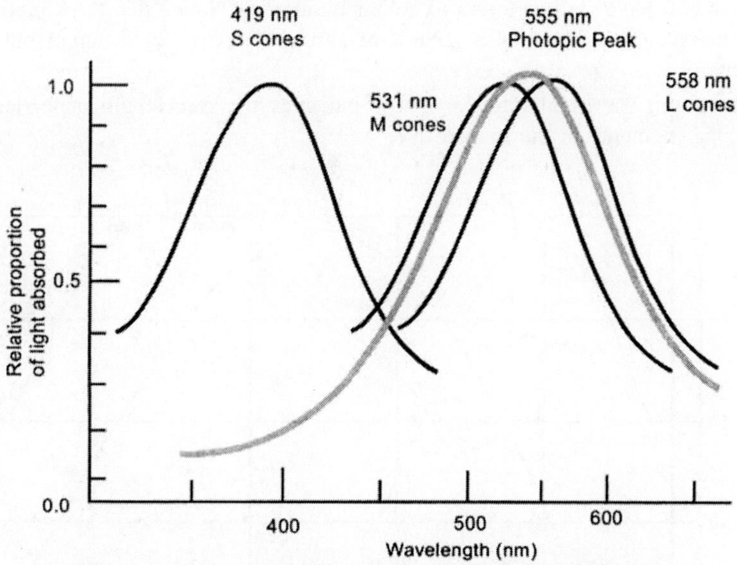

Fig. 5.5. Relative response of the eye to different colours.

5.6. THREE COLOUR THEORY

If you ask someone why red, green and blue are used in computer monitors — the immediate answer is "Because these are the primary colours". If you then ask, "But why are these primary colours?" — The answer you get is that "If you mix light of these colours together you can make any colour". However, there are some very fascinating reasons why RGB are the primary colours — and they have nothing to do with either light ... or Fourier series. They lie in the nature of the human eye.

The major light sensing element in the human eye is the retina. The retina consists of a number of different and important layers of cells. Notice however, that from an engineering viewpoint, the retina is constructed upside down. The major light receptors (the rods and cones) are at the bottom of the retina — not at the top. Thus, light must pass through nerves, blood, and tissue before reaching the main photoreceptors. This means that some serious signal processing is required by the retina and the brain to edit out these structures in your visual perception.

There are two types of photoreceptors, **rods and cones**. **Rods** are responsible for low level light detection and are most sensitive in the blue/green. They are very light sensitive and motion sensitive, but at the cost of resolution. Rods are virtually missing in the fovea (the center of the visual field) but are scattered elsewhere throughout the eye — thus governing peripheral vision as well as nighttime vision. At night, the fovea is very insensitive and most of the visual information is being carried by rods in the periphery of your eye. **Cones** carry the colour information and provide higher resolution — but at the cost of sensitivity. Cones are concentrated in the fovea, providing high resolution central daytime vision.

Rods contain a blue/green pigment called rhodopsin. **Cones** contain three pigments, a blue-sensitive pigment (447 nm), a green sensitive pigment called chlorolabe (540 nm), and a red sensitive pigment called erythrolabe (577 nm). These three pigments are the pigments responsible for "primary colours". Individuals missing one or two of these pigments will have various forms of colour blindness. (Notice that three pigments is by no means standard. There is a species of shrimp that has 20 different photoreceptor pigments.)

The graph shown in Fig. 5.6 below indicates the wavelength dependence of the three cone pigments in the human eye.

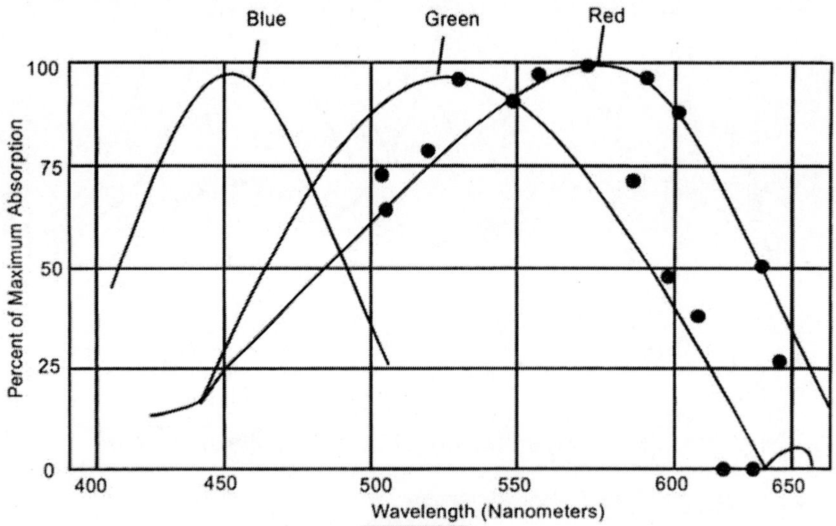

Fig. 5.6.

The idea of trireceptor vision (three colour theory) was worked out far before the physical mechanism of retinal pigments was understood. A common diagram for describing human colour perception was developed by the International Commission on Illumination (CIE). The CIE diagram is an attempt to precisely quantify the trireceptor nature of human vision.

Color perceptions were measured by giving subjects various combinations of the three standard CIE primary colours (435.8 nm, 546.1 nm and 700 nm) and measuring their perceptions. These perceptions are plotted on an *x-y* diagram called the **CIE Chromaticity diagram.**

The "pure" colours lie along the outer locus of the diagram, and the center of the diagram (CIE Illuminant C) is the CIE defined "white". Thus, any point on this diagram can be uniquely indicated by an *x* and *y* value. Therefore, in some fundamental way, only two basic functions are really require to describe any particular colour or hue. This fact will become very important to us in colour TV.

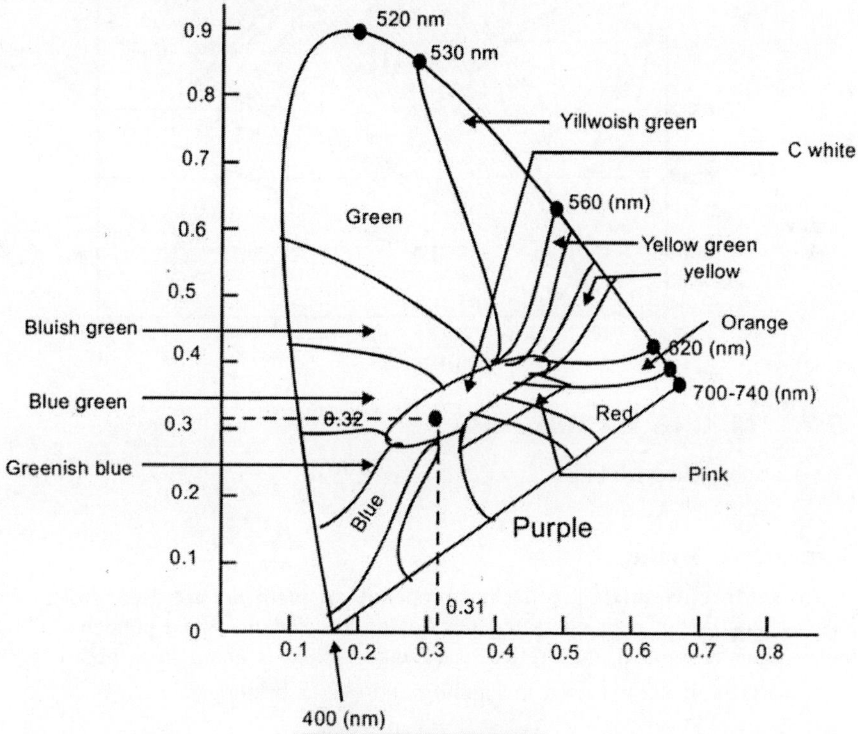

Fig. 5.7. CIE diagram.

It is interesting to compare the CIE Chromaticity diagram against commonly observed colours. In the following graph shown in Fig. 5.8, the CIE Chromaticity diagram is overlaid against surface colours of common paints and dyes (the gray blob) as well as the primary colour triangle of the American NTSC (National Television Systems Committee) colour television system and the European PAL (Phase Alternation Line rate) and SECAM (Sequential Couleur avec Memoire) systems.

All light sensation to the eye is divided in to 3 main groups.

1. Optic Nerve System—Integrate the different colour impressions.
2. Additive Mixing—Basis of colour TV system.
3. A white colour —additive mixing of the sensations from all the three sets of cones.

Any colour other than red, green and blue will excite different sets of cones to generate the cumulative sensation of that colour.

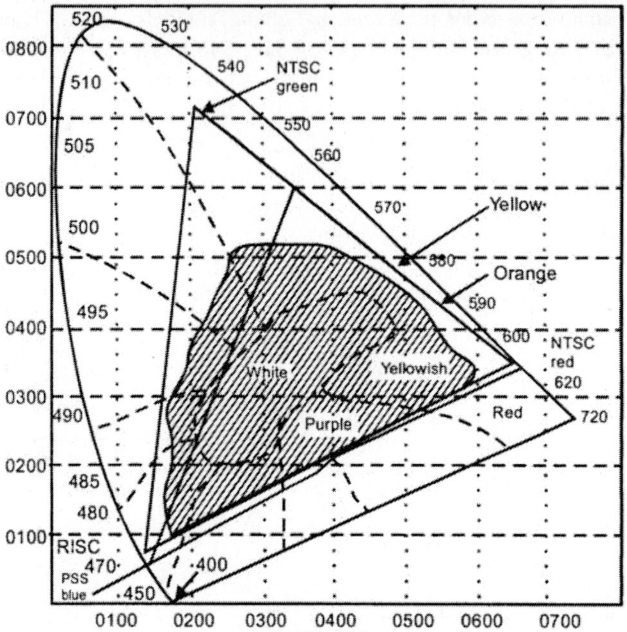

Fig. 5.8.

5.7. MIXING OF COLOURS

1. Subtractive Mixing,
2. Additive Mixing

Subtractive Mixing

In **subtractive mixing**, reflecting properties of pigments are used, which absorbs all wavelengths but their characteristics colour wavelength. When pigments of two or more colour are mixed, they reflect wavelength which is common to both.

This type of mixing used in Painting and colour printing.

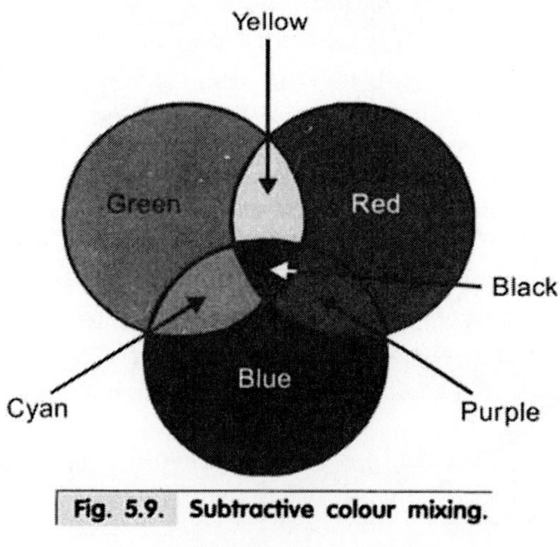

Fig. 5.9. Subtractive colour mixing.

Additive Mixing

Additive mixing is a method of producing various colours. When lights of two or more different colours are mixed and projected on a white screen, the eye perceives it as a combined colour.

Different colours are created by mixing pure colours and not by subtracting parts from white.

The additive mixing of three colours:-RGB in adjustable intensities can create most of the colours.White colour can be created by choosing suitable intensities of RGB.

It is used in **colour TV system.**

Complementary Colours

Red + Green = Yellow

Red + Blue = Magneta (purplish red shade)

Blue + Green = Cyan (greenish blue shade),

They are known as **complementary** colours because when combine with the remaining primary colours, they give white. Like yellow is complementary colour of blue because yellow plus blue will give white.

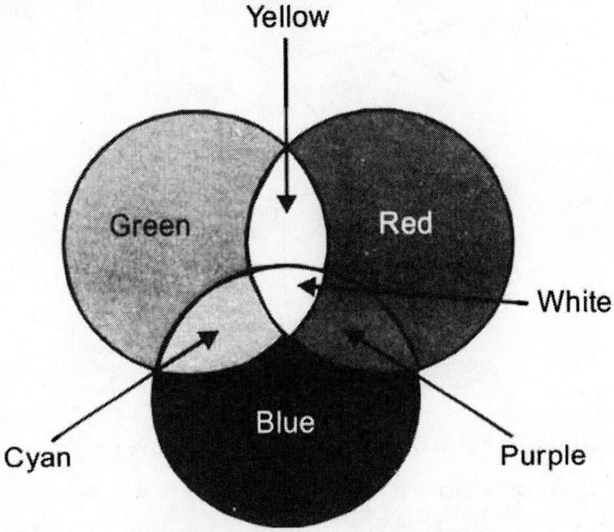

Fig. 5.10. Additive colour mixing.

Grassman's Law

The property of the eye of producing a response which depends on the algebraic sum of the RGB is known as **Grassman's law**.

The eye is not able to distinguish each of colours that mix to form a new colour but instead perceives only the resultant colours.

5.8. TRISTIMULES VALUE OF SPECTRAL COLOURS

Based on the spectral curve and extensive tasks, the primary spectral colours and their intensities required to produce different colours by mixing have been standardized.

The red green and blue have been fixed at wavelength of

$$R = 700 \text{ nm},$$
$$G = 546.1 \text{ nm}$$
$$B = 438.8\text{nm}$$

The components values of the three primary colours to produce various others colour have been standardized are called **Tristimules values** of the different spectral colours. These percentages of light values are based on the sensitivity of eye to different colours.

The reference white 30% (Red)
59% (green)
11% (Blue)

One lumen of white light = 0.3ln R +0.59ln G + 0.11ln B

As per the law of additive colour mixing, one lumen of white light is produced by 0.70 ln of cyan + 0.41ln of magenta + 0.89ln of yellow.

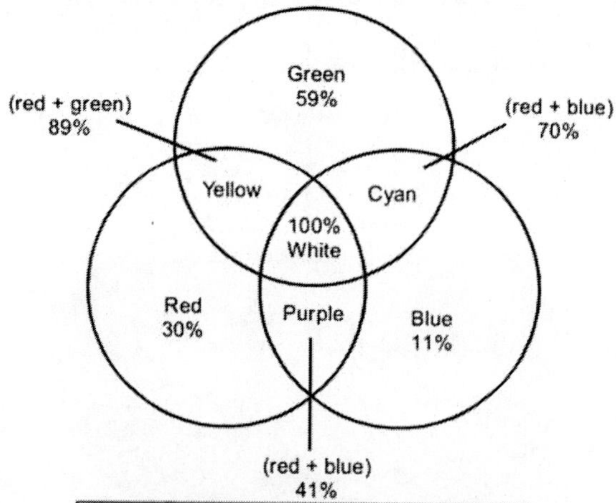

Fig. 5.11. Distribution of colour in %.

5.9. LUMINANCE, HUE AND SATURATION

A colour light has 3 characteristics to specify its visual information:

Luminance

This is the amount of light intensity as perceived by the eye regardless of the colour. It is different from the average brightness which determines the background level of illumination. Lighted parts in black and white picture have more luminance than black area. The output of a monochrome television camera is proportional to the luminance for each pixel. If the pixel is having a colour or hue, the output of monochrome television camera will be automatically reduced to its brightness value perceived by the eye.

Hue

It represent wavelength of the light reaching the eye. The eye cannot distinguish between the closely space wavelength. This is the predominant spectral colour of the received light. Thus the colour of any object is distinguished by its Hue or tint. Green leaves have green hue and red tomatoes red hue.

Saturation

This is the spectral purity of the colour light and is an indication of how the colour is diluted by the white. A fully saturated colour has no white. It represents how much white is present in a colour. White light dilutes a colour. Absence of white in a colour makes it fully saturated colour. If a colour is not fully saturated, it means white is present to some extent in the colour. A light green is not fully saturated; it contains white whereas a vivid green is fully saturated.

The hue and saturation of a colour put together is known as **Chrominance or chroma**. It does not contain brightness information.

5.10. COLOUR TV CAMERA

To reproduce a colour image of the pattern, additional information on the colour content, *i.e.* the "chromaticity" of the individual picture elements, must be transmitted together with the brightness or luminance distribution. This requires first the extraction of the colour information and then a possibility of reproducing the colour image.

In line with the basic principle of colour transmission, three pickup tubes are required; the image to be televised is projected, using the primaries red, blue and green, onto three photosensitive semiconductor targets via an optical beam splitter, called colour splitter, and via correcting filters for matching the signals to the spectral sensitivity of the semiconductor layers. To make the three partial images coincide accurately with their raster's, high mechanical and electro-optical precision is required.

It consists of three camera tubes with each tube receives relatively filtered primary colours as shown in Fig. 5.12. Each camera tube develops a signal voltage proportional to the respective colour intensity received by it. Light from the scene is processed by the objective lens system. The image formed is split in to three images by means of glass prism. Prisms are designed as diachroic mirrors. Diachroic mirrors passes one wavelength and reject other wavelengths. Thus RGB images are formed. The precise colour images which are converted in to video signals by image orthicon or vidicon camera tubes are also passes through colour filter or trimming filters. Thus the three colour signals are generated.

The three video signals produced by the camera represent three primaries of the colour diagram. By selective use of three signals, all colour in the visible spectrum can be reproduced on the screen of a colour picture tube. The luminance signal which represents the brightness of the elements being scanned must also be generated and transmitted along with the colour signals.

Coincidence errors of the colour rasters would cause a loss in definition for the luminance signal. For this reason, colour television cameras with a separate pickup tube for the luminance signal are also used. Progressive developments, already partly implemented in portable colour TV cameras, point to a single-tube colour TV camera producing the tristimulus signals in the red, green and blue channels by a multiplex method.

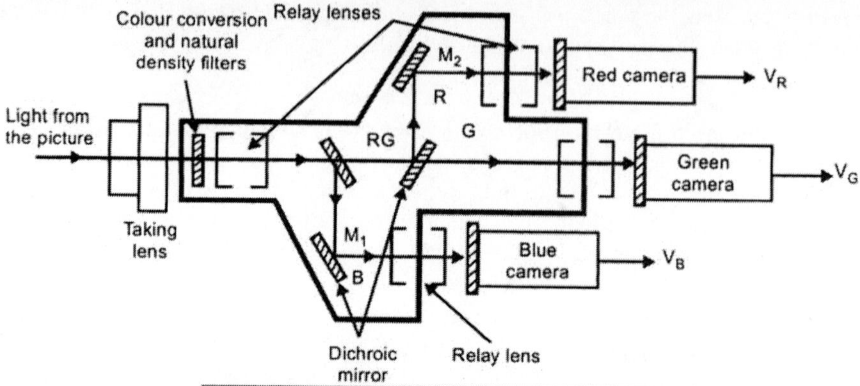

Fig. 5.12. Basic Colour Camera Arrangement.

Gamma Corrections

To compensate for the non-linearity of the system including TV camera and picture tubes, a correction is applied to the voltage produced by the three camera tubes, known as **Gamma corrections**. The camera tube output is corresponding to maximum intensity (100%) of standard white light.

5.11. LUMINANCE, CHROMINANCE SIGNALS AND COLOUR DIFFERENCE SIGNALS

For reasons of compatibility, the colour camera has to deliver the same signal from a coloured pattern, *i.e.* the luminance signal to the monochrome receiver, as the black-and-white camera.

The spectral sensitivity of a black-and white camera corresponds to the brightness sensitivity curve of the human eye to ensure that the black-and-white picture tube reproduces the different colour stimuli as grey levels of the same brightness as perceived by the eye.

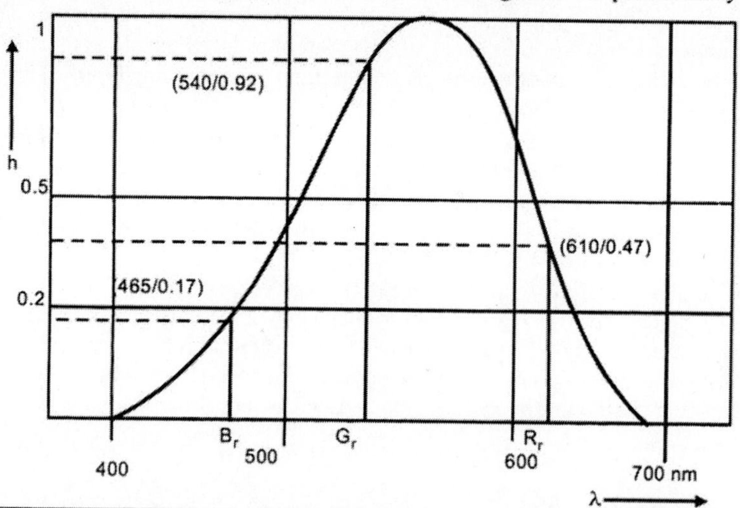

Fig. 5.13. Brightness sensitivity of human eye to receiver primaries.

The colour camera, however, delivers three signals with spectral functions matching the colour mixture curves. To obtain one signal whose signal spectral function corresponds to the sensitivity curve of the eye, coding is required. To this end, the three colour values represented by the functions $r_r(\lambda)$, $g_r(\lambda)$ and $b_r(\lambda)$ are multiplied by the relative luminosity coefficients h_r, h_g and h_b and then added up. Except for the proportionality constant k, the result must be identical to the sensitivity function of the human eye $h(\lambda)$.

$$h(\lambda) = k[h_r \times r_r(\lambda) + h_g \times g_r(\lambda) + h_b \times b_r(\lambda)]. \qquad ...(5.1)$$

The relative luminosity coefficients h_r, h_g and h_b are obtained by normalization from the corresponding values $h(R_r)$, $h(G_r)$ and $h(B_r)$ of the eye (Fig. 5.14).

$$h(R_r) = 0.47$$
$$h(G_r) = 0.92$$
$$h(B_r) = 0.17$$

$$h = 1.56$$

The following figures are calculated for the relative luminosity coefficients:

$$h_r = \frac{h_{(Rr)}}{\Sigma h} = \frac{0.47}{1.56} = 0.30$$

$$h_g = \frac{h_{(Gr)}}{\Sigma h} = \frac{0.92}{1.56} = 0.59$$

$$h_b = \frac{h_{(Br)}}{\Sigma h} = \frac{0.17}{1.56} = 0.11$$

Due to normalization :

$$h_r + h_g + h_b = 1 \qquad ...(5.2)$$

Thus for equation

$$h(\lambda) = k\,[0.30 \times r_r(\lambda) + 0.59 \times g_r(\lambda) + 0.11 \times b_r(\lambda)].$$

is obtained or, written in a simplified way, for the luminance signal Y corresponding to the sensitivity curve of the eye

$$\boxed{Y = 0.30 \times R + 0.59 \times G + 0.11 \times B} \qquad ...(5.3)$$

This equation is one of the most important relations of colour TV engineering.

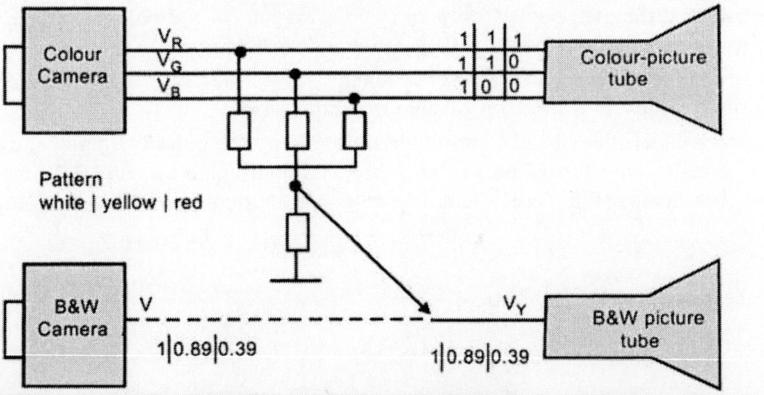

Fig. 5.14. Producing luminance signal V_Y from tristimulus voltages V_R, V_G, V_B and compability relationship.

Technically the luminance signal V_Y is obtained from the tristimulus signals V_R, V_G and V_B via a matrix (Fig. 5.14).

The signals listed in Table 1 below result for a pattern consisting of eight colour bars (standard colour bar sequence; ?the three primaries plus the associated complementary colours and the colourless stripes white and black.

Pattern	R	G	B	Y
White	1	1	1	1.00
Yellow (R + G)	1	1	0	0.89
Cyan (G + B)	0	1	1	0.70
Green	0	1	0	0.59
Purple (R + B)	1	0	1	0.41
Red	1	0	0	0.30
Blue	0	0	1	0.11
Black	0	0	0	0

Fig. 5.15. Signals of standard colour bar sequence.

To reproduce a colour pattern, the three tristimulus signals R, G and B are required. For compatibility, however, colour transmission uses the luminance signal Y and the chrominance signal F. The latter cannot be obtained directly from the tristimulus values but only by way of the colour difference signals:

R-Y, B-Y & G-Y

These are the colour values minus the luminance component.

The chrominance signal carries information on hue and saturation. Therefore two colour difference signals are sufficient to describe the chrominance component.

For this purpose, the quantities R-Y and B-Y were selected.

Referring to the voltage of the luminance signal derived from the coder yields these two colour difference signals as:

The equation

$$VY = 0.30 \times VR + 0.59 \times VG + 0.11 \times VB \qquad ...(5.10)$$

$$VR - VY = 0.70 \times VR - 0.59 \times VG - 0.11 \times VB \qquad ...(5.11)$$

$$VB - VY = -0.30 \times VR - 0.59 \times VG + 0.89 \times VB$$

The colour difference signals only carry information on the colour content. For a monochrome scene ($V_R = V_G = V_B$) they are equal to zero.

The amplitude of the colour difference signals shows the departure of the hue from the colourless, which is a measure of colour saturation.

The hue is determined by the amplitude ratio and by the polarity sign of the colour difference signals. Transformation of the orthogonal coordinate system (B-Y and R-Y) into polar coordinates (Fig. 5.16) yields the colour saturation from the vector length A:

$$A = \sqrt{(B-Y)^2 + (R-Y)^2} \qquad ...(5.13)$$

and the hue from the angle α

$$\alpha = \arctan \frac{(R-Y)}{(B-Y)} \qquad ...(5.14)$$

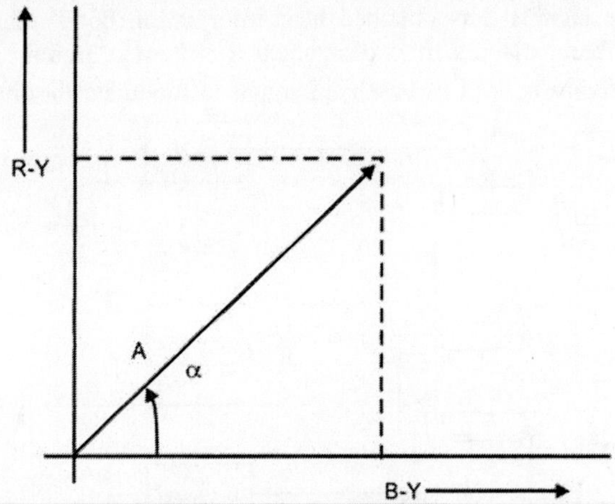

Fig. 5.16. **Representation of chromaticity as a function of colour difference signal.**

5.12. PRODUCTION OF COLOUR DIFFERENCE VOLTAGES

Investigations have shown that the resolution of the eye is lower for coloured pattern details than for brightness variations. Therefore it is sufficient to transmit only the luminance signal with the full bandwidth of 5 MHz. The bandwidth of the chrominance signal can be reduced to about 1.5 MHz by taking the two colour difference signals via low pass filters.

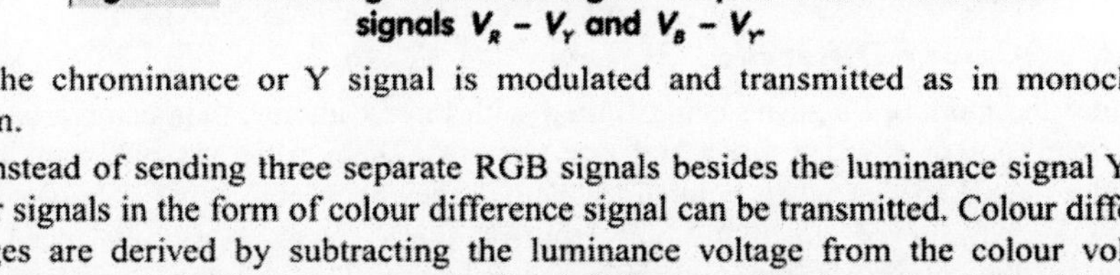

Fig. 5.17. **Producing luminance signal VY plus colour difference signals $V_R - V_Y$ and $V_B - V_Y$.**

The chrominance or Y signal is modulated and transmitted as in monochrome system.

Instead of sending three separate RGB signals besides the luminance signal Y, only colour signals in the form of colour difference signal can be transmitted. Colour difference voltages are derived by subtracting the luminance voltage from the colour voltages. Only (R-Y) and (B-Y) are produced. Third difference signal can be derived from the other two. In the coder, the signals V_R, V_G and V_B produced by the colour camera are converted into the luminance component V_Y and the colour difference signals $V_R - V_Y$ and $V_B - V_Y$ (Fig. 5.18) and, in this form, applied to the reproducing system.

By using the arrangement shown in Fig. 5.17, colour difference voltages are derived by subtracting the luminance voltage from the colour voltages. Only $(R - Y)$ and $(B - Y)$ are produced. Voltage Vy is low because Rc chosen is very small to avoid cross talk. Hence it is amplified. Amplified Y signal is inverted to obtain $-Y$ as the output.

One adder circuit adds the red camera output to $-Y$ to obtain the $[R + (-Y)]$ or $(R - Y)$ signal. Similarly the second adder combines the blue camera output to $-Y$ and delivers $(B - Y)$ as its output.

The difference signals thus obtained bear information both about the 'Hue' and 'Saturation' of different colours. It is designated as **Chroma signal.**

The Chroma signal is used as baseband signal to modulate the main video carrier.

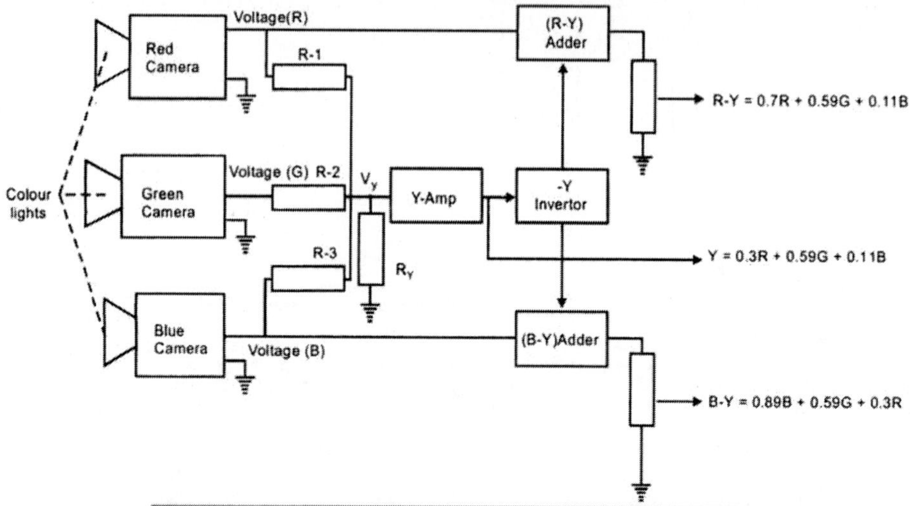

Fig. 5.18. Production of three colour difference signals.

5.13. COLOUR DIFFERENCE SIGNALS ACTUALLY TRANSMITTED

Out of three colours difference signals if only two colour difference signals are transmitted, the third one can be obtained locally in the receiver from the remaining two colour difference signals. At the receiving end all the three difference signals are needed.

The circuit would have become complex if a single carrier was required to be modulated by three signals, hence only two colour difference signals [(B-Y) and (R-Y)] are used for modulating the colour subcarrier. The colour difference signal G-Y is not used in modulation process.

Encoding of Colour Difference Signals

Consider the model of a scene being filmed with three cameras. One camera has a red filter, one camera a green filter and one camera a blue filter. Assume that the cameras all adjusted so that when pointed at "white" they each give equal voltages. To create the Y signal, the red, green and blue inputs to the Y signal must be balanced to compensate for the colour perception misbalance of the eye. The governing equation is:

$$Y = 0.3R + 0.59G + 0.11B \quad ...(5.14)$$

Two colour difference signals [(B-Y) & (R-Y)] are used for modulating the colour subcarrier. Each colour-difference signal modulaies an assigned subcarrier. The two subcarriers are of identical frequency (about 4.43MHz) but of different phase. The phase difference between the two subcarriers is 90 degrees, so the original signals modulating the two subcarriers can be demodulated without crosstalk. The two subcarriers are obtained from a common crystal-controlled oscillator. The type of modulation used is suppressed-carrier amplitude modulation. Because the subcarrier is suppressed, only the sidebands are obtained at the output of the modulators. This results in complete cancellation of the chrominance signal when no colours are present.

The colour video signal contains two independent information, that of Hue and Saturation. It is difficult matter to modulate them to one and the same carrier in such a way that there can be easily recovered at the receiver without affecting each other.

Two signals are then created to carry the chrominance information. One of these signals is called "Q" and the other is called "I".

The positive polarity of Q is purple, the negative is green. The positive polarity of I is orange, the negative is cyan. Thus, '**Q**' is often called the "green-purple" or "purple-green" axis information and '**I**' is often called the "orange-cyan" or "cyan-orange" axis information.

It turns out that the human eye is more sensitive to spatial variations in the "orange-cyan" than it is for the "green purple". Thus, the "orange-cyan" or I signal has a maximum bandwidth of 1.5 MHz and the "green purple" only has a maximum bandwidth of 0.5 MHz.

5.14. THE COLOUR DIFFERENCE SIGNAL (G-Y) WHICH IS NOT TRANSMITTED

Why (G-Y) signal is not transmitted?

(G-Y) colour difference signal is not transmitted, as it can be produced locally in the receiver easily by passing the available signals B-Y and R-Y through a resistive matrix. The other colour difference signals cannot be produced by resistive matrix.

To understand this, let us determine the relationship between the colour difference signals.

$$0.11B + 0.30R + 0.59G = Y$$

Or $$0.11B + 0.30R + 0.59B = (0.11 + 0.30 + 0.59)\ Y$$

$$0.11\ (B - Y) + 0.30\ (R - Y) + 0.59\ (G - Y) = 0 \qquad ...(5.15)$$

$$G - Y = -0.11\ (B\text{-}Y)/0.59 - 0.30\ (R\text{-}Y)/0.59 \qquad ...(5.16)$$

$$R - Y = -0.11\ (B\text{-}Y)/0.30 - 0.59\ (G\text{-}Y)/0.30 \qquad ...(5.17)$$

$$B - Y = -0.30\ (R\text{-}Y)/0.11 - 0.59\ (G\text{-}Y)/0.11 \qquad ...(5.18)$$

A resistive matrix cannot amplify the inputs but can reduce them. For getting (*G-Y*), the coefficients of inputs are 0.11/0.59 and 0.30 / 0.59, which are only a fraction and so can be achieved by a resistive matrix.

R-Y and *B-Y* cannot be obtained from a resistive matrix because the coefficient 0.59/ 0.30 (*R-Y*) and the coefficient 0.30/0.11 and 0.59/0.11 for (*B-Y*) are greater than 1.

This is the reason why only (*G-Y*) can be obtained locally from the known values of *R-Y* and *B-Y* with the help of a simple resistive matrix, and hence it is not transmitted.

5.15. TRANSMISSION OF CHROMINANCE SIGNAL WITH COLOUR SUBCARRIER

As explained in the preceding sections, the colour TV signal is transmitted in the form of the luminance signal *Y* and the two colour difference signals *R-Y* and *B-Y* for reasons of compatibility. To transmit the complete picture information; luminance plus chrominance, a triple transmission channel would be required. Here one might think of making multiple use of the TV transmission channel either by frequency or time multiplex. However, neither method is compatible with the existing black-and-white transmission method.

To solve this problem, This system has been so designed that's its signals occupies not more than 7 MHz (both video and sound signals) and it carries not only the full black and white (monochrome) signal but in addition, the colour information.

Now the question is how is all this information compresses into a 7MHz band spread?

The answer is that since a video signal does not occupy every cycle of 5 MHz assigned to it, colour information can be inserted within 5 MHz as shown in Fig. 5.19.

A decisive thought for the colour transmission method actually selected is derived from spectral analysis of the luminance signal or Composite Video Signal. It becomes apparent that only certain frequency components occur in the CVS spectrum, these components being mainly multiples of the line frequency due to the periodic scanning procedure.

The varying picture content amplitude modulates the line-repetitive pulse sequence producing sidebands spaced at multiples of the field frequency from the spectral components of the line pulse. It is to be noted that the energy content progressively decreases as the harmonic number increases, being quite high at f_h, $2f_h$ and negligibly small at $319f_h$, $320f_h$.

Figure 5.19 shows a detail of the CVS spectrum.

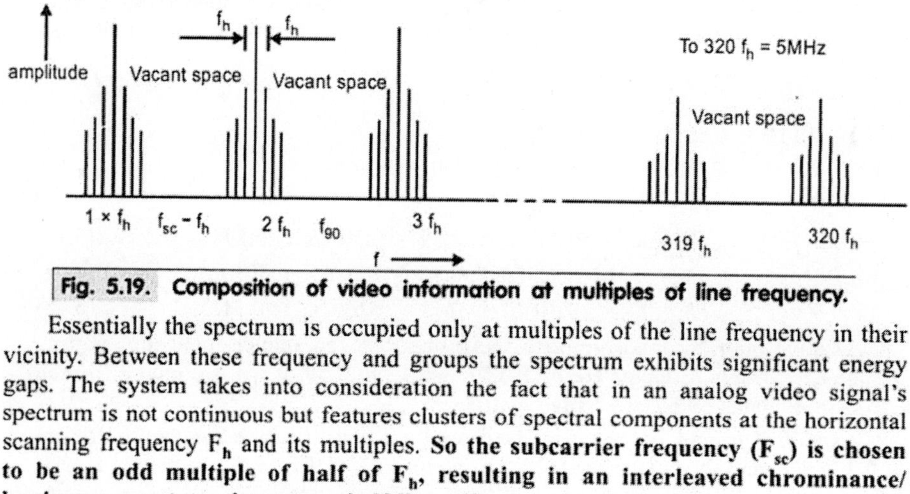

Fig. 5.19. Composition of video information at multiples of line frequency.

Essentially the spectrum is occupied only at multiples of the line frequency in their vicinity. Between these frequency and groups the spectrum exhibits significant energy gaps. The system takes into consideration the fact that in an analog video signal's spectrum is not continuous but features clusters of spectral components at the horizontal scanning frequency F_h and its multiples. **So the subcarrier frequency (F_{sc}) is chosen to be an odd multiple of half of F_h, resulting in an interleaved chrominance/luminance spectrum known as half-line offset.**

This is done by modulating the chrominance signal onto a colour subcarrier whose frequency f_{SC}, and thus also the spectrum of the line-repetitive modulation products, is located between the frequency components of the CVS (Figure 5.20).

Since the colour information is also line repetitive, the spectrum of the chrominance signal consists only of multiples of the line frequency and the corresponding sidebands. Therefore it is appropriate to insert the additional colour information into the gaps of the CVS frequency spectrum. Between any two consecutive harmonics, there are vacant spaces or slots where there is no monochrome energy. The vacant spaces are occupied by the colour sub-carrier (modulated by colour difference signals). The process of placing the energy of one signal between the clusters of energy of another is called **frequency interleaving**.

The energy of the chroma signal is also concentrated in bundles, but the highest energy bundles is at the colour subcarrier frequency, near the upper end of video bandwidth, and as one moves towards the lower integers of scanning frequencies, the energy decreases and becomes negligibly small as f_h is approached. High colour subcarrier frequency makes the interleaving process free of cross interference because the energy of luminance and is high, the energy of chroma signals is low and vice versa as shown in Fig. 5.20(*a*).

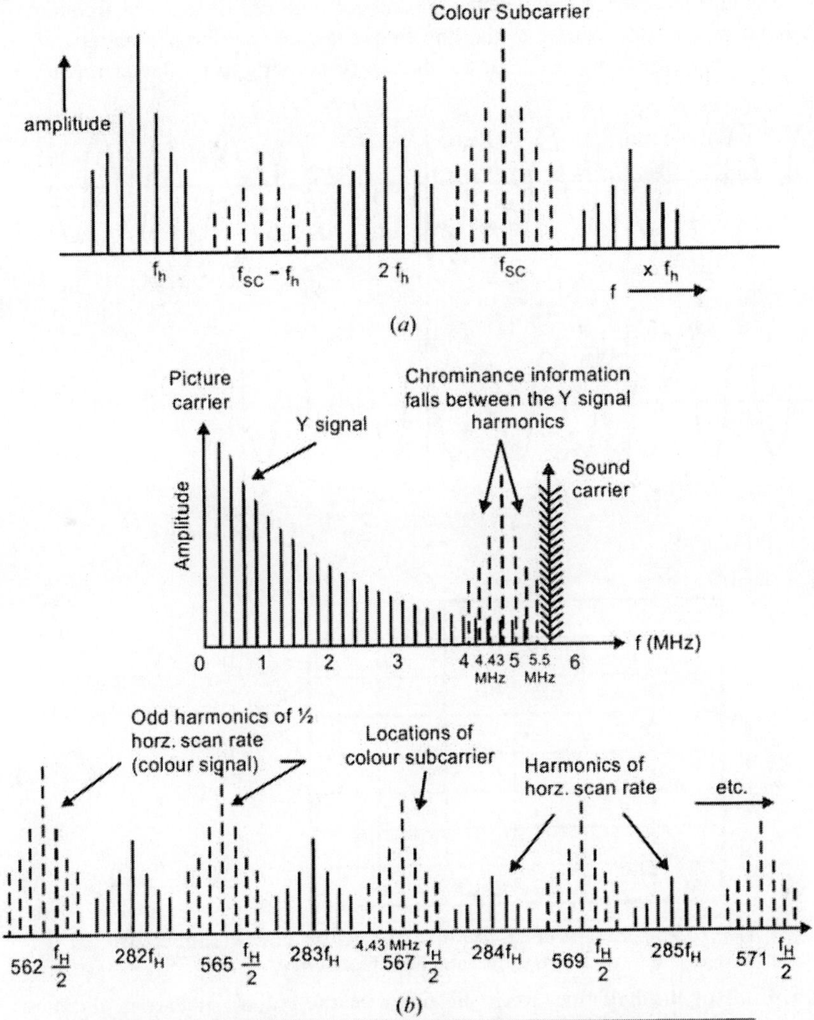

Fig. 5.20. **Interleaving of the of luminance and chroma signals.**

5.16. DETERMINING THE COLOUR SUBCARRIER FREQUENCY

One condition for determining the colour subcarrier frequency results from the symmetrical interleaving of the CVS and chrominance signal spectra: the frequency f_{SC} should be an odd multiple of half the line frequency f_h :

$$f_{SC} = (2n+1) \times \frac{f_h}{2}$$

In this way the half-line offset is obtained.

Generally, the interference which a colour subcarrier produces on the black-and white picture can be explained as follows, if it is assumed that a sine wave in the frequency region of the CVS produces a bright-dark interference pattern on the screen.

If the colour subcarrier frequency is an integer multiple of the line frequency, *i.e.* if there is no offset with respect to the line frequency, an interference pattern of bright and dark vertical stripes appears, their number corresponding to the factor n (Fig. 5.21).

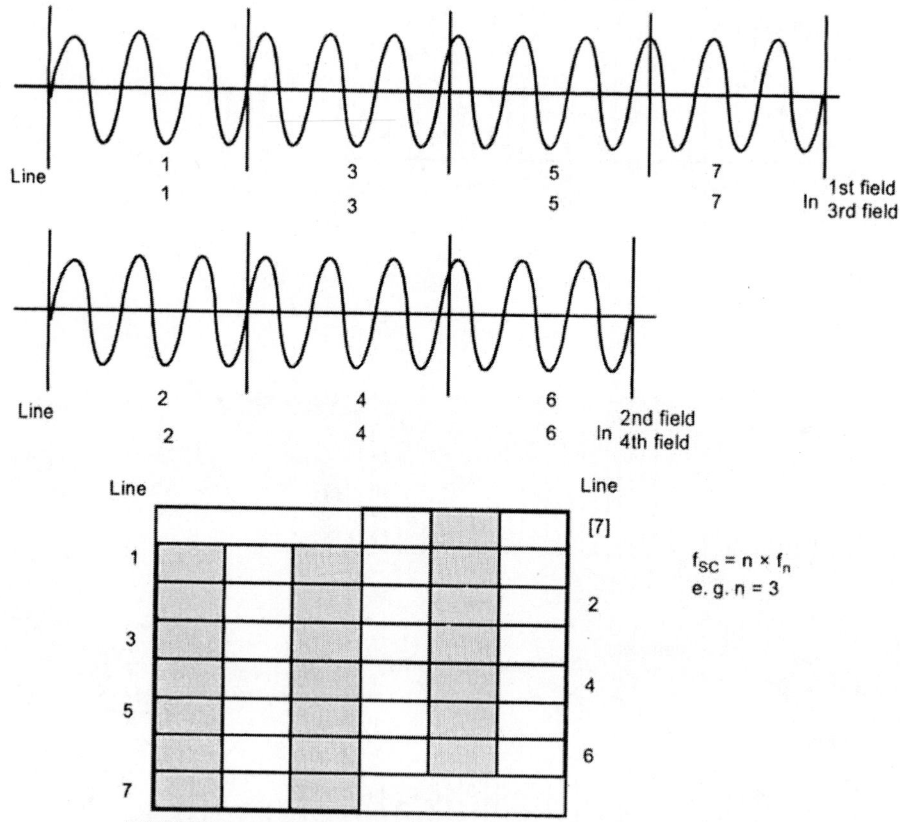

Fig. 5.21. Interference pattern caused by colour subcarrier frequency and line frequency.

As a result of the half-line offset the phase of the colour subcarrier alternates by 180° from line to line of a field. However, because of the odd number of lines, bright and dark dots coincide after two fields. The interference pattern occurring in a rhythm of $f_v/4 = 12.5$ Hz would thus be compensated over four fields (Fig. 5.21). Nevertheless, the compensation of the interference pattern on the screen is not perfect due to the nonlinearity of the picture tube characteristic and the inadequate integrating capability of the human eye.

If a half-line offset is assumed between the colour subcarrier frequency and the line frequency, the subjective annoyance can be further reduced by selecting the colour subcarrier frequency as high as possible. In this way, the interference pattern takes on a very fine structure.

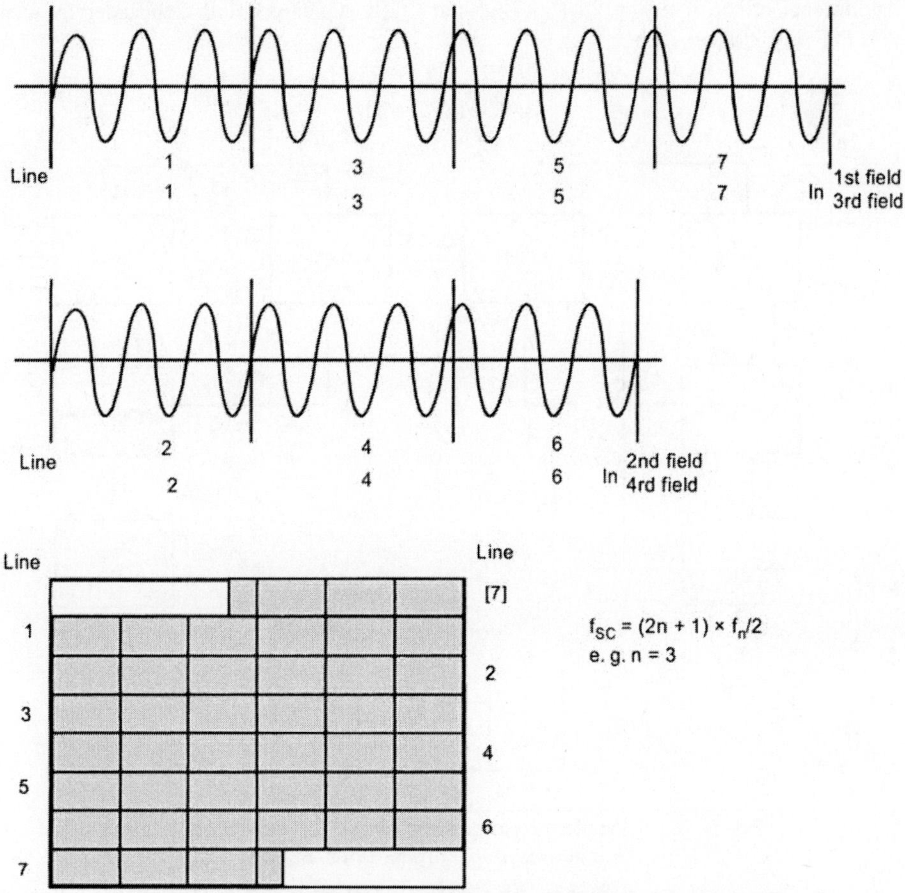

Fig. 5.22. Compensation of interference pattern with half-line offset of colour subcarrier frequency.

However, the colour difference signals modulate the colour subcarrier so that for transmitting the upper sideband a certain minimum spacing of the colour subcarrier frequency from the upper frequency limit of the CVS has to be maintained.

The best compromise has proved to be a colour subcarrier frequency of about 4.4 MHz. Thus the principle of compatible colour TV transmission is found using the luminance signal and the chrominance signal modulated onto the subcarrier; this is the basis of the NTSC system and its variants (Fig. 5.22). For the CCIR-modified NTSC method, a colour subcarrier frequency of has been fixed.

$$f_{SC} = 567 \times \frac{f_h}{2} = 283.5 \times f_h$$

$$= 4.4296875 \text{ MHz}$$

A further development of the NTSC method is the PAL method, which is being widely used today. In the PAL system, one component of the subcarrier is switched by 180° from line to line. However, this cancels the offset for this subcarrier component so that a pronounced interference pattern would appear in the compatible black-and-white picture. This is avoided by introducing a quarter-line offset of the subcarrier plus an additional offset of $fv/2$ = 25 Hz. Thus the CCIR standard colour subcarrier frequency in the PAL system is

$$f_{SC} = 283.75 \times f_h + 25 \text{ Hz}$$
$$f_{SC} = 4.43361875 \text{ MHz}$$

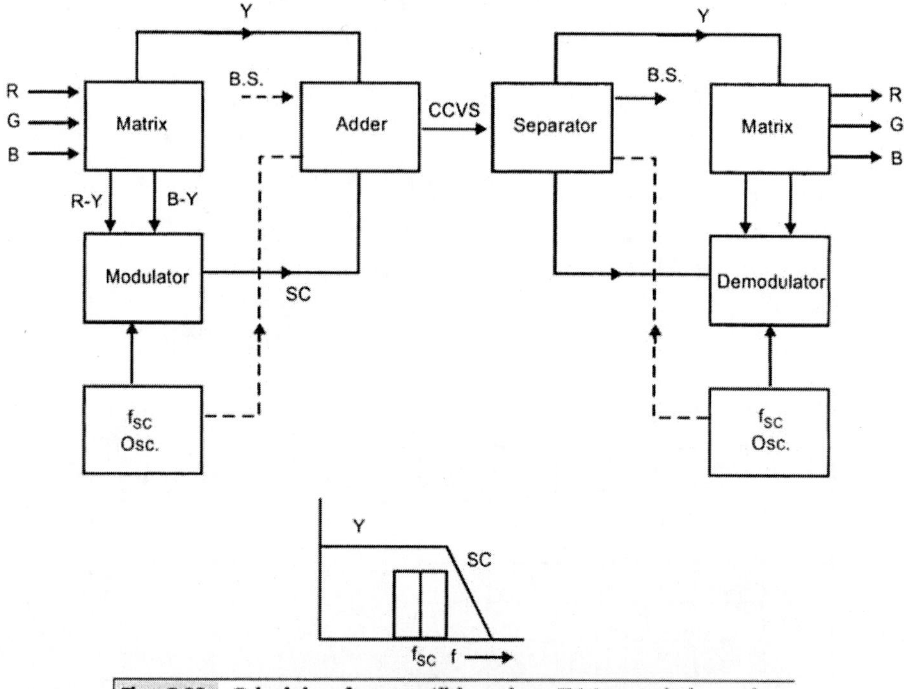

Fig. 5.23. Principle of compatible colour TV transmission using luminance and chrominance signals.

5.17. MODULATION OF COLOUR SUBCARRIER

The chrominance signal is transmitted by modulating the colour subcarrier with the two colour difference signals. The modulation method must permit the colour difference signals to be extracted separately at the receiver end.

In the NTSC and PAL methods, double amplitude modulation is used. The 0° component of the subcarrier is amplitude- modulated by the $(B - Y)$ signal and its 90° component by the $(R - Y)$ signal, the subcarrier being suppressed at the same time. This results in quadrature modulation (Fig. 5.22). The modulation product is a subcarrier frequency whose amplitude and phase are determined by the two colour difference signals. Amplitude and phase modulation take place at the same time. Compared to the chrominance signal represented in Fig. 5.23, the colour saturation now corresponds to the amplitude SC and the hue to the phase angle ϕ_{SC} of the modulated colour subcarrier.

For this reason, the modulated colour subcarrier is also called the chrominance signal.

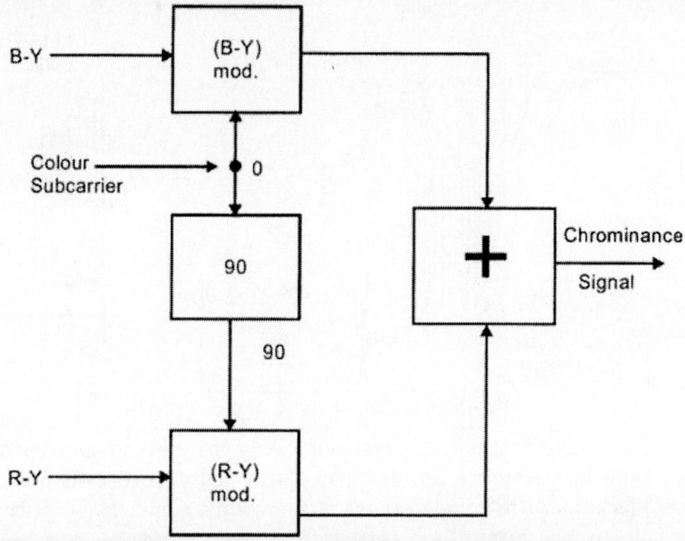

Fig. 5.24. **Producing chrominance signal by quadrature modulation of colour subcarrier with two colour difference signals.**

$$SC = \sqrt{(B-Y)^2 + (R-Y)^2} \qquad ...(5.21)$$

$$\varphi_{SC} = \arctan \frac{R-Y}{B-Y} \qquad ...(5.22)$$

A vector diagram of the chrominance signal shows the position of the different colours on the colour circle (Fig. 5.25). The complementary colours are located on opposite sides of the coordinate zero (colourless). When transmitting a colourless picture element, the colour difference signals and the amplitude of the colour subcarrier equal zero. In this case, the subcarrier does not cause any interference in the black-and-white picture.

To demodulate the chrominance signal, the unmodulated carrier of correct phase is required.

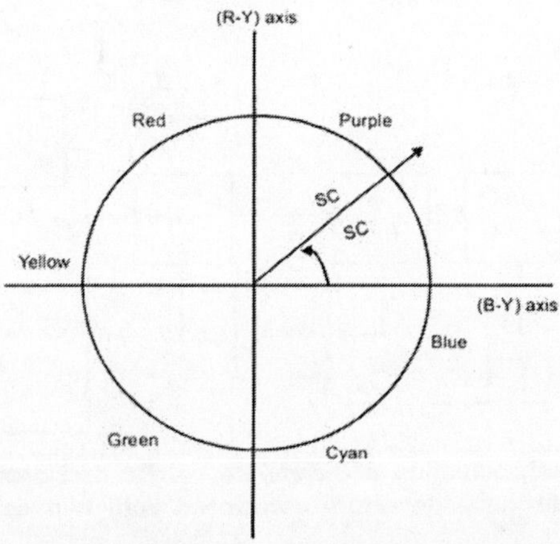

Fig. 5.25. **Vector representation of chrominance signal: colour circle.**

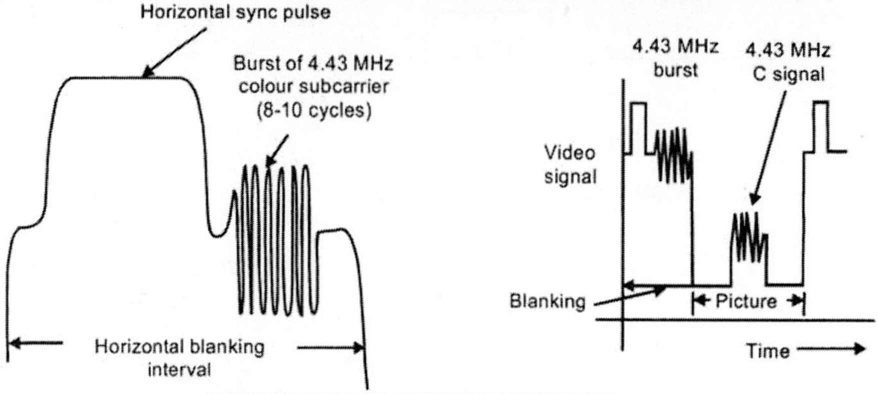

Fig. 5.26. Colour sync signal (burst).

Synchronous detection is used, evaluating only the chrominance component which is in phase with the reference carrier. Since the actual subcarrier is not transmitted, it must be produced as a reference carrier at the receiver end. Now, since the I and Q signals are clearly phase sensitive; some sort of phase reference must be supplied. For synchronization with the subcarrier at the transmitter end, a reference signal is inserted into each line in the H blanking interval, *i.e.* the colour sync signal or burst. The phase reference consists of 8-10 cycles of the 4.43 MHz signal. It is called the "colour burst" and looks something like shown in Fig. 5.26.

In the NTSC system, the burst phase is at 180° compared to the 0° reference phase of the colour subcarrier. In the receiver the burst is separated from the chrominance signal by blanking. A phase comparator produces a control voltage from the departure of the reference carrier phase from the burst phase; this voltage controls the frequency and phase of the reference carrier oscillator via an integrating network. The control voltage becomes zero if the phase difference is 90°. Coming from the reference carrier oscillator, the 90° component is taken directly to the $(R - Y)$ synchronous detector and, after a negative 90° phase shift, as the 0° component to the $(B - Y)$ synchronous detector (Fig. 5.27).

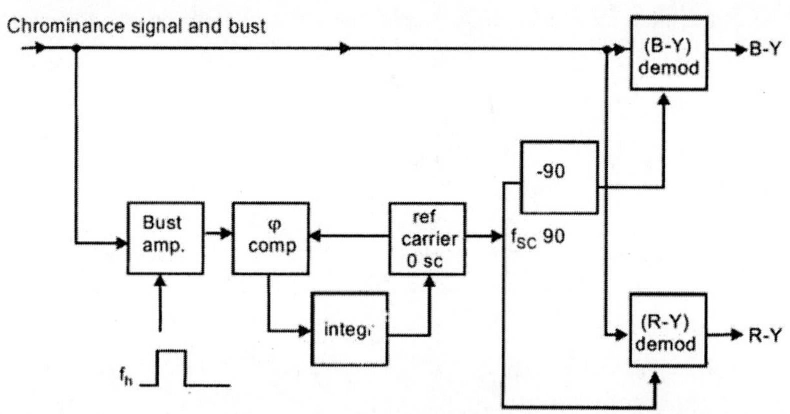

Fig. 5.27. Synchronization of reference carrier oscillator at receiver end and synchronous detectors associated with two colour difference signals.

5.18. DEMODULATION OF COLOUR SUBCARRIER

Two synchronous demodulators (and the associated low-pass filters) are needed to extract these signals.

An example of a circuit to do the demodulation is given below:

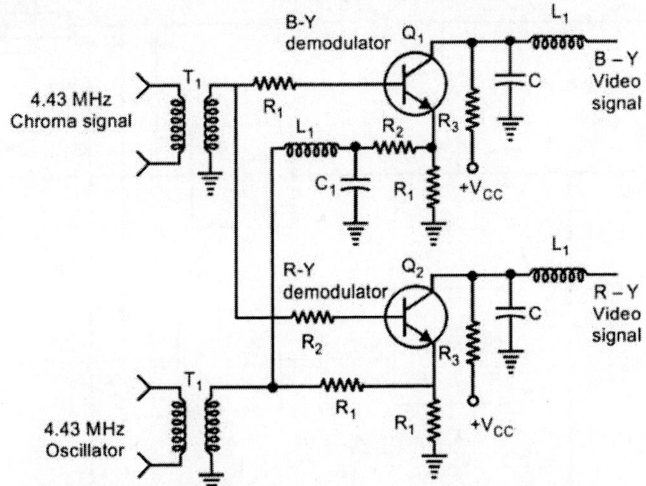

Fig. 5.28. **Demodulation of colour difference signals.**

However, the tristimulus values are required for unbalancing the red, green and blue beams. Two different methods are commonly used for restoring these colour values:

1. Driving colour picture tube with RGB voltages (Fig. 5.29)

The tristimulus voltages V_R, V_G and V_B are produced from the luminance component V_Y and the two colour difference signals $V'_R - V_Y$ and $V'_B - V_Y$ via matrices and applied directly to the control grids of the colour picture tube, the cathodes being at fixed potential.

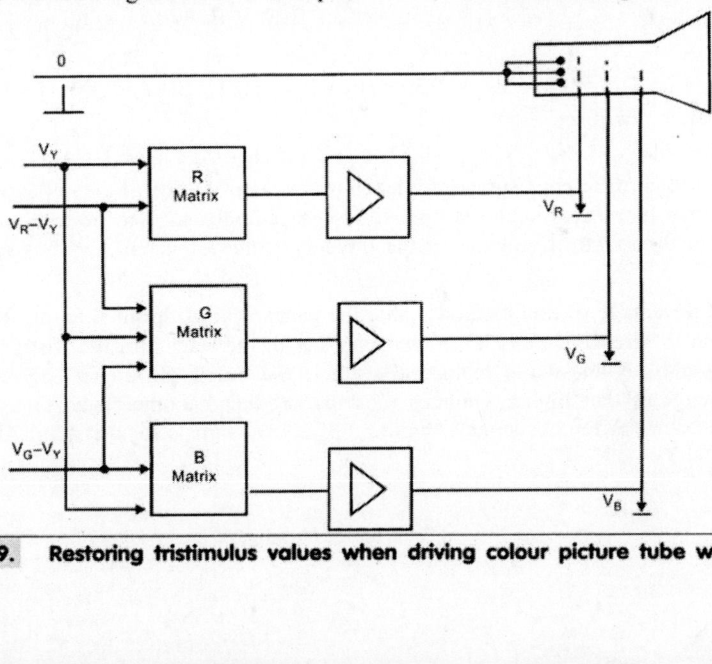

Fig. 5.29. **Restoring tristimulus values when driving colour picture tube with RGB.**

2. Driving Colour Picture Tube with Colour Difference Signals (Fig. 5.30)

The third colour difference signal $V'_G - V_Y$ is obtained in a matrix from the two quantities

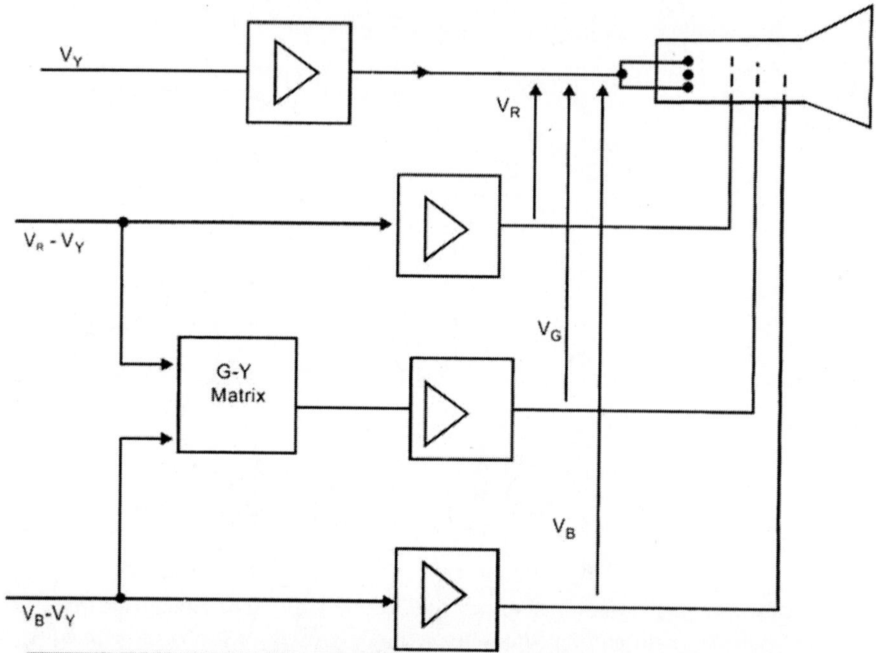

Fig. 5.30. **Restoring tristimulus values when driving colour picture tube with colour difference signals**

$V'_R - V_Y$ and $V_B - V_Y$, based on the following equations :

$$V_Y = 0.30 \times V_R + 0.59 \times V_G + 0.11 \times V_B$$
$$V_Y = 0.30 \times V_Y + 0.59 \times V_Y + 0.11 \times V_Y$$

$$V_Y - V_Y = 0.30 \times (V_R - V_Y) + 0.59 \times (V_G - V_Y) + 0.11 \times (V_B - V_Y) = 0$$

or, after rewriting,

$$V_G - V_Y = -0.51 \times (V_R - V_Y) - 0.19 \times (V_B - V_Y)$$

The colour difference signals are taken to the control grids of the deflection systems; the negative luminance signal is applied to the cathodes so that the tristimulus signals are obtained as control voltages at the three systems, for instance:

$$U_{\text{cont } R} = (U_R - U_Y) - (-U_Y) = U_R \qquad ...(5.21)$$

The advantage of this method is that the bandwidth of the final amplifier stages for the colour difference signals is smaller than that of the stages for the tristimulus signals and that a black-and-white picture appears if the colour difference signals fail. The disadvantage is that higher voltages must be produced in the final stages associated with the colour difference signals. Figure 5.31 shows signals for the standard colour bar pattern.

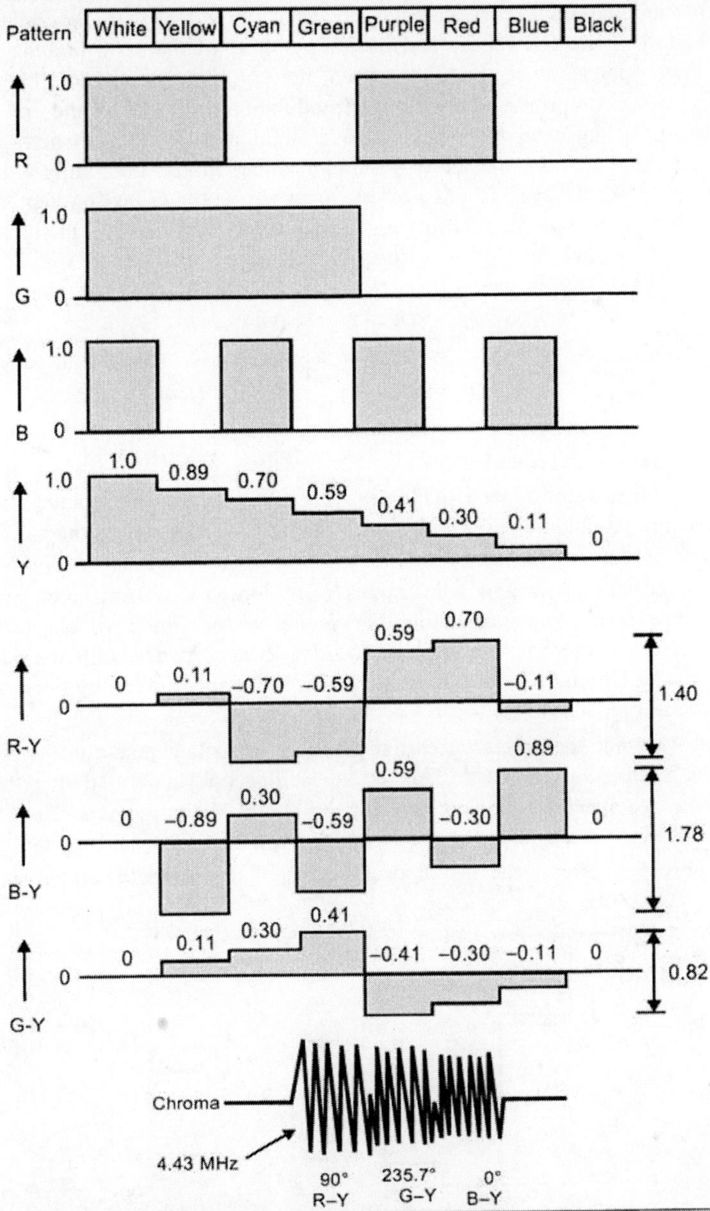

Fig. 5.31. Trstimulus values, luminance component, chroma and colour difference signals for standard colour pattern.

5.19. COMPOSITE COLOUR VIDEO SIGNAL

The chrominance signal is combined with the composite video signal (CVS) to form the composite colour video signal (CCVS). The CCVS is amplitude-modulated onto the RF vision carrier. The full level of the colour difference signals would cause overmodulation of the RF vision carrier by the chrominance signal for certain coloured patterns. This is shown for the signals of the standard colour bar sequence (Table 5.1).

Overmodulation occurs in both directions (Fig. 5.30). In particular, the periodic suppression of the RF carrier and its falling short of the 10% luminance level would cause heavy interference. For this reason, the chrominance signal amplitude has to be reduced. As a compromise between overmodulation on the one hand and degradation of signal-to-noise ratio on the other an overmodulation of 33% in both directions with fully saturated colours has been permitted since, in practice, fully saturated colours hardly ever occur. This is ensured by using different reduction factors for the two colour difference signals, multiplying them by 0.49 for the $(B - Y)$ signal and 0.88 for the $(R - Y)$ signal. In this way the reduced colour difference signals U and V are obtained:

$$\begin{aligned} U &= (B - Y)_{red} = 0.49 \times (B - Y) \\ &= 0.15 \times R + 0.29 \times G + 0.44 \times B \quad \text{...(5.23)} \\ V &= (R - Y)_{red} = 0.88 \times (R - Y) \\ &= 0.61 \times R + 0.52 \times G + 0.10 \times B \quad \text{...(5.24)} \end{aligned}$$

(The values are rounded off.)

The signal values listed in Table 5 are obtained for the colour bar pattern with 100% saturated colours. Fig. 5.31 shows the line oscillogram of the CCVS for a 100% saturated colour bar sequence.

For measurements and adjustments on colour TV transmission systems the test signal used is the standard colour bar sequence for which all chrominance signals, except in the white bar, are reduced to 75% in accordance with the EBU (European Broadcasting Union) standard. In this way, 33% overmodulation by the colour subcarrier is avoided (Fig. 5.33).

To determine the colour subcarrier phase or the colour points in the $(B - Y)$ $(R - Y)$ plane, a vectorscope is used. This is an oscilloscope calibrated in polar coordinates including two synchronous detectors for the U and V components. The detected colour difference signals are taken to the X and Y inputs of the oscilloscope.

Figure 5.32 shows the vector oscillogram of the standard colour bar sequence.

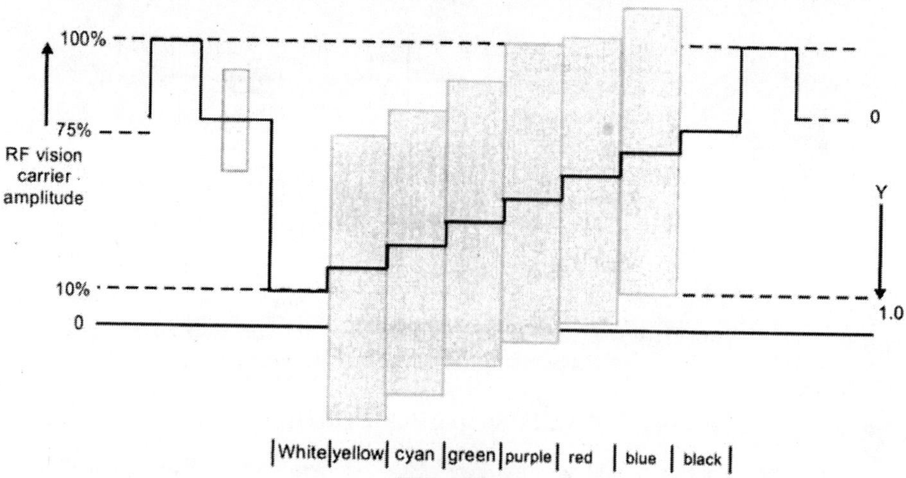

Fig. 5.32.

Amplitude modulation of RF vision carrier by CCVS without reducing colour difference signals.

Table 5.1. Overmodulation of RF vision carrier by standard colour bars.

Colour bar	Y	B – Y	R – Y	SC	Y + SC	Y – SC
White	1	0	0	0	1	1
Yellow	0.89	–0.89	+0.11	0.89	1.78	0
Cyan	0.70	+0.30	–0.70	0.76	1.46	–0.06
Green	0.59	–0.59	–0.59	0.83	1.42	–0.24
Purple	0.41	+0.59	+0.59	0.83	1.24	–0.42
Red	0.30	–0.30	+0.70	0.76	1.06	–0.46
Blue	0.11	+.89	–0.11	0.89	1.00	–0.78
Black	0	0	0	0	0	0

Table 5.2. Modulation of RF vision carrier by standard colours bars with reduced colour difference signals.

Colour bar	Y	U	V	SC_{red}	$Y + SC_{red}$	$Y - SC_{red}$
White	1	0	0	0	1	+1
Yellow	0.89	–0.44	+0.10	0.44	1.33	+0.45
Cyan	0.70	+0.15	–0.62	0.63	1.33	+0.07
Green	0.59	–0.29	–0.52	0.59	1.18	0
Purple	0.41	+0.29	+0.52	0.59	1.00	–0.18
Red	0.30	–0.15	+0.63	0.63	0.93	–0.33
Blue	0.11	+0.44	–0.10	0.44	0.55	0
Black	0	0	0	0	0	0

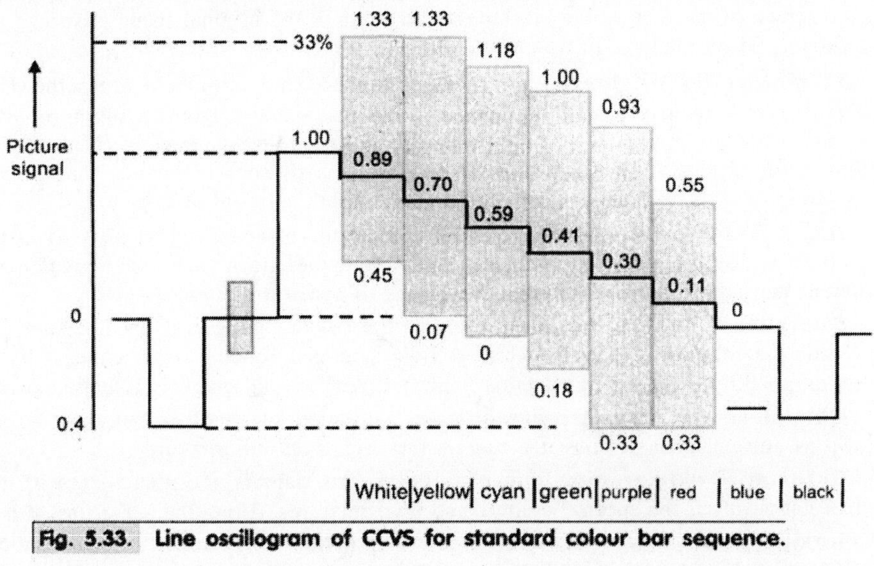

Fig. 5.33. Line oscillogram of CCVS for standard colour bar sequence.

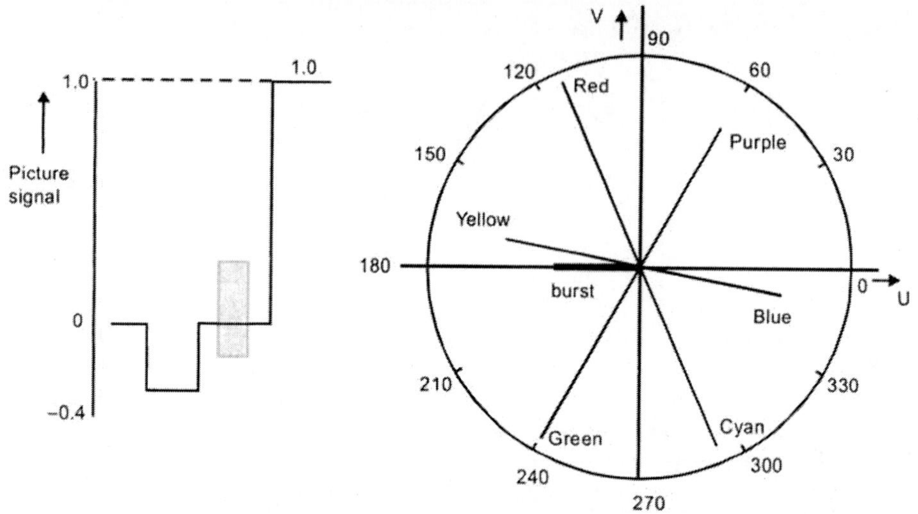

Fig. 5.34. Vector oscillogram of standard colour bar sequence.

5.20. FEW DEFINITIONS

Contrast : It is the difference in intensity between black and white parts of the picture over and above the brightness level.

Contrast Ratio : The ratio of maximum to minimum brightness relative to the original picture is called contrast ratio.

Gamma of the Picture : Ratio of brightness variations in the reproduced picture on the screen of the tube to the brightness variations in the original scene is known as the Gamma of the picture. Its value is close of 0.5.

Luminance (or Brightness) : In fact, luminance and brightness are somewhat different. For a given colour, luminance is the fundamental brightness of the colour impression and is the amount of light intensity as perceived by the eye regardless of the colour. Lighted parts in black and white picputre have more luminance than black areas. Colours also, though equally illuminated appears as more or less bright.

Hue : This is the prominent spectral colour of the received light. The colours of any object is distinguished by its hue or tint. For example, green leaves have green hue. Different hues result from different wavelengths of spectral radiation.

Saturation : This is the spectral purity of the colour light and is an indication of how little the colour is diluted by white. A fully saturated colour has no white. A light green is not fully saturated; it contains white whereas a vivid green is fully saturated.

Chrominance (Chroma) : The hue and saturation of a colour put together is known as chrominance. It does not contain the brightness information.

Chromaticity diagram : It is a diagram that depicts the space coordinate representation of all the spectral colours and their mixtures. Along the perimeter of the curve, various saturated pure spectral colours are represented but as the central area of the triangular is approached, the colours become desaturated representing mixing of colours. On the central point C, (x = 0.31, y = 0.32) lies white which is formed by combining hues having wavelength : 700 nm (red), 546.1 nm (green) and 438.8 nm (blue) with proper intensities. This shade of white corresponds to eye response to a shade of mixture of 30% of red colour, 59% of green colour and 11% of blue colour at the wavelengths specified above. This diagram is useful in determing the results of additive mixing of any two or more colours by geometric construction.

Gamma (γ) correction : For distortion free reproduction of pictures on the receiver screen, the overall TV system should be linear. For this the light emitted from the picture tube (receiving end) should be directly proportional to the light falling on the camera target plate (transmitting end). But this is not so, because the beam current of a picture tube is not linearly related to the driving voltage applied between its grid and cathode. In order to compensate for this non-linearity at the receiver (at picture tube), an opposite distortion (*i.e.*, non-linearity), called **Gamma correction**, is introduced at the trnasmitting end. This is done by passing camera output voltages through corrective networks which have input/output characteristics just opposite to that at the receiving end.

SUMMARY

- *Monochrome compatibility:* A monochrome receiver must reproduce the brightness content of a colour signal correctly in black and white, with no visible interference from the colour information.
- *Reverse compatibility :* A colour television receiver must reproduce a monochrome signal correctly, with no spurious colour components.
- *Scanning compatibility :* The scanning system used for colour transmissions must be identical to the one used by the existing monochrome service.
- *Transmission channel compatibility :* The colour system must fit into the existing monochrome television channel and use the same spacing between the vision and sound carriers.
- To reproduce a colour image of the pattern, additional information on the colour content, *i.e.* the "chromaticity" of the individual picture elements, must be transmitted together with the brightness or luminance distribution.
- A light is a visible form of EM energy and it has two visible characteristics; intensity and colour. Light consists of a spectrum of frequencies of wavelength ranging from 700 nanometer (for red light) to 380 nanometer (for violet light) visible to human eye.
- Visible Spectrum
  - Red –700 nm
  - Orange—600 nm
  - Yellow—550 nm
  - Green—500 nm
  - Blue—400 nm
  - Violet—380 nm
- According to the Helmholtz, there are two light sensitive organs:
  1. Rods – Brightness sensation (Gray from Black)
  2. Cones – Sensitive to colour (RGB) 3 groups.

- Mixing of Colours
  1. Subtractive Mixing,
  2. Additive Mixing
- In subtractive mixing, reflecting properties of pigments are used, which absorbs all wavelengths but their characteristics colour wavelength. When pigments of two or more colour are mixed, they reflect wavelength which is common to both.
- Additive mixing is a method of producing various colours. When lights of two or more different colours are mixed and projected on a white screen, the eye perceives it as a combined colour.
- **Grassman's Law :** The property of the eye of producing a response which depends on the algebraic sum of the RGB is known as Grassman's law.
- The components values of the three primary colours to produce various others colour have been standardized are called Tristimules values of the different spectral colours.
- One lumen of white light = 0.3ln R + 0.59ln G + 0.11ln B
- **Luminance :** This is the amount of light intensity as perceived by the eye regardless of the colour.
- **Hue :** It represent wavelength of the light reaching the eye. The eye cannot distinguish between the closely space wavelength. This is the predominant spectral colour of the received light.
- **Saturation :** This is the spectral purity of the colour light and is an indication of how the colour is diluted by the white.
- **Gamma Corrections :** To compensate for the non-linearity of the system including TV camera and picture tubes, a correction is applied to the voltage produced by the three camera tubes, known as **Gamma corrections.**
- To reproduce a colour pattern, the three tristimulus signals R, G and B are required.
- The chrominance signal carries information on hue and saturation. Therefore two colour difference signals are sufficient to describe the chrominance component. The quantities R-Y and B-Y were selected.
- The bandwidth of the chrominance signal can be reduced to about 1.5 MHz by taking the two colour difference signals via low pass filters.
- Only two colour difference signals [(B-Y) and (R-Y)] are used for modulating the colour subcarrier. The colour difference signal G-Y is not used in modulation process.
- (G-Y) colour difference signal is not transmitted, as it can be produced locally in the receiver easily by passing the available signals B-Y and R-Y through a resistive matrix.
- To transmit the complete picture information; luminance plus chrominance, a triple transmission channel would be required. Here one might think of making multiple use of the TV transmission channel either by frequency or time multiplex. However, neither method is compatible with the existing black-and-white transmission method.

- The answer is that since a video signal does not occupy every cycle of 5 MHz assigned to it, colour information can be inserted within 5 MHz.

$$f_{sc} = (2n+1) \times \frac{f_h}{2}$$

- For the CCIR-modified NTSC method, a colour subcarrier frequency of has been fixed.

$$f_{SC} = 567 \times \frac{f_h}{2} = 283.5 \times f_h$$
$$= 4.4296875 \text{ MHz}$$

- The CCIR standard colour subcarrier frequency in the PAL system is

$$f_{SC} = 283.75 \times f_h + 25 \text{ Hz}$$
$$f_{SC} = 4.43361875 \text{ MHz}$$

- The chrominance signal is transmitted by modulating the colour subcarrier with the two colour difference signals.
- In the NTSC and PAL methods, double amplitude modulation is used. The 0° component of the subcarrier is amplitude- modulated by the $(B - Y)$ signal and its 90° component by the $(R - Y)$ signal, the subcarrier being suppressed at the same time. This results in quadrature modulation.

$$SC = \sqrt{(B-Y)^2 + (R-Y)^2}$$

$$\varphi_{SC} = \arctan \frac{R-Y}{B-Y}$$

- For synchronization with the subcarrier at the transmitter end, a reference signal is inserted into each line in the H blanking interval, *i.e.* the colour sync signal or burst. The phase reference consists of 8-10 cycles of the 4.43 MHz signal.

REVIEW QUESTIONS

1. Draw and explain the block diagram of a colour televisions systems.
2. Differentiate between a colour receiver and black and white receiver. Explain why separate colour sync pulses are needed in a colour television systems.
3. Why three camera pick up tubes are provided in a colour camera ? Explain it.
4. What do you understand by compatibility in TV systems ?
5. Describe how compatibility between colour and monochrome television systems is achieved.
6. Explain the working of a three camera tubes for producing of colour signals. With the help of diagram explain that how luminance signal and colour difference signals are obtained.
7. Distinguish between luminance and chrominance signals.
8. Why G-Y signal is not transmitted alongwith B-Y and R-Y signals ?
9. Why is a separate sub-carrier used for encoding colour signals ?
10. Why quadrature modulation technique is used in colour encoder ?

11. Explain the phenomenon of interleaving of colour signals. How does it help in ensuring compatibility.

12. Explain how synchronous demodulation is used to drive the chroma components from quadrature modulated colour subcarrier.

13. G-Y signal is not chosen as one of the chroma signals for transmission. Why ?

14. Discuss the factors which determine the value of sub-carrier frequency. Show that the colour sub-carrier frequency in CCIR NTSC syntax is 3.579 MHz and in CCIR B PAL system 4.43361 MHz.

15. Why is colour burst signal transmitted alongwith the video signal in colour TV system.

16. Write short notes on

(*i*) Additive mixing

(*ii*) Subtractive mixing

(*iii*) Frequency interleaving

(*iv*) AMSC modulation

MULTIPLE CHOICE QUESTION

1. Bandwidth of the colour signal is about

(*a*) 1.5 MHz (*b*) 3.5 MHz

(*c*) 5 MHz (*d*) 7 MHz

2. Bandwidth of the luminance signal is about

(*a*) 1.5 MHz (*b*) 3.5 MHz

(*c*) 5 MHz (*d*) 7 MHz

3. Chroma signal requires

(*a*) Envelope detector (*b*) Balanced detector

(*c*) Ratio detector (*d*) Foster selley detector

4. Luminance signal is represented by

(*a*) G (*b*) L

(*c*) X (*d*) Y

5. The phase angle between (R-Y) and (B-Y) signal is

(*a*) 0° (*b*) 270°

(*c*) 90° (*d*) 180°

6. Which of the following is used for colour separation in a colour camera

(*a*) Colour conversion filter

(*b*) Image filter

(*c*) Dichroic mirrors

(*d*) None of these

7. The luminance energy is uniformally distributed through the whole video bandwidth. True/False

8. Colour burst is used to

(*a*) Boom intensity of colours in the picture tube

(*b*) Dilute the vivid colours

(*c*) Synchrononise generation of sub-carrier in the receiver

(*d*) Ensure the correct modulation of colours

9. The value of colour sub-carrier in CCIR B system is

(*a*) 3.58 MHz (*b*) 4.43 MHz

(*c*) 4.5 MHz (*d*) 5 MHz

10. The value of colour subcarrier in CCIR M system is about

(*a*) 3.58 MHz (*b*) 4.43 MHz

(*c*) 4.5 MHz (*d*) 5 MHz

11. Video carrier frequency is the same on monochrome and colour receivers.

True/False

12. Colour TV transmitter transmits the colour differences signals are

(*a*) R-Y and B-Y (*b*) G-Y and R-Y

(*c*) B-Y and G-Y (*d*) None of these

13. SECAM system was developed in

(*a*) France (*b*) Japan

(*c*) Germany (*d*) India

ANSWERS

1. (*a*)	**2.** (*c*)	**3.** (*b*)	**4.** (*d*)	**5.** (*c*)
6. (*c*)	**7.** False	**8.** (*c*)	**9.** (*b*)	**10.** (*a*)
11. True	**12.** (*a*)	**13.** (*a*)		

6
CHAPTER

COLOUR TELEVISION SYSTEMS

INSIDE THIS CHAPTER

6.1. INTRODUCTION

There are three Monochrome systems which are discussed in earlier chapters :

1. 525 – Lines (American)
2. 625 – Lines (European)
3. 819 – Lines (French)

In India TV transmission started in 1959 and 625-B monochrome system has been adapted.

Correspondingly there are three Colour systems have been developed which are as follows:

1. National Television System Committee (NTSC) used in America in 1953.
2. Phase Alteration by Lines (PAL) used in German in 1967.
3. Sequential Couleruses A Memoirs (SECAM) used in France.

In India PAL system CCIR (B Standard) is used for colour television transmission and reception.

We will discuss each colour television system in brief in this chapter.

6.2. NTSC METHOD

The work of the first NTSC (National Television System Committee) was built on the basis of earlier standardization attempts. The first NTSC laid the foundations that made monochrome television a practical reality in the United States. The standards it endorsed in 1941 are still in use today. At the dawn of colour television it seemed at first that it was not going to be compatible with monochrome television and would need special receivers. A tug of war of compatible versus incompatible colour television systems ensued. The compatible system won. The second NTSC set up the compatible standards that have been adopted by the U.S. and by the rest of the world in many major respects.

The NTSC, PAL and SECAM methods, which are mainly used for colour TV transmission, differ only with respect to the modulation of the colour subcarrier.

The NTSC method constitutes the basis for the improved variants PAL and SECAM.

The principle of the NTSC method has basically been described in the chapters on modulation of the colour subcarrier and on the composite colour video signal.

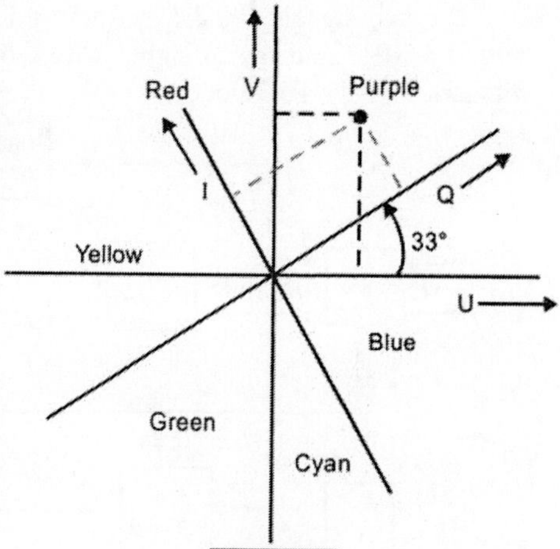

Fig. 6.1.

However, the original NTSC system (US standard) does not transmit the reduced colour difference signals U and V but instead the I and Q components, which are referred to a coordinate system rotated counter-clockwise by 33° (Fig 6.1).

In the colour triangle, the I axis corresponds to the axis for which the eye has maximum colour resolution (Fig. 6.2). This ensures better transmission of colour transitions.

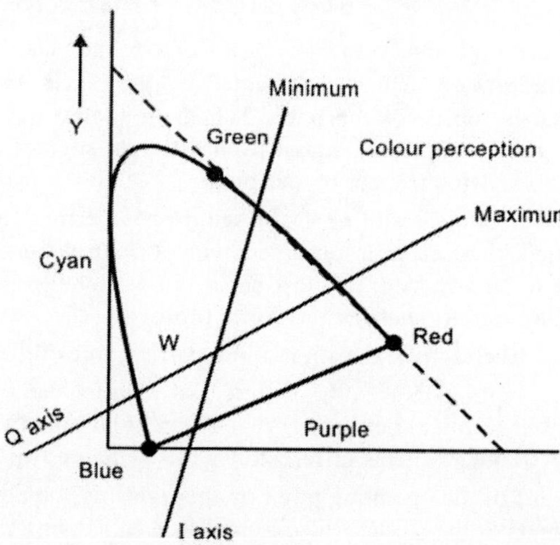

Fig. 6.2.

The modulation signals are thus

$$I = V \times \cos 33° - U \times \sin 33° \quad ...(6.1)$$

$$Q = V \times \sin 33° + U \times \cos 33° \quad ...(6.2)$$

or, using the tristimulus matrix equations,

$$I = 0.60 \times R - 0.28 \times G - 0.32 \times B \quad ...(6.3)$$

$$Q = 0.21 \times R - 0.52 \times G + 0.31 \times B \quad ...(6.4)$$

The two signals I and Q are transmitted with different bandwidth, *i.e.* the I signal with 1.5 MHz and the Q signal with 0.5 MHz. Figure 6.3 shows the complete block diagram of an NTSC coder.

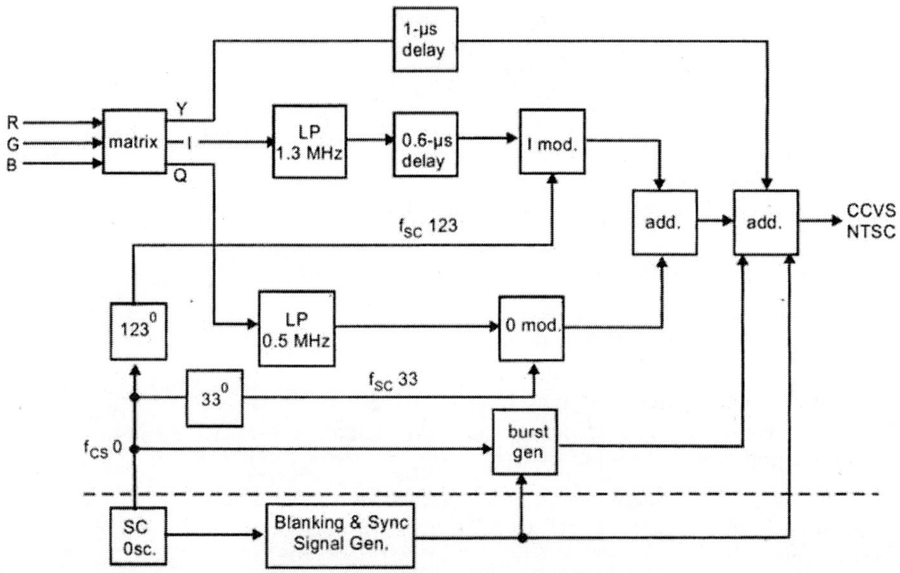

Fig. 6.3. Block diagram of NTSC coder.

The human eye reacts very strongly to incorrect hue. The hue of the colour image reproduced by the picture tube is determined by the phase angle of the chrominance signal referred to the phase of the burst. When producing the CCVS in the studio, it may happen that the chrominance signal from different sources has different delays and thus different phases with respect to the burst.

To correct wrong hue resulting from static phase errors in the transmission path, the NTSC colour TV receiver is provided with a control permitting the phase of the reference carrier to be adjusted. In most cases this is done by referring to the hue of a well-known picture detail, such as the **flesh tone.**

However, this **hue control** does not allow correction of differential phase distortion.

In accordance with DIN 45 061 differential phase means the difference of phase shifts through a four-terminal network at two different points of the transfer characteristic at the subcarrier frequency. The differential gain is defined in a similar way.

Due to the shift of the operating point on the transmission characteristic as a function of the Y components of the CCVS, the chrominance signal suffers a gain change (because of the change in slope of the characteristic) and a phase change (because the transistor input capacitance is dependent on the emitter current and thus on the operating point) when passing, for instance, through an amplifier stage with preceding tuned circuit. While the differential gain can be eliminated to a large extent by negative feedback, the differential phase can be reduced only by limiting the driving level. Figure 6.4 shows this influence on the CCVS and the effect in the vector scope representation.

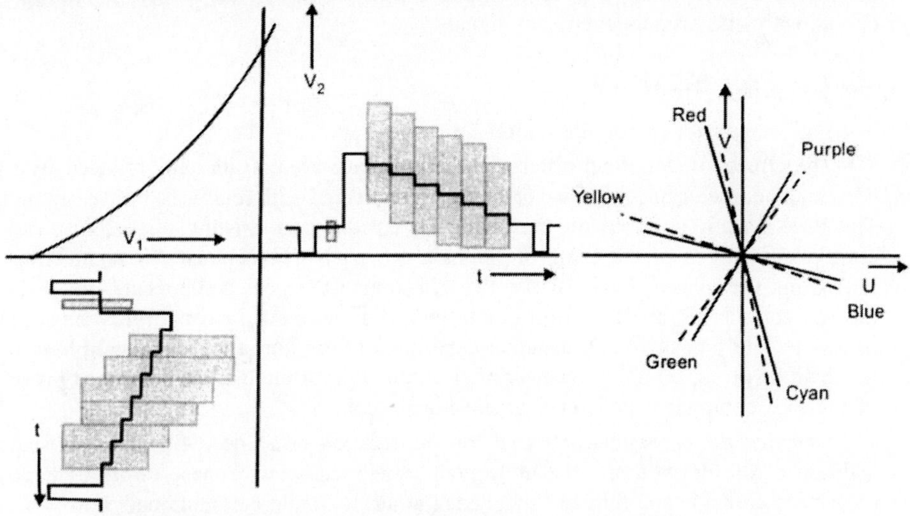

Fig. 6.4. Generation of differential gain and phase distortion.

Advantages of NTSC

1. Eyehas good resolution because of the higher bandwidth in the colour region, saving in bandwidth gains interleaving without colour cross interference.
2. It is less costly than PAL system.
3. Studio mixing is quite easy.
4. Their vertical resolution is double in comparison with PAL and SECAM systems.

Problems with NTSC

NTSC was a compromise aimed at squeezing color information into a 6MHz monochrome television channel. The solution was brilliant, but the electronic tube technologies of the time were at best inadequate and at worst unacceptable. Solid-state technologies have removed many of the early problems. Here are a few of the problems:

1. Transmitter overmodulation : The color-difference signals are scaled to avoid transmitter overmodulation. The video transmitter is modulated in amplitude and uses negative modulation. In a negative modulation system, increasing the brightness of the transmitted picture produces a decrease of the modulation envelope amplitude.

The picture modulation envelope has four reference levels: peak white level, black level, blanking level and sync level. White level (100 IRE) reduces the carrier to 12.5 percent. The carrier cancels at 120 IRE. Saturated yellow and cyan colors can produce video signal levels of 130.8 IRE, resulting in carrier cancellation and inter-carrier "buzz." Because saturated yellow and cyan colors do not normally exist in nature, camera-generated video signals would not create problems. However, synthetically generated signals (100 percent color bars or character generators) could create problems.

2. Video signal distortions : Analog composite video signals are subjected to various types of cumulative distortions and noise. Separate distortions of the luminance and the chrominance components as well as inter-modulation between them are likely to occur. PAL is more tolerant of luminance/chrominance inter-modulation. Such distortions can be reduced by performing all, or at least a major part of, production operations using component video signals.

6.3. PAL METHOD

PAL stands for phase alternation line.

The effects of static and differential phase errors are considerably reduced with the PAL method, the occurrent phase errors being corrected with relatively little extra outlay. The PAL system is based on the following concept: an existing phase error can be compensated by a phase error of opposite polarity. This is realized technically by alternating the phase of one of the two chrominance signal components, for instance the SC_V component, by 180° from line to line. If a phase error exists in the transmission path, alternately positive and negative departures of the chrominance signal phase from nominal are produced in the receiver after elimination of the line-to line polarity reversal of the SC_V component generated at the transmitter end.

Delaying the chrominance signal for the duration of a line (64 μs) and subsequent addition of the delayed and the undelayed signals cause two phase errors of opposite polarity to coincide and thus to cancel each other. It should be mentioned, however, that this method is based on the assumption that the chromaticity does not change within two consecutively transmitted lines. If horizontal colour edges exist, the eye hardly perceives a falsification of the colour transition even in this case.

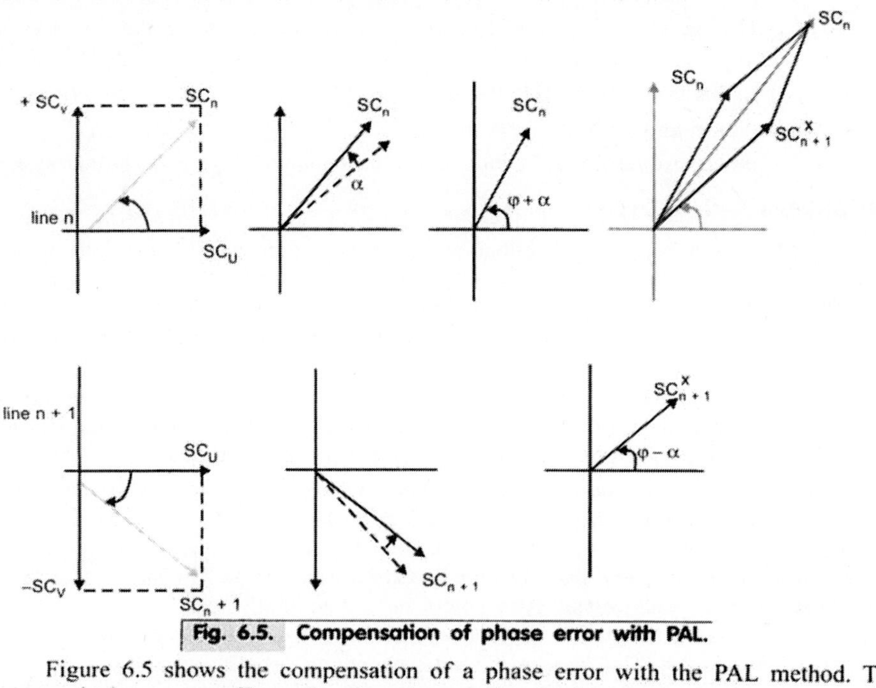

Fig. 6.5. Compensation of phase error with PAL.

Figure 6.5 shows the compensation of a phase error with the PAL method. The assumed phase error-affects the chrominance signal with respect to the burst on the transmission link.

After elimination of the SC_V polarity reversal (PAL switchover) and addition of the chrominance signals in two successive lines, the phase angle of the resulting signal SCres is equal to that of the transmitted chrominance signal, and the original hue is thus maintained. After reducing the resulting signal to half its amplitude, this signal exhibits only slight desaturation.

An additional identification is transmitted with the burst to ensure correct phase reversal to the SC_V component in the receiver or of the reference carrier for the $(R - Y)$ sync detector. To this effect, the burst is split into two components, one being transmitted at 180° and the other at –90° alternating from line to line in phase with the S_{CV} reversal. This yields a swinging burst of 180° ±45° (Fig. 6.6). The actual burst reference phase (180°) is obtained by averaging.

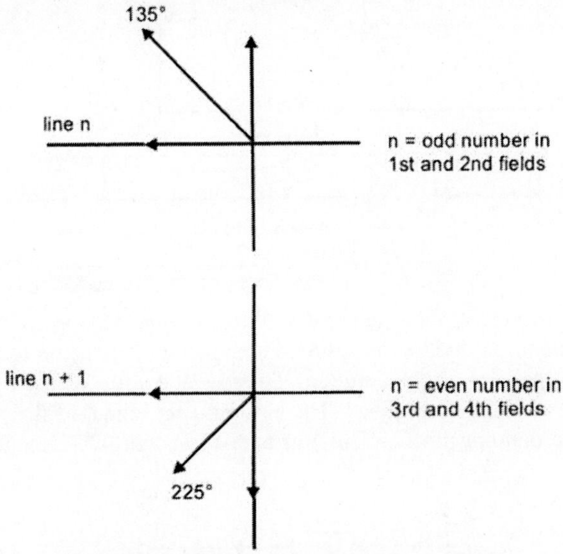

Fig. 6.6. Swinging burst with PAL method.

The reference carrier oscillator in the receiver adjusts, via the phase control, to 90° with respect to the mean burst phase. The identification signal (burst flag) for synchronizing the PAL switch is derived from the burst phase discriminator. With PAL the reduced colour difference signals U and V are directly transmitted, the bandwidth being 1.3 MHz. Limiting the sidebands of the modulated colour subcarrier to different widths no longer has an annoying effect thanks to the phase error compensation. Figure 6.7 is a block diagram of a PAL coder.

Compared with the NTSC coder, the 33° phase shift of the colour subcarrier components is omitted, but reversal of the subcarrier component for the $(B - Y)$ modulator and generation of the swinging burst are added.

The technical realization of the PAL error compensation requires special explanation as against Fig 6.5. For this purpose it is best to start with the group delay decoder included in the PAL decoder. In contrast to the NTSC decoder, the chrominance signal is not simultaneously applied to the two sync detectors in the PAL decoder but is first split into the SC_U **and** SC_V **components.**

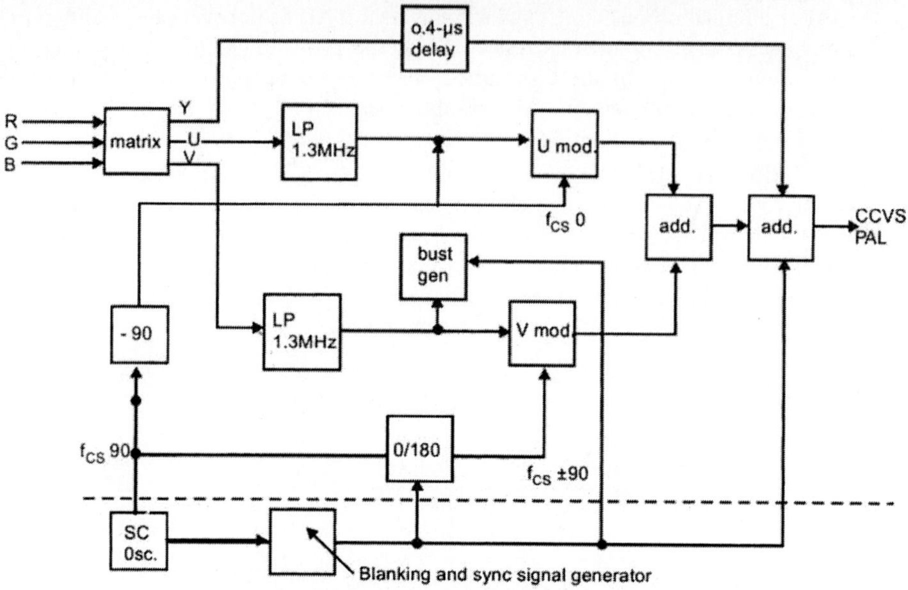

Fig. 6.7. Block diagram of PAL coder.

This is performed in the group delay decoder (Fig. 6.8). At its output, the incoming chrominance signal is divided into three components. It is taken to the two outputs via a 64 µs delay network (line duration) directly and after a 180° phase shift. At the outputs, signal addition takes place. The chrominance signal of the preceding line (SC_n) and that of the ongoing line (SC_{n+1}) are added at the SCU output.

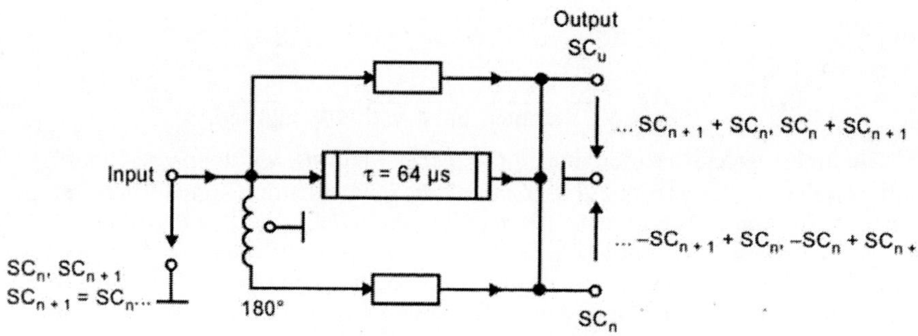

Fig. 6.8. PAL group delay decoder.

Successive lines contain the SC_V component with a 180° phase alternation so that the SC_V component is cancelled every two lines. Thus the SC_U component of the chrominance signal is constantly available at this output.

The input signal is taken to the SC_V output with a 180° phase shift. Addition of the delayed chrominance signal cancels the SC_U component, and the SC_V component appears at this output, although with a 180° phase alternation from line to line. Based on the vector diagrams in Fig. 6.9, the functioning of the group delay decoder can be explained very easily.

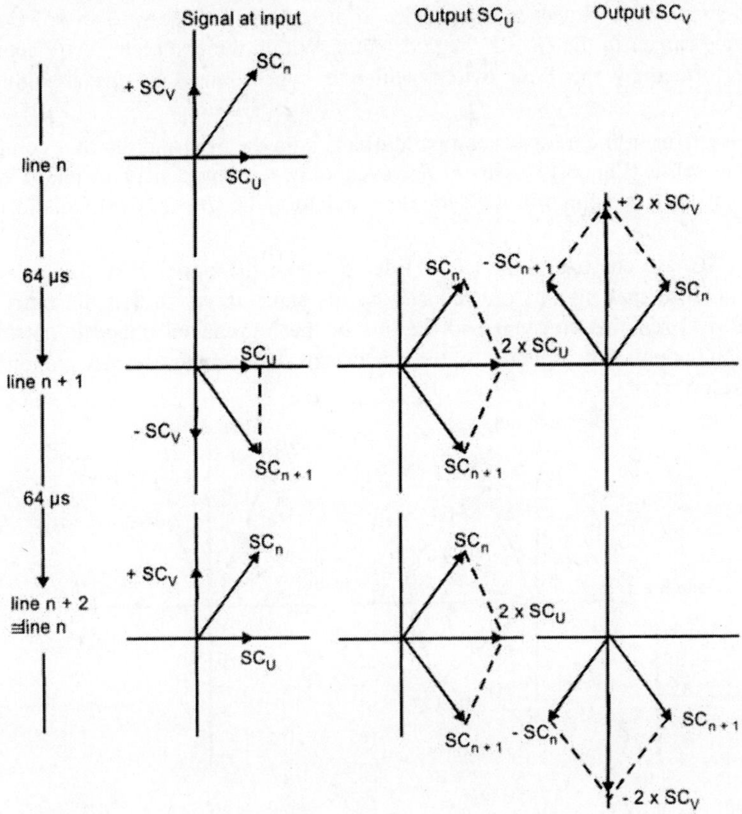

Fig. 6.9. Division of chrominance signal into SC_U and SC_V components in PAL group delay decoder.

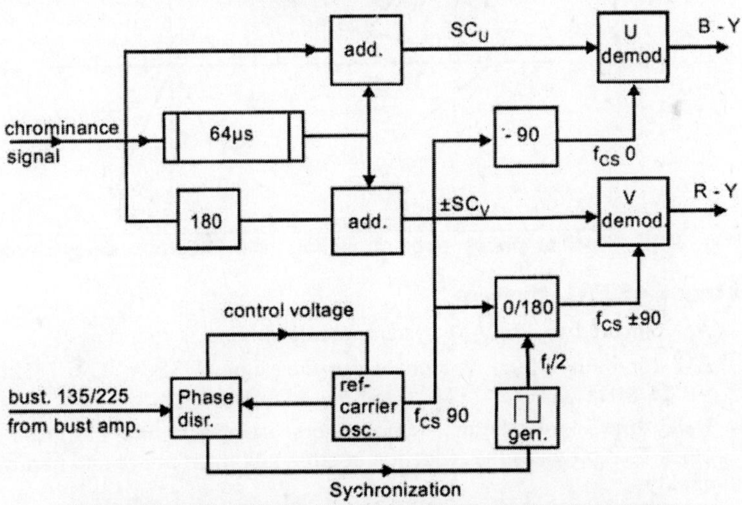

Fig. 6.10. PAL decoder with reference carrier generation.

The line-to-line phase alternation of the SC_V component can be disabled by a controlled switchover. However, it is easier to provide for line-to-line phase reversal of the reference carrier in the $(R - Y)$ sync detector. Within the complete PAL decoder this task is performed by the PAL switch, which is synchronized by the swinging burst (Fig. 6.11).

A phase error in the transmission path affects both the SC_U and the SC_V components in the same sense (Fig. 6.11). Since, however, only the component in phase with the reference carrier is weighted in the sync demodulators, the $(B - Y)$ demodulator delivers the signal

$U' = | SC_U | \times \cos \alpha$ and the $(R - Y)$ demodulator the signal $V' = | SC_V | \times \cos \alpha$. Both colour difference signals are reduced by the same factor so that the ratio V/U or $(R - Y)/(B - Y)$ remains constant and the hue of the reproduced image is not affected. Desaturation, which corresponds to the factor cos θ, becomes significant only with large phase errors.

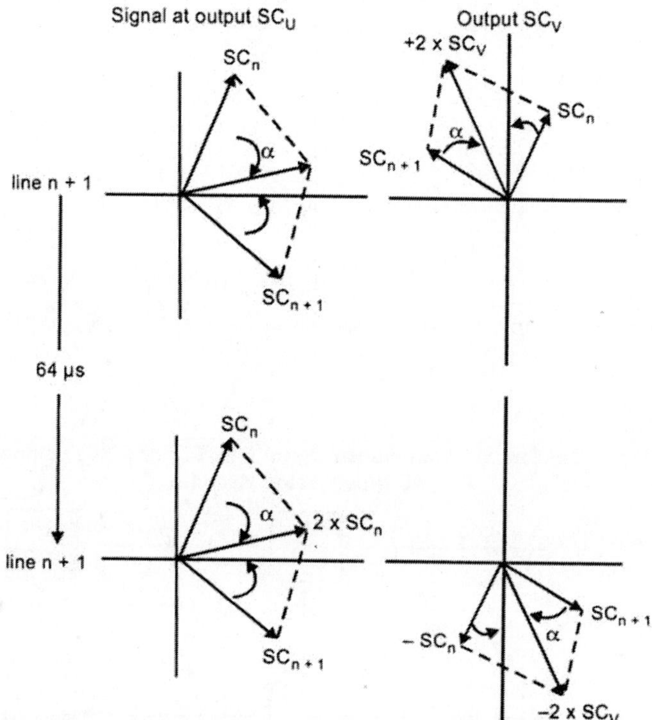

Fig. 6.11. Effect of phase error on signals with PALgroup delay decoder.

Advantages of PAL System

1. As compared to SECAM system, studio mixing is quick easy.
2. BW for both U and V signals are the same, USB = 0.56 MHz and LSB = 1.15 MHz.
3. Delay line is used before demodulators' isolates U and V signals from each other. That provides the reduction in cross talk interference and improves picture quality.
4. Phase over is automatically eliminated, thus a separate hue control is not needed.

Disadvantages of PAL System

1. This system is complex, the use of electronic switching and indent signals make PAL receivers are the costliest.
2. In magnetic recording of pal transmission, complete colour coding needs 8 fields, while NTSC needs only four.

6.4. SECAM METHOD

The short form SECAM derived from "sequential à mémoire" indicates that this is a sequential colour system with memory.

Work on SECAM began in 1956. The technology was ready by the end of the fifties, but this was too soon for a wide introduction. Notably, SECAM did not work with the 819-line television standard then used by the then sole French TV network. France had to start the conversion by switching over to a 625-line television standard, which happened at the beginning of the sixties with the introduction of a second network.

SECAM was inaugurated in France on October 1st, 1967, on *la seconde cha* (the second network), currently called France 2. A group of four men, all dressed in suits, presumably presenters and network officials, were shown standing in a studio. The image was originally black and white and suddenly switched to colour; one of the people said something along the lines of "*now you can see us as we really are*".

As against the NTSC method, the SECAM method too brings an improvement with respect to wrong hues caused by phase errors in the transmission path. Like the PAL method it is based on the assumption that the colour information does not essentially vary from line to line or that the human eye does not perceive any annoyance if the vertical colour resolution is reduced to a certain extent.

Therefore, the colour difference signals $(B - Y)$ and $(R - Y)$ characterizing the colour information need not be transmitted simultaneously. They can be sent separately in successive lines. In the receiver, the signal content of one line is stored for 64 μs via a delay line and processed together with the signal of the next line.

As the two colour difference signals are transmitted separately, the type of modulation can be freely selected. SECAM uses frequency modulation, which is not very interference-prone. However, the reference frequency of the FM demodulator must be kept very stable so that the demodulated colour difference signals are not falsified.

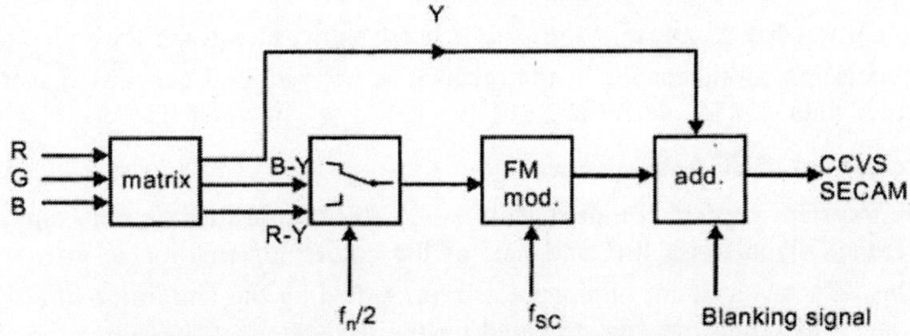

Fig. 6.12. Simplified block diagram of SECAM Coder.

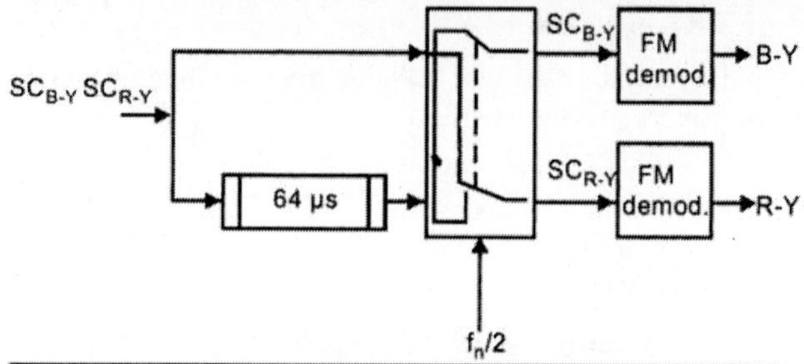

Fig. 6.13. Simplified block diagram of SECAM Decoder.

Fig. 6.13 is a simplified block diagram of a SECAM coder and decoder. In the coder, the $(B - Y)$ and the $(R - Y)$ signals are applied to the frequency modulator in alternate lines. To ensure that in the decoder the demodulated colour difference signals are in synchronism with the transmitter end, identification pulses in the form of the modulated colour subcarrier are transmitted during nine lines of the field blanking interval.

When frequency-modulating the colour subcarrier, the latter is not suppressed. In particular with colours of low saturation, this would produce an interference pattern on a black-and-white receiver in spite of the colour subcarrier offset.

Therefore the colour subcarrier is attenuated by pre-emphasis at the transmitter end and boosted by de-emphasis at the receiver end. The effect of noise is reduced by video frequency pre-emphasis and de-emphasis.

SECAM has gone through several phases of development. Its latest variant, SECAM IIIb or SECAM III opt. is based on slightly different colour subcarrier frequencies for the $(B - Y)$ and $(R - Y)$ signals, further reducing the interference pattern caused by the colour subcarrier.

As against PAL, SECAM features some system-dependent weak points since frequency modulation is utilized at its physical limits.

Advantages of SECAM System

1. Regeneration of subcarrier is not needed, so it is low cost.
2. Hue control is not needed.
3. Picture quality is better since there is no crosstalk interference, as the two phasor do not exist on the same line.
4. Saturation control is also no needed because the carrier amplitude in the Frequency modulation remains same.
5. It is not needed to transmit the colour burst signal alongwith the video signal.
6. Regeneration of subcarrier in the receiver is not required and FM detection is simpler than AMSC detection,

Disadvantages of SECAM system

1. In this system vertical colour resolution is half because there is only one colour difference signal on a line and half of the colour information is lost.
2. In this SECAM system, luminance is represented by the amplitude of a voltage but hue and saturation are indicated by the derivation of the subcarrier. When a composite signal consists of chrominance and luminance is faded out in operation, then luminance signal is readily alternated but not chromimance.

This makes the colour more saturated during fade to black; a pink colour changes to red during fade out.

Overview of PAL and SECAM in relation to NTSC

There are three major TV standards used in the world today. These are the American NTSC (National Television Systems Committee) color television system, the European PAL (Phase Alternation Line rate) and the French-Former Soviet Union SECAM (Sequential Couleur avec Memoire).

The largest difference between the three systems is the vertical lines. NTSC uses 525 lines (interlaced) while both PAL and SECAM use 625 lines. NTSC frame rates are slightly less than 1/2 the 60 Hz power line frequency, while PAL and SECAM frame rates are exactly 1/2 the 50 Hz power line frequency.

Table 6.1. Summary table for existing interlaced colour TV standards

Colour system	Lines	Active lines	Vertical Resolution	Aspect Ratio	Horizontal Resolution	Frame rate
NTSC	525	484	242	4/3	427	29.94
PAL	625	575	290	4/3	425	25
SECAM	625	575	290	4/3	465	25

NTSC

	y	x
R	0.67	0.33
G	0.21	0.71
B	0.14	0.08
White	0.310	0.316

PAL/SECAM

	y	x
R	0.64	0.33
G	0.29	0.60
B	0.15	0.06
White	0.313	0.329

All three systems use the same definition for luminance:

$$Y = 0.299R + 0.587G + 0.114B$$

However, the three systems do not use the same definitions for primary R, G and B colors from the CIE diagram. Summarizing the differences:

As you can see from the CIE drawing below, these numbers essentially translate into a slightly small color triangle — with the largest differences in the green.

The color encoding principles for the PAL system are the same as those of the NTSC system — with one minor difference. In the PAL system, the phase of the R-Y signal is reversed by 180 degrees from line to line. This is to reduce color errors that occur from amplitude and phase distortion of the color modulation sidebands during transmission.

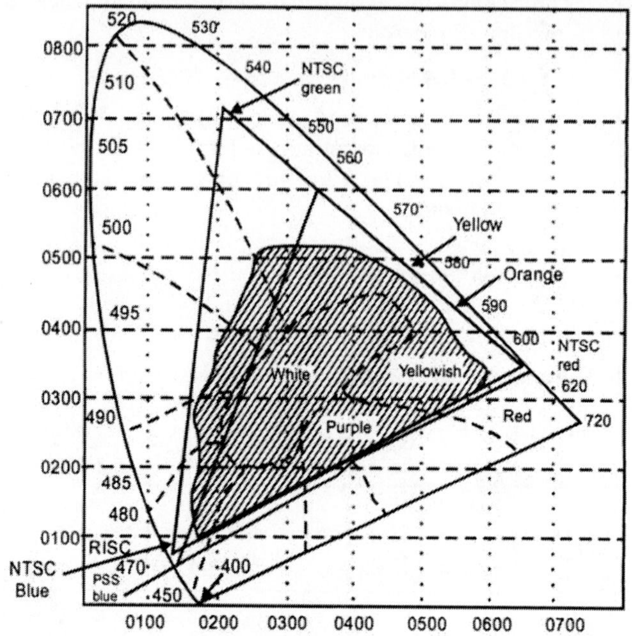

Fig. 6.14.

Saying this more mathematically, the chrominance signal for NTSC transmission can be represented in terms of the *R-Y* and *B-Y* components as.

$$C_{NTSC} = \frac{B-Y}{2.03}\sin(w_c t) + \frac{R-Y}{1.14}\cos(w_c t)$$

The PAL signal terms its *B-Y* component *U* and its *R-Y* component *V* and phase-flips the *V* component (line by line) as:

$$C_{PAL} = \frac{U}{2.03}\sin(w_c t) \pm \frac{V}{1.14}\cos(w_c t)$$

The *U* and *V* components in PAL have roughly the same bandwidth as the *I* and *Q* components in NTSC. Both *U* and *V* are approximately 1.3 MHz. (Recall that *I* is 1.5 MHz and *Q* is 0.5 MHz.)

In order to synchronize this phase flipping, the color burst swings ±45 degrees about the axis. Since the sign of the *V* component tracks with the phase of the color burse, this then produces the required 180 degree line by line phase flip. The *U* signal remains untouched.

SECAM is quite different from both PAL and NTSC. Although all three systems share a common luminance signal *Y* and difference signals *R-Y*, and *B-Y* — the SECAM system differs very strongly from PAL and NTSC in the manner in which the signals are modulated.

In SECAM the *R-Y* and *B-Y* signals are transmitted alternately every line. (The *Y* signal remains on for each line). Since there is an odd number of a line on any given scan, any line will have *R-Y* information on the first frame and *B-Y* on the second.

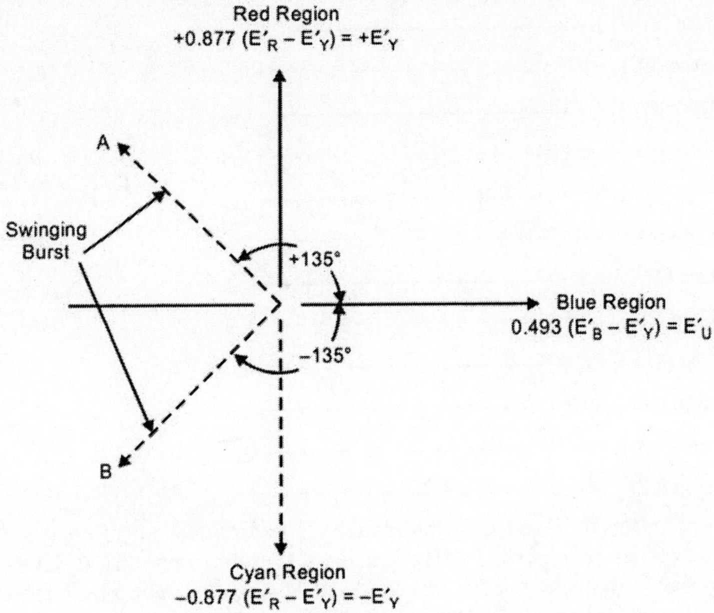

Fig. 6.15.

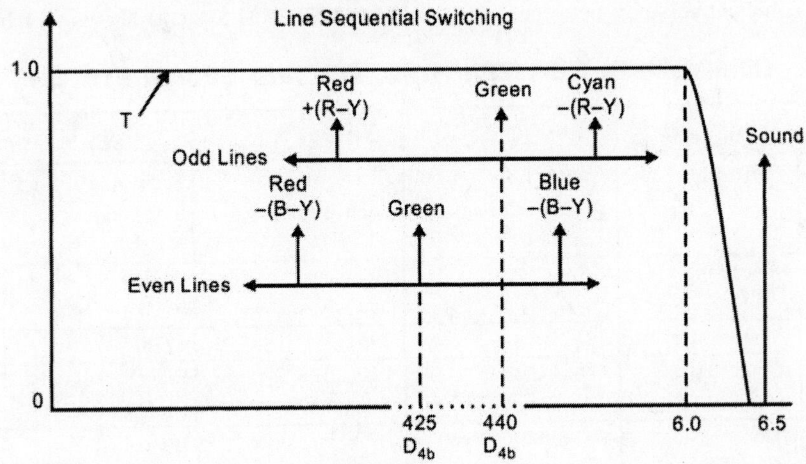

Fig. 6.16.

Field	Line #	Colour	Subcarrier #
Odd (1)	n	f_{ON}	0°
Even (2)	n + 313	f_{ON}	180°
Odd (3)	n + 1	f_{ON}	0°
Even (4)	n + 314	f_{ON}	0°
Odd (5)	n + 2	f_{ON}	180°
Even (6)	n + 315	f_{ON}	180°
Odd (7)	n + 3	f_{ON}	0°
Even (8)	n + 316	f_{ON}	180°
Odd (9)	n + 4	f_{ON}	0°
Even (10)	n + 317	f_{ON}	0°
Odd (11)	n + 5	f_{ON}	180°
Even (12)	n + 318	f_{ON}	180°

Furthermore, the R-Y and B-Y information is transmitted on different subcarriers. The B-Y sub-carrier runs at 4.25 MHz and the R-Y subcarrier runs at 4.4 MHz. The maximum possible frequency deviations for the R-Y signal is - 506 kHz and + 350 kHz and for the B-Y signal is –350 kHz and +506 kHz.

In order to synchronize the line switching, alternate R-Y and B-Y sync signals are provided for nine lines during the vertical blanking interval following the equalizing pulses after the vertical sync.

A comparative study between NTSC, PAL and SECAM systems shown in table-.

6.5. COMPARISON BETWEEN NTSC, PAL AND SECAM SYSTEMS

S.No.	Item name	NTSC	PAL	SECAM
1.	Full form	National TV systems Committee of USA	Phase Alteration by line	Sequential Colour a Memoire
2.	Inventing country	USA	Germany	France
3.	Year	1957	1967	1970
4.	Countries where used	USA, Canada, Japan etc.	India, U.K., Australia etc.	France, Russia Hungary etc.
5.	Transmission of colour	By colour difference signals	By colour difference signals	By colour difference signals
6.	Chroma signal of colour difference signals (weighted)	$I = 0.74\ (R-Y) - 0.27\ (B-Y)$, $Q = 0.48\ (R-Y) + 0.41\ (B-Y)$	$U = 0.493\ (B-Y)$, $V = 0.877\ (R-Y)$	$D_B = +\ 1.5\ (B-Y)$ $D_R = -\ 1.9\ (R-Y)$

continued on next page

S.No.	Item name	NTSC	PAL	SECAM
7.	Modulation sequence	I and jQ both are on each line. Operator j indicates that they are perpendicular to each other.	U and jV both are in each line but phase of V is reversed to $-jV$ every alternate line.	D_R and D_B signals are not present simultaneously but appear alternating on successive lines.
8.	Colour sub-carrier MHz)	3.579545 MHz (3.58 MHz) (Shifted to 57° from reference axis)	4.436 1875 MHz (4.43 MHz)	1.4062 MHz for D_R and 4.250 MHz for D_B. Frequency deviation is 500 MHz in one direction and 3.50 MHz in other direction.
9.	Reference axis	$-(B-Y)$	$-(B-Y)$	Reference axis is not required.
10.	Modulation	Phasors , I and Q modulate the colour subcarrier, using AMSC.	Phasors U and V modulate the colour subcarrier, using AMSC	Colour phasors D_R and D_B modulate the colour subcarriers, using FM (and not AM).
11.	Bandwidth of modulated subcarrier	For I, 1.5 MHz for LSB and 0.5 MHz for USB. For Q, 0.5 MHz for both LSB and USB	USB = 0.56 MHz LSB = 1.3 MHz for U and V both	50 KHz
12.	Polarity of final modulation	Negative	Negative	Positive
13.	Colour burst	In phase colour subcarrier	Switching between +45° and –45° relative to reference axis	In phase with colour subcarrier.
14.	Video bandwidth	4 MHz	5 MHz	6 MHz
15.	Y-signal	$Y = 0.3R + 0.59G + 0.11B$	$Y = 0.3R + 0.59G + 0.11B$	$Y = 0.3R + 0.59G + 0.11B$
16.	Demodulation	I and Q phasors are demodulated by synchronous demodulation inserting subcarrier in phase in one balanced demodulator and in quadrature phase in the other.	U and V phasors are detected by using delay line, causing delay of 64 μs (one line) so as to being $U + jV$ & $U - jV$ together and detect them by synchronous demodulation.	A delay line of 64 μs is used as a one line memory device to produce decoded output of both signals simultaneously. FM detector is used for detection.

continued on next page

S.No.	Item name	NTSC	PAL	SECAM
17.	Intercarrier frequency	4.5 MHz	5.5 MHz	6.5 MHz
18.	Identification signal	Not needed	Needed	Needed
19.	Colour resolution Vertical Horizontal	 Ful Full	 Half Full	 Half Half
20.	Phase distortion	Differential phase errors occur, which are corrected by a manual hue control	Differential phase errors are averaged between $U + jV$ and $U - jV$ signals and hence are automatically corrected	No error.
21.	Noise immunity	High	High	Very high
22.	Receiver controls for picture	Brightness contrast, saturation, hue	Brightness, contrast, saturation,	Brightness, contrast
23.	Studio mixing	Easy	Easy	Difficult as Y and chroma signals have different types of modulations.
24.	Cost	Less than PAL but higher than SECAM	Costliest	Cheapest

Table 6.3. World Television Systems

System	Lines /field	Frame rate	Channel BW	Visual b/w	Sound offset	Vestigial sideband	Vision mod.	Sound mod.	Notes
A	405	25	5	3	-3.5	0.75	Pos.	AM	Old UK VHF system (B/W only)
B	625	25	7	5	+5.5	0.75	Neg.	FM	VHF only in most countries. VHF & UHF in Australia (see systems G and H)
C	625	25	7	5	+5.5	0.75	Pos.	AM	Old VHF system used only in Belgium
D	625	25	8	6	+6.5	0.75	Neg.	FM	VHF only (see system K)

E	819	25	14	10	11.15	2	Pos.	AM	Old French VHF system
F	819	25	7	5	+5.5	0.75	Pos.	AM	Old VHF system used only in Belgium and Luxembourg
G	625	25	8	5	+5.5	0.75	Neg.	FM	UHF only (see system B)
H	625	25	8	5	+5.5	1.25	Neg.	FM	UHF only (see system B)
I	625	25	8	5.5	+5.996	1.25	Neg.	FM	UK, Ireland, South Africa & Hong Kong
K	625	25	8	6	+6.5	0.75	Neg.	FM	UHF only (see system D)
L	625	25	8	6	+6.5	1.25	Pos.	AM	France: audio -6.5 MHz on VHF Band 1 only
M	525	29.97	6	4.2	+4.5	0.75	Neg.	FM	Americas, Japan, Philippines, South Korea (all NTSC-M), and Brazil (PAL-M)
N	625	25	6	4.2	+4.5	0.75	Neg.	FM	Argentina, Bolivia, Paraguay, Uruguay

Systems A, C, E, and F are no longer in use.

SUMMARY

- There are three Colour systems have been developed which are as follows:
  1. National Television System Committee (NTSC) used in America in 1953.
  2. Phase Alteration by Lines (PAL) used in German in 1967.
  3. Sequential Couleruses A Memoirs (SECAM) used in France.

  In India PAL system CCIR (B Standard) is used for colour television transmission and reception.
- The NTSC, PAL and SECAM methods, which are mainly used for colour TV transmission, differ only with respect to the modulation of the colour subcarrier. The NTSC method constitutes the basis for the improved variants PAL and SECAM. However, the original NTSC system (US standard) does not transmit the reduced colour difference signals U and V but instead the I and Q

components, which are referred to a coordinate system rotated counter-clockwise by 33°.

- NTSC colour TV receiver is provided with a control permitting the phase of the reference carrier to be adjusted.
- However, this **hue control** does not allow correction of differential phase distortion.
- **Advantages of NTSC**
  1. Eye has good resolution because of the higher bandwidth in the colour region, saving in bandwidth gains interleaving without colour cross interference.
  2. It is less costly than PAL system.
  3. Studio mixing is quite easy.
  4. Their vertical resolution is double in comparison with PAL and SECAM systems.
- **Problems with NTSC :**
  1. Transmitter overmodulation
  2. Video signal distortions
- PAL stands for phase alternation line.
- PAL system is based on the following concept: an existing phase error can be compensated by a phase error of opposite polarity. This is realized technically by alternating the phase of one of the two chrominance signal components, for instance the SCV component, by 180° from line to line. If a phase error exists in the transmission path, alternately positive and negative departures of the chrominance signal phase from nominal are produced in the receiver after elimination of the line-to line polarity reversal of the SCV component generated at the transmitter end. Delaying the chrominance signal for the duration of a line (64 μs) and subsequent addition of the delayed and the undelayed signals cause two phase errors of opposite polarity to coincide and thus to cancel each other.
- An additional identification is transmitted with the burst to ensure correct phase reversal to the SC_V component in the receiver or of the reference carrier for the $(R - Y)$ sync detector. To this effect, the burst is split into two components, one being transmitted at 180° and the other at –90° alternating from line to line in phase with the S_{CV} reversal.
- With PAL the reduced colour difference signals U and V are directly transmitted, the bandwidth being 1.3 MHz.
- **Advantages of PAL System**
  1. As compared to SECAM system, studio mixing is quick easy.
  2. BW for both U and V signals are the same, USB= 0.56 MHz and LSB= 1.15MHz.
  3. Delay line is used before demodulators' isolates U and V signals from each other. That provides the reduction in cross talk interference and improves picture quality.
  4. Phase over is automatically eliminated, thus a separate hue control is not needed.

- **Disadvantages of PAL System**
  1. This system is complex , the use of electronic switching and indent signals make PAL receivers are the costliest.
  2. In magnetic recording of pal transmission, complete colour coding needs 8 fields, while NTSC needs only four.
- The short form SECAM derived from "sequential à mémoire" indicates that this is a sequential colour system with memory.
- SECAM was inaugurated in France on October 1st, 1967, on *la seconde cha* (the second network), currently called France 2. A group of four men, all dressed in suits, presumably presenters and network officials, were shown standing in a studio. The image was originally black and white and suddenly switched to colour; one of the people said something along the lines of "*now you can see us as we really are*".
- NTSC method, the SECAM method too brings an improvement with respect to wrong hues caused by phase errors in the transmission path. Like the PAL method it is based on the assumption that the colour information does not essentially vary from line to line.
- The colour difference signals $(B - Y)$ and $(R - Y)$ characterizing the colour information need not be transmitted simultaneously.
- SECAM uses frequency modulation, which is not very interference-prone.
- the colour subcarrier is attenuated by pre-emphasis at the transmitter end and boosted by de-emphasis at the receiver end. The effect of noise is reduced by video frequency pre-emphasis and de-emphasis. SECAM has gone through several phases of development.
- **Advantages of SECAM System**
  1. Regeneration of subcarrier is not needed, so it is low cost.
  2. Hue control is not needed.
  3. Picture quality is better since there is no crosstalk interference, as the two phasor do not exist on the same line.
  4. Saturation control is also no needed because the carrier amplitude in the Frequency modulation remains same.
  5. It is not needed to transmit the colour burst signal alongwith the video signal.
  6. Regeneration of subcarrier in the receiver is not required and FM detection is simpler than AMSC detection,
- **Disadvantages of SECAM System**
  1. In this system vertical colour resolution is half because there is only one colour difference signal on a line and half of the colour information is lost.
  2. In this SECAM system, luminance is represented by the amplitude of a voltage but hue and saturation are indicated by the derivation of the subcarrier.
- The largest difference between the three systems is the vertical lines. NTSC uses 525 lines (interlaced) while both PAL and SECAM use 625 lines. NTSC frame rates are slightly less than 1/2 the 60 Hz power line frequency, while PAL and SECAM frame rates are exactly 1/2 the 50 Hz power line frequency.

- All three systems use the same definition for luminance:

$$Y = 0.299R + 0.587G + 0.114B$$

- the chrominance signal for NTSC transmission can be represented in terms of the R-Y and B-Y components as

$$C_{NTSC} = \frac{B-Y}{2.03}\sin(w_c t) + \frac{R-Y}{1.14}\cos(w_c t)$$

The PAL signal terms its B-Y component U and its R-Y component V and phase-flips the V component (line by line) as:

$$C_{PAL} = \frac{U}{2.03}\sin(w_c t) \pm \frac{V}{1.14}\cos(w_c t)$$

The U and V components in PAL have roughly the same bandwidth as the I and Q components in NTSC. Both U and V are approximately 1.3 MHz. (Recall that I is 1.5 MHz and Q is 0.5 MHz.)

REVIEW QUESTIONS

1. What do you understand by I and Q signals ?
2. Draw the block diagram of an NTSC colour transmitter and describe each block.
3. Draw the block diagram of NTSC decoder and explain each block in brief. Explain how chroma signal is processed to obtain the original R, G and B colour signals.
4. Discuss advantages and disadvantages of NTSC system.
5. Describe the main features of the PAL system.
6. What do you understand by weighted colour signals ? How do you produce weighted U and V signals in PAL transmitter ?
7. Write down the similarities and differences of NTSC and PAL system.
8. Draw the block diagram of a PAL transmitter and explain functions of each block.
9. Briefly explain the salient features of SECAM-III.
10. With the help of block diagram, explain the operation of SECAM receiver. Why is 64 μs delay is used in receiver.
11. What is the functions of delay line in colour receiver ?
12. What is the functions of synchronous detector ?
13. What is the main advantage of PAL switching ?
14. Describe briefly the basic difference between SECAM and other colour systems.
15. Compare the performance and features of the NTSC, PAL and SECAM systems.
16. How the chrominance signals transmitted on one carrier in the
    (*a*) NTSC (*b*) PAL and (*c*) SECAM systems.
17. G-Y is not chosen as one of the chroma signals for transmission, why ?
18. Describe the various controls used in PAL receiver.
19. Why is colour burst signal made swinging in PAL system ?
20. Why does SECAM colour encoder require two separate but closely spaced sub-carrier ?

MULTIPLE CHOICE QUESTIONS

1. In NTSC standards, the total number of equalizing pulses per field is
   (*a*) 10 (*b*) 12
   (*c*) 5 (*d*) 6
2. In the term NTSC TV, C stands for
   (*a*) Colour (*b*) Committee
   (*c*) Camera (*d*) Code
3. Weighting factor for U is
   (*a*) 0 (*b*) 1
   (*c*) 0.493 (*d*) 0.877
4. Saturation control is related to
   (*a*) Video amplifier (*b*) Colour killer circuit
   (*c*) Gain of BPA (*d*) Intercarrier frequency
5. Automatic correction of colour error is possible in
   (*a*) NTSC (*b*) PAL
   (*c*) SECAM (*d*) None of these
6. Sound system uses in PAL transmitter
   (*a*) AM (*b*) FM
   (*c*) AMSC (*d*) AMVSB
7. Colour killer circuit kills the colour when
   (*a*) Colour burst is excessive
   (*b*) Colour burst is present
   (*c*) Colour burst is absent
   (*d*) Received used in monochrome
8. The luminance signal modulation in PAL system is
   (*a*) FM (*b*) AM (positive)
   (*c*) AM (negative) (*d*) None of these

ANSWERS

1. (*b*) **2.** (*b*) **3.** (*c*) **4.** (*c*) **5.** (*b*)
6. (*b*) **7.** (*c*) **8.** (*c*)

7

CHAPTER

TELEVISION DISPLAY SYSTEMS

INSIDE THIS CHAPTER

7.1. INTRODUCTION

The picture tube or **kinescope** that serves as the screen for a TV receiver is a specialized form of Cathode Ray Tube (CRT). In the television receiver, the original image is reconstructed essentially by reversing the operation of the video camera. The final image is typically displayed on the face of a cathode-ray tube, where an electron beam scans the fluorescent face, called the "screen," line for line with the pickup scanning. The fluorescent deposit on the tubes inside face glows when hit by the electrons, and the visual image is reproduced.

> **DO YOU KNOW ?**
>
> **Inventions: The Television Tube : Farnsworth developed the vacuum tube television display, an idea he conceived at age 14 and developed at age 21. During a patent law-suit against RCA his high school teacher redrew a drawing Farnsworth had made on the blackboard when he was 14. Farnsworth won the suit and was paid royalties.**

Some cathode-ray tubes can produce multiple beams of electrons and have phosphor screens that are capable of displaying more than one colour. Cathode-ray tubes are used in television sets, computers, automated teller machines, oscilloscopes, and radar displays.

Liquid crystal displays have also been used, mainly on small, portable sets; they are also finding increasing use as light valves on large-screen projectors. Although LCD technology is advancing rapidly, video projectors that use electron tubes can still produce better pictures. Other devices in the receiver extract the crucial synchronization information from the signal and demodulate separate the information signal from the carrier wave) it.

7.2. MONOCHROME TELEVISION DISPLAY TUBE

7.2.1. Introduction to Cathode Ray Tube or CRT

A **monochrome picture tube** has one electron gun and a continuous phosphor coating that produces a picture in black & white. For coloured picture tube the screen is formed of three different phosphors and there are three electrons beams, one for each colour phosphor. Three colours RED, GREEN & BLUE produced by three phosphors combine to produce different colours.

To reproduce the picture at the receiver (the television set) a picture tube (or CATHODE RAY TUBE-CRT) is employed. CRT is a device that can produce an image on a screen with electrical impulses. It is a special-purpose electron tube in which electrons are accelerated by high-voltage anodes, formed into a beam by focusing electrodes, and projected toward a phosphorescent screen that forms one face of the tube. The beam of electrons leaves a bright spot wherever it strikes the phosphor screen. To form a display, or image, on the screen, the electron beam is deflected in the vertical and horizontal directions either by the electrostatic effect of electrodes within the tube or by magnetic fields produced by coils located around the neck of the tube.

In case of a television, the entire front area of the tube is scanned in a fixed pattern called a **raster**, and a picture is created by modulating the intensity of the electron beam according to the programme's video signal. The beam in all modern TV sets is scanned with a magnetic field applied to the neck of the tube with a "magnetic yoke", a set of coils driven by electronic circuits. This use of electromagnets to steer the electron beam is referred to as "magnetic deflection".

The Picture tube screen has a uniform coating of a material which emits light when struck by the beam, and if the beam density is controlled by the picture signal from the camera, a reproduction of the original scene is built up on the picture tube screen. Complete pictures are received at the rate of 25 per second and hence movement is the original scene is portrayed as effectively as in the movies.

Working of CRT

Fig. 7.1 shows the construction of the monochrome CRT.

Cathode-ray tubes have one function that cannot be duplicated by any other tube or transistor; namely, they have the ability to convert electronic signals to visual displays, such as pictures, radar sweeps, or electronic wave forms.

Parts of CRT

All monochrome CRT's used as picture tube in television system, have following main elements:

1. Electron Gun (consists of Cathode, Control grid)
2. Accelerating or screen grid
3. Focusing grid
4. Deflection coils
5. Final Anode (Aquadag coating)

6. Phosphorus coating over screen
7. Screen

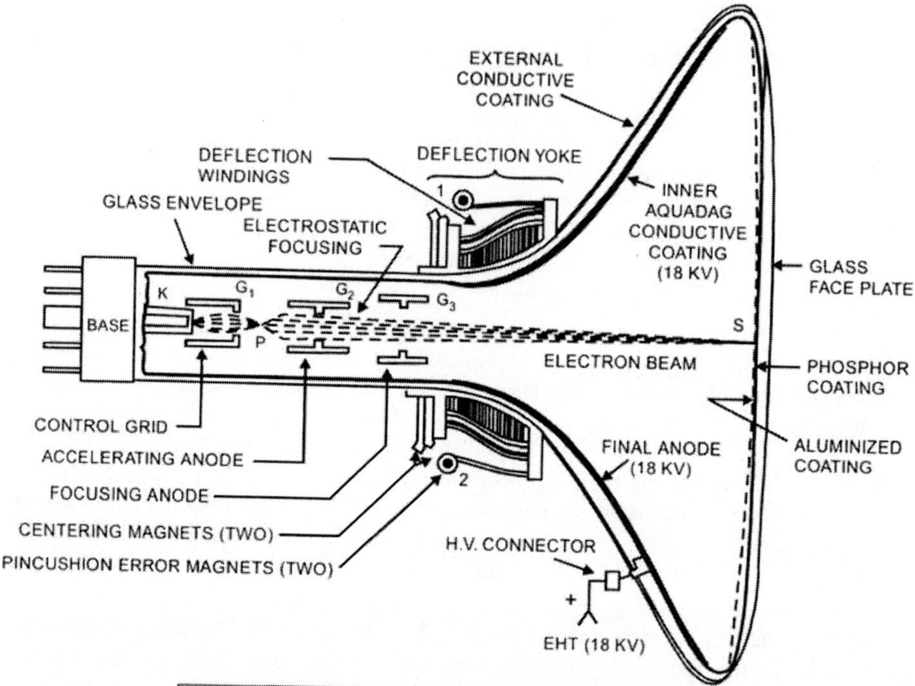

Fig. 7.1. Elements of Monochrome Picture Tube.

The Electron Gun

The electron gun provides an electron beam, which is a highly concentrated stream of electrons. The deflection system positions the electron beam on the screen, and the screen displays a small spot of light at the point where the electron beam strikes it. The *electron gun* is roughly equivalent to the cathodes of conventional tubes. The cathode of the electron gun in the CRT is required not only to emit electrons, but also to concentrate emitted electrons into a tight beam. Generally in the electron tubes, the cathode is cylindrical and emits electrons in all directions along its entire length. This type of cathode is not suitable for producing a highly concentrated electron-beam. The cathode of the CRT consists of a small diameter nickel cap. The closed end of the cap is coated with emitting material. This is shown in Fig. 7.2. Because of this type of construction, electrons can only be emitted in one direction. Notice that the emitted electrons shown in Fig. 7.2 are leaving the cathode at different angles. If these electrons were allowed to strike the screen, the whole screen would glow. Since the object of the electron gun is to concentrate the electrons into a tight beam, a special grid must be used. This special grid is in the form of a solid metal cap with a small hole in the centre. The grid is placed over the emitting surface of the cathode and charged negatively in relation to the cathode, referred as **control grid**. The dotted lines represent the direction of cathode emitted electron repulsion, as shown in figure 3. Since all emitted electrons leave the cathode (point C), their paths can be identified. An electron attempting to travel from point C to point B (downward) will instead follow the path from point

C to point E to point P. Consider an electron leaving from C in the direction of point A (upward). Its path will be curved from point C to point P by electrostatic repulsion. These curving electron paths are due to the negative potential of the grid coupled with the high positive potential of the anode. The potential of the anode attracts electrons out of the cathode-grid area past point P toward the screen. The grid potential may be varied to control the number of electrons allowed to go through the control-grid opening. Since the brightness or intensity of the display depends on the number of electrons that strike the screen, *the control grid is used to control the brightness of the CRT.*

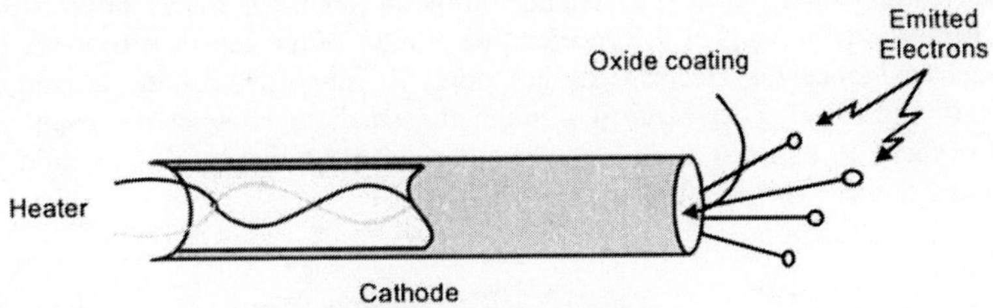

Fig. 7.2. CRT Cathode.

The proper name, ***brightness control***, is given to the potentiometer used to vary the potential applied to the control grid. The control grid actually serves as an electron lens. It is this electronic lens that you adjust when you turn up the brightness control on your TV set. Notice that the effect of the grid is to focus the electron beam at point *P* in Fig. 7.3.

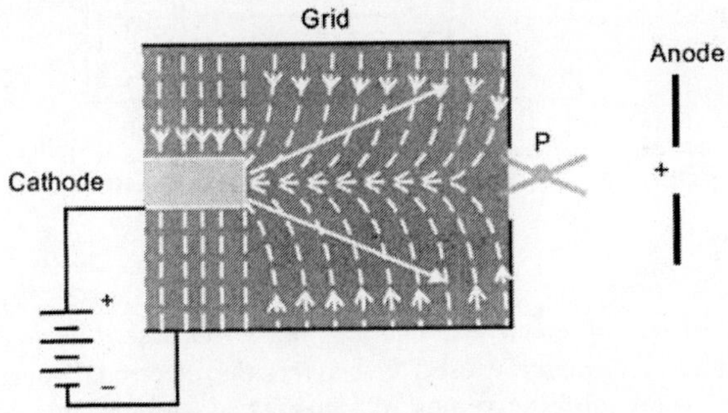

Fig. 7.3. Operation of the CRT grid.

Electrostatics Focusing

After passing point P, the electrons start to spread out, or diverge, again. Therefore, it becomes necessary to provide some additional focusing to force the electrons into a tight beam again. Electrostatic focusing is achieved by the combined efforts of accelerating anode and focusing anode, so that electrons get focused over the screen.

Electrostatic focusing is preferred over magnetic focusing because it is not much affected by change in the line voltage and need no ion-spot correction.

This is done by two additional positively-charged electrodes as shown in figure 4. The first electrode is commonly called the ***focusing anode***. Generally, the focusing anode is charged a few hundred volts positive with respect to the cathode. Electrons emitted by the cathode are attracted to the focusing anode. This is the reason that they travel through the small hole in the grid. The second electrode, called the **accelerating**

anode, *is charged several thousand volts positive in relation to the cathode.* Any electrons approaching the focusing anode will feel the larger electrostatic pull of the accelerating anode and will be bent through the opening in the focusing anode and will travel into the area labelled D. You might think that once an electron is in this region, it is simply attracted to the accelerating anode and that is the end of it. This does not happen. Because the accelerating anode is cylindrical in shape, the electrostatic field radiating from it is equal in all directions. Thus, an electron is pulled in all directions at once, forcing the electron to travel down the centre of the tube. Then, the electron is accelerated into the accelerating anode. Once it passes the mid-point (point E), it feels the electrostatic attraction from the front wall of the accelerating anode, which causes it to move faster toward the front. Once the electron reaches point F, equal electrostatic attraction on either side of the opening squeezes it through the small opening in the front of the anode. From there, it is joined by millions of other electrons and travels in a tight beam until it strikes the screen (point S).

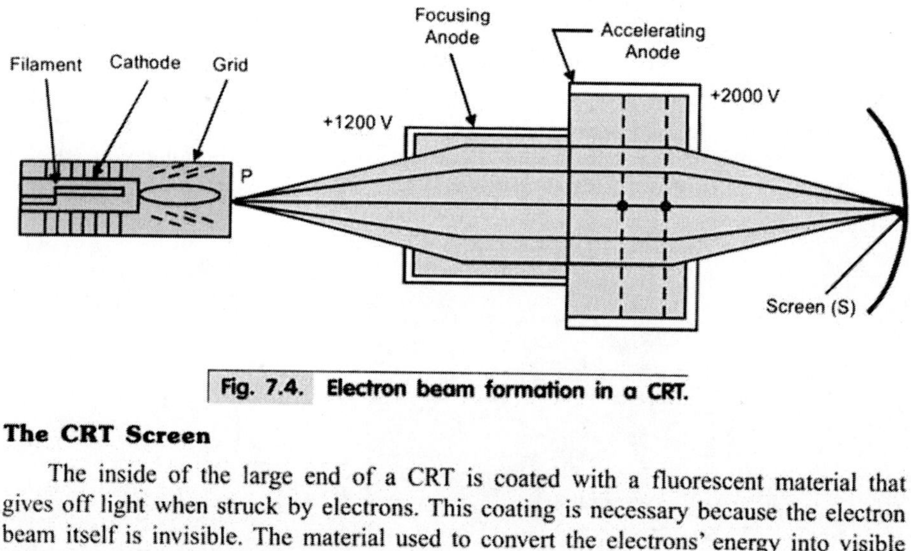

Fig. 7.4. Electron beam formation in a CRT.

The CRT Screen

The inside of the large end of a CRT is coated with a fluorescent material that gives off light when struck by electrons. This coating is necessary because the electron beam itself is invisible. The material used to convert the electrons' energy into visible light is a phosphor. Many different types of phosphor materials are used to provide different colours displays and displays that have different lengths of persistence (duration of display). In order to reach the screen, electrons from the cathode are accelerated to relatively high velocities. When these electrons strike the screen, they dislodge other electrons from the material of the screen. If these secondary emission electrons are allowed to accumulate, they will form a negatively-charged barrier between the screen and the electron beam, causing a distorted image on the CRT screen. In order to give the electron stream sufficient velocity to reach to screen material with proper energy to cause it to fluoresce, a second anode is included within the tube. This is conductive coating with colloidal graphite on the inside of the wide bell of the tube. A special coating called an aquadag coating is applied to the inside of the tube usually extends from almost halfway into the narrow neck to within 3cm of the fluorescent screen, as shown in Fig. 7.5. This coating is composed of a conductive material, such as graphite, and has the same high-positive potential applied to it that is applied to the accelerating anode.

This allows the aquadag to perform two functions.

1. First, since the aquadag coating is positive, it attracts the secondary emitted electrons and removes them.
2. Second, because the aquadag is operated at a high-positive potential and is mounted in front of the accelerating anode, it aids in the acceleration of electrons toward the screen.

The voltage applied to the Aqudag coating or Final anode is depend upon on the tube size and it is about 18 kv for a 48 cm monochrome tube.

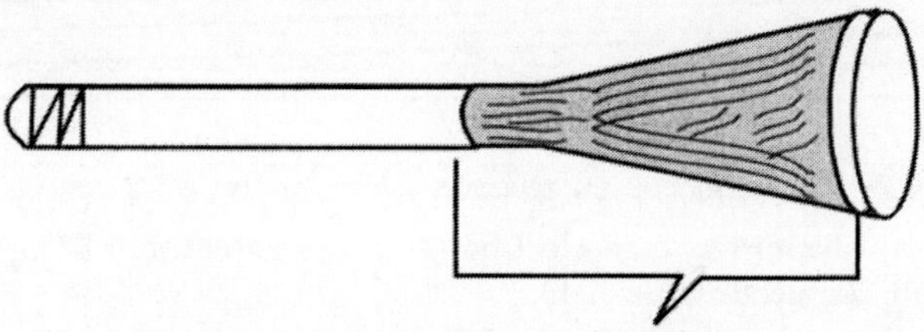

Fig. 7.5. **Aquadag coating in a CRT.**

Deflection system

Beam Deflection

At this point, you have a bright spot in the centre of the CRT screen as shown in Fig. 7.6. Having watched TV, you know that a TV picture consists of more than just a bright spot in the centre of the picture tube. Obviously, something is necessary to produce the picture. That something is called *deflection.* For the CRT to work properly, the spot must be moved to various positions on the screen. In TV set, the spot is moved horizontally across the CRT face to form a series of tightly packed lines. As each line is displayed, or traced, the electron beam is moved vertically to trace the next line as shown in Fig. 7.7. This process starts at the top of the tube and ends when the last line is traced at the bottom of the CRT screen. Because the beam is swept very quickly across the CRT and the phosphor continues to glow for a short time after the beam has moved on, you do not see a series of lines, but a continuous picture.

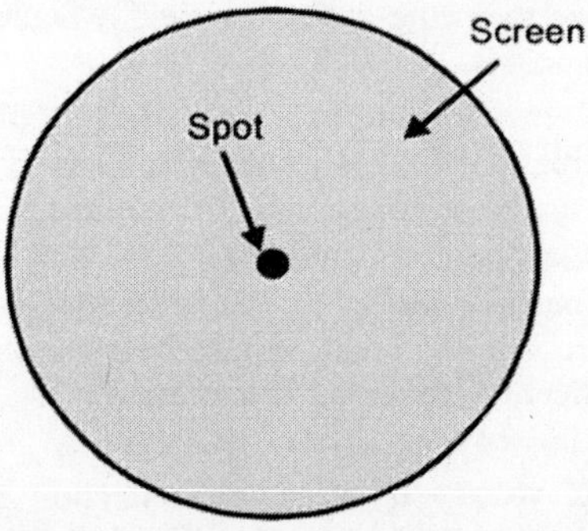

Fig. 7.6. **Impact of an electron beam on a CRT screen.**

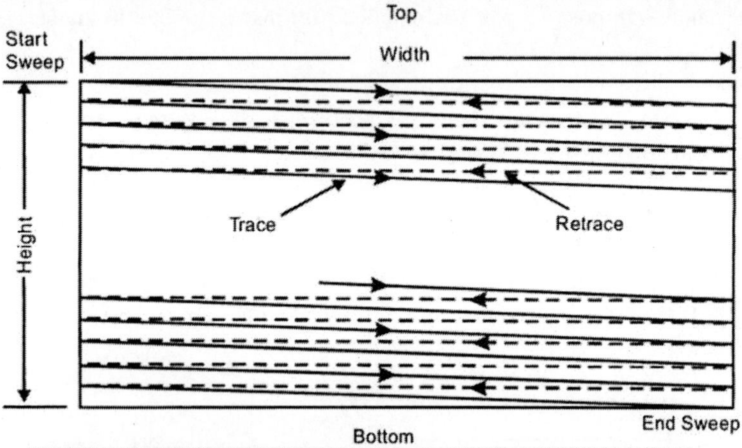

Fig. 7.7. Deflection of an electron beam across a TV screen.

There are two ways to move an electron (and thus an electron beam): either with a magnetic or with an electrostatic field.

Because of this, there are three possible ways to move or deflect an electron beam in a CRT :

1. Magnetically,
2. Electromagnetically, and
3. Electrostatically.

All three ways are used in electronics. In general, though, electrostatic and electromagnetic deflections are used most often.

In TV picture tube Electromagnetic deflection is preferred for the following reasons:

(*i*) Due to shortage of deflecting time, it is necessary to have high deflecting field to achieve the desired deflection. For example: *In picture tube with 15kv at the final anode, about 7500v would be needed for 50cm screen.* It is difficult to achieve the desired high voltage at the deflection frequencies. On the other end with magnetic deflection it is large that would be necessary to achieve the same deflection. It is easy to generate large current than large voltages.

(*ii*) With *electrostatic deflection* the electrons beam gain energy. Thus large deflection angle tend to defocus the beam. The deflection plates need to be placed further apart as the deflection is made larger, thus requiring higher voltages to produce the same deflection field. Magnetic deflection is free from both the shortcomings.

(*iii*) For ES deflection, two delicate pairs of deflecting plates are needed inside the picture tube, whereas for magnetic deflection two pairs of deflecting coils are mounted outside and close to the neck of the tube. In combination, the vertical and horizontal deflection coils are called deflection yoke and it is fitted outside and close to the neck of the tube. The tube neck is made narrow so that the deflecting coils are near the beam axis to be more effective. Figure 7.8 (*a*) and (*b*) shows the deflection yoke and orientation of magnetic field in vertical and horizontal deflection coils.

(*iv*) If the electrostatic voltage between the deflection plates is high, that causes sparking between the plates. In magnetic deflection there is no reason of sparking.

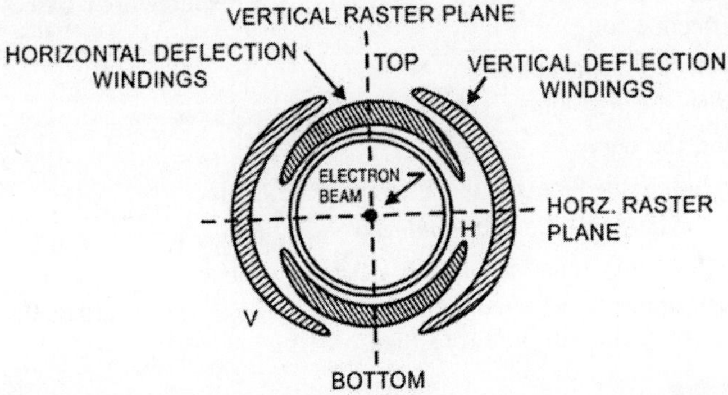

(a) Position of vertical and horizontal deflection coils (cross-sectional view).

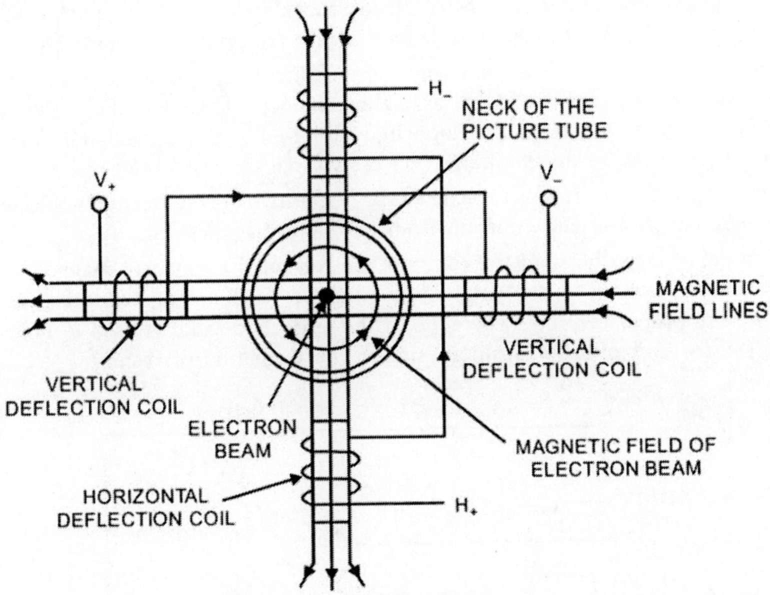

(b) Orientation of magnetic field in vertical and horizontal deflection coils.

Fig. 7.8.

To generate uniform or linear magnetic field for linear deflection, a special type of winding is used. In this arrangement the thickness of the deflection winding varies as the cosine of the angle from a central reference line, such a winding is known as **cosine winding.**

Yoke Selection Parameters

The selection of a yoke for a given application involves the consideration of a number of parameters, some of which require design tradeoffs. The typical selection procedure includes the following steps:

- Select a CRT and determine the maximum deflection angle
- Determine yoke inside diameter (ID) dimensions from the CRT neck size

- Find the yoke energy constant from the yoke manufacturer, based on the ID and deflection angle
- Establish the anode voltage
- Determine the half-angle deflection for the two axes
- Calculate the energy constants
- Set the minimum time for the half-angle deflection
- Find the maximum allowable induced voltage
- Calculate the maximum allowable yoke inductance

For any given application, a number of choices usually exist, although some may be more practical to realize in hardware than others.

Deflection Angle

A deflection angle is the maximum angle through which the beam can be deflected without striking the side of the bulb. Most picture tubes have a deflection angle of ±45 degree (90 degree). Typical values of deflection angles are 70, 90,110. and 114 degree.

The advantage of large deflection angle is that for equal picture size the picture tube is shorter and can be installed in a smaller cabinet. However a large deflection angle requires more power from the deflection circuit. For this reason the tubes are made with narrow neck to put the deflection coils to closer to the electron beam.

For portable television receiver, picture tube has a narrower deflection angle which makes the tube longer but current drain from the battery remain low.

To keep the electron beam at the centre, a pair of small permanent magnets in the form ferrite rings mounted on the yoke cover. The two rings are rotated with the help of tabs to ensure proper centring. The edge of the deflection yoke is used to hold two small magnets for correcting pincushion error in the formation of raster.

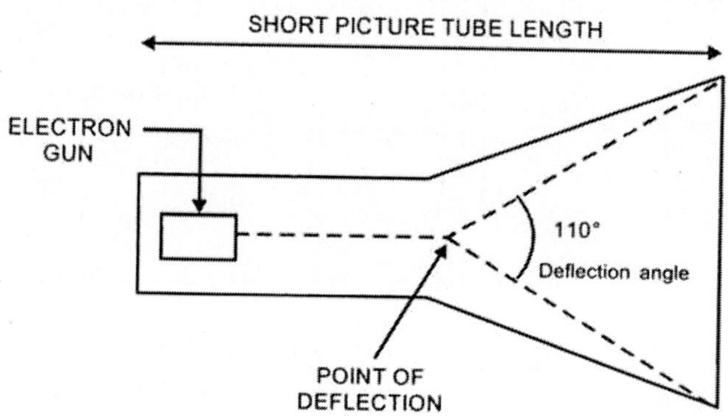

Fig. 7.9. Deflection angles.

Screen brightness

There are some shortcomings in CRT:

1. It is estimated that about 50% of the light emitted at the screen where the electron beam strikes it, travels back in the tube. Another 20% or so is lost in the glass of the tube because of internal reflection and about 20% of it reaches to the viewer.

2. Any ions in the beam, which do exist despite best precautions while degaussing, damage the phosphor material on hitting it and thus cause a dark brownish patch on the screen.

How to Overcome These Shortcomings ?

Practically all modern picture tube employ a very **thin coating of aluminum** on the back surface of the screen phosphor. The aluminized coating is very thin and with a final anode voltage of 10kv or more, the electrons of the beam have enough velocity to penetrate this coating and excite the phosphor. This aluminized coating is connected to the high voltage anode coating.

Spark-gap protections are used to protect the receiver circuit due to any arcing. They are may be metallic spark gap or neon bulbs, to provide the shunt path for the surge current.

High Voltage Filter Capacitor

A grounded coating of aquadag material is used at the outer surface of the picture tube. This provides shielding from stray fields and acting as one plate of a capacitor, the other plate being inner aquadag material coating. Glass between the two aquadag coatings plates acts as dielectric.

For EHT voltage rectifier this capacitor used as filter capacitor. This capacitor holds the charge even if power supply of the tube is removed and it must be discharged by shorting the external anode button to the ground before handling the tube for servicing.

Picture Tube Transfer Characteristics

The picture tube converts the electrical video signal into light information. The transfer characteristics plotted as light output against electrical signals on log-log scale is not linear but follows a square law. The mutual transfer characteristics of a picture tube is plotted as the beam current i_b in μA against the control grid voltage V_{gk} as shown in Fig. 7.10. It is similar to the grid plate characteristics of vacuum tubes. The grid

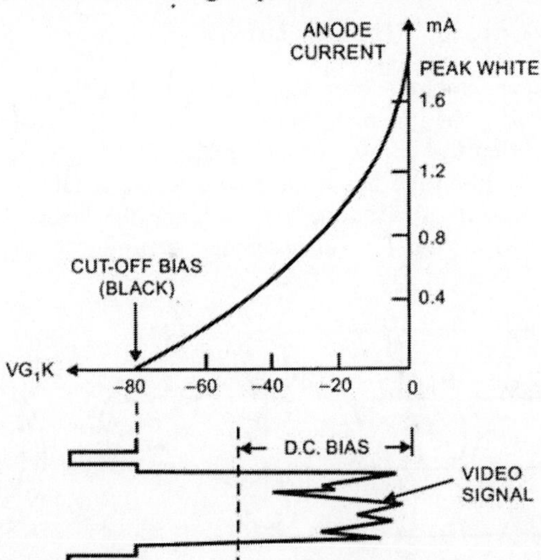

Fig. 7.10. Transfer characteristics of a picture tube.

of a picture is a field bias and is set is set with brightness control for optimum average brightness of the picture on the screen. Finally the video signal that controls the variation in brightness on the screen may be applied either at grid or cathode of the picture tube. The process of varying the beam current to control instantaneous brightness is known as z-axis modulation.

The peak to peak value of the ac video signal determines the contrast in the picture. At cut-off the grid voltage is negative and is reduces the beam current to a value low enough to extinguish the beam and shows the black level in the picture. The smaller the value of voltage at the screen grid, the smaller the beam current and this characteristics curve is in-between the beam current against the control grid voltage.

7.3. COLOUR PICTURE TUBES

In colour picture tubes, voltages pertaining to three primary colours give rise to three electron beams. These three beams activate three phosphor elements (one each), closely placed in a group, either in the form of dots or in the form of stripes. Thus there are lights of three primary colours coming out of these three phosphor elements about a point. The eye integrates the three colours and perceives the single resultant colour.

Types of Colour Picture Tubes

Shadow mask picture tube was first developed by Dr Goldsmith and his team in 1950, at the Radio Corporation of America (RCA) in the USA. The picture tube embodied all the basic features of a modern colour picture tube. This tube evolved into many derivatives that improved upon its low beam efficiency and convergence adjustments. These shadow mask picture tube are classified on the basis of electron gun configurations and the manner in which phosphors are arranged on the screen.

1. Delta Gun colour picture tube
2. Precision- In- Line (PIL) picture tubes
3. Trinitron

7.3.1. Delta Gun Colour Picture Tube

Delta Gun picture tubes employs three separate Guns, equally spaced at 120 degree with respect to each other, generates three high velocity (25 kv acceleration) electrons beams. They are mounted on the corners of an equilateral triangle, forming the Greek letter delta (D) that is why the tube has been named as Delta Gun picture tubes. The guns are slightly tilted inwards to the tube axis to make the beam converge near the faceplate. BR & G represents guns for video signals pertaining to blue, red and green colours, respectively as shown in Fig. 7.11.

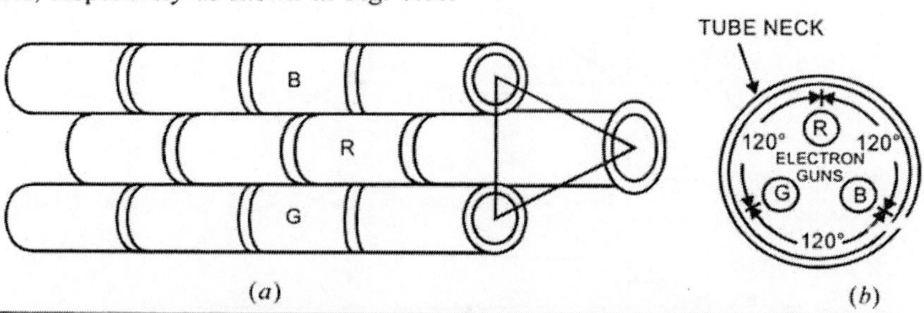

Fig. 7.11. Arrangement of three electron guns in Delta colour picture tube.

Each gun in the three gun delta picture tube has its own separate set of cathode, control grid, accelerating grid and focus grid so that individual colour signals can be coupled to the grid cathode circuit to modulate beam intensity.

Shadow Mask

On the screen, there are about one million dots of phosphor in groups of three, corresponding to red, green and blue light. These are deposited on the back of the face plate. Each group is called **Triads.** Thus there are 333000 triads.

Each dot is of 0.42 mm diameter with a spacing of 0.72 mm between the respective dots of the adjacent triads.

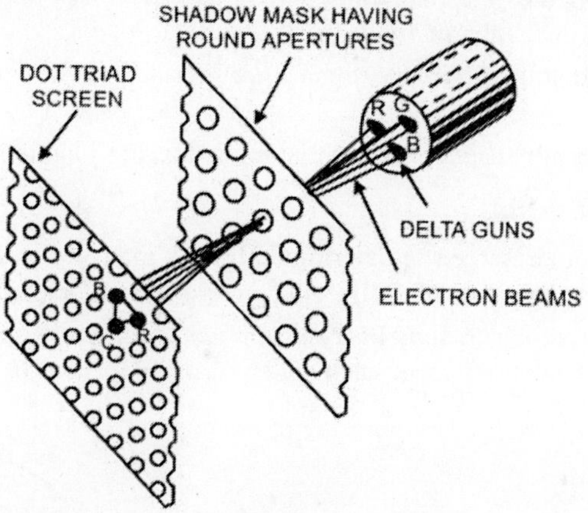

(a)

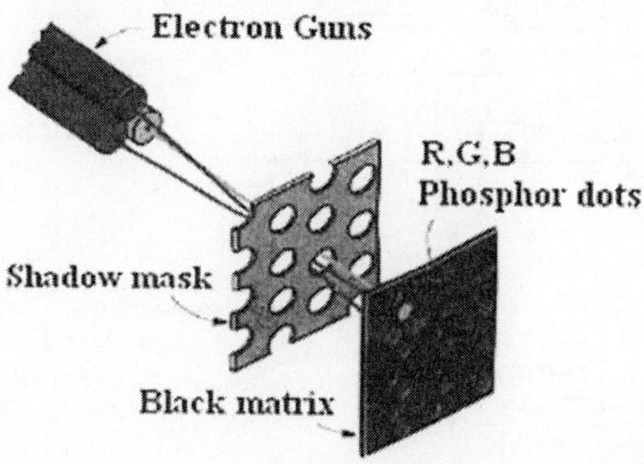

(b)

Fig. 7.12. Picture tube with shadow masks and screen phosphor.

A thin perforated metal sheet is placed about one cm behind the tube screen, to ensure the correct landing of beam on phosphor, is known as **Shadow mask.** It is supported by a heavy internal frame for rigidity. The mask has one conical hole for every phosphor dot triad on the screen as shown in Fig. 7.12. The hole is conical to prevent emission of secondary electrons. Shadow mask is used to ensure that the beams land on the correct phosphor. Three electron beams converge on a hole of the mask, and from there they diverge to strike the respective dots of triads. The shadow mask thus blocks an appreciable portion of the beam current dissipating several watts of power. The transparency of the shadow mask is only 20%. The remaining 80% of the total beam current energy is dissipated as heat loss in the shadow mask. **This is why colour picture tubes use higher anode voltage and need larger beam currents and cathode currents than monochrome tubes.**

Phosphor used for colour tubes are :

Red : Zinc sulphide

Green : Zinc silicate

Blue : Europium and yttrium

Persistency : 1 ms

Space between the phosphor dots filled with the black pigments to improve the contrast by absorbing ambient light.

All the three colour difference signals are then fed to the three grids of the colour picture tubes.

The inverted luminance signal (–*Y*) is applied at the junction of the three cathodes.

Principle of Working

The overall colour screen is determined by the intensity of the each beam and the phosphors. The arrangement of Delta gun picture tube is shown in Fig. 7.13.

The inverted luminance signal (-*Y*) is applied at the junction of the three cathodes. The signal voltage subtract from each other to develop control voltages for the three electron guns.

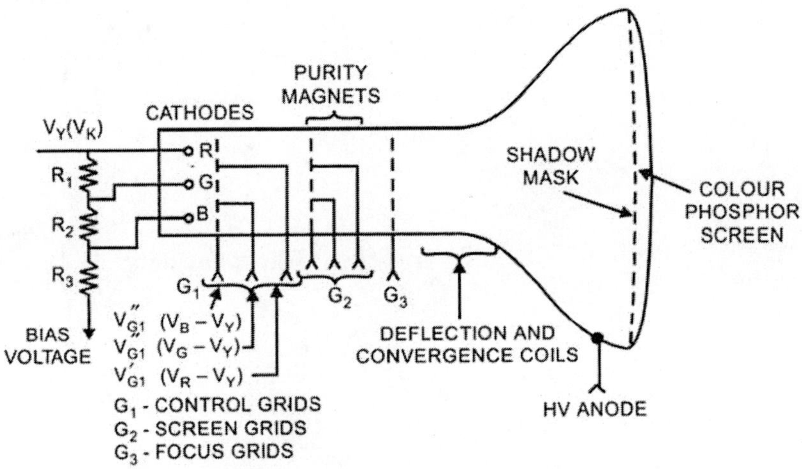

Fig. 7.13. Delta gun picture tube.

The electron beams from the three electron guns strikes three phosphor dots of red, green and blue (R, G, and B primary colours) phosphor in a triad glow simultaneously. The intensity of glow is proportional to the intensity of video signal of the respective colour. The eye integrates the three colours emitted by the phosphor dots at a time and perceives the resultant colour of the concerned pixel as in the original picture. Triads glow one after another in quick succession due to deflection of the beams and thus the whole picture is reproduced in its original forms.

In the absence of two electron beams, dots of only one colour phosphor get excited and raster will have only one of the primary colours. In case when all the three electron beams are off than the picture will be black in colour.

If the amplitude of colour difference signal drops to zero, the only Y signal control the three guns and a monochrome picture is produced on the screen.

$$V_{G1} - V_k = (V_R - V_Y) - (-V_Y) = V_R$$
$$V_{G2} - V_k = (V_G - V_Y) - (-V_Y) = V_G$$
$$V_{G3} - V_k = (V_B - V_Y) - (-V_Y) = V_B$$

Purity and Convergence

- While deflecting the three beams by vertical and horizontal deflecting coils it is necessary to ensure that each produces a pure colour and all the three colour raster overlap each other.
- For obtaining purity in colour each beam should land at the center of the corresponding phosphor dot irrespective of location of the beam on the raster.
- This needs precise alignment of the colour beams and is carried out by a circular magnet assembly known as the **purity magnet.**
- Purity magnets mount on the neck of the tube and close to the deflection yoke.

Yoke Position

The position of the yoke on the tube neck determines the location of the deflection center of the electron beams. A wrong setting will result in poor purity due to improper entry angles of the beam in to the mask opening.

Convergence

The technique of bringing the beams together so that they hit the same part of the screen at the same time to produce three coincident raster is referred as **convergence**.

Convergence errors are caused by:

1. Non-coincident convergence plane
2. Non-uniformity of the deflection field.
3. Flat surface of the picture screen.

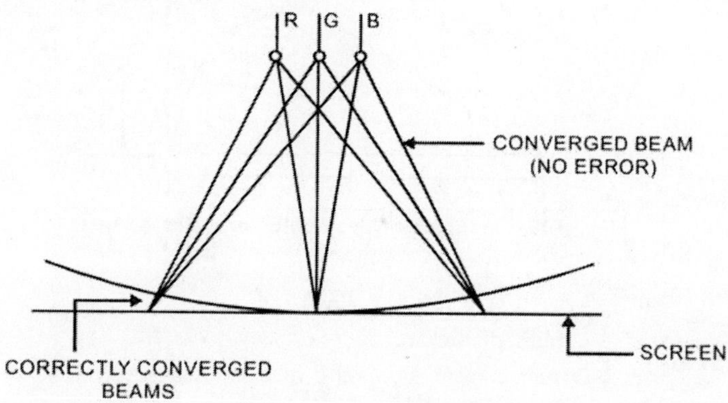

(*a*) Correct convergence.

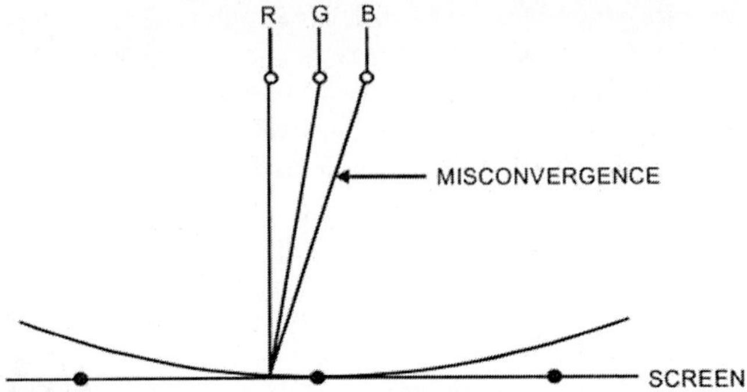

(*b*) Incorrect convergence.

Fig. 7.14. Correct and incorrect convergence of the beam.

Static and Dynamic convergence

Static convergence : It involves movement of the beam by permanent magnetic fields, one correctly set; bring the beam in to convergence in the central area of the screen. The permanent ring magnets are adjusted for static convergence in the central area of the screen.

Dynamic convergence : Apart from the static convergence at the centre it is necessary to ensure dynamic convergence when the beams are deflected to form a raster. Dynamic convergence over the rest of the screen is achieved by continuously varying (dynamic) magnetic field.

Pin–cushion Effect

We know that the screen of a picture tube is rectangular, it is not circular, and the magnetic deflection would not be uniform. The distance of corners is more as compared to the centre point. Deflections of electron beam under magnetic field along the circular arc and on a rectangular surface are illustratrated in Fig. 7.15.

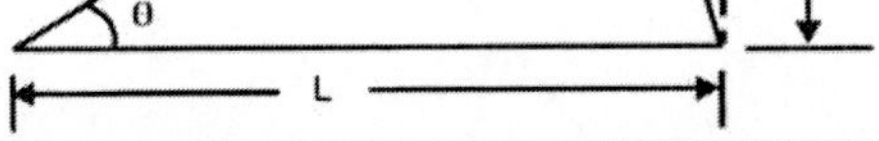

Fig. 7.15. Deflection for circular and flat screens.

We get

$$d = L.\theta = L \sin \theta$$

where d = Length of the arc

L = Distance from the centre of a circular screen

θ = deflection angles in radians

Practically for rectangular screen, the same deflection angle θ will have to move the beam to longer distance d and that will be equal to $L \tan \theta$. This causes the raster to be stretched at the corners and shape of the raster is shown in Fig. 7.16.

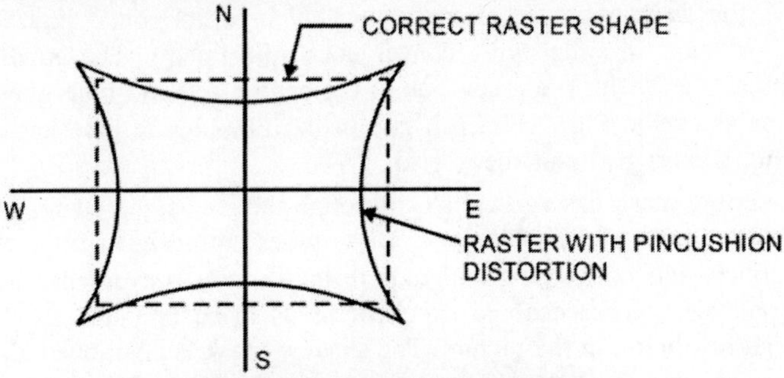

Fig. 7.16. Pincushion distortion.

Dynamic pin cushion correction is used with colour picture tubes. Such a correction automatically increases horizontal width and vertical size in those regions of the raster that are shrunken because of pin cushion distortion.

Degaussing

Degaussing means demagnetizing iron and steel parts of the picture tube mounting. The main cause of poor purity is the susceptibility of the mask and its mounting frame to become magnetized by the earth magnetic field and / or by any other strong magnetic field. The effects of those localized magnetic field is the deviation of electrons beams from their normal path. To prevent such effect the picture are magnetically shielded. It is done by placing thin silicon steel housing around the belly of the tube.

Drawbacks of Delta Gun Tubes

1. As the beams are not in the same line, focus is not sharp all over the screen.
2. Electrons transparency of the shadow mask is very low *i.e.* 20%. Thus 80% of the beams are stopped from striking the phosphor dots, resulting in poor efficiency.
3. Convergence of the three beams on the respective dots is difficult. It required 4 static and one dynamic convergent magnets assembly, resulting 12 preset controls.

These tubes are not used and are being replaced by PIL and Trinitron picture tube.

7.3.2. Precision In-Line Picture Tube (PIL Tube)

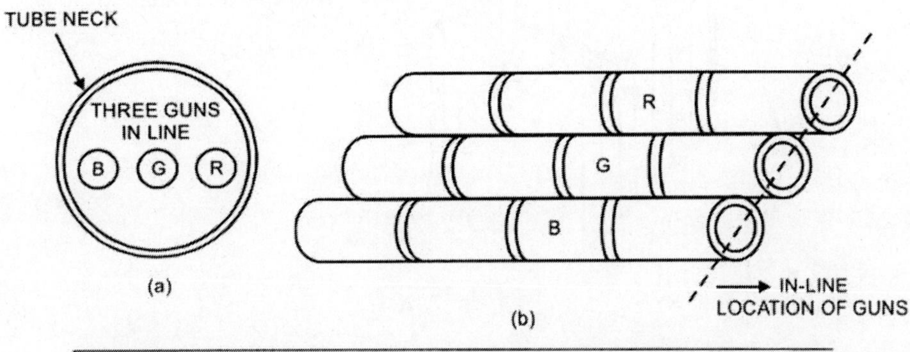

Fig. 7.17. Arrangement of three guns in precision In-Line Picture tube.

In this the three guns are aligned precisely in a horizontal line. The in line gun configuration helps in simplifying convergence adjustments. The guns are mounted in the horizontal plane with the green gun at the centre, red and blue guns are on the left and right as shown in Fig. 7.17 Each gun in the PIL picture tube has its own cathode control grid, screen grid and focus grid.

The aperture mask has vertical slots through the converged beoms pass to strike the corresponding colour phosphor stripes. These tubes are design with such precision that they are almost self converging and require hardly any convergence adjustments. The improved gun design has enabled excellent focus to be maintained with a small spot size for high resolution in the picture. The shadow mask is positioned about 1cm behind the tube screen. Fig. 7.18 shows an arrangement of vertical stripe phosphor screen and Fig. 7.18 illustrates the gun and the mask structure of PIL picture tube.

The PIL picture tube carries with it the deflection yoke and magnet assembly for purity and convergence, factory adjusted and bonded on to the tube neck. The deflection yoke used a torroidal yoke having precision wound coils and is less bulky.

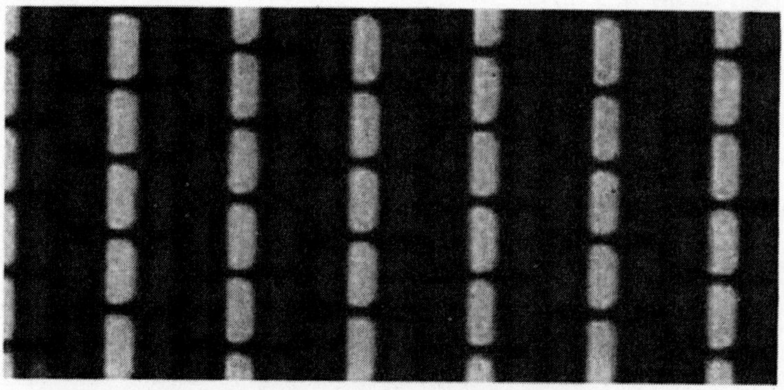

Fig. 7.18. Arrangement of vertical stripe phosphor screen.

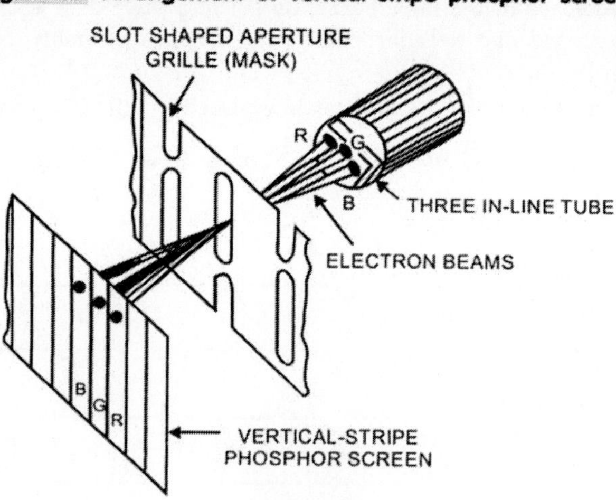

Fig. 7.19(a)

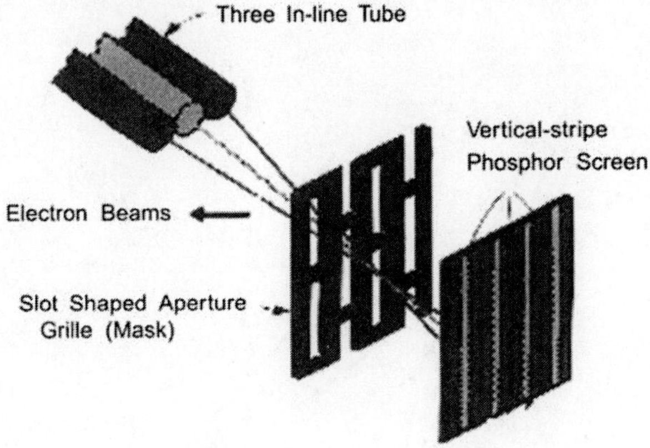

Fig. 7.19(b). PIL-Picture tube with shadow masks and screen phosphor.

Advantages

1. The PIL tubes are very efficient that mean higher electron transparency (50%) and needs fewer convergence adjustments.
2. No needs of vertical convergence, only horizontal convergence is required at the stage of manufacturing.
3. Focus is sharp all over the screen.

Convergence Errors

Since the three raster should be accurately superimposed on each other, it's give some constraints on the distribution of the deflection. Non optimum distribution of field produces the convergence error as the screen is nearly flat surfaces.

There are two kinds of colour errors are:

1. Astigmatism effect
2. Comma effect

Astigmatism Effect

It is clear from the Fig. 7.20, that with the uniform field, the rays from a vertical row of points (AGB) converge to produce a vertical focal line on the screen but the rays from a horizontal point (RGB) converge beyond the screen. This effect is referred as **astigmatism** and results convergence error. In PIL picture tube, beams RGB emerge only from a horizontal line and so vertical astigmatism have no effect. To eliminate horizontal astigmatism, the horizontal focus should be a point on the screen but not a line for any deflection. Because of the change in the shape of deflection field, opposite changes in horizontal and vertical astigmatism occurs. Suitable field adjustments are required to make sure that the beams coming from in line guns in horizontal line coverage should be at the same point on the screen irrespective of the deflection angle.

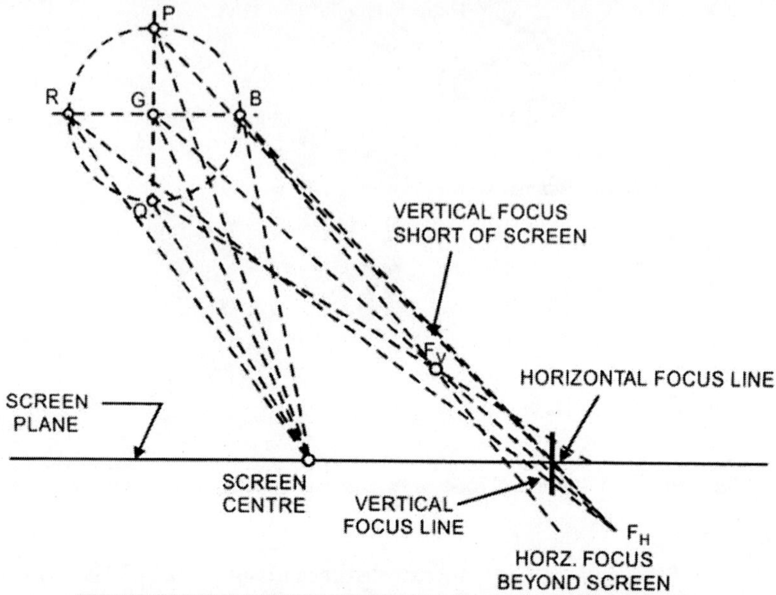

Fig. 7.20. **Astigmatism effect in uniform deflection field.**

Comma Effect

Due to the non-uniformity of the deflection field, all the beams are not deflected by the same amount. The green beam deflects less as compare to other two beams. Such a distortion is known as **Comma effect** and results in misconvergence of the beams as shown in Fig. 7.21.

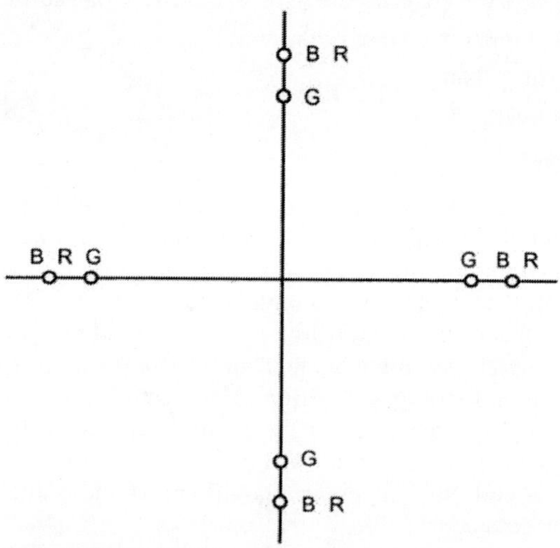

Fig. 7.21. **Comma effect in PIL picture tube.**

7.3.3. Trinitron Picture Tube

Trinitron picture tube is developed by Sony in 1968, and employing, for the first time in PIL beam construction as shown in Fig. 7.22. It is three in line cathode colour picture tube. It employs single gun having three in line cathodes for each of the red, green and blue beams. This simplifies the construction problem because one electron gun can easily be accommodated. The control grid is a single cup type electrode with three holes for three beams. The screen and focus grid are common for all three beams.

The beams cross over to converge within the focus field of the single-gun lens before leaving the gun to cross again at the slots in the shadow mask. This ensure of passing all the three beams through the central portion of the relatively large common electrostatic lens, minimizing spherical aberrations in the outer beams. All the three beams are in focus together at the centre of the main focus electrode.

Three phosphors triads are arranged in vertical strips as in the PIL tube. A metal aperture grille or mask is provided very close to the screen. The shadow mask has long slots, one slot per vertical line, increasing transparency to about 80%. Figure 7.22 shows the schematics of Trinitron picture tube.

The latest version of Trinitron picture tube is developed in 1978. It consists of low magnification electron gun assembly, long focusing electrodes and larger aperture lens system. A high quality picture with good resolution all over the screen is obtained by new high precision deflection yoke with minimum convergence adjustments.

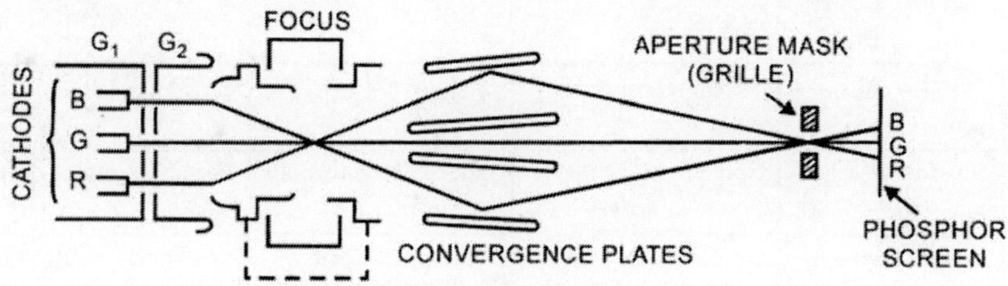

Fig. 7.22. Trinitron picture tube.

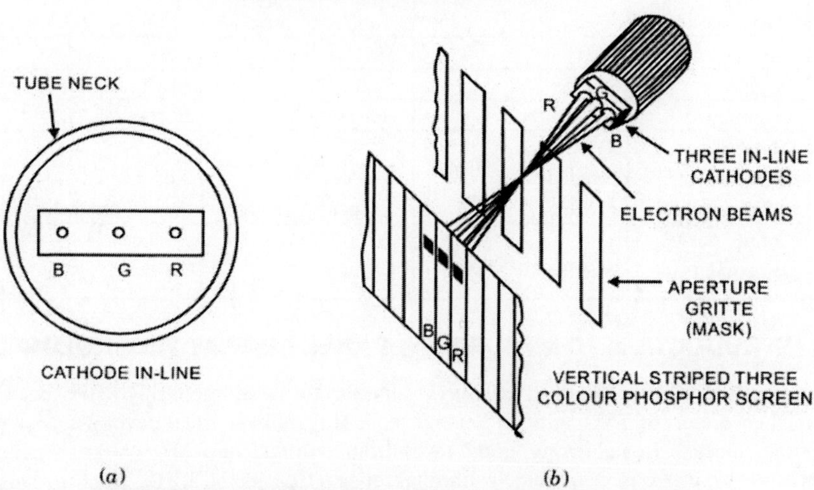

Fig. 7.23. Arrangement of cathodes and vertical striped, three colour phosphor screen.

Advantages

- Greater transparency about 80%.
- Sharper focus and brighter picture.
- No problem of convergence adjustment during servicing.
- Improved vertical resolution.

7.3.4. Comparisons Between Three Colour Picture Tubes

Table 7.1

Sl. No	Description	Delta gun	PIL	Trinitron
1.	Construction of three guns	Guns are equally spaced at 120° in the shape of delta	Guns are precisely aligned in a horizontal line in the same plane	Three cathodes are aligned in a line producing three beams
2.	Number of cathodes	3	3	3
3.	Number of control grids	3	3	1
4.	Number of focus grids	3	3	1
5.	Number of screen grids	3	3	1
6.	Shadow mask aperture	Conical holes	Short vertical slots	Vertical slots of full length
7.	Screen phosphors on which three electron beams strikes	Dots arranged in triad	Vertical stripes arranged in group of three	Vertical stripes arranged in group of three
8.	Transparency	20%	50%	80%
9.	Focus	Sharp	Sharp	Sharpest
10.	Brightness	Good	Better	Best
11.	Convergence a. Vertically b. horizontally	 Needed Needed	 Not needed Needed at factory	 Not needed Needed at factory
12.	Colour purity adjustment	Needed	Needed	Needed

7.4. INTRODUCTION TO MAJOR FLAT PANEL DISPLAY TECHNOLOGIES

Flat panel matrix-addressed displays suitable for wall mounting have been a long-cherished dream of TV engineer. Several technologies have been develoveped with only limited success in achieving good resolution, contrast and brightness. Most of these technology are based on electroluminescence (EL) as in LED, gas discharge plasma (PL), liquid crystals (LC) and vacuum fluorescence (VF) using phosphor coating anodes. Some of the technologies are being discussed in further sections.

Various Flat Panel Display Systems

1. LCD
2. Plasma
3. OLED, etc.

7.4.1. LCD Technology

Liquid crystal displays (*LCDs*) are the most common type of flat panel displays (*FPDs*) and has been widely used since the early 1970's. Like popular LCD monitors you commonly see used with computers, LCD TVs have a slim design and a flat viewing surface, but have been fine tuned for video display. Recent advances in flat panel LCD technology now allow for larger screens, wider viewing angles, and higher-quality video images. LCD TVs are also competition for trendy-but-heavy, plasma display technology. They are several times lighter than comparably sized plasmas, and are far more durable.

Working Principle of LCD

Two sheets of polarized transparent material, one with a special polymer coating that holds liquid crystals, are adhered together. Electric current is passed through individual crystals, which interpret the information from the broadcast signal to allow or disallow light through them to create an image. The crystals themselves do not produce light, so the technology is "non-emissive" and therefore does not give off radiation like an older television does. Fluorescent tubes housed behind the transparent material illuminate the image. They require less power to operate than CRT and plasma televisions which require powering hundreds of electrodes to stimulate the phosphorous.

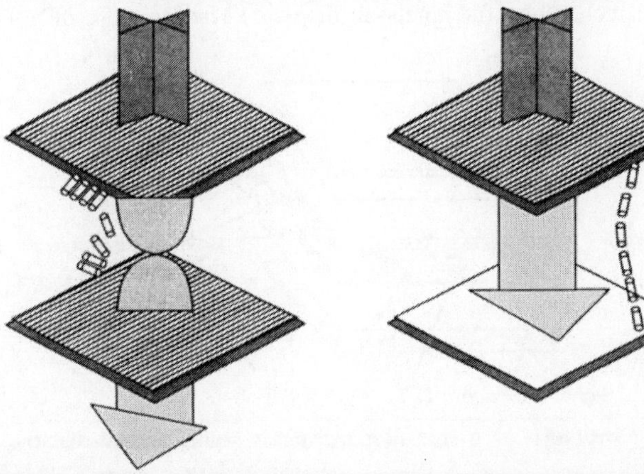

Fig. 7.24. Principle of a passive-matrix twisted nematic liquid crystal display.

All *LCDs* utilize the fact that certain organic molecules (liquid crystals, *LC*) can be reoriented by an electric field. As these materials are optically active, their natural twisted structure can be used to turn the polarization of light by, for example, 90 degrees. Two crossed polarizer's normally do not transmit any light but if a 90°-twisted *LC* is inserted in between, light will be transmitted as shown to the left in Fig. 7.24. On the other hand, applying an electric field will unwind the helical structure and the *LC* therefore loses its polarization-rotating characteristics. As a result, the display turns dark as shown to the right in Fig. 7.24. An *LCD* consists of an array of picture element ("pixels") which can be individually addressed according to the principle below.

Direct-view *LCD* s can largely be categorized into *reflective* and *transmissive* displays which utilize ambient light and light from a fluorescent backlight tube, respectively.

The **twisted nematic (TN)** type of *LC* s shown in Fig. 7.24 are in the passive matrix configuration (*i.e.* simple electrodes for applying the electric field) primarily used in products such as wrist-watches, handheld calculators, pagers, pocket games, and other inexpensive devices requiring low power consumption and a small form factor. Its response speed is; however, insufficiently fast for high resolution and/or high frame-rate displays.

In the late 1980's, this problem was solved by the introduction of **super twisted nematic (STN)** *LC*s which are twisted 270° instead of 90°. With a response speed of around 200 ms, *STN* enabled screen sizes of 8-10 inch with *VGA* resolution (640 × 480 pixels). It is no exaggeration to say that *STN* was the enabling technology for notebook computers.

Despite this progress, though, the response speed is still not high enough for displaying fast moving images such as mouse movements or video. Moreover, increased screen-size and pixel counts negatively affects *STN* parameters such as contrast, grey scale capability, and noise because of a large capacitance and limited conductivity of the electrodes.

To solve this, a sample-and-hold circuit can be attached to each pixel which maintains the voltage during one frame scan. Such a circuit was practically realized by the advent of **thin-film transistors (TFTs).** At first, *TFT*s were extremely expensive to manufacture and the price of a notebook with *STN* and *TFTLCD* s could differ more than Rs 40000 approximately. Today, however, production technology has caught up and *TFTLCDs* are now the mainstream technology for notebook displays. The structure of a *TFTLCD* is shown in Fig. 7.25.

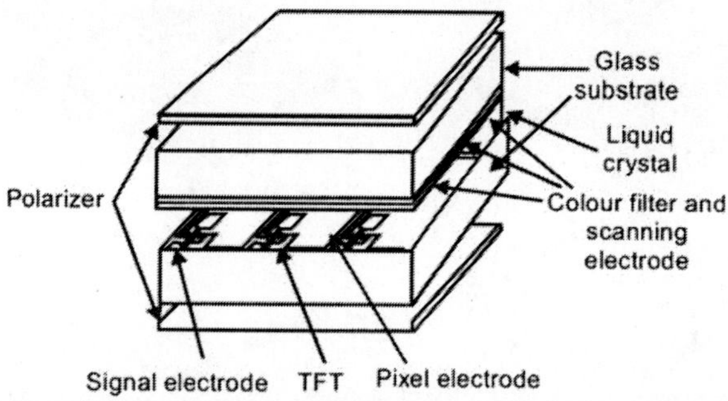

Fig. 7.25. Structure of a thin-film transistor liquid crystal display.

The first *TFT*s were made from cadmium selenide (CdSe) in 1972 but an investment momentum in the solar cell industry convinced the Japanese to move towards amorphous silicon (a-Si) although CdSe both has higher electron mobility and handles higher current densities.

The electron mobility is a crucial parameter increasing the frame rates and/or pixel count. It is also important when downsizing *TFT* devices for fine-pitch displays such as 120 ppi (pixels per inch) or more. Polycrystalline silicon (p-Si), with a higher mobility than a-Si, is expected to play an important role in this context. Indeed, projection displays using tiny *TFTLCD* shutters (typically 2-inch diagonal) often employ p-Si because of a pixel pitch of only a few tens of micrometers. Whereas these devices are grown at high temperatures on expensive quartz substrates, direct-view displays require low-temperature processes which are compatible with conventional glass. Low temperature processing of p-Si is therefore attracting extremely much attention at the moment.

In 1998, Sharp Corp. and the Semiconductor Energy Laboratory announced a new technology called continuous grain silicon (*CGS*) which could potentially revolutionize *TFTLCDs*. With mobility close to crystalline silicon, a 2.6-inch projection display device for high-definition *TV* was successfully demonstrated. In addition to enabling higher resolutions, the *CGS* and p-Si technologies allow driver circuitry - and eventually complete LSIs and CPUs - to be integrated monolithically.

Apart from being employed in notebook computers and, recently, in desktop monitors with diagonal sizes up to 30 inch, *TFTLCD* s are used in reflective displays for mobile terminal applications requiring video speed.

LCD TVs at a Glance

- Use non-emissive technology; do not emit radiation.
- Use fluorescent tubes which require less power to operate.
- Use non-emissive technology; do not emit radiation.

7.4.2. Plasma Display Panels

A **plasma display panel (PDP)** is essentially a matrix of tiny fluorescent tubes which are controlled in a sophisticated fashion. There are two main types, *DC* - and *AC* of which the latter has become main stream because of simpler structure and longer lifetime. This section treats the *AC* -type.

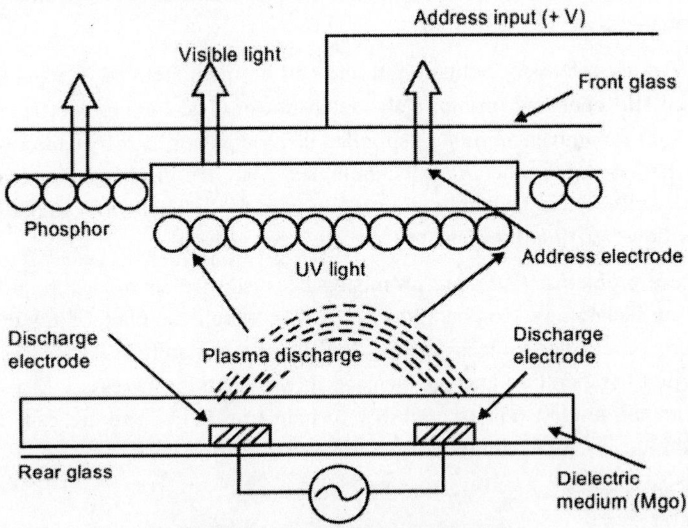

Fig. 7.26. Principle of an AC PDP.

A plasma discharge is first induced by the positive period of an *AC* field (see Fig. 7.26) and a layer of carriers is shortly thereafter formed on top of the dielectric medium. This causes the discharge to stop but is induced again when the voltage changes polarity. In this way, a sustained discharge is achieved. The *AC* voltage is tuned just below the discharge threshold so the process can be switched on/off by adding a relatively low voltage at the address electrode.

The discharge creates plasma of ions and electrons which gain kinetic energy by the electric field. These particles collide at high speed with neon and xenon atoms, which thereby are brought to higher-energy states. After a while, the excited atoms return to their original state and energy is dissipated in the form of ultraviolet radiation. This radiation, in turn, excites the phosphors which glow in red, green, and blue (***RGB***) colors, respectively. Since each discharge cell can be individually addressed, it is possible to switch on and off picture elements (pixels).

To generate color shades, the perceived intensity of each *RGB* color must be controlled independently. While this is done in *CRTs* by modulating the electron beam current, and therefore also the emitted light intensities, *PDP* s accomplish shading by pulse code modulation (*PCM*). Dividing one field into eight sub-fields, each with pulse weighted according to the bits in an 8-bit word, makes it possible to adjust the widths of the addressing pulses in 256 steps. Since the eye is much slower than the *PCM*, it will integrate the intensity over time. Modulating the pulse widths will therefore translate into 256 different intensities of each color. The number of color combinations is therefore $256 \times 256 \times 256 = 16{,}777{,}216$.

At first, *PDPs* had problems with disturbances caused by interference between the *PCM* and fast moving pictures. By fine-tuning the *PCM* scheme, however, this problem has been eliminated.

While *PDP* s are relatively light weight and can be manufactured at a thickness of 3-4 inches, they still consume prohibitively much power. The luminous efficiency, *i.e.* the amount of light for a given amount of supplied electric power, is still at approximately 1 lm/W, about 10% of some other *FPD* technologies. Also, the discharge process causes sputtering of the cells which inevitably reduces life time. With a new protective dielectric layer of MgO, however, this problem has largely been solved.

Despite these problems, *PDPs* are promising because of their modest requirements on manufacturing technology. Compared to *TFTLCDs*, which use photolithographic and high-temperature processes in clean rooms, *PDPs* can be manufactured in less clean factories using low-temperature and inexpensive direct printing processes. Also, *PDP* s feature wide viewing angles, no susceptibility to magnetic fields, and are easy to scale up for wall-hanging *TV* applications.

7.4.3. Comparisons between LCD and Plasma Television

Table 7.2.

Description	Plasma televisions	LCD televisions	Advantage
1. Screen Size	Screen sizes range from 32 inches to 60 inches. Larger plasmas are in production, but are still prohibitively expensive for consumer use.	Sizes range from 13 inches to 45 inches. Larger screens are expected soon, and will likely also be expensive.	Plasma TVs are the current size champions. Larger LCD TVs and plasmas are already in development.
2. Viewing Angle	Up to 160°	Up to 170°	LCD TVs.
3. Screen Refresh Rates	Plasma displays refresh and handle rapid movements in video about as well as CRT televisions.	LCD TVs were originally designed for data display, and not video. Therefore refresh rates had to be improved. LCD TVs with refresh rates of 16 ms or lower show very little noticeable artifacts.	Slight edge to plasma technology.
4. Burn-in or Stuck Pixels	Plasma TVs can suffer from burn-in produced by static images. After extended periods, stationary images "burn in" and produce an after-image ghost which remains permanently on the screen. With technologies such as 'pixel orbiter,' new plasma TVs have addressed burn-in and significantly reduced the issues of older models.	LCD TVs do not suffer from burn-in, but can have a "retained pixel charge" which may also produce ghosting. Stuck pixels are also possible with an LCD display.	With the latest plasma technology this would be even. The "pixel orbiter" is automatic and does not require additional efforts by the end-user.
5. Product Life-span	Typical plasma TVs have a life span of 30,000 to 40,000 hours, which equates to at least three years, three months of 24/7 usage before the TV fades to half the original brightness.	LCD TVs also have replaceable backlights, but the expense of replacing one when the time comes may be greater than simply replacing the entire TV.	LCD TVs lamps will last a bit longer than plasmas, but both are likely to meet general consumer expectations.
6. Weight	Plasma displays are fairly heavy, and may need additional supports to be mounted onto a wall.	LCD TVs weigh less than comparably sized plasma TVs.	LCD TVs are considerably lighter, more portable, and cheaper to ship.

Description	Plasma televisions	LCD televisions	Advantage
7. Durability	Plasmas are very fragile making them tricky to ship and install. Unlike the commercials where plasmas are mounted on the ceiling, plasmas are best installed by a professional, and should be installed on a wall that can bear a good deal of weight.	Much more durable then plasmas. End users can easily mount an LCD TV themselves if desired.	LCD TVs are far less fragile than plasmas.
8. Shipping	Due to their fragile nature, plasma TVs needs to be shipped by specialty carriers. Overnight or fast delivery options are not recommended. Special shipping methods and their heavier weight add to higher shipping costs.	Shipping LCD TVs is not difficult, and is not as expensive as shipping plasma displays.	LCD TVs are lighter and far less fragile than plasma displays making shipping easier and less expensive.
9. Installation	Plasmas are heavier, use more power, and run hotter than LCD TVs, and therefore require more planning when mounting them. Plasmas are generally best installed by professionals.	End users can easily install LCD TVs themselves, or can use them just as they use a traditional TV using a stand.	LCD TVs are much easier to install than plasma TVs.
10. Brightness	Plasma TVs range from 800-1000 cd/m^2, but are measured based on a different standard than LCD TVs. When compared under "real world" circumstances, plasma TVs brightness is typically closer to 100 cd/m^2.	Measured under the more stringent "real world" standards, LCD TVs average a brightness rating of 500-600 cd/m^2.	Both LCD and plasma TVs will meet the brightness expectations of most consumers. However, in 'real world' situations with ambient light, LCD TVs will generally look a little brighter.
11. Thickness	As thin as 3 inches deep.	As thin as 2 inches deep.	LCDs TVs are just a bit thinner.
12. Contrast Ratios	Current plasmas measure contrast ratios of up to 3000:1. However, when compared to LCD TVs in "real world" situations, contrast ratios for plasma TVs drop to approximately 200:1.	LCD TV contrast ratios are measured using "real world" standards. Typical contrast ratios range from 350-450:1.	LCD TVs contrast ratios measured in real world situations double typical plasma TVs.

7.4.4. Organic Light Emitting Diode (OLED)

Introduction

An **OLED** is a thin film light emitting diode in which the emissive layer is an organic compound. It promises to reduce thickness drastically. It can be used at varied places very effectively. It will allow 40"+ size televisions to be made very thin that can be rolled up and pushed away for storage.

An OLED consists of following parts shown in Fig. 7.27.

1. Substrate
2. Anode
3. Organic Layers
4. Cathode

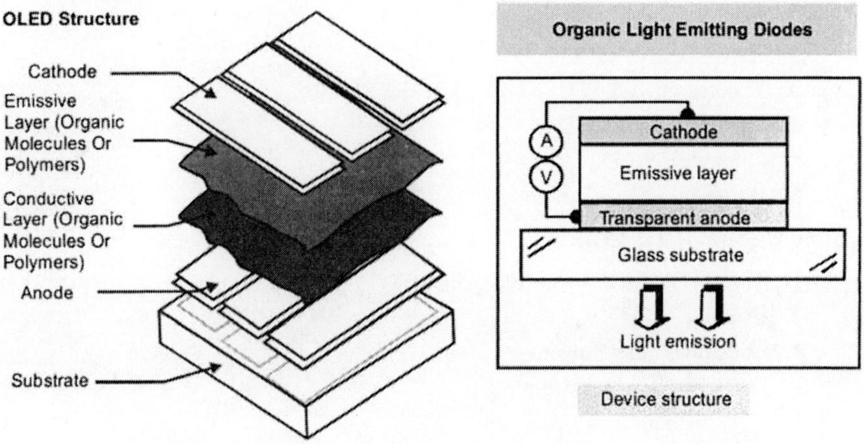

(*a*) (*b*)

Fig. 7.27. Structure of OLED.

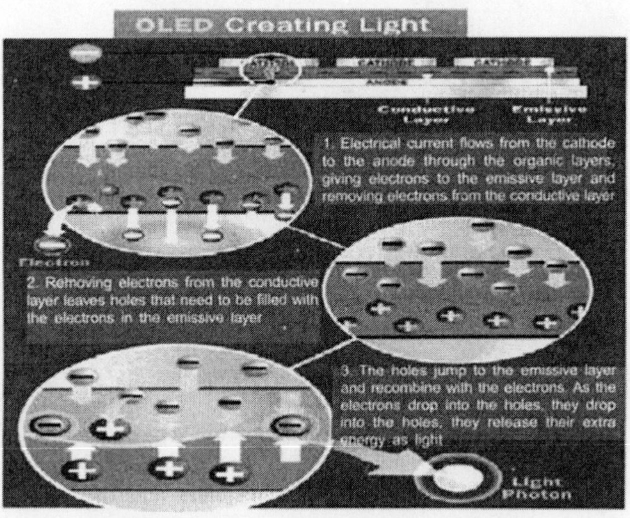

Fig. 7.28. The working principle of OLED.

How it works...?

To understand the working principle of OLED refers Fig. 7.28. In Fig. 7.28, three steps are illustrated which are as follows:

1. Electrical current flows from the cathode to the anode through the organic layers, giving electrons to the emissive layer and removing electrons from the conductive layer.
2. Removing electrons from the conductive layer leaves holes that need to be filled with the electrons in the emissive layer.
3. The holes jump to the emissive layer and recombine with the electrons. As the electrons drop into the holes, they release their extra energy as light.

Types of OLEDs

There are several types of OLEDs:

1. Passive-matrix OLED.
2. Active-matrix OLED.
3. Transparent OLED.
4. Top-emitting OLED.
5. Foldable OLED.
6. White OLED.

Advantages of OLEDs

1. Thinner, lighter & more flexible.
2. Brighter.
3. Consumes less power.
4. Don't require backlighting.
5. Easy to manufacture.
6. Have large fields of view.

Disadvantages of OLEDs

1. Problems regarding its lifetime.
2. Manufacturing cost is quite high at present.
3. Water can damage it very easily.

Let's compare it with latest spearhead technologies like LCDs :

1. Increased brightness.
2. Wide viewing angle, up to 160 degrees
3. Lighter weight.
4. Require very less power, only 2-10 volts.
5. Greater durability.
6. Broader operating temperature ranges.

SUMMARY

- A monochrome picture tube has one electron gun and a continuous phosphor coating that produces a picture in black & white. For coloured picture tube the screen is formed of three different phosphors and there are three electrons beams, one for each colour phosphor. Three colours RED, GREEN & BLUE produced by three phosphors combine to produce different colours.

- CRT is a device that can produce an image on a screen with electrical impulses.
- The entire front area of the tube is scanned in a fixed pattern called a *raster.*
- All monochrome CRT's used as picture tube in television system, have following main elements:
  1. Electron Gun (consists of Cathode, Control grid)
  2. Accelerating or screen grid
  3. Focusing grid
  4. Deflection coils
  5. Final Anode (Aquadag coating)
  6. Phosphorus coating over screen
  7. Screen
- The electron gun provides an electron beam, which is a highly concentrated stream of electrons.
- This special grid is in the form of a solid metal cap with a small hole in the centre. The grid is placed over the emitting surface of the cathode and charged negatively in relation to the cathode, referred as control grid.
- Since the brightness or intensity of the display depends on the number of electrons that strike the screen, *the control grid is used to control the brightness of the CRT.*
- The proper name, *brightness control*, is given to the potentiometer used to vary the potential applied to the control grid.
- Electrostatic focusing is achieved by the combined efforts of accelerating anode and focusing anode, so that electrons get focused over the screen. Electrostatic focusing is preferred over magnetic focusing because it is not much affected by change in the line voltage and need no ion-spot correction.
- THE CRT SCREEN The inside of the large end of a CRT is coated with a fluorescent material that gives off light when struck by electrons.
- The material used to convert the electrons' energy into visible light is a phosphor.
- A special coating called an aquadag coating is applied to the inside of the tube usually extends from almost halfway into the narrow neck to within 3cm of the fluorescent screen.
- This allows the aquadag to perform two functions.
  1. First, since the aquadag coating is positive, it attracts the secondary emitted electrons and removes them.
  2. Second, because the aquadag is operated at a high-positive potential and is mounted in front of the accelerating anode, it aids in the acceleration of electrons toward the screen.
- There are two ways to move an electron (and thus an electron beam): either with a magnetic or with an electrostatic field.
- In TV picture tube Electromagnetic deflection is preferred for the following reasons:
  (*i*) Due to shortage of deflecting time, it is necessary to have high deflecting field to achieve the desired deflection. With magnetic deflection it is large

that would be necessary to achieve the same deflection. It is easy to generate large current than large voltages.

(*ii*) With *electrostatic deflection* the electrons beam gain energy. Thus large deflection angle tend to defocus the beam.

(*iii*) For ES deflection, two delicate pairs of deflecting plates are needed inside the picture tube, whereas for magnetic deflection two pairs of deflecting coils are mounted outside and close to the neck of the tube. In combination, the vertical and horizontal deflection coils are called deflection yoke and it is fitted outside and close to the neck of the tube. The tube neck is made narrow so that the deflecting coils.

(*iv*) If the electrostatic voltage between the deflection plates is high, that causes sparking between the plates. To generate uniform or linear magnetic field for linear deflection, a special type of winding is used known as cosine winding.

- A deflection angle is the maximum angle through which the beam can be deflected without striking the side of the bulb. The advantage of large deflection angle is that for equal picture size the picture tube is shorter and can be installed in a smaller cabinet.
- There are some shortcomings in CRT:
  1. It is estimated that about 50% of the light emitted at the screen where the electron beam strikes it, travels back in the tube. Another 20% or so is lost in the glass of the tube because of internal reflection and about 20% of it reaches to the viewer.
  2. Any ions in the beam, which do exist despite best precautions while degaussing, damage the phosphor material on hitting it and thus cause a dark brownish patch on the screen.
- Spark-gap protections **are used to protect the receiver circuit due to any arcing.**
- A grounded coating of aquadag material is used at the outer surface of the picture tube.
- The transfer characteristics plotted as light output against electrical signals on log-log scale is not linear but follows a square law.
- The process of varying the beam current to control instantaneous brightness is known as z-axis modulation.
- In colour picture tubes, voltages pertaining to three primary colours give rise to three electron beams. These three beams activate three phosphor elements (one each), closely placed in a group, either in the form of dots or in the form of stripes.
- **Types of Colour Picture Tubes**
  1. Delta Gun colour picture tube
  2. Precision- In- Line (PIL) picture tubes
  3. Trinitron
- Delta Gun picture tubes employs three separate Guns, equally spaced at 120 degree with respect to each other, generates three high velocity (25 kv acceleration) electrons beams. Each gun in the three gun delta picture tube has its own separate set of cathode, control grid, accelerating grid and focus grid

so that individual colour signals can be coupled to the grid cathode circuit to modulate beam intensity.

- A thin perforated metal sheet is placed about one cm behind the tube screen, to ensure the correct landing of beam on phosphor, is known as Shadow mask.
- Phosphor used for colour tubes are;

  Red : Zinc sulphide

  Green : Zinc silicate

  Blue : Europium and yttrium

  Persistency : 1 ms
- All the three colour difference signals are then fed to the three grids of the colour picture tubes. The inverted luminance signal (-Y) is applied at the junction of the three cathodes.
- For obtaining purity in colour each beam should land at the center of the corresponding phosphor dot irrespective of location of the beam on the raster. This needs precise alignment of the colour beams and is carried out by a circular magnet assembly known as the purity magnet.
- The technique of bringing the beams together so that they hit the same part of the screen at the same time to produce three coincident raster is referred as convergence. Convergence errors are caused by:
  1. Non-coincident convergence plane
  2. Non-uniformity of the deflection field.
  3. Flat surface of the picture screen.
- *Static convergence* : It involves movement of the beam by permanent magnetic fields, one correctly set; bring the beam in to convergence in the central area of the screen.
- Dynamic convergence over the rest of the screen is achieved by continuously varying (dynamic) magnetic field.
- Dynamic pin cushion correction is used with colour picture tubes. Such a correction automatically increases horizontal width and vertical size in those regions of the raster that are shrunken because of pin cushion distortion.
- Degaussing means demagnetizing iron and steel parts of the picture tube mounting.
- **Drawbacks of Delta Gun Tubes**
  1. Focus is not sharp all over the screen.
  2. Electrons transparency of the shadow mask is very low *i.e.* 20%.
  3. Convergence of the three beams on the respective dots is difficult.
- In this the three guns are aligned precisely in a horizontal line. The in line gun configuration helps in simplifying convergence adjustments.
- **Advantages :**
  1. The PIL tubes are very efficient that mean higher electron transparency (50%) and needs fewer convergence adjustments.
  2. No needs of vertical convergence, only horizontal convergence is required at the stage of manufacturing.
  3. Focus is sharp all over the screen.

- Non optimum distribution of field produces the convergence error as the screen is nearly flat surfaces. There are two kinds of colour errors are:
  1. Astigmatism effect
  2. Comma effect
- With the uniform field, the rays from a vertical row of points (AGB) converge to produce a vertical focal line on the screen but the rays from a horizontal point (RGB) converge beyond the screen. This effect is referred as *astigmatism* and results convergence error.
- Due to the non-uniformity of the deflection field, all the beams are not deflected by the same amount. Such a distortion is known as *Comma effect* and results in misconvergence of the beams.
- Trinitron picture tube is developed by Sony in 1968. It is three in line cathode colour picture tube. It employs single gun having three in line cathodes for each of the red, green and blue beams. This simplifies the construction problem because one electron gun can easily be accommodated.
- The shadow mask has long slots, one slot per vertical line, increasing transparency to about 80%.
- **Advantages :**
  - Greater transparency about 80%.
  - Sharper focus and brighter picture.
  - No problem of convergence adjustment during servicing.
  - Improved vertical resolution.
- Most of these technology are based on electroluminescence (EL) as in LED, gas discharge plasma (PL), liquid crystals (LC) and vacuum fluorescence (VF) using phosphor coating anodes. Various Flat Panel Display Systems
  1. LCD
  2. Plasma
  3. OLED, etc.
- Liquid crystal displays (*LCDs*) are the most common type of flat panel displays (*FPDs*) and has been widely used since the early 1970's. LCD TVs have a slim design and a flat viewing surface, but have been fine tuned for video display.

REVIEW QUESTIONS

1. Draw the diagram of monochrome picture tube. Explain the construction and working of monochrome picture tube.
2. Explain the focussing action achieved in monochrome picture tube.
3. What is the function of an aquadag coating on the inner side of the tube ? A grounded coating is provided on the outer surface of the picture, why ?
4. Why magnetic deflection is preferred over electrostatic deflections ?
5. Why is the neck of picture tube made narrow ?
6. Discuss the relative merits of magnetic deflection over electrostatic deflections.
7. What do you understand by convergence ? Describe briefly now static convergence and dynamic convergence are adjusted.
8. Define the term deflection angle of a picture tube. Give the advantages of providing a large deflection angle yoke.

9. Define the construction and working of a delta gun picture tube. Write the disadvantges of delta gun picture tube.
10. Explain the working of precision-in-line picture tube. How it is different from a delta gun colour picture tube ?
11. Describe trintron picture tube with the neat diagram and write the advantages of this.
12. What do you meant by colour purity ? Explain the procedure of colour purity adjustment.
13. What do you understand by pincushion effect ?
14. What do you meant by degaussing ? Why is it necessary ?
15. What is the function of shadow mask ?
16. What are the holes in shadow mask of a deltagun picture tube conical ?
17. Explain the working principle of LCD. Write the main features of LCD.
18. Explain the working principle of plasma picture display system. Compare the LCD and Plasma display system.
19. Describe the working principle of OLED. List the types of OLED and their application.
20. Compare the LCD, OLED and plasma display system.

MULTIPLE CHOICE QUESTIONS

1. Aquadag coating form a
   (*a*) Capacitor (*b*) Resistor
   (*c*) Inductor (*d*) Transformer
2. A TV set specified as 50 cm size has the screen whose width is
   (*a*) 30 cm (*b*) 40 cm
   (*c*) 50 cm (*d*) 60 cm
3. Greater the screen size
   (*a*) Larger the voltage required for picture tube anode
   (*b*) Lesser the voltage required for picture tube anode
   (*c*) Voltage required is independent of screen size
   (*d*) None of these
4. Phosphors glow if electron hitting them are of
   (*a*) High velocity (*b*) Low velocity
   (*c*) Zero velocity (*d*) None of these
5. The number of guns in a colour picture tube are
   (*a*) 2 (*b*) 4
   (*c*) 3 (*d*) 5
6. Which colour tube has the maximum transparency for the beam to reach the screen ?
   (*a*) Delta gun (*b*) PIL
   (*c*) Trintron (*d*) None of these
7. Pincushion effect in a colour picture tube can be corrected by
   (*a*) Magnetic ring (*b*) Current through deflection coils
   (*c*) Colour purity adjustments (*d*) Currents through convergence coils
8. A trintron colour picture tube employs
   (*a*) Three-in-line cathode beam
   (*b*) A common electron gun

(*c*) Three separate electron gun
(*d*) (*a*) and (*b*)

9. Slotted aperture shadow mask is used in
(*a*) Trinton tube (*b*) PIL tube
(*c*) Delta gun (*d*) (*a*) and (*b*)

10. The telavion picture tube has
(*a*) Magnetic focus and electrostatic deflection
(*b*) Magnetic focus and magnetic deflection
(*c*) Electrostatic focus and electrostatic deflection
(*d*) Electrostatic focus and magnetic deflection

11. In picture tube the phosphor used has a persistence of
(*a*) 5 ms (*b*) 50 μs
(*c*) 500 ms (*d*) 100 ms

12. OLED stands for
(*a*) Organic light emitting diode
(*b*) Organge light emitting diode
(*c*) Organic light energy displacement
(*d*) Organic level energy diode

13. LCD displays more commonly employ
(*a*) Nematic crystal
(*b*) Sematic crystals
(*c*) Cholesteric crystal
(*d*) Twisted nematic crystals.

ANSWERS

1. (*a*)	2. (*b*)	3. (*a*)	4. (*a*)	5. (*c*)
6. (c)	7. (*b*)	8. (*a*)	9. (*d*)	10. (*d*)
11. (*a*)	12. (*a*)	13. (*d*)		

8 CHAPTER

TELEVISION APPLICATIONS

INSIDE THIS CHAPTER

8.1. INTRODUCTION

The TV broadcasting means transmission of electromagnetic waves in all the direction using a transmitting antenna from the transmitting station. This type of system essentially needs an antenna at the receiver. In earlier days the television transmission was limited to the VHF band only but nowadays many numbers of channels are allocated from the UHF band. The distance of transmission is restricted to line of sight between the transmitting and receiving antennas. For VHF channels it is 120 km and for UHF channels is about 90 km. There are many applications of television in the fields of science, industry, education, entertainment etc.

8.2. CABLE TELEVISION

A television distribution system in which station signals, picked up by elevated antennas, are delivered by cable to the receivers of subscribers. It is called **cable TV** or community *antenna television*. The transmission of TV programs into the home and office via coaxial cable. Cable TV organizations have tremendous potential for new services since they are already wired into so many homes.

Cable television means the transmission of televised images to viewers by means of coaxial cables. Cable systems receive the television signal, which is sent out over cables to individual subscribers, by a common antenna (CATV) or satellite dish. Early cable systems developed in the late 1940s to improve reception of commercial programming in rural areas. In the 1960s, cable systems expanded to large urban areas, where reception can also be poor, and the cable television industry began introducing its own networks, such as Home Box Office (HBO), founded in 1972, to provide programming exclusively to subscribers. Beginning in 1975, cable networks began distributing their shows to local cable operators via satellite, thus increasing the amount of programming available nationally.

New technologies, such as fiber optics, digital compression, and interactive television, allow cable operators to offer more programming choices and services. The cable lines installed by cable operators are also to use to provide broadband Internet access to the homes of subscribers.

Technically, both cable TV and CATV involve distributing a number of television channels collected at a central location (called a head end) to subscribers within a community by means of a network of optical fibers and/or coaxial cables and broadband amplifiers.

As in the case of radio broadcasting, the use of different frequencies allows many channels to be distributed through the same cable, without separate wires for each. The tuner of the TV, VCR or radio selects one channel from this mixed signal.

Traditional cable TV systems worked strictly by way of analog system (*i.e.* using standard radio waves) but many modern cable TV systems also employ the use of digital cable technology, which uses compressed digital signals, allowing them provide many more channels than they could with analog alone.

Basic Elements of Typical Cable TV System

Figure 8.1 shows a block diagram of typical cable TV system. Major elements of cable TV system are namely:

1. Head end or combining network
2. Trunk line cables
3. Branch line cable
4. Drop line
5. Amplifiers
6. Directional coupler

Brief descriptions of these blocks are as follows:

1. Head end or combining network : Head end unit is to receive the signals from various VHF/UHF antennas, satellites, and local studio and process them. They are amplified by a low noise amplifier (LNA) and adjusted for correct level. A frequency translator is used to convert UHF frequencies to VHF frequencies, so that transmission loss through the cable is minimised because at UHF frequencies losses are more comparatively VHF frequencies. The signal from the head-end then goes on to the trunk lines.

2. Trunk line : Generally trunk line is a coaxial cable which is an efficient wideband line has the advantage of shielding. A good coaxial cable for trunk line is about 19 mm in diameter. In future optical fibre cable may be used with reduced cost because of advantages of very high capacity, low loss and no electromagnetic interference.

3. Branch line : A coaxial cable of about 10 mm in diameter is used because of short length.

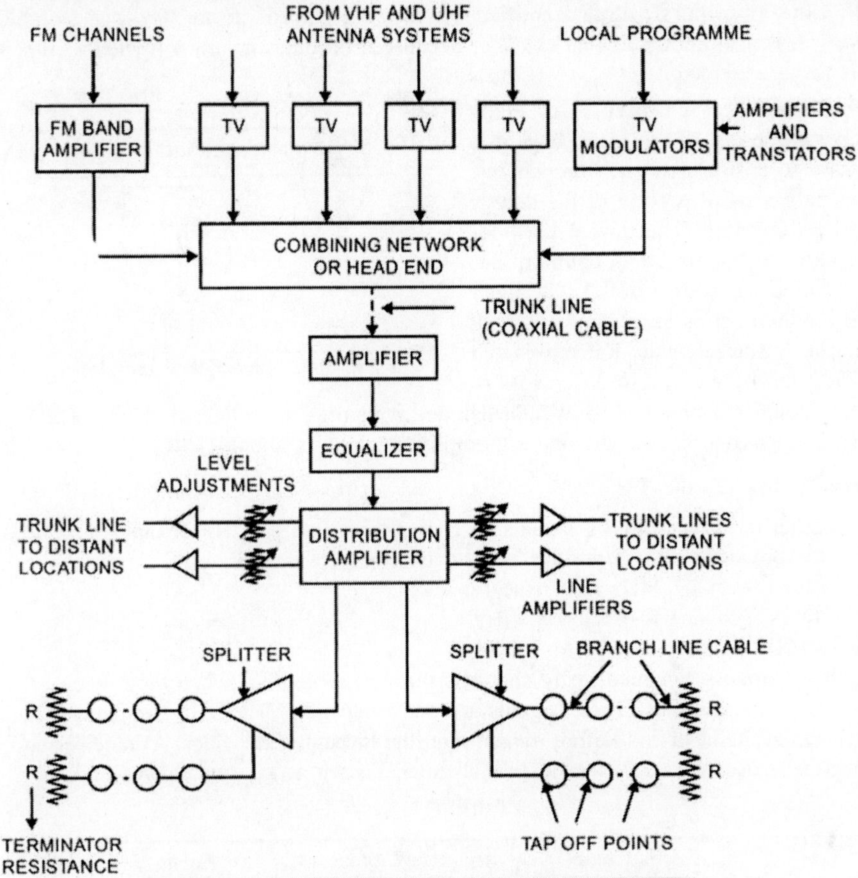

Fig. 8.1. Block diagram of Cable system Network.

4. Drop line : The line running from the branch to the subscriber's is known as Drop line. It is normally a coaxial cable of 6 mm diameter which is covered with the weatherproof polythene jacket. Cable network requires having good impedance matching so that a good strength and quality of signal can be achieved. Usually the cables used have the characteristics impedance of 75 Ω.

5. Amplifiers : Amplifiers are required at several points along the line to progressive signal attenuation which occurs due to various losses such as DC loss (I^2R), skin effect loss and dielectric loss. These losses increase in proportion to the square root of the frequency. At high VHF, these losses may be doubled in comparison to the losses low VHF. Thus the trunk amplifiers with a gain 20 to 30 dB are placed at suitable interval along the route. Amplifiers keep the signal level between 1 and 3 mV.

6. Equaliser : An equaliser consists of a bandpass filter arrangement with an adjustable frequency response. The purpose of this is that the outputs from the splitters should have uniform relative amplitude response across the entire VHF band.

7. Distribution amplifier : This amplifier take care of the impedance mismatch caused by the connection with the trunk line and compensates the loss in the trunk line up to the point connection, gain of this amplifier is 20 to 30 dB.

8. Line amplifiers : Line amplifier provides the gain of 20 to 30 dB at suitable intervals. It is also known as line extender because it becomes necessary when a branch line is to be extended.

9. Directional coupler : As many branches come out of the trunk line, it is necessary to ensure that each branch line extract only a small portion of the energy from the trunk line. The device used for that is known as direction coupler, has three terminals as shown in 8.2. The inner conductor loop act as capacitance as well as inductance and internally it is terminated in a 75 Ω resistance. The insertion loss of direction coupler is about 1 dB at 300MHz and it ignores the reflected energy. The tap loss is compensated by the distributing amplifier which is about 13dB.

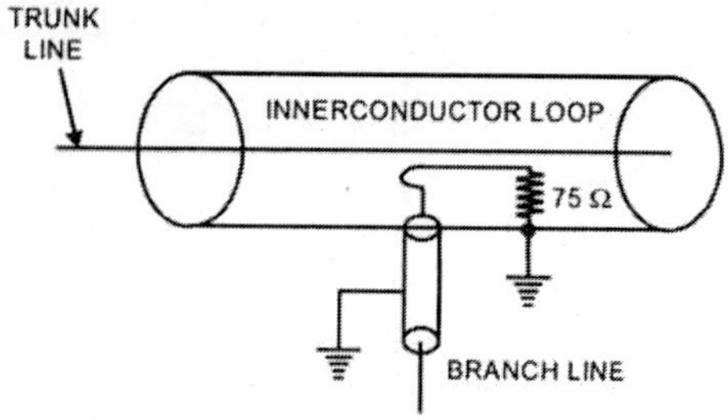

Fig. 8.2. **Directional coupler.**

Channels for Cable TV

In earlier days of cable TV signals on cables were sent on VHF channels and used for the TV broadcasting. Earlier we had eleven channels in VHF band.

Ist channel—41-47 MHz (not used for TV)

IInd to IVth channels—47 to 68 MHz

Vth to XIIth channel— 174 to 230 MHz

Cable operators can send many channels through cable lines when they receive the signals from the satellites. These signals are converted into VHF/UHF frequencies and multiplexed at the head end before transmitting the through trunk lines. At receiver end, a converter is used to select the desired channel, known as **set-top converter.**

Table 8.1.

Band	Channel name	Frequency range	Purpose
VHF band I	1st channel 2nd - 4th channel	41–47 MHz 47–68 MHz	Not used for TV Used for TV broadcast
VHF band II	X channel Y channel Z channel 2 + 1 channel 2 + 2 channel (S_1 to S_{10}) 10 channels	68–75 MHz 75–82 MHz 82–89 MHz 89–96 MHz 96–103 MHz 104–174 MHz	Not used for TV broadcast but used in defence, police and FM broadcast for ATV network
VHF band III	E_5 to E_{12} 8 channels	174–230 MHz	Allows for TV broadcast.
Superband	S_{11} - S_{12} channels 10 channels	230–300 MHz	Used for CATV
Hyperband	S_{21} - S_{41} 21 channels	302–470 MHz	Cable TV
UHF band IV	E_{13} - E_{28} 16 channels	470–502 MHz	Used for cable TV
VHF band V	S_{47} - S_{71} 33 channels	606–870 MHz	Allocated for TV broadcast.

Cable Based Services

1. The CATV service provider can receive signals from satellites and make a large number of channels available to the viewers through cables.
2. Premium channels as Pay- TV channels, which are exclusive to those who can pay additional fees.
3. Cable operators can generate their own TV program and send it on cable.
4. Internet service through cable TV also provided recently.
5. Coaxial cables are capable of bi-directional carriage of signals as well as the transmission of large amounts of data. Cable television signals use only a portion the bandwidth available over coaxial lines. This leaves plenty of space available for other digital services such as broadband internet and digital telephony.
6. *Broadband internet* is achieved over coaxial cable by using cable modems to convert the network data into a type of digital signal that can be transferred over coaxial cable. One problem with some cable systems is the older amplifiers placed along the cable routes are unidirectional thus in order to allow for uploading of data the customer would need to use an analog modem to provide for the upstream connection. This limited the upstream speed to 56k and prevented the always-on convenience broadband internet typically provides. Many large cable systems have upgraded or are upgrading their equipment to allow for bi-directional signals, thus allowing for greater upload speed and always-on convenience, though these upgrades are expensive.
7. Another service being added to many cable systems is *digital telephone service*. Digital telephony is accomplished by installing a special telephone interface at the customer's premises that coverts the analog signals from the customer's in-home wiring into a digital signal, which is then sent on the local loop (replacing the analog last mile, or POTS) to the company's switching center, where it is connected to the PSTN. The biggest obstacle to digital telephone service is the need for nearly 100% reliable service for emergency calls. One of the standards available for digital cable telephony, Packet Cable, seems to be the most promising and able to work with the Quality of Service demands of traditional analog POTS service. The biggest advantage to digital cable telephone service is similar to the advantage of digital cable TV, namely that data can be compressed, resulting in much less bandwidth used than a dedicated analog circuit-switched service. Other advantages include better voice quality and perhaps future integration to a VoIP network providing cheap or unlimited nationwide and international calling. Note that in most cases, digital cable telephone service is separate from cable modem service being offered by many cable companies and does not rely on IP traffic or the Internet.

Cable television is facing increasing competition from satellite television.

8.3. CLOSED CIRCUIT TELEVISION (CCTV)

Closed Circuit Television (CCTV) is a system in which video signal obtained by one or more cameras tubes is sent to one or more monitors through coaxial cables.

CCTV as the name implies is a TV system in which the circuit is closed and all the elements are directly connected. CCTV is the use of video cameras to transmit a signal to a specific place, limited set of monitors. It differs from broadcast television in that the signal is not openly transmitted, though it may employ point to point wireless links. It utilizes CCD video cameras (to produce the video), cable or wireless transmitters/ receivers or Internet (to transmit the video), and monitors (to see the video).

The type of link depends on distance between the two locations, numbers of receivers and mobility of the camera or the receiver. Various types of links used in CCTV are shown in Fig. 8.3.

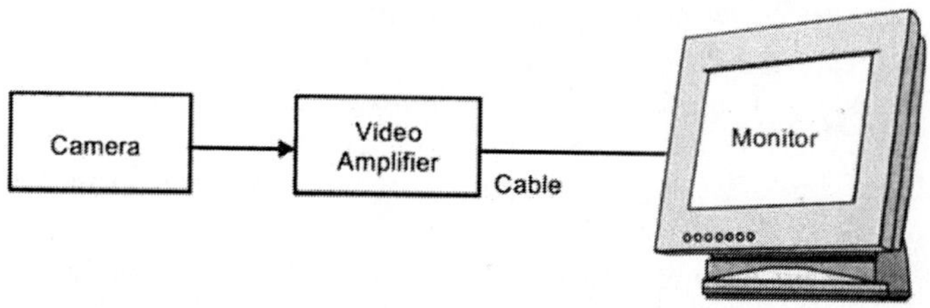

Fig. 8.3(a). Direct camera link to one monitor.

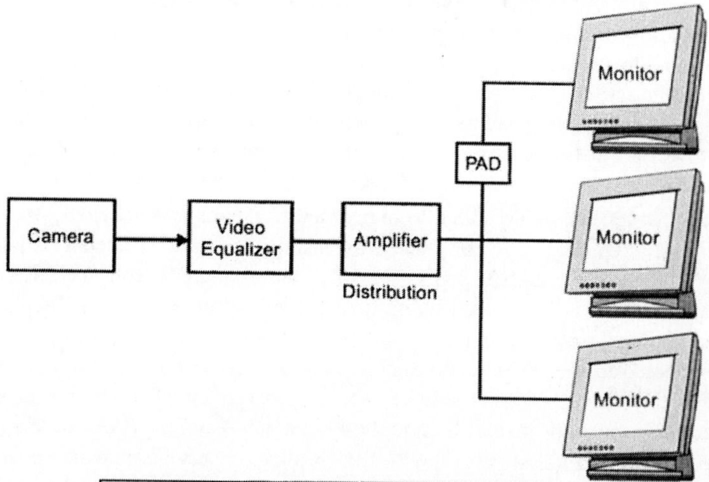

Fig. 8.3(b). Direct link to several monitors.

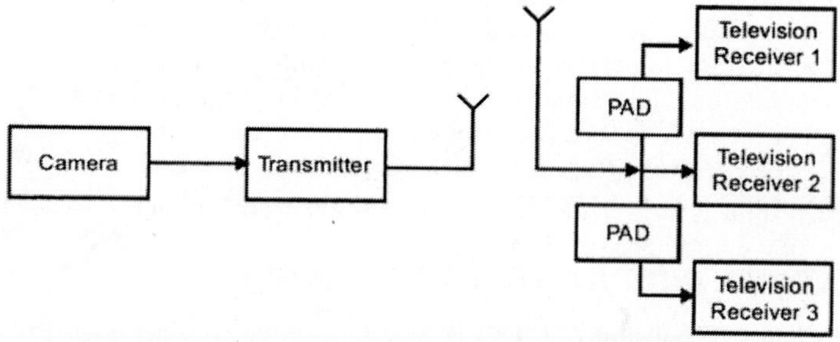

Fig. 8.3(c). Wireless link to several receivers.

Block Diagram of CCTV

Figure 8.4 shows the simplified block diagram of CCTV. Essential elements of CCTV are as follows:

1. Camera tube
2. Scanning circuits
3. Blanking and sync pulses
4. Video amplifier (each for transmitter & receiver)
5. Coaxial cable
6. Video monitor

1. **CCTV cameras** can produce images or recordings for surveillance purposes, and can be either video cameras, or digital stills cameras. CCTV camera tube is used to convert optical signal into video signal. It consists of scanning circuits to deflect the electron beams horizontally and vertically, synchronized with blanking and sync pulse generators. Usually vidicon type camera is used in CCTV.

   The first CCTV cameras used in public spaces were crude, conspicuous, low definition black and white systems without the ability to zoom or pan. Modern CCTV cameras use small high definition colour cameras that can not only focus to resolve minute detail, but by linking the control of the cameras to a computer, objects can be tracked semi-automatically. For example, they can track movement across a scene where there should be no movement, or they can lock onto a single object in a busy environment and follow it. Being computerized, this tracking process can also work between cameras.
2. **Scanning circuits** are required so that the electron beam scan the complete raster of the camera tube and video monitor.
3. **Blanking and sync pulses :** Blanking pulses are required to blank during retrace period of horizontal and vertical scanning process in the transmitter and receiver. Sync pulses are required for the correct synchronization between transmitter and receiver. Sync pulses are transmitted for the horizontal and vertical scanning process along with the video signal.
4. **Video amplifier (transmitter end):** At the transmitter end the output of the camera tube is amplified by using the video amplifier along with the blanking and sync pulses, known as composite video signal. Since the high frequency components of video signal are attenuated more in the coaxial cable than low frequency components, Pre-emphasis of high frequency is needed before the transmission.
5. **Coaxial cable :** Normally coaxial cable is used to carry the signals from transmitter to the monitoring room. The characteristics impedance of coaxial cable is 75 Ω.
6. **Video amplifier (Receiver end) :** The composite video signal get attenuated when they travels from transmitter to monitoring room through the coaxial cable. The voltage level of the signal drops below the required level of the monitor. The input impedance of the amplifier matches with the impedance of the cable. In case of more than one monitor a distribution amplifier is used which feeds signal to individual monitor through matching pad.

7. **Video monitor :** A video monitor also called a broadcast monitor, broadcast reference monitor or just reference monitor, is a device similar to a television, used to monitor the output of a video-generating device, such as a media play out server, IRD, video camera, VCR, or DVD player. It may or may not have audio monitoring capability. Unlike a television, a video monitor has no tuner and, as such, is unable to independently tune into an over-the-air broadcast. One common use of video monitors is in Television stations and outside broadcast vehicles, where broadcast engineers use them for confidence checking of signals throughout the system. Video monitors are used extensively in the security industry with Closed-circuit television cameras and recording devices.

Common display types for video monitors

- Cathode ray tube
- Liquid crystal display
- Plasma display

A video monitor is a TV receiver without RF, IF and detector stages. Video monitors consist of video amplifiers, deflection stages and picture tube. The scene at which the camera tube was focused is displayed on the picture tube. In case of many cameras are installed but monitoring is done by a single person, than the facility of switching of the video cameras is used to monitor each camera video signal. Normally CCTV is used for the video display so there is no sound processing.

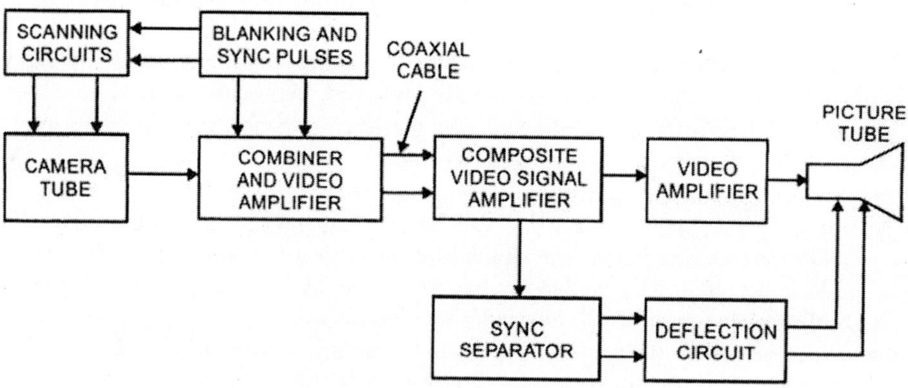

Fig. 8.4. Block diagram of CCTV network.

A more advanced form of CCTV, utilizing Digital Video Recorders (DVRs), provides recording for possibly many years, with a variety of quality and performance options and extra features (such as motion-detection and email alerts). In industrial plants, CCTV equipment may be used to observe parts of a process from a central control room; when, for example, the environment is not comfortable for humans. CCTV systems may operate continuously or only as required to monitor a particular event.

CCTV is commonly used for the security and surveillance purpose and it improves security, operations and risk management of any business. CCTV is often used for surveillance in areas that may need monitoring such as banks, casinos, airports, military installations and convenience stores.

Applications of CCTV

1. Surveillance
2. Education
3. Medical care
4. Industry
5. Safety
6. Traffic control
7. Crowd control
8. Aerospace and oceanography

8.4. TV VIA SATELLITE

Satellite television is television delivered by the means of communications satellites, as compared to conventional terrestrial television - which is broadcast over radio waves originating from several independent, ground-based television stations - and cable television - which is distributed by underground coaxial cables. In many areas of the world, satellite television services supplement older terrestrial signals, providing a wider range of channels and services, including subscription-only services.

Satellite television is television delivered by way of orbiting communications satellites located 37,000 km (22,300 miles) above the earth's surface. The first satellite television signal was relayed from Europe to the Telstar satellite over North America in 1962. The first domestic North American satellite to carry television was Canada's Anik 1, which was launched in 1973.

Satellites used for television signals are generally in either naturally highly elliptical (with inclination of +/-63.4 degrees and orbital period of about 12 hours) or geostationary orbit 37,000 km (22,300 miles) above the earth's equator.

The processing of the received signal from the satellite to the picture tube and loudspeaker is illustrated in the Fig. 8.5.

Satellite television, like other communications relayed by satellite, starts with a transmitting antenna located at an uplink facility. Uplink satellite dishes are very large, as much as 9 to 12 meters (30 to 40 feet) in diameter. The increased diameter results in more accurate aiming and increased signal strength at the satellite. The uplink dish is pointed toward a specific satellite and the uplinked signals are transmitted within a specific frequency range, so as to be received by one of the transponders tuned to that frequency range aboard that satellite. The transponder 'retransmits' the signals back to Earth but at a different frequency band (a process known as translation, used to avoid interference with the uplink signal), typically in the C-band (4–8 GHz) or K_u-band (12–18 GHz) or both. The leg of the signal path from the satellite to the receiving Earth station is called the downlink.

A typical satellite has up to 32 transponders for Ku-band and up to 24 for a C-band only satellite, or more for hybrid satellites. Typical transponders each have a bandwidth between 27 MHz and 50 MHz. Each geo-stationary C-band satellite needs to be spaced 2 degrees from the next satellite (to avoid interference). For K_u the spacing can be 1 degree. This means that there is an upper limit of 360/2 = 180 geostationary C-band satellites and 360/1 = 360 geostationary K_u-band satellites. C-band transmission is susceptible to terrestrial interference while K_u-band transmission is affected by rain (as water is an excellent absorber of microwaves at this particular frequency).

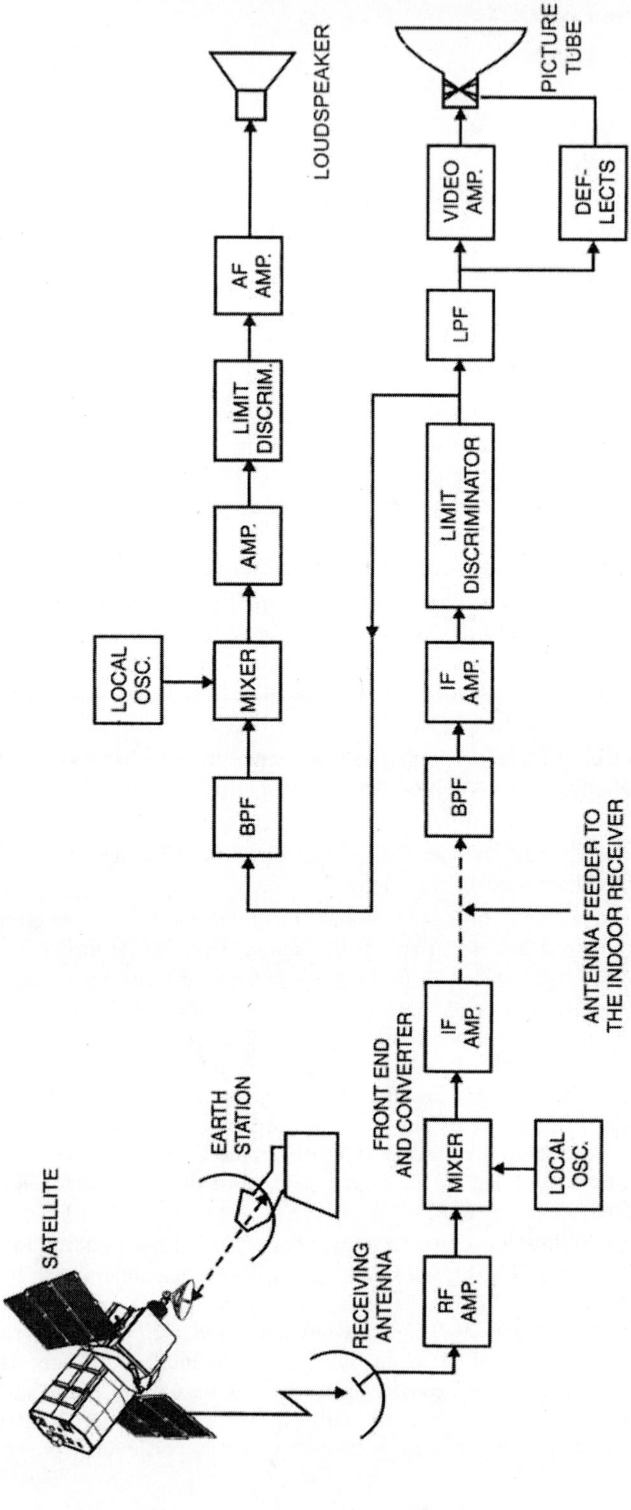

Fig. 8.5. A simple block diagram of satellite television.

The down linked satellite signal, quite weak after travelling the great distance, is collected by a parabolic receiving dish, which reflects the weak signal to the dish's focal point. Mounted on brackets at the dish's focal point is a device referred as feed horn. This feed horn is essentially the flared front-end of a section of waveguide that gathers the signals at or near the focal point and 'conducts' them to a probe or pickup connected to a low-noise block down converter or LNB. The LNB amplifies the relatively weak signals, filters the block of frequencies in which the satellite TV signals are transmitted, and converts the block of frequencies to a lower frequency range in the L-band range. The evolution of LNBs was one of necessity and invention.

The original C-Band satellite TV systems used a Low Noise Amplifier connected to the feed horn at the focal point of the dish. The amplified signal was then fed via very expensive 50 Ohm impedance coaxial cable to an indoor receiver or in other designs fed to a down converter (a mixer and a voltage tuned oscillator with some filter circuitry) for down conversion to an intermediate frequency. The channel selection was controlled, typically by a voltage tuned oscillator with the tuning voltage being fed via a separate cable to the head end. But this simple design evolved.

Designs for micro strip based converters for Amateur Radio frequencies were adapted for the 4 GHz C-Band. Central to these designs was concept of block down conversion of a range of frequencies to a lower, and technologically more easily handled block of frequencies (intermediate frequency).

The advantages of using an LNB are that cheaper cable could be used to connect the indoor receiver with the satellite TV dish and LNB, and that the technology for handling the signal at L-Band and UHF was far cheaper than that for handling the signal at C-Band frequencies. The shift to cheaper technology from the 50 Ohm impedance cable and N-Connectors of the early C-Band systems to the cheaper 75 Ohm technology and F-Connectors allowed the early satellite TV receivers to use, what were in reality, modified UHF TV tuners which selected the satellite television channel for down conversion to another lower intermediate frequency centred on 70 MHz where it was demodulated. This shift allowed the satellite television DTH industry to change from being a largely hobbyist one where receivers were built in low numbers and complete systems were expensive (costing thousands of Dollars) to a far more commercial one of mass production.

Direct broadcast satellite dishes are fitted with an LNBF, which integrates the feed horn with the LNB.

The satellite receiver demodulates and converts the signals to the desired form (outputs for television, audio, data, etc.). Sometimes, the receiver includes the capability to unscramble or decrypt; the receiver is then called an integrated receiver/decoder or IRD. The cable connecting the receiver to the LNBF or LNB must be of the low loss type RG-6, quad shield RG-6 or RG-11, etc. It cannot be standard RG-59.

Special on-board electronics in the receiver box help tune the signal and then convert it to a frequency that a standard television can use.

There are two primary types of satellite television distribution:

1. Direct broadcast satellite (DBS) and
2. Television receive-only (TVRO).

8.4.1. Direct Broadcast via Satellite

Direct broadcast satellite (DBS) is a term used to refer to satellite television broadcasts intended for home reception, also referred to more broadly as direct-to-home signals. The expression *direct-to-home* or DTH was, initially, meant to distinguish the transmissions directly intended for home viewers from cable television distribution services that sometimes carried on the same satellite. The term predates DBS satellites and is often used in reference to services carried by lower power satellites which required larger dishes (1.7m diameter or greater) for reception.

"Direct broadcast satellite" can either refer to the communications satellites themselves that deliver DBS service or the actual television service. DBS systems are commonly referred to as "minidish" systems. DBS uses the upper portion of the K_u band.

DBS systems rely upon proprietary reception equipment, most often in the form of a television set-top signal descrambling box. This measure assures satellite television providers that only authorized; paying subscribers have access to the content.

8.4.2. Television Receive-Only

Television receive-only, or TVRO, refers to satellite television reception equipment that is based primarily on open standards equipment. This contrasts sharply with direct broadcast satellite, which is a completely closed system that uses proprietary reception equipment. TVRO is often referred to as "big dish" satellite television.

TVRO systems are designed to receive analog satellite signals from both C-band and K_u-band satellite television or audio signals. TVRO systems tend to use larger rather than smaller satellite dish antennas, since it is more likely that the owner of a TVRO system would have a C-band-only setup rather than a K_u-band-only setup. Additional receiver boxes allow for different types of digital satellite signal reception, such as DVB/MPEG-2 and 4DTV.

Parabolic Earth Station antennas receive signals from a single satellite at a time. Simulsat is a quasi-parabolic satellite earth station antenna that is capable of receiving satellite transmissions from 35 or more C and Ku Band satellites simultaneously.

8.5. PICTURE PHONE

Picture phone is the another fascinating application of televisions, in which two persons can see each other while talking over the telephone lines. In this application both persons have a unit which contains a small picture tube for displaying the picture of another persons and a mixture TV camera to send your image through the telephone line. The highest modulating frequency is used in picture phone is limited to about 1.5 MHz.

Picture Phone was developed by the Bell System (AT&T/Western Electric). The Picture Phone was connected to the Central Office via 3 standard wire pairs (for comparison, a regular telephone line uses a single wire pair). One pair carried the 1 MHz Picture Phone video signal in one direction, the other carried video in the opposite direction. These had to be specially equalized to carry the signal but were still otherwise just ordinary telephone wire. The third pair carried the normal 2-way voice call plus carried the Touchtone dialling to set up the call.

A Picture Phone Central Office had a second switch operating in parallel to the regular switch for those making Picture Phone calls. That takes care of local calls. Many calls share the same communications system whether microwave, coaxial cable, or (nowadays) fibre optic is the medium. Even when offices are connected via ordinary wire pairs many calls share the same wires. A voice channel is allotted only 3 kHz for each direction. But a Picture Phone video signal takes 1 MHz. It is 333 times of the audio bandwidth. A few video calls would fill up all available bandwidth. Video compression technique is being used to increase the number of users.

8.6. FACSIMILE

Fax (short for facsimile, from Latin *fac simile*, "make similar", *i.e.* "make a copy") is a telecommunications technology used to transfer copies (facsimiles) of documents, especially using affordable devices operating over the telephone network. The word telefax, short for telefacsimile, for "make a copy at a distance", is also used as a synonym. Although *fax* is not an acronym, it is often written as "FAX". The device is also known as a telecopier in certain industries. When sending documents to people at large distances, faxes have a distinct advantage over postal mail in that the delivery is nearly instantaneous, yet its disadvantages in quality have relegated it to a position beneath email as the prevailing form of electronic document transferral.

Facsimile Transmission, communication system that copies, sends, and receives documents by way of telephone networks or over Internet connections. Also called faxing, this method of communication allows people to share exact copies of important papers by duplicating and sending them on one end, and then receiving and reproducing them on the other.

There are several different indicators of fax capabilities: Group, class, data transmission rate, and conformance with ITU-T (formerly CCITT) recommendations. Fax machines utilize standard PSTN lines and telephone numbers.

A "fax machine" usually consists of an image scanner, a modem, and also offered as options for many high-volume workgroup printers and photocopiers.

How Facsimile Machines Work

The standard facsimile machine works like a combination telephone and photocopier. The user places the documents into a document feeder on the sending machine, and then dials the telephone number of the receiving fax machine. A gear mechanism pulls the original document over an optical scanner. The scanner records variation between light and dark areas of the document as dots arranged in a series of rows or columns. A photoelectric cell converts the dots into electronic impulses, which are then transmitted to the receiving fax machine via telephone lines. A simple block diagram of facsimile system is shown in Fig. 8.6.

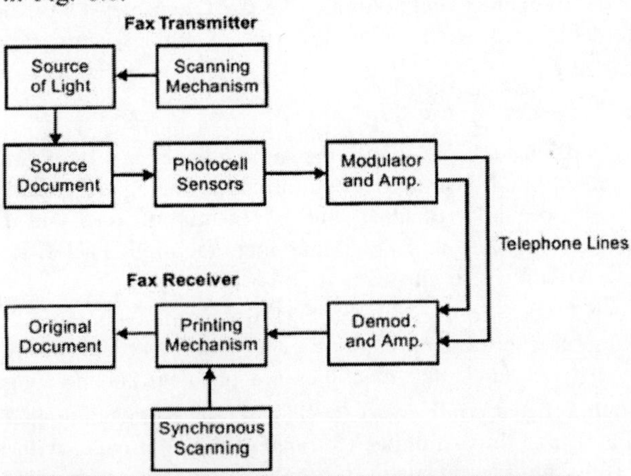

Fig. 8.6. A simple block diagrma of FAX system.

The receiving fax machine decodes the electrical impulses into a series of dots. It sends the decoded signal to a print mechanism built into the fax machine, which prints a duplicate of the original document. International standards ensure that fax machines around the world are compatible with each other.

Before a fax machine can send a document, it must first convert the document into a digital image. Although some advanced fax machines can actually recognize certain characters, most facsimile machines simple scan one page of the document at a time, converting the page into a digital image in much the same way as a photocopier takes a picture of the document to be copied.

Once the fax machine has converted the document into a digital image, it needs to get the image ready to send across a telephone line. The machine prepares to send the document by modulating the digital image into sound that can be recognized by another computer or fax machine.

When the document is ready to be sent, the fax machine sends an "off hook" signal to the attached telephone line followed by the appropriate dual-tone multi-frequency (DTMF) sounds necessary to dial the receiving fax machine. The receiving machine answers with a special tone known as a line acceptance tone (LAT); this tone indicates to the sending machine that another machine has answered. The sending machine then initiates a data transfer session with the fax machine or computer that answered the phone.

Once the sending and receiving fax machines have connected to one another, the sending machine sends the fax as a series of audible tones. The receiving machine listens to these tones, demodulates the data into a digital picture identical to the one scanned by the sending machine, then either stores or prints the received images. When the sending fax machine signals that the last page has been sent, the receiving machine acknowledges that the fax was fully received and disconnects the telephone call. Some sending fax machines subsequently print a report indicating that the fax was sent successfully.

Although devices for transmitting printed documents electrically have existed, in various forms, since the 19th century, modern fax machines became feasible only in the mid-1970s as the sophistication increased and cost of the three underlying technologies dropped. Digital fax machines first became popular in Japan, where they had a clear advantage over competing technologies like the teleprinter. Over time, faxing gradually became affordable, and by the mid-1980s, fax machines were very popular around the world.

Analogue Formats

Group 1 and 2 faxes were sent in the same manner as a frame of analogue television, with each scanned line transmitted as a continuous analogue signal. Horizontal resolution depended upon the quality of the scanner, transmission line, and the printer. Analogue fax machines are obsolete and no longer manufactured. ITU-T Recommendations T.2 and T.3 were withdrawn as obsolete in July 1996.

- Group 1 faxes conform to the ITU-T Recommendation T.2. Group 1 faxes take six minutes to transmit a single page, with a vertical resolution of 98 scan lines per inch. Group 1 fax machines are obsolete and no longer manufactured.
- Group 2 faxes conform to the ITU-T Recommendations T.30 and T.3. Group 2 faxes take three minutes to transmit a single page, with a vertical resolution of 100 scan lines per inch. Group 2 fax machines are almost obsolete, and are

no longer manufactured. Group 2 fax machines can interoperate with Group 3 fax machines.

Digital Formats

Group 3 and 4 faxes are digital formats, and take advantage of digital compression methods to greatly reduce transmission times.

Group 3 faxes conform to the ITU-T Recommendations T.30 and T.4. Group 3 faxes take between six and fifteen seconds to transmit a single page (not including the initial time for the fax machines to handshake and synchronize). The horizontal and vertical resolutions are allowed by the T.4 standard to vary among a set of fixed resolutions :

1. Horizontal: 100 scan lines per inch
   Vertical: 100 scan lines per inch
2. Horizontal: 200 or 204 scan lines per inch
   (*i*) Vertical: 100 or 98 scan lines per inch ('Standard')
   (*ii*) Vertical: 200 or 196 scan lines per inch ('Fine')
   (*iii*) Vertical: 400 or 391 (note not 392) scan lines per inch ('Superfine')
3. Horizontal: 300 scan lines per inch
   (*i*) Vertical: 300 scan lines per inch
4. Horizontal: 400 or 408 scan lines per inch
   (*i*) Vertical: 400 or 391 scan lines per inch ('Ultrafine')

Group 4 faxes conform to the ITU-T Recommendations T.563, T.503, T.521, T.6, T.62, T.70, T.72, and T.411 to T.417. They are designed to operate over 64 kbps digital ISDN circuits. Their resolution is determined by the T.6 recommendation, which is a superset of the T.4 recommendation.

Fax Over IP (FOIP) can transmit and receive pre-digitized documents at near real-time speeds. Scanned documents are limited to the amount of time the user takes to load the document in a scanner and for the device to process a digital file. The resolution can vary from as little as 150 DPI to 9600 DPI or more. This type of faxing is not like the e-mail to fax service that still uses fax modems at least one way.

In many corporate environments, standalone fax machines have been replaced by "fax servers" and other computerized systems capable of receiving and storing incoming faxes electronically, and then routing them to users on paper or via an email (which may be secured). Such systems have the advantage of reducing costs by eliminating unnecessary printouts and reducing the number of inbound analog phone lines needed by an office.

Data Transmission Rate

Several different telephone line modulation techniques are used by fax machines. They are negotiated during the fax-modem handshake, and the fax devices will use the highest data rate that both fax devices support, usually a minimum of 14.4 kbps for Group 3 fax.

ITU Standard	Released Date	Data Rates (bps)	Modulation Method
V.27	1988	4800, 2400	PSK
V.29	1988	9600, 7200, 4800	QAM
V.17	1991	·14400, 12000, 9600, 7200	TCM
V.34	1994	28800	QAM
V.34bis	1998	33600	QAM

Note that 'Super Group 3' faxes use V.34bis modulation that allows a data rate of up to 33.6 kbps.

8.7. DIGITAL TELEVISION

As in other areas of electronics, digital technology has made its impact in television system too. In TV transmission system, the input signal at the camera tube, and the output processed to the picture tube are inherently taken as analog continuous with respect to space and time. If televisions signals are digitized in binary sequences, several additional processing of signals are possible. However, to convert video signal requires high speed A/D converters and very high bit rates for transmission. With the advancements in the technology of micro electroics a wide variety of studio equipments are available to produce variety of digital video effects. Digital techniques are mainly employed in time base correction, frame synchronisating digital recording and signal processing for improved picture in television receivers.

The CCIR has unanimously adopted in 1982, Recommendation 601, for encoding parameters of Digital Television for studios, which established an agreement on a digital code compatible with both 525/60 and 625/50 standards.

Analog signals suffers degradation in transmission path due to noise, crosstalk, linear and non-linear distortion in the circuitry. In a digital system, the analog signal is converted into digital signal by using A/D converter. There are three basic functions of A/D converter are sampling, quantization and encoding.

(*i*) **Sampling :** According to the sampling theorem, given by the Nyquist (also known as Nyquest criteria) for any band limited signal (or TV signal), is sampled at rate of twice the maximum band limit frequency. The spectrum or the fourier transform of a band limited signal sampled at the interval T seconds, is a periodic signal consisting of the original spectrum of the signal scaled by $1/T$ and shifted on the frequency domain by $n\omega_0$, repeated at ω_0 interval given by $w_0 = 2\pi/T$. If the sampling rate is less than the Nyquist frequency, the analog signal cannot be reconstructed fully.

In case of TV, the composite video signal is sampled at 10 million samples per second to effectively cover the video bandwidth of 5 MHz.

(*ii*) **Quantization :** The sampled output has to be assigned a discrete numerical value as the A/D converter maps a range the input amplitude in the minimum resolution range, onto each single digital output value.

(*iii*) **Encoding :** After quantization, the sampled numerical value is assigned a binary code along with additional coding like parity bits for error detections, scrambling etc, depending on the nature and error characteristics of the system handling the digital signal.

For displaying the picture, the digital pulses must be reconverted into analog signal, using D/A (digital to analog) converter. The analog signal so produced is faithfully the same as the video signal of the original picture. The output would be free of noise and other spurious effects.

The basic block diagram of production and reproduction of digital video is shown in Fig. 8.7.

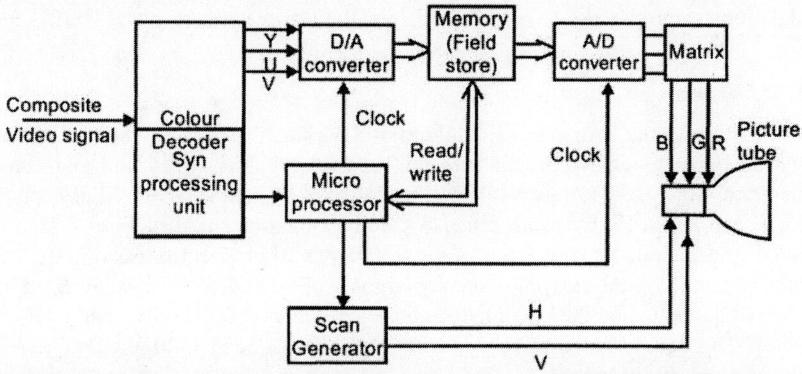

Fig. 8.7. Block diagram of production and reproduction of digital video.

The analog video signal output of the camera tube and matrix includes luminance signal (Y) and colour difference signals (U and V). The CVS is converted in to digital pulse by using analog to digital converter. The CVS is sampled at 10 million samples per second to effectively convert the video bandwidth of 5 MHz. The digitized video signals are recorded into the memory at 50 Hz field rate. For playback, bits are read from the memory at 100 Hz and then converted from digital to analog by digital to analog converter unit for display on the screen of a monitor at double the field and line scan rates to have high resolution. By using micro-processor digital pulses are processed to have many various special featrues.

Digital TV Transmission and Reception

Composite video signal from the camera pickup is digitized, it requires of high bit rate to define the details of luminance and colours of every pixels. For normal quality picture, the bit rate is 216×10^6 bits per second. It is quite difficult to handle such a high bit rate because there may be problem while broadcasting. So it is necessary to reduce the bit rate and adopt a suitable modulation system.

Data compression technique is used to reduced the number of bits by removing the redundant bits in a picture, because most of the TV picture contains redundant information. This type of information needs to be send only once. It can be readout from the memory.

Modulation for Digital Video and Sound

Modulation for digital video : The digital data consisting of video and audio informations alongwith the control signals, are interleaved in time. The multiplexed pulses modulate a suitable carrier for transmission. The method of modulation used is orthogonal frequency divisions multiplex (OFDM). In OFDM technique a large number of closely spaced carriers are modulated individually and separately, and then transmitted in a slot in a frequency spectrum. The carrier power requried is quite low and the systems is almost free from noise and interference. For the transmission of digital signals by long distances. Optical fibre cable is more suitable. Satellite links may also be used because the microwaves are capable of carrying such signals.

Modulation for digital sound : In digital televisions system, the sound is also to be digitized. Stereo sound is used in digital televisions. The upper limit for audio frequency is 15 KHz and samples are taken at an intervals of 31.25 μs. Each channel is sampled to 14 bits which give 16238 levels. Then there bits are compressed to 10 per samples to save bandwidth. One parity bit is added to each 10 bit word to have even parity. The modulation technique used for this is called digital quadrature phase shift keying (DQPSK).

Digital TV Receiver

Figure 8.8 shows the simple block diagram of a digital TV receiver which is basically a superhetrodyne receiver. Receiving antenna, receiver the EM signal and and convert it into electrical signals. The tuner selects the desired channel. A sharp filter separates the video IF and audio IF, by using either SAW filter or a mechanical filter. Video bits are detected and decoded to give the original samples after decompression. By using DAC, converts the digital bits into analog signals. For audio, IF is converted into intercarrier frequency and the sampled audio pulses are obtained by using DQPSK demodulator. The demodulated pulses pass through the decoder and then DAC circuits to get the original audio signal.

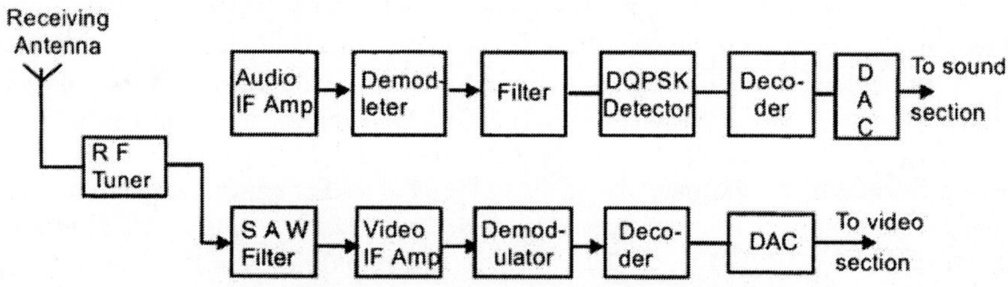

Fig. 8.8. Block diagram of a digital receiver.

Advantages of Digital TV

(*i*) A good quality of picture is obtained in digital televisions because noise and cross colour effect can be minimized and flicker is removed totally.

(*ii*) In digital TV, the signal can be easily stored, delayed and manipulated by means of microcomputer chips, integrated in the system.

(*iii*) Some special features like picture in picture, freezing of frame, teletext and TV games and become available in digital TV.

(*iv*) Some of the special effects like zooming, editing, blasting, wide dimension TV can be made available in digital TV.

(*v*) The problem of component aging on the alignment, tunning etc. and consequent instability encountered in analog system is eliminated with digital technology which can have self tunning and adjustments routines.

Disadvantages of Digital TV

(*i*) For the long distance transmission it requires the use of optical fibre or microwave satellite links because of high bit rate so high bandwidth.

(*ii*) High bit rate requires high speed logic families, like ECL or schottky TTL, which are costly.

(*iii*) In analog system, the quality degrades gradually whereas in dighital system the picture quality remain good till a certain C/N ratio but below that it degrades very rapidly.

8.8. TV GAMES

An application of digital electronics and IC technology to TV products is TV games. Earlier days, solid state games used TTL and provides the logic for playing questions answer games on the TV screen. Later on paddle type game were developed include generation of sounds. With the developments of microprocessor many advanve features are added in the TV game. Mainly there are two sections of a TV game, namely,

(*i*) Game unit

(*ii*) TV receiver

Game Unit : The game unit includes complete electronic circuitry necessary for generating signals. These signals are connected to the televisions receiver to display the current status of the game on the TV scree. A basic block diagram of TV game is shown in Fig. 8.9.

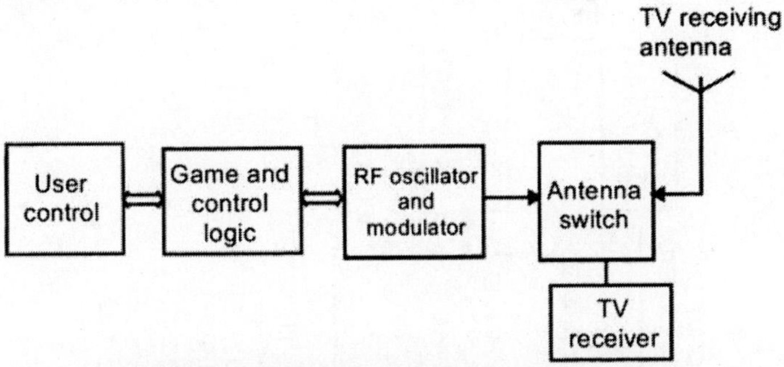

Fig. 8.9. Simple block diagram of a TV game.

The functions of each block is explained in brief.

(*i*) **User Control Unit.** This unit has various controls required to the user for playing the game. It contains associated circuitry to generate the command signals to initiate various actions.

(*ii*) **Game and Control Logic Unit.** This unit is the heart of any TV game. It produces video signals required to display the game character and field on the screen. It also provide a logical circuit for game playing rules score display etc. It also inserts the sound at any point in the game by pulse detection gating. It has sync generation section which produces the horizontal and vertical sync pulses to correct the composite video signals.

An RF oscillator is used which is set for channel 3 or 4. Its output is modulated and provides input signal to the TV receiver.

Modern TV games uses a microprocessor microcontroller chip which coordinate and control functions of various elements. A software is used to play the game, so many types of games can be played with a single hardware unit, which can process many number of softwares easily. That provides the more flexibility on the system. Figure 8.10 shows the block diagram of microprocessor controlled TV games.

User Control and Interface : This units provides a link between the system and player. The player uses many input peripherals to give commands to the microprocessor, such as keyboard, joystick, or suitable switches. The output from the input peripherals. For example, a keyboard, is analog in nature and being converted into digital form by using ADC as interface unit in built in the input device.

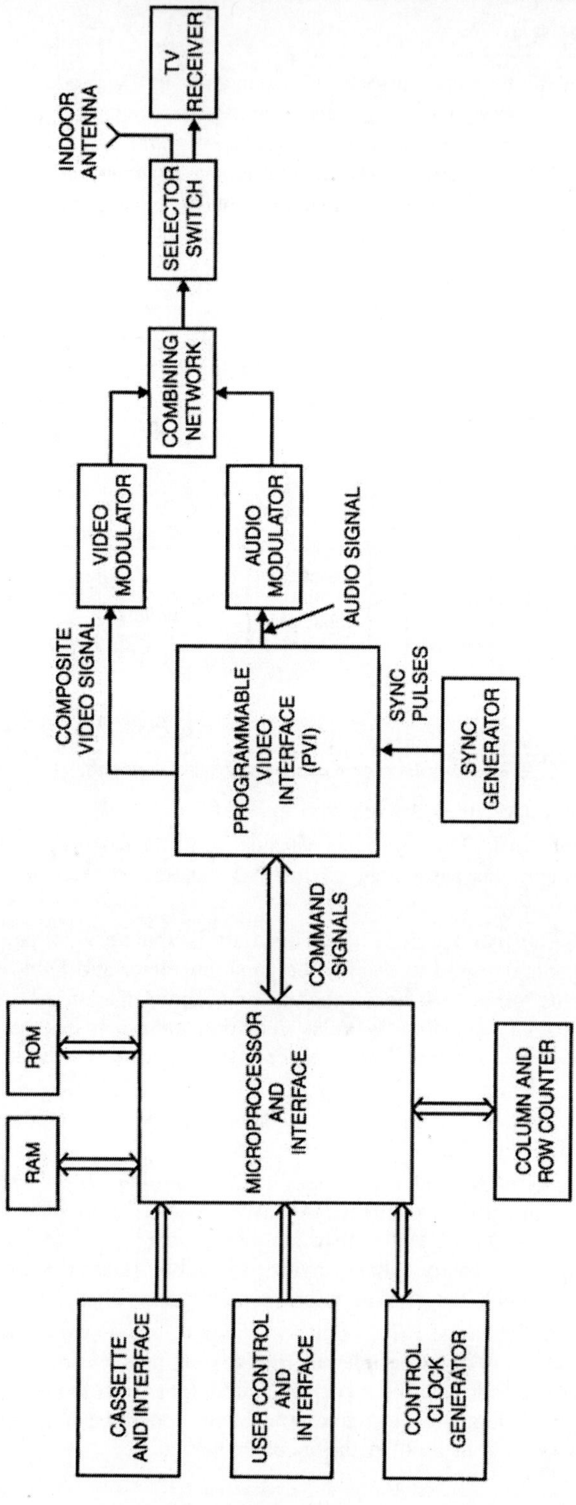

Fig. 8.10. Basic block diagram of the microprocessor controlled TV game.

Cassette and Interface : This unit contains the software for one or more games. These are fed to the memory through suitbale interface. The cassette output is analog in nature and it is converted into digital by the interface section. Random Access Memory (RAM) is used to store the contents of the required game.

Memories : There are two types of memories are used here, which are RAM (Random Access memory) and ROM (Read only memory). RAM stores information temporarily which is changed by the command given by the player or from the microprocessor. Whereas ROM stores permanent information which can be accessed by sending appropriate address from the microprocessor.

Counters : There are two counters, row and column. The receiver screen has 150 rows and 250 columns. There are used to determine the row and column positions of any object. The position and velocity of the objects in game is determined by the counters with the help of micrporocessor. For generating corresponding video signals, the counters feed the digital signals to the programmable video interface (PVI) unit.

Programmable Video Interface (PVI) : This unit generates video and audio signals. By receiving the digital commands from the microprocessor both directly and indirectly. It receives the timing signals from the clock to generate appropriate sound at the correct movements. Sync and colour burst pulse are also added here to the video information to forms colour composite video signals.

Sound Generator : Different types of sounds are generated in a game. The audio signals are processed within the PVI. The microprocessor is controlled such that it can control a square wave generator whose frequncy is equal to 78 KHz. Generation of sounds makes the game more ralistic.

Microprocessor : Microprocessor is the heart of the TV games. it processes the data received from the player or game cassette or RAM or ROM and generates the digital output which is used by the PVI to produce the proper video and audio signals simultaneously continously.

Modulators : The video and audio signals received from the PVI together with the sync pulses can be fed directly to the receiver through a cable. Video modulator uses AM VSB technique to modulate the video signal and frequency modulation is used for audio signals.

Selector Switch : It is a two way switch which select either TV or game at a time. The modulated signals from the game console are fed to one terminal. The second terminal is connected to the TV receiving antenna. The common point of this switch is connected to the receiver at the antenna terminal.

8.9. TELETEXT

The composite video signal has some portion which is not used for TV programme. Such a portion is the back porch of the vertical blanking pulse, covering 15 scanning lines in CCIR B and in CCIR M systems. Eight of these lines are used by a new service, designated as **teletext** which provides one-way communications of text. It is transmitted during the vertical blanking pulse and is displayed on the screen of picture tube in the receiver. A teletext service transmits information in the form of pages. Each page gives a screenful of information.

Data for each row is transmitted serially as 8 bit words. The word consists of data bits and some parity bits. The parity bits allow detections of error which is automatically corrected by using hamming code. The bit rate for 625 line system is 6.9375 Mbps and for 525 lines system is 15.72 Mbps. It corresponds to maximum bandwidth of 3.46075 MHz for 625 lines and 2.86 MHz for 525 line system.

A typical schematic block diagram of teletext decoder in a TV receiver to extract information from the data transmitted, is shown in Fig. 8.11.

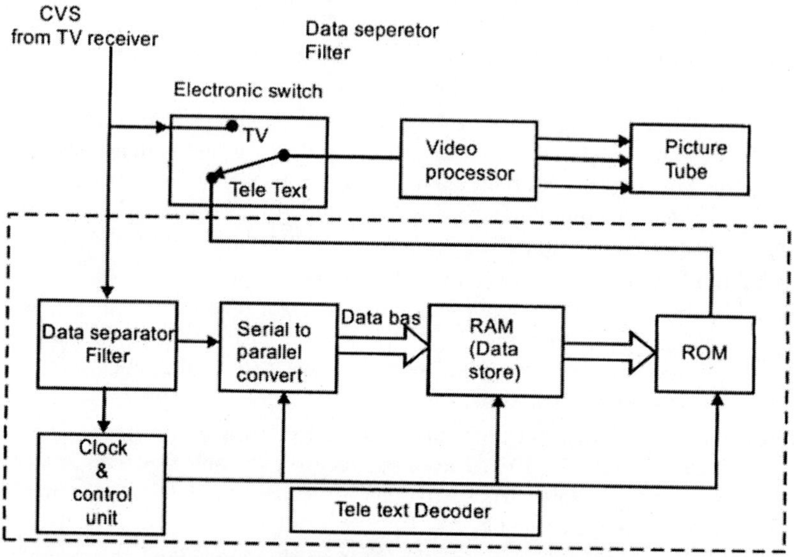

Fig. 8.11. Block diagram of teletext system.

At the transmitter, the information is converted into a digital form and is transmitted sequentially (during the vertical blanking) with the normal picture signal. Thus it is a continous transmission of the data alongwith the channel programme.

At the TV receiver, a decoder separates the digital signals from the normal picture signal, and stores them into RAM. Storing is done on receipt of an instruction bit, generated by the control unit, when it gets a command from the user. The control unit will decode the users command and will therefore enable RAM to store the required page number which can be displayed on the screen. The normal picture would remain suppressed so long as the teletext page is being displayed.

8.10. BASIC IDEAS FOR HDTV

HDTV stands for High Definition Television. HDTV enhances the quality of the picture on your screen and the quality of the sound. In the future HDTV will allow the connectivity of computers and televisions. Some of the features of HDTV are :

- Higher picture sharpness
- Larger picture
- More colours
- Wider picture
- Multi-channel digital CD quality sound

The basic concept behind high-definition television is actually not to increase the definition per unit area but rather to increase the percentage of the visual field contained by the image.

The majority of proposed analog and digital HDTV systems are working toward approximately a 100% increase in the number of horizontal and vertical pixels. (Proposals are roughly 1 MB per frame with roughly 1000 lines by 1000 horizontal points). This typically results in a factor of 2-3 improvement in the angle of the vertical and horizontal fields. The majority of HDTV proposals also change the aspect ratio to 16/9 from 4/3 — making the image more "movie-like".

Under the proposed HDTV standard pictures will have 6 times the resolution of current TV and it will be a full 60 frames per second temporal resolution which is twice the current NTSC images. Motion will be seen smooth and the picture will be clear enough to sit very close to a very large screen. The picture will be displayed in a panoramic 16:9 horizontal-to-vertical aspect ratio to be more like movies and add the feeling of realism to TV.

High Definition Television (HDTV) is actually a subset of the Digital Television (DTV) family of formats, as defined by the Advanced Television Systems Committee (ATSC). DTV uses digital data (1's and O's) transmission of the picture and sound information, as opposed of the traditional *analog* signals used for what we know as analog television, devised by the National Television System Committee (NTSC). The relatively new DTV picture formats are generally characterized by the horizontal and vertical resolutions, aspect ratio, interlaced or progressive scanning, and refresh rate.

The Table 8.1 below summarizes all 18 of the ATSC Digital Television formats. There are a total of six (6) HDTV formats, of which 720p/30 and 1080i/30 are the most common. Again, It is important to realize that HDTV is only a subset of the DTV standards, and so DTV is the more general term, while HDTV specifically references the six high definition formats of the 18 DTV formats. The DTV formats are most frequently referred by their horizontal lines of resolution and whether they scan in progressive or interlace (e.g., 480p, 720p, 1080i). The suffix "p" stands for progressive scan, while the suffix "i" stands for interlaced scan. Sometimes, they are further distinguished by their refresh rate, as designated with a slash ("/"), followed by the refresh rate. For example, "1080i/30" refers to 1080 horizontal lines of resolution with interlaced scanning at 30 Hz refresh rate.

Standard Definition Television (SDTV) consists of the first DTV format of 480i/30. It is equivalent to interlaced video output of DVD-Video in 4:3 aspect ratio. This format is used for when bandwidth is a bigger consideration than absolute picture quality. SDTV uses a data rate of about 4-7 Mbps, so three to six SDTV channels can be crammed into the same bandwidth as a HDTV channel.

Enhanced Definition Television (EDTV) is a step up from SDTV, but not quite as good as HDTV. EDTV consists of some 11 formats as shown in the table 8.1. The vertical resolution is limited to 480 lines, but horizontal resolution varies 640 to 704 vertical lines. It encompasses both 4:3 and 16:9 aspect ratios, a number of refresh rates, and both interlaced and progressive scanning. EDTV is used when better picture quality is desired, but without the full bandwidth of HDTV.

Table 8.2. Summary of the 18 Digital Television formats, including 6 HDTV formats.

Format Index	Vertical Resolution	Horizontal Resolution	Aspect Ratio	Scan Type	Refresh Rate [Hz]	Type
1.				Interlaced	30	SDTV
2.		640	4 : 3		24	
3.				Progressive	30	
4.					60	
5.				Interlaced	30	
6.					24	
7.	480	704	4 : 3	Progressive	30	EDTV
8.					60	
9.				Interlaced	30	
10.					24	
11.		704	16 : 9	Progressive	30	
12.					60	
13.					24	
14.	720	1280	16 : 9	Progressive	30	
15.					60	HDTV
16.				Interlaced	30	
17.	1080	1920	16 : 9	Progressive	24	
18.					30	

High Definition Television (HDTV) uses a data rate of 25 - 27 Mbps for the best possible picture. All HDTV formats are in 16:9 aspect ratios. The 720 vertical resolutions only use progressive scanning, but at various refresh rates. The highest resolution is commonly used in interlaced scanning mode (1080i), due to limitations of current broadcast and consumer equipment. But the format includes 1080p; to accommodate future growth as imaging and display technologies catches up. HDTV is used for premium programming when picture quality is of utmost priority, and bandwidth is less of a concern. This includes select prime time shows, major sporting events, and premium movies.

At its highest resolution, HDTV offers 2,116,800 pixels (picture elements). This is over a sixfold improvement in picture detail of standard definition television which only has 307,200 pixels. Colour resolution is also improved by a factor of two. All of the DTV formats use MPEG-2 as the video compression standard, just like DVD-Video. MPEG-2 is a flexible video encoding algorithm and scales up nicely for the higher resolutions of DTV. With digital transmission, there are no analog transmission artifacts and degradations such as snow due to weak signal, double images or ghosting due to multi-path interference of large buildings and structures, and sparkles due to noise from a vacuum cleaner.

Technical Details

An HDTV-compatible TV usually uses a 16:9 aspect ratio. The high resolution images (1920 pixels × 1080 lines or 1280 pixels × 720 lines) allow much more detail to be shown compared to analog television or regular DVDs. MPEG-2 is currently used as the compression codec. Like NTSC and PAL, 1920 x 1080 broadcasts generally use interlacing to reduce bandwidth demands. Alternating scan lines are broadcast 50 or 60 times a second, similar to PAL's 50 Hz and NTSC's 60 Hz interlacing. This format is entitled 1080i, or *1080i60*. In areas traditionally using PAL 50 Hz *1080i50* is also used. 1080p, "true high-definition", is not currently used for broadcasting due to high bandwidth demands. Progressive scan formats are also used wifh frame rates up to 60 per second. The 1280 × 720 format in practice is always progressive scan (with the entire frame refreshed each time) and is thus termed 720p. FOX, ABC, and ESPN currently broadcast 720p60 content. A mix of interlaced and progressive scan is also encountered, where the 1080i60 (or 50) signal can be deinterlaced by simple weaving of fields to frames, giving in effect a 1080p30 (or 25) image.

- Bandwidth : Current bandwidth is limited to 6MHz on current terrestrial channels. HDTV will require 18MHz for 1050 lines / 600 pixels.

  Options :

- Change channels from 6 to 20MHz
- Compress signal to fit inside the existing 6MHz channels
- Multiple channels for HDTV signal
- Distribution (terrestrial, cable, satellite)
- Interlaced/Non-Interlaced
- Compression.

The TDM (Time Division Multiplexing) used in HDTV is called Multiplexing Analog components (MAC). A basic working of MAC encoder, receiver and decoder are discussed here in brief.

MAC Encoder

A simple block diagram of MAC encoder is illustrated in Fig. 8.12. The colour camera tubes produced the video signals R, G and B which represents the intensity of light of three primary colours (RGB), present in each pixel of the picture. These signals are fed to a resistor matrix to give luminance signal Y and colour weighted difference signals designated as D_B and D_R.

There are three bandpass filters, which allow bandwidth of 5.6 MHz for Y-signal and 2.75 MHz for DB and DR signals. The minimum sampling rate is equal to twice the maximum bandwidth frequency. The samples are coded as 8 bit codes, producing a word of 8 bits for each samples of the analog waveform taken. Digital pulses are to a frame store. The frame store isolates the input and the output and hence synchronization is not needed. The total storage capacity is about 20 to 25 million. Samples per second.

The luminance 'Y' signal is delayed by one line by using two RAM's, one for storing luminance signal for the current line and other for the previous line. With this automatically synchronizes the sequence of luminance signal and the chroma signal and the chroma signal.

Line sequential switch is an electronic switch which allow CB signal on odd numbered lines and CR signal on even numbered lines. The luminance signal samples at 13.5 MHz and chrominance signal at 6.75 MHz. Then they are compressed by clocking at a higher frequency of 20.25 MHz. There is a time multiplex switch, the Y and C signals are multiplexed so that odd line contains Y and DB signal and the even line Y and DR signal. The Y and C bits are divided into time slots.

In frequency modulator stage, the multiplexed signal modulates a sine wave are carrier using frequency modulation. Thus we get a modulated signal duly multiplexed for Y and C. Power amplifier amplifies the power and deliver it to the transmitting antenna to radiate in the space.

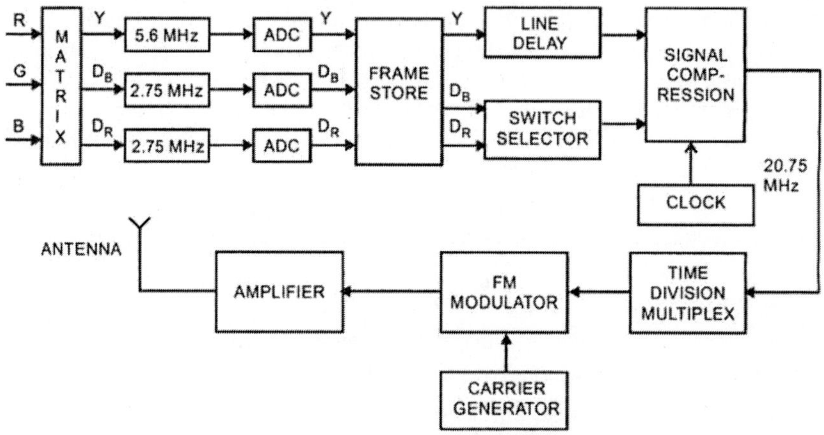

Fig. 8.12. Block diagram of MAC encoder.

MAC Receiver

Figure 8.13 shows the block diagram of MAC receiver. In the receiver section, the received information is detected by FM detector. The output of FM detector consists of words of 8 bits which are demultiplexed to give Y and C words in separate time slots. There words are decompressed to offset the effect of compression which was done in the encoder. Thus after expansion Y + D_B signal occupies one full line of Y + DR signal another full line. Now these multiplexed signals are combined in a matrix to recover R, G and B signals, which are fed to the picture tube to reproduce the orignal picture.

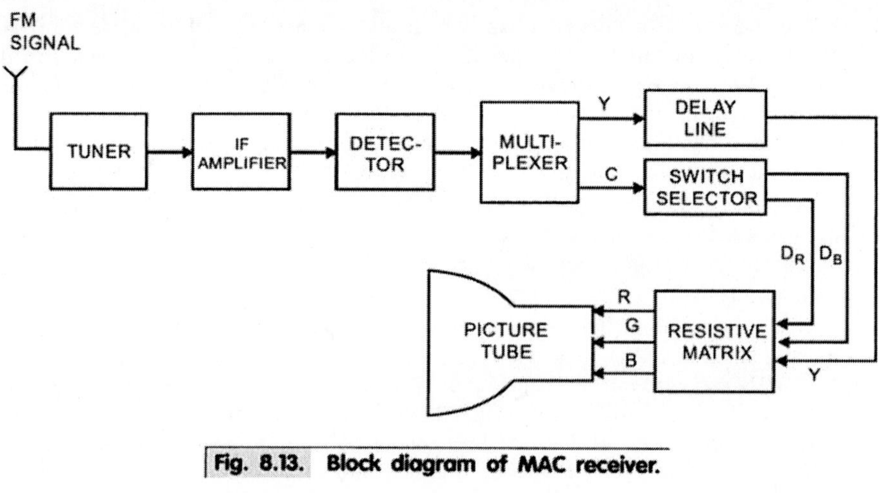

Fig. 8.13. Block diagram of MAC receiver.

Thus the important function of the multiplexing analog components (MAC) system is to eliminate the need of interleaving of chroma signal in between luminance signals within common frequency spectrum. This eliminate crosstalk interferences and allows wide screen to be used.

Japan has developed a different systems of HDTV which has been adopted by USA, canada and many other countries. This system offers 1125 lines, 60 field frequency and interlaced scanning 2 : 1 ratio for the new HDTV system. The technique used is known as multiple sub-nyquist encoding (MUSE). It is based on compressions of the bandwidth.

8.11. TV REMOTE CONTROL

The first TV remote control, called "Lazy Bones," was developed in 1950 by Zenith Electronics Corporation (then known as Zenith Radio Corporation). Lazy Bones used a cable that ran from the TV set to the viewer. In 1956, Zenith's Dr. Robert Adler suggested using "ultrasonic," that is, high-frequency sound, beyond the range of human hearing.

Hard Wired : The first TV remote control used a cable that ran from the TV set to the viewer. A motor in the TV set operated the tuner through the remote control. By pushing buttons on the remote control, viewers rotated the tuner clockwise or counter clockwise, depending on whether they wanted to change the channel to a higher or lower number. The remote control included buttons that turned the TV on and off.

Flashmatic : Operated by means of four photo cells, one in each corner of the TV cabinet around the screen. The viewer used a highly directional flashlight to activate the four control functions, which turned the tuner clockwise and counter-clockwise and which turned the sound and power on and off.

Ultrasonic : The "ultrasonic's," that is, high-frequency sound, beyond the range of human hearing. The original Space Command remote control was expensive because an elaborate receiver in the TV set, using six additional vacuum tubes, was needed to pick up and process the signals. In the early 1960s, solid-state circuitry (*i.e.*, transistors) began to replace vacuum tubes. Hand-held, battery-powered control units could now be designed to generate the inaudible sound electronically.

Infrared : The IR remote works by using a low frequency light beam, so low that the human eye cannot see it, but which can be detected by a receiver in the TV.

Remote control consists of a small portable handset having a key pad on the front, as shown in Fig. 8.14. With the development of infrared LED's and infrared photodiodes, infrared light has been widely used as the link for remote control of TV receivers. For the transmission of remote control commands, infrared light is being modulated using pulse code modulation. The information of the remove control command is being defined by the varying time intervals between a sequence of very short IR pulses. This short IR pulses enables the IR LED to the driven with high current pulses of over one ampere. A large transmission distance and high interference immunity while ensuring a long battery life for the control unit.

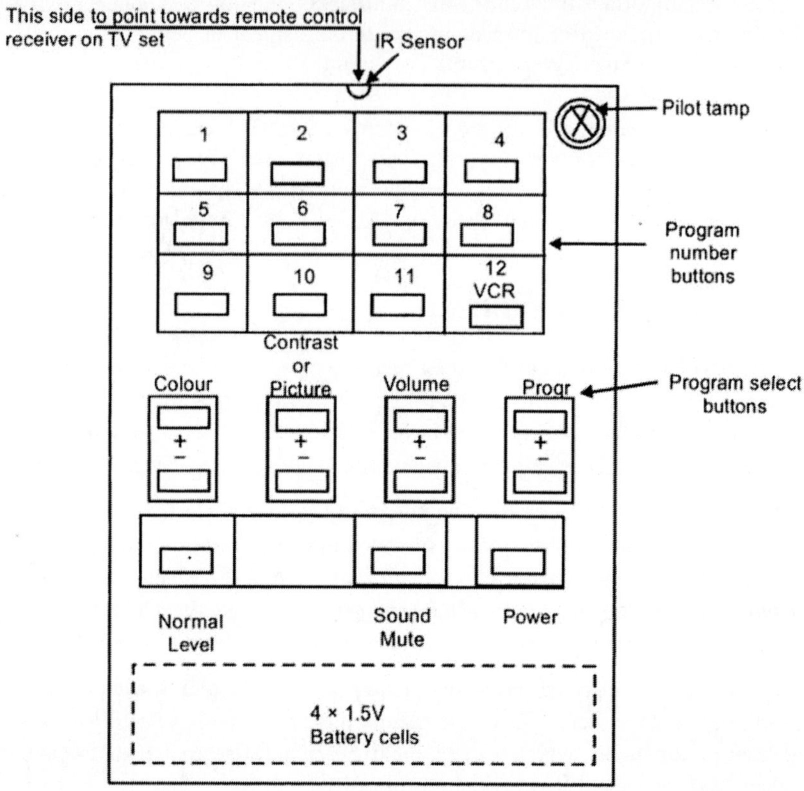

Fig. 8.14. Front view of a typical infrared wave remote control transmitter.

Whenever a key is pressed on the remote control, the IR light generated by LED is chopped in a predetermined sequence. Thus the chopped IR light forms a string of a coded pulses in serial bits form. The code is unique for each command signal given by pressing the relevent key of the remote control. There is no on/off switch on the remote control device, as there is very less current drain from the battery in the absence of a pressed key.

The basic functions performed by the unique codes are as follows :

(*i*) Channel selections

(*ii*) Volume control

(*iii*) Fine tunning

(*iv*) Brightness control

(*v*) Contrast control

(*vi*) Colour saturation.

All the above functions have two positions : Up and down. In down position, value of the parameter decreases, while in up position, it increases.

The controlling function of the signals sensed from the remote control unit and decoded by the microcomputer in the television receiver as illustrated in Fig. 8.15.

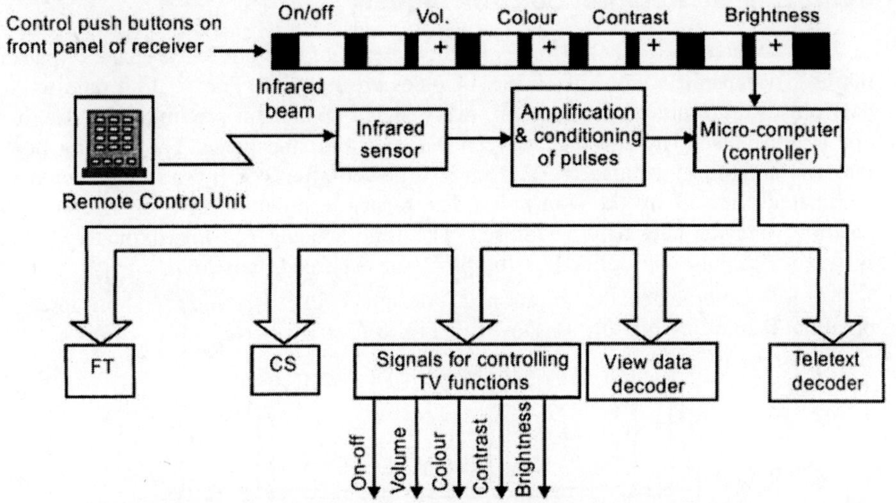

Fig. 8.15. Controlling in TV receiver acting on remote signal.

The modern remote control units may also have bass control, treble control, teletext selection and so on.

A typical IR transmitted IC SAA is IC developed for infrared remote control of colour television receiver. As CMOS IC, it consumes minimum current and is highly flexible in applications. It can transmit up to 1024 commands with the single SAA/IRT 1250 transmitter IC. An IR remote control receiver IC, SAA 1251/SAA 1290 is meant to be a receiver. A block diagram of an IR remote control using IRT 1250 at the transmitter and SAA 1251 or CCU 20 or MAA 4000 or SAA 1290 or SAA 1293 as IR receiver, shown in Fig. 8.16.

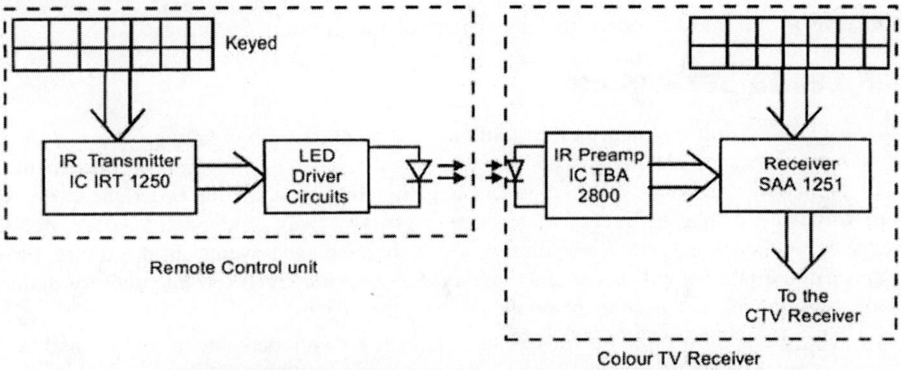

Fig. 8.16. Block diagram of an IR remote control.

At the colour TV receiver, a photodiode converts the received IR transmission into electrical signals. A pre-amplifier such IC TBA 2800, amplifies the electrical signal. The pulse code modulated signals are decoded and converted into respective command. The TV set ON/OFF is effected by a TRIAC in its main circuit, switched ON/OFF by mains flip-flop through an optocoupler for isolation. The circuits for colour, contrast, volume etc. are provided analog signals from the digital to analog converters to set the desired values.

Synthesis of Remote Control Signals

The command signals are transmitted by means of IR light in the shape of packaged pulses. To transmit a word of 10 bit, 14 pules are required. For a 10 bit data word, 11 data pulses are required. In addition, every signal contains a preliminary pulse, a start and a stop pulse. The spacing between the start and stop pulse. The spacing between the start and stop bit pulse is 3T. This is followed after 1 T by the eleven pulses and terminated after 3T by the stop pulse. The binary information of a bit is contained in the time interval between two pulses. The time spacing (T) approximately 1000 μs between two pulses is defined as the basis for each code employed.

A long interval of the duration 2T designate bit '1', while a short interval of duration T signifies bit '0', as shown in Fig. 8.17.

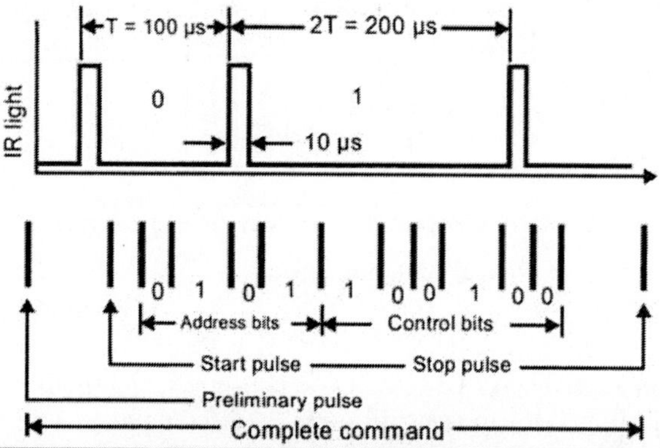

Fig. 8.17. Representing bit 0 and bit 1 by time interval (b) command word 0101100100, as an example.

In order to minimised interference from IR sound transmissions, a preamplifier such as TBA Z800 is used which equipped with AGC in the receiver. The gain is set relative to the interference signal by means of the preliminary pulses.

8.12. 3D TELEVISION

3D Television augments the traditional TV technology by showing the viewer not only sequences of 2D images but streams of three-dimensional scene representations. To the viewer at home this will mean a completely new media experience. He will perceive the displayed events in a more immersive way, and he may even get the chance to choose his own viewpoint to watch the displayed events. In the future, three-dimensional movies will become a standard and provide enhanced interactivity options, e.g. by allowing the user to navigate through the scenes.

The production pipeline for 2D television has developed into a mature and well-understood process over many years. Scenes are recorded with cameras from single view points; captured image streams are post processed, transferred to receivers, and displayed on planar screens. In contrast, the production process for 3D television requires a fundamental rethinking of the underlying technology. Scenes have to be recorded with multiple imaging devices that may be augmented with additional sensor technology to capture the three-dimensional nature of real scenes.

In addition, the data format used in 3D television is a lot more complex. Rather than normal video streams, time-varying computational models of the recorded scenes are required that comprise of descriptions of the scenes' shape, motion, and multi-view appearance. The reconstruction of these models from the multi-view sensor data is one of the major challenges that we face today. Finally, the captured scene descriptions have to be shown to the viewer in three-dimensions which requires completely new display technology.

Three-dimensional TV is expected to be the next revolution in the history of television. It has only recently become feasible to deal with the high processing and bandwidth requirements for real-time acquisition, transmission, and display of high-resolution 3D TV content. In our past work, we have built a complete end-to-end 3D TV system that performs real-time acquisition, transmission, and 3D display of dynamic scenes. More recently, we have been participating in standardization on multi view video coding, and we have also developed an anti-aliasing method for 3D displays.

Advances in optical technology and computing power are bringing life-like 3DTV closer, with potential applications not only in entertainment, but also in education, scientific research, industry, medicine, and many other areas.

3DTV will require the integration of a diversity of key technologies from computing to graphics, imaging to display, and signal processing to communications.

3D television employs some form of stereoscopy and a special viewing device to project a television program into a realistic three-dimensional field. Occasional 3D episodes became moderately popular in the late 1990s when several shows used the technique to attract viewers and increase ratings.

Until a few years ago, 3D films required special glasses which had one red lens and one of either blue or green. Phillips, however, has released a prototype of a new type of 3D display which allows for 3D perception without special virtual glasses.

Today, digital broadcast networks carry hundreds of channels and will presumably be capable of carrying even more as the most advanced video codec's become deployed. This makes it plausible that a number of them will be dedicated to 3D TV. Similar to HDTV, the introduction of 3D TV can proceed gradually, with one 3D channel at first and more to follow, depending on market demand. Our initial system demonstrated that 3D TV offers a richer, more immersive experience than regular TV. It increases entertainment value and realism without the encumbrance of special glasses. Recent work has centred on improving the quality of the experience further.

In our initial 3D TV system, image acquisition consisted of an array of hardware-synchronized cameras that captured multiple views of the scene. To deal with the high processing and bandwidth requirements, the system used a fully distributed architecture with clusters of PCs. A multi-projector 3D display with horizontal parallax was used on the output side. The system was scalable in the number of acquired, transmitted, and displayed real-time video streams. More recently, we have concentrated research efforts on anti-aliasing techniques that aim to improve the rendering of multi-view video on 3D displays. View interpolation techniques are utilized to achieve an oversampling of the multi-view signal in the view dimension. The oversampled signal is then filtered to suppress high frequency portions of the signal that contribute to aliasing, and finally sub-sampled to match the display characteristics.

8.13. VIDEO RECORDING FUNDAMENTALS

8.13.1. Basic Elements of a Magnetic Tape Recorder

A magnetic tape is moved in the direction indicated by a tape-drive device or *transport*. The magnetic coating of the tape contacts the magnetic heads in a prescribed sequence, starting with the erase head and ending with the reproduce head. The basic elements of a magnetic tape recorder are illustrated in Fig. 8.18.

The erase head demagnetizes the tape coating by exposing the magnetic particles to a high frequency field that is several times greater in strength than the coercivity of the particles. As the tape is drawn past the erase head, the erasing field gradually decays, leaving the magnetic coating in a demagnetized state.

The tape then moves into contact with the record head, which consists of a ring-shaped core made of a relatively high-permeability material, and having a nonmagnetic gap. A magnetic field fringes from the gap, varying in accordance with the magnitude of the current signal flowing in the head coil. With low-level signals the field is small, and some magnetic particles in the tape coating will be forced into alignment with the field. As the signal field is increased, a larger number of particles will become oriented in the direction of the recording field. As the tape is moved past the record gap, the magnetic coating acquires a net surface magnetization having both magnitude and direction. This magnetization is a function of the recording field at the instant the tape leaves the *recording zone,* a small region in the vicinity of the trailing edge of the gap.

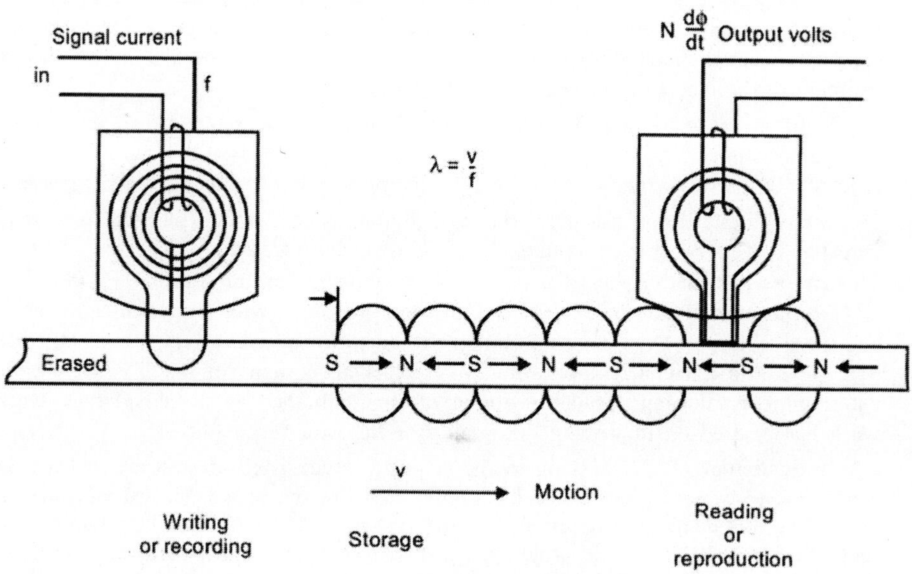

Fig. 8.18. Fundamental of recording and reproduction process.

The magnetization of the fundamental recording system just described is not necessarily linear with respect to the head current. Linear magnetization can be achieved by adding a high-frequency ac bias current to the signal current. Audio recorders use such a scheme to linearized the tape and reduce the distortion. In video recorders, the signal information in the form of a frequency-modulated carrier is recorded directly, without ac bias.

When the tape approaches the nonmagnetic gap of the reproduce head, the flux $d\Phi/dt$ from the magnetized particles is forced to travel through the high-permeability core to link the signal windings and produce an output voltage. The output voltage is proportional to $d\Phi/dt$, the rate of change of the inducted flux, and therefore will rise at the rate of 6 dB per octave until a wavelength is reached where the gap and spacing losses begin to reduce the head output.

Recording Signal Parameters

The physical distance that one cycle of the recorded signal occupies along the tape is called the *wavelength*, which is directly proportional to the relative velocity between the head and the tape, and inversely proportional to the frequency of the recorded signal. It may be expressed as

$$\lambda = v/f \qquad ...(8.1)$$

where: λ = wavelength, in

v = velocity, in/s

f = frequency, Hz

The *linear packing density* is the number of flux reversals per unit length along the recording medium. Because there are two flux reversals, or bits, per cycle, the linear packing density may be expressed as

$$\text{bits/in} = 2\left(\frac{2}{\lambda}\right) = \frac{2}{\lambda} \qquad ...(8.2)$$

The ***area packing density*** is the number of bits per unit area and is, therefore, equal to the number of recorded tracks per inch times the linear packing density, or

$$\text{bits/in}^2 = \left(\frac{2}{\lambda}\right)\frac{\text{tracks}}{\text{in}}$$

When the magnetization is oriented in the direction of relative motion between the head and tape, the process is referred to as ***longitudinal recording*****. If the magnetization is aligned perpendicular to the surface of the tape, it is called *vertical* or *perpendicular* recording.**

Transverse recording exists when the magnetization is oriented at right angles to the direction of relative head-to-tape motion. From these definitions, longitudinal magnetization patterns are produced by both rotary- and stationary-head recorders.

The Recording Process

The recording process consists of applying.a temporally changing signal voltage to a record head as the tape is drawn by the head. The magnetic field that results from the energized head records a magnetization pattern that spatially approximates the voltage waveform. In *saturation* or *direct recording*, the signal consists of polarity changes with modulated *transition times* or zero crossings.

Strict linear replication of this signal is not required because the information to be recovered depends only on knowledge of when the polarity transitions occur. Examples are digital recording, where the transitions are synchronized with a bit time interval and occur at bit positions depending upon the coded pattern, and FM video recording, where a modulated sine wave is applied so that the transitions occur not regularly but according to the signal information contained in the modulation.

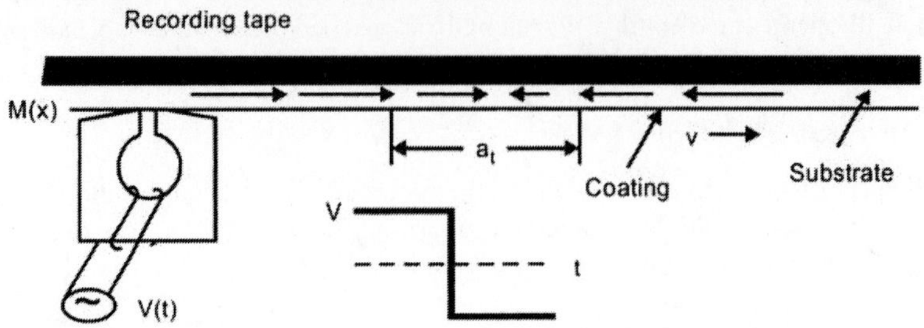

Fig. 8.19. Resultant magnetization from a step change in head voltage.

The essential process in direct recording, therefore, is the writing of a transition or polarity change of magnetization. In Fig. 8.19, the resulting magnetization from a step change in head voltage is shown. In saturation recording, the spatial variation of magnetization will not be a perfect replica of the time variation of signal voltage. Even if the head field change is perfectly abrupt, the magnetization will gradually change from one polarity to another. In Fig. 8.19, this is indicated by a gradual change in vector lengths; the notation *at* denotes an estimate of the distance along tape over which the magnetization reverses. The nonzero distance between polarity changes of magnetization is due to the finite-loop slope at the coercivity, combined with the gradual decrease of the head field away from gap center. This process is illustrated in Figures 8.20 and 8.21. In Figure 8.20, the *M-H* remanence loop is shown for a well-oriented tape sample.

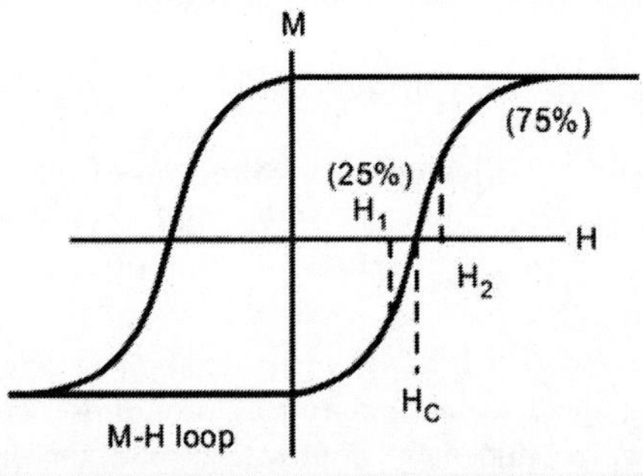

Fig. 8.20. M-H hysteresis loop for a typical tape.

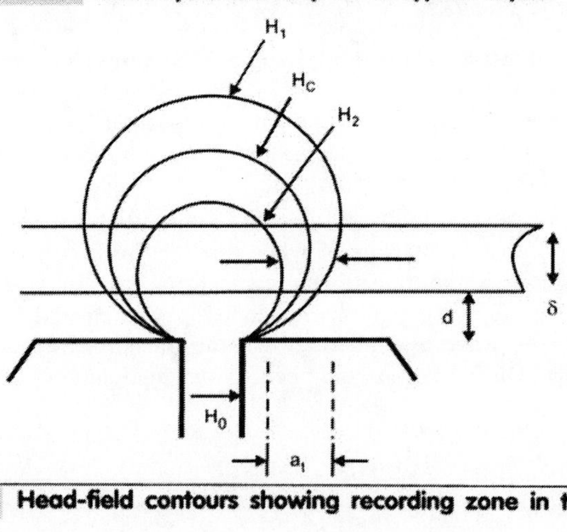

Fig. 8.21. Head-field contours showing recording zone in tape.

The magnetization M is the remanence magnetization that results from the application of a field H, which is subsequently removed. If the tape is saturated in one direction, for example, M, and a positive field is applied, then the magnetization will start to switch toward the positive direction when the field is close to the remanent coercivity Hcr. Because the slope of this M-H loop is not infinitely steep for fields near $_{Hc}r$, the switching will take place gradually. $H1$ denotes the field that switches 25 percent of the particles to leave the magnetization at $-M/2$; $_{Hc}r$ is, in fact, the 50 percent reversing field that leaves $M = 0$. H_2 denotes the 75 percent switching field that leaves the magnetization halfway to positive saturation ($+M/2$). During recording, a finite transition width will occur, depending on how H_1 and H_2 are spatially separated. In Fig. 8.21, three contours of recording field are plotted for the three fields H_1, H_cr, and H_2. In plots of head fields, larger fields are closer to the surface of the head and toward the gap center. Thus, along the mid plane of the tape the field magnitudes H_2, H_cr, $H1$ are in decreasing order away from the gap centerline. Therefore, if the tape is initially magnetized negatively, a positively energized head (H_0) will switch the magnetization according to Fig. 8.20 following the spatial change of the fields. This yields a finite transition width *at*. The transition width can be narrowed by using tape with a steeper loop gradient, making H_1 closer to H_2 in magnitude (a narrower spread in switching fields) and decreasing the head-to-tape spacing, which moves H_1 and H_2 closer together spatially (a larger head field gradient) as indicated in Fig. 8.21. In addition, spatial changes in magnetization cause demagnetization fields in the tape that further broaden the transition.

This demagnetization broadening can be reduced by increasing the tape coercivity.

In saturation recording, the signal current is held fixed for all wavelengths. The current level is set to optimize the short-wavelength output, and complete saturation of the tape does not occur. In Fig. 8.22 reproduce voltage versus input current is shown for two different wavelengths in square-wave recording for video tape (Ampex 196). If true saturation were to occur, the curves would increase initially with current as the tape is recorded, and then level, representing a magnetization saturated to full remanence and recorded fully through the tape thickness of 200 μm (5 μm). However, at short wavelengths these curves are peaked, and the current that yields the maximum output represents recording only a very small distance into the tape. For video recording on a type C format machine optimized at 10 MHz ($\lambda \approx 100$ μm), this is a record depth of approximately 50 μm. A mechanism for optimization of this parameter can be seen by considering the change in transition with record current. As the current is increased, the point of recording shifts continuously downstream from the center of the gap. The transition width depends upon the head-field gradient at the recording point. This field derivative, $H_2 - H_1$, divided by the separation between them, increases with distance along the head surface, as shown in Fig. 8.21, reaching a maximum near the gap edge and thereafter decreasing. Because the reproduce voltage increases with decreasing transition width, a maximum voltage will occur as the current is increased. This peaking becomes more pronounced as the wavelength is reduced.

A form of linearity known as *linear superposition* is found in saturation recording. For constant- current recording (strictly, *constant-field amplitude*) the reproduce voltage from a complicated pattern can be shown to closely resemble the linear superposition of isolated transition voltage pulses, according to the timing and polarity change of the series of transitions. The lack of complete linear superposition is believed due to demagnetization fields. This accompanies large head-to-medium separations, as in rigid-disc applications, where the increase in the demagnetization fields can cause significant nonlinearities.

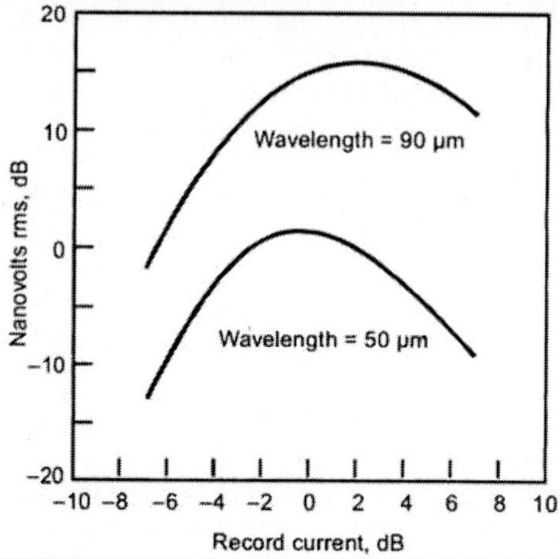

Fig. 8.22. **Reproduce voltage versus record current for typical video tape.**

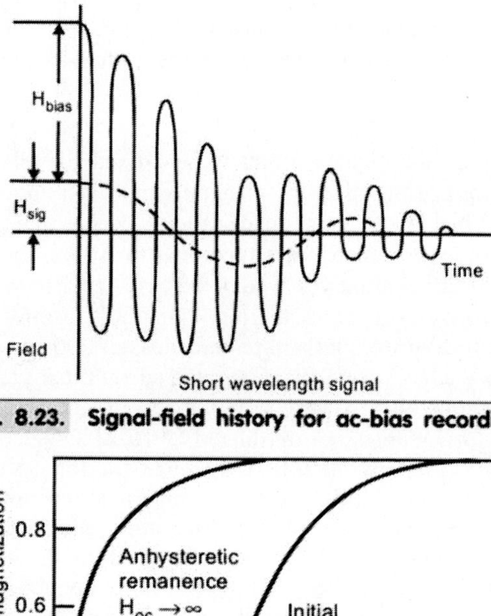

Fig. 8.23. **Signal-field history for ac-bias recording.**

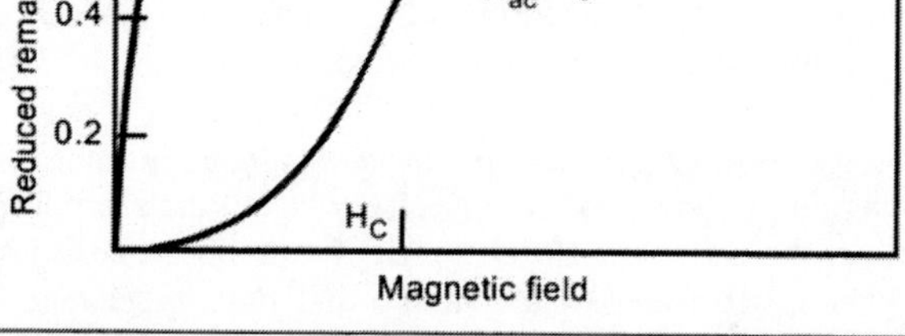

Fig. 8.24. **Comparison of sensitivities of ac-bias and direct recording.**

Bias Recording

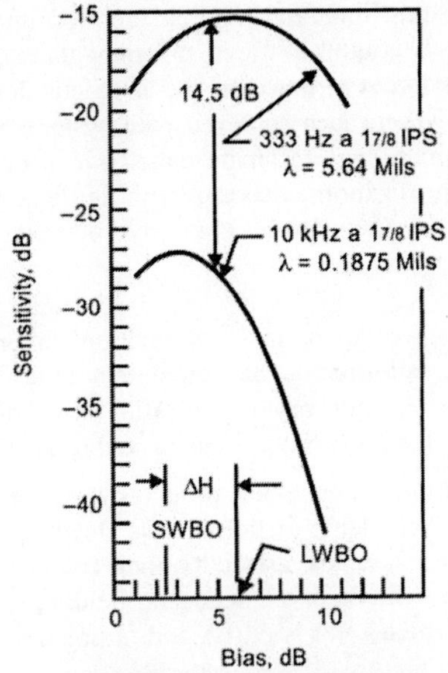

Fig. 8.25. Low-level output sensitivity versus bias for long and short audio wavelengths.

In some applications, predominantly audio recording, strict linearity is required between the reproduced voltage and the input signal. This may be achieved by superimposing a high-frequency, large-amplitude bias current on the signal (Fig. 8.25). The physical process is called *anhysteresis*. The bias field supplies the energy to switch the particles while the resulting remanent magnetization is a balance between the signal field and the inter particle magnetization interactions. In Fig. 8.26 a comparison is shown between the magnetic sensitivity of ac bias recording and direct recording. Alternating-current bias or anhysteresis results in an extremely linear characteristic with sensitivity an order of magnitude greater than that for unbiased recording. In typical audio applications, the bias current is somewhat greater than that of the signal in direct recording. The signal current is approximately an order of magnitude less than the bias current and is set to maintain the harmonic distortion below 1 to 2 percent. For complete anhysteresis there should be many field reversals as the tape passes the recording point where the bias field equals the tape coercivity. This is achieved if the bias wavelength is substantially less than the record-gap length. In fact, to avoid reproducing the bias signal itself, the bias wavelength is usually less than the reproduce-gap length.

As in direct recording, a current optimization occurs, but in bias recording it is with respect to the bias current. In Figure 8.25, reproduce voltage is shown versus bias at short and long wavelengths.

At long wavelengths, the optimum bias occurs approximately when the bias field has recorded through to the back of the medium. This is often taken to be the usable bias current since close to this optimization a minimum in the distortion occurs. For shorter-wavelength machines, the bias is chosen as a compromise between *short-wavelength bias optimization* (SWBO) and *long-wavelength bias optimization* (LWBO).

Particle Orientation

The previous discussion applies generally to all types of recording tape; the most common is that composed of uniaxial elliptical particles oriented *longitudinally* in the direction of head-tape motion. However, *isotropic* tape does exist and is composed of particles of cubic (threefold) symmetry that exhibit high remanence in all directions. In addition, it is conceptually possible to *vertically* orient the grains to result in a tape isotropic in the plane, but capable of recording signals perpendicular to the surface. These last two would be advantageous for transverse recording since the difficult process of orienting elliptical particles along the tape cross direction could be avoided.

During the tape-coating process, elliptical particles will naturally orient along the tape-coating direction. A field applied during coating improves the orientation even further. It is quite difficult to orient these particles vertically because the hydroscopic coating forces overwhelm the magnetic force from a vertical-orienting field. Success has, however, been achieved with systems that *inherently* yield vertical orientation. As an example, barium-ferrite platelets can be coated to yield perpendicular media since the magnetization anisotropy axis is perpendicular to the plane of the particles. Other vertical medium can be made by sputtering CoCr on either a tape or substrate.

Erasure

The writing of new information on previously recorded media requires that the previous information be completely removed. Erasure requirements, in terms of the previous-signal to new-signal ratio, vary from –30 dB for digital systems to as much as –90 dB for professional analog audio. Analog video recorders require about –60 dB.

Erasure is the ac-bias or anhysteretic process with zero-signal field. If a reel of tape is placed in a large ac field that is slowly reduced so that many field cycles occur when the field is near the coercivity, then complete erasure is easily obtained, In addition, the largest amplitude of the acerasing field must be sufficient to reverse at least 99.9 percent of the particles (for –60 dB), and in practice that field is about three times the coercivity.

Most tape recorders utilize erase heads to remove old information before recording new data.

Similar to bias recording, the requirement of the erase frequency is that the wavelength be much less than the erase-head gap to provide sufficient reversals of the particles. However, one important problem occurs with an erase head. As the erase current increases, the erasure level does not continuously increase as more of the *M-H*-loop tail is switched. There is an erasure plateau of about –40 dB for erase-gap lengths of 1 to 2 mils (0.0254 to 0.0508 mm) and tape thicknesses of 200 to 400 μm (5.08 to 10.16 μm). This levelling is believed to be due to the phenomenon of rerecording. As the tape passes the erase head, the field from the portion yet to be erased (entering the gap region) acts as a signal for the bias-erase field to record a residual signal at the recording zone on the far side of the gap. This effect is seen only at long wavelengths where the field is high. The problem is eliminated with double erasure by using a double-gap erase head.

The erasure level can be decreased by decreasing the ratio of the tape thickness (or recording depth) to erase-gap length; this reduces the rerecording field.

8.13.2. Video Cassette Recording and Reproduction

Introduction

A video cassette recorder (VCR) is a device for recording and reproducing a video signal and an audio signal recorded on a recording medium (*i.e.*, a video cassette tape). Development of a semiconductor technology has enabled digital signal processing to be applied to fields of wide-band video and audio recording and reproducing, among others. Video cassette recorders have gained wide acceptance in the consumer market place where video information is recorded on magnetic video tape stored in a video cassette. A video cassette recorder can record image and audio signs onto a magnetic tape and reproduce the signals from the magnetic tape on which the signals are recorded. The video cassette recorder can reproduce the recorded video signal at a normal, low or high speed, or can stop reproducing the signal. Video tape cassette recorders are widely used to record and/or reproduce video signals on a recording medium (tape). The development of a recording and reproduction medium and recording and reproduction technology has enabled general home video cassette recorders to utilize digital signal processing. The integration of a VCR with a camera to produce what is commonly called a camcorder has made possible the recording of home movies on magnetic tape in a true portable system.

8.13.3. Video Tape Recording

Videotape machines can be categorized using several criteria, including tape size and style, scanning method, recording format, and input/output connections. Tape widths range from 1/4-in up to 2-in, and styles include both open reel and cassette. The cartridge style, in which a single reel is contained in a removable shell, has found little acceptance in the video industry, despite widespread adoption by the audio industry. Scanning methods include *transverse* and *helical scan*.

Like cartridges, the *arcuate scanning* method has not been utilized successfully. Recording formats include: *composite* (direct colour), *component*, and Y/C (*colour-under*) methods in both analog and digital designs. The 2-in quadruplex format is the only commercially successful format to utilize transverse scanning. The 2-in format is also one of a limited number of open-reel machines; others include the 1-in type B and C helical-scan machines. Other than the formats mentioned, nearly all other commercially successful videotape machines have been of the helical-scan, cassette variety. Figure 8.26 shows the two basic types of helical tape wrap geometries.

The classic type C format tape path is illustrated in Fig. 8.27.

The basics of today's VTRs are in many ways similar to the machines produced in the mid 1960s. The 2-in VTRs are relatively straightforward and easy to understand and having a basic knowledge of these formats makes it much simpler to grasp the concepts behind the newer machines.

Fundamental Concepts

Basic building blocks of VTRs include the transport and related control systems, servos, video and audio circuits, and in many cases, additional circuitry for editing or other special functions, such as *dynamic tracking*. The *transport* is an electromechanical assembly that provides a precise path through which the tape moves. Also included are the various motors and guides used to move and position the tape. Magnetic tape transports require mechanical alignments to be maintained precisely. Tape must be capable of moving at any speed and in forward or reverse without experiencing edge damage. This is especially true in editing machines. Control of the early VTRs was accomplished using simple relay circuits, whereas today's units rely heavily on microprocessor-based systems. Servos are used to control the head *drum assembly*, capstan, and in many units the reel motors. The basic analog video circuits consist of input, modulator, demodulator, and output stages. Audio in most analog VTRs is comparable to professional audio recorders.

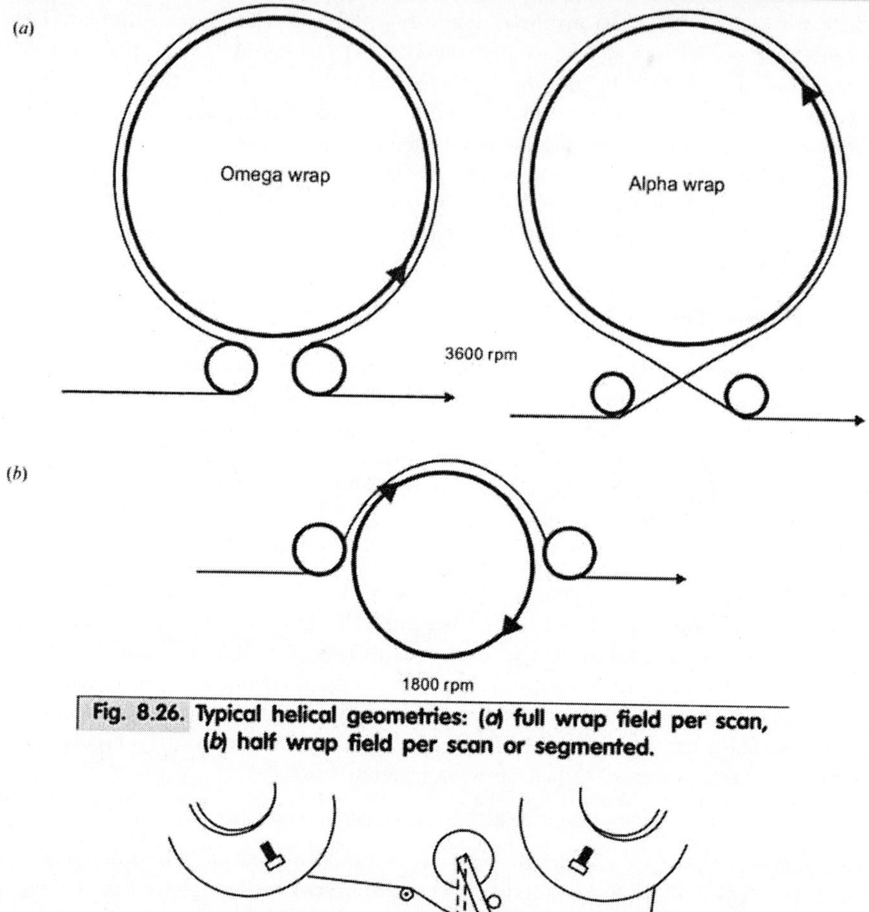

Fig. 8.26. **Typical helical geometries: (a) full wrap field per scan, (b) half wrap field per scan or segmented.**

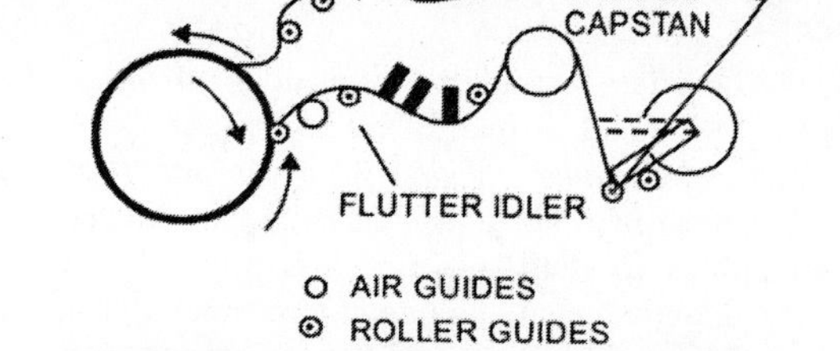

Fig. 8.27. **Type C format transport layout.**

Tape Transport and Control

Fundamental to any videotape machine is the mechanical transport assembly. The transport provides the mechanical interface between the machine electronics and the information recorded on the tape. Depending on the format, the transport will include the following:

- Loading mechanism and reel hubs
- Tape guides and sensors

- Erase head
- Drum assembly containing the video heads, audio heads, and *control track* head
- The capstan

Additional transport parts found in editing machines include:

- Time code heads
- Audio confidence playback heads
- Audio-only erase heads
- Flying erase heads mounted in the drum assembly for video editing

Flying erase heads are used during *insert edits* and at the beginning of *assemble edits*, until sufficient time has passed to allow the full track head to begin erasing the tape.

Constant torque motors linked to the reel hubs prevent slack tape and minimize tape stretch resulting from excessive tension. Brakes are engaged when the tape is stopped to prevent spillage.

On many units, the brakes are also used to slow the tape during movement; however, more sophisticated machines rely on eddy currents within the motors to provide gentler tape deceleration.

Precision rotating and fixed guides are used to route the tape around, over, and through the various assemblies of the transport. Some machines also make use of air (either pressure or vacuum) to position, handle, or buffer the tape. The 2-in machines used a *vacuum guide* to control and position the tape at the head drum assembly. Several formats have used *vacuum capstans* to precisely control tape movement without a pinch roller. Pressurized air escaping through small holes in fixed guide assemblies provides an *air bearing*, reducing both tape and guide wear.

Loading and unloading of tape transports has also been accomplished quickly and accurately by using air pressure or vacuum.

Head drum assemblies contain one or more video heads. Relative tip-to-tape movement is typically 1000 inches per second (ips) in professional machines, and around 250 ips in consumer machines. A capstan/pinch roller combination or a vacuum capstan assembly is used to precisely control tape movement. Tracking controls are used to electronically change the mechanical timing relationship between control track pulses and the recorded video tracks. A nonstandard recording could have these pulses delayed or advanced with reference to the video tracks. Either way, this causes the RF signal recovered from the tape to be less than ideal, as the head is not centered on the recorded tracks. By adjusting the tracking control, the relative timing between the control track and video tracks can be changed, centering the heads on the recorded tracks and allowing as much RF as possible to be recovered from the tape.

Control circuitry provides the proper signals to the various motors involved. Control systems were rather primitive on early machines, whereas today's machines rely almost exclusively on microprocessor-based control. Relays, and later transistor drivers, were used for motor control.

Modes included what has become the standard transport controls—stop, play, record, fast forward, and rewind. Many newer machines also include forward and reverse search modes, which combine viewable pictures with rapid tape movement.

Servo Systems

Servos used in video tape machines include the capstan and drum servos. Nearly all professional units allow the capstan and drum servos to be referenced and locked to external sync. Once a transport can be locked to external sync, it then becomes relatively easy to lock two transports for editing purposes. Reel servos are used in many VTRs, but early models relied on constant torque motors. The constant torque method is also used on many types of consumer VCRs. Some machines tie the reel servos to the motion of the capstan, assuming that if the capstan is stationary, tape is not moving. These systems usually include vacuum capstans. Capstan motion, along with pulse generators located in the reel tables, can be used to calculate tape pack diameters precisely.

These calculations, along with tape tension detector(s), provide sufficient information to gently and precisely handle tape in a manner that allows rapid high-speed acceleration and deceleration.

Machines using this type of tape handling can provide prerolls as short as a single frame (with the head in *ready* mode). Microprocessor control of the servo system adds considerable flexibility to a video recorder. A servo system based on microprocessor control is shown in Fig. 8.28.

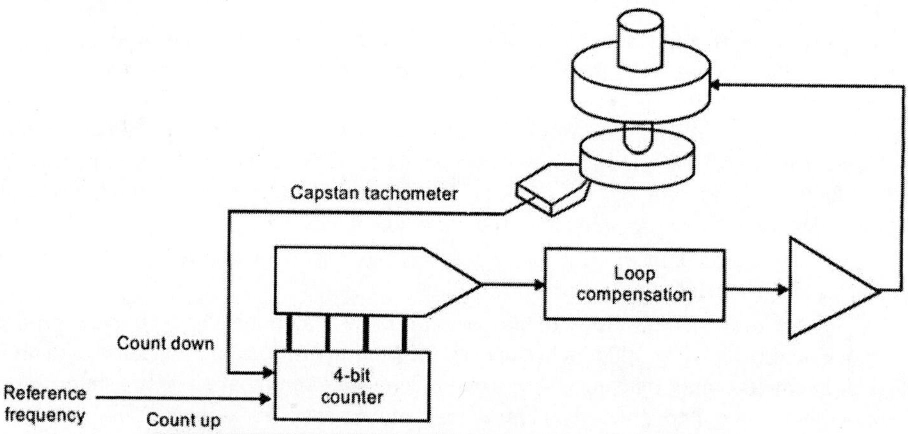

Fig. 8.28. Microprocessor-control servo block diagram.

Capstan servos are responsible for tape movement during most modes that provide viewable pictures. Drive oscillators typically have a wide range to provide the tape speeds required. Error correction signals are provided through a feedback loop that includes the control track head. The capstan servo must position the tape such that recorded tracks are physically located where they can be read by properly positioned video heads.

Video Signal Path

On the record side, 1 V peak-to-peak composite video is applied to an input amplifier and sync separator. Sync pulses are routed to the servos so that the transport mechanisms can be properly phased and standard recordings can be made. Having the servos locked to an external signal during record can cause the input video to be recorded in such a way that upon playback it will not be positioned properly with respect to the reinserted sync pulses. A worst-case scenario involves recording a nonsynchronous source, such as a satellite feed onto a machine locked to house reference.

Upon playback, the recording will appear nonsynchronous and be unusable. Most machines automatically switch from external reference during playback to incoming video reference during record.

After the tape is in motion and the servos are locked, a record lockout circuit verifies that the tape can be recorded on. Then the erase heads are energized, erasing any previous recordings and preparing the tape for the new information. Video goes through a pre-emphasis circuit and is frequency modulated before being applied to the record amplifier and record heads. Many machines provide inputs and outputs for the record amplifiers and playback preamplifiers. In this manner, the modulated RF signal can be dubbed from one machine to another. This process does not address time base errors, but because the demodulation/modulation steps are avoided, the duplicate tape is of higher quality than when the signal is returned to baseband video. Figure 8.29 shows a basic block diagram of a typical analog composite video recorder.

When a properly recorded tape is played back, the capstan and drum servos position the tape within the transport such that the reproduce heads are properly aligned with the recorded tracks.

Relative movement of the tape and head assemblies causes an induced current in the heads proportional to the recorded information on the tape. Preamplifiers located close to the video heads are used to increase signal levels, thereby minimizing noise. Once amplified, the reproduced signal is equalized to produce a flat frequency response. Rotary transformers provide a signal path from the spinning heads to the stationary transport. A limiting amplifier and demodulator complete the playback signal path. Because the recorded signal is frequency-modulated, amplitude variations in the reproduced signal do not affect the reproduced signal. Passing it through a limiting amplifier provides the demodulator with a constant-level signal. The demodulator circuitry, rather than responding to amplitude changes, is sensitive to zero crossings. In the demodulator, the carrier frequency is removed from the signal, and only baseband video remains. The baseband video is applied to a de-emphasis circuit to remove frequency distortions added to improve the recording process. Finally, an output amplifier is used to set the signal level and provide sufficient power to drive the output.

Generally, the output amplifiers are considered the final stage in the playback circuitry. However, because of time base errors caused by both the playback and record processes, additional circuitry is required to stabilize and correct the signal to meet the requirements of the RS- 170(A). In the Ampex *quads* of the early 1960s, this was accomplished with three or four stabilizing units—the Amtec, Colourtec, velocity compensator (optional), and a processing amplifier (proc amp). Later, a buffer type unit was used, followed by actual timebase correction units.

Time base correctors (TBCs) have reached a level of sophistication such that today complete units are available as plug-in cards for personal computers. In many videotape machines not intended for broadcast use, stabilizer circuitry is incorporated to provide output signals suitable for display on a colour monitor or television.

With analog recording systems, recording and playback errors are cumulative. Machine-to -machine dubs using baseband video can be performed without a TBC; however, signal quality degrades quickly. The reason is twofold:

The time base errors in the active picture area become evident because of uncorrected timing and phase errors in the video signal, especially the chroma.

Errors in horizontal and vertical sync pulse timing cause instabilities in the recorder servos.

These servo instabilities compound the errors within the active picture area, causing much larger errors when the recording is played back.

Dropout compensation is accomplished by repeating previous lines whenever a loss of RF occurs. Dropouts are fairly common on video recordings and are caused by either airborne particulates or small sections of tape with poor response. Particulates that end up between the head and tape during record will prevent RF from being recorded, thus causing a permanent loss of signal. During the playback process, particulates will cause random and generally non-repeatable dropouts. Tape problems are caused either by physical damage or a imperfections in the magnetic coating of the tape. Manufacturing techniques have improved considerably over the years and high-quality tape is readily available.

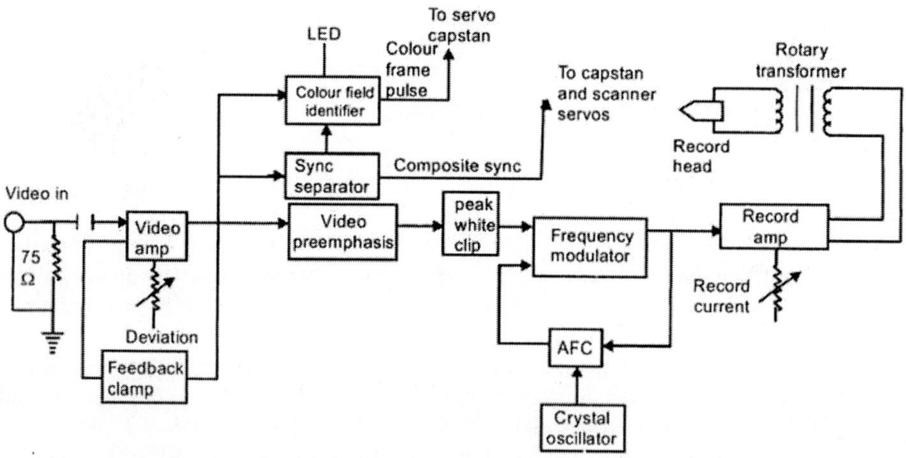

(*a*) **video record circuits**

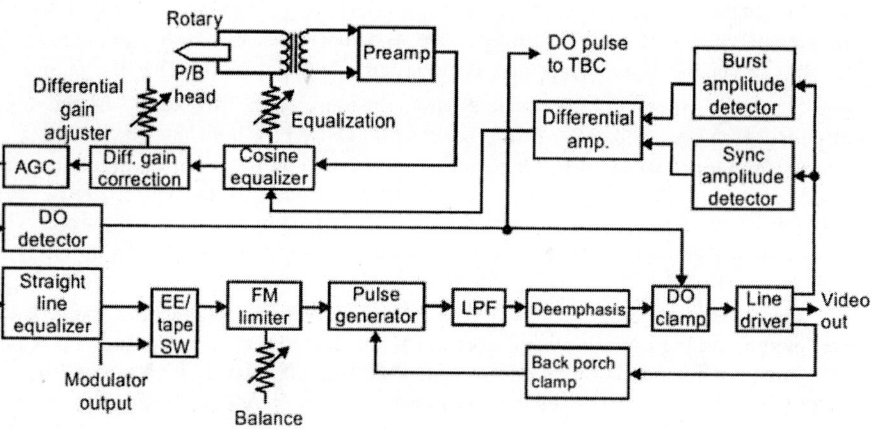

(*b*) **reproduce circuits**

Fig. 8.29. Single-head VTR: (*a*) video record circuits, (*b*) reproduce circuits.

Today, airborne particulates are the primary cause of dropouts not caused by physical damage; maintaining a clean environment is the best way to reduce dropouts of this nature.

The previous description covers the composite recording process used in the 1-in type C format.

Both the 2-in quads and the 1-in type B format used similar circuits; the main difference is the segmenting of the video signal so that it can be recorded to tape using separate heads. For *colour-under recordings*, including the 3/4-in U-matic and the VHS and 8 mm consumer formats, two parallel circuit paths are used, one for luminance and the other for chrominance. (See Figure 8.30). These units typically provide composite and Y/C inputs and outputs. Many decks also provide dubbing connections. Although the distribution of separate Y/C signals throughout a facility may be considered "component distribution," the colour-under recording process is not considered a component format. In the colour-under process, composite input signals are fed through a low-pass filter to separate the luminance from the chrominance and limit high-frequency luminance components to approximately 3 MHz. The chrominance subcarrier is bandpass limited to provide a 0.5-MHz-wide double sideband signal. This signal is amplitude modulated onto a carrier near 650 kHz, which is added to the FM luminance carrier for recording on tape. The recording spectrum is shown in Figure 8.31.

Colour-under systems use helical-scanning. Normally, two heads are used to record the combined signals, one for each field. Typically the tape wraps around the head, allowing the heads to be switched by a symmetrical square wave locked to incoming vertical sync or, during playback, generated by the control track signal.

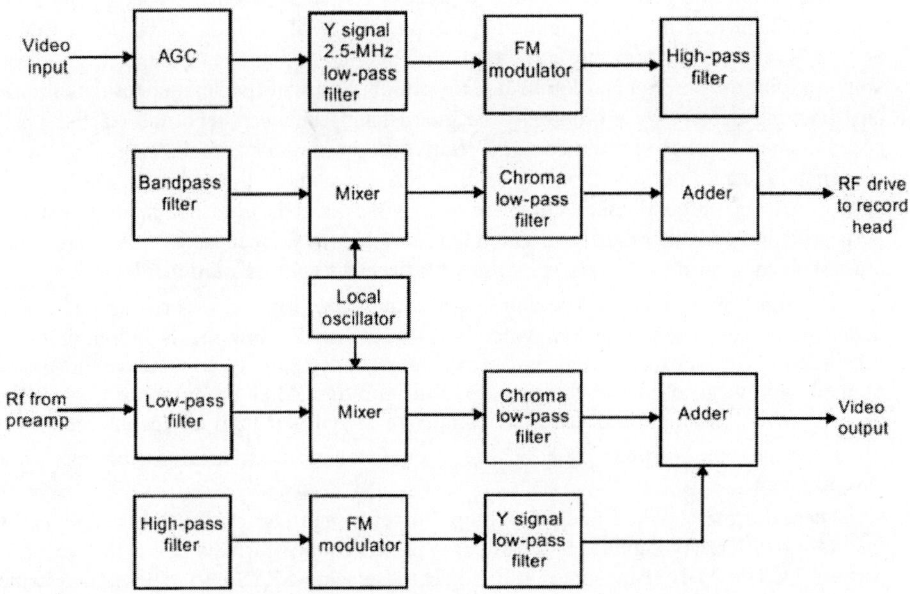

Fig. 8.30. Block diagram of a colour-under system.

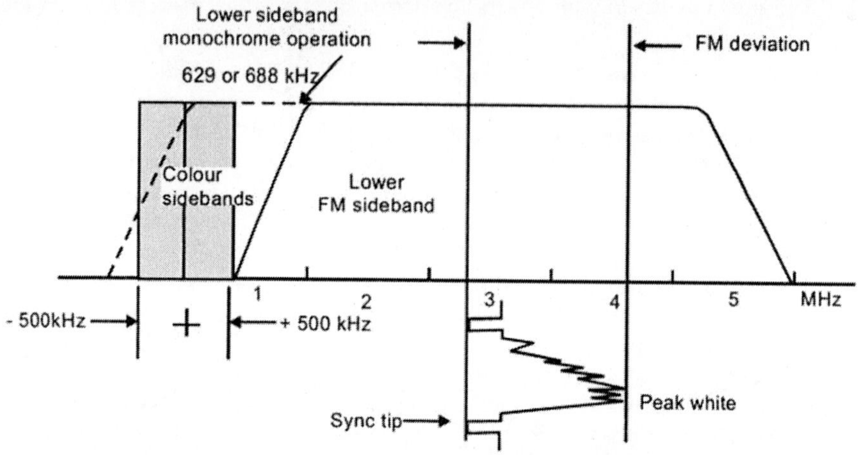

Fig. 8.31. The recording spectrum used by video tape machines.

(The recording spectrum used by video tape machines that use the colour-under recording process. The AM chrominance subcarrier is added to the FM luminance carrier and recorded on tape. Part of the lower sideband of the FM luminance signal is removed to allow for the chrominance signal.)

One advantage of the colour-under process is that time-base errors in the colour signal are kept to a minimum, allowing the output signal to be displayed on a monitor or television with little or no correction circuitry. Disadvantages include high-chroma noise and limited chroma and luminance resolution. Many of the smaller formats improve tape packing density by eliminating the guard bands between recorded tracks. This is accomplished by angling the head gap relative to the recorded track. Typically, the gap is perpendicular to the track; on the smaller formats without guard bands, the gap is up to ± 15° from perpendicular. On these units, one head is angled one direction from perpendicular, and the other head is angled the other way. Because of this, guard bands are not needed as the off-axis response of adjacent tracks is minimized.

The most straightforward method of accomplishing timebase-correction of colour-under signals is to treat the luminance and chrominance components individually (see Fig. 8.32). After separating the two components, independent error measurements can be made and used to correct the respective instabilities. After the time-base errors have been removed, a stable and coherent composite signal can then be reconstructed.

Like the colour-under process, the component formats separate luminance and chrominance.

These formats take an additional step, however, and separate the chrominance into the colour difference signals, R–Y and B–Y. The primary formats using this recording process include M-II (Panasonic) and Betacam/Betacam-SP (Sony). These two formats are similar but not interchangeable.

Although the same component signals are used for input and output (Y, R–Y, and B–Y), signal levels are different. Care must be taken to adjust input/output (I/O) levels when decks based on these different formats are interconnected. Most decks provide composite and component inputs and outputs as well as dubbing connectors. These units provide reasonably high-quality encoders and decoders for composite-to-component signal conversion. Three parallel signal paths exist; the luminance path typically has a

4.2–4.5 MHz bandwidth, whereas the paths used for the colour difference signals have only a 1.5-MHz bandwidth each. On these machines, one head records the FM luminance signal on a single track per field. A second track, recorded by a second head, contains the chrominance signal, as well as FM audio. The chrominance signal is a *compressed time division multiplexed* (CTDM) signal in which the analog R–Y and B–Y components are time compressed and then multiplexed into a single signal. Delaying the B–Y signal approximately one-half of a field allows the R–Y information to be recorded first, followed by the B–Y. FM audio is placed at the beginning of the track, before the colour difference information.

Upon playback, timing differences between the three component video signals are corrected and the components are time base-corrected separately. The component signals are output a specified signal levels and also fed to an encoder. The encoder combines the signals to provide a composite output that can be used if desired.

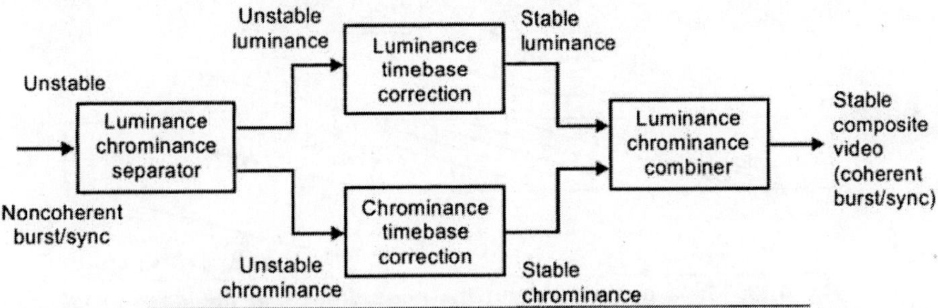

Fig. 8.32. Time-base correction of colour-under signals.

The Helical Scan System

In an audio cassette deck, which only registers audio signals, the tape passes over a static recording/playback head at constant speed. The higher the speed of the tape, the more tape particles pass the head opening and the higher the frequencies that can be registered. Thanks to the extremely narrow head opening, it is possible to record and play back the entire tone range, up to 18,000 or 20,000 Hz, despite a slow tape speed of no more than 4.75 centimeters per second.

However, to register video signals, a range of 3.2 MHz is required and so a tape speed of approximately 5 meters per second is a prerequisite. This is over 100 times as fast as the tape speed for an audio cassette deck. The required high recording speed for video recorders is realized by the helical scan system without such high tape speeds. The system basically consists of a revolving head drum that has a minimum of two video heads.

The head drum has a diameter of approximately 5 cm and rotates at a speed of 1500 revolutions per minute. The 1/2" (12.65 mm) wide videotape is guided around half the surface of this drum, in a slightly oblique manner. This is achieved by positioning the head drum at a slight angle. The tape guidance mechanism then ensures that the tape is guided through the device at a speed of approximately 2 cm per second (half of the low tape speed that is used in audio cassette decks).

In the meantime, the rapidly revolving video heads write narrow tape tracks of no more than 0.020 to 0.050 mm wide on the tape, next to each other, diagonally. Every half revolution, each of the two heads writes one diagonal track which equals half an image. The first head writes one track, *i.e.*, the first field (the odd numbered scanning lines). The second head writes a second track, *i.e.*, the other half of the image (the second field: the even numbered scanning lines), which precisely fits in the first image. This corresponds to the interlacing principle, as applied in television. One full revolution of both heads results in two diagonal tracks right next to each other, together forming one entire image scan (a frame). This means that two apparently contradictory requirements can be realized simultaneously: low tape speed of only 2 cm per second and at the same time a high registration speed (relative tape speed) of no less than 5 meters per second. These two requirements make it possible to record the high video frequencies up to 3.2 MHz. At the same time, the low tape speed gives a time capacity up to three hours.

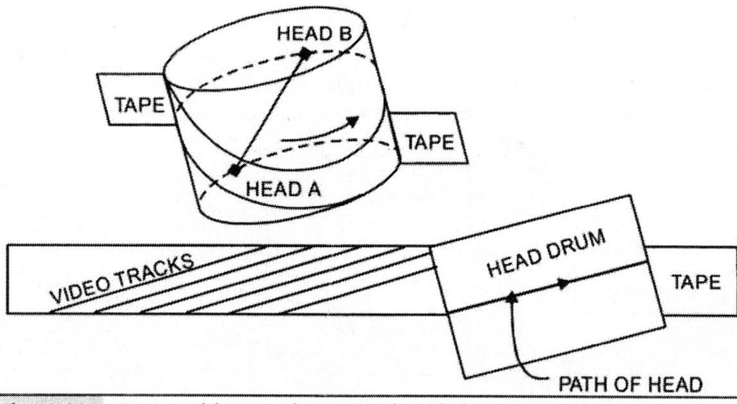

Fig. 8.33. Tape guidance along the head drum with the video heads writing tracks on the tape.

Azimuth Settings

Compared with early video recorders, modern day video recorders have their video tracks lying right next to each other. To avoid interference, the two video heads are angled slightly away from each other. As a result, the video head openings that transmit the magnetic tracks to the tape create an angle between them. The heads are 15 degrees angled in opposite direction, making a total angle of 30 degrees. This diverted registration angle ensures no problems are caused if the heads slightly lose track when playing back and touch the next track. The heads only register tape information at an angle that precisely corresponds to the position of the head opening. This system is called the **azimuth recording system**. If the video heads stray too far from the track, which could lead to distorted images, tracking control can correct this.

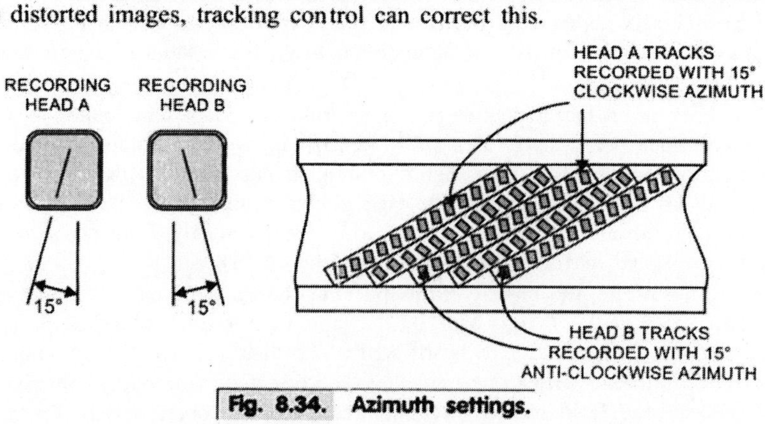

Fig. 8.34. Azimuth settings.

Synchronization Track

The revolutionary speed of the head drum and the video heads needs to maintain constancy within strict parameters. Moreover, the tracks must be scanned during playback in precisely the same way as they were recorded. Each tape track is synchronized at the recording stage by means of field synchronization pulses. These pulses are generated in the video recorder by a separate head which are recorded on a separate narrow track at the side of the video tape. This is called the **synchronization, servo or control track.**

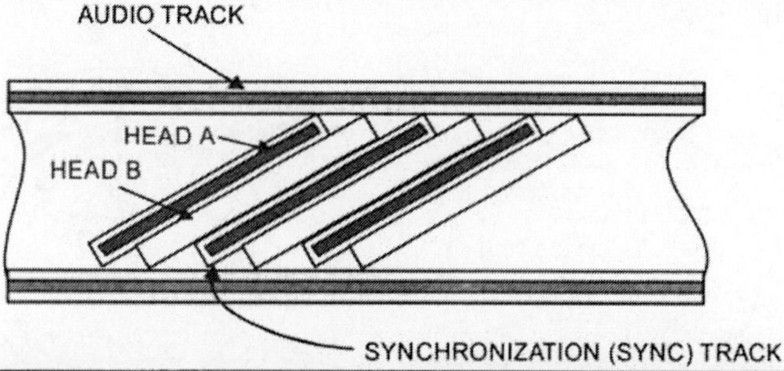

Fig. 8.35. Position of the video, audio and synchronization tracks on the tape.

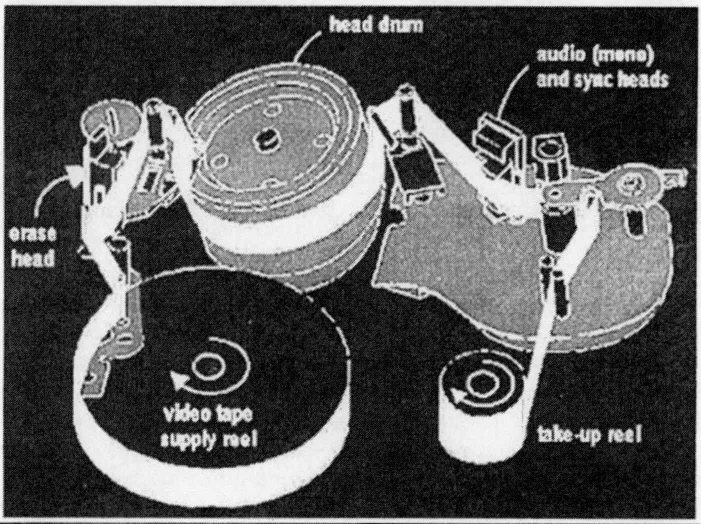

Fig. 8.36. Position of the audio, sync and erase heads inside the VCR.

8.13.3. Video Systems

There are three major video systems in use today:

1. Video Home System (VHS)
2. Betamax
3. Video Hi8

When the video recorders were first introduced, Philips also developed a system called V2000. Despite the fact that is a high quality system, it was not successful in the market. Although Betamax was reasonably successful at first, its popularity waned and VHS was adopted as the world standard.

Betamax

The Sony Betamax System, launched in 1975, was based on the pre-existing professional Sony U-matic-system. In the Betamax system, the video tape is guided along the head drum in a U-shape for all tape guidance functions, such as recording, playback and fast forward/backward. When the cassette is inserted, the tape is guided around the head drum (called threading). Threading the tape takes a few seconds, but once the tape is threaded, shifting from one tape function to another can be achieved rapidly and smoothly.

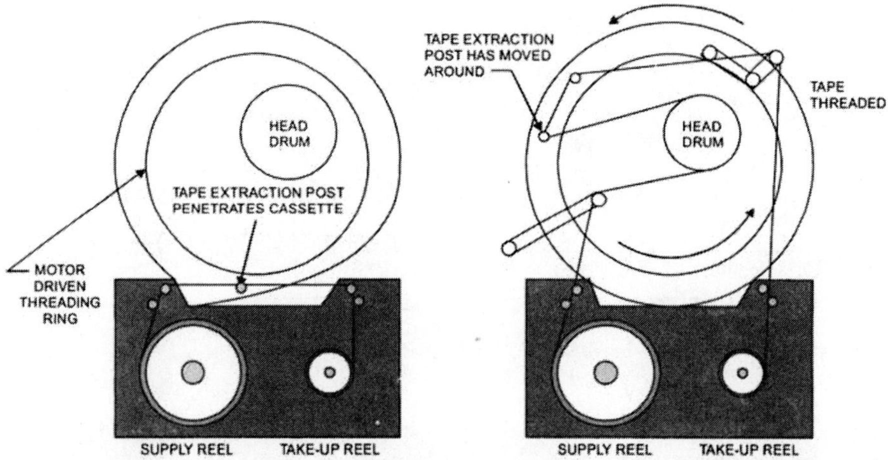

Fig. 8.37. The Betamax U-system before and after threading.

VHS

JVC's VHS System was introduced one year after the launch of Betamax. In VHS, the tape is guided through in an M-shape; the so-called **M-tape guidance system.** It is considered simpler and more compact than the U-system. Threading is faster and is done every time the tape guidance function is changed. It is therefore somewhat slower and noisier than the U-system. This problem is being solved by "Quick-start" VHS video recorders, which allow fast and silent changes in tape guidance functions. To avoid excessive wear, M-tape guidance system recorders are provided with an automatic switch-off feature, activated some minutes after the recorder is put on hold, which automatically unthreads the tape. An improvement of the basic VHS system is HQ (High Quality) VHS.

In the VHS system different starting points were used than in Betamax, such as track size and relative speed. VHS has rather wide video tracks, but a slightly lower relative tape speed, and that also counts for the audio track. In general, the advantages of one aspect are tempered by the disadvantages of the other. The end result is that there is not too much difference between the sound and image qualities of both systems.

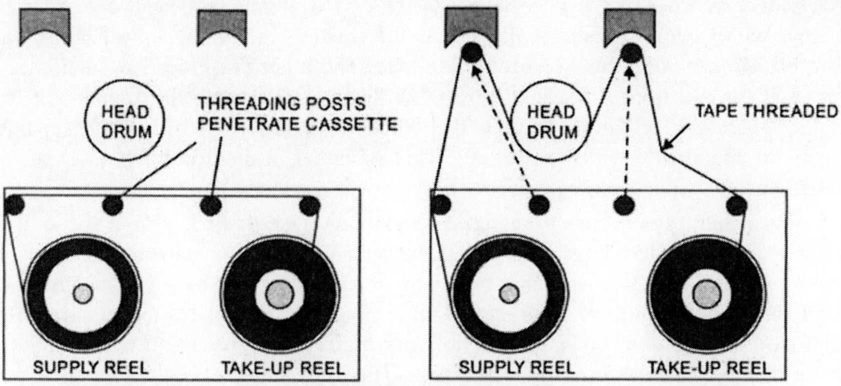

Fig. 8.38. The VHS M-system before and after threading.

Video Hi8

As a direct addition to the Video-8 camcorders, there is a third system: Video Hi8, which uses a smaller cassette than VHS and Betamax. The sound recording takes place digitally, making its sound quality very good. When using the special Hi8 Metal Tape, the quality of both image and sound are equivalent to that of Super-VHS. The Video-Hi8-recorder can also be used to make audio recordings (digital stereo) only. Using a 90 minute cassette, one can record 6 × 90 minutes, making a total of 18 hours of continuous music. The video Hi8-system also allows manipulating digital images, such as picture-in-picture and editing. Video Hi8 uses a combination of the M- and U-tape guidance system.

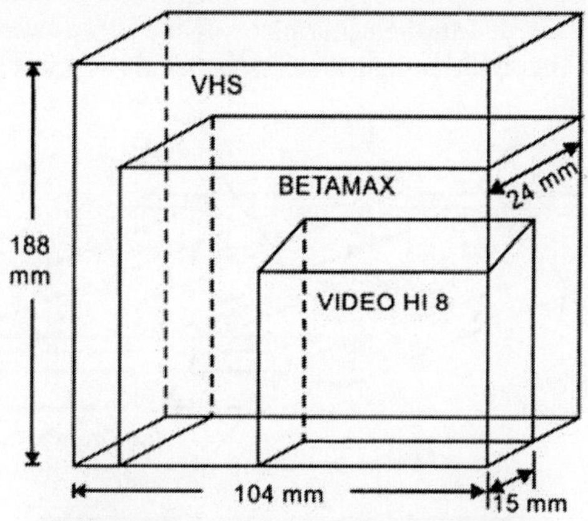

Fig. 8.39. Cassette sizes compared.

8.13.4. Sound Recording

Mono

In case of a mono video recorder, the audio signal which corresponds with the image is transferred to a separate, fixed audio head. As in an audio cassette deck, this head writes an audio track in longitudinal direction of the tape. This is called **linear** or **longitudinal track recording.**

The video recorder has two erase heads. One is a wide erase head covering the whole tape width which automatically erases all existing image, synchronization and sound information when a new recording is made. The other erase head is smaller and positioned at the position of the audio track. With this erase head, the soundtrack can be erased separately, without affecting the video information. In this way, separate audio can be added to a video recording. This is called **audio dubbing**, and can be particularly useful when making your own camera recordings.

The linear audio track does have some restrictions. Due to its low tape speed, it is not suitable for hi-fi recordings. Moreover, the audio track is so narrow (0.7 mm for VHS and 1.04 mm for Betamax) that not even stereo sound can be recorded properly. The frequency range is limited as is the dynamic range (which relates to the amount of decibels), and the signal-to-noise ratio is not very high.(The signal-to-noise ratio relates to amount of noise compared to the total signal. The higher this ratio, the less noise and the better the signal will be). The sound quality of the mono track can be improved by a noise reduction system. There is a way to get superior hi-fi stereo sound quality on a videotape (used in hi-fi video recorders,) which will be discussed later.

Hi-fi Stereo Sound

Hi-fi video recorders were developed for improved sound quality. The most common quality of video images is HQ. (The recorder is labeled 'VHS High Quality Hi-fi Stereo'). Conventional mono video recorders use linear audio registration, which does not allow hi-fi recordings. A special method was therefore devised to record stereo sound with hi-fi quality.

In the case of hi-fi, the audio signal is also put on tape via revolving heads similar to the video signal, not on the linear track. As there is no space between the video tracks, as the video tracks lie right next to each other with no space in between, the audio tracks need to be recorded in the same place as the video tracks. The way this is realized is by recording the audio signal under (deeper than) the video signal.

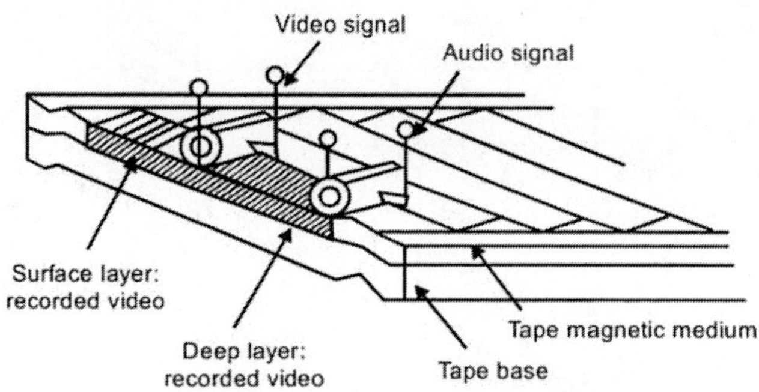

Fig. 8.40. Hi-fi video recording, where the audio signal is recorded at a deeper level, after which the video signal is recorded on top.

In hi-fi video recorders, the audio signal is modulated to a high carrier frequency. This is realized via FM modulation, with the right channel stereo signal at a slightly higher frequency than the left channel. The corresponding video and audio signals are written to tape immediately after each other. First the FM audio signal is registered at a deep level in the tape's magnetic coating. Straight after the audio signal, the video signal is recorded. As the frequency of the video signal is higher than the audio signal, it will not register as deep in the tape coating as the audio signal. The video signal erases the audio signal in the top layer and records the video signal instead. Thus, the audio and video signal tracks are written in the same magnetic layer, separately, one on top of the other. The entire magnetic coating is only 0.004 mm thick. To ensure that the two do not interfere, the audio and video tracks are written on tape from a different angle, by means of a different head with a different azimuth setting.

To guarantee compatibility with cassettes not recorded in stereo hi-fi, the fixed audio recording/playback heads remain in place. So, a hi-fi video recorder always has two audio registration systems installed. This offers possibilities for amateur video makers to do audio dubbing using an audio mixer to combine the sound of the hi-fi track with other sounds and to write the mix to the linear audio track. In this way synchronized recordings will be left intact.

Quality Audio Recorders

A hi-fi video recorder is also suitable as a high-quality audio recorder, not only because of the professional recording quality, but also because of the long play possibilities and the low recording costs.

The specifications of hi-fi video sound registration systems equal those of professional tape decks and compact discs. The entire sound spectrum can be covered without any problems, and the dynamic range is 80 dB, close to the 90 dB that compact discs can cover. (As the video recorder is a recording medium, a couple of good microphones can actually cover the whole 80 dB range.) Recordings made on a hi-fi video recorder result in almost immeasurable wow and flutter and very little harmonic distortion. The low tone quality of a hi-fi video recorder is remarkably good compared to tape recordings of cassette decks. A disadvantage is that sound editing is not possible via a VCR. Instead, the required tape segments can be copied unto another tape without hardly any loss of sound quality.

A hi-fi-video recorder needs to be tuned very accurately. As the two rotating audio heads function alternately, the recorded sound consists of successive particles and need to fit together perfectly. If they do not, the result is rumble, which is a humming sound. In high quality, well-tuned hi-fi video recorders you will not hear this sound.

Super-VHS

Super-VHS or S-VHS (for Betamax: ED-Beta) is a major step forward in the field of video registration. It is a recording-playback system of such high quality that its recordings are equal to the quality of direct TV broadcast signals. S-VHS offers better image quality than normal VHS, fuller colours, more sharpness, clearer colour separations and colour fields, and eliminate moire effects. Details not visible on normal VHS become visible on S-VHS, such as fine fabric patterns and eyelashes. As in all video recording systems, recording image and sound on magnetic tape involves the actual image, the colours, the horizontal and vertical synchronization pulses for perfect image building, and finally the sound. S-VHS requires so much information that it takes a frequency band of 7 million Hz (7 MHz) to store all the information. As this would be too much, in S-VHS the 7 MHz bandwidth is reduced to 5 MHz, without seriously reducing the image quality.

8.13.5. Analog versus Digital Video Cassette Recorders

Analog video cassette recorders adopt a consecutive recording method, while digital video cassette recorders adopt a non-consecutive recording method.

The digital video cassette recorder has a search mode and a normal reproduction mode.

The digital video cassette recorder cannot reproduce a picture having the quality of a normal picture in the search mode, since it scans more tracks at a time than in the normal reproduction mode. In the VCR which is capable of recording and playing back only the analog video signal, the analog video signal of about 5 MHz bandwidth is recorded on slant tracks of a tape at an azimuth angle of 6.degree. In a recording mode, each of the slant tracks having a track pitch at a predetermined distance.

In a digital recording format of a digital video cassette recorder, an input video signal is pulse code-modulated (PCM) and pixel values thereof are then coded in the unit of a desired bit. Namely, the input video signal is converted into digital video data, which is then compressed by a variable length coding which varies a length of bit being coded every pixel. The digital video data compressed by the variable length coding is partitioned in the unit of a desired sync-block and then recorded on a video tape in a predetermined order. Digital video cassette recorders for public use have better picture quality and are able to record longer programs in comparison with conventional analog system VTR.

A digital video cassette recorder can be operated in the same way as a conventional analog video cassette recorder is operated, and in particular it is needed that in the cases of a normal mode (SD mode) and a long play mode having a 1/2 tape speed of the normal tape speed, a digital video cassette recorder can be operated in the same way as an analog video cassette is operated. A high definition digital video cassette recorder (HD-DVCR) is a system using an inner error correction code and an outer error correction code for channel coding. The HD-DVCR is used as a data storage in which data compressed according to the MPEG II standard can be stored and from which the stored data is reproduced. The HD-DVCR uses the same recording format and channel processing portion as those of SD-DVC (standard digital video camcorder). A video cassette recorder is operable in various modes, such as play mode, review mode, rewinding mode, fast forward mode, through the selective operation of the driving mechanism. Generally, a VCR is capable of recording and playing back only an analog video signal. A high definition VCR is capable of recording and playing back only a digital video signal or both the analog and digital video signals compatibly.

8.13.6. Video Server Systems

Video servers, with their unique set of features and functions, have reshaped the way program segments are stored and played to air. These systems have emerged from the realm of limited use, special-purpose devices to mainstream video production. Among the many attributes of video servers are:

- They permit material from a single storage source to be used simultaneously by multiple users
- Provide a migration path to the all-digital facility that is not necessarily format-limited
- Result in a reduction in lost or misplaced materials
- Reduction in the size and space requirements relative to a tape environment
- Complete computer-control capabilities
- Near-instant access and playback of video segments

The end-result of these attributes is an environment where multiple applications and/or services can be generated from a single system. This reduces the amount of playback equipment required, reduces tape consumption, and generally permits more efficient use of human resources.

Basic Architecture

The basic video server architecture consists of three elements:

- A multiple hard disk drive system capable of fast and simultaneous data access, with sufficient capacity and redundancy for the contemplated application. A disk array controller manages data distribution and communications among all drives.
- Fast data communication interfaces among disk drives and networks. Several approaches can be used to perform fast data transfer such as very fast CPUs, multiple CPUs and buses, and routing switchers. Interfaces may also include data compression encoding and decoding.

Video Recording Systems

An operating system capable of handling multiple digital audio and/or video data streams in any combination of record and playback modes, while ensuring correct file management and easy access.

The design and performance of the server involves trade-offs among various parameters, including the quality of compressed signals, storage capacity, data speed, play time, number of channels, access speed, and reliability. Table 8.1 shows the storage space requirements for different audio and video signals.

All video servers have in common a large storage capacity and multiple channel capability; differences involve the basic architecture and performance targets, because different systems may be designed to meet specific requirements. The most common application groups for broadcasting include the following:

- **Transmission.** High compression ratios can be used for server-to-air applications, typically commercial and short-length program replay. The resulting bit rates and number of channels are usually low, therefore, with relatively low bandwidth requirements.
- **Data cache.** A caching system is basically a temporary random access disk buffer. It is commonly used in conjunction with tape library systems in on-air applications. Caches are well suited to commercial on-air insertion applications where elements are repeated several times a day and last-moment changes occur frequently. Bandwidth and channel requirements are usually not critical.
- **News.** For most news materials, moderate overall quality levels are acceptable. Video compression is used to reduce the data file size and increase the transfer rate. Multiple accesses to multiple segments must be possible and a moderate bandwidth is necessary. Guaranteed availability of output ports might be required for direct on-air programming of news materials.
- **Production.** Wide bandwidth must be provided to meet simultaneous transfer demands for large uncompressed video files with real-time and random access capabilities. In a production facility, a server stores all compressed or uncompressed audio and video data files for use in post-production and distribution. The server is central to the production operation and, thus, determines the overall performance of the facility.

- **Video on demand (VOD).** This server application must deliver a large number of channels, each of relatively low video quality, such as MPEG-1 (1.5 Mbps). A high overall bandwidth may be required to satisfy all demands. Short access times are necessary for VOD, whereas long access times permit *near-VOD* (NVOD) only.

Table 8.3. The storage space requirements for different audio and video signals.

Media Signals	Specifications	Data Rate
Voice-grade audio	1 ch; 8-bit @ 8 kHz	64 kbits/s
MPEG audio Layer II	1 ch; 16-bit @ 48 kHz	128 kbits/s
MPEG audio Layer III	1 ch; 16-bit @ 48 kHz	64 kbits/s
AC-3	5.1 ch; 16-bit @ 48 kHz	384 kbits/s
CD	2 ch; 16-bit @ 44.1 kHz	1.4 Mbits/s
AES/EBU	2 ch; 24-bit @ 48 kHz	3.07 Mbits/s
MPEG-1 (video)	352 × 288, 30 f/s, 8-bit	1.5 Mbits/s
MPEG-2 (MP @ ML)	720 × 576, 30 f/s, 8-bit	15 Mbits/s, max.
MPEG-2 (4:2:2 P@ML)	720 × 608, 30 f/s, 8-bit	50 Mbits/s, max.
ITU-R Rec. 601	720 × 480, 30 f/s, 8-bit	216 Mbits/s
HDTV	1920 × 1080, 30 f/s, 8-bit	995 Mbits/s

Video servers have redefined many of the common applications of video tape recorders. The

VTR is being displaced in direct-to-air uses, complex post production, and desktop editing. As new VTR formats evolve, they will increasingly be aimed at acquisition and long-form storage applications. Among the technical frontiers for video servers is the ability to transfer files at *faster than real-time*. Such features place considerable demands on the bandwidth and throughput of the system; however, they provide many operational benefits, such as accelerated non-linear editing and reduced program load-up times.

Video Server Design

A number of operational scenarios can be implemented with a video server system. The simplest involves multiple streams of one or two video inputs, as illustrated in Figure 8.41. All the material is stored within the server and plays out to multiple channels as required.

Fig. 8.41. Simple application of a video server in a one-in, many-out configuration.

In Fig. 8.42, the basic system is expanded to allow for some material, particularly long form programs, to be held in an external storage area or device (most likely on tape) and loaded into the server on an as-needed basis to coexist with the short-form (typically commercial) library that is stored in the server. The server then manages the entire play-out process to one or more outputs.

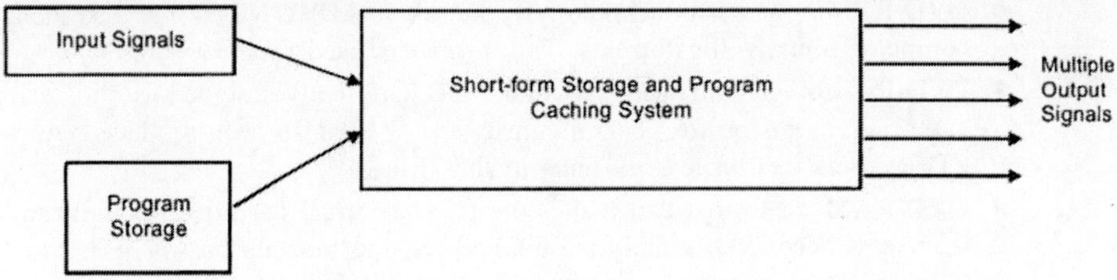

Fig. 8.42. A server system expanded to utilize long-form program storage.

DVD Devices

The compact disc has been an unqualified success. Beginning with the CD-Audio format in 1982 and the subsequent development of CD-ROM, CD-R, and CD-RW (and other less popular variations) the CD revolutionized data storage particularly at the consumer level. However, the CDs limited capacity and slow throughput bit rate made it unsuitable for high bandwidth or large volume applications such as high quality digital video. For example, the White Book Video CD format and its MPEG-1 coding were deemed unsuitable. Thus, the motion picture industry sought to develop a small optical disc holding a feature film coded with high quality digital audio and digital video. In December 1994, Sony and Philips proposed the MultiMedia Compact Disc (MMCD). In January 1995, Toshiba and Time Warner proposed the Super Density disc (SD). The DVD Technical Working Group subsequently outlined the following criteria for the new video disc:

- A single standard that is fully interchangeable for TV and PC applications
- Fully backward compatible with existing CD media
- Forward compatible with write-once and rewritable media
- Low cost and high quality disc replication
- A single universal file structure
- Support for both linear and non-linear applications
- High capacity that is expandable for high definition media

The **"Digital Versatile Disc"** provides exceptionally efficient video, audio, and data storage and access in a convenient form factor. The DVD is high capacity multimedia data storage medium that can accommodate a complete motion picture on a single disc, content rich multimedia or high quality multi-channel audio. Most DVD hardware will play audio CDs and CD-ROMs. The physical dimensions are identical to compact discs.

Like CDs, DVDs store data in microscopic grooves running in a spiral around the disc. All DVD drive types use laser beams to scan these grooves. Minuscule reflective bumps (*lands*) and non reflective holes (*pits*) aligned along the grooves represent the zeros and ones of digital information.

DVDs use smaller tracks (0.74 microns wide, compared to 1.6 microns on CDs) as well as new modulation and error correction methods. These technologies allow them to store seven times the data of a CD.

There are a number of DVD formats, the most common being:

- DVD Video—for viewing movies and other visual entertainment. The total capacity is 17 GB if two layers on both sides of the disk are utilized.

- DVD-ROM—the basic technology is the same as DVD Video, but also includes computer friendly file formats. This format is used to store data.
- DVD-R—storage with a capacity of 4.7 GB. Originally designed for professional authoring, a version for general consumer use has also been produced. As with CD-R, users can write only once to this disk.
- DVD-RAM—a format that makes the DVD a virtual hard disk, with a random read-write access. Originally a 2.6-GB drive, the capacity has increased to 4.7-GB-per-side. It can be rewritten more than 100,000 times.
- DVD-R—similar to DVD-RAM except that its technology features a sequential read-write access more like a phonograph than a hard disk. The read-write capacity is 4.7 GB per side. It can be re-written up to about 1,000 times.
- DVD Audio—an audio format that more than doubles the fidelity of a standard CD.
- DVD Multi is a set of specifications that define which drives will read and write which disks for the various DVD consumer and computer applications. DVD Multi is designed to provide broader compatibility across DVD disks, and embraces all existing format versions.

SUMMARY

- A television distribution system in which station signals, picked up by elevated antennas, are delivered by cable to the receivers of subscribers. It is called **cable TV** or community *antenna television.*
- Traditional cable TV systems worked strictly by way of analog system (*i.e.* using standard radio waves) but many modern cable TV systems also employ the use of digital cable technology, which uses compressed digital signals, allowing them provide many more channels than they could with analog alone.
- Figure 8.1 shows a block diagram of typical cable TV system. Major elements of cable TV system are namely:
    1. Head end or combining network
    2. Trunk line cables
    3. Branch line cable
    4. Drop line
    5. Amplifiers
    6. Directional coupler
- Channels for Cable TV

  Ist channel—41-47 MHz (not used for TV)

  IInd to IVth channels—47 to 68 MHz

  Vth to XIIth channel— 174 to 230 MHz
- Cable Based Services
    1. The CATV service provider can receive signals from satellites and make a large number of channels available to the viewers through cables.
    2. Premium channels as Pay- TV channels, which are exclusive to those who can pay additional fees.
    3. Cable operators can generate their own TV program and send it on cable.
    4. Internet service through cable TV also provided recently.

5. *Broadband internet* is achieved over coaxial cable by using cable modems
6. Another service being added to many cable systems is *digital telephone service*.

- **Closed Circuit Television (CCTV)** is a system in which video signal obtained by one or more cameras tubes is sent to one or more monitors through coaxial cables.
- **Block Diagram of CCTV :** Essential elements of CCTV are as follows:
  1. Camera tube
  2. Scanning circuits
  3. Blanking and sync pulses
  4. Video amplifier (each for transmitter & receiver)
  5. Coaxial cable
  6. Video monitor
- **Applications of CCTV**
  1. Surveillance
  2. Education
  3. Medical care
  4. Industry
  5. Safety
  6. Traffic control
  7. Crowd control
  8. Aerospace and oceanography
- **Satellite television** is television delivered by the means of communications satellites, as compared to conventional terrestrial television - which is broadcast over radio waves originating from several independent, ground-based television stations - and cable television - which is distributed by underground coaxial cables.
- Picture phone is the another fascinating application of televisions, in which two persons can see each other while talking over the telephone lines.
- The highest modulating frequency is used in picture phone is limited to about 1.5 MHz.
- Fax (short for facsimile, from Latin *fac simile*, "make similar", i.e. "make a copy") is a telecommunications technology used to transfer copies (facsimiles) of documents, especially using affordable devices operating over the telephone network.
- A "fax machine" usually consists of an image scanner, a modem, and also offered as options for many high-volume workgroup printers and photocopiers.
- In a digital system, the analog signal is converted into digital signal by using A/D converter. There are three basic functions of A/D converter are sampling, quantization and encoding.
- A digital TV receiver which is basically a superhetrodyne receiver.
- An application of digital electronics and IC technology to TV products is TV games.
- **Modern TV games** uses a microprocessor microcontroller chip which coordinate and control functions of various elements.

- A teletext service transmits information in the form pages. Each pages give a screenful of information.
- HDTV stands for High Definition Television. HDTV enhances the quality of the picture on your screen and the quality of the sound.
- The basic concept behind high-definition television is actually not to increase the definition per unit area but rather to increase the percentage of the visual field contained by the image.
- An HDTV-compatible TV usually uses a 16:9 aspect ratio. The high resolution images (1920 pixels × 1080 lines or 1280 pixels × 720 lines) allow much more detail to be shown compared to analog television or regular DVDs.
- The TDM (Time Division Multiplexing) used in HDTV is called Multiplexing Analog components (MAC).
- Remote control consists of a small portable handset having a key pad on the front.
- 3D Television augments the traditional TV technology by showing the viewer not only sequences of 2D images but streams of three-dimensional scene representations.
- **3DTV** will require the integration of a diversity of key technologies from computing to graphics, imaging to display, and signal processing to communications.
- When the magnetization is oriented in the direction of relative motion between the head and tape, the process is referred to as ***longitudinal recording*****. If the magnetization is aligned perpendicular to the surface of the tape, it is called *****vertical***** or *****perpendicular***** recording.**
- *Transverse recording* exists when the magnetization is oriented at right angles to the direction of relative head-to-tape motion.
- A video cassette recorder (VCR) is a device for recording and reproducing a video signal and an audio signal recorded on a recording medium (*i.e.*, a video cassette tape).
- Servos used in video tape machines include the capstan and drum servos.
- There are three major video systems in use today:
  1. Video Home System (VHS)
  2. Betamax
  3. Video Hi8
- In case of a mono video recorder, the audio signal which corresponds with the image is transferred to a separate, fixed audio head.
- Hi-fi video recorders were developed for improved sound quality. The most common quality of video images is HQ. (The recorder is labeled 'VHS High Quality Hi-fi Stereo').
- Video servers, with their unique set of features and functions, have reshaped the way program segments are stored and played to air.
- The "Digital Versatile Disc" provides exceptionally efficient video, audio, and data storage and access in a convenient form factor.
- All DVD drive types use laser beams to scan these grooves. Minuscule reflective bumps (*lands*) and non reflective holes (*pits*) aligned along the grooves represent the zeros and ones of digital information.
- DVDs use smaller tracks (0.74 microns wide, compared to 1.6 microns on CDs) as well as new modulation and error correction methods. These technologies allow them to store seven times the data of a CD.

REVIEW QUESTIONS

1. Explain the close circuit television system with the help of block diagram in brief.
2. How is CCTV system different from regular TV broadcasts ?
3. Draw the block diagram of cable television systems and explain the working in brief.
4. Explain the different types of tapping used to extract signal from a branch line in a cable TV systems.
5. Describe various cable TV channels used.
6. Discuss the main merits and applications of the CCTV and CATV system.
7. Draw a neat block diagram of a digital TV receiver and explain the function of each block.
8. Describe with the help of a block diagram the working of a HDTV transmitter.
9. Explain the working of HDTV receiver with the help of block diagram.
10. Discuss the merits of a HDTV.
11. Draw and discuss the block diagram of a TV games consist of microprocessor for processing input data and generating digital outputs for the TV.
12. List the various standards of HDTV.
13. What is teletext ? How is it used in modern TV receiver ?
14. Explain the functioning of a remote control unit for TV.
15. Explain how the documents are reproduced at the facsimile receiver.
16. Explain the satellite TV with the help of a block diagram.
17. What do you understand by DTH ?
18. Explain the working principle of fax.
19. Describe the principle of video recording and reproduction.
20. Draw block diagram of VCR (video recording and reproduction system) and explain the function of each block.
21. Compare audio cassette recording with video cassette recording.
22. What is the principle of magnetic tape ?
23. Write short notes on :
    (*i*) 3D Television (*ii*) Anhysteresis
    (*iii*) Video tape recording (*iv*) The Helical scan system
    (*v*) Azimuth recording system (*vi*) Video systems
    (*vii*) VHS (*viii*) Betamax
    (*ix*) Video Hi8 (*x*) Sound recording
    (*xi*) Super-VHS
    (*xii*) Analog versus digital video casette recorders
    (*xiii*) Video server
    (*xiv*) DVD devices (*xv*) Video phone

MULTIPLE CHOICE QUESTIONS

1. In digital TV, audio signal being fed to the loud speaker can be
    (*a*) Analog only (*b*) Digital only
    (*c*) Both digital and analog (*d*) None of these
2. Modulation method used for digital video transmission is
    (*a*) AM (*b*) FM
    (*c*) TDM (*d*) OFDM

3. MAC system of wide definition TV was developed in
   (*a*) Europe (*b*) Japan
   (*c*) USA (*d*) India
4. Aspect ratio for WDTV for european MAC system is
   (*a*) 4 : 3 (*b*) 5 : 3
   (*c*) 16 : 9 (*d*) 2 : 1
5. In facsimile electronic scanning uses
   (*a*) Light beam (*b*) Electron beam
   (*c*) Electrothermal beam (*d*) Ultra sonic rays
6. Which of the following system does not use EM fields at any stage of acquiring the TV signals
   (*a*) CATV (*b*) CCTV
   (*c*) VCR (*d*) MATV
7. Video programmes sent on CATV are in the form of
   (*a*) AM VSB modulation (*b*) Unmodulated carrier
   (*c*) FM modulation (*d*) Video base band signals
8. The field frequency of HDTV is
   (*a*) 15 (*b*) 30
   (*c*) 60 (*d*) 120
9. Which of the following function is not performed by head-end in cable TV
   (*a*) Translation (*b*) Mixing
   (*c*) Sync generation (*d*) Amplification
10. Which of the following stage is not a part of CCTV monitor
    (*a*) Video detector (*b*) Picture tube
    (*c*) Sync separator (*d*) Video amplifier
11. Video recording is based on the principle of
    (*a*) Electromagnetism
    (*b*) Piezo electric effect
    (*c*) Faraday's law of electrolysis
    (*d*) EM induction
12. Video reproducing or playback is based on the principle of
    (*a*) Electromagnetism (*b*) Piezo electric effect
    (*c*) Faraday's law of electrolysis(*d*) EM induction
13. The duration recording is increased by
    (*a*) High speed (*b*) Slant
    (*c*) FM (*d*) Straight
14. Relative speed of tape is increased by using
    (*a*) Rotating head (*b*) Stationary head
    (*c*) Small gap-width (*d*) Large gap width
15. In DVD, laser beam is reflected from a
    (*a*) Flat (*b*) Pit
    (*c*) Glass (*d*) Subtrate

ANSWERS

1. (*a*)	2. (*d*)	3. (*a*)	4. (*c*)	5. (*a*)
6. (*b*)	7. (*a*)	8. (*c*)	9. (*c*)	10. (*a*)
11. (*a*)	12. (*d*)	13. (*b*)	14. (*a*)	15. (*a*)

PART 2

RADAR ENGINEERING

1
CHAPTER

INTRODUCTION TO RADAR

INSIDE THIS CHAPTER

1.1. INTRODUCTION

Radar is an acronym for **Radio Detection and Ranging**. The term "radio" refers to the use of electromagnetic waves with wavelengths in the so-called radio wave portion of the spectrum, which covers a wide range from 10^4 km to 1 cm. It is a system used to detect, range (determine the distance) and map objects such as aircraft and rain. Strong radio waves are transmitted, and a receiver listens for reflected echoes. By analysing the reflected signal, the reflector can be located, and sometimes identified. Although the amount of signal returned is tiny, radio signals can easily be detected and amplified.

It can operate in darkness, haze, fog rain and snow. It has ability to measure distance with high accuracy in all weather conditions.

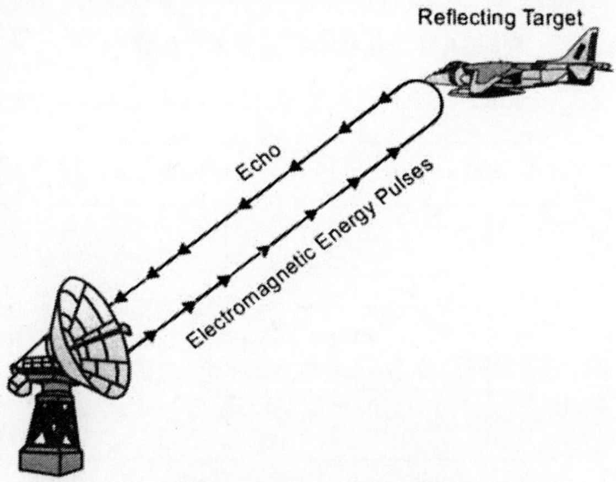

Fig. 1.1. Radar Principle.

The electronics principle on which radar operates is very similar to the principle of sound-wave reflection. If you shout in the direction of a sound-reflecting object (like a rocky canyon or cave), you will hear an echo. If you know the speed of sound in air, you can then estimate the distance and general direction of the object. The time required for a return echo can be roughly converted to distance if the speed of sound is known.

Radar uses electromagnetic energy pulses in the same way, as shown in Fig. 1.1. The radio-frequency (rf) energy is transmitted to and reflects from the reflecting object. A small portion of the energy is reflected and returns to the radar set. This returned energy is called an **ECHO**, just as it is in sound terminology. Radar set use the echo to determine the direction and distance of the reflecting object.

Fig. 1.2. An Air Traffic Control Radar.

Figure 1.2 shows the real look of an Air Traffic Control Radar, which is being used as Ground Control Approach Radar at air bases in civil as well as in defense.

1.2. RADAR BLOCK DIAGRAM

A practical radar system requires seven basic components as illustrated below :

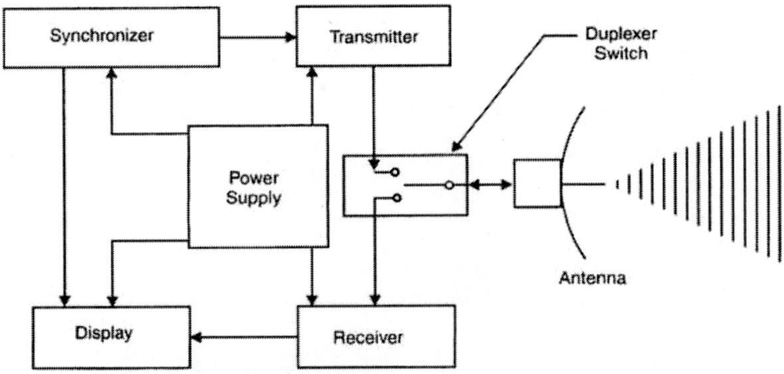

Fig. 1.3. A Practical Simple Radar Block Diagram.

1. Transmitter. The transmitter creates the radio wave to be sent and modulates it to form the pulse train. The transmitter must also amplify the signal to a high power level to provide adequate range. The source of the carrier wave could be a Klystron, Traveling Wave Tube (TWT) or Magnetron. Each has its own characteristics and limitations.

2. Receiver. The receiver is sensitive to the range of frequencies being transmitted and provides amplification of the returned signal. In order to provide the greatest range, the receiver must be very sensitive without introducing excessive noise. The ability to discern a received signal from background noise depends on the signal-to-noise ratio (S/N).

3. Power Supply. The power supply provides the electrical power for all the components. The largest consumer of power is the transmitter, which may require several kW of average power. The actually power transmitted in the pulse may be much greater than 1 kW. The power supply only needs to be able to provide the average amount of power consumed, not the high power level during the actual pulse transmission. Energy can be stored, in a capacitor bank for instance, during the rest time. The stored energy then can be put into the pulse when transmitted, increasing the peak power.

4. Synchronizer. The synchronizer coordinates the timing for range determination. It regulates that rate at which pulses are sent (*i.e.* sets PRF) and resets the timing clock for range determination for each pulse. Signals from the synchronizer are sent simultaneously to the transmitter, which sends a new pulse, and to the display, which resets the return sweep.

5. Duplexer. This is a switch that alternately connects the transmitter or receiver to the antenna. Its purpose is to protect the receiver from the high power output of the transmitter. During the transmission of an outgoing pulse, the duplexer will be aligned to the transmitter for the duration of the pulse, PW. After the pulse has been sent, the duplexer will align the antenna to the receiver. When the next pulse is sent, the duplexer will shift back to the transmitter. A duplexer is not required if the transmitted power is low.

6. Antenna. The antenna takes the radar pulse from the transmitter and puts it into the air. Furthermore, the antenna must focus the energy into a well-defined beam, which increases the power and permits a determination of the direction of the target. The antenna must keep track of its own orientation, which can be accomplished by a synchro-transmitter. There are also antenna systems, which do not physically move but are steered electronically (such as phased array antenna, in these cases, the orientation of the radar beam is already known a priori).

7. Display. The display unit may take a variety of forms but in general is designed to present the received information to an operator. The most basic display type is called an A-scan (amplitude vs. Time delay) as shown in Fig. 1.4. The vertical axis is the strength of the return and the horizontal axis is the time delay, or range. The A-scan provides no information about the direction of the target.

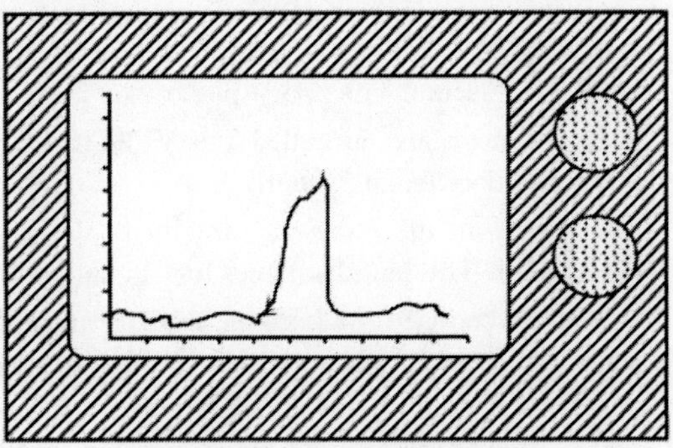

Fig. 1.4. An "A"-Type Scope.

The most common display is the PPI (plan position indicator) as shown in Fig. 1.5. The A-scan information is converted into brightness and then displayed in the same relative direction as the antenna orientation. The result is a top-down view of the situation where range is the distance from the origin. The PPI is perhaps the most natural display for the operator and therefore the most widely used. In both cases, the synchronizer resets the trace for each pulse so that the range information will begin at the origin.

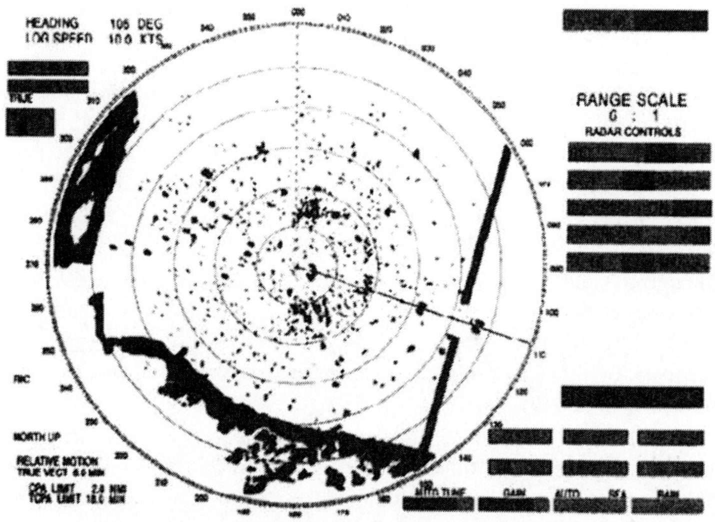

Fig. 1.5. Display In Plan Position Indicator (PPI).

1.3. DETERMINATION OF CO-ORDINATES IN RADAR

Radar requires a more precise reference system. Radar surface angular measurements are normally made in a clockwise direction from TRUE NORTH, as shown in Fig. 1.6, or from the heading line of a ship or aircraft. The surface of the earth is represented by an imaginary flat plane, tangent (or parallel) to the earth's surface at that location. This plane is referred to as the HORIZONTAL PLANE. All angles in the up direction are measured in a second imaginary plane that is perpendicular to the horizontal plane.

This second plane is called the VERTICAL PLANE. The radar location is the center of this coordinate system.

The line from the radar set directly to the object is referred to as the **LINE OF SIGHT (LOS)**. The length of this line is called **RANGE.**

The angle between the horizontal plane and the LOS is the ELEVATION ANGLE. The angle measured clockwise from true north in the horizontal plane is called the **TRUE BEARING or AZIMUTH** angle.

These three coordinates of range, bearing, and elevation describe the location of an object with respect to the antenna.

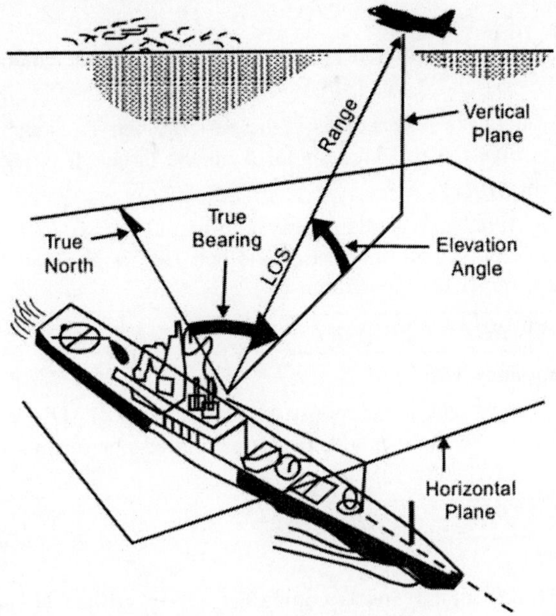

Fig. 1.6. Radar Reference Coordinates.

1.3.1. Range

The detection and ranging part of the radar is accomplished by timing the delay between transmission of a pulse of radio energy and its subsequent return. If the time delay is Δt, then the range may be determined by the simple formula :

$$R = c \times \left(\frac{\Delta t}{2}\right) \qquad \text{...(1.1)}$$

where $c = 3 \times 10^8$ m/s, the speed of light at which all electromagnetic waves propagate. The factor of two in the formula comes from the observation that the radar pulse must travel to the target and back before detection, or twice the range.

A radar pulse train is a type of amplitude modulation of the radar frequency carrier wave, similar to how carrier waves are modulated in communication systems. In this case, the information signal is quite simple, a single pulse repeated at regular intervals. The common radar carrier modulation, known as the **pulse train** is shown below. The common parameters of radar as defined by referring to Fig. 1.7.

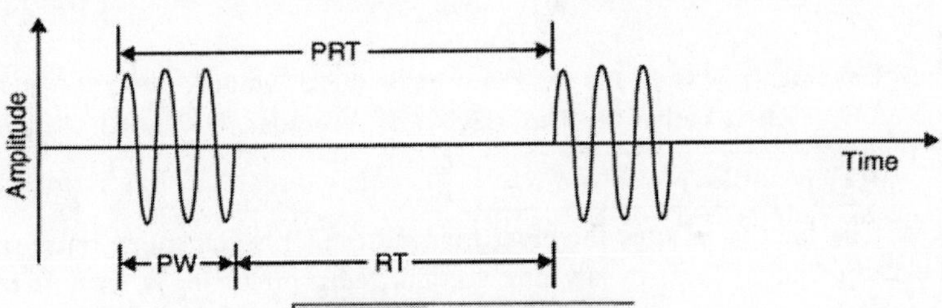

Fig. 1.7. A Radar Pulse.

Common Parameters of Radar Pulse

Pulse Width (PW)

PW has units of time and is commonly expressed in ms. *PW* is the duration of the pulse.

Rest Time (*RT*)

RT is the interval between pulses. This is the period when Tx is silence (not firing) and Rx is ready to receive the reflected signal from the target. It is measured in ms.

Pulse Repetition Time (*PRT*)

PRT has units of time and is commonly expressed in ms. *PRT* is the interval between the start of one pulse and the start of another. *PRT* is also equal to the sum of pulse width and rest time. It can be written as

$$\boxed{PRT = PW + RT}$$

Pulse repetition frequency (*PRF*)

PRF has units of time^{-1} and is commonly expressed in Hz (1 Hz = 1/s) or as pulses per second (pps). *PRF* is the number of pulses transmitted per second and is equal to the inverse of *PRT*.

$$\boxed{PRF = 1/PRP}$$

Radio frequency (RF)

RF has units of time^{-1} or Hz and is commonly expressed in GHz or MHz. *RF* is the frequency of the carrier wave which is being modulated to form the pulse train.

Peak power (*Pt*)

The power *Pt* in the radar equation is called by the radar engineer, the Peak Power. The **peak power** as used in the radar equation is not the instantaneous peak power of a sine wave. It is defined as the power averaged over that carrier frequency cycle which occurs at the maximum of the pulse power. Peak power is usually equal to one-half of the maximum instantaneous power.

Average Power (*Pav*)

The **average power *Pav*** is defined as the average transmitted power over the pulse repetition time or period. If the transmitted waveform is a train of rectangular pulses of width and pulse repetition period or time.

$$\boxed{PRT = \frac{1}{PRF}} \quad \text{...(1.2)}$$

$$\boxed{Pav \text{ (Average Power)} = P_t \times \frac{PW}{PRT} = P_t \times PW \times PRF} \quad \text{...(1.3)}$$

(Average Power = Peak Power × Pulse Width/*PRT* = Peak Power × Pulse width × PRF)

Duty Cycle

The ratio of average power to the Peak power or pulse width to the *PRT* or Pulse width multiplied by PRF is called the Duty Cycle of the radar. It may be written as,

$$\boxed{\text{Duty cycle} = PW \times PRF} \quad \text{...(1.4)}$$

Example. The fraction of time the radar transmitter (T_x) is generating pulse power is termed the Duty-Cycle. Thus, in any one second, radar producing 1 μs pulses at a rate of 1000 pulse per sec is "on". The duty cycle may be calculated by using Eqn. 1.4.

$$\text{Duty cycle} = PW \times PRF = 1 \times 10^{-6} \times 1000 = 0.001 \text{ s}$$

The duty cycle is then 0.001, and the average power is given by the product of duty cycle and peak power; thus if the peak power is 500 kW, the average power is than

$$\text{Average Power } (Pav) = Pt \times \text{Duty Cycle}$$

$$Pav = 500 \times 10^{3} \times 0.001$$

$$Pav = 500 \text{ W}$$

1.3.2. Bearing

The **TRUE BEARING** (referenced to true north) of a radar target is the angle between true north and a line pointed directly at the target. This angle is measured in the horizontal plane and in a clockwise direction from true north. The bearing angle to the radar target may also be measured in a clockwise direction from the centerline of your own ship or aircraft and is referred to as the **RELATIVE BEARING**. Both true and relative bearing angles are illustrated in Fig. 1.8.

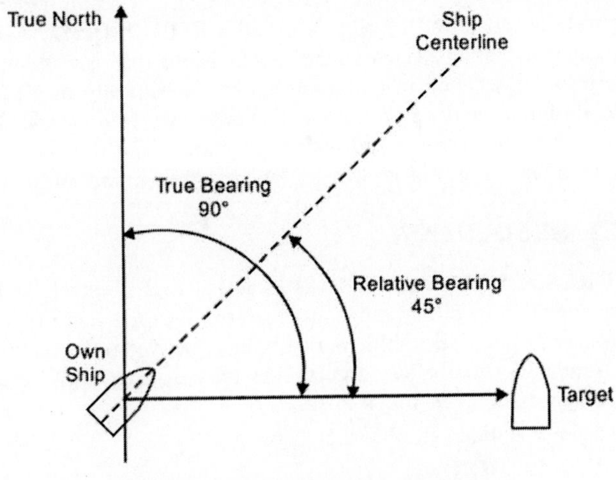

Fig. 1.8. True and Relative Bearings.

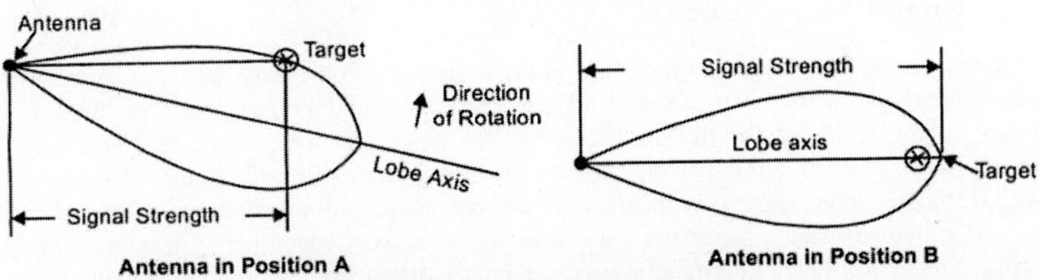

Fig. 1.9. Determination of Bearing.

The antennas of most radar systems are designed to radiate energy in a one-directional lobe or beam that can be moved in bearing simply by moving the antenna. As you can see in Fig. 1.9, the shape of the beam is such that the echo signal strength varies in amplitude as the antenna beam moves across the target. At antenna position A, the echo is minimal;

at position B, where the beam axis is pointing directly at the target, the echo strength is maximum. Thus, the bearing angle of the target can be obtained by moving the antenna to the position at which the echo is strongest. In actual practice, search radar antennas move continuously; the point of maximum echo return is determined by the detection circuitry as the beam passes the target or visually by the operator.

Weapons-control and guidance radar systems are positioned to the point of maximum signal return and maintained at that position either manually or by automatic tracking circuits.

1.3.3. Altitude

Many radar systems are designed to determine only the range and bearing of an object. Such radar systems are called **TWO-DIMENSIONAL (2D) radars**. In most cases these systems are further described as **SEARCH RADAR SYSTEMS** and function as early-warning devices that search a fixed volume of space. The range and bearing coordinates provide enough information to place the target in a general area with respect to the radar site and to determine distance, direction of travel, and relative speed.

However, when action must be taken against an airborne target, altitude must be known as well. An altitude is height of the target from the ground plane, sometimes it is considered as the height from the Mean Sea Level (MSL). A search radar system that detects altitude as well as range and bearing is called a **THREE-DIMENSIONAL (3D) radar**. Altitude- or height-finding search radars use a beam that is very narrow in the vertical plane. The beam is scanned in elevation, either mechanically or electronically, to pinpoint targets. Height-finding radar systems that also determine bearing must have a beam that is very narrow in both the vertical and horizontal planes. The display system use for indicating the height of the target is known as **Height Measuring Indicator (HMI).**

1.4. TARGET RESOLUTION

The **TARGET RESOLUTION** of a radar is its ability to distinguish between targets that are very close together in either range or bearing. Weapons-control radar, which requires great precision, should be able to distinguish between targets that are only yards apart. Search radar is usually less precise and only distinguishes between targets that are hundreds of yards or even miles apart.

Resolution is usually divided into two categories :

1. RANGE RESOLUTION and
2. BEARING RESOLUTION.

1.4.1. Range Resolution

Range resolution is the ability of a radar system to distinguish between two or more targets on the same bearing but at different ranges. The degree of range resolution depends on the width of the transmitted pulse, the types and sizes of targets, and the efficiency of the receiver and indicator. Pulse width is the primary factor in range resolution. A well-designed radar system, with all other factors at maximum efficiency, should be able to distinguish targets separated by one-half the pulse width time. Therefore, the theoretical range resolution of a radar system can be calculated from the following formula :

$$\boxed{R_{RES} = c \times PW/2}$$

The above formula is often written as :

$$\boxed{R_{RES} = c/2\beta} \qquad (PW = 1/\beta)$$

where β is the bandwidth of transmitter pulse.

For example, if a radar system has a pulse width of 5 microseconds, the range resolution is calculated as follows :

$$R_{RES} = c \times PW/2$$
$$R_{RES} = 3 \times 10^8 \times 5 \times 10^{-6}/2$$
$$= 3 \times 5 \times 10^2/2 = 7.5 \times 100$$
$$R_{RES} = 750 \text{ m}$$

In the above example, targets on the same bearing would have to be separated by more than 750m to show up as two targets on your indicator.

1.4.2. Bearing Resolution

Bearing, or azimuth, resolution is the ability of a radar system to separate objects at the same range but at different bearings. The degree of bearing resolution depends on radar beam width and the range of the targets.

Range is a factor in bearing resolution because the radar beam spreads out as range increases.

A **RADAR BEAM** is defined in width in terms of HALF-POWER POINTS. All the points off the centerline of the beam that are at one-half the power level at the center are plotted to define beam width. When the half-power points are connected to the antenna by a curve, such as that shown in Fig. 1.10, the resulting angular width of the curve is called the **ANTENNA BEAM WIDTH**. The physical size and shape of the antenna determines beam width. Beam width can vary from about 1 degree up to 60 degrees. In Fig. 1.10, only the target within the half-power points will reflect a useful echo. Two targets at the same range must be separated by at least one beam width to be distinguished as two objects.

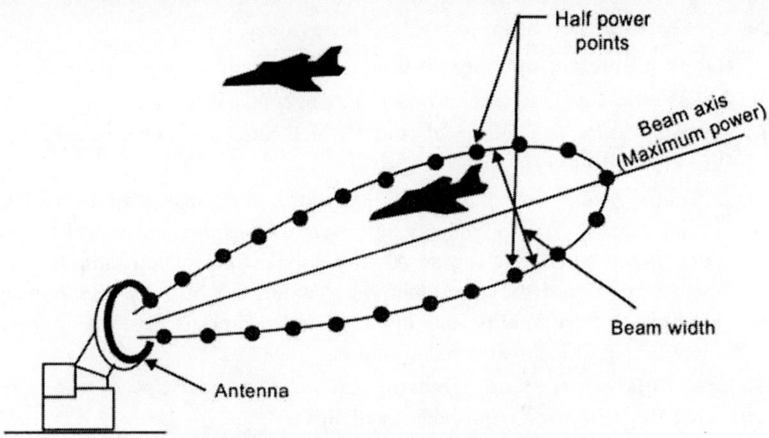

Fig. 1.10. Beam Half-power Points.

1.5. APPLICATIONS OF RADAR

Radar has been employed to detect target on the ground on the sea, in the air in space and even below ground. The major areas of radar applications are briefly described below :

1. Military applications. In military radar has wide application. Radar is used by the military specially for surveillance purpose of the enemy aircraft. The surveillance radat detect, locate and identified the enemy targets. It is also being used for providing navigational aids to military or as well as civil aircraft. It is also used for tracking the targets and controlling of weapons.

2. Targeting radars use the same principle but scan a much smaller area far more often, usually several times a second or more, where a search radar might scan a few times per minute. Some targeting radars have a range gate that can track a target, to eliminate clutter and electronic counter-measures.

3. Navigational radars resemble search radar, but use very short waves that reflect from earth and stone. They are common on commercial ships and long-distance commercial aircraft.

4. Weather radars can resemble search radars. These radar use radio waves with horizontal, dual (horizontal and vertical), or circular polarization. The frequency selection of weather radar is a performance compromise between precipitation reflectivity and attenuation due to atmospheric water vapour. Some weather radar uses doppler to measure wind speeds

5. General purpose radars are increasingly being substituted for pure navigational radars. These generally use navigational radar frequencies, but modulate the pulse so the receiver can determine the type of surface of the reflector. The best general-purpose radars distinguish the rain of heavy storms, as well as land and vehicles. Some can superimpose sonar and map data from GPS position.

6. Radar Proximity Fuses are attached to anti-aircraft artillery shells or other explosive devices, and detonate the device when it approaches a large object. They use a small rapidly pulsing omnidirectional radar, usually with a powerful battery that has a long storage life, and a very short operational life. The fuses used in anti-aircraft artillery have to be mechanically designed to accept fifty thousand gravities of acceleration, yet still be cheap enough to throw away.

7. Radar altimeters measure an aircraft's true height above ground.

8. Air Traffic Control uses Primary and Secondary Radars

(*a*) *Primary radar* is a "classical" radar which reflects all kind of echoes, including aircraft and clouds.

(*b*) *Secondary radar* emit pulses and listen for special answer of digital data emitted by an Aircraft Transponder as an answer. Transponders emit different kind of data like a 4 octal ID (mode A), the onboard calculated altitude (mode C) or the Callsign (not the flight number) (mode S). Military use transponders to establish the nationality and intention of an aircraft, so that air defenses can identify possibly hostile radar returns.

9. Law Enforcement and Highway safety. The radar speed meter, familiar to many is used by police for enforcing speed limit.

10. Radar has also found application in industry for the non contact measurement of speed and distance. It has been used for oil and gas exploration.

1.6. RADAR FREQUENCIES

Conventional radar generally operates in what is called the microwave region (a term not rigidly defined). Microwaves are since they are defined in terms of their wavelength (λ) in the sense that micro refers to tinyness. In short a microwave is a signal that has a wavelength of 1 foot or less ie $\lambda \leq 30.5$ cms. So all the frequencies above approximately 1000 Mhz (1 Ghz) to about 1000 Ghz are microwave frequencies. IEEE standard Radar Frequencies Letter–Band nomenclature given in Table 1.1.

EM spectrum frequency range and their general applications are given in Table 1.2.

Table 1.1. I E E E STANDARD RADAR FREQUENCIES

Band Designation	*Nominal Frequency range*	*Specific radar frequency range Based on ITU*
HF	3-30 Mhz	
VHF	30-300 Mhz	138-144 Mhz
		216-225 Mhz
UHF	300-1000 Mhz	420 –450 Mhz
		850-942 Mhz
L	1-2 Ghz	1215-1400 Mhz
S	2-4 Ghz	2300-2500 Mhz
		2700-3700 Mhz
C	4-8 Ghz	5250-5925 Mhz
X	8-12 Ghz	8500-10680 Mhz
KU	12-18 GhZ	13.4-14 Ghz
		15.7-17.7 Ghz
K	18-27 Ghz	24.05-24.25 Ghz
Ka	27-40 Ghz	33.4-36 Ghz
V	40-75 Ghz	59-64 Ghz
W	75-110 GhZ	76-81 Ghz
		92-100 Ghz
mm	110-300 Ghz	126-142 Ghz
		144-149 Ghz
		231-235 Ghz
		238-248 Ghz

Table 1.2. EM spectrum frequency range and their general applications

Frequency	*Description*	*Application*
Upto 300 Hz	Extremely low frequency (ELF)	Special Communications
300 Hz – 3 KHz	Voice frequency	
3 KHz – 30 KHz	Very low frequency (VLF)	Shore-based communications, experimental
30 KHz – 300 KHz	Low frequency (LF)	Shore-based communications, navigation
300 KHz – 3 MHz	Medium frequency (MF)	Commercial broadcast band (550 KHz to 1700 KHz), communications on either side of the broadcast band
3 MHz – 30 MHz	High frequency (HF)	Ship and shore long-range communications
30 MHz – 300 MHz	Very high frequency (VHF)	Communications, navigation
300 MHz – 3 GHz	Ultra high frequency (UHF)	Line-of-sight communications to 400 MHz, above his frequency, radar and special equipments
3 GHz – 30 GHz	Superhigh frequency (SHF)	Radar and special equipments
30 GHz – 300 GHz	Extremely high frequency (EHF)	Radar and special equipments

1.7. RADAR DEVELOPMENT

Fig. 1.11. An Imagination Picture of Radar.

History of Radar

In 1887 the German physicist Heinrich Hertz began experimenting with radio waves in his laboratory. He found that radio waves could be transmitted through some materials, and were reflected by others. Although this was predicted earlier by James Clerk Maxwell, his work was not widely known as the time and much of the research became known through Hertz. In fact today we use his name as the SI unit of frequency, the hertz, often abbreviated to Hz.

Some years later a German engineer, Chistian Huelsmeyer, proposed the use of radio echoes to avoid collisions in marine navigation. In 1904 he completed construction of a device he called the **telemobiloscope**, which consisted of a simple spark gap aimed using a funnel-shaped metal antenna. When a reflection was seen by the two straight antennas attached to the receiver, a bell sounded. Although very simple, the system could detect shipping accurately up to about 3 km. Nevetheless the naval world seemed uninterested in his invention, and it was not put into production.

Invention of Modern Radar

The invention of modern radar is generally credited to Sir Robert Watson-Watt. In 1915 he joined the Royal Aircraft Factory at Ditton Park as a meteorologist, where he attempted to use radio signals generated by lightning strikes to map out thunderstorms. The difficulty in pinpointing the direction of these high-speed signals led to the use of rotating directional antennas, and in 1923 the use of oscilloscopes in order to display them in 2-D. At this point the only missing part of a functioning radar was the broadcaster.

Nikola Tesla, in August 1917, first established principles regarding frequency and power level for the first primitive RADAR units in 1934. In the 1917 *The Electrical Experimenter*, Tesla stated the principles of modern military radar in detail. Tesla's study of high voltage, high frequency alternating currents led to this development. Tesla had formed the concept of using radio waves to detect objects at a distance.

Tesla stated,

"For instance, by their [standing electromagnetic waves] use we may produce at will, from a sending station, an electrical effect in any particular region of the globe; [with which] we may determine the relative position or course of a moving object, such as a vessel at sea, the distance traversed by the same, or its speed."

Tesla proposed to use electromagnetic waves to determine the relative position, speed, and course of a moving object and other modern concepts of radar. Tesla had proposed that it might help find submarines (for which it isn't well-suited), though it was first applied successfully to find aircraft (after their later proliferation) and surface ships during World War II. Emil Girardeau, working with the first French radar systems, stated he was building radar systems "*conceived according to the principles stated by Tesla*".

In 1934, Watson-Watt was well established in the area of radio, and was approached by H.E. Wimperis from the Air Ministry, who asked about the use of radio to produce a 'death ray'. While he knew this to be unlikely, he did mention Meanwhile attention is being turned to the still difficult, but less unpromising, problem of radio detection and numerical considerations on the method of detection by reflected radio waves will be submitted when required. Watson-Watt and his assistant Arnold Wilkins published a report on the topic in February 1935, titled *The Detection of Aircraft by Radio Methods*.

Meanwhile in Germany, Hans Eric Hollmann had been working for some time in the field of microwaves, which were to later become the basis of almost all radar systems. In 1935 he published *Physics and Technique of Ultrashort Waves*, which was then picked up by researchers around the world. At the time he had been most interested in their use for communications, but he and his partner Hans-Karl von Willisen had also worked on radar-like systems.

In the autumn of 1934 their company, GEMA, built the first commercial radar system for detecting ships. Operating in the 50 cm range it could detect ships up to 10 km away, similar in purpose to Huelsmeyer's earlier device. In the summer of 1935 a pulse radar was developed with which they could spot the ship the *Königsberg* 8 km away, with an accuracy of up to 50 m, enough for gun-laying. The same system could also detect an aircraft at 500 m altitude at a distance of 28 km. The military implications were not lost this time around, and construction of land and sea-based versions took place as Freya and Seetakt.

Development in England and Germany

At this point both England and Germany knew of the ongoing efforts by their "competition". Both nations were intensely interested in the other's developments in the field, and engaged in an active campaign of espionage and false leaks about their respective equipment. But it was only in Britain that the usefulness of the system became obvious, so while the German systems had the edge technologically (operating on much shorter wavelengths) only Britain started true mass deployment of both the radars and the control systems needed to support them.

Invention of Chain Home in England

Shortly before the outbreak of World War II several radar stations known as **Chain Home** (or *CH*) were constructed in the south of England. As one might expect from the first radar to be deployed, CH was a simple system. The broadcast side was formed from two 300′ (100 m) tall steel towers strung with a series of cables between them.

The output of a powerful 50 MHz radio of about 200 kW (up to 800 kW in later models) was fed into these cables, pulsed at about 50 times a second. A second set of 240′ (73 m) tall wooden towers were used for reception, with a series of crossed antennas at various heights up to 215′ (65 m). In fact most stations had more than one set of each antenna, tuned to operate at different frequencies.

The CH radar was read with an oscilloscope. When a pulse was sent out into the broadcast towers, the scope was triggered to start its beam moving horizontally across the screen very rapidly. The output from the receiver was amplified and fed into the vertical axis of the scope, so a return from an aircraft would deflect the beam upward. This formed a spike on the display, and the distance from the left side-measured with a small scale on the bottom of the screen - would give the distance to the target. By rotating the receiver antennas to make the display disappear, the operator could determine the direction (this is the reason for the cross shaped antennas), the size of the vertical displacement indicated something of the number of aircraft involved, and by comparing the strengths returned from the various antennas up the tower, you could determine the height.

CH proved highly effective during the Battle of Britain, and are often credited with allowing the RAF to defeat the much larger Luftwaffe forces. Whereas the Luftwaffe had to hunt all over to find the RAF fighters, the RAF on the other hand knew exactly where the Luftwaffe bombers where, and could converge all of their fighters on them.

Very early in the battle the Luftwaffe made a series of small raids on a few of the stations, but they were eventually returned to operation in a few days. In the meantime the operators took to broadcasting radar-like signals from other systems in order to fool the Germans into believing that the systems were still operating. Eventually they gave up trying to bomb them. The Luftwaffe apparently never understood the importance of radar to the RAF's efforts, or they would have assigned them a much higher priority — it is clear they could have knocked them out continually if they wished.

In order to avoid the CH system the Luftwaffe adopted other tactics. One was to approach Britain at very low levels, below the sight line of the radar stations. This was countered to some degree with a series of shorter range stations built right on the coast, known as Chain Home Low (*CHL*). These radars had originally been intended to use for naval gun-laying an known as Coastal Defense (*CD*), but their narrow beams also meant they could sweep an area much closer to the ground without seeing the reflection of the ground (or water) itself. Unlike the larger CH systems, CHL had to have the broadcast antenna itself turned, as opposed to just the receiver. This was done manually on a pedal-crank system run by WAAFs until more reliable motorized movements were installed in 1941.

Later Adaptions

Similar systems were later adapted with a new display to produce the **Ground Controlled Intercept stations** starting in late 1941. In these systems the antenna was rotated mechanically, followed by the display on the operators console. That is, instead of a single line across the bottom of the display from left to right, the line was rotated around the screen at the same speed as the antenna was turning.

The result was a 2-D display of the air around the station with the operator in the middle, with all the aircraft appearing as dots in the proper location in space. These so-called **Plan Position Indicators** (*PPI*) dramatically simplified the amount of work needed to track a target on the operator's part. Such a system with a rotating, or sweeping, line is what most people continue to associate with a radar display.

Rather than avoid the radars, the Luftwaffe took to avoiding the fighters by flying at night and in bad weather. Although the RAF was aware of the location of the bombers, there was little they could do about them unless the fighter pilots could see the opposing planes. However, just this eventuallity had already been forseen, and Watson-Watt (likely at the urging of Tizzard) had already started work on a miniaturized radar system suitable for aircraft, the so-called **AI** (*airborne interception*) set. Initial sets were available in 1941 and fitted to Bristol Blenheim aircraft, replaced quickly with the better performing Bristol Beaufighter, which quickly put an end to German night- and bad-weather bombing over England.

The next major development in the history of radar was the invention of the **cavity magnetron** by Randall and Boot of Birmingham University in early 1940. This was a small device which generated much more powerful microwaves than previous devices, which in turn allowed for the detection of much smaller objects and the use of much smaller antennas. The secrecy of the device was so high that it was decided in 1940 to move production to the USA, which resulted in the creation of the MIT Radiation Lab to develop the device further.

German Developments

German developments mirrored those in England, but it appears radar received a much lower priority until later in the war. The Freya was in fact much more sophisticated than its CH counterpart, and by operating in the 1.2 m wavelength (as opposed to ten times that for the CH) the Freya was able to be much smaller and yet offer better resolution. Yet by the start of the war only eight of these units were in operation, offering much less coverage.

However the Germans did not have an airborne system of any sort deployed until 1942, leaving them with the problem of having to get their fighters into that 300m range solely with ground-based equipment. To fill this need another system known as Wuerzburg was deployed, starting in 1941.

Unlike other systems, the Wuerzburg was mounted on a highly directional parabolic antenna that was sensitive in only one direction. This made it useless for finding the targets, but once guided to one by an associated Freya it could track it with extreme accuracy: later models were accurate to 0.2 degrees or less. In order to do this the radar sent out two lobes and the return of each was shown on the display. By keeping the returns from both the same strength, the operator kept the Wuerzburg pointed directly at the target.

The downfall of the German radar network was that it could only track a single aircraft per Wuerzburg. In fact the system required two Wuerzburgs per interception, one for the target, and one for the fighter. This meant that as a raid developed, only a few night fighters could be directed at any one time, as only a small number of the eventual 5,000 Wuerzburgs would be within their 25 km range at any one time.

1.8. DIFFERENT TYPES OF RADAR

Radar systems may be divided into types based on the designed use. This section presents the general characteristics of several commonly used radar systems:

- Air-Defense Radars
- Air Traffic Control Radars
- Fire-control radar or tracking radar
- Speed gauges
- Airborne radar
- Mortar locating radar
- Radar Satellites
- Weather Radar
- Ground Penetrating radar

Depending the desired information radar units must have different qualities and technologies

On reason of these different qualities and techniques radar units are classified is:

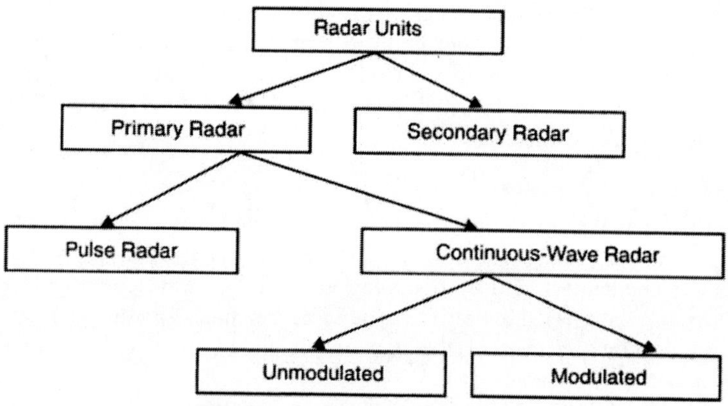

Fig. 1.12. Radar Classifications.

Primary Radar

A primary radar transmits high-frequency signals which are reflected at targets. The arisen echoes are received and evaluated. This means, unlike secondary radar units a primary radar unit receive its own emitted signals as an echo again.

Secondary Radar

At these radar units the airplane must have a transponder (transmitting responder) on board and receives an encoded signal of the secondary radar unit. An active also encoded response signal, which is returned to the radar unit then is generated in the transponder. In this response can be contained much more information, as a primary radar unit is able to acquire (E.g. an altitude, an identification code or also any technical problems on board such as a radio contact loss...). Example of secondary radar is IFF (Identification of Friend and Foe).

Pulse Radar

Pulse radar units transmit a high-frequency impulsive signal of high power. After this a longer break in which the echoes can be received follows before a new transmitted signal is sent out. Direction, distance and sometimes if necessary the altitude of the target can be determined from the measured antenna position and propagation time of the pulse-signal.

Continuous-Wave Radar (CW Radar)

CW radar units transmit a high-frequency signal continuously. The echo signal permanently is received and processed. The receiver needn't be mounted at the same place as the transmitter absolutely. Every firm civil radio transmitter can work as a radar transmitter at the same time, if a remote receiver compares the propagation times the direct one with the reflected signal. Tests are known that the correct location of an airplane can be calculated from the evaluation of the signals by three different television stations.

Unmodulated CW-Radar

The transmitted signal of these equipments is constant in amplitude and frequency. These equipments are specialized in speed measuring. Distances cannot be measured. E.g. these equipments are used as speed gauges of the police.

Modulated CW-Radar

The transmitted signal is constant in the amplitude but modulated in the frequency. This one gets possible after the principle of the propagation time measurement with that again. It is advantage of these equipments that an evaluation is carried out without reception break and the measurement result is therefore continuously available. These radar units are used everywhere there where the measuring distance isn't too large and it's necessary a continuous measuring (e.g. an altitude measuring in airplanes or as weather radar/wind profiler).

Radar units whose transmitting impulse is too long to get a well distance resolution also use a similar principle of obtaining. Often these equipments modulate its transmitting pulse additional, to obtain a distance resolution also within the transmitting pulse with help of the pulse-compression.

1.9. BASIC PULSE RADAR SYSTEM

A very basic radar block diagram of pulsed radar system was shown in Fig. 1.3. A more detailed block diagram of high–power pulsed radar set is shown in Fig. 1.13.

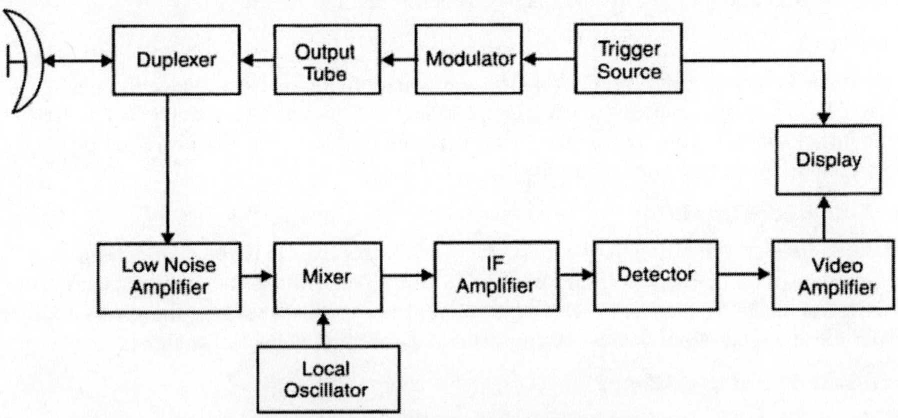

Fig. 1.13. Pulsed Radar Block Diagram.

Duplexer. This is a switch, which alternately connects the transmitter or receiver to the antenna. Its purpose is to protect the receiver from the high power output of the transmitter. During the transmission of an outgoing pulse, the duplexer will be aligned to the transmitter for the duration of the pulse, PW. After the pulse has been sent, the duplexer will align the antenna to the receiver. When the next pulse is sent, the duplexer will shift back to the transmitter. A duplexer is not required if the transmitted power is low.

Antenna. The antenna takes the radar pulse from the transmitter and puts it into the air. Furthermore, the antenna must focus the energy into a well-defined beam which increases the power and permits a determination of the direction of the target. The antenna must keep track of its own orientation, which can be accomplished by a synchro-transmitter. There are also antenna systems, which do not physically move but are steered electronically.

Transmitter (Output Tube). The transmitter creates the radio wave to be sent and modulates it to form the pulse train. The transmitter must also amplify the signal to a high power level to provide adequate range. The source of the carrier wave could be a Klystron, Traveling Wave Tube (TWT) or Magnetron. Each has its own characteristics and limitations.

Radar Modulator

The radar modulator is a device, which provides the high power to the transmitter tube to transmit during transmission period. It makes the transmitting tube ON and OFF to generate the desired waveform. Modulator allows storing the energy in a capacitor bank during the rest time.

The stored energy then can be put into the pulse when transmitted. It provides rectangular voltage pulses which acts as the supply voltage to the output tube such as magnetron, thus switching it ON and OFF as required. Commonly a thyratom valve is being used as switching device, which is a gas filled valve.

Trigger Source (Synchronizer)

The synchronizer coordinates the timing for range determination. It regulates that rate at which pulses are sent (*i.e.* sets PRF) and resets the timing clock for range determination for each pulse. Signals from the synchronizer are sent simultaneously to the transmitter, which sends a new pulse, and to the display, which resets the return sweep.

Receiver

Receiver is usually of superhetrodyne type whose function is to detect the desired echo signal in the presence of noise, interference and clutter.The receiver in pulsed radar consists of the Low Noise RF Amplifier, mixer, local oscillator IF amplifier, detector, video amplifier and radar display.

Low Noise RF Amplifier

Low Noise RF Amplifier is the first stage of the receiver.It is low noise transistor amplifier or a parametric amplifier or a TWT amplifier. Silicon bipolar transistor is used at lower radar frequencies (below L-band 1215 to 1400 Mhz) and the GaAs FET is preferred at higher frequencies. It amplifies the received weak echo signal.

Mixer and Local Oscillator

These convert RF signal output from RF amplifier to comparatively lower frequency levels called Intermediate Frequency (IF). The typical value for pulse radar is 30 Mhz or 60 Mhz.

IF Amplifier

Its consists of a cascade of tuned amplifiers, these can be synchronous, that is, all tuned to the same frequency and having identical band pass characteristics. If a really large bandwidth is needed, the individual IF amplifiers may be stagger tuned. The typical value for pulse radar is 30 Mhz or 60 Mhz.

Detector

Detector is often a Schottky-barrier diode which extracts the pulse modulation from the IF amplifier output. The detector output is then amplified by the Video Amplifier to a level where it can be properly displayed usually on CRT (Cathode Rays Tubes) directly or via digital signal processors. Synchronizing pulses are applied by the trigger source to the display unit.

Display Unit

The received video signals are displays on the CRT (Cathode Rays Tube) for further observations and actions. Different types of Display system which are used in radar, discussed in Chapter 8 in this book.

1.10. INTRODUCTION TO RADAR EQUATION

The radar equation relates the range of radar to the characteristics of the Tx (transmitter), Rx (receiver), Ae (antenna), target and the environment. It is useful to determine the maximum range at which particular radar can detect a target and its makes to understand the factor affecting radar performance. In this section, the simple form of equation is derived.

The radar equation is an important tool for following aspect:

1. Assessing the performance of radar.
2. Designing of new radar systems.
3. Assessing the technical requirement for new radar procurement.

1.11. SIMPLE FORM OF RADAR EQUATION

If the power of radar Tx is denoted by Pt and if an isotropic antenna (which will radiate energy uniformly in all direction) then the power density at a distance "R" from the radar is equal to the radiated power divided by the surface area $4\pi R^2$ of an imaginary sphere of radius R, or

i.e., power density at range R from an isotropic antenna

$$= Pt/4\pi R^2 \text{ watt/m}^2 \quad \text{...(1.1)}$$

However, we employ directive antenna (with narrow beam width) to concentrate the radiated power Pt in a particular direction.

The gain of an antenna is a measure of the increased power density radiated in some direction as compared to the power density that would appear in that direction from an isotropic antenna. The maximum gain "G" of an antenna may be defined as

$$G = \frac{\text{Maximum power density radiated by a directive antenna}}{\text{Power density radiated by loss less isotropic antenna}}$$

Power density at a distance R from directive antenna of power Gain "G",

$$= Pt\ G/4\pi R^2 \text{ watt/m}^2 \quad \text{...(1.2)}$$

The target intercepts a portion of the incident energy and re radiates it in various directions. It is only the power density re radiated in the direction of the radar (echo signal) that is of interest. The radar cross section of the target determines the power density returned to the radar for a particular power density incident on the target. It is denoted by 'σ' and is often called, for short, target cross section or radar cross section. The radar cross section is defined by the following :

$$\text{Reradiated power density back at the radar} = \frac{Pt.G.\sigma}{4\pi R^2.4\pi R^2} \quad \text{...(1.3)}$$

The radar cross section has units of area, but it can be misleading to associate the radar cross section directly with the targets physical size. Radar cross-section more dependent on the targets shape than on its physical size.

The radar antenna received a portion of the echo power. If the effective area of receiving antenna is denoted Ae, the power Pr received by the radar is,

$$Pr = \frac{Pt.G.\sigma.Ae}{4\pi R^2.4\pi R^2} \text{ watts} \quad \text{...(1.4)}$$

The maximum range of radar Rmax is the distance beyond which the target can not detected. It occurs when the received signal power Pr just equals the minimum detectable signal (Smin).

Substituting, $S_{min} = Pr$ in equation (1.4)

$$S_{min} = \frac{Pt.G.\sigma.Ae}{4\pi R^2.4\pi R^2} \text{ watts} \quad \text{...(1.5)}$$

i.e.,

$$R^4_{max} = \frac{Pt.G.\sigma.Ae}{(4\pi).(4\pi)(S_{min})} \quad \text{...(1.6)}$$

$$R^4_{max} = \frac{Pt.G.\sigma.Ae}{(4\pi)^2.(S_{min})} \quad \text{...(1.7)}$$

$$\boxed{\mathbf{R_{max} = \sqrt[4]{\frac{Pt.G.\sigma.Ae}{(4\pi)^2.S_{min}}}}} \quad \text{...(1.8)}$$

where,
Pt = Transmitter Power,
G = Maximum Gain of Antenna
Ae = Aperture area of receiving Antenna
σ = Cross section area of the target
S_{min} = Minimum detectable signal (mds)
R_{max} = Maximum range of radar

This is the fundamental form of radar range equation. The important parameters are the transmitting gain and the receiving effective area. The transmitter power Pt has not been specified as either the average or the peak power.

If the same antenna is used for both transmitting and receiving, as it usually is in radar, antenna theory gives the relationship between transmit gain G and the receive effective area Ae as

$$G = \frac{4\pi . Ae}{\lambda^2} \quad ...(1.9)$$

where λ = wavelength ($\lambda = c/f$ where c = velocity of propagation and f = frequency) equation (1.8) can be substituted, first for Ae and the G, to give two other forms of the radar equation,

$$R_{max} = \sqrt[4]{\frac{Pt . G^2 \sigma . \lambda^2}{(4\pi)^3 S_{min}}} \quad ...(1.10)$$

or

$$R_{max} = \sqrt[4]{\frac{Pt . Ae^2 \sigma}{4\pi\lambda^2 S_{min}}} \quad ...(1.11)$$

These three forms of the radar equations (1.8), (1.10) and (1.11) are basically the same.

1.12. PREDICTION OF RANGE PERFORMANCE

The simple form of radar equation is not adequately predicting the range performance of actual radars. There are some drawbacks of simple form of radar equation appended below:

1. The fluctuation and uncertainties in targets radar cross section
2. The nature of minimum detectable signal, which is affected by the receiver noise.
3. The losses experienced in the radar system.
4. Propagation effect by earth surface and atmosphere.

The range of radar, therefore, will be a function of the Probability of detection (*Pd*) and the Probability of false alarm (*Pfa*).

1.12.1. Radar Performance

All of the parameters of the basic pulsed radar system will affect the performance in some way. Here we find specific examples and quantify this dependence where possible.

Transmitter Power

The power *Pt* in the radar equation is called by the radar engineer, the Peak Power. The peak power as used in the radar equation is not the instantaneous peak power of a sine wave. It is defined as the power averaged over that carrier frequency cycle which occurs at the maximum of the pulse power. Peak power is usually equal to one-half of the maximum instantaneous power.

The average power *Pav* is defined as the average transmitted power over the pulse repetition period. If the transmitted waveform is a train of rectangular pulses of width (*T*) and *PRP* (*Tr*).

$$PRP = 1/PRF \quad ...(1.12)$$

$$Pav(\text{Average Power}) = Pt \times T/Tr = Pt \times T \times PRF \quad ...(1.13)$$

(Average Power = Peak Power × Pulse Width/*PRP* = Peak Power × Pulse width × PRF).

The ratio of average power to the Peak power or pulse width to the *PRP*

or Pulse width × *PRF*, is called the **Duty Cycle** of the radar.

The fraction of time the radar transmitter (Tx) is generating pulse power is termed the Duty-Cycle. Thus, in any one second, a radar producing 1 μ sec pulses at a rate of 1000 per sec is "on" for

$$(1 \times 10^{-6}) \times (1000) = 0.001 \text{ s}$$

The duty cycle is then 0.001, and the average power is given by the product of duty cycle and peak power; thus if the peak power is 500 kw, the average power should be 500 w.

Typical values of radar power lies within the range 100 to 500 Kw.

Writing the equation in terms of the average power rather than the peak power, we get

$$R^4_{max} = \frac{Pav.G.\sigma.Ae}{(4\pi)^2 S_{min} T.fr} \quad ...(1.14)$$

where $fr = PRF$

Pav = Average power

If the radar range is to be doubled, we have to increase the transmitter power 16 times since $R_{max} \propto$ ¼ root of peak power.

Pulse Width (PW)

The duration of the pulse and the length of the target along the radial direction determine the duration of the returned pulse. In most cases the length of the return is usually very similar to the transmitted pulse. In the display unit, the pulse (in time) will be converted into a pulse in distance. The range of values from the leading edge to the trailing edge will create some uncertainty in the range to the target. Taken at face value, the ability to accurately measure range is determined by the pulse width.

If we designate the uncertainty in measured range as the range resolution, R_{RES}, then it must be equal to the range equivalent of the pulse width, namely:

$$R_{RES} = c\ PW/2 \quad ...(1.15)$$

Now, you may wonder why not just take the leading edge of the pulse as the range which can be determined with much finer accuracy? The problem is that it is virtually impossible to create the perfect leading edge. In practice, the ideal pulse will really appear like:

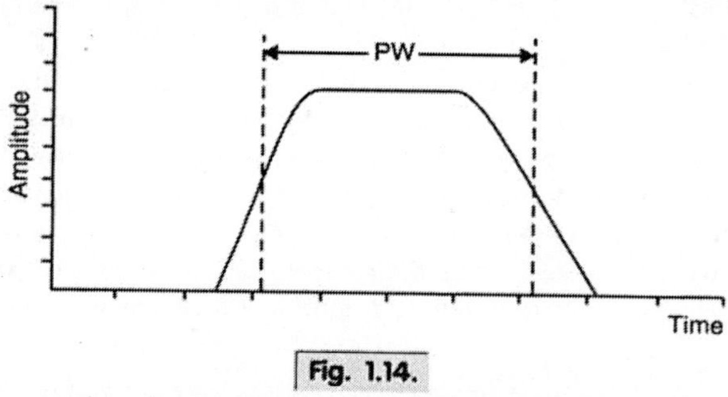

Fig. 1.14.

To create a perfectly formed pulse with a vertical leading edge would require an infinite bandwidth. In fact you may equate the Bandwidth, β, of the transmitter to the minimum pulse width, PW by :

$$PW = 1/\beta \quad ...(1.16)$$

Given this insight, it is quite reasonable to say that the range can be determined no more accurately than $cPW/2$ or equivalently

$$\boxed{R_{RES} = c/2\beta} \quad ...(1.17)$$

In fact, high-resolution radar is often referred to as wide-band radar, which you now see as equivalent statements. One term is referring to the time domain and the other the frequency domain. The duration of the pulse also affects the minimum range at which the radar system can detect. The outgoing pulse must physically clear the antenna before the return can be processed. Since this lasts for a time interval equal to the pulse width, *PW*, the minimum displayed range is then:

$$\boxed{R_{MIN} = c \times \frac{PW}{2}} \quad ...(1.18)$$

The minimum range effect can be seen on a *PPI* display as a saturated or blank area around the origin.

Increasing the pulse width while maintaining the other parameters the same will also affect the duty cycle and therefore the average power. For many systems, it is desirable to keep the average power fixed. Then the *PRF* must be simultaneously changed with *PW* in order to keep the product *PW* × *PRF* the same. For example, if the pulse width is reduced by a factor of ½ in order to improve the resolution, then the *PRF* is usually doubled.

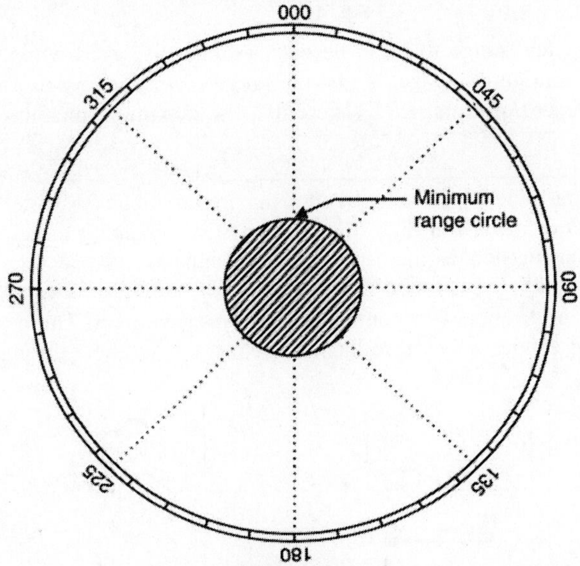

Fig. 1.15.

Pulse Repetition Frequency (PRF) and Range Ambiguities

The frequency of pulse transmission affects the maximum range that can be displayed. Recall that the synchronizer resets the timing clock as each new pulse is transmitted. Returns from distant targets that do not reach the receiver until after the next pulse has been sent will not be displayed correctly. Since the timing clock has been reset, they will be displayed as if the range where less than actual. If this were possible, then the range information would be considered ambiguous. An operator would not know whether the range were the actual range or some greater value.

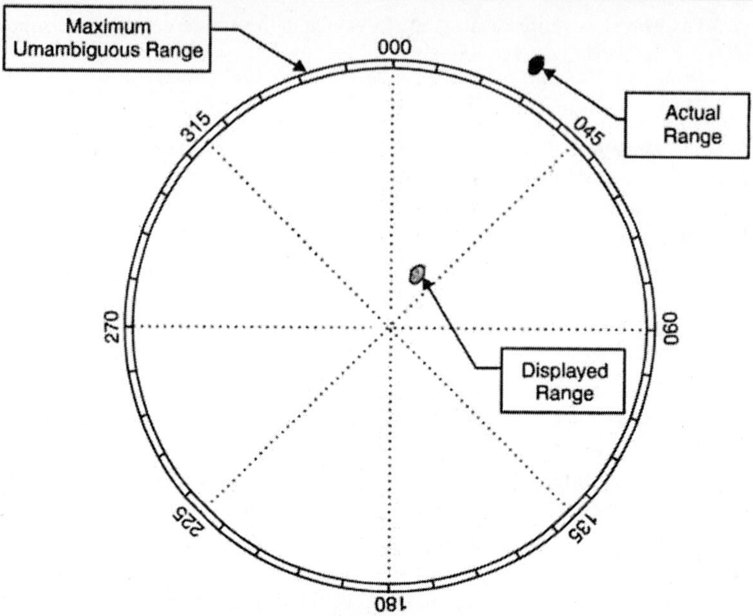

Fig. 1.16.

The maximum actual range that can be detected and displayed without ambiguity, or the maximum unambiguous range, is just the range corresponding to a time interval equal to the pulse repetition time, *PRT*. Therefore, the maximum unambiguous range,

$$R_{UNAMB} = c\ PRT/2 = c/(2PRF) \quad ...(1.19)$$

The *PRF* is determined primarily by the maximum range at which targets are expected. If the *PRF* is made too high, the likelihood of obtaining target echoes from the wrong pulse transmission is increased. Echoes signals received after an interval exceeding the *PRP* (Pulse Repetitation Period) are called multiple times around echoes.

They can result in erroneous or confusing range measurement. The nature of some multiple-time-around echoes may cause them to be labeled as ghost or angel targets or even flying saucers.

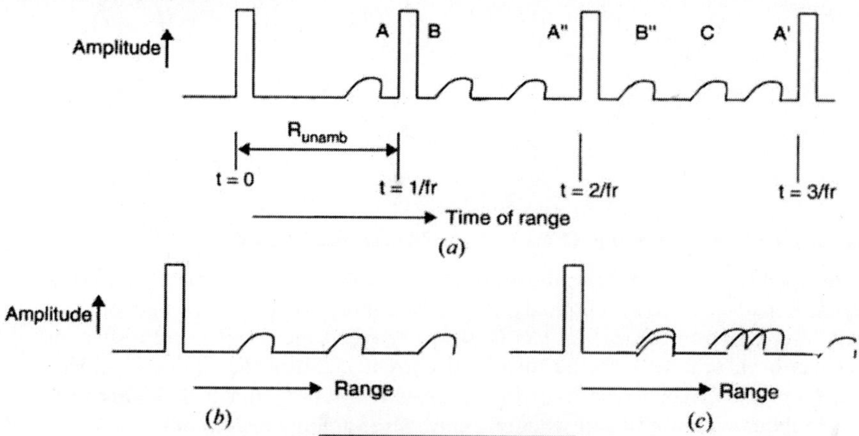

Fig. 1.17. (a, b and c)

(*a*) Three targets *A*, *B* & *C* where *A* is within Runamb and *B* & *C* are multiple time around echoes (targets); (*b*) appearance of the three targets on *A*–scope (*c*) appearance of three targets on the *A*-scope with a changing *PRF*.

To minimize the multiple time around targets, a staggered *PRF* can be used.

When a radar is scanning, it is necessary to control the scan rate so that a sufficient number of pulses will be transmitted in any particular direction in order to guarantee reliable detection. If too few pulses are used, then it will more difficult to distinguish false targets from actual ones. False targets may be present in one or two pulses but certainly not in ten or twenty in a row. Therefore to maintain a low false detection rate, the number of pulses transmitted in each direction should be kept high, usually above ten.

For systems with high pulse repetition rates (frequencies), the radar beam can be repositioned more rapidly and therefore scan more quickly. Conversely, if the *PRF* is lowered the scan rate needs to be reduced. For simple scans it is easy to quantify the number of pulses that will be returned from any particular target. Let τ represent the dwell time, which is the duration that the target remains in the radar's beam during each scan. The number of pulses, *N*, that the target will be exposed to during the dwell time is:

$$\boxed{N = \tau PRF} \qquad ...(1.20)$$

We may rearrange this equation to make a requirement on the dwell time for a particular scan

$$\boxed{\tau_{min} = N_{min}/PRF} \qquad ...(1.21)$$

So it is easy to see that high pulse repetition rates require smaller dwell times. For a continuous circular scan, for example, the dwell time is related to the rotation rate and the beam-width.

$$\boxed{\tau = \theta_B/\theta_s} \qquad ...(1.22)$$

where θ_B = beam-width [degrees] θ_s = rotation rate [degrees/sec] which will give the dwell time (τ) in seconds. These relationships can be combined; giving the following equation from which the maximum scan rate may be determined for a minimum number of pulses per scan:

$$\boxed{\theta_s = \theta_B \; PRF/N} \qquad ...(1.23)$$

Scan rate may be calculated by using following equation,

$$\boxed{\text{Scan rate } (\theta_s) = \frac{RPM \times 360°}{60 \text{ s}}} \qquad ...(1.24)$$

Radar Operating Frequency

Finally, the frequency of the radio carrier wave will also have some affect on how the radar beam propagates. At the low frequency extremes, radar beams will refract in the atmosphere and can be caught in "ducts" which result in long ranges. At the high extreme, the radar beam will behave much like visible light and travel in very straight lines. Very high frequency radar beams will suffer high losses and are not suitable for long range systems.

The frequency will also affect the beam-width. For the same antenna size, low frequency radar will have a larger beam-width than a high frequency one. In order to keep the beam-width constant, low frequency radar will need a large antenna.

1.13. MINIMUM DETECTABLE SIGNAL

The receiver is sensitive to the range of frequencies being transmitted and provides amplification of the returned signal. In order to provide the greatest range, the receiver must be very sensitive without introducing excessive noise. The ability to discern a received signal from background noise depends on the signal-to-noise ratio (*S/N*).

The background noise is specified by an average value, called the noise-equivalent-power (*NEP*). This directly equates the noise to a detected power level so that it may be compared to the return. Using these definitions, the criterion for successful detection of a target is

$$P_r > (S/N)\ NEP,$$

where P_r is the power of the return signal. Since this is a significant quantity in determining radar system performance, it is given a unique designation, S_{min}, and is called the **Minimum Signal for Detection.**

$$\boxed{S_{min} = (S/N)\ NEP} \quad ...(1.25)$$

Since S_{min}, expressed in Watts, is usually a small number, it has proven useful to define the decibel equivalent, *MDS*, which stands for **Minimum Discernible Signal**.

$$\boxed{MDS = 10 \text{ Log } (S_{min}/1 \text{ mW})} \quad ...(1.26)$$

When using decibels, the quantity inside the brackets of the logarithm must be a number without units. In the definition of MDS, this number is the fraction S_{min}/1 mW. As a reminder, we use the special notation dBm for the units of *MDS*, where the "*m*" stands for 1 mW. This is shorthand for decibels referenced to 1 mW, which is sometimes written as db/1 mW.

In the receiver, *S/N* sets a threshold for detection which determines what will be displayed and what will not. In theory, if *S/N* = 1, then only returns with power equal to or greater than the background noise will be displayed.

Detection is based on establishing a threshold level at the output of the receiver.

1.13.1. Threshold Detection

The target is said to be present when the receiver output crosses the threshold level and if the level of the receiver output signal is low and not able to cross the threshold level, it may be considered as noise. So the detection of target or noise may be decided by the crossing of threshold level and this type of detection is known as **threshold detection.**

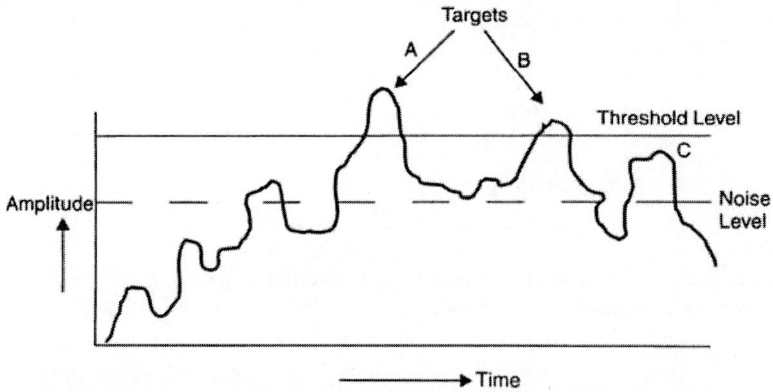

Fig. 1.18. Typical Output of the Radar as a Function of Time.

In the figure A, B and C represent signal pulse noise. A and B are valid detection, but C is missed detection. If the threshold level is set too low, noise might exceed it and be mistaken as target. This is called **False Alarm**. If the threshold were set too high, noise might not be large enough to cause false alarm, but weak targets echoes might not exceeds the threshold and would not be detected, when this occurs it is called **Missed Detection.**

1.14. RADAR RECEIVER NOISE AND SIGNAL-TO-NOISE RATIO

"**Noise** is unwanted Electromagnetic energy which interfere with the ability of the receiver to detect the wanted signal". Noise may be generated within the receiver it self or it may enter through the receiving antenna along with the reflected signals.

Thermal Noise or Johnson Noise is the noise, which is generated by the thermal motion of the conducted electrons in receiver input stages. The magnitude of the noise is directly proportional to the bandwidth and the absolute temperature of the resistive input circuit. The thermal noise power of a radar receiver may be written as

Thermal noise power (watt) $\infty \; T \,.\, Bn$

Or $= k.T \,.\, Bn$ ...(1.27)

Where, K = Boltzman's Constant = 1.38×10^{-23} Joules/degree

of a in Superhetrodyne receiver the bandwidth of the IF Amplifier which may be called as **matched filter**.

We can defined bandwidth noise (Bn) as

$$Bn = \frac{\int_0^{\infty} |H(f)|^2 \, dt}{|H(f_0)|^2} \qquad ...(1.28)$$

In the above Eqn. (1.28),

f_0 = Frequency of the maximum response

$H(f)$ = The IF amplifier function of frequency response

The Noise Figure (F_n) in practical receiver may be defined as the ratio of noise out of a practical receiver to that noise out of an ideal receiver assuming that no other noise is present except thermal noise in the receiver.

Noise figure (F_n) may be written as

$$F_n = \frac{\text{Practical receiver Noise output}}{\text{Ideal receiver Noise output}}$$

$$F_n = \frac{N_0}{K.T_0.Bn.Ga} \qquad ...(1.29)$$

Where, N_0 = noise output

Ga = gain of the receiver

T_0 = Standard temperature is to be taken 290 degree Kelvin.

The noise N_0 is measured over the linear portion of the receiver input-output characteristics, usually at the output of the IF amplifier.

The receiver bandwidth B_n is that of the IF amplifier in most radar receiver. The Available gain Ga is the ratio of the signal out (So) to the signal in (Si), and kT_0Bn is the input noise (Ni) in ideal receiver.

$$F_n = \frac{Si / Ni}{So / No} \quad ...(1.30)$$

$$= \left(\frac{Si}{So}\right) \times \left(\frac{No}{Ni}\right)$$

$$= \frac{Si}{(Ga . Si)} \times \left[Ga \times \frac{(Ni + Nr)}{Ni}\right]$$

$$F_n = \left[\frac{1 + Nri}{Ni}\right]$$

Thus,

$$\left(\frac{Nri}{Ni}\right) = (F_n - 1)$$

$$Nri = Ni(F_n - 1)$$

Or

$$\boxed{Nri = k . T_0 . B(F - 1) = S_{(min)}}$$

$$\boxed{k . T_0 . B(F - 1) = S_{(min)}}$$

Where Nri = Noise power generated at the receiver input.

Substitute the value of $S_{(min)}$ in equation (1.11),

$$\boxed{R_{max} = \sqrt[4]{\frac{Pt . A_e^2 . \sigma}{4\pi\lambda^2 . k . T_0 . B . (F - 1)}}} \quad ...(1.31)$$

By substituting the capture in terms of antenna diameter and for the various constants. From antenna theory, we know that

$$\boxed{A_e = \frac{0.65\pi(D^2)}{4}} \quad ...(1.32)$$

We may further simplify the radar range equation and may be written as

$$\boxed{R_{max} = (48)\sqrt[4]{\frac{Pt . D^4 . \sigma}{\lambda^2 . B . (F - 1)}} \text{ Km}} \quad ...(1.33)$$

or

$$R_{max} = 48\left[\frac{(Pt . D^4 . \sigma)}{\lambda^2 B(F - 1)}\right]^{1/4} \quad ...(1.34)$$

or

$$\boxed{R_{max} = 26\left[\frac{(Pt . D^4 . \sigma)}{\lambda^2 B(F - 1)}\right]^{1/4} \text{ Nm}} \quad \text{(since 1 Nm = 1853 m)}$$

where, R_{max} = Maximum radar range,

Pt = Peak power

D = Antenna diameter in meters

σ = Radar cross section of the target (m^2)

λ = wavelength in meter

F = Noise figure in ratio

1.15. SYSTEM LOSSES

In the radar equation one of important features has not being considered that is losses occurred in the radar system it self such as,

(a) The antenna beam shape loss,

(b) The collapsing loss,

(c) The losses in the microwave plumbing etc.

The losses are very significant while prediction of radar performance. In a complete radar system, there are so many losses, which are very negligible as an individual but if combine them that may be more significant.

1.15.1. Plumbing Loss

The RF Plumbing losses in the radar are generally considered the losses in microwave components, which are being used in the radar to convey the energy from antenna to the receiver and from transmitter to the antenna. These microwave components are circulator, duplexer, directional coupler, wave-guides and bends etc.

Transmission Lines Loss

The losses occurs in transmission line are not one-way loss; it is always two-way loss because these lines are used for reception and transmission both. The theoretically two ways plumbing loss in decibels in radar operating in the S band is approximately 6.8 dB.

Duplexer Loss

The loss of signal power due to duplexer normally depends upon the type of the duplexer. The duplexer loss is not same as during transmission as during reception of the energy. Manufactures provides the detail of losses occurred in the duplexer.

Antenna losses

There are losses, which are being treated as antenna losses. Some losses of antenna are given below:

1. Antenna beam shape loss,
2. Scanning losses,
3. Radome losses.

1.16. PROPAGATION EFFECT ON EM WAVES IN ATMOSPHERIC CONDITIONS

Electromagnetic wavefronts travel through empty space in straight lines at the speed of light, but the **REFRACTIVE INDEX** of the atmosphere affects both the travel path and the speed of the electromagnetic wavefront. The path followed by electromagnetic energy in the atmosphere, whether direct or reflected, usually is slightly curved; and the speed is affected by temperature, atmospheric pressure, and the amount of water vapor present in the atmosphere, which all affect the refractive index. As altitude increases, the combined effects of these influences, under normal atmospheric conditions, cause a small, uniform increase in signal speed. This increase in speed causes the travel path to curve slightly downward, as shown in **Fig. 1.19.** The downward curve extends the radar horizon beyond a line tangent to the earth, as illustrated in **Fig. 1.20.**

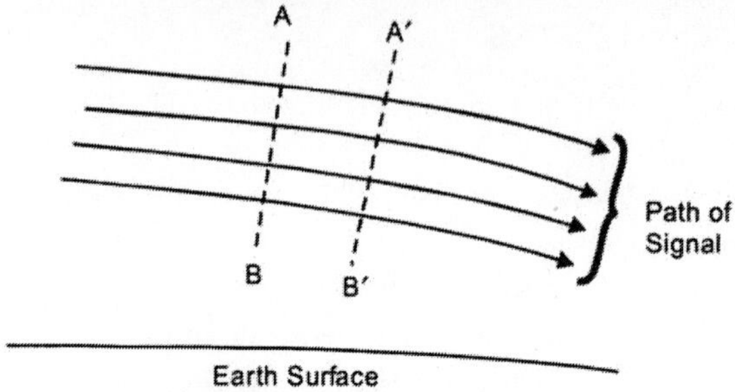

Fig. 1.19. Wavefront Path.

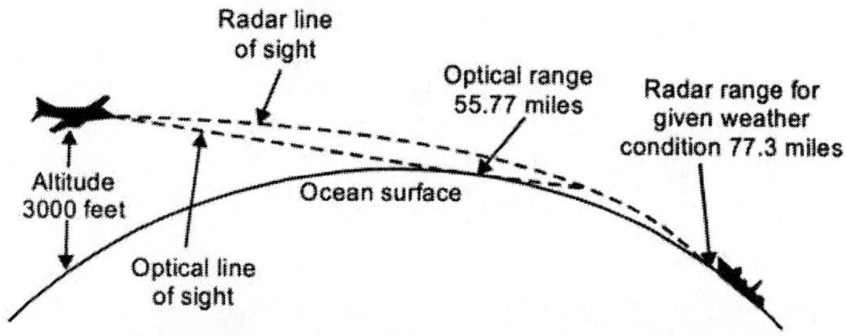

Fig. 1.20. Extension of the Radar Horizon.

The reason for the downward curve can be illustrated using line **AB** in **Fig. 1.19.** Line ***AB*** represents the surface of a wavefront with point *A* higher in altitude than point *B*. As wavefront *AB* moves to the point represented by *A′B′*, the speed at *A* and *A′* is faster than the speed at *B* and *B′* since *A* and *A′* are at a greater altitude. Therefore, in a given time, the upper part of the wavefront moves farther than the lower part. The wavefront leans slightly forward as it moves. Since the direction of energy propagation is always perpendicular to the surface of a wavefront, the tilted wavefront causes the energy path to curve downward.

Refraction

Refraction is the bending of electromagnetic waves caused by a change in the density of the medium through which the waves are passing. A visible example of electromagnetic refraction is the apparent displacement of underwater objects caused by the bending of light as it passes from the atmosphere into the water.

Index of Refraction

Index of Refraction has been established which indicates the degree of refraction, or bending, caused by different substances. Because the density of the atmosphere changes with altitude, the index of refraction changes gradually with height.

Temperature Inversion

The temperature and moisture content of the atmosphere normally decrease uniformly with an increase in altitude. However, under certain conditions the temperature may first increase with height and then begin to decrease. Such a situation is called a **temperature inversion.**

Moisture Lapse

An even more important deviation from normal may exist over the ocean. Since the atmosphere close to the surface over large bodies of water may contain more than a normal amount of moisture, the moisture content may decrease more rapidly at heights just above the sea. This effect is referred to as **MOISTURE LAPSE**.

Ducting

Either temperature inversion or moisture lapse, alone or in combination, can cause a large change in the refraction index of the lowest few-hundred feet of the atmosphere. The result is a greater bending of the radar waves passing through the abnormal condition. The increased bending in such a situation is referred to as **DUCTING** and may greatly affect radar performance. The radar horizon may be extended or reduced, depending on the direction the radar waves are bent. The effect of ducting on radar waves is illustrated in **Fig. 1.21.**

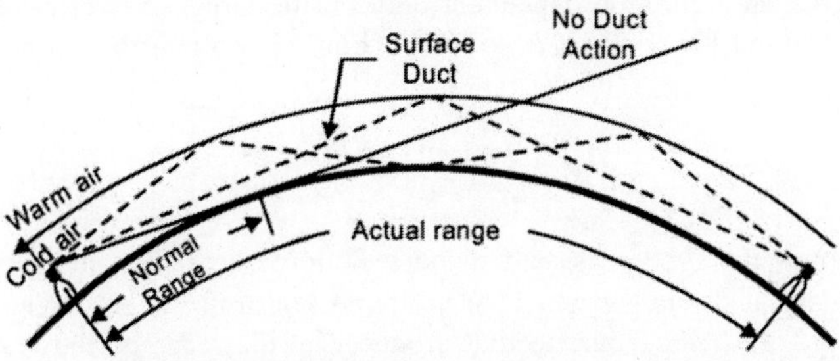

Fig. 1.21. Ducting Effect on the Radar Wave.

Another effect of the atmosphere on radar performance is caused by particles suspended in the air. Water droplets and dust particles diffuse radar energy through absorption, reflection, and scattering so less energy strikes the target. Consequently, the return echo is smaller. The overall effect is a reduction in usable range that varies widely with weather conditions. The higher the frequency of a radar system, the more it is affected by weather conditions such as rain or clouds. In some parts of the world, dust suspended in the air can greatly decrease the normal range of high-frequency radar.

1.17. RADAR CROSS-SECTION OF TARGETS

The radar cross-section σ is the property of a target or scattering object, which is included in the radar range equation to represent the magnitude of echo signal returned to the radar by the target. Radar cross-section RCS is not equal to the physical cross-sectional area of the target. But it is an equivalent area that can be used to relate incident power density at the target to the reflected power density that results at the reciever.

The radar cross-section was defined in the simple radar range equation in terms of following equation

$$\text{Reradiated power density back at the radar} = \frac{P_t G}{4\pi R^2} \cdot \frac{\sigma}{4\pi R^2}$$

In another way, the RCs may also be defined as

$$\sigma = \frac{\dfrac{\text{Power reflected toward source}}{\text{Unit solid angle}}}{\dfrac{\text{Incident power density}}{4\pi}} \qquad \text{...(1.35)}$$

$$\sigma = 4\pi R^2 \frac{|E^r|^2}{|E^i|^2} \qquad ...(1.36)$$

where R = Range to the target

E^r = Electric field strength of the echo signal back at the radar

E^i = Electric field strength incident on the target.

In the Eqn. (1.36), it is assumed that the target is far enough from the radar that the incident wave can be considered to be planar rather than spherical.

The equation (1.36) is usually written in terms of electric field amplitude and in order to make the definition dependent only on the target characteristics, we eliminate range R by taking the limit as $R \to \infty$. So Eqn. (1.36) may be rewritten as

$$\sigma = 4\pi \lim_{R\to\infty}\left[4\pi R^2 \frac{|E^r|^2}{|E^i|^2}\right] \qquad ...(1.37)$$

Sometimes the $RCS\sigma$ is said to be a fictional area that intercepts a part of the power incident at the target which, if scattered uniformly in all direction, produces an echo power at the radar equal to that produced at the radar by the real target. But the real targets do not scatter the incident energy uniformly in all directions.

The power scattered from a target in the direction of the radar receiver *i.e. RCS*, can be calculated by solving Maxwell's equations using the proper boundary conditions.

The *RCS* is just defined as a single real scalar number. The above definition of *RCS* uses a single polarization for the transmitted and receiving wave usually the same. The polarization state of a transverse EM plane wave is a two dimensional vector and two orthogonal polarization basis vector are required to fully describe the wave. The commonly there are two types of polarization basis one is linear (horizontal and vertical polarization) and circular (right and left circular polarization). To understand the effect of polarization, *RCS* must be generalized to the polarization scattering matrix (PSM) S that relates the complex amplitudes of the incident and backscattered fields.

The following relation for a radar using a linear polarization basis is given by Knott, Shaeffer and Tully, 1985; Mott, 1986; Holm, 1987.

$$\begin{bmatrix} E^r_{(H)} \\ E^r_{(V)} \end{bmatrix} = \begin{bmatrix} S_{HH} & S_{HV} \\ S_{VH} & S_{VV} \end{bmatrix} \begin{bmatrix} E^i_H \\ E^i_V \end{bmatrix}$$

$$= S\begin{bmatrix} E^r_H \\ E^r_V \end{bmatrix} \qquad ...(1.38)$$

The target back scattering characteristics are described by four complex number rather than of a single real number. If we assume that both radar transmitter and receiver have only the vertical components, than the *RCS* may be expressed as follows :

$$\sigma = |S_{VV}|^2 \qquad ...(1.39)$$

In general radars area designed to measure the full Polarization Scattering Matrix (PSM). There are other designs measure the magnitude of the elements of the PSM, but not the phases or the magnitudes of two of the PSM elements, which are known as polarimetric measurements. The polarimetric measurements can be used for the analysis of various target.

SOLVED EXAMPLES

Example 1. Calculate the range of a target if the time taken by the radar signal to travel to the target and back is 100 μs.

Solution.

$$\text{Range } (R) = c \times \left(\frac{\Delta t}{2}\right)$$

$$c = 3 \times 10^8 \text{ m/s}$$

$$\Delta t = 100 \text{ μs}$$

By putting the values in equation, we get

$$= (3 \times 10^8) \times \left(\frac{100 \times 10^{-6}}{2}\right)$$

$$\boxed{R = 15 \text{ km.}} \quad \textbf{Ans.}$$

Example 2. If the transmitted peak power of radar is 100 KW, pulse repetition frequency is 1000 pps and the pulse width is 1 μs than calculate the average power in dbs.

Solution. We have

$$Pt = 100 \text{ KW}$$

$$PRF = 1000 \; pps$$

$$PW = 1 \text{ μs}$$

We know that $\text{Duty Cycle} = PW \times PRF$

By putting the value in the above equation, we get,

$$\text{Duty cycle} = (10^{-6}) \times (1000) = 0.001$$

We know, $\text{The average power} = Pt \times \text{Duty cycle}$

$$= (100 \times 10^3) \times 0.001$$

$$\boxed{\text{Average power} = 100 \text{ w}} \quad \textbf{Ans.}$$

or we can write in dbs, $100 \text{ w} = 20 \text{ dbs}$

Example 3. A typical pulse waveform of radar is shown below. In which some parameters of radar is shown. Calculate the (*a*) Average Power, (*b*) Duty Cycle and (*c*) Maximum range of radar.

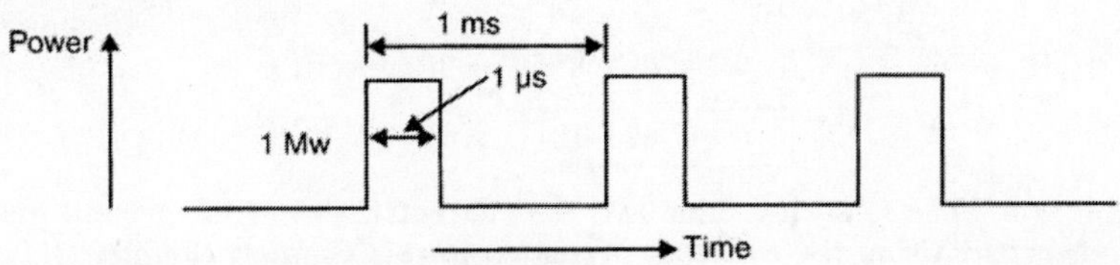

Fig. 1.22.

Solution. We have,

$$\text{Peak power } (Pt) = 1 \text{ Mw}$$

$$\text{Pulse Width } (T) = 1 \text{ μs}$$

$$\text{Pulse Repetition Period } (Tr) = 1 \text{ ms}$$

We know,

$$PRf\ (fr) = 1/Tr$$
$$= 1/(1 \times 10^{-3})$$
$$PRF = 1000 \text{ hz}$$

(*a*)

$$Pav = Pt \times T \times fr$$
$$= (10^6) \times 1 \times 10^{-6} \times (1000)$$

$$\boxed{Pav = 1000 \text{ w} = 1 \text{ KW}}$$ **Ans.**

(*b*)

$$\text{Duty Cycle} = T \times fr$$
$$= (10^{-6}) \times (1000) = 0.001$$

$$\boxed{\text{Duty Cycle} = 0.001}$$ **Ans.**

(*c*) Maximum Unambiguous Range (Rmax) = $c/(2 \times fr)$

$$= \frac{(3 \times 10^8)}{(2 \times 1000)}$$
$$= (3/2) \times 10^5$$

$$\boxed{R\text{max} = 150 \text{ Km}}$$ **Ans.**

or

$$R\text{max} = \frac{150 \times 10^3}{1852} = 81 \text{ nmi} \qquad (1 \text{ nmi} = 1853 \text{ m})$$

Example 4. Determine the maximum unambiguious range and range resolution of a pulse radar having pulse width is 5 μs at a PRF of 1000 hz.

Solution.

$$R\text{max (unamb)} = c/\ (2 \times fr)$$
$$= \frac{(3 \times 10^8)}{(2 \times 1000)} = 150 \text{ km}$$

$$\boxed{R\text{max (unamb)} = 150 \text{ Km}}$$ **Ans.**

$$\text{Range Resolution} = c \times \frac{T}{2} = \frac{(3 \times 10^8) \times (5 \times 10^{-6})}{2}$$
$$= 7.5 \times 100 = 750 \text{ m}$$

$$\boxed{\text{Range Resolution} = 750 \text{ m}}$$ **Ans.**

Example 5. A radar is to have a maximum range of 250 km. Determine the maximum allowable PRF for unambiguous reception.

Solution.

$$R\text{max (unamb)} = 250 \text{ km} = c/(2 \times fr)$$
$$(250 \times 10^3) = (3 \times 10^8)/2 \times fr$$
$$fr = (3 \times 10^8)/2 \times (250 \times 10^3)$$

$$\boxed{fr = 600 \text{ hz}}$$ **Ans.**

Example 6. The following table lists the characteristics of the ground pulse-echo type search radar. Using the concepts presented in this chapter, complete this table.

frequency, f	**5600 MHz**
wavelength, λ	____
pulse width, PW	**1.3 μ sec**
pulse repetition frequency, PRF	____

pulse repetition time, *PRT*	——-
peak power	——
average power	——
duty cycle	**8.3×10^{-4}**
Antenna rotation rate, θ	**16 RPM**
Horizontal beamwidth, θ_B	——
vertical beam width, ϕ_B	**4°**
effective aperture, A_e	**$0.9\ m^2$**
power gain, *G*	**3940**
directive gain, G_D	——
number of returns per sweep, N_B	**9.9**
minimum discernible signal, *MDS*	**–83 dBm**
receiver sensitivity, S_{min}	**-5.012×10^{-12}**
maximum unambiguous range, R_{unamb}	——
maximum theoretical range, R_{max}	**50 km**
minimum range, R_{min}	——
range resolution, R_{res}	——
radar cross-section, σ	**$5\ m^2$**

Solution.

(*a*) Wavelength (λ)

$$\text{Operating frequency } (f) = 5600 \text{ MHz} = 5600 \times 10^6 \text{ Hz}$$

$$\lambda = c/f$$

The value of $c = 3 \times 10^8$ m/s

$$\lambda = (3.0 \times 10^8 \text{ m/s}) / (5600 \times 10^6 \text{ hz})$$

$$\lambda = 0.0536 \text{ m} \quad \text{or} \quad \boxed{\lambda = 5.36 \text{ cm}}$$ **Ans.**

(*b*) Pulse Repetitions Time (PRT)

We know, $\text{duty cycle} = \dfrac{PW}{PRT}$

$$PRT = PW/\text{Duty cycle} = (1.3\ \mu s) / (8.3 \times 10^{-4})$$

$$\boxed{PRT = 1.57 \times 10^3\ \mu s}$$ **Ans.**

(*c*) Pulse Repetition Frequency (PRF)

We know, $PRF = 1/PRT$

$$PRF = 1/(1.57 \times 10^{-3})$$

$$\boxed{PRF = 637 \text{ hz}}$$ **Ans.**

(*d*) Peak Power (*Pt*)

We know, by eqn. (2.5), $S_{min} = Pt\,.\,G.Ae.\ \sigma/(4\pi R^2)^2$

$$Pt = (S_{min})\ (4\pi R^2)^2/G.Ae.\ \sigma$$

$$= (5.012 \times 10^{-12})\ [(4\pi)\ (50 \times 10^3)^2]^2/(3940)\ (0.9)\ (5.0)$$

$$\boxed{Pt = 279 \text{ KW}}$$ **Ans.**

(*e*) Average Power (Pav)

$$\text{Average Power } (Pav) = Pt \times \text{Duty Cycle}$$
$$= (279 \times 10^3)\ (8.3 \times 10^{-4})$$

$$\boxed{\text{Average Power} = 231.6 \text{ W}} \quad \textbf{Ans.}$$

(*f*) Directive Gain (G_D)

$$\text{Scan Rate} = RPM \times 360°/60 \text{ s} \qquad \text{(By using eqn. 2.24)}$$
$$= 16 \times 360/60$$

$$\boxed{\text{Scan Rate} = 96°/\text{s}}$$

We know,

$$\theta_B = (N_B \times \text{scan rate})/PRF \qquad \text{(By using eqn. 2.23)}$$
$$= 9.9 \times 96/637$$
$$\theta_B = 1.49°$$
$$G_D = 41253/\text{horizontal beamwidth } (\theta_B) \times \text{vertical beam width } (\Phi_B)$$

(From Antenna theory)

$$= 41253/(1.49 \times 4)$$

$$\boxed{G_D = 6921} \quad \textbf{Ans.}$$

(*g*) Range unambgious (R_{unamb})

$$R_{unamb} = (c \times PRT)/2 \qquad \text{(By using eqn. 2.17)}$$
$$= [3 \times 10^8 \times 1.57 \times 10^{-3}]/2$$
$$= (4.71 \times 10^5)/2$$

$$\boxed{R_{unamb} = 235.5 \text{ km}} \quad \textbf{Ans.}$$

(*h*)

$$R_{min} = R_{res} \qquad \text{(By using equation 2.18)}$$
$$R_{min} = (c \times PW)/2 = (3 \times 10^8)\ (1.3 \times 10^{-6})/2$$

$$\boxed{R_{min} = 195 \text{ m}} \quad \textbf{Ans.}$$

Example 7. Calculate the minimum receivable signal in a radar receiver which has an IF BW of 1 Mhz and noise figure of 6 db.

Solution. We have,

$$BW = 1 \text{ Mhz}$$
$$Fn = 6 \text{ db} \qquad (F = \text{Antilog } 6/10 = 4 \text{ Ratio})$$

Minimum detectable noise power (P_{min}) = $k \,.\, T_r \,.\, B_n$

Noise temperature of receiver (T_r) = $(F - 1)\,.\,T$

Therefore,

$$P_{min} = k(F - 1)\,.\,T\,.\,B_n$$
$$k = 1.38 \times 10^{-23}$$
$$T = 290° \text{ K}$$
$$P_{min} = (1.38 \times 10^{-23}) \times (4 - 1) \times (290) \times (1 \times 10^6)$$
$$= 4 \times 10^{-21} \times 3 \times 10^6$$

$$\boxed{P_{min} = 12 \times 10^{-15} \text{ W}} \quad \textbf{Ans.}$$

Example 8. A radar is operating at 100 Ghz, if the antenna diameter is 2 m, calculate the beamwidth of the antenna. (Assume $\theta = 70 \times \lambda/D$ degree)

Solution. We know that $\theta = 70 \times \lambda/D$ Degree (from antenna theory)

$$\lambda = c/f$$
$$= (3 \times 10^8)/100 \times 10^9$$
$$\lambda = 0.003 \text{ m}$$

By putting the value of λ in above formula,

$$\theta = 70 \times (0.003)/2$$

$$\boxed{\theta = 0.105 \text{ degree}} \quad \textbf{Ans.}$$

Example 9. What is the peak power of a radar whose average power is 200 w, pulse width (PW) is 1 μs and has PRF of 1000 hz ? Also calculate the range of this ground–based air surveillance radar if it has to detect a target with a radar cross section of 2m² when it operates at a frequency of 2.9 Ghz with a rectangular – shaped antenna that is 5 m wide, 2.7 m height, antenna aperture efficiency of 0.6 and mds is 10^{-12} w.[M.D. University, 2004]

Solution. We have following parameters,

$P_{av} = 200$ w, $Pw = 1$ μs $= 1 \times 10^{-6}$ s, $PRF = 1000$ hz, $\sigma = 2$ m², $S_{min} = 10^{-12}$ w.

Calculation of Peak power(*Pt*)

$$Pt = Pav/PW \times PRF \quad \text{(By using eqn. 2.13)}$$
$$Pt = 200/(1 \times 10^{-6})\ (1000)$$

$$\boxed{Pt = 200/0.001 = 200 \text{ kw}} \quad \textbf{Ans.}$$

Calculation of effective antenna area (*Ae*)

$$\text{physical Area of antenna } (A) = 5 \text{ m} \times 2.7 \text{ m}$$
$$= 13.5 \text{ m}^2$$

By antenna theory, we know that

$$Ae = \rho_a \cdot A$$

Where, Ae = Effective area of antenna

ρ_a = Antenna aperature efficiency

$$\boxed{Ae = 0.6 \times 13.5 = 8.10 \text{ m}^2} \quad \textbf{Ans.} \quad (8.0 \text{ m}^2 \text{ App})$$

Calculation for wavelength (λ)

$$\lambda = c/f = (3 \times 10^8)/2.9 \times 10^9 = 3/2.9 \times 10$$

$$\boxed{\lambda = 0.103 \text{ m}} \quad \textbf{Ans.}$$

We know maximum range of radar is given by following equation,

$$R^4 = [Pt \times (Ae)^2 \times \sigma]/4\pi \times \lambda^2 \times S_{min} \quad \text{(By using eqn. 2.11)}$$

$$R^4 = \frac{[(200 \times 10^3) \times (8)^2 \times (2)]}{[(4 \times 3.14) \times (0.103)^2 \times 10^{-12}]}$$

$$= 1921.2 \times 10^{17}$$

$$R = \sqrt[4]{1921.2 \times 10^{17}} = 117731.63 \text{ m}$$

$$R = 117.73 \text{ km}$$

or $R = 118$ km

or $R = 63.68$ nmi **Ans.** (1 nmi = 1853 m)

Example 10. Calculate the maximum range of a radar system which operates at 3 cm wavelength with a peak power of 500 kw, if its P_{min} is 10^{-12} w, the capture area of its antenna is 5 m square and radar cross section area of target is 20 m square. **[M.D. University, 2003]**

Solution. We have,

$$\lambda = 3 \text{ cm} = 0.03 \text{ m}$$

$$Pt = 500 \text{ kW} = 500 \times 10^3 \text{ W}$$

$$S_{min} = 10^{-12} \text{ W}$$

$$Ae = 5 \text{ m}^2$$

$$\sigma = 20 \text{ m}^2$$

$$R^4 = [Pt \times Ae^2 \times \sigma]/4\pi \times \lambda^2 \times S_{min} \quad \text{(By using eqn. 2.11)}$$

By putting the values, $R^4 = [(500 \times 10^3) \times (5)^2 \times (20)]/[(4 \times 3.14) \times (0.03)^2 \times (10^{-12})]$

Solve the equation for R, We get

$$R^4 = 25 \times 107/1.1304 \times 10^{-14}$$

$$R^4 = 2.2116065 \times 10^{22}$$

$$R = 3.85 \times 10^5 \text{ m}$$

$R = 385$ Km or $R = 208$ nmi **Ans.**

Example 11. A radar operating at 10 Ghz with the peak power of 500 kW, The power gain of antenna is 5000 and minimum power of receiver is 10^{-14}. Calculate the maximum range of radar if the effective area of antenna is 10m^2 and radar cross section is 4 m^2.

Solution. Given,

$$Pt = 500 \text{ kW}$$

$$G = 5000$$

$$S_{min} = 10^{-14}$$

$$Ae = 10 \text{ m}^2$$

$$\sigma = 4 \text{ m}^2$$

$$\text{Operating frequency} = 10 \text{ Ghz}$$

By using equation for radar range,

$$R_{max} = \{(Pt \times G \times Ae \times \sigma)/(4\pi^2 \times S_{min})\}^{1/4}$$

$$R_{max} = \{(500 \times 10^3 \times 5000 \times 10 \times 4)/(4\pi^2 \times 10^{-14})\}^{1/4}$$

$$R_{max} = \{(10^3 \times 10^8 \times 10^{14})/(157.75)\}^{1/4}$$

$$R_{max} = (0.06339 \times 10^{24})^{1/4} = 282167.93 \text{ m}$$

$R_{max} = 282.167$ kM **Ans.**

Example 12. Calculate the average power when the peak power is 200kW, pulse width is 2 is and rest time is 2000s.

Solution. We know,

$$P_{av} = Pt \times \text{duty cycle}$$

$$\text{Duty cycle} = PW/PRT \qquad \text{(Since } PRT = PW + RT\text{)}$$
$$= (2 \times 10^{-6})/(2 \times 10^{-6} + 2000) = 1/2000$$
$$= 0.5 \times 10^{-3}$$
$$P_{av} = (200 \times 10^{3}) \times (0.5 \times 10^{-3})$$
$$P_{av} = 200 \times 0.5$$
$$\boxed{P_{av} = 100 \text{ W}} \quad \textbf{Ans.}$$

IMPORTANT RELATIONS

1. $R = C \times \dfrac{\Delta t}{2}$
2. $PRT = PW + RT$
3. $PRF = \dfrac{1}{PRT}$
4. $P_{av} = P_t \times \dfrac{PW}{PRT} = P_t \times PW \times PRF$
5. $\text{Average Power} = \text{Peak Power} \times \dfrac{\text{Pulse Width}}{\text{Pulse Repetition Time}}$

   $= \text{Peak Power} \times \text{Pulse Width} \times \text{Pulse Repetition Frequency}$
6. $\text{Duty cycle} = PW \times PRF$
7. $R_{RES} = C \times \dfrac{PW}{2}$
8. $R_{RES} = \dfrac{C}{2} \times \dfrac{1}{\beta} \qquad \left(\beta = \text{Bandwidth of transmitter} \times \text{Pulse; } PW = R_{RES} = \dfrac{1}{\beta}\right)$

   where, R = Range of the target

   C = Speed of light or propagation speed of EM wave = 3×10^8 m/s

   Δt = Delay time between the transmission of radio energy and the return of reflected energy or round trip propagation time.

   PRT = Pulse Repetition Time or Period

   PRF = Pulse Repetition Frequency

   P_{av} = Average Power

   R_{RES} = Range Resolution

   β = Bandwidth of Transmitter Pulse
9. Various forms of Radar Range Equations :

   (i) $R_{max} = \sqrt[4]{\dfrac{P_t.G.\sigma.Ae}{(4\pi)^2.S_{min}}}$

   (ii) $R_{max}^4 = \dfrac{P_t.G.\sigma.A_e}{(4\pi)^2.S_{min}}$

   (Simple form of Radar range equation)

   (iii) $R_{max} = \sqrt[4]{\dfrac{P_t.G^2.\sigma.\lambda^2}{(4\pi)^3.S_{min}}} \qquad \left(A_e = \dfrac{G.\lambda^2}{4\pi}\right)$

(iv) $R_{max} = \sqrt[4]{\dfrac{P_t . A_e^2 . \sigma}{4\pi\lambda^2 . S_{min}}}$ $\left(G = \dfrac{4\pi . A_e}{\lambda^2}\right)$

(v) $R_{max} = \sqrt[4]{\dfrac{P_{av} . G . \sigma . A_e}{(4\pi)^2 . S_{min} . T . fr}}$ $\left(P_{av} = P_t \times \dfrac{T}{T_r} = P_t \times T \times PRF\right)$

(vi) $R_{max} = \sqrt[4]{\dfrac{P_t . A_e^2 . \sigma}{4\pi\lambda^2 K . T_0 . B . (F-1)}}$ $(S_{min} = K . T_0 . B(F-1))$

(vii) $R_{max} = (48)\sqrt[4]{\dfrac{P_t . D^4 . \sigma}{\lambda^2 . B . (F-1)}}$ Km $\left(A_e = \dfrac{0.65\pi (D^2)}{4}\right)$

(viii) $R_{max} = 26\left[\dfrac{(P_t . D^4 . \sigma)}{\lambda^2 . B . (F-1)}\right]^{1/4}$ Nm (1 Nm = 1853 m)

10. $R_{RES} = \dfrac{C}{2\beta}$

11. $R_{min} = C \times \dfrac{PW}{2}$

12. $R_{4\,NAMD} = C\dfrac{PRT}{2} = \dfrac{C}{2PRF}$

13. $N = t . PRF$

14. $\tau_{min} = \dfrac{N_{min}}{PRF}$

15. $\tau = \dfrac{\theta_B}{\theta_S}$

16. $\theta_s = \theta_B . \dfrac{PRF}{N}$

17. Scan Rate $(\theta_s) = \dfrac{RPM \times 360°}{60\ s}$

18. $S_{min} = \dfrac{S}{N} . NEP$

19. MDS $= 10 \log\left(\dfrac{S_{min}}{1\ mW}\right)$

where, R_{max} = Maximum radar range
P_t = Transmitter Peak Power,
G = Maximum Gain of Antenna
A_e = Aperture Area of Receiving Antenna
σ = Cross-section Area of Target
S_{min} = Minimum Detectable Signal (MDS)
λ = Wavelength
f_r = PRF
T = Pulse width

T_2 = PRT
P_{av} = Average Power
R_{min} = Minimum displayed range
R_{UNAMB} = Maximum Unambiguous Range
N = Number of Pulses During Dwell Time
τ = Dwell Time
θ_B = Beamwidth in Degrees
θ_s = Rotation rate (degree/Scan)
RPM = Rotation per minute
MDS = Minimum Discernible Signal
F = Noise Figure in Ratio
D = Antenna Diameter in Meters
T_0 = Standard Temp. (290 °K)
B = IF Bandwidth

SUMMARY

- Radar is an acronym for Radio Detection and Ranging.
- Echo is a small portion of the energy is reflected and returned to the radar.
- Seven Basic components of simple radar system.
  (*a*) Transmitter used to generate the high power rf pulses to send to the space through antenna.
  (*b*) Receiver used to received the echo signals from the target and process them.
  (*c*) Power supply provided the electrical power for all the components. Transmitter needed maximum electrical power, which may be several KW.
  (*d*) Synchroniser coorrdinates the timing for range determination.
  (*e*) Duplexer act as a switch that alternately connects the transmitter or receiver to the antenna.
  (*f*) Antenna radiate the *RF* energy into the space and received the reflected signal from the target.
  (g) Display unit used to represent the target location, in range, azimuth and elevation as per the radar applications.
- Range determination is accomplished by measuring the time taken by *RF* energy EM wave to travel from radar to target and back. It can be determined by using following formula, $R = c \times \Delta t/2$ where $c = 3 \times 10^8$ m/s.
- Common parameters of Radar pulse :
  (*a*) *PW* is the duration for which radar transmitter is ON.
  (*b*) Rest time (*RT*) is the interval between the pulses and to receive echo signal from the target, during the silence period.
  (*c*) *PRT* is the interval between the start of one pulse and the start of another. It may written as $PRT = PW + RT$.

(*d*) *PRF* is the number of pulses transmitted in one second and is equal to inverse of *PRT*. *PRF* = *1/PRF*.

(*e*) *Pt* (Peak power) is defined as the power averaged over that carrier frequency cycle which occurs at the maximum of the pulse power.

(*f*) Average Power (*Pav*) is defined as the average transmitted power over the pulse repetition period. It my be written as

$$Pav = Pt \times \frac{PW}{PRT} = Pt \times PW \times PRT$$

(*g*) Duty cycle is the ratio of average power to the peak power or *PW* to the *PRT* is called the duty cycle. It may be defined as the energy transmitted in one second by the transmitter is also as duty cycle.

$$\text{Duty cycle} = PW \times PRF$$

or

$$\text{Duty cycle} = \frac{Pav}{Pt}$$

- Applications of Radar :
  - (*a*) Surveillance search radar in military.
  - (*b*) Navigational radar in military as well as in civil.
  - (*c*) Targetting radar or tracking radar
  - (*d*) Weather radar
  - (*e*) Radar proximity fuses
  - (*f*) Radar altimeter
  - (*g*) Air traffic control radar
  - (*h*) Traffic radar
  - (*i*) Monitoring of space vehicle
  - (*j*) In industry for oil and gas exploration.
- Radar generally operates in microwave region. The wave length $\lambda \leq 30.5$ cm.
- History of radar.
- Different types of radar such as
  - (*a*) Air defence radar
  - (*b*) Air traffic control radar
  - (*c*) Fire control radar or tracking radar
  - (*d*) Airborne radar
  - (*e*) Mortar locating radar
  - (*f*) Ground penetrating radar.
- The radar equation relates the range of radar to the characteristics of the Tx (transmitter), Rx (receiver), Ae (antenna), target and the environment. It is useful to determine the maximum range at which particular radar can detect a target and its makes to understand the factor affecting radar performance.
- Radar range equation may be written as below :

$$R_{max} = \sqrt[4]{\frac{P_t G \sigma A_e}{(4\pi)^2 S_{min}}}$$

Or

$$R_{max} = \sqrt[4]{\frac{P_t G^2 \sigma . \lambda^2}{(4\pi)^3 S_{min}}}$$

Or

$$R_{max} = \sqrt[4]{\frac{P_t . A_e^2 \sigma}{4\pi \lambda^2 S_{min}}}$$

where,
P_t = Transmitter Power,
G = Maximum Gain of Antenna
A_e = Aperture area of receiving Antenna
σ = Cross section area of the target
S_{min} = Minimum detectable signal (mds)
R_{max} = Maximum range of radar
λ = Wavelength

- **Prediction Of Range Performance**

  The simple form of radar equation is not adequately predicting the range performance of actual radars. There are some drawbacks of simple form of radar equation appended below:

  1. The fluctuation and uncertainties in targets radar cross section
  2. The nature of minimum detectable signal, which is affected by the receiver noise.
  3. The losses experienced in the radar system.
  4. Propagation effect by earth surface and atmosphere.

- The uncertainty in measured range as the **range resolution, R_{RES}**, then it must be equal to the range equivalent of the pulse width, namely:

$$R_{RES} = c\ PW/2$$

- $PW = 1/\beta$
- Given this insight, it is quite reasonable to say that the range can be determined no more accurately than $cPW/2$ or equivalently

$$R_{RES} = c/2\beta$$

- The minimum displayed range is

$$R_{MIN} = c \times PW/2$$

- The maximum actual range that can be detected and displayed without ambiguity, or the **maximum unambiguous range,** is just the range corresponding to a time interval equal to the pulse repetition time, PRT. Therefore, the maximum unambiguous range,

$$R_{UNAMB} = c\ PRT/2 = c/(2PRF)$$

- If the PRF is made too high, the likelihood of obtaining target echoes from the wrong pulse transmission is increased. Echoes signals received after an interval exceeding the PRP (Pulse Repetition Period) are called **multiple times around echoes**.

- The duration that the target remains in the radar's beam during each scan is known as Dwell-Time. The number of pulses, N, that the target will be exposed to during the dwell time is:

$$N = \tau\ PRF$$

- $$\tau = \theta_B / \theta_s$$

  Where θ_B = beam-width [degrees] θ_s = rotation rate [degrees/sec] which will give the dwell time (τ) in seconds.

- Scan rate may be calculated by using following equation,

$$\text{Scan rate } (\theta_s) = (RPM \times 360^\circ)/60 \text{ s}$$

$$\theta_s = \theta_B\ PRF/N$$

- $$P_r > (S/N)\ NEP,$$

  Where P_r is the power of the return signal and NEP is Noise Equivalent Power. Since this is a significant quantity in determining radar system performance, it is given a unique designation, S_{min}, and is called the **Minimum Signal for Detection.**

$$S_{min} = (S/N)\ NEP$$

  Since S_{min}, expressed in Watts, is usually a small number, it has proven useful to define the decibel equivalent, MDS, which stands for **Minimum Discernible Signal.**

$$MDS = 10 \text{ Log } (S_{min}/1 \text{ mW})$$

- If the receiver output is not sufficient level to cross the threshold, only noise is said to be present. This is called the **Threshold Detection**.
- If the threshold level is set too low, noise might exceed it and be mistaken as target. This is called False Alarm. If the threshold were set too high, noise might not be large enough to cause false alarm, but weak targets echoes might not exceeds the threshold and would not be detected, when this occurs it is called **Missed Detection**.
- "**Noise** is unwanted E M energy which interferes with the ability of the receiver to detect the wanted signal". It may generate within receiver it self or it may enter through the receiving antenna along with the desired signal.
- The noise generated by thermal motion of the conduction electron in receiver input stages is called the **Thermal Noise or Johnson Noise**. Its magnitude is directly proportional to the bandwidth and the absolute temperature of the resistive potion of the input circuit. The available thermal noise power (watt) generated at input of a radar receiver of bandwidth Bn (hertz) at a temperature T (degree Kelvin) is

  Available thermal noise power = $K \cdot T \cdot Bn$

  where, K = Boltzmans Constant = 1.38×10^{-23} Joules/degree

- Different forms of radar range equations in terms of noise figure,

- $$R_{max} = \sqrt[4]{\frac{P_t \cdot A_e^2\ \sigma}{4\pi\lambda^2\ k \cdot T_0 \cdot B(F-1)}}$$

- $$R_{max} = 48\sqrt[4]{\frac{P_t D^4 .\sigma}{\lambda^2 B(F-1)}} \text{ Km}$$
- $R_{max} = 48\{(P_t . D^4 . \sigma) / [\lambda^2 B(F-1)]\}^{1/4}$ Km
- $R_{max} = 26\{(P_t . D^4 . \sigma) / [\lambda^2 B(F-1)]\}^{1/4}$ Nm (since 1 Nm = 1853 m)

where R_{max} = Maximum radar range
P_t = Peak power
D = Antenna diameter in meter
σ = Radar cross section of the target in m^2
λ = Wavelength in meter
F = Noise figure in ratio

- **System Losses**

  In the radar equation one of important features has not being considered that is losses occurred in the radar system it self such as,

  (*a*) The antenna beam shape loss,
  (*b*) The collapsing loss,
  (*c*) The losses in the microwave plumbing etc.

- **Propagation Effect**

  In most cases of, the earth's surface and the medium in which radar wave propagates can have a significant effect on radar performance. In some cases the propagation effect might be important enough to overshadow all other factors that contribute to abnormal radar performance.

  The effects of non-free space propagation on radar are of three categories.

  (*a*) Attenuation
  (*b*) Refraction
  (*c*) Lobe

UNSOLVED PROBLEMS

1. For a maximum unambiguous range of 1000 km, calculate the PRF required for the radar. [1500 hz]
2. Determine the maximum unamb. range and range resolution of a pulse radar whose pulse duration is 10 μs at a PRF 1000 hz. [150 km, 1.5 km]
3. A radar receiver having BW of 3 Mhz, calculate the highest resolution in range. [50 m]
4. A pulse radar measure the round trip propagation time with an error of 2 μs. What will be the corresponding error in the measurement of range. [300 m]
5. Calculate the duty cycle of a radar which transmits a 1.5 ms pulse at a PRF of 8 kHz. If the peak power of this radar is 500 kilowatts, what is the average power ? What is the resting time ? [2×10^{-3}, 6 kw]
6. A pulsed radar has a duty cycle of 0.016. If the resting time is 380 ms, what is the pulse width ? What is the PRF ? What is the minimum range, in meters, of this radar ? [6.18 μs]

7. If a CRT is designed so that the spot traverses the indicator face in 900 ms, what range could be indicated by a spot deflection half way across the indicator ? [67 km]
8. What is the minimum range for radar with a PW of 5 ms ? [750 m]
9. What should be the PRF of radar in order to achieve a maximum unambiguous range of 50 nmi ? [1.61 Khz]
10. A radar transmitter has a peak power of 400 kw, a PRF of 1500 PPS and pulse width of 0.8 μs. Calculate : (*a*) maximum unamb. range; (*b*) duty cycle; (*c*) average power; (*d*) suitable band width. [(*a*) 100 km (*b*) 0.0012 (*c*) 480 w (*d*) 1.25 Mhz]
11. If the radar is designed for operation at 5 Ghz with an antenna of diameter 4 m, calculate the peak power required to have a maximum range of 500 km with a target of cross-sectional area 20 m^2 and S_{min} is 10^{-12} W. [8.94 MW]
12. What is the peak power of a radar whose average power 400 W, pulse width of 1 μs and pulse repetition frequency of 1000 hz. It has to detect the target of cross-section 4 m^2 when it operates at a frequency of 3 Ghz with a antenna diameter of 2 m and mds is 10^{-10} W. Calculate the maximum range in nmi. [7.212 nmi]
13. Calculate the maximum range of a radar system which operates 2 mm wavelength with a peak power of 1000 kW, if its Smin is 10^{-12} W, the capture area of its antenna is 5 m^2 and radar cross-section area of target is 10 m^2. [1257 km]
14. A radar has IF bandwidth of 3 Mhz, calculate the minimum detectable power, if the noise figure of that receiver is 6 db. [3.6×10^{-16} W]
15. If the operating frequency of a surveillance radar is 10 Ghz, the diameter of the radar antenna is 2 m. The maximum range of the radar is 1000 km with a target of cross-sectional area 20 m^2 and S_{min} of that receiver is 3.6×10^{-16} W. Calculate the peak power of the transmitter. [2000 kW]
16. A radar transmitter has a peak power of 400 kW, a PRF of 1500 PPS and a PW of 0.8 μs, calculate
    (*a*) Maximum unambiguous range,
    (*b*) The duty cycle,
    (*c*) The average transmitted power. [(*a*) 120 km, (*b*) 0.0012, (*c*) 480 W]
17. Calculate the minimum receivable signal in radar receiver which has an intermediate bandwidth of 1.9 MHz and a 8 db noise figure. **[UPTU, 2004]** (4.106×10^{-14} W)
18. Calculate the maximum range of a radar operating at 3 GHz and using a peak pulse power of 28 MW. The antenna diameter is 60 m, the target cross-section is 1 m^2, the receiver noise figure is 1.1 (Power factor) and receiver BW is 6 KHz. **[UPTU, 2004]** (133.85 Km) (**Hint.** Refer equation 2.33)

REVIEW QUESTIONS

1. What do you understand by terms RADAR and explain the echo pulse principle.
2. What are the basic functions of radar ? In indicating the position of target, what is the difference between azimuth and elevation ? **[M.D. University, 2004]**
3. Explain the working of pulse radar with the help of block diagram. **[M.D. University, 2001, 2003]**
4. Explain the various applications of radar. **[M.D. University, 2004]**
5. What are the radar frequencies, explain in brief. **[M.D. University, 2003, 2005, 2007]**

6. What is the maximum unambiguous range ? How it is related with pulse repetition rate? **[M.D. University, 2003]**
7. Explain a Radar system with diagram. **[UPTU, 2003]**
8. Discuss Radar applications. **[UPTU, 2003]**
9. Explain the principle of radar system. **[UPTU, 2004]**
10. Explain the limitations and application of radars. **[PTU, 2004]**
11. What is Radar ? Describe the principle operation of a basic radar with the help of neat block diagram. What are the various applications of Radar ? **[M.D. University, 2005]**
12. What is the difference between pulse interval and PRF ? **[M.D. University, 2005, 2007]**
13. What are the salient factors governing the choice of the PRF for a particular radar ? What are difference between the pulse interval and the PRF ?
14. Draw block diagram of a pulse radar and explain each block briefly. **[UPTU, 2006]**
15. Explain the block diagram of radar. Also discuss the radar frequencies according IEEE standard. Also discuss its applications. **[M.D. University, 2006]**
16. Define the following terms in brief :
    (*i*) Pulse Width
    (*ii*) Pulse Repetition Time
    (*iii*) Average Power **[M.D. University, 2007]**
    (*iv*) Duty Cycle **[M.D. University, 2007]**
    (*v*) Range Resolution
    (*vi*) Azimuth Resolution
    (*vii*) Radar frequencies **[M.D. University, 2004]**
17. Draw the block diagram of a basic Radar set and explain the essentials of its operations. What are the basic functions of Radar ? **[M.D. University, 2007]**
18. Derive the simple form of radar range equation and relates the P_t (Transmitter peak power) to the maximum range of the radar.
19. What are the factors affecting the radar range ? **[M.D. University, 2001]**
20. What do you understand by prediction of range performance ?
21. What is mds to noise ratio? Why it replaces minimum detectable signal in radar system ? **[M.D. University, 2003]**
22. What is the maximum unambiguous range ?
23. How noise affects the detection process of radar ?
24. What are the different types of system losses? Explain them briefly. **[M.D. University, 2003]**
25. Derive an radar range equation considering internal noise of receiver. **[UPTU, 2003]**
26. How to calculate dwell time ? **[PTU, 2004]**
27. How to find false alarm ?
28. What are PRF and range ambiguities ? **[PTU, 2004]**
29. What is a search radar equation ? **[PTU, 2004]**
30. Explain the production of radar range in terms of S/N ratio. **[PTU, 2004]**
31. Explain what do you mean by ambiguous reception with the help of an example. **[UPTU, 2004]**

32. Derive the basic Radar range equations as governed by the minimum receivable echo power P_{min} ? **[M.D. University, 2007]**
33. Derive radar range equation and comment on it. **[UPTU, 2006]**
34. Calculate the maximum radar range in terms of signal to noise ratio. **[M.D. University, 2006]**
35. Explain the PRF and range ambiguities. **[M.D. University, 2006]**
36. What do you understand by radar cross-section ? Explain its effects on radar performance.
37. Write short technical notes on :
    (*a*) Propagation effect
    (*b*) Multiple around echoes
    (*c*) Noise and clutter
    (*d*) Probability of detection
    (*e*) Minimum detectable signal **[M.D. University, 2007]**
    (*f*) Antenna loss
    (*g*) Range ambiguities and PRF
    (*h*) Plumbing loss
    (*i*) Ducting effect on radar wave
    (*j*) Signal to noise ratio **[M.D. University, 2007]**
    (*k*) Minimum unambiguous range. **[M.D. University, 2007]**
    (*l*) Radar cross-section.

2

CHAPTER

CW AND FREQUENCY MODULATED RADAR

INSIDE THIS CHAPTER

2.1. INTRODUCTION

A pulse radar, which is transmitting the burst of pulses of EM energy and received back the reflected signal from the target. The time taken by the EM energy to travel up to the target and came back to the radar antenna will be considered. In other words we can say that the separation of signal of the echo and the energy transmitted by the transmitter is made on the basis of difference in time. If the radar transmitter transmits the energy continuous than the separation of transmitted energy and the received energy from the target is not possible as it is done in pulse radar. We have to use separate antenna for transmission and reception both and separating of the received signal from the transmitted signal in terms of frequency not in the time as in pulse radar.

The working of CW radar is based upon the Doppler effect. In this chapter we will be discussing about the Doppler effect and the radar based on Doppler principle such as CW radar, FM CW radar and multiple frequency radar.

2.2. DOPPLER EFFECT

Everyday life has multiple examples of the Doppler phenomenon with sound; the whistle from a moving train is a good example. As the train approaches a stationary listener, the pitch (frequency) of the whistle sounds higher than when the train passes by, at which time the pitch sounds the same as if the train were stationary. As the train recedes from the listener, the pitch decreases. Car horns exhibit the same phenomenon, as does all sound. Note that in the above example if a car horn is stationary and a listener is on the train, the Doppler principle still applies. As the listener on the train approaches the stationary horn, the pitch of the horn sounds higher; as the train recedes from the stationary horn the pitch sounds lower (to anyone on the train).

Electromagnetic waves radiated by radar, as well as sound waves, obey the Doppler principal, although electromagnetic waves travel at the speed of light and audio waves travel at the speed of sound. The Doppler effect is a frequency shift that results from relative motion between a frequency source and a listener. If both source and listener are not moving with respect to each other (although both may be moving at the same speed in the same direction), no Doppler shift will take place. If the source and listener are moving closer to each other, the listener will perceive a higher frequency — the faster the source or receiver is approaching the higher the Doppler shift. If the source and listener are getting farther apart, the listener will perceive a lower frequency — the faster the source or receiver is moving away the lower the frequency. The Doppler shift is directly proportional to speed between source and listener, frequency of the source, and the speed the wave travels (speed of light for electromagnetic)

It is well known in the fields of optics and acoustic that if either the source of oscillation or the observer of the oscillation is in motion, an apparent shift in frequency will result. This is the **Doppler effect** and is the basis of **continuous wave (cw) radar**.

If, suppose the range to the target is r and the wavelength λ. The energy travel from the radar antenna to the target and from target to the radar than the total wavelength is given by $2R/\lambda$ and there is 2π radian is the phase change. The unit for R and the λ is considered same. As the energy travel two-way so than the total phase change (Φ) is

$$\Phi = 2\pi \times 2R/\lambda \text{ rad} = 4\pi R/\lambda \text{ rad.} \quad ...(2.1)$$

The R will change continuously as the target is in motion so the phase will also change accordingly. A change in phase with respect to the time may be called as angular frequency (ω) may be given by

$$\omega = 2\pi fd = d\Phi/dt = (4\pi/\lambda) \cdot (dR/dt)$$

Therefore, $$\boxed{\omega = 4\pi Vr/\lambda} \qquad (dR/dt = Vr) \quad ...(2.2)$$

Where, fd = Doppler frequency shift

Vr = dR/dt is the radial velocity (meter/sec) or rate of change of range with time.

If the angle between the target and the radar antenna is θ, then the redial velocity may be written as $v \cos \theta$. As shown in Fig. 2.1.

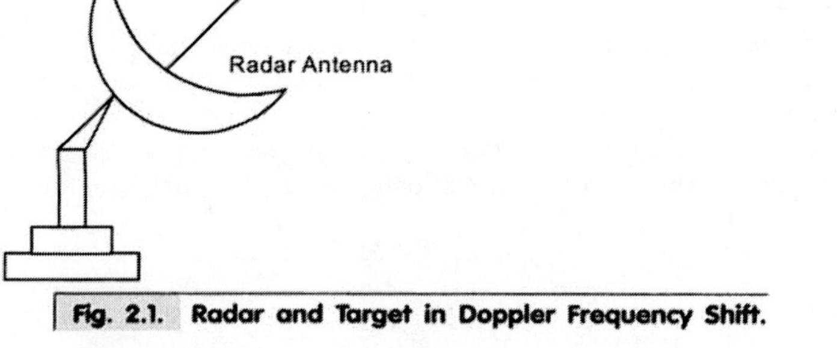

Fig. 2.1. Radar and Target in Doppler Frequency Shift.

v is the speed of the target

We know that $\quad d\Phi/dt = w$

or $\quad w = 2\pi fd$

We can write $\quad 2\pi fd = 4\pi Vr/\lambda$

$$\boxed{fd = 2Vr/\lambda} \quad (\lambda = c/f) \qquad ...(2.3)$$

$$\boxed{fd = 2fVr/c} \qquad ...(2.4)$$

Where f = transmitted frequency and c = velocity of EM wave = 3×10^8 m/sec, fd = Doppler frequency shift.

If the Doppler frequency (fd) is in Hz, Vr in knots and wavelength (λ) in cm than fd may be given as

$$\boxed{fd = 103 Vr(kt)/\lambda(\text{cm})} \qquad ...(2.4)$$

Or if the wavelength (λ) is in m, Vr in knots than the above equation may be rewritten as

$$\boxed{fd = 1.03\ Vr(kt)/\lambda(\text{m})} \qquad ...(2.5)$$

2.3. CW RADAR

2.3.1. Simple CW Radar

The block diagram of a simple CW Radar is shown in Fig. 2.2. A CW radar works on the principle of Doppler frequency shift due to the motion of the target. In CW radar, transmitter radiates the EM energy continuously unlike a pulse radar transmitter which transmits in pulse. Even the transmitter signal are very strong but still received signal from the target, may be separated easily. Because in this radar change in the frequency is observed instead of the considering the time taken by the receiver as in pulse radar. A simple working of CW radar block diagram is explained below.

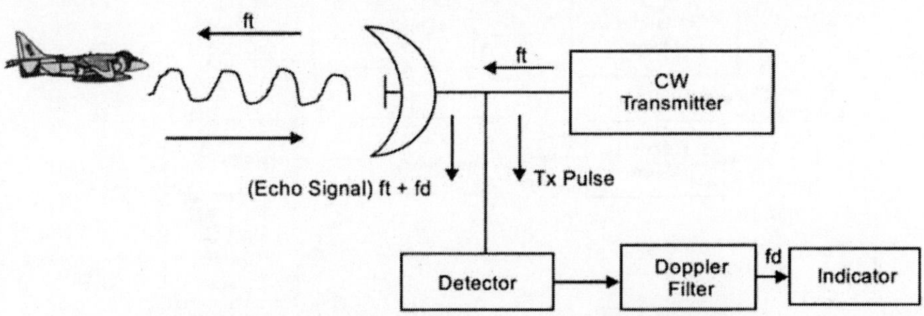

Fig. 2.2. The Block Diagram of a Simple CW Radar.

Transmitter. In CW radar, transmitter radiates the continuous energy in the form of unmodulated sinusoidal oscillation at frequency F_r. A common antenna is used in the block diagram, which will transmit the energy and as well receives the reflected energy.

According to the movement of the target the reflected energy frequency will vary. If the target is closing at that received energy will have frequency (f_t + fd) and if the target is moving away than the received energy frequency will have (f_t – fd). In other words we can say that if the target is approaching to the radar the frequency will increase and if it is moving away from the radar, the received frequency will be less then the transmitted frequency.

Detector. The purpose of detector or mixer is to multiplies the received signals (f_t + fd) (f_t – fd) frequency with the reference signal (f_t) frequency which is taken from the transmitter directly.

Doppler filter. Detector will produce the difference in frequency and further pass it on to the Doppler filter, which is a low pass filter. Doppler filter will allow the frequencies from the detector and reject the other frequencies. It will stop the low frequency from the transmitter to ever as it uses the low frequencies cut off filter. The upper frequency cut off level is decided by the maximum expected radial velocity of the target.

The Doppler filter passes only the difference of frequencies it does not indicates the sign(+ or –). So you cannot identify the direction of the target movements. The output from the Doppler filter is further amplified up to that level, so that we can indicates over to the frequency meter or some audio phone.

2.3.2. CW Radar with IF Amplification

The B/D of CW radar with IF amplification is shown in Fig. 2.3. The receiver operates with a non-zero IF. Separate antenna is shown for transmission and reception. The noise generated in the crystal detectors or in the vacuum tubes may be called as **flicker effect** and it will vary as $1/f^{\alpha}$, the α is approximately unity. The flicker noise formed in the CW radar is considerable as; it is generated at the low frequencies range in which generally CW radar operate. To eliminate this noise in the superhetrodyne receiver, IF (Intermediate Frequency) stage is used.

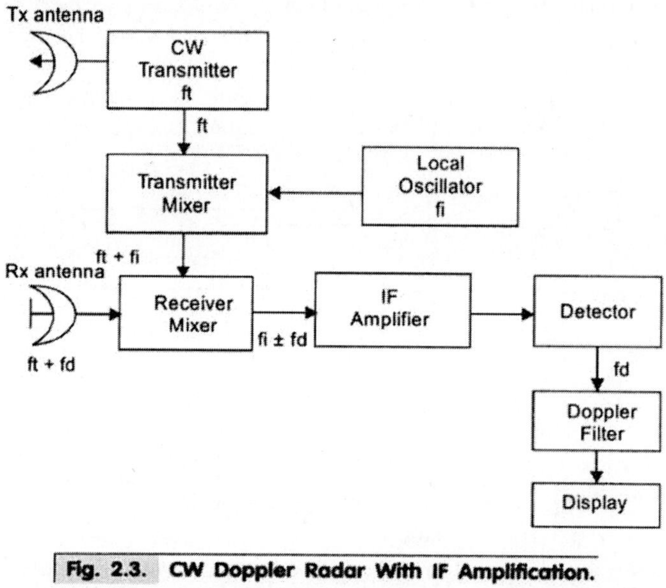

Fig. 2.3. CW Doppler Radar With IF Amplification.

A small portion of the TX output is mixed with the output at a local oscillator, and the sum is fed to the receiver mixer. This also receives the doppler shifted signal from its antenna and producing an output difference frequency that is typically 30 Mhz, plus or minus the Doppler frequency. The output of this mixer is amplified and demodulated again, and the signal from the second detector is just the Doppler frequency. Its sign is lost, so that it is not possible to tell whether the target is approaching or receding. The overall receiver system is rather similar to the superhetrodyne receiver. Extra sensitivity is provided by the lowered noise, because the output of the diode mixer is now in the vicinity of 30 Mhz, at which FM noise is disappeared.

The problem of isolation is the main determining factor; as a consequence the CW power from such a radar seldom exceeds 100 W and is often very much less. Gunn or IMPATT diodes or, for the highest powers CW magnetrons are used as power oscillators in the transmitter.

The CW radar can be used to measure the speed of the automobiles; shells guided missiles .It can be used to detect movement of troops vehicles in the dark and even in the bad weather. It can be used as radio proximity fuse. However, practical application of CW radar is limited by the fact that several targets at a given bearing tend to cause confusion. Further it is limited in the maximum power it transmits and this places a limit on its maximum range. Also, it is not capable of indicating the range of the targets and can show only its velocity. CW radar has other advantages likes its uses low transmitting power, low power consumption, simple circuitry and small size. Hence it can be used for mobile applications. It can be used by traffic police to monitor the speed of vehicles also it can be used in aircraft navigation for speed measurement and as rate climb meter.

2.4. FREQUENCY MODULATED CW (FMCW) RADAR

The transmitter in a CW radar is not modulated. Therefore it can neither provide range of the target nor sense which particular cycle of oscillation is being received at any instant. This major drawback can be eliminated by frequency modulating the transmitted signal even it increases the bandwidth. The spectrum of a CW transmission can be broadened by the application of modulation, amplitude, frequency, or phase. An example of an amplitude modulation is the pulse radar.

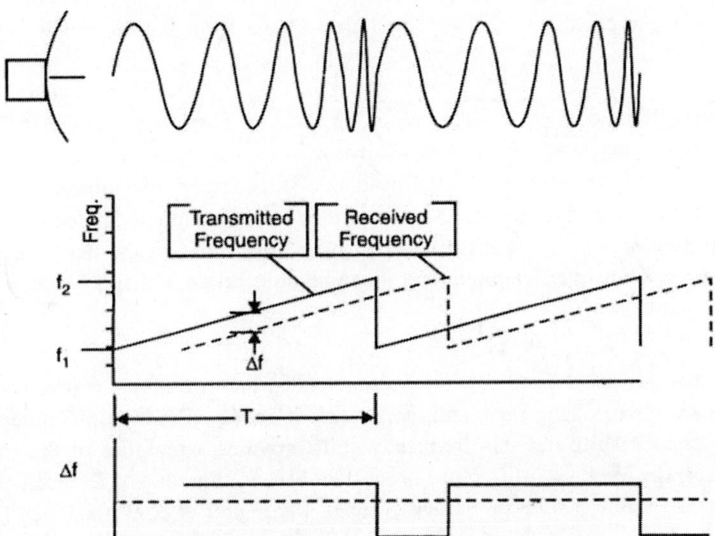

Fig. 2.4. FMCW Radar Theory of Operation.

Theory of Operation

A widely used technique to broaden the spectrum of CW radar is to frequency - modulate the carrier. The timing mark is the changing frequency. The transit time is proportional to the difference in frequency between the echo signal and the transmitter signal. The greater the transmitter frequency deviation in a given time interval, the more accurate the measurement of the transit time and the greater will be the transmitted spectrum. By measuring the frequency of the return signal, the time delay between transmission and reception can be measure and therefore the range determined as before. Of course, the amount of frequency modulation must be significantly greater than the expected doppler shift or the results will be affected.

The simplest way to modulate the wave is to linearly increase the frequency. In other words, the transmitted frequency will change at a constant rate.

The FMCW system measures the instantaneous difference between the transmitted and received frequencies, Δf. This difference is directly proportional to the time delay, Δt, which is takes the radar signal to reach the target and return. From this the range can be found using the usual formula,

$$R = c \times \Delta t/2 \qquad ...(2.5)$$

The time delay can be found as follows :

$$\boxed{\Delta t = T \times \Delta f/(f_2 - f_1)} \qquad ...(2.6)$$

where: f_2 = maximum frequency

f_1 = minimum frequency

T = period of sweep from f_1 to f_2,

and Δf = the difference between transmitted and received.

There is a slight problem, which occurs when the sweep resets the frequency, and the frequency difference becomes negative (as shown in the plot of Δf vs. time). The system uses a discriminator to clip off the negative signal, leaving only the positive part, which is directly proportional to the range.

Combining these equations (2.5 & 2.6) into a single form for the range

$$\boxed{R = 2cT\Delta f/(f_2 - f_1)} \qquad ...(2.7)$$

where Δf is the difference between the transmitted and received frequency (when both are from the same sweep, *i.e.* when it is positive).

Another way to construct a FMCW system is to compare the phase difference between the transmitted and received signals after they have been demodulated to receiver the sweep information. This system does not have to discriminate the negative values of Δf. In either case however, the maximum unambiguous range will still be determined by the period, namely

$$R_{unamb} = c \times T/2$$

FMCW radar generally used for radar altimeters in the aircraft. If it is considered that the relative velocity of the earth and the aircraft is not equal to zero than, there will be doppler frequency shift and this frequency shift provides a measure of the relative velocity. The average frequency difference is constant because the time difference between transmission and reception of a particular cycle of the signal is constant.

The rate of change of the frequency is given by

$$\left(\frac{df}{dt}\right) = \frac{2\Delta f}{T/2}$$

where, Δf = The range of frequency

T = Modulation period

The beat frequency (f_b) is corresponding to the range (R) may be given by

$$\left[\frac{df}{dt}\right]\left[\frac{2R}{c}\right] = \left[\frac{2\Delta f}{T/2}\right]\left[\frac{2R}{c}\right]$$

$$\boxed{f_b = \frac{8R.4\Delta f}{f_m}} \qquad \left(\because \frac{1}{T} = f_m\right) \quad ...(2.8)$$

where f_m = minimum beat frequency

$$\boxed{R_{min} = \frac{c}{8\Delta f}} \quad ...(2.9)$$

In case of altimeter R_{min} is the fixed error in the range.

FMCW systems are often used for radar altimeters, or in radar proximity fuses for warheads. These systems do not have a minimum range like a pulsed system. However, they are not suitable for long-range detection, because the continuous power level they transmit at must be considerably lower than the peak power of a pulsed system. You may recall that the peak and average power in a pulse system were related by the duty cycle,

$$P_{av} = DC \times Pt$$

For a continuous wave system, the duty cycle is one, or alternatively, the peak power is the same as the average power. In pulsed systems the peak power is many times greater than the average.

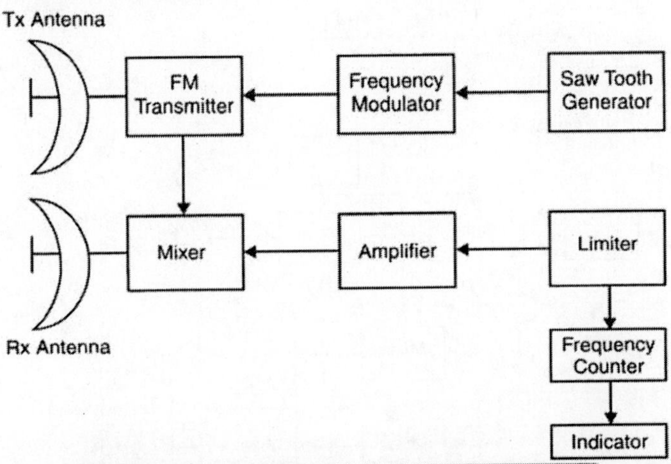

Fig. 2.5. Block Diagram of Simple FM-CW Radar.

A block diagram illustrating the principle of FM-CW radar is shown in Fig. 2.5. A portion of the transmitter signal acts as the reference signal required to produce the beat frequency. It is introduced directly into the receiver via a cable. Ideally, the isolation between transmitting and receiving antenna is made sufficiently large so as to reduce to a negligible level the transmitter leakage, which arrives at the receiver via the coupling between the antennas. The beat frequency is amplified and limited to remove any amplitude fluctuations. The frequency of amplitude limited beat note is usually measured with a cycle–counting frequency meter calibrated in distance.

FM-CW radar application. One of the major applications of the FM-CW radar principle has been as an altimeter on board aircraft to measure height above the ground. The large target cross section and the relatively short range required of altimeters permit low transmitter power and the low antenna gain. Since the relative motion, and hence the doppler velocity between the aircraft and ground, is small the effect of the doppler frequency shift may usually be neglected.

2.5. MULTIPLE FREQUENCIES CW RADAR

The variation of phase with frequency is the fundamental basis of radar measurement of time delay or range measurement. It is easier to analysis the pulse radar and FM CW radar in the term of time do main. The principle used in multiple frequencies CW radar is the measurement of range by computing the phase difference. A measurement of the range 'R' of stationary target by employing continuous wave radar transmitting sine waves ($2\pi f_t$). The time taken by the sine wave is $T = 2R/c$. c is the velocity of light or propagation (3×10^8 m/s). The output given by the phase detector, which will compare the transmitted signal on the received signal, is may be written as $\Delta\varnothing = 2\pi ft$, by replacing the value of T ($T = 2R/c$), we get

$$\Delta\varnothing = 2\pi ft \times 2R/c \quad ...(2.10)$$

$$\Delta\varnothing = 4\pi fR/c \quad ...(2.11)$$

The corresponding range may be given by the formula given below.

$$\boxed{R = c\Delta\varnothing/4\pi f} \quad ...(2.12)$$

Or we may write it in terms of λ ($\lambda = c/f$)

$$\boxed{R = (\lambda/4\pi) \times \Delta\varnothing} \quad ...(2.13)$$

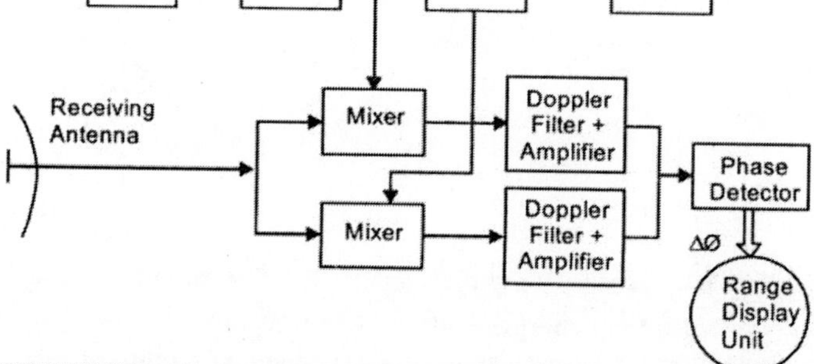

Fig. 2.6. A Simplified Block Diagram of Multiple (Two) Frequencies CW Radar.

The maximum error occurs in measure net of phase difference is 2π radian. If we put the value of $\Delta Ø = 2\pi$. The maximum ambiguity, in range is $R = \lambda/4\pi \times 2\pi = \lambda/2\pi$, it is considerably very small in practical interest.

A simple block diagram of multiples frequencies CW Radar is shown in Fig. 2.6. It is almost as CW radar except it has got one more channel and a measuring device. The better accuracy in range measurement may be provided by the large frequency difference between the two transmitted signals.

Transmitting three or more frequencies instead of just two can make both more accurate and unambiguous range measurement.

This technique by using pulse compression (long pulse) might be useful for satellite or space tracking. It may be used for missile guidance and surveying.

SOLVED EXAMPLES

Example 1. With a CW transmitter frequency of 5 GHz, calculate the doppler frequency seen by a stationary radar when the target radial velocity is 100 Km/h.

[M.D. University, 2001]

Solution. Given $RF = 5$ GHz, $Vr = 100$ Km/h

$$Vr = \frac{100 \times 10^3}{60 \times 60} \text{ m/s} = 27.8 \text{ m/s}$$

$$fd = \frac{2Vr}{\lambda} \quad \text{(By using equation 2.3)}$$

$$\lambda = \frac{c}{f} = \frac{3 \times 10^8}{5 \times 10^9} = \frac{3}{50} \text{ m}$$

$$\lambda = 0.06 \text{ m.}$$

By putting the value in equation above, we get

$$fd = \frac{2 \times 27.8}{0.06} = 926.66 \text{ Hz}$$

$$\boxed{fd = 927 \text{ Hz (approximately)}} \quad \textbf{Ans.}$$

Example 2. A target is closing on a radial of a radar with a relative velocity of 200 knots. The radar transmits continuous wave energy at a wave length of 5 cm. What will be the doppler shift of the target ? What will the doppler shift if the target alters its course by 45° ?

Solution. We have,

$$Vr = 200 \text{ knots}$$

$$\lambda = 5 \text{ cm} = \frac{5}{100} = 0.05 \text{ m}$$

We know,

$$fd = \frac{2Vr}{\lambda} \quad \text{(By equation 2.3)}$$

Radial velocity Vr to be calculated in mtr

Since, 1 knot = 0.508 m/s

So, 200 knots = 101.60 m/s

$$Vr = 101.60 \text{ m/s}$$

By putting the radar in above equation in fd, we get

$$fd = \frac{2\times 101.60}{0.05}$$

$$\boxed{fd = 4064 \text{ Hz or } 4.06 \text{ Khz}}$$ **Ans.**

The velocity component (V) in the close direction can be calculated by using,

$$V = Vr \cos\theta$$

$$V = 101.60 \times \cos(45°) \qquad (\text{since } \cos 45° = \sqrt{1/2})$$

$$V = 101.60\times\sqrt{1/2} = 101.60 \times 0.707$$

$$V = 71.84 \text{ m/s}$$

Frequency shift,

$$\Delta fd = \frac{2\times 71.84}{0.05} = 2873 \text{ Hz} = 2.873 \text{ KHz}$$

$$\boxed{\Delta fd = 2.873 \text{ Khz}}$$ **Ans.**

Example 3. A frequency modulated radar sweeps from 400 Hz to 800 MHz in 10 μs. What is the maximum unambiguous range which can be measured by this radar ?

Solution. By using equation (1.1)

$$Rmax = \frac{c.\Delta t}{2}$$

We have,

$$\Delta t = 10 \ \mu s$$

$$c = 3 \times 10^8 \text{ m/s}$$

By putting the values in above equation, we get

$$Rmax = \frac{(3\times 10^8)\times(10\times 10^{-6})}{2}$$

$$Rmax = \frac{3\times 10\times 10^8\times 10^{-6}}{2} = 3\times 5\times 10^2$$

$$Rmax = 1500 \text{ m.}$$

$$\boxed{Rmax = 1.5 \text{ Km}}$$ **Ans.**

Example 4. A CW transmit frequency of 10 GHz and doppler frequency is 1000 Hz. Calculate the radial velocity of the target in knots.

Solution. We have, $RF = 10 \text{ GHz} = 10 \times 10^9$ Hz

$$fd = 1000 \text{ Hz}$$

We know,

$$fd = \frac{2Vr}{\lambda} \qquad \text{[By equation (2.3)]}$$

or

$$Vr = \frac{\lambda.fd}{2}$$

$$\lambda = \frac{c}{f} = \frac{3\times 10^8}{10\times 10^9} = 0.03 \text{ m}$$

By putting the values of λ and fd, we can get Vr

$$Vr = \frac{0.03\times 1000}{2} = 15 \text{ m/s} \qquad (1 \text{ knot} = 0.508 \text{ m/s})$$

$$\boxed{Vr = \frac{15}{0.508} = 29.52 \text{ knots}}$$ **Ans.**

IMPORTANT RELATIONS

1. $f_d = \frac{2Vr}{\lambda} = \frac{2fVr}{c}$
2. $f_d = \frac{103\,Vr(\text{knots})}{\lambda(\text{cm})}$
3. $f_d = \frac{1.03\,Vr(\text{knots})}{\lambda(\text{cm})}$
4. $\Delta t = \frac{T \times \Delta f}{(f_2 - f_1)}$ (In FMCW Radar)
5. $R = 2cT\Delta f\,(f_2 - f_1)$ (In FMCW Radar)
6. $f_b = \frac{8R.4\Delta f}{f_m}$ (In FMCW Radar)
7. $R_{\min} = \frac{c}{8\Delta f}$ (In FMCW Radar)
8. $R = \frac{c\Delta\phi}{4\pi f}$ (Multiple Frequency CW Radar)
9. $R = \left(\frac{\lambda}{4\pi}\right) \times \Delta\phi$ (Multiple Frequency CW Radar)

where, f_d = Doppler Frequency Shift (Hz)
V_r = Radial Velocity (m/s)
f = Transmitted Frequency (Hz)
c = Velocity of EM wave = 3×10^8 m/s
λ = Wavelength (m)
Δt = Time delay
T = Period of Sweep from f_1 to f_2
f_2 = Maximum frequency
f_1 = Minimum frequency
Δf = Frequency Difference Between Transmitted and Recevied
f_m = Minimum Beat Frequency
f_b = Beat Frequency

SUMMARY

- It is well known in the fields of optics and acoustic that if either the source of oscillation or the observer of the oscillation is in motion, an apparent shift in frequency will result. This is the Doppler effect and is the basis of continuous wave (CW) radar.

$$fd = 2V_r/\lambda$$
$$fd = 2fV_r/c \quad \text{(since } \lambda = c/f\text{)}$$
$$fd = 1.03\ V_r(kt)/\lambda(\text{m})$$

- The FMCW system measures the instantaneous difference between the transmitted and received frequencies, Δf. This difference is directly proportional to the time delay, Δt, which is takes the radar signal to reach the target and return. From this the range can be found using the usual formula.

$$R = c \times \Delta t/2.$$

The time delay can be found as follows:

$$\Delta t = T \,.\, \Delta f/(f_2 - f_1)$$

Where, f_2 = maximum frequency

f_1 = minimum frequency

T = period of sweep from f_1 to f_2,

and Δf = the difference between transmitted and received.

$$R = 2cT\Delta f/(f_2 - f_1)$$

- $$f_b = \frac{8R.4\Delta f}{f_m}$$

where, f_b = Beat frequency

R = Range

f_m = Minimum beat frequency

- $$R_{min} = \frac{C}{8\Delta f}$$

- For a continuous wave system, the duty cycle is one, or alternatively, the peak power is the same as the average power. In pulsed systems the peak power is many times greater than the average.
- The major applications of the FM-CW radar principle has been as an altimeter on board aircraft to measure height above the ground

UNSOLVED PROBLEMS

1. A CW radar operating at 5 cm wavelength and target radial velocity is 200 knots, calculate the doppler frequency of the radar. [4.06 KHz]
2. A target with a radial velocity of 100 Km/h closes to the CW radar, which is operated at 20 GHz. Calculate the doppler frequency of the target. [3.706 KHz]
3. A target is closing on a radial of a radar with a relative velocity of 200 Km/h. The CW radar is operating at 5 GHz. What will be the Doppler shift of the target ? What will be the doppler shift, if the target changes its course by 30° ? [1853 Hz, 1605 Hz]
4. For FMCW radio altimeter, operating at 4 GHz, calculate the beat frequency f_b between the transmitted and received signals when Δf = 30 KHz, R = 2.5 Km and f_m = 10 KHz. [20 KHz] **[UPTU, 2003]**

REVIEW QUESTIONS

1. What do you understand by term Doppler Effect ? **[M.D. University, 2004]**
2. Explain the working of FMCW radar with the help of block diagram. **[M.D. University, 2001, 2004, 2007]**
3. How do you find out the relative velocity of a target using CW radar ? How does flicker noise affects its working ? **[M.D. University, 2001]**

4. What do you understand by Multiple Frequency CW radar?
5. Write the application of CW radar.
6. Explain the working of CW radar with help of block diagram. **[UPTU, 2004]**
7. What is the difference between Pulse radar and CW radar ?
8. Write short notes on
   (*a*) Limitation of CW radar (*b*) Advantage of Doppler effect
   (*c*) Multi frequency CW radar (*d*) Doppler effect. **[M.D. University, 2001]**
9. Will the doppler shift for an opening target indicate an increase or decrease in frequency ? Why ?
10. Why can't a CW radar measure range ? What is the main advantage of a CW radar ?
11. Explain, with neat diagram, how target velocity can be obtained by CW radar. **[UPTU, 2003]**
12. Explain in brief how FMCW radar is used for measurement of range. **[M.D. University, 2003]**
13. What is doppler Principle ? **[PTU, 2004]**
14. Draw the block diagram and explain the operations of a CW Doppler radar using an intermediate frequency in the receiver. How the drawbacks of the basic CW radar have been overcome ? **[M.D. University, 2005]**
15. Draw block diagram of CW doppler radar with non-zero I.F. receiver and explain it. **[UPTU, 2006]**
16. With the help of a block diagram, explain the working of multiple frequency CW radar. **[M.D. University, 2007]**

3
CHAPTER

MTI AND PULSE DOPPLER RADAR

INSIDE THIS CHAPTER

3.1. INTRODUCTION

The radars discussed till now were required to detect targets in the presence of noise. But in practical radars have to deal with more than receiver noise when detecting targets while they can also receives echoes from the natural environment such as land, sea and weather. These echoes are called **clutter**, since they tend to clutter the radar display with unwanted information's. Clutter echoes signal has greater magnitude then echo signal receives from the aircraft. When an aircraft echo and a clutter echo appear in the same radar resolution cell, the aircraft might not be detected. But the Doppler effect permits the radar to distinguish moving targets in the presence of fixed target even the echo signal from fixed targets has comparatively greater magnitude than the moving targets such as aircraft.

3.2. MOVING TARGET INDICATOR RADAR

Echo signals from fixed targets are not shifted in frequency, but the echo from a target moving with relative velocity Vr will be shifted in frequency by an given by the Doppler formula, $fd = 2vr/\lambda$, when λ is wave length of the transmitted signal.

Pulse radar, which makes use of the Doppler shift for detecting moving targets is either, MTI radar or a pulse Doppler radar. The MTI radar has a PRF low enough to not have any range ambiguities $Run = c/fp$ (such as multiples around echo). It does however have many ambiguities in the Doppler domain (such as blind speed).

The pulses Doppler radar on the other hand have a PRF large enough to avoid doppler ambiguities but it can have numerous range ambiguities. These are also a medium PRF pulse doppler that accepts both range and doppler ambiguities.

There is a type of MTI which does use the informations directly from the doppler but it observing the area map scan to scan. Only moving objects which are changes their position scan to scan from mapping operations are displayed. This type of MTI is known as **Area MTI**.

Figure 3.1 shows the five successive return on MTI radar. This sweep shown several fixed targets and a moving target. On the basis of single sweep. Moving targets cannot be distinguished from fixed targets. Echoes from fixed targets remain constant throughout, but echoes from moving targets varies sweep to sweep at a rate corresponding to the doppler frequency.

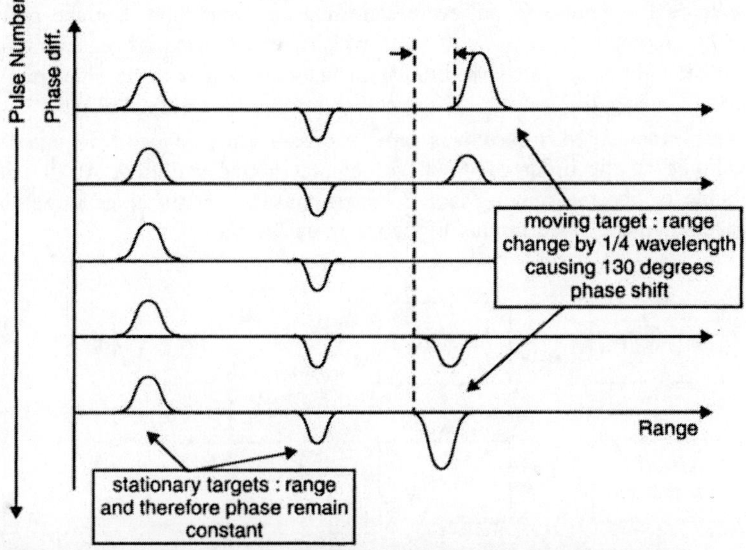

Fig. 3.1. Five Sequential Returns from Pulse Comparison Output.

3.2.1. A Simple Block Diagram of MTI Radar

A schematic diagram of MTI radar is shown in Fig. 3.2. This diagram has got two channels, one is receiver channel and other is coherent oscillator (COHO Channel). A receiver channel is identical to the standard radar receiver except the local oscillator which is designed to have good frequency stability and a limiter is included in the IF stage so that the amplitude of echo signal fed to the phase detector should have the same amplitude. The reference signal is provided by the COHO oscillator (Coherent Oscillator) and is a stable oscillator (STALO) operating at a frequency that is near to the difference between transmitter and Stalo frequencies.

The phase of each transmitter pulse is randomly related to the phase of the last pulses in the radar using magnetrons as transmitter. It is essential that the phase of the echo oscillator be matched with the phase of the transmitter pulse is injecting in to Coho channel. If the frequency of the Coho oscillator is stable, its phase in the time interval between pulses is then related to the phase of the oscillations of the last outgoing pulse in exactly the same way for successive pulses. The phase detector used in MTI radar compare the phases of the Coho oscillator and the amplitude limited intermediate frequency echo signal. The phase detector output for successive pulses may be subtracted from each other by employing a single delay line. A delay line provides the delay of one PRP, it may employ a mercury delay line or quartz transmission line, upon which are impressed mechanical vibrations lying in the frequency range 10 to 20 Mhz, modulated by the video signals. The line length is such as to introduce a delay corresponding exactly to the period between successive pulses. It may be ensured by using the delayed outgoing pulse to trigger off the next outgoing pulse. The delayed and undelayed echoes are subtracted by first reversing the polarity of one of them and adding.

The phase comparison is carried out at IF stage instead of doing at the much higher transmitted frequency. The output of the phase detector is fed as the input to the delay line canceller, which is acting as a high pass filter. It will separate the moving target from the echo received from the stationary target. The fixed targets with the constant amplitude from pulse to pulse are cancelled on subtraction. But the moving target has varying amplitude from pulse to pulse which are not cancelled on subtraction. The output from the subtractor unit is in bipolar in nature and it is to be converted in to unipolar video by using full wave rectifier before presenting on display unit.

By Practical MTI system is capable of canceling echoes from fixed targets having amplitude of the order of 40 dB or greater above visibility. At the same time it is possible to observe moving targets superimposed over the fixed targets and which are weaker than the fixed targets by as much as 25 dB.

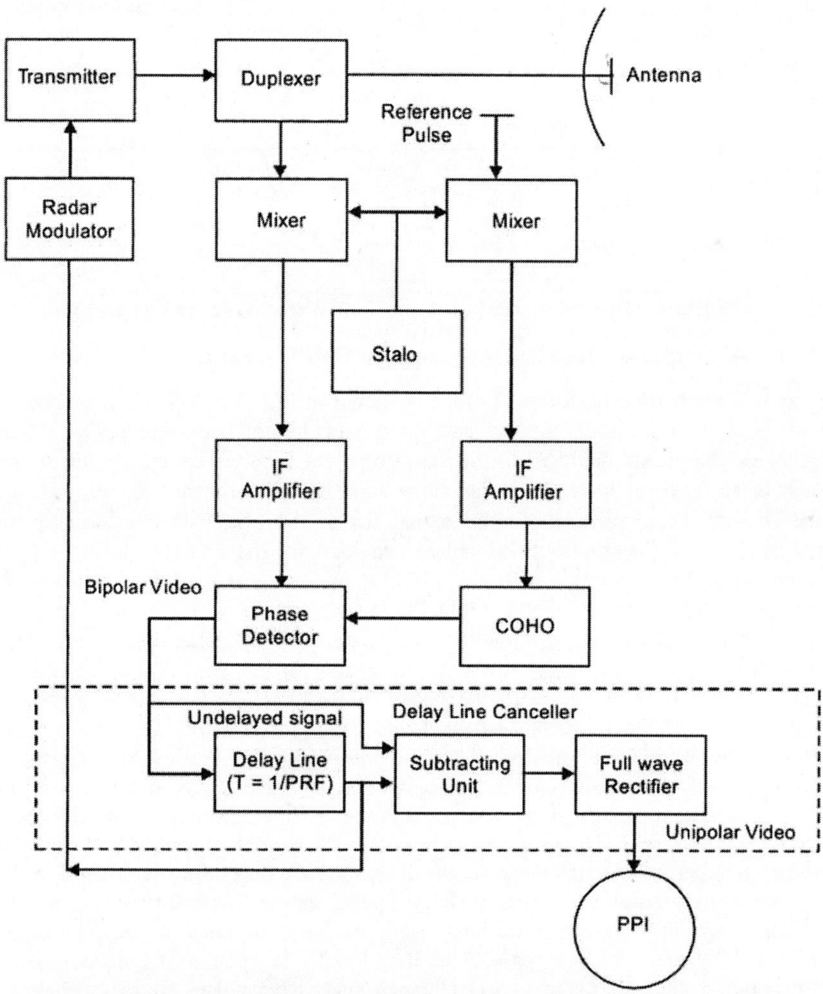

Fig. 3.2. Block Diagram of MTI Radar with the Delay Line Canceller.

3.2.2. MTI Radar using Power Oscillator as Transmitter

A block diagram of MTI radar using a power oscillator is shown in Fig. 3.3. A portion of the transmitted signal is mixed with the STALO output to produce an IF beat signal whose phase is directly related to the phase of the transmitter. This IF pulse is applied to the coherent oscillator (COHO) and causes the phase of the COHO CW oscillation to "lock" in step with the phase of the IF reference pulse. The phase of the COHO is then related to the phase of the transmitted pulse and may be used as the reference signal for echoes received from the particular transmitted pulse. Upon the next transmission another IF locking pulse is generated relock the phase of CW COHO until the next locking pulse comes along. This type of MTI radar has wide applications.

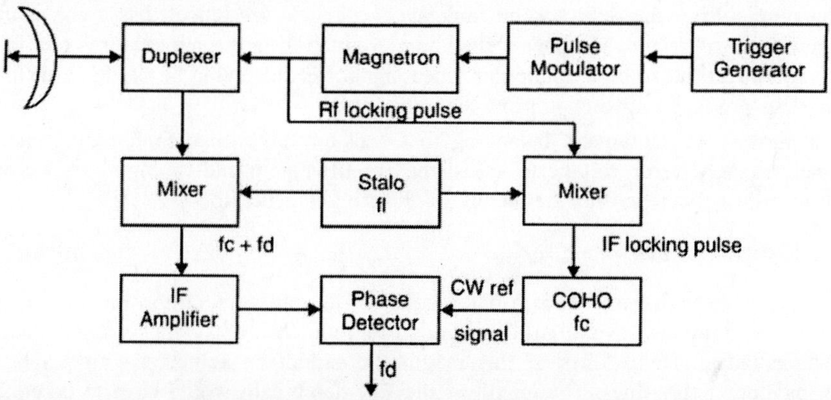

Fig. 3.3. Block Diagram of MTI Radar (Using Power Oscillator As Transmitter).

3.3. DELAY LINES AND CANCELLERS

A block diagram of a typical delay lines cancellation network is shown in Fig. 3.4. The bipolar video from the phase detector modulates a carrier before being applied to

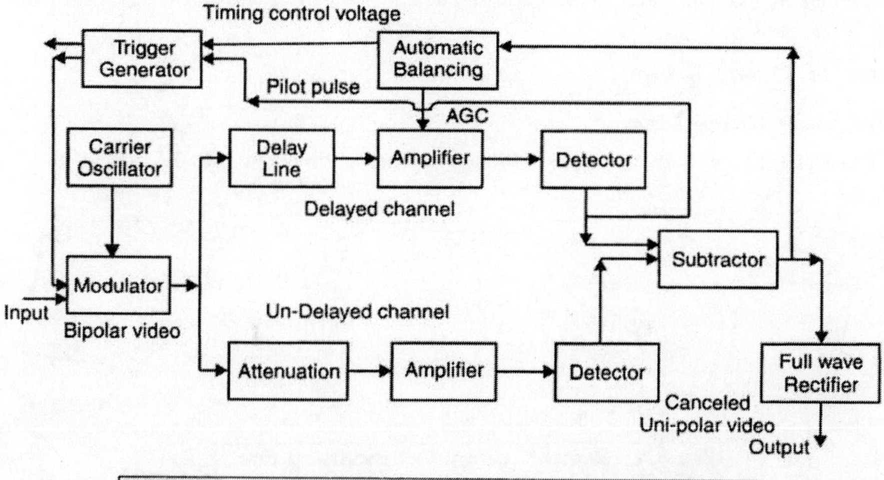

Fig. 3.4. A Block Diagram of Single Delay Line Canceller.

the delay lines. The carrier frequency is normally 15 or 30 Mhz. The radar output is not directly applied to the delay lines as a video signal since it would be differentiated by the crystal transducer that convert the electromagnetic energy into acoustic energy, and vice versa. The modulated bipolar video is divided between two-channel .In one channel the signal is delayed by a PRP, while in the other channel it reaches directly ie undelayed. There is considerable attenuation in the signal introduced by the delay line and must be amplified in order to bring it back to its original level. Since the introduction of an amplifier into the delay channel can alter the phase of the delayed waveforms and introduce a line delay, an amplifier with the same delay characteristic is also used in the direct (un delayed) channel. An attenuator might also be inserted in the direct channel to make equalizing the gain. Good cancellation in a typical application might result in an uncancelled voltage residue of the order of 1% or 40db.

The output from the delayed and undelayed channels are detected to remove the carrier and than subtracted. The uncancelled bipolar video from the sub tractor is rectified in a full wave rectifier to obtain unipolar video signal for displaying on the PPI. Ideally, only moving produced an output from the subtractor.

The purpose of automatic balancing to detect any amplitude timing differences and generates AGC error voltage to adjust the amplifier gain and timing control error voltage to adjust the repetition frequency of the trigger generator.

3.3.1. Delay Lines

The delay line must introduce a delay equal to the pulse–repetition interval. Thus delay lines as long as several milli seconds are required for typical ground based surveillance radars. Delay lines of this magnitude cannot be achieved with practical electro magnetic delay lines. The length of the EM delay path would have to be equal to twice the unambiguous range of the radar. Converting the EM wave in to acoustic waves and accomplishing the delay in an acoustical delay line overcome this difficulty. After the delay the acoustic wave are converted back to EM waves. The velocity of acoustic wave depends on the delay medium, but it is of the order of magnitude of 10^{-5} that of EM waves; both liquids and solids have been used as the acoustic delay media. Mercury & fused quartz are the two media most widely employed for MTI radar application, but water, water –glycol mixture, aluminum, magnesium and glass have also been used in delay lines.

Types of Delay Lines

(*i*) Acoustic Delay Line

The basic elements of an acoustic delay line are outlined in Fig. 3.5 below.

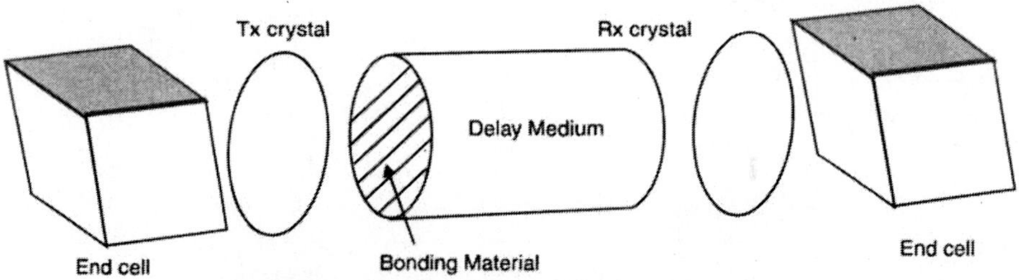

Fig. 3.5. Elements of an Acoustic Delay Line.

The EM energy is converted into acoustic energy by piezoelectric transmitting crystal. A similar transducer (the receiving crystal) at the o/p of the line converts the acoustic energy back to EM energy.

(*ii*) Quartz Crystal

The **Quartz crystal** transducers are normally a high *Q* device with an inherently small bandwidth. However, when transducer is coupled to a delay medium, the medium has a dumping effect, which broadens the bandwidth. Consequently, acoustic delay lines are relatively broadband devices.

(*iii*) Liquid Mercury

One of the simplest acoustic delay lines consists of a straight cylindrical tube filled with **liquid mercury**. The transit time of acoustic waves in mercury at room temperature is approximately 17.5 μs/inch. To produce a delay of 1000 μs the line must be 57 inch in length exclusive of end cells. This is manageable size in ground-based radar.

A more compact configuration may be had by **folding the line** back on it itself one or more times. Another method of obtaining a more compact delay line is to make use of **multiple reflection** in a tank filled with liquid. The alignment of the reflecting surface is a problem, and it has been difficult to obtain a leak proof construction.

One of The disadvantages of either solid or liquid delay is the large insertion loss. The insertion loss of a typical folded mercury delay line with 1000 μ sec is as follows:

Impedance mismatch loss of the crystal	36 dB
Free space attenuation in mercury	12.8 dB
Tubular attenuation	2.7 dB

The total insertion loss is 52db.

3.3.1.1. Modern Trends in Delay Line using Fiber Optics in MTI Radar

In the advanced version of the radar, nowadays we are using Fiber Optics as a delay-line media, which is an excellent propagating medium for time delays. It has low loss, low dispersion, lightweight, and is inexpensive. (Details of this is beyond the scope of this book). (Reference : IEEE Transaction on Microwave theory and techniques, Vol. 45, No. 8 August 19971473).

In MTI radar, the small echo from the target has to be distinguished from the large echo from fixed objects (clutter) The filtering of the clutter in a typical MTI radar system is performed by delaying a returned pulse by the pulse period, and subtracting it from the following pulse, as shown in Fig. 3.6.

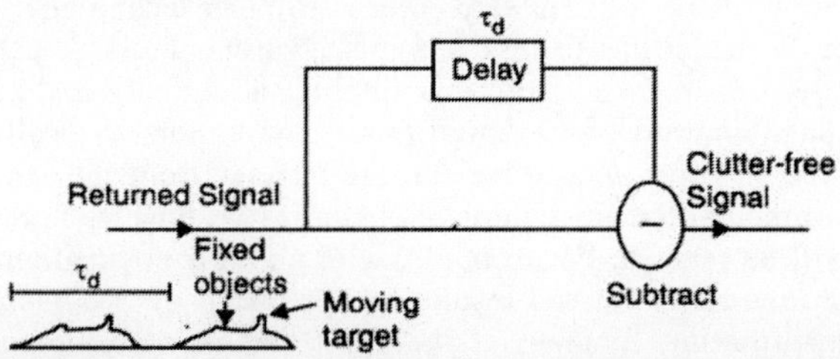

Fig. 3.6. Schematic of an MTI Radar in which the Pulse is Delayed by the Pulse Period and Subtracted from the Following Pulse.

Signal processing using optic-fiber delay lines is a very powerful technique for processing high bandwidth signals. Moreover, the signal carried by the intensity modulated optical beam is insensitive to electromagnetic radiation. Various optical delay-line configurations have been proposed to achieve different radio-frequency (RF) filter functions. These configurations can be either optically coherent or incoherent. Coherent optical processing allows more flexibility in shaping of the filter-transfer function. Nevertheless, the phase of the optical beam has to be precisely controlled since the filter is highly sensitive to any phase variation. Moreover, the longest time delay in the filter has to be shorter than the coherence time of the optical source. Incoherent processing does not require any control of the optical phase. The filter is, therefore, less sensitive to external perturbation (e.g., variation of temperature). Long time delays are made possible, but the shortest delay line has to be longer than the coherence length of the optical source. In this approach, optical signals are detected using a pair of matched photo detectors and combined electrically. Another unique advantage of the filter is that the two signals, which recombine inside the photo detector, are at two different wavelengths avoiding problems of optical coherence, which appear in traditional delay-line filters.

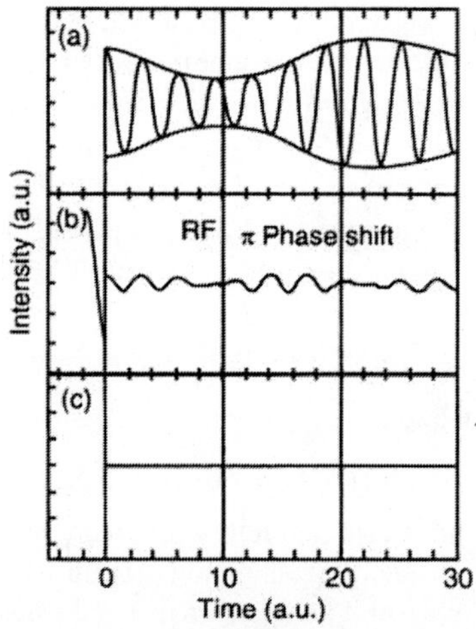

Fig. 3.7.

(*a*) Time-domain representation of a radar pulse. (*b*) Resulting filtered signal after filtering using p phase shift of an RF carrier—due to the p phase shift the cancellation is incomplete. (*c*) Resulting filtered signal after filtering using inversion inside an semiconductor operational amplifier (SOA)—the cancellation is theoretically complete.

A major application of the feed forward filter is in radar technology. In MTI radar, the small echo from the target has to be distinguished from the large echo from fixed objects(clutter). In a typical radar system, this filtering is digitally realized after down conversion to base band and analog-to digital (A/D) conversion. Typically, 14–18 b of resolution are required in order to resolve the faint target from the strong clutter. If some filtering is performed in the analog domain before A/D conversion, the requirement on A/D resolution will be relaxed. For example, with a 30-dB attenuation of the clutter in the analog domain, the required A/D resolution is reduced by 5 b. Optical filters are ideal to perform this filtering function at the front end of the antenna as they are compatible with the emerging technology of optically controlled phased-array antennas, and long low cost time delays are easily achieved. The clutter, which has the same phase from pulse to pulse, will be removed, while the doppler shifted-target echo will be preserved. All-optical subtraction is, therefore, a key operation of the filtering function.

In principle, subtraction in the analog domain can also be made by shifting one of the signals by half the RF period out of phase) and then combining them. However, for MTI radar this technique does not lead to a perfect cancellation of the clutter as analytically shown in Fig. 3.7(*a*) represents the simulated pulse of a radar, which consists of an RF signal amplitude modulated. Fig. 3.7(*c*) shows the filtered signal after recombination of the signal represented in Fig. 3.7(*a*) with a RF phase shifted version of itself. Although strong attenuations achieved, part of the signal still remains. The no perfect cancellation stems from the fact that a phase shift is not necessarily equivalent to an inversion. In contrast, Fig. 3.7(*c*) shows the cancellation when the signal represented in Fig. 3.7(*a*) is recombined with an inverted version of itself. A theoretically perfect cancellation is obtained. The difference between these two techniques will depend on the relative frequency of the RF signal compared to the slow modulation on top of the return pulse. At high RF frequencies, the small extra time delay introduced to perform the phase shift will become negligible compared to the typical period of the intensity modulation. Therefore, the difference between inversion and phase shift will not be significant. However, at lower frequencies, the extra time delay introduced to perform the shift may become comparable to the typical period of the intensity modulation, leading to a non-ideal cancellation of the clutter.

All-Optical Filter Approach

In Fig. 3.8 describes the schematic of the filter. An external cavity tunable laser operating at λ_1 =1525 nm is directly RF intensity modulated. The coherence length of the laser is on the order of 1 km. This beam is then split into two parts; one part goes to the long time delay consisting of 500 m of optic fiber. The other part is combined with a continuous wave (CW) beam from a **distributed feedback (DFB)** laser operating at λ_2 =1526.6 nm. These beams are then amplified in an erbium doped fiber amplifier (EDFA) and sent into the SOA. Due to cross-gain modulation in the SOA, the emerging signal λ_2 at represents the inverted version of the RF signal.

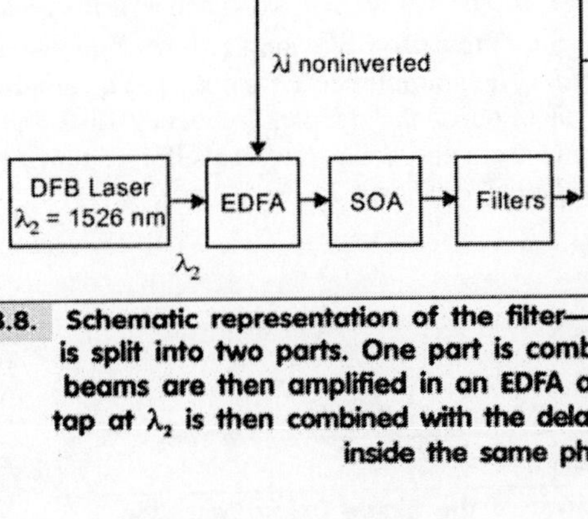

Fig. 3.8. Schematic representation of the filter—a directly modulated laser beam is split into two parts. One part is combined with a CW beam at λ_2 both beams are then amplified in an EDFA and sent to an SOA. The negative tap at λ_2 is then combined with the delayed positive tap at λ_1 and detected inside the same photo detector.

Using a differential configuration of the filter as shown in Fig. 3.8. This all-optical notch filter can be used in MTI technology to filter out the clutter signal. It can be used to filter the clutter at Intermediate frequency (IF) in the analog domain prior to A/D conversion, thus alleviating the requirement for high-resolution A/D converters. Moreover, as the signal stays in the optical domain, cascading of delay-line processors is made possible.

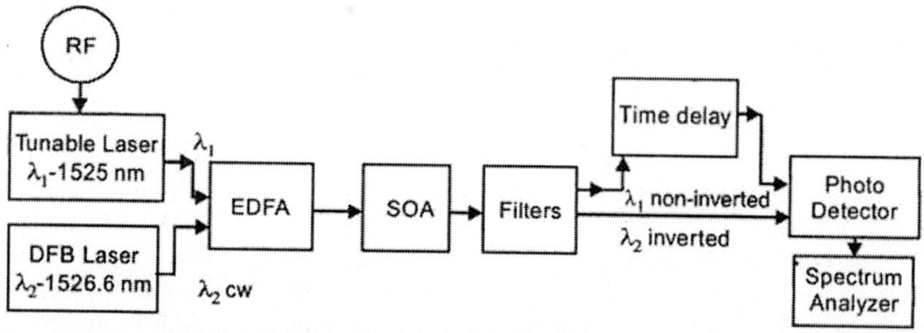

Fig. 3.9. Schematic representation of the differential configuration of the filter, showing improved performance. Beams at λ_1 and λ_2 go through the SOA. The positive tap at λ_1 is delayed and is recombined with the negative tap at λ_2. Both beams are then detected in the same photo detector.

3.3.1.2. Response of The Delay Lines Cancellers

The video signal received from a particular target at a range R can be expressed as,

$$V_1 = K \sin (2\pi \, fd \, t - \phi_0)$$

Where K is the amplitude of video signal & ϕ_0 =Phase shift ($4\pi ft \, R/c$)

At a time t-T, where T is the Pulse repetition time, the voltage from the same target is

$$V_2 = K \sin (2\pi fd(t - T) - \phi_0$$

The output from the subtractor is given below, provided rest every thing remain constant over the time interval T.

$$V = V_1/2 - V_2/2 = K \sin \pi fdT \cos \{2\pi fd(t - T/2) - \phi_0\} \quad ...(3.1)$$

The each video voltage is divided by the factor ½ because it is assumed that voltage of video, are equally divided between the delayed and the un delayed channel of the receiver.

Above equation (3.1) indicates that the output from the single–delay line canceller is a cosine wave at a doppler frequency fd with an amplitude $2K \sin \pi fdT$. The amplitude of the canceller video output is thus a function of the doppler frequency shift and the PRT. The ratio of amplitude of the output from the delay line canceller, $K \sin (\pi \, fd \, T)$ to the amplitude of the normal radar video, K is shown in Fig. (3.10).

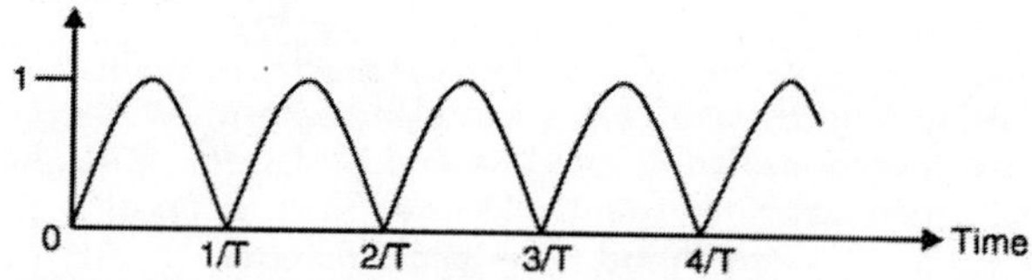

Fig. 3.10. Frequency Response of the Single Delay Canceller.

3.3.1.3. Double Cancellation

A single delay line canceller is not sufficient to reject all the clutter as desired in the vicinity of DC components or at doppler frequencies corresponding to the different PRF and its harmonics. The frequency response function of double delay line canceller is much wider than the single delay line canceller. To have the widened clutter rejection, the output must be pass through the second delay line canceller.

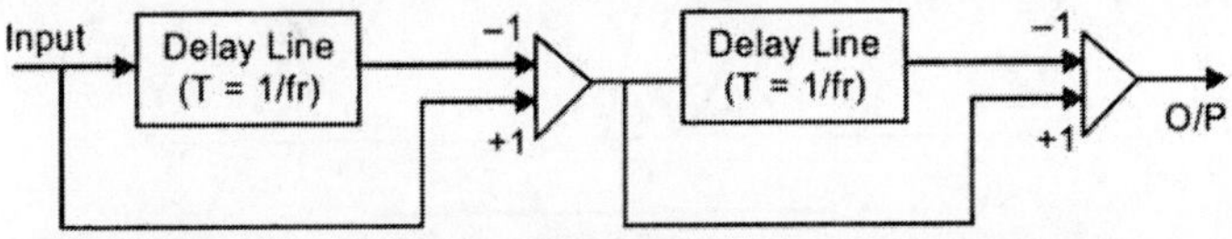

Fig. 3.11. A Diagram of Double Delay Line Canceller.

The sine wave response of the double delay line canceller is a squared of the single delay line characteristics. Two single delay line canceller are configured in cascaded form to have double delay line canceller.

A double delay line canceller shown in the Fig. 3.11. Two delay lines are being arranged in the cascaded form alongwith the two adders. A input signal $f(t)$ is inserted in the Ist delay line canceller and to Ist adder simultaneously. The output from the Ist delay line canceller after delaying of one PRF ($T = 1/fr$) feed to the Ist adder. In the next stage, the output of adder is bifurcated and one is given to the IInd adder directly and the second input is feed through the IInd delay line canceller. The output of double delay canceller is as follows :

$$f(t) - f(t + T) - f(t + T) + f(t + 2T) \quad ...(3.2)$$

There is one more configuration commonly known as the three pulse comparison as shown in Fig. 3.12. A three-pulse comparison canceller configuration has the same frequency response characteristics as the double delay canceller. The operation of this configuration is explained here. In this configuration, we use two delay line canceller and a adder. A input signal $f(t)$ is bifurcated and one is fed to the Ist delay line canceller and second is given to the adder, which is connected at the output end of IInd delay line canceller. The output from the Ist delay line canceller (after introducing delay of 1 PRP) is again divided in two parts and one is fed to the IInd delay line canceller and other part is given to the adder directly. The output of IInd delay line canceller is finally fed to the adder. The adder get three signal, one is directly from the input signal, second from the IInd delay line canceller after getting delay of 2 PRP and the third input is from the output end of 1st delay line.

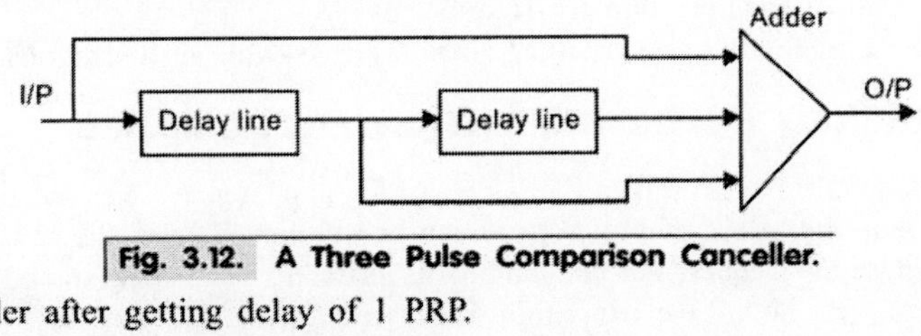

Fig. 3.12. A Three Pulse Comparison Canceller.

Canceller after getting delay of 1 PRP.

The output of the adder as follows :

$$f(t) - 2f(t + T) + f(t + 2T) \quad ...(3.3)$$

The three-pulse comparison canceller is same as the double delay canceller until in pefect adjustment. The frequency response characteristics curve is shown in the Fig. 3.13.

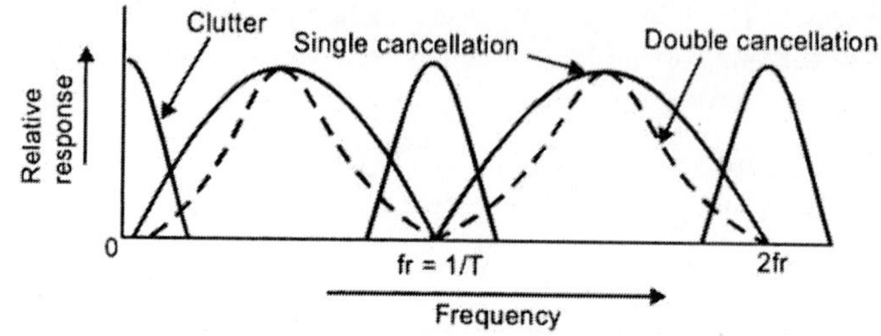

Fig. 3.13. Frequency Response of Single Delay Line Canceller and Double Delay Line Canceller.

3.3.1.4. Blind Speed

Blind speed is a very serious problem in MTI radar. When the doppler shift equals an integer multiple of PRF (Pulse Repetition Frequency), the delay line canceller not only eliminate the DC components caused by clutter but it also rejects moving targets. In other words, blind speed can be explained as the relative velocities at which the MTI response is zero called **blind speeds.**

The response from the single delay line canceller will be zero, When

$$\boxed{fd = n/T = nfr}$$

Where $n = 0, 1, 2, 3, ...; T = PRT$ and fr is the PRF.

The nth blind speed Vn can be expressed as

$$\boxed{Vn = n\lambda/2\ T = n\lambda fr/2}$$

Where, $n = 1, 2, 3, ...$

If λ is in meters, fr in hz and the relative velocity in knots,

The blind speeds are written as

$$Vn = n\lambda fr/1.02 = n\lambda fr \qquad ...(3.4)$$

This limitation of blind speed can be reduced up to some extent by employing following steps :

1. Operate the radar at long wavelength (means low frequency),
2. Operate with a higher PRF,
3. Operate with more than one PRF,
4. Operate with more then one RF frequency.

Choice of these methods has particular advantages as well as limitations.

3.4. STAGGERED PULSE REPETITION FREQUENCIES

If a radar is operating at multiples PRFs or its PRF is changed either pulse to pulse or scan to scan, than the affect of blind speed can be eliminated from the radar. If two radar operating at same frequencies but having its different PRF then if one radar is blind to moving target. So, if we use single radar but having different PRF than the same affect can be achieved. When the PRF is changing pulse to pulse than it may be called as staggered PRF. Staggering of PRF is generally employed in Air Traffic Control Radar such as Surveillance Radar Element (SRE).

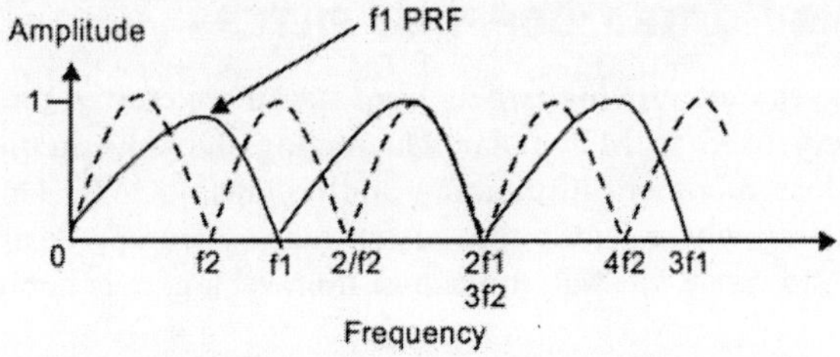

Fig. 3.14. Frequency Response of Two PRFs F_1 and F_2, where $F_1 = 3F_2/2$.

In the Fig. 3.14, frequency response of two PRF is shown. Suppose the first PRF is F_1 shown in bold line and the blind speed of second PRF is F_2 shown in dotted lines. If we observe the Fig. 3.14, it is clear that at particular position when $2f_1 = 3f_2$, both the PRFs have the same blind speed.

The multiples PRFs can be obtained by using several methods. Using following techniques may vary the PRFs :

1. Pulse to pulse (known as staggered PRF)
2. Scan to scan
3. Dwell to dwell.

The problems occurs in using staggered PRF is that residual of uncancelled echoes of clutters, which are due to second time around echoes. So to minimize the second time around echoes affect, if we use unstaggered PRF in the sector where second time around are expected more and rest of the sector used staggered PRFs.

3.4.1. Sub Clutter Visibility

Sub clutter visibility may be defined as the ability of MTI radar to detect the moving target, if the target is superimposed over the clutters. SCV defined the performance of an MTI radar and it is measured in dbs. It may be defined as

SCV = The gain in signal-to clutter power ratio.

Suppose an MTI radar has SCV 20 dB than it means that a moving target can be detécted in the presence of clutter even though the clutter echo power is 100 times the target echo power.

3.4.2. The Cancellation Ratio

The cancellation ratio may be defined as the ratio of a fixed target signal voltage using MTI cancellation to the signal voltage without MTI cancellation. We may write

$$\text{Cancellation ratio (CR)} = \frac{\text{Fixed targets signal voltage after MTI cancellation}}{\text{Signal voltage without MTI cancellation}}$$

The CR is a number that is always less than One (1) and may be expressed in dB.

3.4.3. The Target Visibility Factor

The target visibility factor is the ratio of the signal strength from a target, which is moving at a specified radial velocity to the signal strength from the same target when it is moving at an optimum radial velocity.

3.5. RANGE GATED DOPPLER FILTERS

In order to separate moving targets from stationary clutter, the delay line canceller has been widely used in MTI radar. Quantizing the time in to small interval can eliminate the loss of range information and collapsing loss. This process is known as the range gating where width depends on range accuracy desired. After quantizing the radar return to range interval, the output from each gate is applied to a narrow band filter.

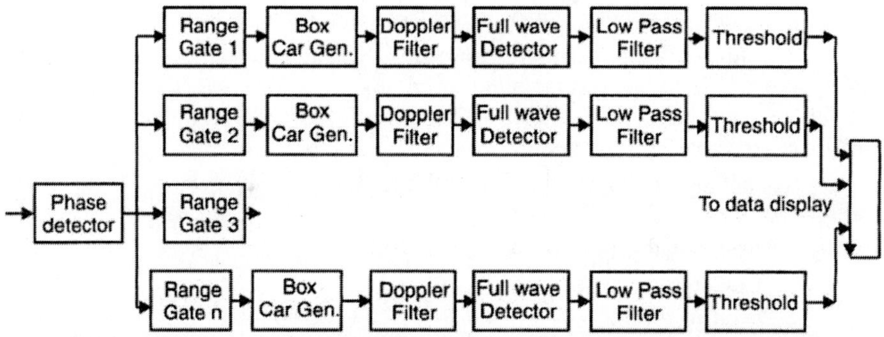

Fig. 3.15. Block Diagram of MTI Radar Using Range Gates and Doppler Filters.

A block diagram of the video of an MTI radar using multiple range gates followed by clutter rejection filter is shown in Fig. 3.15. Here the range gates sample the output of the phase detector sequentially range interval. Each range open in sequence just long enough to sample the voltage of the video waveform corresponding to a different range interval in space or it acts as a switch/gate which open and close at the proper time. The output of the range gate is given to a circuit known as box car generator. Its function is to aid in the filtering and detection process enhancing the fundamental of the modulation frequency and eliminating harmonics of the PRF. The clutter rejection filter is nothing but a band pass filter whose bandwidth depends on the extent of the expected clutter spectrum. The filtered output from the Doppler filter is further feed to a full wave linear detector which convert the bipolar video to unipolar video. A low pass filter or integrator passes these unipolar video to the threshold detection circuit. Any signal crosses the threshold level is treated as a target. The outputs from each range channels are combined for display on the PPI or any other display unit.

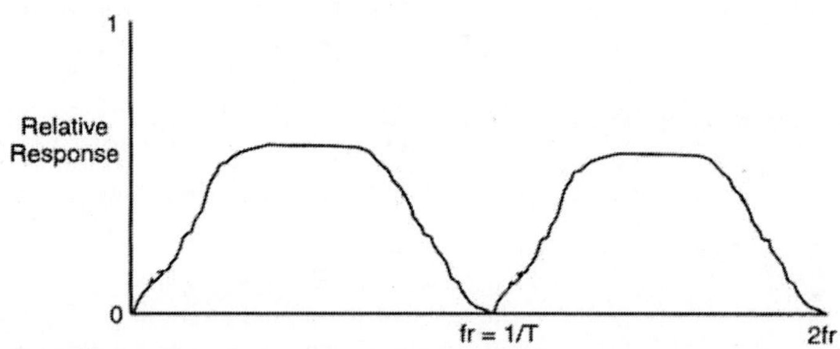

Fig. 3.16. Frequency Response Characteristics of an MTI using Range Gates and Dopplers Filters.

The presentation of this type of MTI radar is far better than the display from normal MTI radar.

The frequency – response characteristics of an MTI radar using range gates and filter is shown in Fig. 3.16, the shape of the rejection band is mainly determined by the shape of the band pass filter.

It must be pointed out that the MTI radar using range gates and filters is more complex then an MTI with single delay line canceller a better MTI performance is achieved from a better match between clutter filter characteristics and clutter spectrum.

3.6. NON-COHERENT MTI

The echo signal received from a moving target or from clutter fluctuates both in amplitude and phase. Where the MTI makes use of the phase fluctuation than it is called coherent MTI and where the amplitude fluctuation is being considered than it is called non-coherent MTI. In non-coherent MTI, the amplitude fluctuation is used to recognize the doppler components produced by a moving target. It is also called externally coherent.

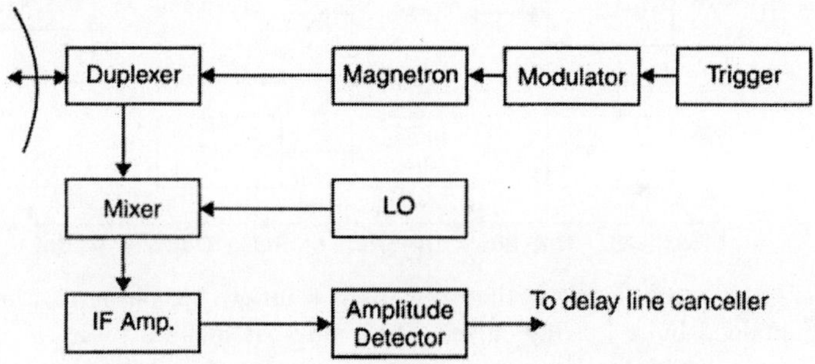

Fig. 3.17. Block Diagram of Non-Coherent MTI Radar.

The block diagram of non–coherent MTI is shown as Fig. 3.17. In non-coherent MTI amplitude limiter can not be used otherwise desired amplitude fluctuation would be lost. Instead of using phase detector we are using amplitude detector. Therefore, IF amplifier should be linear and should have large dynamic range. A logarithm amplifier may be used as IF amplifier to have logarithm gain characteristics, such as protection from saturation and have uniform output with variations in the clutter input amplitude. The output of IF amplifier to be detected through the amplitude detector. The output of amplitude detector may be further processed either using delay line canceller or it may be detected over the A-scope. A butterfly effect can be observed on the A-scope due to the doppler in amplitude fluctuation. The Transmitter should be stable over the pulse duration to prevent beat from the overlapping ground clutter.

The main advantage of non-coherent MTI is that it is very simple and may be used where space and weight are limited.

3.7. PULSE DOPPLER RADAR

Pulse radar is a combination of pulse radar and CW radar. It works on the principal of Doppler shift as MTI radar follows. As per the Nyquist Criterion the sampling rate (*i.e.* PRF) should be greater and equal to the twice of the doppler shift frequency but in MTI due to use of low frequency its became under sampled. It will leads to ambiguous estimation of target speed and occurrence of blind speed, where target appears stationary and unresolvable against back ground clutter. Pulse Doppler radar being high PRF radar, it can remove the doppler ambiguities. To extract the doppler shift information of the carrier the pulse radar should be modified by introducing a coherent oscillator (COHO) for frequency stability in he transmitter and receiver chain. It employs the coherent radar system.

Pulse Doppler radar is classified as high PRF and as medium PRF. In high PRF pulse radar there is ambiguity in the range but unambiguities in the velocity. In the medium PRF pulse radar there is ambiguities in range and velocity both.

The block diagram of Pulse Doppler Radar is shown as Fig. 3.18.

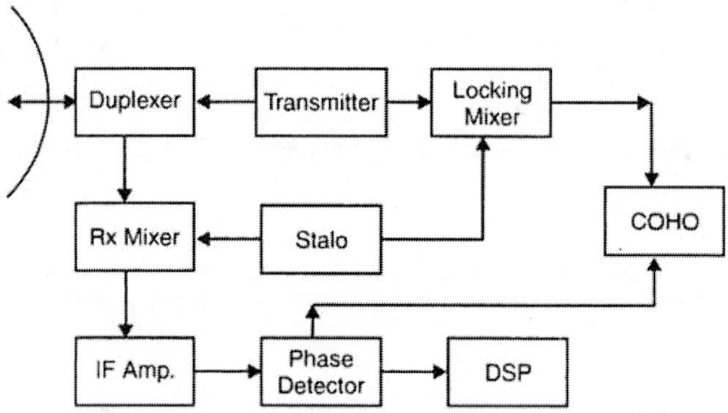

Fig. 3.18. The Block Diagram of Pulse Doppler Radar.

A STALO (stable local oscillator) is used to allow the phase of transmitter signal to be maintained by a locking mixer. The output of locking mixer given to lock the COHO phase and inurn it serves as reference phase for the detector at intermediate frequency. Now the, phase detector measures the difference in phase between two RF signal. Due to the target motion the phase path of the echo changes pulse to pulse and by the same amount phase difference will vary.

Applications. Pulse radar has many applications; some of those are listed below:

1. It is being used as weather warning radar at the airbases to detect and measure thunderstorm, turbulence in the air.
2. It is very useful in detecting and estimating the target motion, locking of particular target out of a group.
3. To observe thunderstorm, rain and hail, a double polarization doppler radar is being used.

Advantages :

1. A pulse doppler radar has got the ability to reject the unwanted echoes by using doppler filters or by a range gating.
2. It can measure the range and velocity over predetermined limits, even in presence of multiples target.
3. Signal to noise ratio can be increased by using coherent integration.

SOLVED EXAMPLES

Example 1. A CW radar (MTI) is operating at a PRF of 1 KHz, find the lowest blind speed. If it is operating at 2 cm wave lengths.

Solution. Lowest blind speed $= n.\dfrac{\lambda}{2} \times \text{PRF}$

We have,

$$\text{PRF} = 1 \text{ KHz} = 1000 \text{ Hz}$$
$$\lambda = 2 \text{ cm} = 0.02 \text{ m}$$

for the calculation of lowest blind speed, the value of $n = 1$,

By puttting the values in above equation, we get

$$\text{Lowest blind speed} = 1 \times \frac{0.02}{2} \times 1000 = 10 \text{ m/s}$$

or

$$= \frac{10 \times 60 \times 60}{1000} = 36 \text{ Km/hr}$$

$\boxed{\text{Lowest Blind Speed} = 36 \text{ Km/hr}}$ **Ans.**

Example 2. If two MTI radar system are operating at the same PRF but with the different operating frequencies. And the 2nd blind speed of one is equal to fourth speed of the other than find the ratio of their operating frequencies.

Solution. Let the operating frequency of first radar $= f_a$

Operating frequency of second radar $= f_b$

Now, second blind speed of first radar $= \left[2 \times \dfrac{C}{2f_a}\right] \times PRF$

and, fourth blind speed of second radar $= \left[4 \times \dfrac{C}{2f_b}\right] \times PRF$

Therefore,

$$\left[2 \times \frac{C}{2f_a}\right] \times PRF = \left[4 \times \frac{C}{2f_b}\right] \times PRF$$

or

$$\frac{2}{f_a} = \frac{4}{f_b}$$

or,

$$\frac{f_a}{f_b} = \frac{2}{4} = \frac{1}{2}$$

Their operating frequency ratio will be,

$\boxed{\dfrac{f_a}{f_b} = \dfrac{1}{2}}$ **Ans.**

Example 3. A pulse doppler radar has a carrier frequency of 9 GHz and PRF of 400 Hz. Find its blind doppler frequencies and the radial velocity of target which would be undetected by the radar.

Solution. Doppler frequencies $(fd) = n \,.\, PRF$

We have,

$$PRF = 4000 \text{ Hz}$$

Than

$$\text{Doppler frequencies} = n(4000)$$
$$= 1 \times 4000 = 4000 \text{ Hz}$$
$$= 2 \times 4000 = 800 \text{ Hz}$$
$$= n \times 4000 = n \,.\, 4000 \text{ Hz.}$$

What radial target velocities would be undetected by the radar ?

We have, operating frequency = 9 GHz = 9 × 10^9 Hz

$$\lambda = \frac{c}{f} = \frac{3\times10^8}{9\times10^9} = 3.33 \text{ cm}$$

$$\lambda = 0.03 \text{ m.}$$

Now, we know

$$Vr \text{ (blind speed)} = \frac{n\,.\,PRF\,.\,\lambda}{2} = \frac{n\times(4000)\times(0.03)}{2}$$

$$\boxed{V_r = n\,.\,66.67 \text{ m/s}} \quad \textbf{Ans.}$$

Any integer multiple of 66.67 m/s is undetected by this radar system.

Example 4. What modification can be made to the radar system (used in problem 3) that would eliminate blind speeds below 2000 m/s and find out the maximum unambigious range of that modified radar system ?

Solution. We have $Vr = 2000$ m/s

We know that $$Vr = \frac{n\,.\,PRF\times\lambda}{2} \qquad \text{(since } n = 1)$$

By putting the values, $$2000 = \frac{1\times PRF\times 0.033}{2}$$

$$PRF = \frac{2\times2000}{0.033} = \frac{4000}{0.033} = 121212 \text{ Hz}$$

$$\boxed{PRF = 121212 \text{ Hz} = 121.21 \text{ KHz}} \quad \textbf{Ans.}$$

Maximum unambigious range is given by formula,

$$R_{unam} = \frac{C\,.\,\Delta t}{2}$$

where $$\Delta t = PRP = \frac{1}{PRF}$$

$$\Delta t = \frac{1}{121212} = \frac{1}{121.21\times10^3}$$

By putting the values, we get

$$R_{unam} = \frac{3\times10^8}{2}\times\frac{1}{121.2\times10^3}$$

$$= \frac{3\times10^8\times10^{-3}}{2\times121.21} = \frac{3\times10^5}{242.42}$$

$$= 0.0123752 \times 10^5 \text{ m}$$

$$R_{unam} = 1237.52 \text{ m}$$

or $$R_{unaml} = 1.237 \text{ Km.} \quad \textbf{Ans.}$$

Example 5. The MTI radar is used by a traffic control police to measure the speed of vehicles. If the dopper frequency shift measured from the moving vehicle is 1.5 KHz. Calculate the speed of vehicle, if radar is operating at 1 GHz with PRF of 1000 Hz.

Solution. The speed of the vehicle = $Vr = \frac{f_d \times \lambda}{2}$ (where f_d = dopper frequency)

$$Vr = \frac{1500 \times 0.3}{2} \text{ m/s} \qquad \left(\text{since } \lambda = \frac{C}{f} = \frac{3 \times 10^8}{1 \times 10^9} = 0.3 \text{ m}\right)$$

$$Vr = 225 \text{ m/s}$$

$$Vr = \frac{225 \times 60 \times 60}{1000} \text{ Km/hr}$$

$$\boxed{V_r = 810 \text{ Km/hr}} \quad \textbf{Ans.}$$

IMPORTANT RELATIONS

1. $V = \frac{V_1}{2} - \frac{V_2}{2} = K \sin \pi f_d T \cos \left\{2\pi f_d \left(t - \frac{T}{2}\right) - \phi_0\right\}$

   (The output from the subtractor)

2. $f(t) - 2f(t+T) + f(t+2T)$ (The output of the adder)
3. $f(t) - f(t+T) - f(t+T) + f(t+2T)$ (The output of double delay canceller)
4. $f_d = \frac{n}{T} = nfr$
5. $V_n = n\frac{\lambda}{2}T = \frac{n\lambda f_r}{2}$
6. $V_n = \frac{n\lambda f}{1.02} = n\lambda fr$
7. $f_d = 2\left(\frac{v}{\lambda}\right)\cos\theta$ (Moving platform MTI)
8. $\Delta f_d = \frac{2v}{\lambda}\sin\theta\,\Delta\theta$ (Moving platform MTI)
9. $2A\sin\left(\frac{\Delta\phi}{2}\right) = A\Delta\phi$
10. $W(f) = |g(f)|^2 = |g_0|^2 \exp\left[-a\left(\frac{f}{f_0}\right)^2\right]$
11. $t_0 = \frac{n_B}{f_r} = \frac{\theta_B}{\theta_S}$

where, V = Video signal output from the subtractor
K = Amplitude of video signal
ϕ_0 = Phase shift
V_n = nth blind speed
f_d = Doppler frequency shift
v = Platform speed
θ = Azimuth angle
$W(f)$ = Clutter power spectrum as a function of frequency

$G(f)$ = Fourier Transform of the input waveform

f_0 = Radar carrier frequency

a = Parameter which depends on the clutter

t_0 = Finite time

nB = Number of hits received

f_r = PRF

θ_B = Antenna beamwidth

θ_S = Antenna scanning rate

SUMMARY

- **Principle Of MTI Radar.** When an aircraft echo and a clutter echo appear in the same radar resolution cell, the aircraft might not be detected. But the doppler effect permits the radar to distinguish moving targets in the presence of fixed target even the echo signal from fixed targets has comparatively greater magnitude than the moving targets such as aircraft.
  - There is a type of MTI, which does use the information directly from the doppler but it observing the area map scan to scan. Only moving objects, which are, changes their position scan to scan from mapping operation are displayed. This type of MTI is known as **Area MTI**.
  - The delay line must introduce a delay equal to the pulse – repetition interval.

UNSOLVED PROBLEMS

1. A CW radar is operating at a PRF of 10 KHz and be operating frequency is 2 GHz, find the lowest blind speed of the MTI radar.
2. Two MTI radar system are operating at the same PRF but have different operating frequencies. If the third blind speed of one is equal to sixth blind speed of the other, find the ratio of their operating frequencies.
3. An MTI radar in operating at 1 GHz will a PRF of 1 KHz. Calculate the first three blind speeds.
4. Calculate the two lowest blind speeds for a MTI radar operating at λ = 10 cm and PRF = 1000 Hz. **[M.D. University, 2001]**
5. If a MTI radar is used by a traffic control police to measure the speed of car. If the Doppler frequency shift measurement from the car in 1000 Hz. Calculate the speed of car. The PRF of radar in 1.5 KHz and operating at frequency 10 GHz.
6. A pulse Doppler radar has a carrier frequency of 10 GHz and PRF of 5000 Hz.
   (*a*) What are its blind doppler frequencies ?
   (*b*) What radial target velocities would be undetected by the radar ?
   (*c*) What modification can be made to the radar that would eliminate blind speeds below 2000 Knots (use PRF). (**Hint**. 2000 Knots = 10^{16} m/s)
   (*d*) What would the maximum unambiguous range of the modified radar be ?

REVIEW QUESTIONS

1. What do you understand by term MTI ?
2. Describe the working of Non-coherent MTI radar with the of block diagram. **[M.D. University, 2004]**

3. What are the advantages of digital MTI over analog MTI ? **[M.D. University, 2004]**
4. Explain MTI radar functional block diagram using Power Amplifier.
5. Discuss the limitations of MTI radar. **[M.D. University, 2003]**
6. What is blind speed and how you can avoid it ?
7. What is the necessity of delay line canceller ? Deseribe various types of delay-lines used in MTI radar.
8. Differentiate between MTI and Pulse Doppler radar. **[M.D. University, 2001]**
9. Explain the working of coherent AMTI radar with the help of block diagram.
10. Describe coherent MTI radar with neat diagram. **[UPTU, 2003]**
11. Explain the MTI radar block diagram. **[M.D. University, 2003, 2006]**
12. What is subclutter visibility ? **[PTU, 2004]**
13. Explain the principle and working of MTI radar. **[PTU, 2004]**
14. Give a brief description of MTI radar. **[UPTU 2004]**
15. Draw a block diagram and explain the operations of an MTI Radar. What are delay lines in such radar ? **[M.D. University, 2005]**
16. What is Non-coherent MTI ? **[M.D. University, 2005]**
17. Explain the use of delay line cancellers. Give its filter characteristics. **[UPTU 2006]**
18. Explain MTI radar and give its limitations. **[UPTU, 2006]**
19. Discuss the limitations to MTI performance. **[M.D. University, 2006]**
20. Discuss the pulse Doppler radar. **[M.D. University, 2006]**
21. Write short notes on :
    (*a*) Staggered PRF
    (*b*) Delay lines and cancellers **[M.D. University, 2001, 2004]**
    (*c*) Doppler range gate filters
    (*d*) Digital MTI signal processing
    (*e*) Blind phase
    (*f*) Blind speed **[M.D. University, 2001]**

4

CHAPTER

RADAR TRANSMITTER AND RECEIVER

INSIDE THIS CHAPTER

4.1. INTRODUCTION

Radar transmitter is used to generate the adequate RF power in any radar system. As per the radar range equality, as we have discussed earlier in chapter-2, the transmitting power varies as the fourth root of the range. It means if we wants to double the range the power has be to increased by 16 times. The selection of the radar transmitter should be best suited according to the radar application because it does not only represents a large part of the initial cost of a radar system but it requires a operating cost such as fuel or need of prime power.

In this chapter student will study about the some sub assemblies of the radar-system such as radar receivers, display unit, duplexer and receiver protectors. In the radar receiver, firstly, studying about the different types of receivers used in the radar and than the brief knowledge of receiver noise. Some parts of receiver are also discussed here such as mixer, low noise front ends and receiver protectors. At the last working of duplexer is being discussed in this chapter.

There are some qualities of a radar system in which radar engineer is more interested. The Tx should provide sufficient power to detect a target, be easily modulated to produce the desired waveform, generate a stable signal so that doppler signal processing can be performed easily, provide the proper signal band width, be of high efficiency, be of high relativity, be easy to maintain, have long life, be able to operate with a minimum of personal, be of a size and weight suitable for the intended applications, and be of affordable price. It is not possible to have all of the above properties together in one transmitter, so compromise is necessary.

The radar performance is depending upon the average power of the Tx rather than peak power. Although there have been radar with average power greater than a million watts, power cannot be increased without limit since high power transmitter are heavy, required more space, and can consume much prime power or fuel. An indicator of the performance of radar is the product of the antenna area lines and the average power. Thus minimizing total system cost requires a reasonable balance between the costs of these two sub systems (Tx and antenna).

Radar Txs are based on either a power amplifier or power oscillator such as klystron amplifier and magnetron oscillator respectively. In the earlier day of microwave radar in the 1940 and 1950's, the magnetron oscillator was made because no other microwave tube available at that time. Magnetron had many serious limitations such as magnetron are noisy device that make difficult to use in MT1; they cannot produce the large average power however they were capable of producing large peak power. The disadvantages of the magnetron are overcome by the use of klystron amplifier TWT as the transmitter

A transmitter, however, is more than just the active RF source. It includes the exciter and drives amplifier that provides the signal to be amplified if the power amplifiers are used as source. If the transmitter generates a pulse waveform, a pulse modulator of some type is needed. There must be a D.C. Power supply for generating the necessary voltage and currents to operate the RF power devices, the modulator, cooling for the tube, heat exchange for the liquid cooling system, protection devices (crow bar) for arc discharge, safety interlocks, monitoring devices insulators and X-ray shielding.

The efficiency quoted for most tubes is the **RF conversion efficiency** defined as "the ratio of RF power o/p available from the tube to the D.C. power input of the electron stream." The system engineer, however, is more concerned with the overall transmitter efficiency, which is the ratio if the RF power available from the TX to the total power needs to operate the tube. If, for example the RF efficiency of a microwave tube were 40 to 50 % the transmitter efficiency might be 20 to 25%.

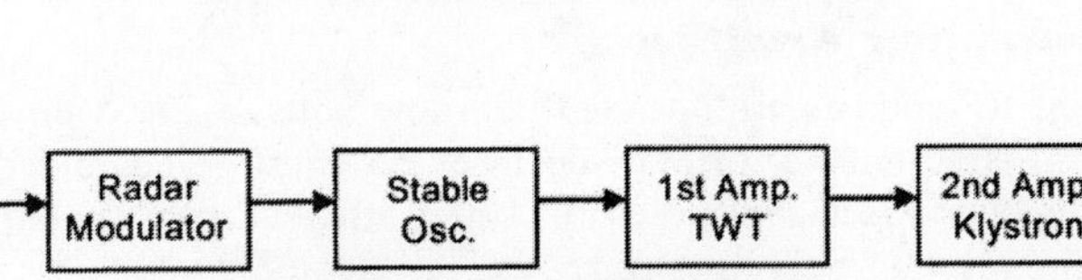

Fig. 4.1. Self Excited Oscillator Tx.

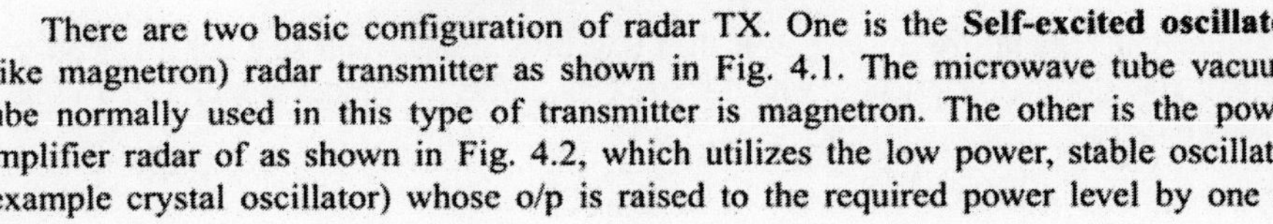

Fig. 4.2. Power Amplifier Tx.

There are two basic configuration of radar TX. One is the **Self-excited oscillator** (like magnetron) radar transmitter as shown in Fig. 4.1. The microwave tube vacuum tube normally used in this type of transmitter is magnetron. The other is the power amplifier radar of as shown in Fig. 4.2, which utilizes the low power, stable oscillator (example crystal oscillator) whose o/p is raised to the required power level by one or more amplifier stages. The tubes used in amplifier stages are multicavity klystron, Traveling wave tube (TWT). The choice between the two types of TX is governed by the particular radar application. The self-excited power oscillators transmitters are usually are smaller in (physical) size than the power amplifier radar transmitter. The various power amplifier transmitters are generally capable of higher power than the self-excited power oscillator transmitter. The power amplifier transmitter has better frequency stability, which is of importance for MT1 and other doppler radar. Self excited power oscillator transmitter, therefore find applications in radars where high frequency stability are not much needed but small size and portability are important (airborne radar). The power amplifier radar transmitter are usually used in radars where high power and frequency stability are important e.g.; long range surveillance radar MT1 and other doppler radar.

4.2. RADAR RF SOURCES

There are different RF power sources available for high power radar application include the klystron, TWT, solid state transistor amplifier, Twystron, magnetron, cross-field amplifier, amplitron, grid controlled vacuum tube, extended interaction amplifier, gyrotron, and others. None provide all the desirable features that might be wanted. Each of there devices are will be briefly summarized below. All are power amplifier except the magnetron. The gyrotron can be either an amplifier or an oscillator.

Klystron

Klystron has high gain and good efficient, and is capable of higher overage and peak power than most other RF power sources. It can have a wide bandwidth when its power is large enough, long life, low interpulse noise, and good stability for doppler processing. It requires very high voltages are X-ray shielding when it has to generate high peak power.

Traveling Wave Tube (TWT)

In comparison to the Klystron, TWTs have slightly less power, slightly less gain, and slightly less efficient but they are capable of wide band width, especially at low-power levels.

Hybrid Klystron

These are similar to klystron, but with one or more of the resonant cavities replaced by multiple cavities similar to what are used in a TWT. They are Twystron, the extended interaction Klystron and the clustered cavity Klystron. Band width can be 15 to 20 percent more.

Solid State Transmitter Amplifier

These types of RF sources are operated with low voltage. They are inherently of low power so that a large number of individual devices must be combined to generate sufficient power for most radar application for better efficient, they should be operated long pulses and employ pulse compression.

Magnetron

The magnetrons utilize lower power than Klystron. It has small size but it has poor noise and stability characteristics, which restrict its ability to cancel clutter when used in an MTI radar.

Cross-field Amplifier

These are capable of high power, good, efficient, and wide band width, but are of relatively low gain (about 10 dB). Cross-field amplifier are generally noises and less stable than other RF power sources.

Gyrotron

They can produce very high power in the millimeter wave region, but they requires large magnetron field. Its use in radar is limited.

4.2.1. Classification of RF Power Sources

The conventional vacuum tubes, such as triodes, tetrodes and pentodes are still used as signal sources of low power output at low microwave frequencies.

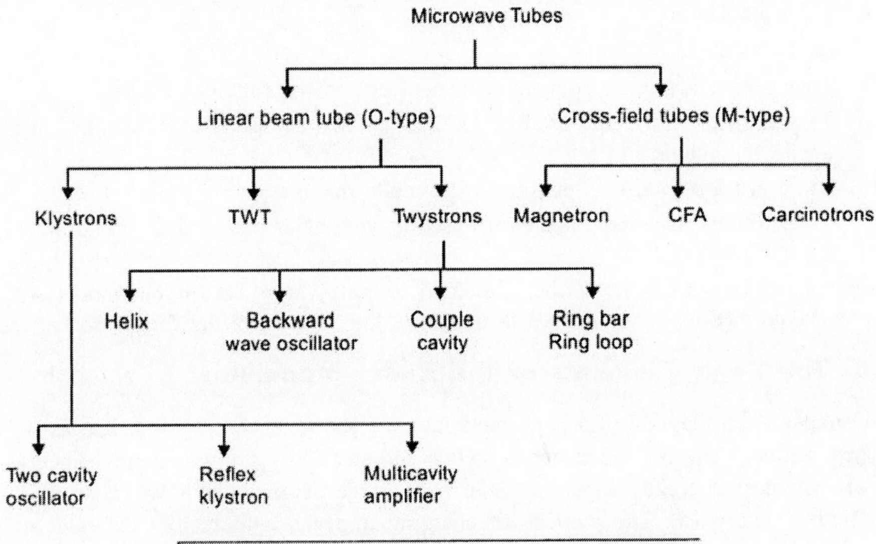

Fig. 4.3. Microwave Vacuum-tube Family Tree.

Microwave tubes are generally categorised in two different groups as shown in Fig. 4.3, one is linear beam tubes and other is cross field tubes. Klystrons, TWT and Twystrons are linear beam or '0' type tubes in which the accelerating electric field is in the same direction as the static magnetic field used to focus the beam of electrons. Magnetron, cross field Amplifier and carinotrons are crossed field type tubes or M-Type, where the static magnetic field is perpendicular to the electric field. The basic principle of operation of the microwave tubes is the transfer of power from a dc voltage source to a source of ac voltage by means of a current density modulated electron beam. This can be achieved by accelerating electron in a static electric field and retarding them in an ac field. The density modulation of the electrons beam facilities more electrons to be retailed by ac field and than accelerated by ac field which therefore make's possible a net energy to passed to the ac electric field.

In this chapter the operating principles of some mostly used microwave tubes such as Klystrons, Reflex Klystrons, Travelling wave tubes (TWT), Magnetrons and Backward oscillator etc.

4.3. RADAR MODULATOR

4.3.1. Introduction

A modulator is the heart in the transmitter .It not only controls the radio frequency energy to be produced in sharp pulses but also controls the repetition frequency and shape of the pulse. The **modulator** is the device which turns the transmitting tube on and off in such a manner as to generate the desired waveform. When the transmitted waveform is a pulse, the modulator is some time called pulser. Each RF power tube has its own particular characteristics, which determine the type of modulator.

4.3.2. Types of Modulators

Pulse modulators are divided into two categories and they are

1. Grid Modulators
2. Anode Modulators

Grid modulators are not employed at present due to the following reasons :

(*a*) A grid modulator requires high H.T. to be maintained longer time than an anode modulator.

(*b*) More precise control is possible with anode modulation.

(*c*) Transmitter valves like magnetrons used at centimeter wavelengths do not have control grid.

Anode modulators can be further classified as **Soft Valve Modulators** and Hard Valve Modulators. Both are discussed in details in this chapter in the further sections.

4.3.3. The Basic Elements of the Radar Modulator

A simple block diagram of a radar modulators is shown in the Fig. 4.4. It consists of energy source, charging impedance, a switch and an energy storage circuits. Energy which is supplied externally accumulates in the storage devices at a slow rate during the period of interpulse. The purpose of charging impedance to restrict the rates at which energy can be delivered to the storage devices. The switch is closed when the pulse is formed and the accumulated energy rapidly discharged through the load. The load may be a magnetron valve. The charging impedance prevents the energy to dissipated from the storage elements to the source, during the discharge period.

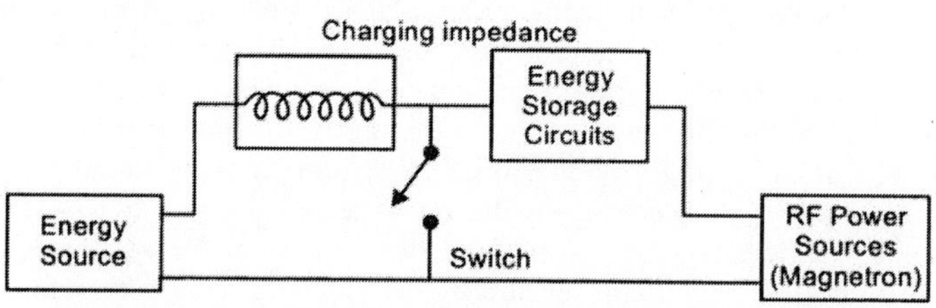

Fig. 4.4. A Simpler Block Diagram of Radar Modulator.

Three basic types of radar modulator used for magnetron oscillator

1. The line–type modulator with a gas tube switch and a delay line storage element.

2. The pulsactor, which uses saturable reactances for both switching and storage purposes and
3. The hard tube modulator incorporating a vacuum tube switch and capacitor storage elements.

4.3.3.1. Line-Type Modulator

The modulator containing a gas tube switch and a delay line as the energy –storage element is called a line type modulator as shown in Fig. 4.5. It is commonly employed with high power magnetron. The delay line of the line type modulator is called the pulse-forming network (PFN). Radio frequency energy in radar is transmitted in short pulses with time durations that may vary from 1 to 50 microseconds or more. A special modulator is needed to produce this impulse of high voltage. The hydrogen thyratron modulator is the most common radar modulator As circuit for storing energy the thyratron modulator uses essentially a short section of artificial transmission line which is known as the pulse-forming network (PFN). Via the charging path this PFN is charged on the double voltage of the high voltage power supply with help of the magnetic field of the charging coil. Simultaneously this charging coil limits the charging current. The charging diode prevents that the PFN discharge its own about the intrinsic resistance of the power supply again.

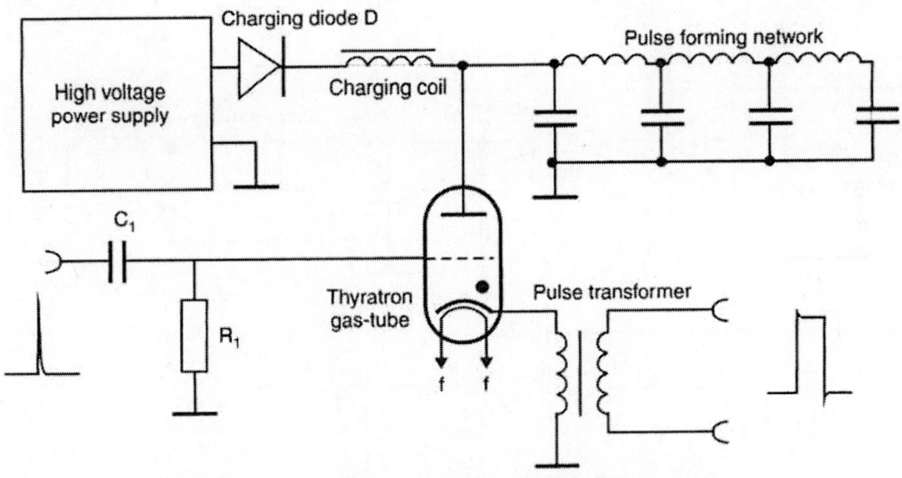

Fig. 4.5. Schematics Diagram of Radar Modulator.

The function of thyratron is to act as an electronic switch, which requires a positive trigger of only 150 volts. The thyratron requires a sharp leading edge for a trigger pulse and depends on a sudden drop in anode voltage (controlled by the pulse-forming network) to terminate the pulse and cut-off the tube. The **R-C Combination** is acting as a DC-shield and protects the grid of the thyratron. This trigger pulse initiates the ionization of the complete thyratron by the charging voltage. This ionization allows conduction from the charged pulse-forming network through pulse transformer. The output pulse is then applied to an oscillating device, such as a magnetron.

Fig. 4.6. Thyratron Modulator of the Russian P-18 Radar.

The Charge Path. The thyratron is an open circuit between the trigger pulses. Therefore in the Fig. 4.7, it is shown as an open switch.

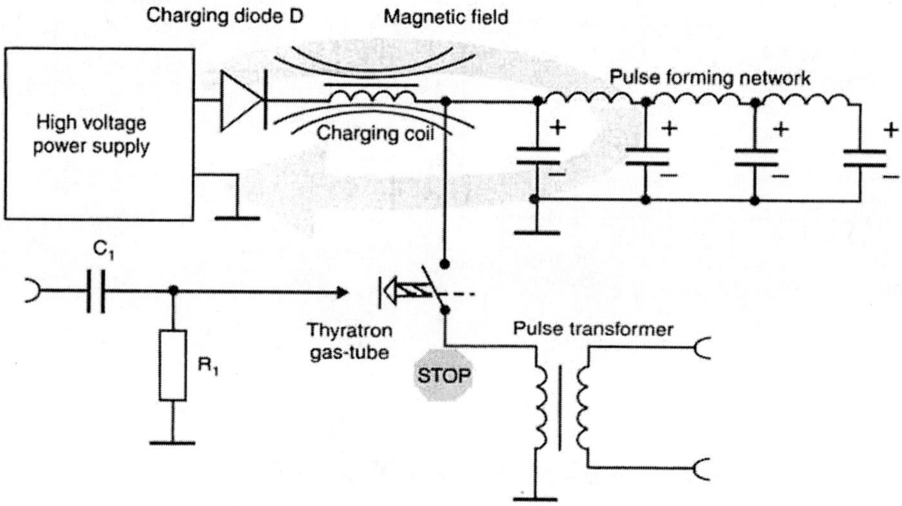

Fig. 4.7. Charging Path of PFN.

After the power supply switching on, the current flows through the charging diode and the charging coil and charges the condensers of the pulse-forming network (PFN). However, the induction of the charging coil offers a great inductive resistance to the current and builds up a strong magnetic field. The charging of the condensers follows an exponential function. The self- induction of the charging coil overlaps for this.

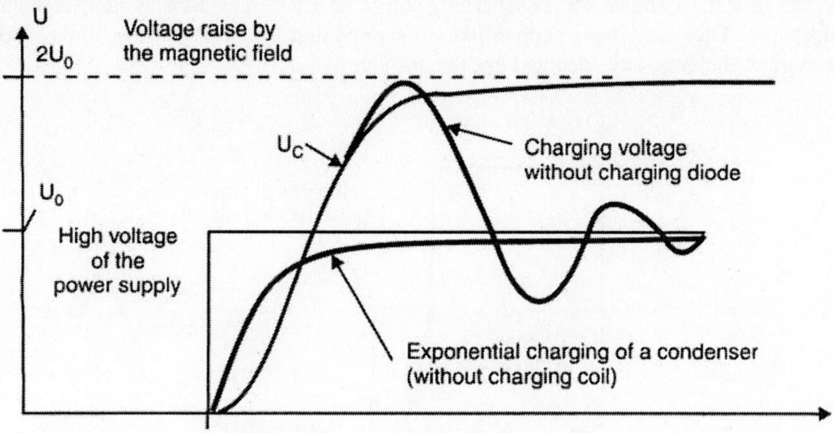

Fig. 4.8. Charging Curve of PFN.

Figure 4.8, shows the Charging curve of PFN. If the condensers are charged with the power supplies voltage, decreases the current and the magnetic field breaks down. The breaking down magnetic field causes an additional induction of a voltage. This one continues the charging of the condensers up to the double voltage of the power supply. Now the condensers would discharged about the power supplies resistance, but the charging diode cut-off this current direction and the energy remains stored therefore in the condensers

The Discharging Path. The Discharging Path of modulator shown in Fig. 4.9. When a trigger pulse is applied to the grid of the thyratron, the tube ionizes causing the pulse-forming network to discharge through the thyratron and the primary of the pulse transformer.

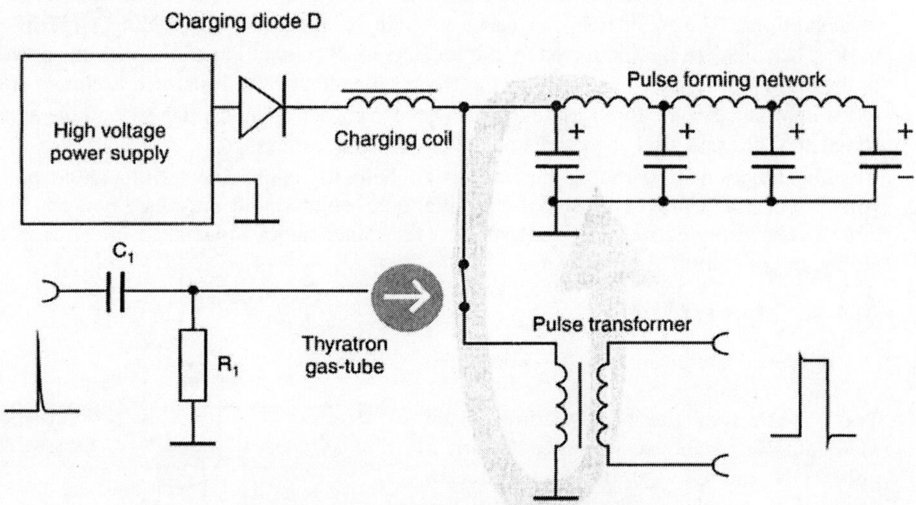

Fig. 4.9. The Discharging Path of PFN.

Figure 4.10 shows the Discharging curve of PFN. The first condenser begins the discharge. This one would happen like an exponential function, but the other condensers recharges the first one, delayed by the PFN-coils.

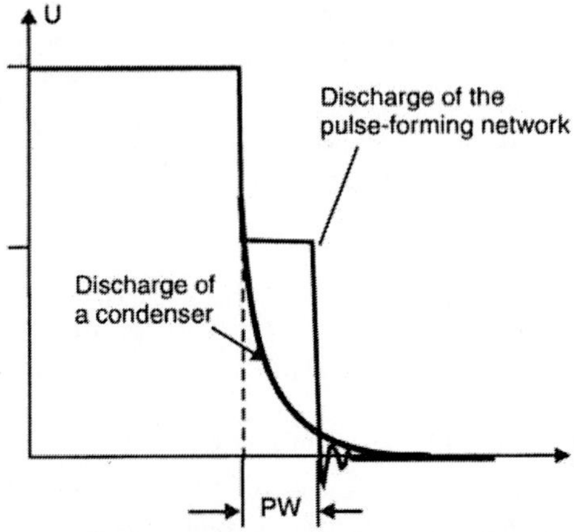

Fig. 4.10. Discharging Curve of PFN.

For the duration PW a current flows through the pulse transformer therefore. The high voltage pulse for the transmitting tube can be taken on the secondary coil of the pulse transformer. Exactly for this time an oscillating device swings on the transmit frequency. Because of the inductive properties of the PFN the positive discharge voltage has a tendency to swing negative. If the oscillator and pulse transformer circuit impedance is properly matched to the line impedance, the voltage pulse that appears across the transformer primary equals one half the voltage to which the line was initially charged.

The magnetron is non-linear impedance and will not be matched to the line under all conditions. The mismatch can cause a spike to appear at the leading edge of the pulse. The spike can be minimized by introducing an RC circuit in parallel with the primary of the pulse transformer. This is called the De-spiking circuit. The resistance is chosen equal to the impedance of the PFN, and the capacitance is chosen small enough so as to be almost completely charged after the oscillator draws full-load current.

The function of damping network is to help to reduce the trailing edge of the voltage pulse and prevent post pulse oscillations, which could introduce noise or false targets. The damping network consists of a resistance and a inductance in series at the secondary of the pulse transformer.

4.3.4. Thyratron

A typical thyratron is a gas-filled tube for radar modulators is shown in Fig. 4.11 The function of the high-vacuum tube modulator is to act as a switch to turn a pulse ON and OFF at the transmitter in response to a control signal. The grid has complete control over the initiation of cathode emission for a wide range of voltages. The

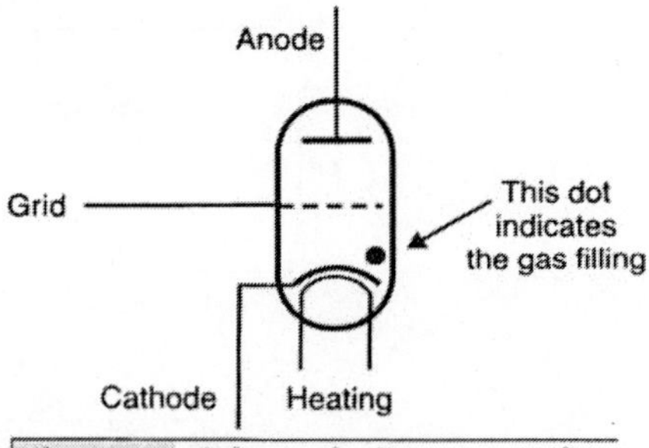

Fig. 4.11. Schematic Representation of Thyratron.

anode is completely shielded from the cathode by the grid. Thus, effective grid action results in very smooth firing over a wide range of anode voltages and repetition frequencies. Unlike most other thyratrons, the positive grid-control characteristic ensures stable operation. In addition, using the hydrogen-filled tube reduces deionization time. A trigger pulse ionizes the gas between the anode and the cathode. Only by removing the plate potential or reducing it to the point where the electrons do not have enough energy to produce ionization will tube conduction and the production of positive ions stop. Only after the production of positive ions is stopped will the grid be able to regain control. Because of the very high anode voltage the anode is attached most on the upper end of the glass bulb. Therefore the tube looks very ancient. By the ionized gas it shines in the ionizated condition like a glow lamp.

Fig. 4.12. A Photograph of Thyratron.

4.3.5. Saturables-Reactor Modulator

The saturable reactor which is an iron core inductance and it is so designed that its magnetic core is driven into saturations for normal values of current in the coil. When the current flowing through the coil is small and the core in unsaturated than the incremental inductance is high. And when the inductances is low for high currents when the core is saturated. The saturable reactor has relatively long life than other radar modulators. It is not using either any mechanical part or any electronic tubes. It has one main draw back that it cannot control the pulse shape results in poor and uncontrolled pulse shape.

A two-stage saturable-reactor modulator or a pulsactor is shown as **Fig. 4.13**.

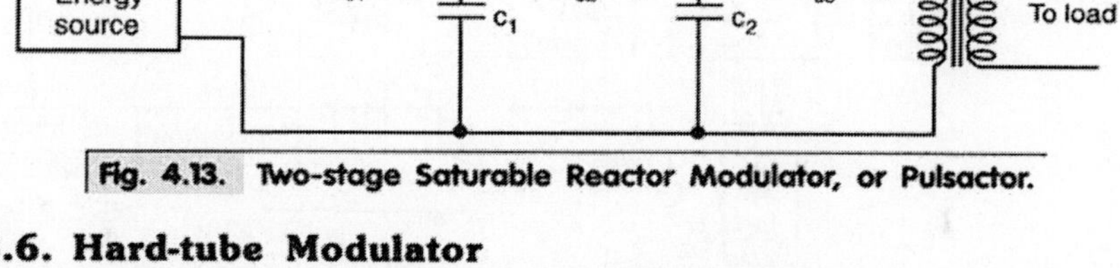

Fig. 4.13. Two-stage Saturable Reactor Modulator, or Pulsactor.

4.3.6. Hard-tube Modulator

In the hard valve modulator the switching is accomplished with hard vacuum tubes instead of soft gas tubes. Basically it is a high power video pulse generator. A block diagram of a particular hard tube modulator is shown as **Fig. 4.14**. The trigger pulses initiate the driver at the proper instants of time. The driver generates the desired pulse wave shape. The power amplifier is being used to amplify the waveform generated by the driver, to the level requires pulsing the transmitter. There are many numbers of pulse-forming circuits, which could be for the driver. The video amplifiers using high power tubes, are being used as power amplifier. A single high power tube working as a blocking oscillator can be used as pulse modulator. A blocking oscillator is a self-excited, overdriven oscillator.

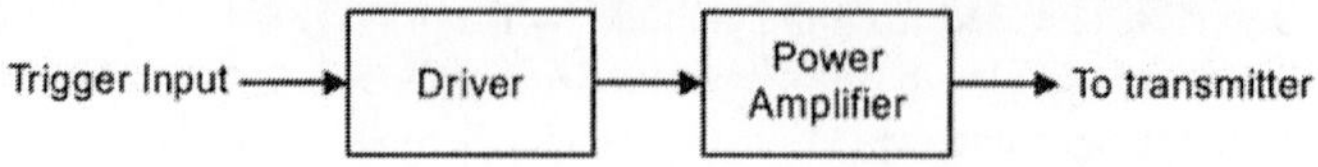

Fig. 4.14. Block Diagram of a Hard Tube Modulator.

A major limitation of the hard tube modulator in the past has been the lack of vacuum tubes capable of handling the large power required for pulsing big radar transmitter.

4.3.7. Comparison of Magnetron Modulators

1. The Line-type modulator is simple and efficient. Due to its small size and light weight, it is desirable in airborne radar. The problem with this modulator is changing of pulse duration from pulse to pulse, requires the switching another pulse forming network.
2. The saturable reactor has the advantage of no active elements, it has long life, it is more difficult to control the shape of the pulse in this modulator.
3. The hard tube modulator can change pulse duration, pulse shape or PRF with little efforts. It may also be used to generate group of pulses. it is less efficient than the others but provides the greatest flexibility in operation.

4.4. A PRACTICAL RADAR TRANSMITTER

A practical radar transmitter block diagram is shown in the Fig. 4.15. The Fig. 4.16 shows the block diagram of practical modulator. The working of the transmitter is explained briefly in the following paragraph.

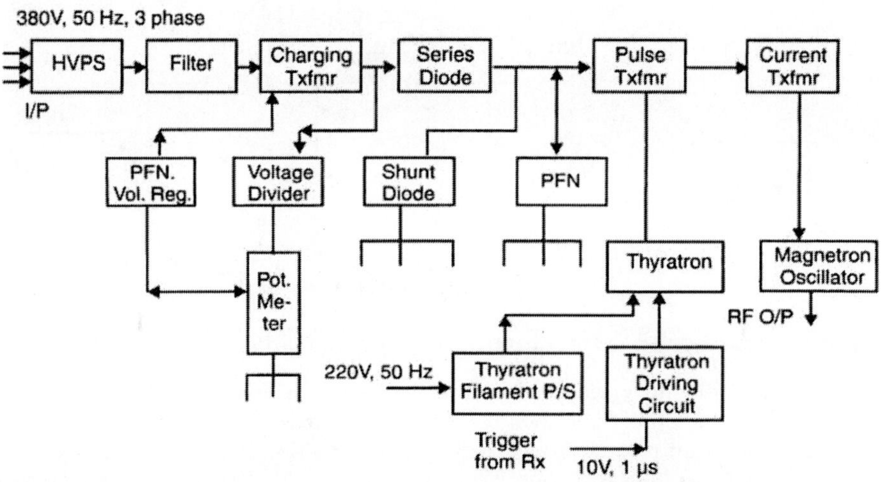

Fig. 4.15. A Practical Radar Transmitter Block Diagram.

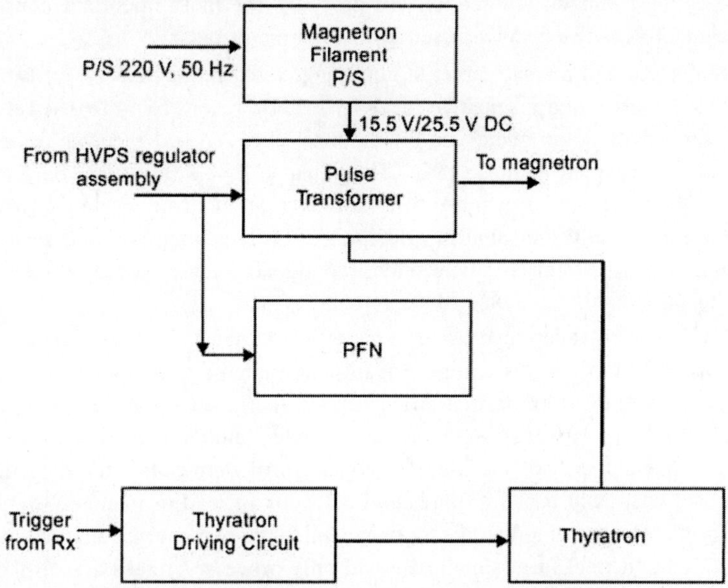

Fig. 4.16. Block Diagram of Modulator Assembly.

The working of the transmitter. 380 V 3 phase 50 hz fed as input to the High voltage power supply (hvps), and its amplified to the level of ±6 kv which is further fed to the charging transformer through the filter network. Charging transformer double up the voltages from +6 kv to +12 kv and its primary acts as a charging choke. A series diode, which is connected between charging and pulse transformer, is an anti-flow diode. It protects the discharge of PFN networks through the source.

The PFN voltage regulator adjust the power output of the charging transformer, Its purpose is to maintain the EHT constant Shunt diode is also a protection circuit to damp out the –ve swing in EHT. It is connected in the reverse connection to the EHT. PFN is a bank of capacitors and it stores the charge up to required level. Whenever thyratron valve get the trigger from the radar modulator through the thyratron driving circuit, the thyratron will act as a short switch and the PFN is discharging through the primary of the pulse transformer. The pulse transformer is a step up transformer and ratio of this is 1:4.5, its raises the input voltages from +6 kv to the ± 28 kv. This 28 kv finally given to the cathode of the magnetron oscillators, the –28 kv is given to the magnetron through the current transformer, which is to sense a sample of transmitter energy approximately 5-10 V and gives to the trigger system in the receiver. The magnetron will generate the high RF pulses for the 1 μs (Pulse Width) and that propagating in to space through RF plumbing with help of the antenna.

4.5. RADAR RECEIVERS

The ability of a radar receiver to detect the presence of the echo and extract information in the presence of noise. There are noises which can enter the receiver through the antenna along with the desired signal are called **external noise** or it may be generated within the receiver itself is called the **internal noise**. The noise can never be completely eliminated, it must be minimized it optimum radar performance is to be obtained. Noise and its elimination are probably the most important consideration in the design of sensitive receiver used with long-range radar.

The detection of a small target at long-range radar requires large Transmitter power, large Antenna area, and a sensitive receiver. A good receiver can be used to reduce the transmitter power requirements or the antenna size. A good receiver design based as maximizing the signal to noise ratio. The signal to noise ratio may be maximized by designing the receiver as a matched filter or as a cross corelator. There are two techniques (which are essentially the same in principle) of receiver improved the desired signal in an optimum manner and eliminate undesired signals such as noise. A good receiver is design to have minimum noise of its own.

In the modern radar receiver by employing a matched filter, maximize the peak-signal-to-noise ratio and discriminate against unwanted signal. When the clutter echoes are large enough to mask desired target echoes, the receiver has to in cooperate means for separating the moving targets from stationary clutter echoes by recognizing the doppler frequency shift of the moving targets. In modern radars the detection decision is made automatically based on threshold detection instead of manually by an operator. The information regarding the targets range and angle may extract automatically. Targets are acquired and tracked automatically and only processed tracks are displayed or sent to some automatic devices such as air traffic control system or weapon control system or computer for future action. Receiver should be designed to have sufficient gain, amplification, stability, dynamic range, proper tuning capability, ruggedness and simplicity. Protection must be provided against overloads or saturation and burnout from nearly interfering transmitter. Timing and reference signals are needed to properly extract target information. If the receiver is of the superheterodyne type than the total gain of a receiver is divided in the IF and video stages. The typical power gain from the input of the output of the display device might vary from 150 to 200 dB.

A receiver should have good dynamic range to be able to detect the weakest signal and yet not saturate with largest expected echo signal.

The receiver bandwidth should be enough to accommodate the transmitted frequency spectrum and the doppler frequency. Radar transmitter such as magnetron tends to drift in frequency. To over come this problem of drifting in frequency, a technique called **Automatic frequency control (AFC)** is being employed to control the magnetron frequency.

4.5.1. Types of Radar Receiver

There are different types of radar receiver, which are employed in radar as per the radar application. Types of radar receiver are as follows :

1. Super heterodyne receiver
2. Super regeneration receiver
3. Crystal video receiver
4. Tuned radio frequency (TRF)

Although each of there has particular areas of application for which it is pretend than other, the **super heterodyne receiver** has been widely used in radars. Because of its sensitivity, high gain, relativity, it is being pretend most.

The **super regeneration receiver** are used in radar becomes applications because its compactness, a single tube may be used for receiving & transmitting.

The **crystal video receiver has** poor sensitivity but it is simple than others. The TRF receiver is similar in many respects to the crystal video receiver but has an RF amplifier of sufficient gain to reduce noise produced by the detector. There is no IF stages in TRF receiver.

4.5.2. Super Heterodyne Receiver

A block diagram of simple super heterodyne receiver is shown in Fig. 4.17. The working of this receiver is explained in brief here.

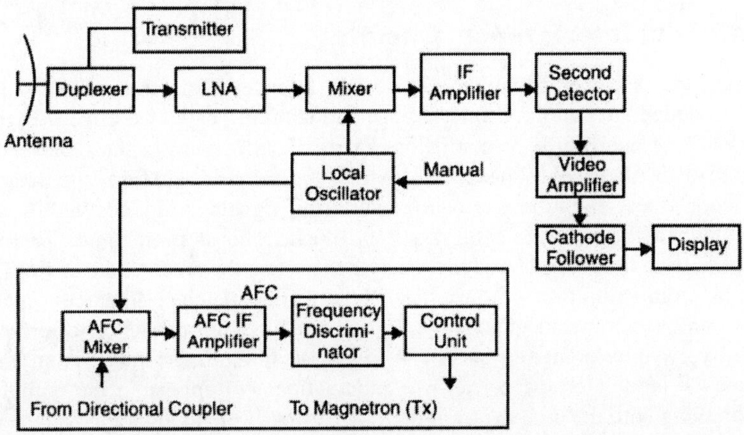

Fig. 4.17. Superhetrodyne Receiver with Automatic Frequency Control (AFC).

The reflected signals which are picked by the antenna and passes through the duplexer. The signals are being feed to the first stage of receiver, that is, low noise RF amplifier for amplification because the received signal strength is very- very low. The noise generated by the input stage of the receiver determines the receiver sensitivity, provided external noise in negligible. The function of mixer stage or first detector is to converts the RF frequency to a lower Intermediate frequency (IF), usually it is 30 or 60 MHz. This is accomplished by heterodyning (mixing) the RF signal with a local oscillator frequency in a non-linear element (mixer) and extracting the signal components at the difference frequency (IF). The Intermediate Frequency is easier to amplified by two or more stages of IF amp that whose overall band width is sufficient to pass all the frequencies in the pulse so that the pulse is not a appreciably distorted. The IF output from the last stage is fed to the second detector, which is either a vacuum tube or crystal diode, extract the video modulation from the carrier. The video signal are being amplified by means of wide band RC coupled voltage amplifier called the video amplifier. This amplifier amplifies the pulse with very little distortion. The Cathode follower is the final stage from which the pulses are fed to indicators. The cathode follower is used for low impedance output. The pulses are not distorted in shape due to this low impedance and making also becomes easy. Finally videos are displayed over the display unit.

4.5.3. Automatic Frequency Control

The AFC system normally employed to keep the receiver in tune with the transmitter is known as a difference frequency system. A sample of transmitted signal is coupled to the AFC mixer and heterodyned with the local oscillator frequency. If there is no drift in the frequency of transmitter and receiver than difference frequency will be same as IF. If the receiver is not in tune with the transmitter than the difference frequency is differ from the IF. So any deviation from the IF is detected by AFC discriminates, which will produce an error voltage whose magnitude is proportional to the deviation from the correct IF and sign determines the direction of the error.

The error signals are further fed to the AFC motor to rotate the plunger of the Magnetron in the respective direction to maintain the transmitter frequency constant.

4.5.4. Anti Interference Circuits

The radar receiver should minimize interference from other radiating equipments and from deliberate jamming. Interference and jamming have two effects on the receiver. First is that of annoyance or confusion. Extra signal present to the receiver make it all the more difficult for the operator to extract the information from the desired signals. The second effect due to power of the interfering signals is large enough to saturate the receiver when the receiver saturates it is blocked for a finite time. There are some techniques used in super heterodyne receiver to protect the receiver either from extraneous signal or from saturation of the receiver. By using matched filter concept, undesired signals can be eliminated effectively. The use of blocking capacitor to remove d-c bias caused by cw interference as shown in Fig. 4.18. If the interference is an amplitude or frequency – modulated cw carrier, the modulation will appear in the output of second detector along with the d-c bias from the CW carrier. If the modulation is to be eliminated, a high pass filter inserted between the second detector and the first stage of video amplifier must replace the simple blocking capacitor. A high pass filter was used for this purpose is called **Fast time constant circuit (FTC).** In most radar the technique employed is called **Instantaneous automatic gain center (IAGC)** to protect the radar receiver from over loading. IAGC is based on negative controlling the gain of the IF amp. The logarithmic amplifier whose response O/p is proportional to the logarithmic of the input.

The like hood of receiver saturation can be reduced further by programming the receiver gain as a function of time so that it is low for targets at close range and high for targets at long – range. This is called **sensitivity time control (STC).** At close ranges the effect of the inverse – fourth power variation of echo signal Vs range results in large target echoes and large clutter echoes.

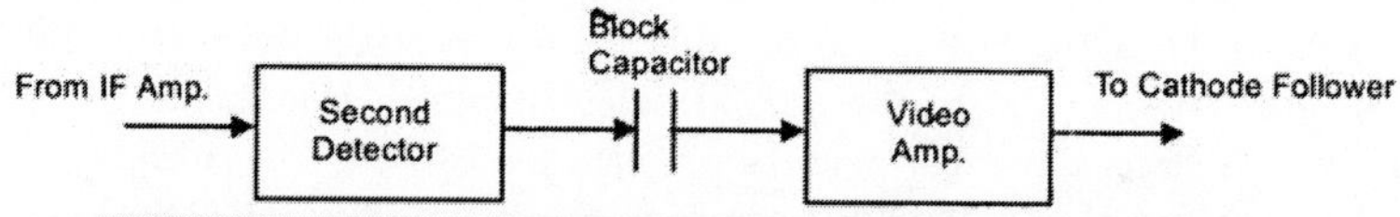

Fig. 4.18. Removal of DC Bias by using Blocking Oscillator.

There are anti interference techniques usually depend upon the operator's judgment as to whether to use or not. In some radar the o/p from the radar receiver is fed directly in to some form of data processor such as digital computer.

4.6. RECEIVER NOISE

4.6.1. Noise

The signal to noise ratio is a very important factor in a radar receiver since it determines the maximum detection range of a radar system. A weak signal from a distant target is received and displayed on an 'A' scope along with noise if the noise amplitude is more than the signal amplitude can be reduced, the signal merges in the noise and the target remain undetected. If the noise amplitude can be reduced considerably, the same weak signal is detectable. Thus, by increasing the signal to noise ratio the maximum detection range of radar set can be increased.

If there is no noise present in the receiver, it would be possible to detect any weak signal, by providing the receiver with sufficient amplification. Practically it cannot be achieved since noise is always present, and unlimited amplification of weak signals amplifies noise as well. Noise is unwanted EM energy, which interfere with the ability of receiver to detect the wanted signal. It may originate within the receiver itself, or it may enter via the receiving antenna along with the desired signal. There are various types of noise generated in side the receiver, which are as follows

1. Thermal noise
2. Shot noise
3. Partition noise
4. Induced grid noise
5. Crystal noise
6. Flicker noise
7. Transistor noise.

4.6.2. Noise Figure

An ideal receiver adds no noise of its own to the signal being amplified. All practical receiver, however generate noise to some extent. A noise figure or noise factor can be explained as a measure of the noise produced by a practical receiver compared with the noise of an ideal receiver. Thus, the noise figure Fn of a linear system may be defined as

$$F_n = S_{in}/N_{in} \times S_{out}/N_{out} \quad ...(4.1)$$

Or

$$F_n = S_{in} \times N_{out}/N_{in} \times S_{out}$$

We know,

$$G = S_{out}/S_{in} \quad ...(4.2)$$

And N_{in} is function of the standard temperature (T) and the band width (B_n), which is directly proportional to T and B_n

$$N_{in} \propto T\,B_n$$

$$N_{in} = K\,T\,B_n \quad ...(4.3)$$

By substituting the value of N_{in} & G. It can be written as

$$\boxed{F_n = N_{out}/K\,T\,B_n G} \quad ...(4.4)$$

Where, F_n = Noise figure
S_{in} = Available input signal power
N_{in} = Available input noise power ($KT\,B_n$)
S_{out} = Available output signal power
N_{out} = Available output noise power

K = Boltzmen constant = 1.38×10^{-23} J/° K
T = Standard temperature (in Kelvin = 290° k)
B_n = Band width
G = Gain

If we express the above equation, it may be considered as the degradation of signal to ratio by the receiver, or it may be interpreted as the ratio of the actual available output noise power (N_{out}) to noise power, which would be available if the receiver merely amplified the thermal noise of the source. The noise figure is commonly expressed in decibels (dB), that is, 10 $\log_{10} F_n$. Some time, it may be called as noise factor.

The noise figure may be also written as

$$F_n = (KTB_nG + \Delta N)/KTB_nG \quad \text{...(4.5)}$$

Or

$$F_n = 1 + \Delta N/KTB_nG \quad \text{...(4.6)}$$

Where ΔN is additional noise added by the receiver itself.

In order to standardize the definition, the institute of Radio Engineers specifies that $T = 290°$ k or 63° F be employed for the calculation of noise figure. It may be convenient to write in the factor of K To is approximately 4×10^{-21}watt/cps.

4.6.3. Noise Figure of Networks in Cascade

If we consider two networks in cascade, with different noise figure and available gain but have the same noise Bandwidth Bn as shown in Fig. 4.19.

F_{n1} and F_{n2} are the noise figure of two networks,

G_1 and G_2 are the available gain of respective networks.

Fo the over all noise figure of the networks

No the output noise of the two circuit in cascade

From the definition of noise figure given by Eqn. (4.1) and (4.4), the output noise (No) of the two networks in cascade is

No = noise from network 1 at the output of network 2 + noise ΔN_2 introduced by network 2

$$No = FoKTB_nG_1G_2 = F_1KTB_nG_1G_2 + \Delta N_2$$

$$No = F_1KTB_nG_1G_2 + (F_2 - 1)KTB_nG_2$$

which result in

$$Fo = F_1 + (F_2 - 1)/G_1 \quad \text{...(4.7)}$$

So, it is necessary that gain of first network should be large enough and have a low noise figure. But second stage must also have a low noise figure. Too large a first stage gain is not always desirable since the dynamic range of the receiver is reduced by the gain G_1 of the low noise amplifier.

This is very important in the designing of multistage receiver. It is not sufficient that only the first stage of a low noise receiver have small noise figure. The succeeding stage must also have a small noise figure. If instead of using amplifiers as the first stage, the mixer (diode) is being used, than the G_1 should be interpreted less than units.

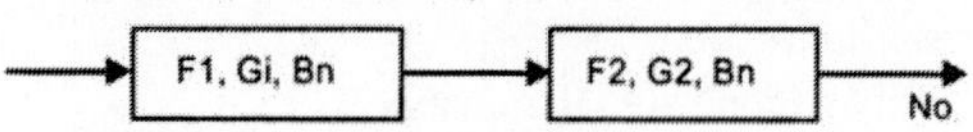

Fig. 4.19. Two Networks in Cascade.

The noise figure of '*N*' networks in cascade form may be represented by

$$Fo = F_1 + (F_2 - 1)/(G_1) + (F_3 - 1)/(G_1G_2) + \dots + F_{n-1}/G_1G_2 - G_{n-1} \quad \dots(4.8)$$

4.6.4. Measurement of Noise Figure

The noise figure of a radar receiver can degrade in operation and cause reduced capability. Therefore, some means for monitoring the noise figure should be provided in operating radars so that worsening of receiver sensitivity can be detected and concerned. The monitoring of the noise figure can be accomplished either automatically or manually by the operator

There are two basic methods of measuring noise figure. In one method of measuring noise figure is with a known source of broadband noise power such as an argon gas tube or solid-state noise source. The noise figure is determined by measuring

1. The noise power N_1 of the receive when a matched impedance at temperature $T = 290°$ K is connected to the receiver input.
2. The noise power output N_2 when a matched noise generator of temperature T_2 is connected to the receiver input. The temperature T_2 is the equivalent noise temperature of the broadband noise of generator. The noise figure can be shown to be

$$F_n = \{(T_2/T_{-1}) - Y(T_1/T_{-1})\}/(y - 1) \quad \dots(4.9)$$

Where $y = N_2/N_1$

T = standard reference temperature of 290°K

The measurement of noise figure can be made during radar operation without degrading the receiver sensitivity by pulse modulating the noise source in synchronize with the radar trigger and injecting the noise in to the receiver during the 'fly back' or dead time of the radar, just prior to the next transmitter pulse. The measurement of the receiver output with the noise source on (N_2) and the noise source of (N_1) can be made on alternate pulse periods.

The second method of measuring the receiver noise figure or sensitivity by using a calibrated signal generator. With a matched resistance at the receiver input, the output power is due to receiver noise alone. The signal generator power is then applied to the receiver input and adjusted until the signal plus noise power is equal to twice the receiver noise power read with the matched resistance. The input signal under this condition is sometimes said to be the minimum discernible signal (mds). It is also proportional to the receiver noise figure.

The sensitivity of a radar may be visually displayed by using the measurement of receiver noise to display the normal range rings on the PPI only within the radar can detect the targets reliably. This provides the operators with a continuous and immediate indication of radar sensitivity. While measuring the receiver noise figure, the noise source or signal generator is usually inserted by a direction coupler ahead of the duplexer and other RF components so that the overall noise figure of the system is measured rather than that of the receiver alone.

4.7. MIXERS

4.7.1. RF Amplifiers

For frequencies up to 700 Mc/s, RF amplifiers are used in the initial stages of receiver. Usually one or two stages of grounded grid amplifiers with special valves are used because the partition noise is minimum and the output and the input circuits are screened from each other by the grounded grid amplifiers. There is damping in the grid tuned circuit as a result of transit time effects at high frequencies, that is, when the travel time of electrons from the cathode to grid is comparable with the period of the input, more induced current flows to the grid. This effect is equivalent to a resistance across the input circuit. The input circuit is damped and its bandwidth is increased. The effective input resistance as a result of the transit time varies inversely as the square of the frequency, that is larger the frequency, less is the value of the resistance and greater is the damping due to the cathode lead inductance and Cgk being in series.

4.7.2. Introduction of Mixer

For frequencies beyond 700 mc/s, the RF amplifier is dispensed with due to the reason mentioned and the received signal is directly fed into the mixer, usually a crystal along with the local oscillator output. Diode mixers are not generally used at high frequencies because of noise consideration. There are two mixer circuits using crystals and they are co-axial lines and wave-guide types shown in Fig. 4.20(*a* & *b*) respectively. In the co-axial type of mixer, the crystal is the termination for the line and the reactance portion of the crystal is tuned out by means of a stub as shown in Fig. 4.20. The condenser C is a radio frequency bypass and is built in the line by having a di-electric washer.

In the wave-guide type of mixer, the crystal is placed parallel to the electric field and the dimensions of the wave-guide are such that the wave-guide impedance is equal to the crystal resistance. The reactive component of the crystal impedance is tuned out by means of a piston shown in the diagram. The condenser C is also of the 'built-in' type for radio frequency bypass.

As we know many radar superhetrodyne receivers do not employ a low noise RF amplifier. Instead, the first stage is simply the mixer. Although, the noise figure of a mixer front-end may not be as low as other devices that can be used as receiver front ends it is acceptable for many radar application when other factor besides low noise are important. The function of the mixer is to convert RF energy to IF energy with minimum loss and without spurious response. Silicon point contact and schottky barrier diodes based on the non-linear resistance characteristic of metal to semi conductor contact have been used as the mixing element. Schottky barrier diodes are made of either silicon or GaAs. GaAs preferred for the higher microwaves frequencies. The schottky barrier diodes have had lower noise figure and lower flicker noise than conventional point contact diodes, but the silicon point contact diode has had better burnout properties. An integral part of the mixer is local oscillator. The IF amplifier is also of importance in mixer design because of its influence of the overall noise figure.

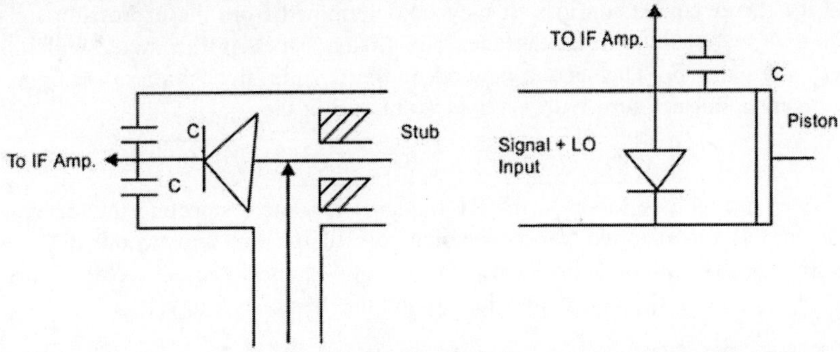

(*a*) Mixer using crystal in Coaxial Lines (*b*) Mixer using crystal in Transmission

Fig. 4.20.

4.7.3. Conversion Loss and Noise Temperature Ratio

The conversion loss of a mixer is defined as

$$LC = \text{(available RF power)/(available IF power)}$$

It is a measure of the efficiency of the mixer in converting RF signal power into IF. The conversion loss of typical microwave crystals in a conventional single ended mixer configuration varies from about 5 to 6.5 dB. A crystal mixer is called "Broad Band" when the signal and image frequencies, both terminated in matched loads. A signal impressed in the RF signal channel of broadband mixer is converted in equal portions to the IF signal and the RF image. Therefore the theoretical conversion loss can never be less than 3 dB with this configuration. The image frequency is defined as that frequency which is displaced from the local oscillator frequency fLo by the IF frequency, and which appears on the opposite side of the local oscillator frequency as the signal frequency fRF. It is equal to 2fLo-fRF.

Short-circuiting or open circuiting the image frequency termination results in a "narrow band mixer". The conversion loss is less in the narrow band than in the broadband mixer. In principle, it can be about 2 dB lower. The design of a broadband mixer has been simpler to achieve and less critical than a narrow band mixer. The noise temperature ratio of a crystal mixer (not to be confused with effective noise temperature) is defined as

$$tr = \text{Actual available IF noise power/ Available noise power from an equivalent resistance.}$$

(or) $$tr = (F_c KTB_n G_c)/(KTB_n) = F_c G_c = F_c/L_c \quad ...(7.10)$$

where F_c = Crystal mixer noise figure

and $L_c = 1/G_c$ = conversion loss.

The noise temperature ratio of a crystal mixer varies approximately inversely with frequency from about 100KHz, (the exact value depends upon the diode), down to a small fraction of a hertz. This is called flicker noise or $1/f$ noise. Above approximately 500 KHz, the noise – temperature ratio approaches a constant value. At a frequency of 30MHz, a typical radar IF, is might range from 1.3 to 2.0.

From equation (1) the noise figure of mixer is $F_C = trL_c$. This, however, is not a complete measure of the sensitivity of a receiver with a mixer front end. The overall noise figure depends not only on the mixer stage but also on the noise figure of the IF stage and the mixer conversion loss. It may be determined from the expression for the noise figure of two networks in cascade. The first network is the mixer with noise figure, trL_c and gain, G_c. The second network is the IF amplifier with a noise figure of FIF. The receiver noise figure with a mixer front end is then

$$Fo = F_1 + (F_2 - 1)/G_1 = L_c(tr + Fif - 1)$$

(This does not include losses in the RF transmission line connecting the receiver in to the antenna) If for example, the conversion loss of the mixer were 6.0 dB, the if noise figure 1.5 dB and the noise temperature ratio diodes, the Rx noise figure is approximately equal to the conversion loss times the IF noise figure.

4.7.4. Types of Mixers

There are various type of radar mixer used in different types of radar as per the application. These are being discussed in the following sections.

4.7.4.1. Balanced Mixers

Noise that accompanies the local oscillator (LO) signal can appear at the IF frequency because of the non-linear of the mixer. The L/O noise must be removed if Rx sensitivity is to be maximized. One method of eliminating L/O noise that interferes with desired signals is to insert a narrow band pass RF filter between the L/O and the mixer. The center frequency of the filter is that of the L/O, and its bandwidth must be narrow so that L/O noise at the signal and the image frequencies do not appear at the mixer. Since the Rx is tuned by changing the L/O frequency, the narrow band filter must be tunable also. A method of eliminating L/O noise without the disadvantage of a narrow bandwidth filter is a balanced mixer. A balanced mixer uses a hybrid junction, a magic *T* or an equivalent. This is a all four-port junction. Figure 4.21 illustrates a magic *T* in which the L/O and RF signals all applies to two ports. Diode mixers are in each of the remaining two arms of the magic *T*. At one of the diode the sum of the RF and L/O signal appears, and at the other diodes the difference of the two is obtained. (In a magic *T*, the L/O signal will be applied to the *H* plane arm and the RF signals will be applied to the *E* plane arm. The diode mixers would be mounted at equal distances in each other co-linear arms). The two-diode mixers should have identical characteristics and be well matched. The IF signals is recovered by subtracting the outputs of the diode mixers.

In Fig. 4.21 (*a*) & (*b*) the balanced diodes are shown reverse so that if output can be added. L/O noise at the two-diode mixers will be phase and will be cancelled at the output. It is only the AM noise of the L/O, which is cancelled. The FM noise inserted by the L/O is unaffected by the balanced mixer.

In a single ended mixer, the mixing action generates all harmonics of the RF and L/O frequency, and combinations there of. The output is designed to filter all the frequency of interest; usually the difference frequency a balance mixer suppresses the even harmonics of the L/O signals. A double balanced mixer is basically two single-ended mixers connecting in parallel and 180 deg out of phase. It suppresses even harmonics of both the RF and L/O signals.

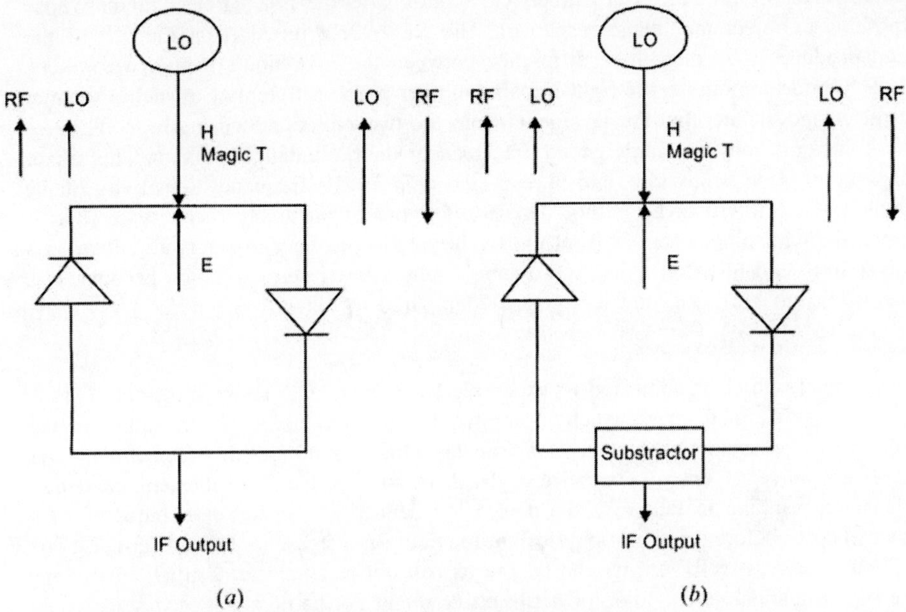

Fig. 4.21. Principle of the Balanced Mixer Based on the Magic-T.

4.7.4.2. Reactive Image Termination

If the image frequency of a mixer is presented with the proper reactive termination (such as an open or a short circuits), the conversion loss and the noise figure can be one to two dB less than with the broad band mixer in, which the image frequency is terminated in matched load. The reactive termination causes energy converted to IF.

Both the sum and the image frequency can be reflected back to the diode mixers to achieve a low conversion loss, but a number of adjustments are required for good results. One method for terminating the image in a reactive load is to employ a narrow band filter, or preselector, at the RF signal frequency. A limitation of this approach is that it is not suitable for very wide bandwidth. Also, the high Q of the filter introduces a loss, which will increase the system noise figure.

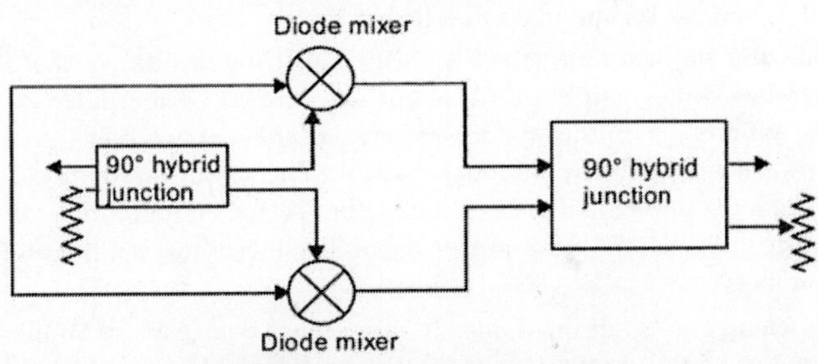

Fig. 4.22. Image Recovery Mixer.

A method for achieving a reactive termination without narrow bandwidth components is an image recovery mixer shown in Fig. 4.22. This is also been called an image enhance mixer or product return mixer (it is similar to the image reject mixer whose purpose is to reject the image response). The RF hybrid junction on the left of the circuit produces a 90 deg phase difference between the L/O inputs to the two mixers. The IF hybrid junction on the right imparts another phase differential in such a manner that the images cancel, but the IF signals from the two mixers added in phase. The two mixers in figure may be single ended, balanced or double balanced mixers. This mixer is capable of wide bandwidth, and is restricted only by the frequency sensitivity of the structure of the microwave circuits. The noise figure of an image recovery circuits is competitive with other receiver front ends. The image recovery mixer is attractive as a receiver front end because of its high dynamic range, low intermodulation product, less susceptibility to burn out, and less cost as compared to other front ends.

4.7.4.3. Diode Burnout

A crystal, which is subjected to excessive RF power, may suffer burn out. This is rather a loose defined term, which is applied to any irreversible deterioration in the direction or conversion properties of a crystals diode as the result of electrical over load. If excessive RF energy is applied to the diode to open circuit or the semi-conductor to puncture, resulting in failure of the device. As defined above, however, burn out of a diode can occur before the one of physical destruction. An increase in the receiver noise due to effects of excessive RF energy can be just as harmful as complete destruction. Perhaps more so, for gradual deterioration of performance might not be notice as readily as would catastrophic failure. It is for this reason that some means of automatic monitoring of receiver noise figure is necessary if the radar is to be maintained in prime operating condition.

Degradation in the noise figure of a pre-determined amount usually is considered as the criterion for diode failure defining burn out. Sometime an increase in noise figure of 3db has been used as the criterion. In other cases, a 1db increase has been used. However, with schottky and point contact diodes, there is an increase in IF noise and a decrease in the breakdown voltage at lower power level that would be indicated by the above criteria.

One of the causes of diode burnout in radar receiver has been the increased RF leakage through a conventional duplexer due to aging of the TR tube. When the transmitter fires, the TR tube breaks down. A finite time usually on the order of several nano seconds must elapse before breakdown if complete. During this time, RF energy leaks into the receiver. This is called the spike- leakage energy. From 1 to 10 Ergs of spike-leakage energy might be required to burn out microwave crystal diodes. The amount of energy contained within the remainder of the pulse after the initial spike is usually small and is not as serious as spike-leakage.

When a solid-state duplexer or when a solid-state limiter follows the TR switch, there need be no initial spikes and burn out is not determined by the pulse energy. Burn out due to pulses without an initial spike, but greater than about one microsecond in duration, is determined primarily by the peak power. (The burn out condition for pulses 1micro seconds or greater is essentially the same as for CW). Crystal diodes can withstand several watts or more seconds, the peak power capability increases, but not at a sufficient rate for constant energy.

In addition to leakage through the duplexer, diode burn can result from the accidental reception of power from nearby radars or from the discharge of static electricity through the diode.

4.8. DUPLEXER

4.8.1. Introduction

Whenever a single antenna is used for both transmitting and receiving, as in a radar system, problems arise. Switching the antenna between the transmit and receive modes presents one problem; ensuring that maximum use is made of the available energy is another. The simplest solution is to use a switch to transfer the antenna connection from the receiver to the transmitter during the transmitted pulse and back to the receiver during the return (echo) pulse. No practical mechanical switches are available that can open and close in a few microseconds. Therefore, ELECTRONIC switches must be used. Switching systems of this type are called duplexer. A simple block diagram of Radar System using Duplexer is shown in Fig. 4.23.

The duplexer enables a radar system to transmit powerful signals and still receive very weak radar echoes. The duplexer acts as a gate between the antenna and the receiver and transmitter. It keeps the intense signals from the transmitter from passing to the receiver and overloading it, and also ensures that weak signals coming in from the antenna go to the receiver. A pulse radar duplexer connects the transmitter to the antenna only when a pulse is being emitted. Between pulses, the duplexer disconnects the transmitter and connects the receiver to the antenna. If the receiver were connected to the antenna while the pulse was being transmitted, the high power level of the pulse would damage the receiver's sensitive circuits. In continuous-wave radar the receivers and transmitters operate at the same time. These systems have no duplexer. In this case, the receiver separates the signals by frequency alone. Because the receiver must listen for weak signals at the same time that the transmitter is operating, high power continuous-wave radar systems use separate transmitting and receiving antennas.

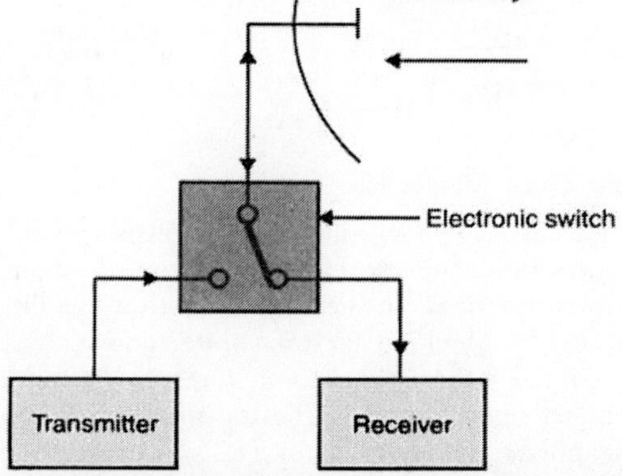

Fig. 4.23. A Simple Block Diagram of Radar System using Duplexer.

4.8.2. Basic Duplexer Operation

In selecting a switch for this task, you must remember that protection of the receiver input circuit is as important as are output power considerations. At frequencies where amplifiers may be used, amplifier tubes can be chosen to withstand large input powers without damage. However, the input circuit of the receiver is easily damaged by large applied signals and must be carefully protected.

An effective radar duplexing system must meet the following four requirements:

1. During the period of transmission, the switch must connect the antenna to the transmitter and disconnect it from the receiver
2. The receiver must be thoroughly isolated from the transmitter during the transmission of the high- power pulse to avoid damage to sensitive receiver components.
3. After transmission, the switch must rapidly disconnect the transmitter and connect the receiver to the antenna. For targets close to the radar to be seen, the action of the switch must be extremely rapid.
4. The switch should absorb an absolute minimum of power both during transmission and reception. Therefore, a radar duplexer is the microwave equivalent of a fast, low-loss, single-pole, double- throw switch. The devices developed for this purpose are similar to spark gaps in which high-current microwave discharges furnish low-impedance paths. A duplexer usually contains two switching tubes.

Generally, Gas- Discharge devices are being used as duplexer but ferrites and solid state varactor diode have also being used. Its make use of fast acting radio– frequency switches known as TR (Transit-Receive) and ATR (Anti-Transmit-Receiver) which are operated by energy from the transmitted, and which protect the sensitive receiver from the damaging effect of the high energy of the transmitter. In a typical application the transmitter power might be 10^6 watts and maximum safe power at receiver might be a few watts. The duplexer must provide isolation between Tx and Rx approximately 60 dB, with the negligible reduction in the transmitted power.

4.8.3. Types of Duplexer

There are different types of duplexer have been used, brief explanation of these duplexer are given below :

1. Branch type duplexer
2. Balanced duplexer
3. Ferrite duplexer

4.8.3.1. Branch Type Duplexer

Branch type duplexer, as shown in Fig. 7.9 consists of ATR switch and an ATR switch, both of which are gas-discharge tubes. During the transmissions, both the TR and ATR tubes ionizes and fired. In the fired condition the TR acts as a short circuit to protect the receiver by blocking the transmitter power. As the TR tube is located at a distance of $\lambda/4$ from the transmission lines so its act as short circuit at the receiver end but as an open circuit at the transmission line and does not limit the flow of trnasmitter power. The purpose of TR is to protect the receiver from excess power by disconnecting the receiver from the transmitter during transmission. The ATR act as a open circuit during the transmission, so it does not effect the flow of transmission.

Both TR and the ATR are not fired.

While receiving the reflected signals from the target. The open circuit of the ATR gap is positioned a $\lambda/4$ away from the transmissions line and act as a short circuit across the line. The short circuit is located a λ/4 from the branch line of receiver than the transmitter is disconnected from the line and reflected received signal power is directed toward the receiver. So, during reception ATR is not fired and the entire energy of echo signal is directed toward the receiver.

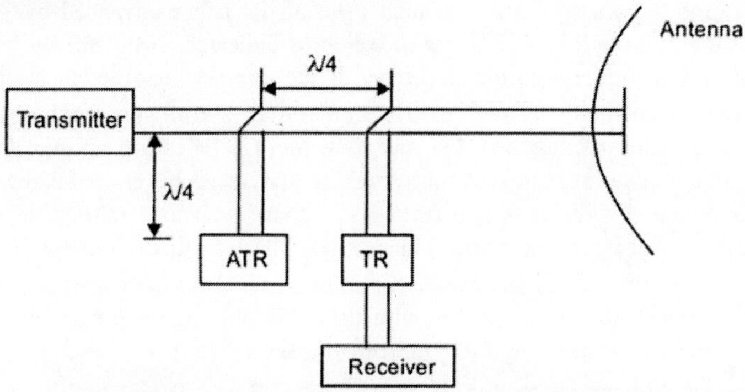

Fig. 4.24. Principle of Branch-Type Radar Duplexer.

4.8.3.2. Balanced Duplexer

Branch duplexer has got a main drawback that its bandwidth is limited, so for the broadband operation cannot be possible. To have more broadband operation, a balanced duplexer arrangement is used. It is shown in Fig. 4.25. Balanced duplexer uses the principle of of hybrid junction. A hybrid junction is a combination of waveguides, coaxial, or other transmissions lines having four branches. According to the principle of hybrid junction that if the energy entering from any one branch or arm is equally transfered into two arms of the remaining. A short-slot hybrid junctions is used in the balanced duplexer which is basically a directions coupler with a coupling ratio of 3 dB. It consists of two wave guides which are coupled through the slots cut in the narrow wall of the waveguides.

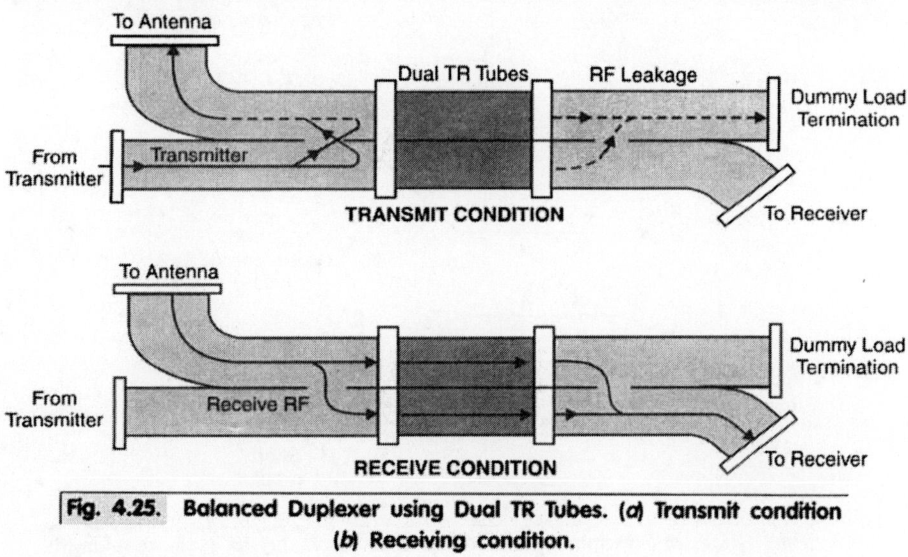

Fig. 4.25. Balanced Duplexer using Dual TR Tubes. (*a*) Transmit condition (*b*) Receiving condition.

During transmissions, the whole energy should be directed towards the antenna and receiver should be isolated at least by 30 dB. When the transmitter is 'on' then the whole transmitted power is divided into two equally path into waveguides by the first short slot hybrid junction. Both the TR tubes are fired which makes waveguide short circuited and the energy is reflected towards the antenna arm. One of the property of the short-slot hybrid that each time the energy passes through the slot, its phase advanced by 90°. The energy travel as shown in the Fig. 7.10 in the direction as indicated. If the energy leaks through the TR tubes than this energy is directed towards the dummy load not to the receiver.

During receptions, the reflected signal power should be reached to the receiver with the minimum attenuations. On reception the TR tubes are unfired and acts as open circuits, so that whole energy should reach to the receiver. First the received energy is divided into two equally parts and due to 90° phase advanced introduced by the slot, the energy joined in the receiving arm. No energy will enter to the dummy load.

There is another form of balanced duplexer which is uses four ATR tubes and two hybrid junction. Balanced duplexer with the ATR has higher power-handling. Capacity and less bandwidth than the TR balanced duplexer.

4.8.3.3. Phase Shift Duplexer

There is one more method to have a duplexer function, which is called phase shift duplexer. In this type of duplexer two hybrid junction and a λ/2 power sensitive phase

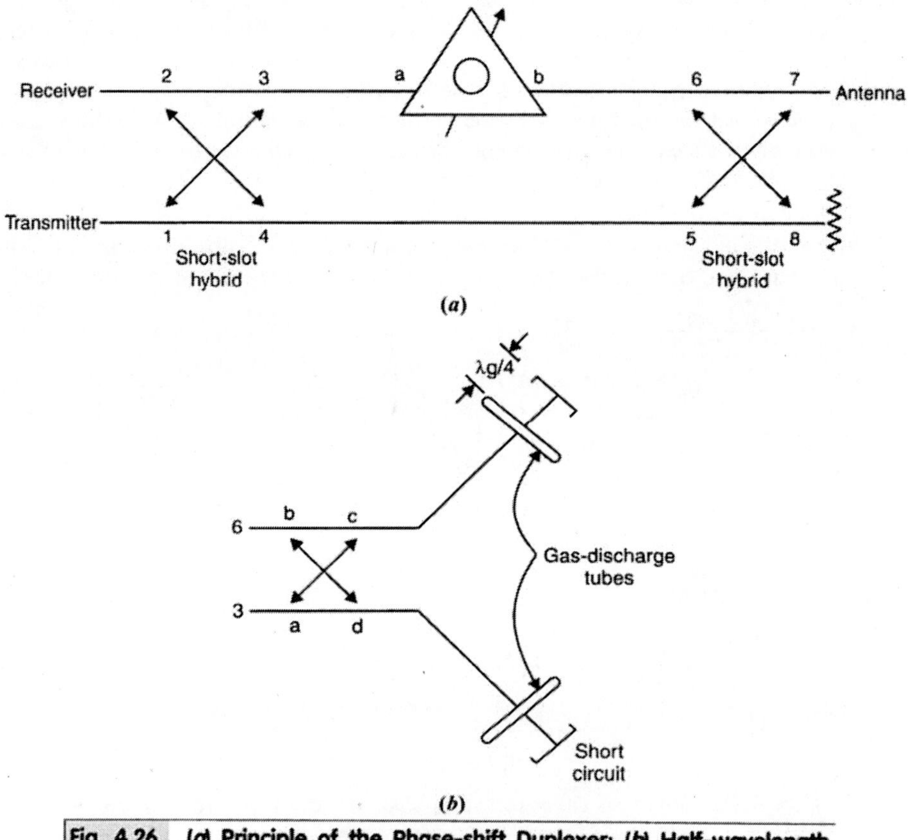

Fig. 4.26. (a) Principle of the Phase-shift Duplexer; (b) Half-wavelength, Power-sensitive Phase Shifter.

shifter are used as shown in the Fig. 4.26. During transmissions, the whole power is given of the arm 1, where hybrid junction divides to its in two equally part between branch 3 and 4. A 90° phase shift is introduced by the hybrid junctions which will make the whole energy to reach to the antenna arm.

During receptions the received energy is received by the antenna arm 7 which is passed through phase shifter. The purpose of power sensitive phase shifter is to change the phase by λ/2. Due to the phase of the received energy, the whole received energy will enter to the receiver arm. The received energy will not go to the transmission arm due to the mismatches of the phase.

The λ/2 power-sensitive phase shifter uses a hybrid junction and two gas-discharge tubes as shown in Fig. 4.26(*b*) to provide the λ/2 phase shift.

4.8.3.4. Ferrite Duplexers

There are some duplexer which are uses the ferrite devices such as ferrites phase shifter, isolators, circulators to perform the duplexer actions. Instead of using power-sensitive phase shifter in the phase shift duplexer. We may use a ferrite phase shifter. The ferrite duplexer uses a magic T, two non-reciprocal ferrite phase shifters and a short slot hybrid junction to perform the duplexer operation as shown in Fig. 4.27. The phase shifter provides the 45° phase shift either it advances the phase or it may retart the phase according to the directions of the energy flow. A TR may be used to protect the energy flow. A TR may be used to protect the receiver against leakage from the Transmitter or reflected energy of transmitter from the antenna. The ferrite duplexer has lot of advantages over gas discharge tubes. It has wide bandwitdh and fast recovery time than the gas discharge tubes. But it has large size and weight and it requires magnetic field for operations.

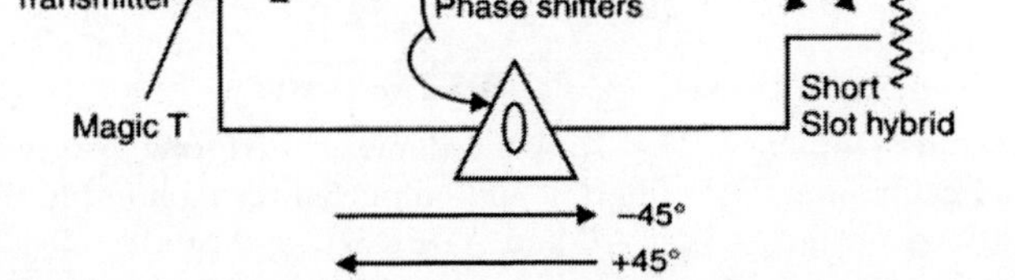

Fig. 4.27. The Non-Reciprocal Differential Phase Shift Ferrite Duplexer.

4.8.3.5. Solid-State Duplexer

As till now, we have discussed about the gas discharge tubes used in the duplexer. These duplexer have disadvantages such as long recovery time, radioactive priming and large size. These tubes may be replaced by the solid-state duplexer which has advantages of fast recovery time and small in size. The only problem with the solid-state duplexers is that it has less power handling capacity. But if PIN diodes are used in solid-state duplexer, it can handle more power.

Generally all solid state duplexer consists of a three port ferrite circulator with a diode limiter at the receiver.

4.8.4. Types of Modern Duplexer Used In Modern Radar

(*a*) **Cassini Duplexer.** This unit performs T/R switching, receiver protection, and calibration for the radar which will map Saturn's moon Titan when the Cassini spacecraft arrives there in 2004. The Cassini radar is capable of imaging, altimetry, scatterometry and radiometry

Fig. 4.28. Cassini Duplexer.

(*b*) **ERS-1 Duplexer.** This Ku-band duplexer, designed for low earth orbiting radar altimetry sensors, includes an EMI filter in the receive path.

Fig. 4.29. ERS-1 Duplexer.

(*c*) **SeaWinds Duplexer.** This Ku-band duplexer performs T/R switching, receiver protection and calibration, EMI filtering and output power monitoring for the SeaWinds instruments flying on the QuikSCAT and ADEOS-II spacecraft

Fig. 4.30. SeaWinds Duplexer.

4.9. ANTENNA PARAMETERS

The purpose of this chapter is to provide introductory in sight in to radar antenna and their characteristics. An antenna has either to receive energy from an electromagnetic field or to radiate electromagnetic waves produced by a high frequency generator. In this section antennae, which are used in radar units, are treated especially. The types of antenna used in the radar system mainly depend upon the application of the radar. For example long-range detection radar (surveillance radar) needs large aperture of the antenna. The Antenna used normally for radar applications different from the antenna used in communication system. Radar antenna with the shaped directive pattern can be scanned either mechanically or electronically.

In general an antenna is a transmission device, or transducer, between a guided wave (transmission line) and a free space wave or vice versa. The basic parameter of an antenna will be discussed in the brief in the following section.

The radar antenna acts as a transducer, which converts electrical pulses from the transmitter to the free space in the form of EM waves and receives the reflected EM signals from the target in free space and converts it in to electrical signals.

In the radar equation we have studied about the antenna gain and the aperture area of the antenna. For the large antenna gain the effective aperture area of the antenna must be large. Both the parameter is proportional to each other.

If we say in simple terms, the function of the radar antenna in transmitting mode should concentrate the radiated energy in to a shaped beam to the desired direction. And in the receiving mode, from the target and deliver it to the receiver.

Mostly radars, as we know, are operated in the microwave frequencies region. So the main advantage of microwave frequencies for radar application in that an aperture of relatively small physical size but it is quite large enough in terms of wavelength can be obtained. The antenna having high gain with the narrow beam widths are possible at microwave frequency. But it is quite difficult to achieve at HF.

In the microwave frequencies region, the most popular radar antenna is, the parabolic reflector with its optics characteristics. Some other types of antenna are also found application in radar such as, microwave lenses, surface wave antennas (for airborne radars).

4.10. PARAMETERS OF RADAR ANTENNA

The important parameters of radar antennas, discussed here in brief. Which are as follows :

1. Directivity or Directive gain.
2. Power gain.
3. Effective receiving aperture or total scattering cross section.
4. Polarizations.
5. Side lobes.

4.10.1. Directivity or Directive Gain

The directivity of an antenna is given by the ratio of the maximum radiations intensity (power per unit solid angle) $u(\theta, \emptyset)$ mass to the average radiations intensity

*U*av (average over a sphere) or it may be defined as at a certain distance from the antenna the directivity may be expressed as the ratio of the maximum to the average polynthing vector.

$$G_D = \frac{\text{Maximum radiation}}{\text{Average radiation}} \quad ...(4.1)$$

$$G_D = \frac{u(\theta, \varnothing)_{Max}}{Uav} \text{ or } \frac{S(\theta, \varnothing)_{Max}}{Sav} \quad ...(4.2)$$

The smaller the beam solid angle, the greater the directivity.

The ability of radar antenna to concentrate energy in a particular direction is known as **gain.**

There are two related terms of antenna gain are the power gain and the directive gain. Both the terms are of interest to the radar system engineer. The directive gain or directivity is related to radiation pattern of the antennas. Whereas the power gain is more suitable related to the radar range equations.

The power radiated from an antenna per unit solid angle is called the radiation intensity u (watts per steradian or per square degree). A radiation intensity pattern may be plot between radiation intensity and the angular co-ordinate, this plot is known as a **radiation-intensity pattern.**

The power density, which may be desired as power per unit area, can be plotted as a function of angle is known as a power pattern. The terms, power density pattern and radiation intensity pattern are identical when plotted on a relative basis that is when the maximum is normalized to up to a value of unity. When both are plotted on a relative basis is called the antenna **radiation pattern.**

The directive gain may be written as the average radiation intensity over a solid angle of 4π radians is equal to the total power radiated divided by 4π

$$G_D = \frac{(\text{Maximum power radiated / unit solid angle}) \times 4\pi}{\text{Total power radiated}} \quad ...(4.3)$$

With the help of above equation. We can find out the directive gain from the radiation pattern.

The directive gain equation can be written as,

$$G_D = \frac{4\pi}{B} \quad ...(4.4)$$

In above equation 4π is denoted the maximum power per unit solid angle approximately and the total power radiated is find by integrating the volume contained under the radiation pattern which is also known as beam area (B).

The angular range of the antenna pattern in which the half of the maximum power at least is still emitted is described as a Beam Width. Bordering points of this may or lobe are therefore the points at which the field strength has fallen in the room around 3 dB regarding the maximum field strength. This angle is then described as an beam width or half width of the antenna.

The beam area (B) is the solid angle from when adds the total power radiated would pass if the power per unit solid angle were equal to P (θ, $\varnothing$) max over the beam area. This solid angle can often be described approximately in term of angles subtended by the half power points of the main to be in the two principles planes as given by

$$B \simeq \theta B \times \varnothing B. \quad ...(4.5)$$

Now by putting the value of B in the above equation, we get,

$$GD\ \emptyset = \frac{4\pi}{(\theta B \emptyset B)} \quad ...(4.6)$$

Where θB and $\emptyset B$ are the half power beam width into principal planes such as horizontal and vertical respectively, minor lobes are neglected. If the half power beams widths are measured in radians.

If the half power beam width are measured in degrees, and minor lobes effects are neglected than this can be written as

$$G_D = \frac{4\pi}{\theta B \emptyset B} \approx \frac{41000}{(\theta B^\circ \times \emptyset B^\circ)} \quad ...(4.7)$$

($4\pi sr$ = 4125 square degree)

Suppose as antenna has a main lobe with both half-power beam widths (HPBWs = 2o°) its directivity gain is approximately

$$G_D = 4\pi sr/B(sr) \sim 41000(\text{deg}^2)/(\theta B^\circ \emptyset B^\circ)$$

$$= 41000(\text{deg}^2)/(2o^\circ \times 2o^\circ)$$

$$G_D \simeq 103$$

Or $\quad G_D \simeq$ 20db (dB above isotrope)

It means that the antennas radiates a power in the direction of the main lobe maximum which is about 100 times as would be radiated by a non directional (isotropic) antenna for the same power point.

4.10.2. Power Gain

The power gain, which is denoted by (G), includes the effect of the antenna losses and any other related losses, which reduces the antenna efficiency. In the directive gain definition we have not considered the losses such as ohmic (resistance) heating, RF heating or mismatched antenna, we have considered only the radiation pattern. But these losses are taken into account while calculating the power gain. It may be defined as the ratio of maximum radiation intensity from particular antenna to the radiation intensity from (loss less) isotropic source with some power input.

This defination must be taken care while using radar range equation since its includes the losses introduced by the antenna. The directive gain is important for resolution consideration, average and the accuracy, which is related to the antenna beam width.

The directive gain is always greater than power gain. If there is no losses, the both the gain (power and Directive) are more or less equal to each other. Both are related by the radiation efficiency factor (ρr) and can be defined as follow:

$$G = \rho r \times G.D \quad ...(4.8)$$

The theory of antenna related to the directivity and power gain is same for the transmission and receiver antenna in radar.

4.10.3. Effective Aperture or Total Scattering Cross Section

One of the important parameter related to the gain is the effective receiving aperture or total scattering cross section or effective area. It may be considered as a measure of the effective area presented by the antenna to the incident wave.

We know from the radar range equation that

$$G = 4\pi Ae/\lambda^2 \quad ...(4.9)$$

Where λ = Wave length

Ae = effective area or effective aperture area

ρ_a = antenna aperture effiency

The effective area (Ae) may be defined as the product of the antenna aperture effiency (ρ_a) and the physical area of antenna.

$$Ae = \rho a \times A \quad \text{...(4.10)}$$

$$G = 4\pi \times (\rho_a \times A)/\lambda^2 \quad \text{...(4.11)}$$

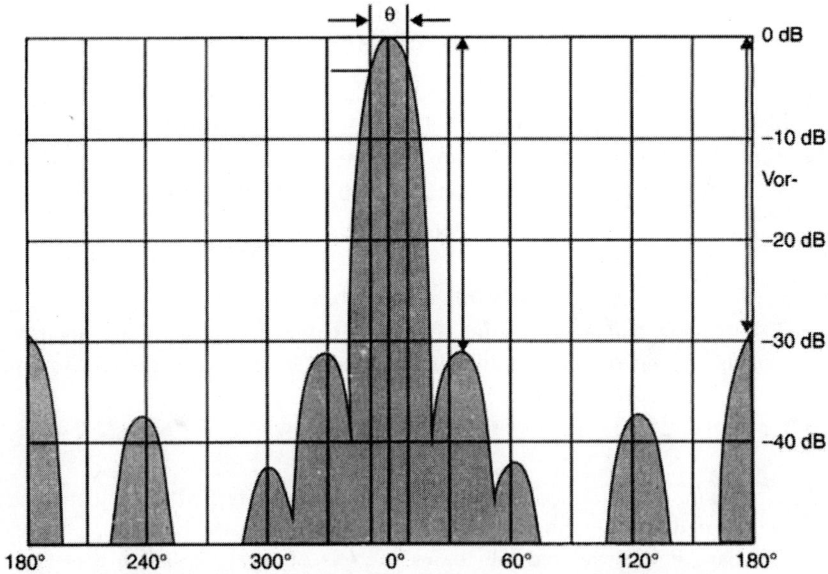

Fig. 4.31. Antenna Pattern in a Rectangular-Coordinate Graph.

4.10.4. Wave Polarization

We have considered the case of a plane electromagnetic wave say propagating in Z-direction. Then the electric vector E is in X-direction while magnetic vector H is in K-direction. There is an orthogonal (perpendicular) relationship between E, H and Poynting vector P. To describe an electromagnetic wave completely, we should not only include the parameters such as its wavelength, phase velocity and power but also the *instantaneous orientation of its field vectors.* It is important to mention here that if the magnitude and direction of E are known then the magnitude and direction of H are automatically specified. Thus, it is sufficient to specify the quantities related to electric vector E.

The ***wave polarization*** *is defined as the orientation of electric field vector E at a given instant of time in space.*

There are three types of polarization :

(1) Linear Polarization : A plane electromagnetic wave in which electric vector vibrates simple harmonically along a fixed straight line perpendicular to the direction of propagation without changing its orientation is called *plane polarized* electromagnetic wave as shown in Fig. 4.32. The wave is travelling out of page in Z-direction.

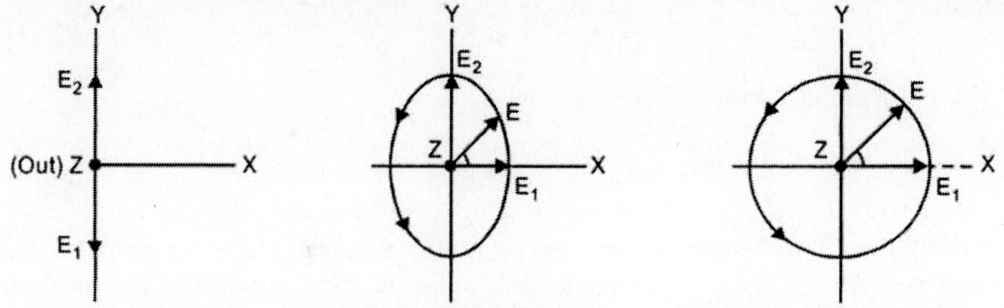

(a) Linear polarization (b) Elliptical polarization (c) Circular polarization.

Fig. 4.32.

(2) Elliptical Polarization : Let us consider the case of two plane polarized electromagnetic waves which superimpose over each other. Now, the resultant vector rotates under certain conditions. If both-the magnitude and orientation of the electric field vector E vary continuously, then the tip of vector traces an ellipse as shown in Fig. 4.32(*b*). The wave is said to be **elliptically polarized.**

(3) Circular Polarization : In the above case, if the magnitude of resultant vector remains constant but the orientation varies continuously then the tip of vector E traces a circle as shown in Fig. 4.32(*c*). The wave is said to be **circularly polarized.**

Mathematical Consideration : A plane polarized wave propagating in Z-direction can be considered to be the sum of two plane polarized components in perpendicular directions but in phase as shown in Fig. 4.33. Here E has been resolved into two components E_1 and E_2. The two components are given by

$$E_x = E_x^0 e^{j(\omega t - kx)}$$

or
$$E_x = \hat{a}_x E_x^0 e^{j(\omega t - kx)} \quad ...(4.12)$$

Similarly,
$$E_y = \hat{a}_y E_y^0 e^{j(\omega t - kx)} \quad ...(4.13)$$

Here, $\hat{a}_x$ and $\hat{a}_y$ are unit vectors along X and Y-axis respectively.

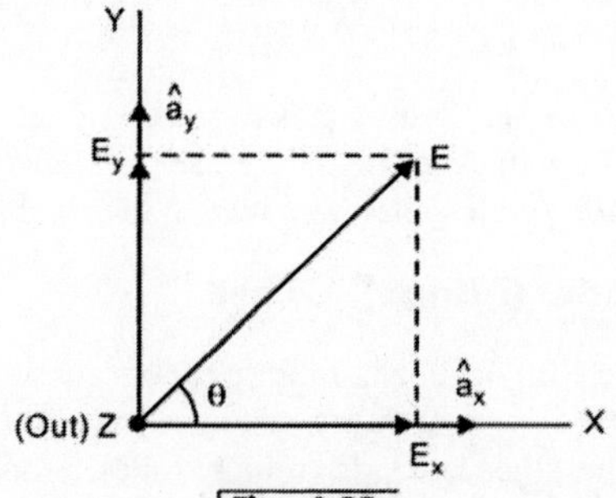

Fig. 4.33.

(1) Plane Polarized Wave : The general solution of plane polarized wave is given by

$$E(r, t) = (\hat{a}_x E_x^0 + \hat{a}_y E_y^0) e^{j(\omega t - kx)} \quad ...(4.14)$$

Here, E_x^0 and E_y^0 must have the same phase if $E(r, t)$ is to be plane polarized. The magnitude E_0 is given by

$$E_0 = \sqrt{[(E_x^0)^2 + (E_y^0)^2]} \quad ...(4.15)$$

The polarization vector E makes an angle

$$\theta = \tan^{-1}\left(\frac{E_y^0}{E_x^0}\right) \text{with } \hat{a}_x \quad ...(4.16)$$

(2) Elliptically Polarized Wave : In elliptically pllarized wave, the two linearly polarized components (one in X-direction and other in Y-direction) are not equal in magnitude. As the magnitude is real, we can expression them as

$$E_x(r, t) = E_1 \cos(\omega t - kz) \quad ...(4.17)$$

and

$$E_y(r, t) = E_2 \sin(\omega t - kz) \quad ...(4.18)$$

where E_1 is amplitude of linearly polarized light in X-direction and E_2 is the amplitude of wave linearly polarized in Y-direction.

The resultant of eqns. (4.17) and (4.18) is given by

$$\frac{E_x^2}{E_1^2} + \frac{E_y^2}{E_2^2} = 1$$

This is equation of ellipse. Therefore, elliptically polarized wave is the resultant of two plane perpendicular polarized waves of unequal amplitudes.

(3) Circularly Polarized Wave : The components of actual electric field obtained by taking real part of Eqn. (4.13) are given as

$$E_x(r, t) = E_0 \cos(\omega t - kz) \quad ...(4.19)$$

and

$$E_y(r, t) = E_0 \sin(\omega t - kz) \quad ...(4.20)$$

In case of circularly polarized wave, the amplitude of two components is the same, *i.e.*,

$$E_1 = E_2 = E_0$$

The resultant of Eqns. (4.19) and (4.20) is given by

$$\frac{E_x^2}{E_0^2} + \frac{E_y^2}{E_0^2} = 1$$

or

$$E_x^2 + E_y^2 = E_0^2 \quad ...(4.21)$$

Eqn. (4.21) represents a circle. Thus, electric vector E, at a fixed point in space, sweeps around a circle at a frequency of ω. Therefore, *circularly polarized wave is the resultant of two perpendicular plane polarized waves with equal amplitude.*

4.10.5. Major and Side (Minor) Lobes

The pattern shown in the Fig. 4.31, has radiation concentrated in several lobes. The radiation intensity in one lobe is considerably stronger than in the other. The strongest lobe is called a major lobe; the other is a side (minor) lobes. Since the complex radiation patterns associated with arrays frequently contain several lobes of varying intensity. In general, major lobes are those in which the greatest amount of radiation occurs. Side or minor lobes are those in which the radiation intensity is least.

In general side lobes are not at all desirable but in radar up to some extent side lobes are permitted. Low side lobes are generally required for radar system but if it is too large than the maximum portion of the radiated energy contained in the side lobes, which result, reduction in main beam energy and in turn reduces the gain also. If the side lobes are high than there are chances of getting false target indication because strong echo signal may enter through side lobes and mislead the radar operator. It is very easy to jam the radar, if its side lobes too high and it is more likely to get interference from nearby other transmitters.

The level of side lobes should of the order of 20 to 30 dB below the main beam and it can be achieved easily with the practical antenna. To achieve very low side lobes is quite difficult practically due to limitation imposed by the nature but theoretically there is no reason why they should not be possible.

4.10.6. Front-to-Back Ratio

The front-to-back ratio of an antenna is the proportion of energy radiated in the principal direction of radiation to the energy radiated in the opposite direction. A high front-to-back ratio is desirable because this means that a minimum amount of energy is radiated in the undesired direction.

4.10.7. Beam Efficiency

Beam efficiency is also an important parameter to judge the quality of transmitting and receiving antenna. In terms of power, the **beam efficiency** can be defined as the ratio of power transmitted (may be received) within cone angle 61 to the power transmitted (or received) by the antenna and it is dimensionless.

$$\text{Beam efficiency} = \frac{\text{Power transmitted (received) within cone angle } \theta_1}{\text{Power transmitted (received) by the antenna}}$$

where the major lobe is directed along the z-axis where $\theta = 0$ and θ_1 is the half angle *of the CMC within which* the percentage of the total power is to be found, as shown in Fig. 4.34.

The beam efficiency should be high usually 90. It is necessary for the antenna used in radar, astronomy and other applications where reception through the side lobes is not desired.

In other terms, the **beam efficiency** can be defined as the ratio of the main beam area to the total beam area. It is denoted by ε_M,

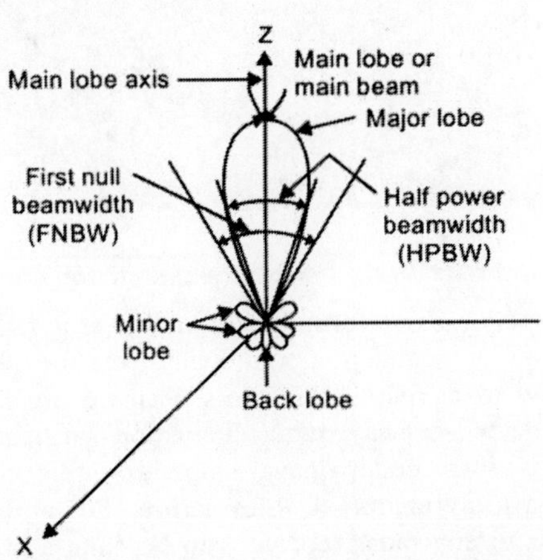

Fig. 4.34. Radiation Lobes and Beamwidth of Antenna Pattern.

$$\varepsilon_M = \frac{\Omega_M}{\Omega_A} = \text{Beam efficiency} \quad ...(4.22)$$

where Ω_A is the total beam area or beam solid angle and consists of the main beam (Ω_M) and the main lobe area (Ω_m).

4.11. INTRODUCTION TO RADAR DISPLAYS UNITS

In this section students will be studying about the different types of the display system used in the radar system to display the received videos signal.

When radar was first used by the military, the information it provided was displayed on a single-unit console. The console included a radar indicator ("scope") and its associated controls, and a number of receiver and transmitter controls. As the development of radar progressed, ships were furnished with more than one type of radar (air search, surface search, etc.). The displaying of radar information began to get complicated. It soon became apparent that information from several different radars had to be available at each of several physically separated consoles. Also, in some cases, information from the same radar needed to be displayed in more than one way at the same time. For example, the information from air-search radar might be needed for both air search and air control at the same time, requiring two different types of display.

Fig. 4.35. A Radar Controller using Radar Equipment for Combat Operation.

The device used to display radar information is known as a radar indicator or radar display. Since indicators can be located at a point away from the other radar equipment, they are frequently referred to as remote indicators. Remote indicators are sometimes referred to as repeaters. The present-day remote indicator can operate with any of the search radars in use today. Since modern naval ships are equipped with severalradars and many indicators for displaying target information, the problem of getting the information from any radar to any radar repeater can be quite a problem. The obvious solution is to run a cable from each radar to every indicator, but this requires a large amount of space for the cables and adds too much weight to be practical. The accepted solution is a centralized distribution system, consisting of a distribution panel with a single input cable from each radar and a single output cable to each indicator. The system operates automatically. When an operator selects particular radar, the switchboard connects the operator's console to the desired radar. Although the change occurs rapidly, it is complicated, in that several electronic connections are required for the inputs (timing, or trigger, pulses from the modulator; video signals from the receiver; and antenna synchronization signals for PPI sweeps).

There are two basic methods of displaying informations on a cathode ray tube.

1. Deflection-Modulated CRT
2. Intensity Modulated CRT

In Deflection –Modulated CRT targets are indicated by the deflection of the electron beam,such as A-Scope.The other display method is the Intensity-Modulated CRT, an example of which is the PPI-Scope.It is normally dark in the absence of the target. A target is indicated by intensifying the beamand presenting a luminous spot on the face of the CRT .In general, defelection modulated scopes have the advantage of simpler circuits than those of intensity modulated scope.

The most common types of displays (indicators) are described in next section.

4.12. A- SCOPE

The A- Scope display, shown in **Fig. 4.36**, presents only the range to the target and the relative strength of the echo. Such a display is normally used in weapons control radar systems. The bearing and elevation angles are presented as dial or digital readouts that correspond to the actual physical position of the antenna. The A-scope normally uses an electrostatic-deflection CRT. Applying a sawtooth voltage to the horizontal deflection plates produces the sweep. The electrical length (time duration) of the saw tooth voltage determines the total amount of range displayed on the CRT face.

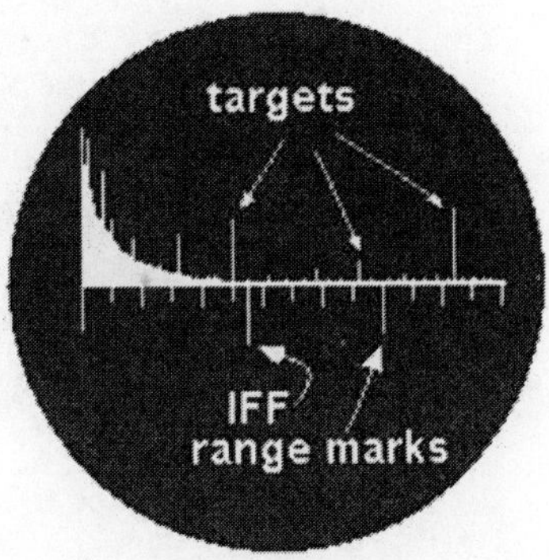

Fig. 4.36(a). A-Scope of the (Russian) P-18.

The A- scope display is using in older radar units only as monitoring oscilloscope. In modern digital radar units don't exist a similar targets video signal. The target messages are transmitted to the displays as a digital word. There isn't any possibility to get a synchronizing signal for these digital signals. Well, the oscilloscope can get an internal trigger only. Therefore it is impossible to analyze the bit sequence with a simple oscilloscope. The only one statement is possible: a digital word is existent that means, obviously the driver module for this line works. In Fig. 4.36. A control-pulse shown at an A-scope of the P-18, in Fig. 4.37. The attempt to show digital words on an oscilloscope. 'A'-scope is a deflection-modulated rectangular presentation in which the vertical deflection is proportional to the amplitude of the received echopower and the horizontal coordinates represents the range as per the time delay.

Fig. 4.36(b). A Control-pulse Shown at an A-scope of the P-18.

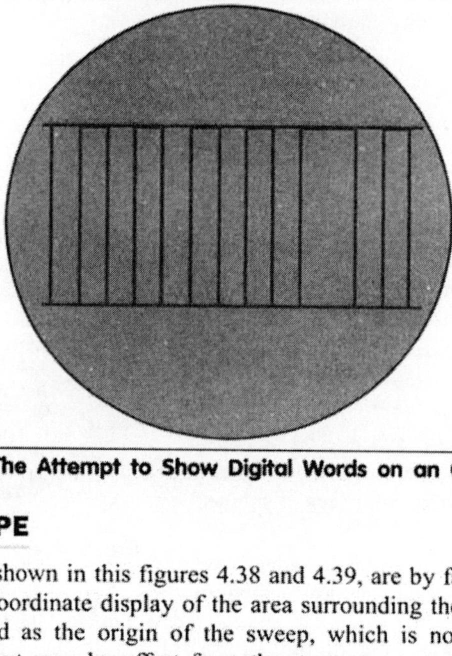

Fig. 4.37. The Attempt to Show Digital Words on an Oscilloscope.

4.13. PPI SCOPE

The **PPI Scopes** shown in this figures 4.38 and 4.39, are by far the most used radar display. It is a polar coordinate display of the area surrounding the radar platform. Own position is represented as the origin of the sweep, which is normally located in the center of the scope, but may be offset from the center on some sets. The PPI uses a radial sweep pivoting about the center of the presentation. This results in a map-like picture of the area covered by the radar beam. A long-persistence screen is used so that the display remains visible until the sweep passes again. Bearing to the target is indicated by the target's angular position in relation to an imaginary line extending vertically from the sweep origin to the top of the scope. The top of the scope is either true north (when the indicator is operated in the true bearing mode) or ship's heading (when the indicator is operated in the relative bearing mode).

The basic PPI screen presentation results from raw (unprocessed) video. Raw video provides only a "blip" on the indicator screen, leaving target interpretation entirely to the operator. PPI is an intensity modulated having circulator representations or display in which echoes are shown in two co-ordinates *i.e.* range and Azimuth in polar coordinates.

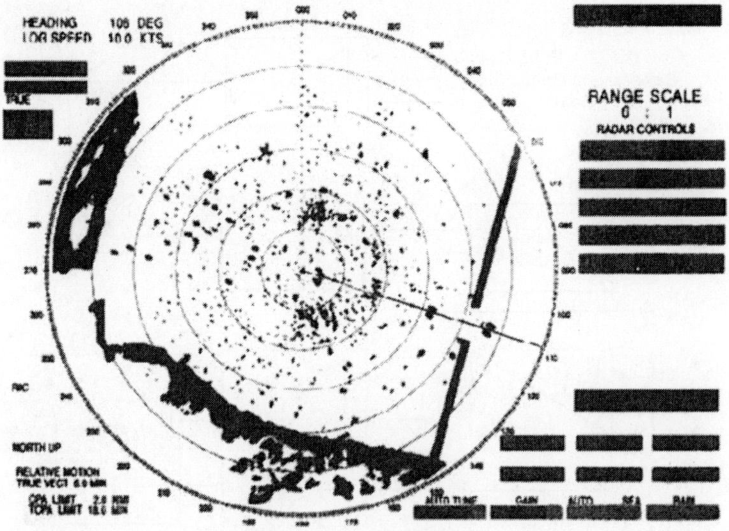

Fig. 4.38. Actual Representation of Radar Information on the PPI.

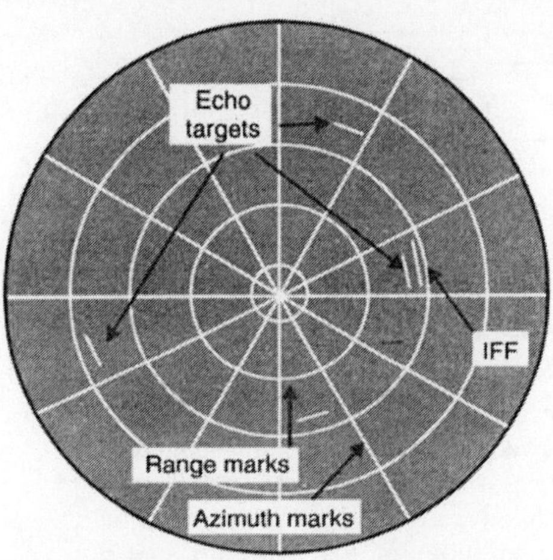

Fig. 4.39. Plan Position Indicator (PPI).

4.13.1. Functional Block Circuit Diagram of a PPI-scope

The brief explanation of functional block diagram of PPI is being discussed here. The function of each block is explained in brief. In Fig. 4.40 shows the functional Block Diagram of PPI.

4.13.1.1. Gate Circuit

The gate circuit develops pulses, which synchronize the indicator with the transmitter. Trigger pulses from the synchronizer synchronize the gate circuit itself. It then provides timing for the intensity gate generator; sweep generator circuit, and the sweep control circuit. While the PPI-Scope at older radar trigger pulses directly from the modulator synchronize units.

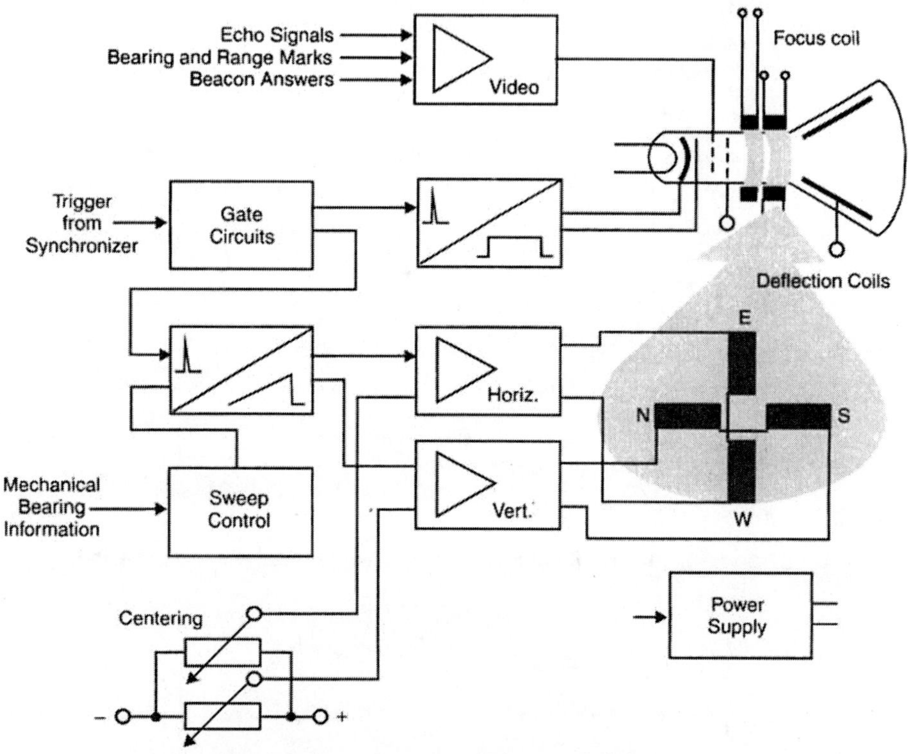

Fig. 4.40. Functional Block Circuit Diagram of a PPI-Scope.

4.13.1.2. Sawtooth-Generator

The sweep generator circuit produces currents, which deflect an electron beam across the CRT. Varying voltages from the sweep control circuit are applied to deflection coils. Gate voltages determine sweep rate, and therefore, the effective distance (range) covered by each sweep. The amplitudes of the sawtooth sweep currents are varied sinusoidally (like a sine wave), corresponding to the rotation of the antenna. Notice that there is a 90 degree phase difference between the amplitude variations of the horizontal and vertical waveforms

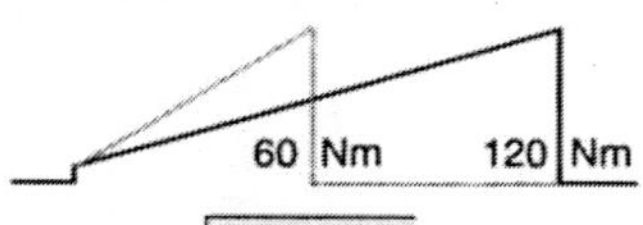

Fig. 4.41.

4.13.1.3. Intensity Gate Generator

The intensity gate generator provides a gate, which unblanks the CRT during sweep periods. The intensity of the trace appearing on the CRT is determined by the DC level of this gate. This circuit is also synchronized by the gate circuit.

4.13.1.4. Deflection Current Amplifier

Electron deflection in the electromagnetic CRT is proportional to the strength of the magnetic fields. Magnetic field strength depends on current in the coils. The sweep circuits associated with electro magnetically deflected cathode-ray tubes must provide currents, rather than voltage, to produce the desired beam deflection.

A sawtooth current is required to produce a linear trace. Because of the inductance of the coil, a trapezoidal voltage must be applied across the coil to produce a sawtooth of current through it. This is illustrated in the following view.

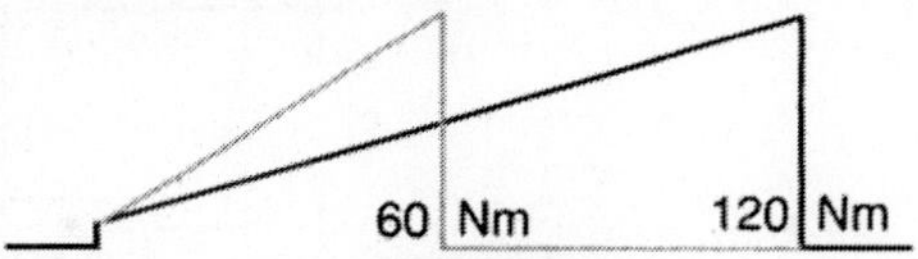

Fig. 4.42.

4.13.1.5. Video Amplifier

The video amplifier circuit amplifies the video signal from the receiver and applies it to the CRT intensity-modulating element (control grid). At this a balanced relationship is necessary between brightness and contrast. The brightness of the grid of the azimuth- and range marks should be adjusted very weakly.

4.13.1.6. Sweep Control Circuit

To synchronize sweep rotation with antenna rotation, you must convert antenna azimuth (bearing) information into electrical signals. These signals, usually provided by synchros, control the amplitudes and polarities of the sawtooth sweep currents applied to the deflection coils.

The amplitudes of the sawtooth sweep currents are varied sinusoidally (like a sine wave), corresponding to the rotation of the antenna.

4.13.1.7. Cathode Ray Tube (CRT)

There are two possibilities for the deflection of the electron beam:

- The electrostatic controlled deflection and focusing, and
- The electromagnetic controlled deflection and focusing of the electron beam.

The primary difference between electromagnetic and electrostatic cathode-ray tubes lies in the method of controlling deflection and focusing of the of the electron beam. Both types employ electron guns and use electrostatic fields to accelerate and control the flow of electrons. The physical construction of a CRT employing electromagnetic deflection is similar to an electrostatic type. A picture of electrostatics CRT is shown as Fig. 4.43 and Fig. 4.44.

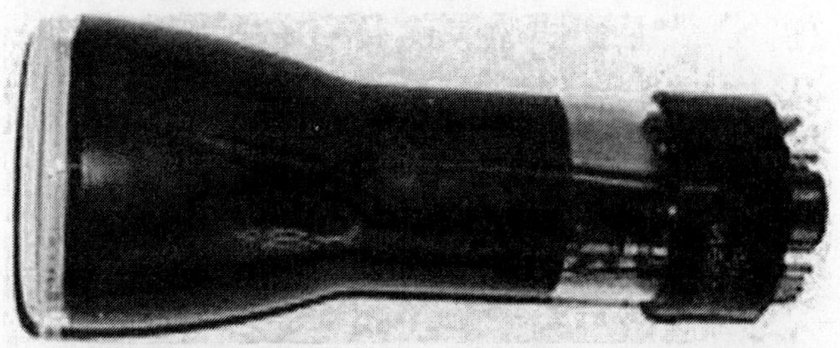

Fig. 4.43. An Electrostatic Cathode Rays Tube (CRT).

A PPI requires a CRT in which the screen is coated with a long-persistence phosphor. This is necessary because each target reflects energy for only a short period of time during each rotation of the antenna. Therefore, the target indication on the face of the CRT must be able to continue to glow during the portion of antenna rotation when the target is not reflecting energy.

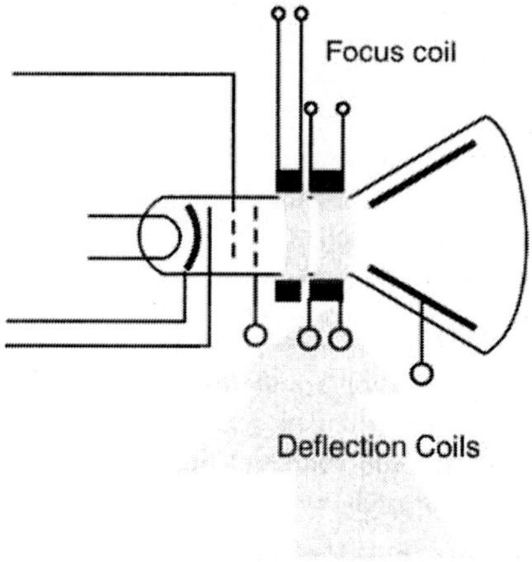

Fig. 4.44. Schematics of Electrostatics CRT.

4.13.2. PPI-Display (Located in the Cockpit)

The PPI Display present a "map style" view of the oncoming weather. It is common to have powerful controls to allow the pilot to zoom in and out of the viewing display screen as he requires.

4.13.3. CAPPI : (Constant Altitude PPI)

This is a horizontal, two-dimensional representation of radar measurement in a certain height (altitude). A CAPPI is calculated from several PPI measuring at different elevation. The different heights are usually represented by color variations on a PPI-Scope. In Fig. 8.11 shows the picture of CAPPI: (Constant Altitude PPI).

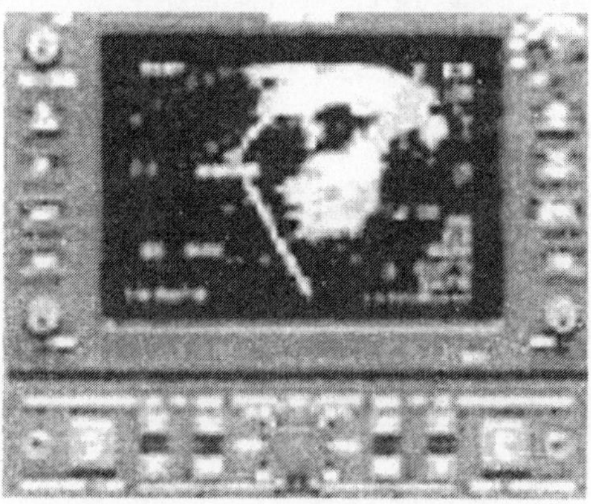

Fig. 4.45. CAPPI: (Constant Altitude PPI).

SUMMARY

- Radar transmitter is used to generate the adequate RF power in any radar system.
- There are some qualities of a radar system in which radar engineer is more interested. The TX should provide sufficient power to detect a target, be easily modulated to produce the desired waveform, generate a stable signal so that doppler signal processing can be performed easily, provide the proper signal band width, be of high efficiency, be of high relativity, be easy to maintain, have long life, be able to operate with a minimum of personal, be of a size and weight suitable for the intended applications, and be of affordable price.
- **There are two basic configuration of radar transmitter:**
  1. Self-excited oscillator
  2. Power amplifier
- There are different RF power sources available for high power radar application include the klystron, TWT, solid state transistor amplifier, Twystron, magnetron, cross-field amplifier, amplitron, grid controlled vacuum tube, extended interaction amplifier, gyrotron and others.
- **Magnetron.** The magnetron was invented by EW Hull of United States in 1921. This was the first microwave high power oscillator.
- **Radar Modulator.** The modulator is the device which turns the transmitting tube on and off in such a manner as to generate the desired waveform.
- When the transmitted waveform is a pulse, the modulator is some time called **pulser**.
- **Types of Modulators :**
  1. Grid Modulators
  2. **Anode Modulators.** Anode modulators can be further classified as Soft Valve Modulators and Hard Valve Modulators
- The modulator consists of a charging impedance, an energy storage element, and a switch.
- Three basic types of radar modulator used for magnetron oscillator :
  1. The line type modulator with a gas tube switch and a delay line storage element.
  2. The pulsactor, which uses saturable reactance for both switching and storage purposes and
  3. The hard tube modulator incorporating a vacuum tube switch and capacitor storage elements.
- The modulator containing a gas tube switch and a delay line as the energy – storage element is called **a line type modulator**.
- As circuit for storing energy the thyratron modulator uses essentially a short section of artificial transmission line, which is known as the **pulse-forming network (PFN)**.
- **Thyratron.** The function of the high-vacuum tube modulator is to act as a switch to turn a pulse ON and OFF at the transmitter in response to a control signal.
- The magnetron is non-linear impedance and will not be matched to the line under all conditions. The mismatch can cause a spike to appear at the leading edge of the pulse. The spike can be minimized by introducing an RC circuit in parallel with the primary of the pulse transformer. This is called the **De-spiking circuit.**

- The function of **damping network** is to help to reduce the trailing edge of the voltage pulse and prevent post pulse oscillations, which could introduce noise or false targets. The damping network consists of a resistance and a inductance in series at the secondary of the pulse transformer.
- **Radar Receivers.** The ability of a radar receiver to detect the presence of the echo and extract information in the presence of noise. There are noises which can enter the receiver through the antenna along with the desired signal are called **external noise** or it may be generated within the receiver itself is called the **internal noise**.
- **Types of Radar Receiver.** There are different types of radar receiver, which are employed in radar as per the radar application. Types of radar receiver are as follow:
  1. Super heterodyne receiver
  2. Super regeneration receiver
  3. Crystal video receiver
  4. Tuned radio frequency (TRF)
- **Automatic frequency control.** The AFC system normally employed to keep the receiver in tune with the transmitter is known as a **difference frequency system**.
- **Anti interference circuits :**
  1. Fast time constant circuit (FTC).
  2. Instantaneous automatic gain center (IAGC)
  3. Sensitivity time control (STC)
- **Noise.** Noise is unwanted EM energy, which interfere with the ability of receiver to detect the wanted signal. It may originate within the receiver itself, or it may enter via the receiving antenna along with the desired signal.
- There are various types of noise generated in side the receiver, which are as follows :
  1. Thermal noise
  2. Shot noise
  3. Partition noise
  4. Induced grid noise
  5. Crystal noise
  6. Flicker noise
  7. Transistor noise.
- **Nose Figure.** A noise figure or noise factor can be explained as a measure of the noise produced by a practical receiver compared with the noise of an ideal receiver. Thus, the noise figure Fn of a linear system may be defined as

$$Fn = Sin/Nin \times Sout/Nout$$

Or

$$Fn = Sin \times Nout/Nin \times Sout$$

$$Fn = Nout/KT\ BnG$$

Or

$$Fn = 1 + \Delta N/KTBnG$$

Where,

Fn = Noise figure

Sin = Available input signal power

Nin = Available input noise power ($KT\ Bn$)

$Sout$ = Available output signal power

$Nout$ = Available output noise power

K = Boltzmen constant = 1.38×10^{-23} J/°K

T = Standard temperature (in Kelvin = 290° K)

Bn = Band width

G = Gain

- **Noise Figure of Networks in Cascade.** The noise figure of 'N' networks in cascade form may be represented by

$$F_o = F_1 + (F_2 - 1)/(G_1) + (F_3 - 1)/(G_1G_2) + \ldots + F_{n-1}/G_1G_2 - GN - 1$$

- **Mixer.** The function of the mixer is to convert RF energy to IF energy with minimum loss and without spurious response.
- The conversion loss of a mixer is defined as

LC = (available RF power) / (available IF power)

It is a measure of the efficiency of the mixer in converting RF signal power into IF.

- The image frequency is defined as that frequency which is displaced from the local oscillator frequency fLo by the IF frequency, and which appears on the opposite side of the local oscillator frequency as the signal frequency fRF. It is equal to $2fLo\text{-}fRF$.
- The noise temperature ratio of a crystal mixer varies approximately inversely with frequency from about 100KHz, (the exact value depends upon the diode), down to a small fraction of a hertz. This is called flicker noise or $1/f$ noise.
- **Types of Mixers :**
  1. Balanced Mixers
  2. Reactive image termination
- **Diode Burnout.** A crystal, which is subjected to excessive RF power, may suffer burn out. This is rather a loose defined term, which is applied to any irreversible deterioration in the direction or conversion properties of a crystals diode as the result of electrical over load.
- **Duplexer.** The duplexer enables a radar system to transmit powerful signals and still receive very weak radar echoes. The duplexer acts as a gate between the antenna and the receiver and transmitter.
- **Types of duplexer.** There are different types of duplexer have been used are given below :
  1. Branch type duplexer
  2. Balanced duplexer
  3. Ferrite duplexer
- **Receiver Protectors.** The duplexer alone cannot always protect the receiver. In addition to the gaseous TR switch, a receiver might require ferrite or diode limiters to limit the amount of leakage that gets by the TR switch. Their limiters are called Receiver Protectors.

- **Modern Trends in Radar Receiver Protectors :**
  1. Wave guide receiver protector.
  2. Coaxial receiver protector
  3. Surface mount receiver protector.
- **Radars Antenna.** The radar antenna acts as a transducer, which converts electrical pulses from the transmitter to the free space in the form of EM waves and receives the reflected EM signals from the target in free space and converts it in to electrical signals.
- The important parameters of radar antennas
  1. Directivity or Directive gain.
  2. Power gain.
  3. Effective receiving aperture or total scattering cross section.
  4. Polarizations.
  5. Side lobes.
  6. Front-to-Back Ratio
- **Directivity or Directive gain.** The directivity of an antenna is given by the ratio of the maximum radiations intensity (power per unit solid angle) $u(\theta, \varnothing)$ mass to the average radiations intensity Uav (average over a sphere).

$$G_D = \text{Maximum radiation/Average radiation}$$
$$G_D = 4\pi/B$$
$$G_D = 4\pi/(\theta B\ \varnothing B)$$

Where θB and $\varnothing B$ are the half power beam width into principal planes such as horizontal and vertical respectively, minor labels are neglected.

If the half power beam width are measured in degrees, and minor labels effects are neglected than this can be written as

$$GD \approx 4\pi/(\theta B\ \varnothing B) \approx 41000\ /(\theta B^{o} \times \varnothing B^{o}) \quad (4\pi sr = 4125 \text{ square degree})$$

- The power radiated from an antenna per unit solid angle is called the **radiation intensity** u (watts per steradian or per square degree).
- The angular range of the antenna pattern in which the half of the maximum power at least is still emitted is described as a Beam Width.

$$B \simeq \theta_B \times \varnothing B.$$

- **Power gain.** The power gain, which is denoted by (G), includes the effect of the antenna losses and any other related losses, which reduces the antenna efficiency. It may be defined as the ratio of maximum radiation intensity from particular antenna to the radiation intensity from (loss less) isotropic source with some power input.
- The directive gain is always greater than power gain. If there is no losses, the both the gain (power and Directive) are more or less equal to each other. Both are related by the radiation efficiency factor (ρ_r) and can be defined as follow:

$$G = \rho r \times G.D$$

- **Effective aperture.** It may be considered as a measure of the effective area presented by the antenna to the incident wave.
- The effective area (Ae) may be defined as the product of the antenna aperture effiency (ρ_a) and the physical area of antenna.

$$Ae = \rho_a \times A$$
$$G = 4\pi \times \rho_a \times A/\lambda^2$$

Where λ = Wave length

Ae = effective area or effective aperture area

ρ_a = antenna aperture effiency

- **Polarization.** The direction of the electric field vector decides the direction of polarization of an antenna. The polarization may be linear, elliptical circular.
- **Major and Side (Minor) Lobes.** The radiation intensity in one lobe is considerably stronger than in the other. The strongest lobe is called a **major lobe**; the other is a side (minor) lobe.
- **Different Types of Radar Antenna :**

1. Half-wave antenna	2. Yagi antennas
3. Parabolic antennas	4. Antennae with cosecant squared pattern
5. Cassegrain Antennas	6. Phased Array Antennas

- **Various types of Parabolic Reflector :**
  1. Truncated parabola or cut parabola
  2. Cylindrical parabola
  3. Cheese antenna or pill box
- The directive gain or power gain of circular aperture with diameter D.

$$G_P = 6.389\,(D\backslash\lambda)^2$$

Where, D = diameter of circular aperture.

- $$HPBW = 70^{\circ}\,\lambda/D$$
- The gain for the rectangular aperture = $7.7(L/\lambda)^2$
- **Radar display.** The device used to display radar information is known as a radar indicator or radar display.
- There are two basic methods of displaying information's on a cathode ray tube.
  1. Deflection–Modulated CRT
  2. Intensity Modulated CRT
- In Deflection–Modulated CRT targets are indicated by the deflection of the electron beam, such as A-Scope. The other display method is the Intensity-Modulated CRT, an example of which is the PPI-Scope. It is normally dark in the absence of the target. A target is indicated by intensifying the beam and presenting a luminous spot on the face of the CRT.
- **The A-Scope.** Display represents only the range to the target and the relative strength of the echo.
- **PPI-Scope.** It is a polar coordinate display of the area surrounding the radar platform. Own position is represented as the origin of the sweep, which is normally located in the center of the scope, but may be offset from the center on some sets.
- **Range-Height Indicator (RHI).** The range-height indicator (RHI) scope is used with height finding search radars to obtain height information. The RHI is a two-dimensional presentation indicating target range and height.
- **Beta Scan Scope.** The Precision Approach Radar found information's are shown in a Beta Scan Scope with two "pictures". The upper picture shows the data determined by the antenna to the height scanning. It is also called Elevation Picture. The lower picture shows the data recorded by the antenna to the horizontal sector scanning. It also is named Azimuth-Picture.
- **Flat Panel Display (FPD).** There are some displays system which are being used for the radar also, such as Flat Panel Display, but all are not suitable for radar. Various types of FPDs are given below :

  (*a*) The Liquid Crystal Display (LCD)

  (*b*) The Plasma Display

REVIEW QUESTIONS

1. What do you understand by Radar Transmitter ?
2. What are the different types of later used for high power radar transmission ?
3. Draw the simple block diagram of Radar transmitter and explain each block in brief.
4. Draw the block diagram of radar modulator pulse type and explain it briefly.
5. Discuss the different types of radar modulator in brief.
6. How the modulator decides the shape of the pulse ?
7. Write the brief note on working of saturables reactor.
8. Explain a line-type modulator with the help of a block diagram. **[UPTU, 2003]**
9. Draw the circuit of line-pulsing modulator and explain its working. **[UPTU, 2004]**
10. With a neat diagram explain the principle and working of a line type modulator. **[UPTU, 2006]**
11. Write short notes on
    (*i*) Pulse forming network (*ii*) Line-type modulator
    (*iii*) Thyratron valve (*iv*) Damping network
    (*v*) Despiking network.
12. What do you understand by radar receiver and explain different types of radar receivers ?
13. Draw the simple block diagram of radar receiver using super hetrodyne as receiver.
14. What is effect of receiver bandwidth over the performance of radar ?
15. What do you understand by radar noise and explain the term noise figure ?
16. Discuss Noise figure of a receiver and noise figure due to RF losses. **[UPTU, 2003]**
17. Explain a balanced Duplexer with diagram. **[UPTU, 2003]**
18. Explain in detail receiver noise figure. **[M.D. University, 2003]**
19. Explain the purpose of mixer and write the different types of mixers used in radar.
20. Explain the principle of branch type radar duplexer with the help of diagram. **[M.D. University, 2003]**
21. How noise effect the detection process of radar ?
22. What are the factors influencing the bandwidth of a radar reciver ? What are the advantages and disadvantages of a very large bandwidth ? **[M.D. University, 2004]**
23. What is a mixer and also gives noise figure of a mixer used as front end ? **[M.D. University, 2003]**
24. What do you understand by ferrite duplexer ?
25. What are the factors influencing the BW (bandwidth) of a radar receiver. **[UPTU, 2004]**
26. Explain the terms "the duplexer cannot always do the entire job of protecting the receiver." **[UPTU, 2004]**
27. Draw the block diagram of a radar receiver and explain it. **[UPTU, 2004]**
28. Explain the operation of Radar receivers and mixer ? **[M.D. University, 2005]**
29. Explain the working of a image recovery mixer and give its limitations. **[UPTU, 2006]**
30. Discuss principle and working of balanced duplexer. **[UPTU, 2006]**
31. Explain the various types of duplexer. **[M.D. University, 2006]**
32. Write short note on :
    (*i*) Advantages of super hetrodyne receiver
    (*ii*) Duplexer **[M.D. University, 2005, 2007]**
    (*iii*) AFC

(*iv*) Balanced mixer

(*v*) Receiver protector **[M.D. University, 2005]**

(*vi*) Noise and clutter **[M.D. University, 2001]**

(*vii*) TR/ATR

(*viii*) Keep alive voltage

(*ix*) Radar mixer **[M.D. University, 2004, 2007]**

(*x*) Noise figure. **[UPTU, 2006]**

(*xi*) Noise temperature **[UPTU, 2006]**

(*xii*) Low noise front ends **[UPTU, 2006]**

(*xiii*) Receiver protections **[UPTU, 2006]**

(*xiv*) Radar receiver **[M.D. University, 2006, 2007]**

33. What do you understand by radar display? **[M.D. University, 2003, 2001]**

34. Explain different types of radar displays in brief. **[M.D. University, 2003]**

35. Explain the working of plan position indicator (PPI) with the help of diagram and why in this method called intensity modulation. **[M.D. University, 2004]**

36. Define CRT and explain its working with the help of sketch diagram.

37. Define the 'A' scope. **[UPTU, 2003]**

38. Write short notes on following :

(*i*) LCD's display (*ii*) PPI (*iii*) A-scope

(*iv*) Beta scan (*v*) Plasma display.

39. What are the various types of radar display? **[PTU, 2004, M.D. University, 2005]**

40. Explain A-scope and PPI display with reference to radars. **[PTU, 2004]**

41. What do you understand by PPI display ? Explain where it is used ? **[UPTU, 2004]**

42. Explain A-scope radar display.

[UPTU, 2006]

5

CHAPTER

RADIO DIRECTION FINDER

INSIDE THIS CHAPTER

5.1. INTRODUCTION

A **radio direction finder (RDF)** is a device for determining the directions of the energy flux of an electromagnetic wave (EM wave). It is the ability of radio wave to travel very long distances "over the horizon", it makes a particularly good navigation system for ships and aircraft that might be flying at long distances from land. Its principle use is in navigation, for determining the angle between the line of bearing connecting a transmitter and a receiver a known direction in space at one or other of those locations.

A direction finder when used at a receiving station consists of a directive array, a receiver having good performance characterstics, and a means for coupling the array to the latter. In case of a transmitting direction finder the receiver is, of course, replaced by a transmitter of suitable design.

RDF's work by pointing a directional antenna in "various directions" and then listening for the directions in which signal from a known station came through most strongly. This sort of system was widely used in 1930s and 1940s.

Certain frequencies are specified and allotted for navigational purposes using RDFs. The allotted frequencies are in the Low frequency (LF)/Medium frequency (MF), High frequency (HF) and VHF bands. The basic principle involved in all the range of frequencies allotted remain same. These will be discussed in this chapter in the following sections.

5.2. CLASSES OF DIRECTION FINDERS

It is commonly supposed that a direction finder always consists of a loop antenna and a receiver. This is very far from being the case, as a direction finder may be either a transmitter or a receiver. Furthermore, the directive antenna system may be installed either on a moving craft or at a fixed location. As far as navigation is concerned, the choice of transmitting or receiving equipment connected to the directive array is dependent only on convenience in relaying the directive informations to the point at which navigation is to be done.

As an example; if a ship desires to learn a line of position from a shore-station equipped with a directive antenna, it may transmit a signal to be picked up by the shore-station's direction finder receiver. Then by means of auxiliary circuits the information of the measurement of the line of bearing is communicated back to the ship for navigation purposes. If the shore-station has a transmitter connected to the array and transmits a signal in various directions on a predetermined schedule, the radio operator abroad ship may identify the characterstic direction of this beam by employing a non-directional receiver. This beam of a rotating radio beam transmitter ("radio light house") has been used extensively in earlier days.

There are different types of information furnished for navigation when the directive system is on the craft or at a fixed location. One gives a line of bearing measured at the craft, and the other gives a similar measurement, establishing a line of position, from the ground or fixed station.

A second difference exists between a directive system at fixed and mobile locations, and that is the accuracy of bearing obtainable. In most cases it is found that a ground or shore-station can use a larger, more carefully chosen site, in which the residual errors are small and can be allowed for by calibration, and this calibration will hold over a long period of time.

5.3. LOOP ANTENNA

The **loop antenna** is a radiating coil of any convenient cross-section of one or more turns carrying radio frequency current. It may assume any shape, rectangular, triangular, square, circle and hexagonal. A loop of more than one turn is often called as **frame**.

When an antenna consisting of one or more turns of a closed loop is considered, important directive properties are found. Consider a rectangular single-turn loop of length '*a*' and width '*b*' which is much less than λ (wavelength) in Fig. 5.1.

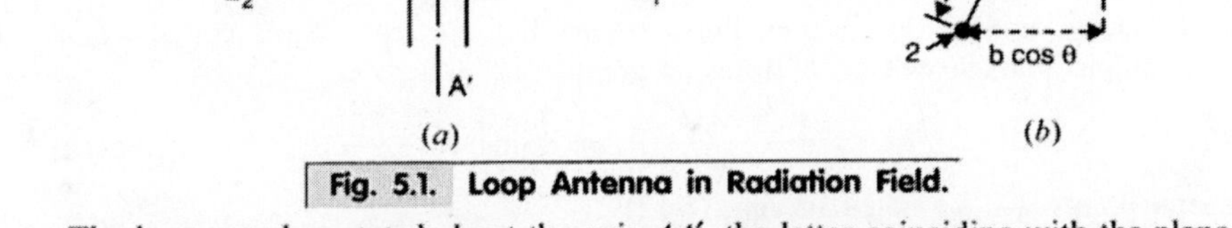

Fig. 5.1. Loop Antenna in Radiation Field.

The loop may be rotated about the axis *AA'*, the latter coinciding with the plane of polarization. The Fig. 5.1(*b*) shows the plan view. The field strength at two points $b \cos \theta$ apart, owing to a wave propagated in the direction shown are

$$E_1 = E_{max} \cos \omega t \quad ...(5.1)$$

$$E_2 = E'_{max} \cos\left(\omega t + \frac{2\pi b \cos \theta}{\lambda}\right) \quad ...(5.2)$$

The vertical conductors 1 and 2 pick up voltages proportional to a :

$$e_1 = E_{1a} \quad \text{...(5.3)}$$

$$e_2 = E_{2a} \quad \text{...(5.4)}$$

This follows because I is constant in the loop for $a = h << \lambda$. These voltages are approximately π out of phase at the terminals where the loop is broken, but there is a resultant e due to the phase change in the distance $b \cos \theta$.

$$e = (E_1 - E_2)a \quad \text{...(5.5)}$$

By putting the values of E_1 and E_2 from equations (5.1 and 5.2), we get

$$e = \left[E_{max} \cos \omega t - E_{max} \cos\left(\omega t + \frac{2\pi b \cos\theta}{\lambda}\right)\right].a$$

$$e = E_{max}\, a\left[\cos \omega t - \cos \omega t \cos\left(\frac{2\pi b \cos\theta}{\lambda}\right) + \sin \omega t \sin\left(\frac{2\pi b \cos\theta}{\lambda}\right)\right] \quad \text{...(5.6)}$$

If one sets $\phi = \dfrac{2\pi b \cos\theta}{\lambda}$

Equation (5.6) becomes

$$e = E_{max}\, a[\cos \omega t(1 - \cos\phi) + \sin \omega t \,.\, \sin\phi] \quad \text{...(5.7)}$$

As much as the loop dimensions are small compared to the wavelength (λ), the quantities $(1 - \cos\phi)$ and $\sin\phi$ may be compared for ϕ quite small.

$$1 - \cos\phi = \frac{\phi^2}{2!} - \frac{\phi^4}{4!} + \ldots$$

$$\sin\phi = \phi - \frac{\phi^3}{3!} + \ldots$$

so that $\dfrac{1 - \cos\phi}{\sin\phi} \cong \dfrac{\phi}{2}$ ...(5.8)

For example if $b = 1$ m, $\lambda = 300$ m and $\phi = 0$ then

$$\phi = \frac{2\pi \times 1}{300} = 0.021$$

and the error of neglecting $1 - \cos\phi$ in comparison with $\sin\phi$ in computing the magnitude of e in eqn. (5.7) cannot exceed 0.0002.

Thus $e = E_{max}\, a \sin \omega t \sin\phi$ ...(5.9)

or making the approximation

$$\phi \cong \sin\phi = \frac{2\pi b \cos\theta}{\lambda} \quad \text{...(5.10)}$$

$$e = \frac{2\pi ab\, E_{max}}{\lambda} \cos\theta \sin \omega t \quad \text{...(5.11)}$$

But ab is the area of the loop A. Furthermore, if there are N turns connected in series, the output voltage will be N times as great.

$$e = \frac{2\pi . N . A . E_{max}}{\lambda} \cos\theta \sin \omega t \quad \text{...(5.12)}$$

Several points will be noted in eqn. (5.12)

1. Since e is proportional to $\sin \omega t$, it is 90° out of phase with the electric field E_1. This is of great importance in direction finders in which the output of a loop is combined with that of a vertical antenna in the proper phase.
2. The voltage 'e' varies as $\cos\theta$, giving the figure of 8 pattern of Fig. 5.2.
3. The voltage pickup varies as the loop area is directly proportional to it. It can be easily demonstrated that for any shape of plane loop, it is the area that enters into the equation corresponding (5.12).

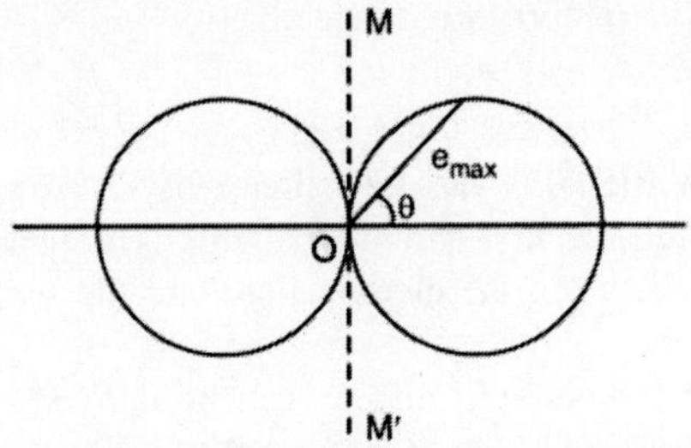

Fig. 5.2. **Directive Pattern (Figure of 8) of Loop Antenna.**

4. The voltage e is directly proportional to the frequency. If the effective height is defined as the voltage amplitude for unit field when $\theta = 0$,

$$h_e = \frac{2\pi NA}{\lambda} = \frac{2\pi NAf}{c} \quad ...(5.13)$$

Example 1. A circular loop 24 in. in diameter with eight turns operates at 1500 kHz (λ = 200 m). The value of h_e is required.

Solution. The loop diameter d = 0.61 m, and

$$A = 0.292 \text{ m}^2$$

$$h_e = \frac{2\pi \times 8 \times 0.292}{200} = 0.0734 \text{ m}$$

$$h_e = 0.0734 \text{ m}$$

5.3.1. Loop and Vertical Antennas

A type of two element array that is of great importance in direction finding is the combination of a vertical loop and an antenna that is non-directional in the horizontal plane. The separation of the two will be assumed to be such a small fraction of a wavelength as to introduce no relative phase shift. This is almost always met in practice with a loop-type direction finder having a vertical "sense antenna".

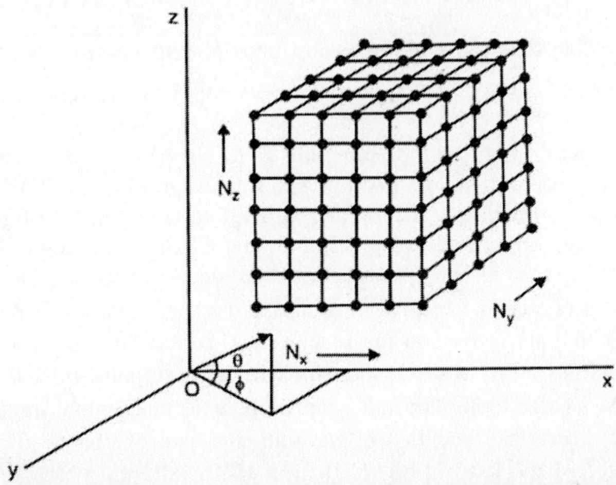

Fig. 5.3. **Three-dimensional array of point sources. Phase shifts between successive sources in multiples of one period are b_x, b_y and b_z. The separations of successive sources are $\alpha_x\lambda$, $\alpha_y\lambda$ and $\alpha_z\lambda$. (Southworth, Proc. I.R.E., 18, 1502, September, 1930).**

In this case the directive pattern are, respectively,

$$E_1 = \cos\theta \cos\omega t \qquad ...(5.14)$$

and

$$E_2 = n\cos(\omega t + \alpha) \qquad ...(5.15)$$

where the constant of proportionality has been ignored. The quantity n is the ratio of the maxima of the second and the first antenna patterns. The phase constant α includes the phase difference of $\pi/2$ between the electric field and the loop terminal voltages as was discussed earlier.

The resultant of the two vectors E_1 and E_2 can be derived as follows.

If exponential notation is used, eqn. 5.14 and 5.15 becomes

$$E_1 = \cos\theta e^{j\omega t}$$

$$E_2 = ne^{j(\omega t + \alpha)}$$

The resultant is $E = E_1 + E_2 = e^{j\omega t}(\cos\theta + ne^{j\alpha})$

Figure 5.4 shows that the magnitude E_0 is

$$E_0 = \cos\theta + ne^{j\alpha}$$

This has a length of $|E| = \sqrt{(\cos\theta + n\cos\alpha)^2 + n^2\sin^2\alpha}$

or

$$|E| = \sqrt{n^2 + \cos^2\theta + 2n\cos\theta\cos\alpha} \qquad ...(5.16)$$

From this result one may proceed to several cases of importance. When $\alpha = 0$, the quantities E_1 and E_2 (representing either radiated field or voltage picked up in receiving antennas) are in phase or π out of phase, depending on θ, and it becomes

$$|E| = \cos\theta + n \qquad ...(5.17)$$

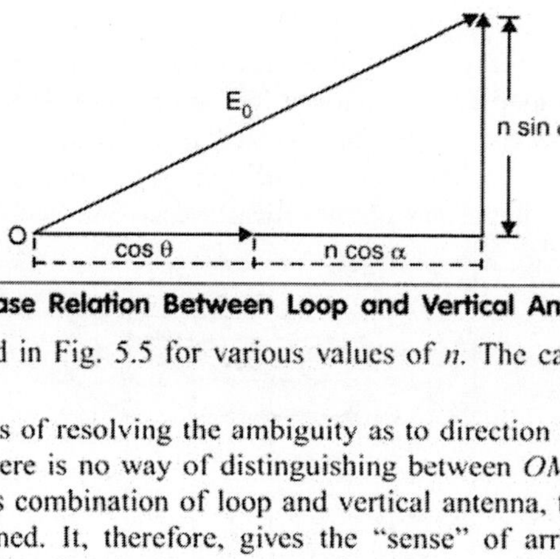

Fig. 5.4. Phase Relation Between Loop and Vertical Antenna Voltage.

This is illustrated in Fig. 5.5 for various values of n. The case for $n = 1$ is that of a cardioid.

It provides means of resolving the ambiguity as to direction of arrival of the waves to the loop alone. There is no way of distinguishing between *OM* and *OM′* of the loop alone, while with this combination of loop and vertical antenna, the minimum direction *ON* is uniquely defined. It, therefore, gives the "sense" of arrival of the signal and accordingly the non-directional antenna used with the loop is called a "**sense antenna**".

The directional property of the loop antenna may be used for finding the directions. The output of loop antenna is feed to the receiver and turning the loop untill a null is obtained. The direction of EM wave is then fix with an ambiguity of 180° as the one normal to the plane of the loop. The null, rather than the maximum signal is chosen because it is more clearly defined that at the null, the rate of change of signal with angular displacement of the loop is higher than at the maximum of the figure of 8.

The two nulls are indistinguishable from each other while considering only the magnitude of the signal, that will arise the ambiguity of 180°. But if we consider the phase of the signal it is distinguishable. Suppose, if the loop is turned clockwise than

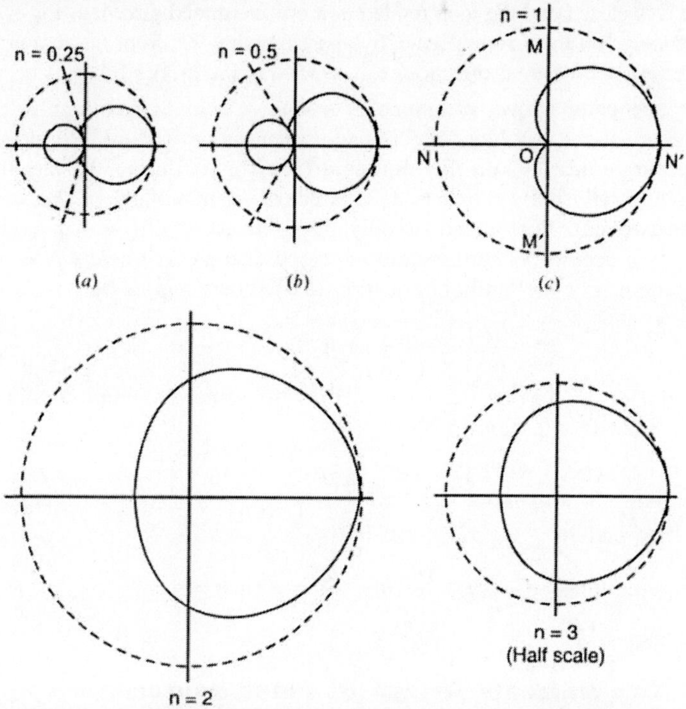

(a) (b) (c) (d) (e)

Fig. 5.5. Directive Patterns of Loop and Vertical Antennas for Various Values of Relative Pickup.

the phase of signal with reference to the field changes from $\pi/2$ leading to $\pi/2$ lagging, and the other null, it changes from $\pi/2$ lagging to $\pi/2$ leading. The process of finding the phase of signal with respect to the reference field is known as sense - finding, which will resolve the ambiguity.

5.3.2. Bilateral and Unilateral Bearings

Applications of this is seen in the circuit of Fig. 5.6, which is adapted for taking aural null bearings with the loop alone ("bilateral") or with loop and sense antenna ("unilateral"). Operation consists in rotating the loop to a minimum signal position with switch S open, thereby establishing accurately the pair of bearings OM and OM'.

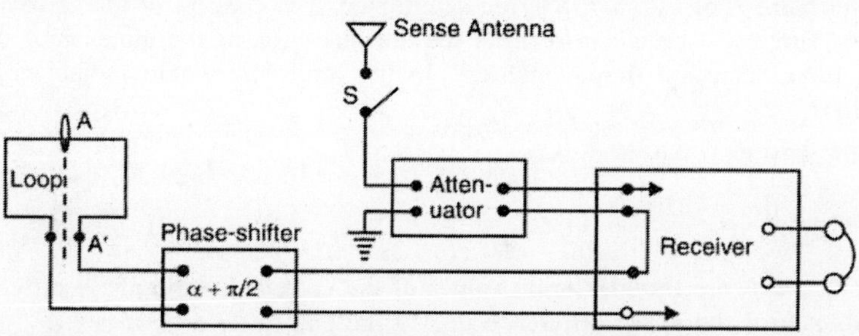

Fig. 5.6. Circuit for Obtaining Bilateral and Unilateral Bearing by Means of an Aural-null Direction Finder.

The S is closed and the loop rotated in a predetermined direction, for example, towards higher readings on the azimuth scale, by approximately 90° from the first null. If the signal has increased, the original direction, e.g., OM, was the proper bilateral bearing.

The attenuator allows variations of n to give a high maximum-to-minimum ratio along the unilateral null like NoN'. Its adjustment is not critical. No attempt is made to secure accurate bearings on the minimum ON. This is due to the fact that it is a null only for n exactly unity. For $n < 1$, it is actualy a maximum of the smaller lobe and there are two null lines symmetrically disposed about ON and diverging by greater amounts as n decreases. Furthermore the slope of the curve near ON is much less than cosine pattern. For the cardioid equation (5.1) shows this to be

$$\frac{d|E|}{d\theta} = -\frac{1}{2}\sin\theta \quad \text{...(5.18)}$$

The factor 1/2 having been included to the same maximum amplitude as for the figure of 8. At $\theta = 0$, eqn. (5.18)

$$\frac{d|E|}{d\theta} = 0 \quad \text{...(5.19)}$$

for the loop pattern

$$E = \cos\theta \quad \text{...(5.20)}$$

and at the minimum $\theta = \pm\pi/2$, so that $\dfrac{dE}{d\theta} = -\sin\theta = \pm 1$

at these angles.

5.3.3. Two-elements Arrays of Point Sources

A fictitious "point source" having equal radiating properties in all directions may be assumed for the purpose of simplifying the analysis of array of a number of elements. When this is done formula may be obtained for directivity as a function of the spacing and phasing of the currents in these point sources. Then superimposed upon this space function will be that due to the directivity of the individual elements if they consist of finite sources.

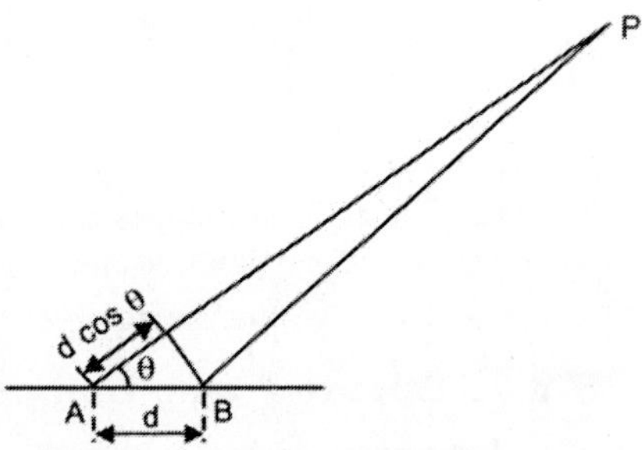

Fig. 5.7. Radiation from Two Point Sources.

If two point sources are separated by a distance d and have flowing in them currents having a phase difference α, the field at a distant point P of Fig. 5.7. may be determined. The concept of the current flowing in a point source will be taken to mean the limiting case, as the dimensions diminish, of a radiator of constant dipole moment. In the limit this would radiate equally in all directions.

The field at P due to A is

$$E_1 = KI_1 \cos\omega t \quad \text{...(5.21)}$$

and due to B,

$$E_2 = KI_2 \cos(\omega t + \alpha + \theta) \quad \text{...(5.22)}$$

where I_1 and I_2 are the maximum values of the currents, α the phase difference in the currents, and θ the phase difference due to the path difference $d\cos\theta$.

$$\theta = \frac{2\pi}{\lambda} d\cos\theta \quad \text{...(5.23)}$$

K is an appropriate constant of proportionaility.

The resultant field is $E = E_1 + E_2$

$$E = Ke^{j\omega t}(I_1 + I^2 e^{j(\alpha+\theta)}) \quad ...(5.24)$$

(It is instructive and more compact in form at this point to utilize exponential notation for equations (5.21 and 5.22)

or if $I_2 = nI_1$

$$E = KI_1 e^{j\omega t} . (1 + ne^{j(\alpha+\theta)}) \quad ...(5.25)$$

5.4. ERRORS IN LOOP DIRECTION FINDER

Loop antennas are subjected to certain errors. Most of these errors can be minimised by proper calibration and construction. The main errors are :

(*i*) Antenna or vertical effect.

(*ii*) Night effect or polarization error.

(*iii*) Error due to abnormal propagation.

5.4.1. Antenna Effect or Vertical Effects

Loop direction finding entirely depends on the fact that when the loop is perpendicular to the direction of propagation, no signal is heard. But anything which causes signal to be heard in this position, is harmful for accurate direction finding. This vertical effect arises due to unequal action of the two vertical sides of the loop antenna. The loop antenna gives spurious bearing if it is not electrically balanced with respect to ground. It can be understood with the help of following figure (5.8)

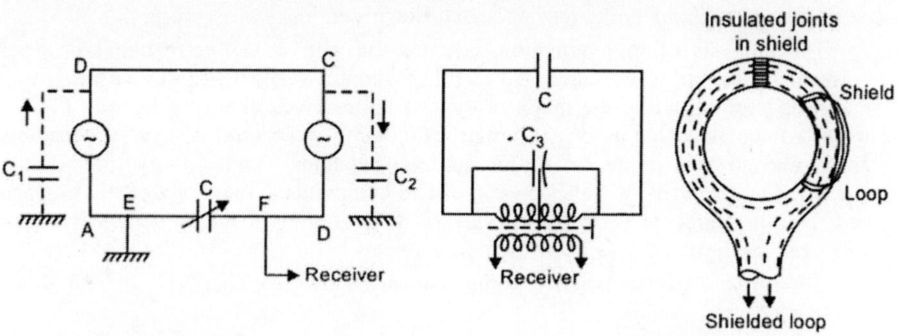

(*a*) Unbalanced loop.

C = Tuning capacitor

C_1 = Capacitor between vertical side AD and ground

C_2 = Capacitor between vertical BC and ground.

(*b*) Balanced loop.

C = Tuning capacitor

C_3 = Compensating capacitor (it is a three plate capacitor having two stators.)

Fig. 5.8.

When the loop is in perpendicular position or zero signal position, emf induced in the vertical sides are having equal magnitude and phase and being directed around the loop in oppossite direction. Thus producing zero resultant voltage around the loop circuit. If one side of the tuning capacitor C is grounded as shown in Fig. 5.9 asymmetry occurs. This results in capacitance current to flow to ground through C_1

and C. However, the corresponding current through C_1 and C does not flow. This shows points E and F are at different impedance as E and F are at different potentials.

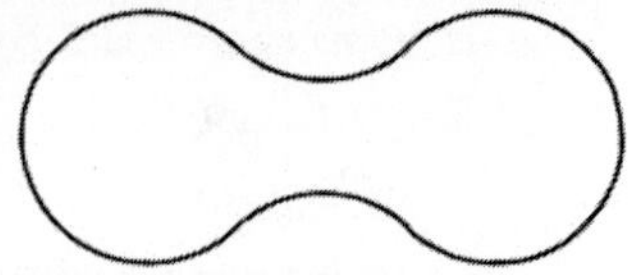

Fig. 5.9. Indistinct Minima Due to Vertical.

It will cause unbalance. Hence, a potential difference is developed across tuning capacitor C and a signal will be heard in the receiver even if the loop is in zero signal position. Due to this vertical effect, either no zero signal position or if it is zero signal will not occur when the loop is at right angles to the transmitter direction. In other words, two zero signal positions will not exactly be spaced 180° apart with respect to each other. The minima may be in distinct as shown in Fig. 5.9. This will introduce errors in the bearing reading.

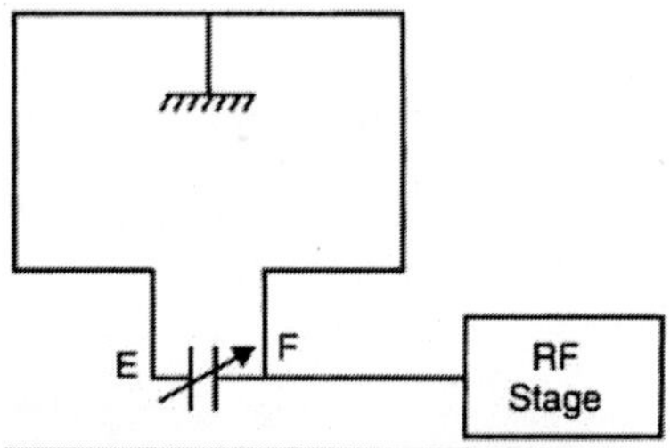

Fig. 5.10. Eliminating Antenna.

The error occurred due to vertical effect may be minimised by rising circuit which is symmetrical w.r.t. ground as shown in Fig. 5.10. Normally in loop antenna a compensating capacitor is employed to compensate for any residual unbalances. The confirmation of balancing arrangement is obtained by varying the compensation untill 180° spaced nulls are obtained.

5.4.2. Polarisation Errors or Night Effect

A loop antenna will give correct bearing only when the passing EM waves are vertically polarised *i.e.* electric vector vertical and magnetic vector horizontal w.r.t. earth and there is no horizontal polarised down coming waves present.

The analysis of loop equation assumes that the waves were being propagated horizontally toward a loop arranged to rotate about a vertical axis AA'. As it is known that over large distances the major portion of the received energy is by way of the sky wave. At the receiving points this may be coming downward at a very considerable angle, and so the simple derivation for loop behavior is not strictly true. One may consider that the arriving waves consist of two components; one horizontally propagated and the other directed vertically downward. It is obvious that any diagonally directed wave can be resolved into these two components.

The horizontally propagated component are shown in Fig. 5.11 in both side and plan views.

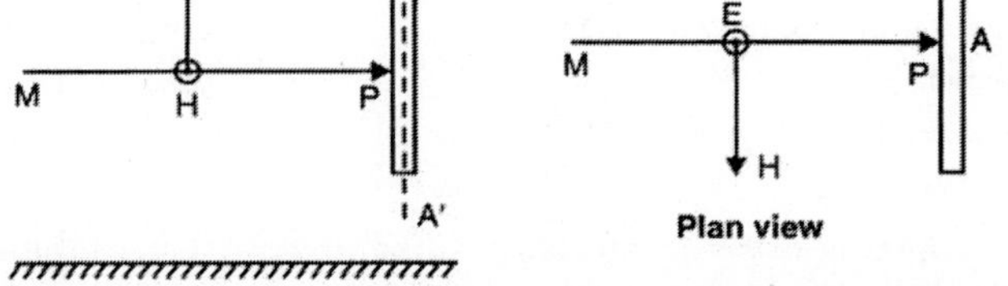

(*a*) Loop in null position (side view). (*b*) Plan view.

Fig. 5.11. Loop and Field Relations-Horizontally Propagated Components.

The electrical field vector E is vertical while the magnetic field H is horizontal for the case of vertical polarisation. If a loop is employed it is obvious that the antenna will be in a position of minimum pickup when its plane is at right angle to the direction of propagation. This is true because all the conductions lie in the plane of the wave front. It is usually simpler to consider the behavior in terms of the magnetic component *H*.

The loop pickup will be zero when *H* is parallel to the plane of the loop. It is evident that the same conclusions will be reached whether the *H* or the *E* components is analysed.

Next, the downward directed ray of Fig. 5.12 may be studied. The direction *MP* is the same as for the case of horizontal propagation and serves to define an azimuth direction in this case. Here the plane of polarisation lies along *MP*. The vector *H* lies in the plane of the loop when the latter is in the position shown. This, therefore, is the null angle, and it will be seen to be the same as for the horizontally propagated component of Fig. 5.12. This case of downward propagated components may be called that of the "**normally polarised wave**".

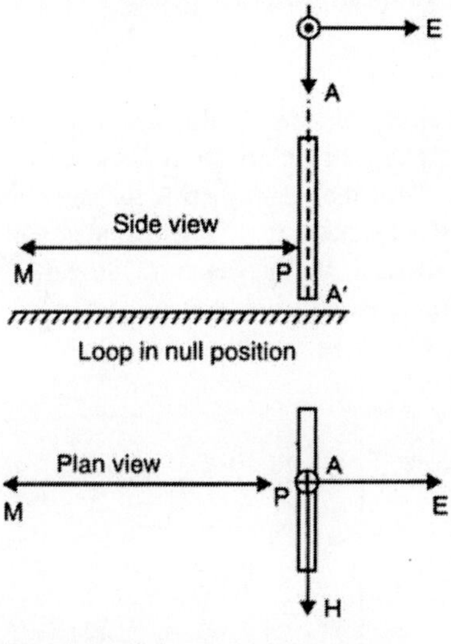

Fig. 5.12. Loop and Field Relations Normally Polarised Components Propagated Vertically Downward.

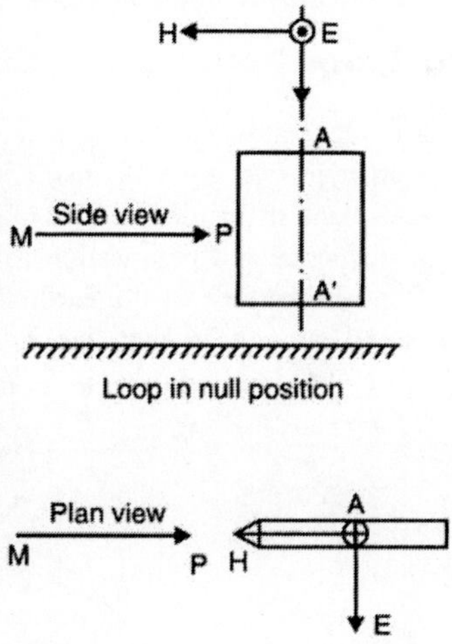

Fig. 5.13. Loop and Field Relations at Abnormally Polarised Components Propagated Vertically Downward.

In contrast, one may take the case where the plane of polarisation of the downward directed ray is rotated 90° w.r.t. direction MP. This is case shown in Fig. 5.13. Since *H* lies along the line MP, the loop will have its positions of minimum when its plane coincides with MP. It will be observed that this component gives an apparent bearing 90° displaced in azimuth from that of the two preceding cases and consequently, in referred to as the "**abnormally polarised wave**". It is this components which accounts for the errors in medium wave direction finding at night known as "**night effect**" and for the major errors at all times on short-wave direction finders.

The horizontally propagated components, whether vertically or horizontally polarised, cannot give rise to night effect or short-wave polarisation errors under any condition. Only the downward directed radiations from the ionosphere, or in certain cases from aircraft transmitters, will produce this effect. The whole class of direction finder derivations of this type, with loops or other types of antenna arrays, may be called "**polarisation errors**".

Night effect may change very greatly in magnitude over a short time of observation because of the rapidly changes in ionosphere conditions.

The characteristics of night effect with an aural null direction finder are :

1. Displaced minima, which may be included in defineness of the minimum or a well defined null point displaced from the proper bearning. In several cases of night effect this displacement or deviation may be 90°.
2. Broad minima, poorly defined but not displaced.
3. Shifting of the null or minimum point during the time of observation or over a few minutes.
4. Abnormal balancer setting in an aural null direction finder, indicating the presence of a large amount of abnormally polarised components.
5. **Fading.** It is due to the interference between the sky wave and ground wave, and, therefore, indicates that the downward coming component is of very appreciable magnitude. Fading may also occur owing to the interference among sky-wave components that have followed different paths.

5.4.3. Loop Polarisation Error

It is evident that when a wave is travelling diagonally downward toward a vertical loop, an error in bearing will result if the plane of polarisation does not coincide with the vertical plane of incidence. The magnitude of the bearing error for any combination of angles of incidence and polarisation may be calculated. It will be assumed that the loop is remote from the surface of the earth. The effect of the presence of the earth's surface is usually small enough so that this assumption is justified.

If the *xz* plane is taken horizontal in Fig. 5.14.

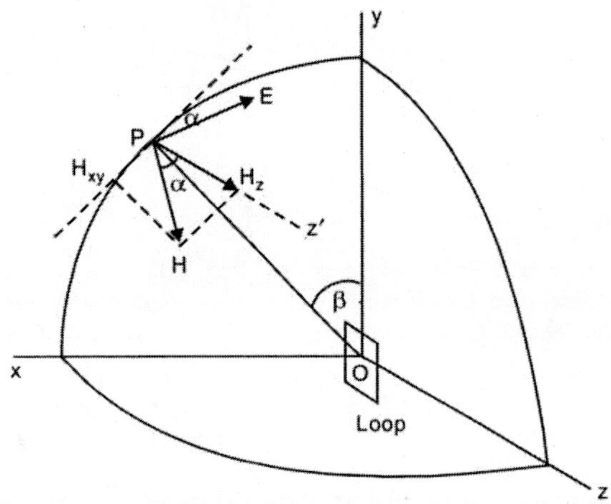

Fig. 5.14. Vector Relations for Wave Propagated Obliquely Downward on Loop Antenna.

With the loop at the origin, the incident wave is propagated along *PO*, with an angle of incidence β. The plane of polarization (*i.e.*, that containing the electric field vector *E*) is assumed to make an angle α with the *xy* plane. For convenience the analysis will be made in terms of the components of the magnetic field *H*, which also makes an angle α with *Pz′* drawn parallel to *Oz*. The *z* component of *H* is

$$H_z = H \cos \alpha \qquad ...(5.26)$$

While the component normal to H_z, lysing in the *xy* plane and tangent to the circle at *P*, is

$$H_{xy} = H \sin \alpha \quad ...(5.27)$$

This quantity may be resolved into x and y components :

$$H_x = H_{xy} \cos \beta$$

$$= H \sin \alpha \cos \beta \quad ...(5.28)$$

$$H_y = H_{xy} \sin \beta$$

$$= H \sin \alpha \sin \beta \quad ...(5.29)$$

The components H_y contributes no voltage to the loop whose axis lies along O_y, since no flux due to this passes through the loop in any position. Hence only H_x and H_z need be considered.

If the wave is propagated horizontally along O_x, $\beta = \pi/2$ and $H_x = 0$, leaving H_z the only component acting on the loop. The plane view of Fig. 5.15 shows the loop in the null position, for which there is no pickup due to H_z. But when $\beta \neq \pi/2$, H_x and H_z give a resultant H_{xz}. Thus the loop must be rotated through the angle δ to the position where its plane is parallel to H_{xz}.

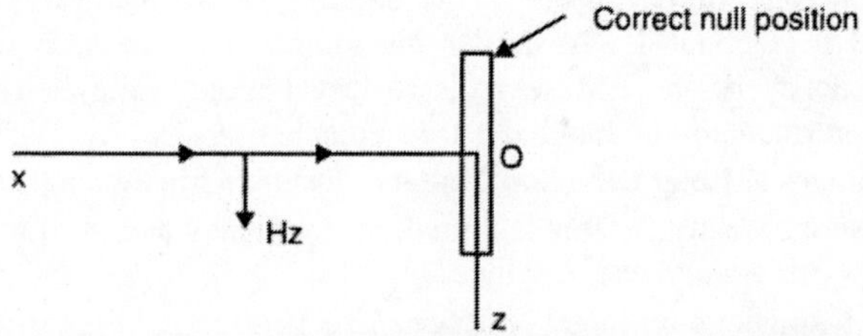

Fig. 5.15. Plan View of Loop in Correct Null Position for Horizontal Propagation.

The angle δ shown in Fig. (5.16) is the deviation error due to polarization effect.

$$\tan \delta = \frac{H_x}{H_z} \quad ...(5.30)$$

$$= \frac{H \sin \alpha \cos \beta}{H \cos \alpha}$$

$$= \tan \alpha \cos \beta \quad ...(5.31)$$

From the above eqn. (5.31), it is evident that waves propagated horizontally will not contribute polarization error for any α, since $\cos \beta = 0$.

Just as is the case in radio receiver measurements, where a "standard signal" is defined in order to give typical measurements, so also may one define a "standard wave" in direction finder. It is a wave downward directed at an angle of 45° ($\beta = \pi/4$) with plane of polarisation at an angle of 45° to the plane of incidence ($\alpha = \pi/4$). The components identified above as H_x and H_z are assumed to be in such time phase as to make δ a minimum.

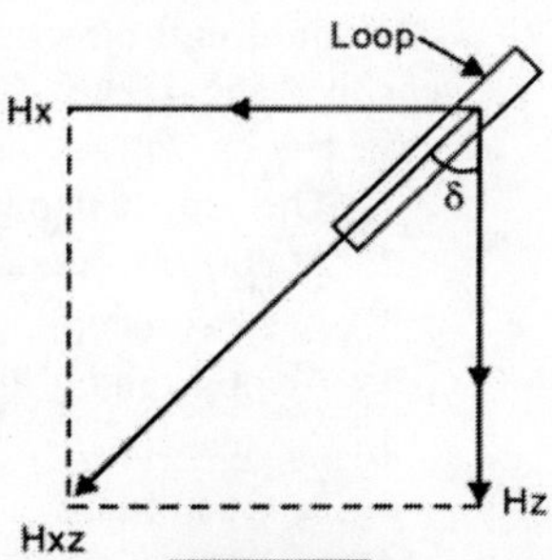

Fig. 5.16.

The standard wave error is defined as the value of δ for such a standard wave from eqn. (5.30) for a loop antenna,

$$\tan \delta = \tan \pi/4 \cos \pi/4 = 0.707$$

$$\delta = 35.2° \quad ...(5.32)$$

This value of standard wave error is almost independent of the presence of the earth of finite σ (conductivity).

The parallel loop direction finder has very small standard wave error. This is true along the line of the principle null. Under given conditions the loops are parallel to H_x, so that this latter gives no pick up. Thus the abnormally polarised component does not cause an error. Proper balancing of the loop circuit is required to make the induced voltage due to H_z be zero in the direction given.

5.4.4. Errors Due to Abnormal Propagation

It is assumed that the EM waves travelled along the great circle path from the transmitter to the direction finder. But sometimes the path deviates from the actual great circle planes, when the propagation is via the ionosphere. Such deviation can occur owing to scattered reflection and tilt of the reflecting region.

These phenomenon are more evident at high frequencies and error may rise upto approximately 1° and can be specified in terms of RMS value.

There are some more errors occurs such as "coastal refractions" and "mountain effect". Coastal refractions caused by the bending of wave towards the land from the coast line and it is about 1°. A similar phenomena may be seen in mountain terrain where the tendency of the EM wave is to travel along valleys. The direction finder shows irregular fluctuations about the true course.

Disadvantages of Loop Direction Finders. It suffers from the following disadvantages :

(*i*) It is suitable only for low and medium frequency and not for higher frequencies except for a very less distance.

(*ii*) There spurious induced voltages are produced if nearly loop, wires and conduction are present.

(*iii*) Its transmission efficiency is very poor and hence cannot be used for transmissions purpose unless the loop designs are modified for UHF uses.

(*iv*) It is subjected to antenna effect and night effect which of course are minimised by balancing and using Adcock antenna. If horizontally polarised down coming waves are absent then bearing accuracy up to 1° to 2° may be achieved.

5.5. AN AURAL NULL DIRECTION FINDER

5.5.1. Components and Types of Operation

An aural null direction finder comprises essentially a directive array confined with a sensitive and highly stable receiver. For mobile installations, as on ship board or aircraft, a loop antenna is almost invariably used. Then the elements consist of

1. The loop antennas
2. A non directional vertical or sense antenna.
3. A sense coupling circuit.
4. A loop tuning and coupling circuit
5. A balancer circuit (usually used in ship board only)
6. A receiver.

Most direction finders of this type are arranged for three types of operation :

1. Communication reception of modulated continuous waves or continuous wave signals. This is often carried out by means of the non-directional vertical antenna alone, or the loop may be used at or near its operations.

2. Bilateral Operations. Here the signal is picked up on the loop, which has of course two directions of null pick up. The result is an accurate bearing but with 180° ambiguity.

If a balancer circuit is used, the signal is also coupled into the loop input circuit in phase quadrature from the nondirectional antenna and is adjusted in amplitude to give the sharpest loop minima. Either modulated CW and CW reception is possible.

3. Unilateral Receiptions. This is also referred to as sense determination or operation with a cardiod pattern. Signals from the loop and the vertical antennas are combined in proper phase relationship to give a single null pattern. This mode of operation permits identification of the appropriate null as determined in bilateral operation, but the accuracy of the unilateral null is insufficient to allow precise bearing determinations by this means alone.

Loop circuits are almost invariably tuned either directly by means of a capacitor or through one or more coupling transformers. Satisfactorily operation in easily obtained at distances upto 35 ft separation between the loop and receiver proper. In modern direction finders the loop antenna is always shielded and is usually balanced to ground. These precautions serve to minimise antenna effect and in the case of aircraft direction finders to reduce precipitation static to a minimum when the vertical antenna is coupled into the receiver at the same.

5.5.2. An example of Shipboard Direction Finder Circuits

An aural-null system when it is used in marine service, it incorporates means for obtaining bilateral bearings including a balancer circuit and means for obtaining "sense". The basic circuit for one band is shown in Fig. 5.18 on shipboard, the presence of metallic objects such as guys, stacks etc. tends to produce, by re-radiating, undesirable voltages in the loop, provision is made in this circuit to introduce a quadrature voltage which can be adjusted to cancel the undesirable picked-up voltage. The operation of this circuit is as follows.

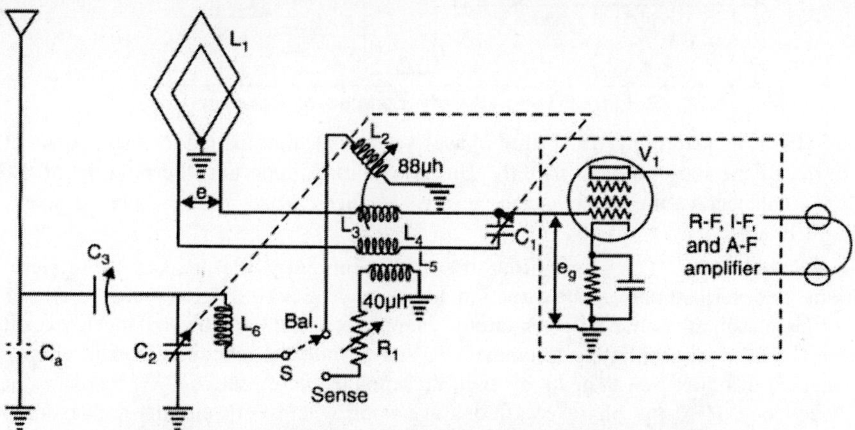

Fig. 5.18. Circuit of Shipboard Direction Finder Showing Balancer and Sense Connections.

Bilateral Operation. Under these conditions this circuit contains the loop inductance L_1 together with two auxiliarly coils L_3 and L_4. All these are resonated by means of capacitor C_1, which is balanced with respect to ground. The phase of the loop terminal voltage is 90° with respect to the incident radiation field. Then the circuit $L_1L_3L_4C_1$, resonant at signal frequency, a phase shift beween input and output of 90° will exit at resonance. The total phase shift due to both these causes is thus 0° or 180°, dependent upon the orientation of the loop. The output of the loop circuit is balanced to ground but is fed to the single-sided amplifier

V_1 in such a fashion that only half of the voltage developed across C_1 is utilized. This diminishes the gain of the system by 6 db but has practically no effect upon signal to noise ratio when gain of V_1 is sufficient high. Time as the loop feeds in, it is important to establish the correct phase relationship beween the signals for use either in balancing or sense determination. In the former the purpose of the antenna signal is to balance out voltage picked up by the loop or circuits and in time quadrature to the desired signal. In the case of sense circuit the antenna should feed a voltage to the receiver in time phase with that of the loop circuit in order to produce a cardioid pattern.

5.5.3. Phase Relations in Senses Resonant Circuits

The loop antenna of a directly tuned input circuit may be coupled to the first RF amplifier by several means, one of the most common circuits being that of Fig. 5.19. Here the loop has the mid point grounded. A shilded transmission line couples the balanced loop output to tuning capacitor C_1, which itself is symmetrical with respect to ground. The RF amplifier is "single sided", *i.e.* not push pull, and is arranged to utilize one-half the voltage appearing across C_1.

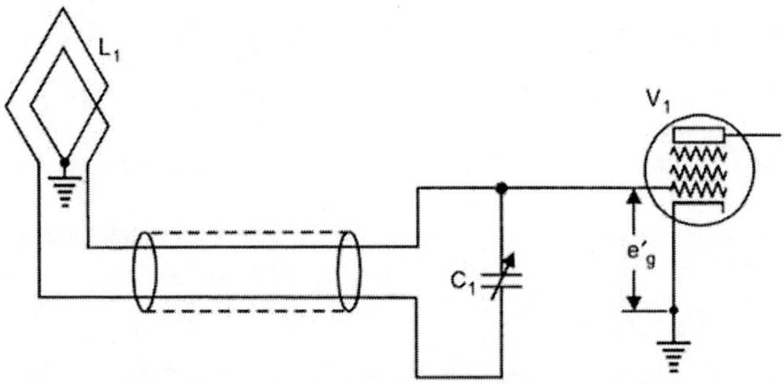

Fig. 5.19. Balanced Loop Directly Coupled to Receiver Unit.

The effect of quadrateral (90° time phase) voltage in diminishing, the sharpness of the minima of the loop pick up pattern. The proper magnitude and the polarity of the introduced balancer voltage together serve to make these broad minima become sharp, well-defined nulls.

Sense Operation. For eliminating the 180° ambiguity that occurs in bilateral operations as described above, the direction finder may be used to give "sense". This is done by means of the same antenna circuit as was described for the balancer, except that switch S is set appropriately to connect L_5 and R_1 into the antenna resonant circuit. L_5 is smaller in inductance than L_2 by such an amount that circuit $L_5L_6C_2$ resonates at signal frequency. Here the phase conditions are same where both circuits are resonant at signal frequency. There is, therefore, a 0° phase shift between radiation field and the grid of tube V_1. The sense antenna voltage thus mixes with loop signal at this grid with either a 0° or 180° phase relationship. These are the conditions under which a cardioid may be obtained. The resistor R_1 perform two functions. In the first place, it broadens out the response curve of the antenna circuit by lowering the Q so that the phase shifts relatively slowly on either side of the resonance frequency. In the second place, R_1 serves to adjust the relative magnitude of the antenna and loop voltages. If R_1 is made an operating adjustment, it may be set to give approximately a cardioid. However, in many direction finders the balancer system consists of a non-directional vertical antenna coupled through a series capacitor C_3 to the circuit $L_2L_6C_2$. This circuit in tuned coupled by a variable mutual inductance to L_3 in the loop circuit. Capacitor C_2, together with the parallel combination of C_3 and C_a, resonates this circuit at a frequency of 15 to 20 percent below signal frequency. Switch 'S' is on position "Bal" for balancing operation. The coupling between L_2 and L_3 may be adjusted from 0 to a rather large value, and its polarity may be reversed. The balancer and loop circuits together constitute two coupled resonant circuits, one is resonating at signal frequency and the other is tuned somewhat below. Because of the small frequency difference between primary and secondary, one may take it to be equivalent to the case in which the both are resonated to a frequency below that of the signal. It will be noted that at resonance the phase shift is 0° beween two resonant circuits, the primary of which is coupled through a reactance from a low impedance generator. At a lower frequency this phase difference approaches 90°. This means that there will be approximately a 90° shift between the antenna and the point at which e_g is measured. The loop and the antenna voltages at e_g at 90° apart in phase. As a consequences, the balancer may bs used to neutralize any pick up in the loop when this pickup voltage is in 90° time phase with respect to the desired signal. This may be due to antenna effect or to a quadrature re-radiated field.

It is set at one frequency only on each band. Then at other frequencies the cardioid is not perfect, but there is an easily distinguishable difference between amplitude at 90° and 270° azimuth positions. This difference is sufficient to established sense uniquely.

A loop type direction finder for marine use generally requires considerable care in installations, in adjustment, and in calibrations in order to give satisfactory operations. Installation includes the choice of loop location with respect to fixed structures and rigging in the ship, and also the addition of compensator to reduce residual bearing errors to a minimum. This is used at frequencies of 1500 KHz and below.

5.6. ADCOCK DIRECTION FINDERS

The horizontal members of a loop contribute nothing to the directional properties for a wave arriving in the horizontal plane. They are responsible for bearing error due to night effect. If these horizontal conductors could be eliminated or their effect nullified, accuracy would be improved. In the case of a single turn loop this can be done by removing the upper conductor, and shielding or balancing out the effect of the lower horizontal wire. Such a directive array is known as an Adcock Antenna.

Two types of Adcock Systems are shown in Fig. 5.20 and Fig. 5.21.

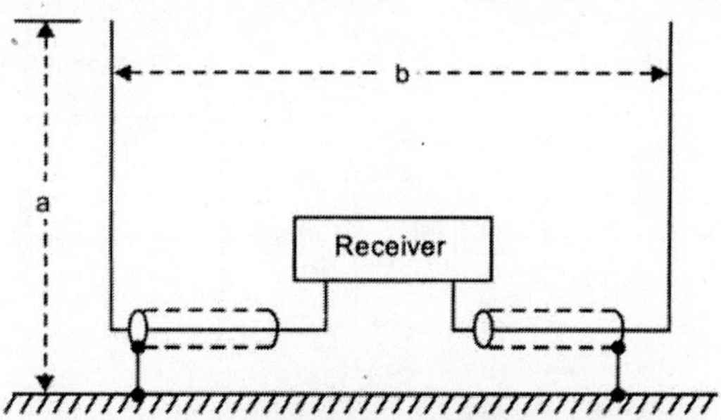

Shielded "U" Adcock

Fig. 5.20. Adcock Direction Finder Employing Shielded Transmission Lines and Direct Coupling.

It can be seen that the pick up is the same that of a loop of one turn of length and width "a" and "b", respectively. Hence the effective height, from equation

$$h_e = \frac{2\pi ab}{\lambda} = \frac{2\pi A}{\lambda} \qquad ...(5.33)$$

Since $N = 1$, a multi turn loop is usually tuned to resonance by means of a capacitor directly or through a coupling transformer. Consequently, it delivers a terminal voltage greater than the open-circuit value due to this resonant rise, the factor for a directly tuned loop being the Q of the loop winding. Furthermore the number of turns N may be considerable. On the other hand, an adcock is limited to $N = 1$, and is usually operated as an untuned antenna. The pick up factor is much less (in the limit by the ratio $1/NQ$) compared to a loop of the same size. As a result, Adcock antennas are made quite large to compensate for this lower efficiency. The dimension 'b' may be as much as $\lambda/5$. Such array are not adapted to shipboard or aircraft use, but are generally employed at fixed ground station for medium and short wave (200 KHz to 20 MHz). At the higher end of this frequency range, however, progress is being made in adapting Adcock Antenna systems for such use abroad mobile craft.

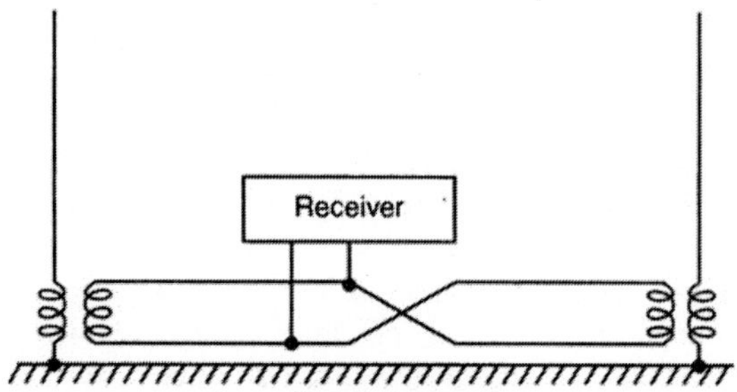

Fig. 5.21. Adcock Direction Finder Employing Coupling Transformers at Bases of Vertical Conductors.

The arrays themselves are offen left fixed in position and are erected in pairs at right angles to each other, with coupling to each obtained by means of a rotating pickup coil system or **Goniometer,** used to read the bearing. This is shown in Fig. 5.22. The Goniometer is ordinarily made a part of the direction finder receiver unit. Its coils are connected by suitable transmission lines to the antenna pairs AA' and BB'.

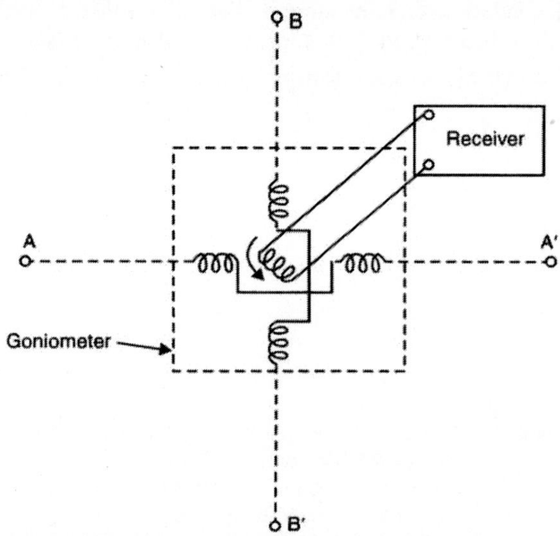

Fig. 5.22. Adcock or Fixed Loop Array with Goniometer (Bellini-Tosi System).

An analysis of several practical Adcock systems was carried out by Barfield. The calculated results were checked experimentally over a wide frequency range. Only the simple case of plane polarisation was actually used in defining the standard wave error and consequently the errors are somewhat low.

The comparison of the various types is still reasonably accurate. This is given in Fig. 5.23 with the case of a single loop included.

An experimental comparision of the observed bearings taken with a loop and with the shielded *u*-adcock and with the balanced-coupled Adcock system was reported by Watson-watt. The results observed the percentage of observations having a certain deviation from the free bearing of the stations being indicated.

Type	*Diagram*	*Dimensions, meters*		*Standard wave error, degrees (at 1000 kc, σ = 1.7 × 10^{-13} emu)*	*Pickup factor, meters*
		a = height of system	*h′ = height above ground of horizontal members*		
Loop		<< λ	<< λ	35 (not affected by σ)	10–20 (1 m sq.) proper number of turns to fit λ used, typical Q, varies as *f*
U	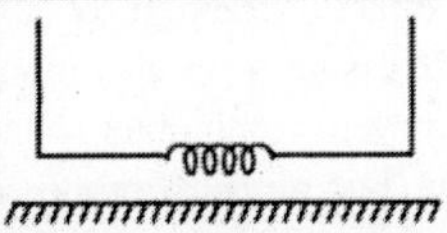	24	0.5	12	100–200, varies as f^3
Shielded U	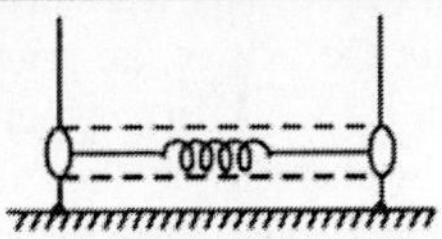	24	0.5	6 (most favourable)	
Elevated or H	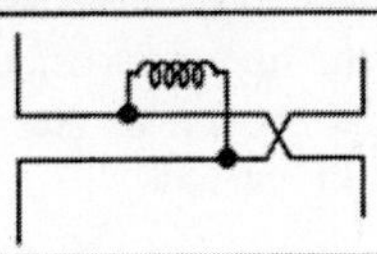	24	15	2 (not greatly affected by σ)	100–200, varies as f^3
Coupled	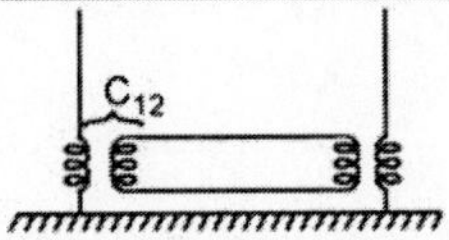	24	0.5	1 (untuned antenna system C_{12} = 10 μμf)	50–100, varies as f^4
Balanced	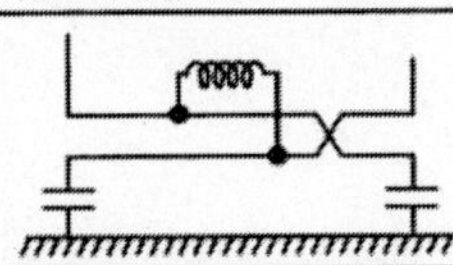	24	2.0	6 (untuned antenna system)	100–200, varies as f^3
Balanced coupled	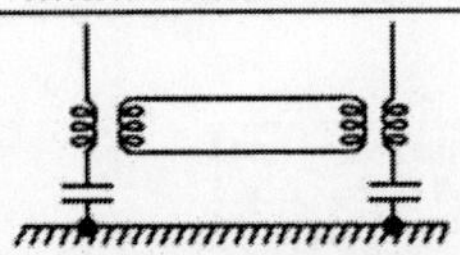	24	2.0	< 1 (untuned antenna system)	50–100

Fig. 5.23. Relative Pickup Factors and Standard Wave Errors for Loops and Adcock Antennas, Based on Experimental Measurements. (Barfield, J. Inst. Elec. Engrs. (London); 76, 423, April , 1935

5.6.1. Comparison between Adcock and Loop Direction Finders

1. The loop antenna direction finder suffers from vertical and polarisation errors but in Adcock direction finder system these errors are minimised.
2. The loop direction finders are suitable at lower frequency, whereas Adcock direction finder is suitable for higher frequency.
3. The induced voltage in Adcock system is less in comparision to loop.
4. A moderate inductive internal impedance is present in the loop direction finder system whereas the Adcock direction finder presents high internal capacitive impedance. Thus for turning the system to resonance, low value capacitance of negligible loss is needed in the loop direction finder whereas high impedance coil is needed in the Adcock direction finder. Due to this the physical structure of Adcock system is increased.
5. Some modified design of loop permits its use in UHF band whereas various circuit arrangement of Adcock permits its use in higher frequencies (UHF/VHF/HF) bands including Automatic direction finding also.
6. Goniometer direction finding is possible with Adcock and loop (crossed loop) both direction finder system.

5.7. GONIOMETER

When it is desirable to avoid the necessity of physically rotating a loop or Adcock antenna, as when their dimensions are very large, it is possible to obtain the effect of rotation by employing two such antenna, at right angles to each other and combining their output in a goniometer as shown in Fig. 5.24. The goniometer consists of two primary coils, one for each antenna, arranged at right angles to each other and coupled to a secondary as shown. The physical proportions are such that the mutual inductance between each primary coil and the secondary is proportional to the cosine of the angle that the axis of the secondary makes with the axis of the primary coil involved. For any positions of the secondary, the relation between the secondary output and the direction from which the waves arrive will then be identical in character to the directional characteristics of a single loop or Adcock, provided that there individual antennas are small enough so that their response is given by Eq. (5.12). However, the orientation of this pattern depends upon the position of the goniometer secondary. Thus the rotation of the goniometer secondary is equivalent electrically to rotating the loop (or Adcock) antenna.

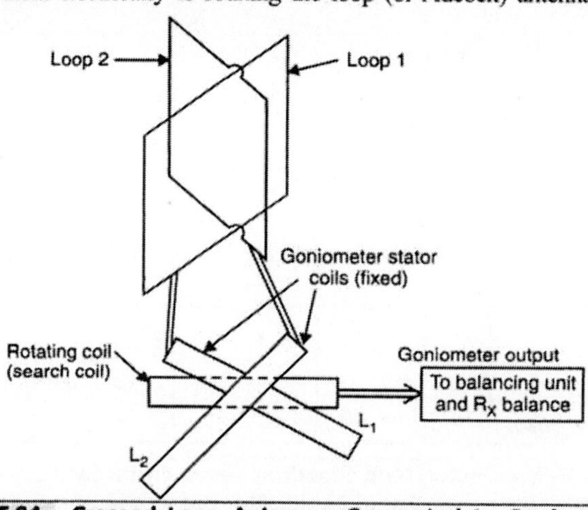

Fig. 5.24. Crossed Loop Antennas Connected to Goniometer.

The goniometer type of direction finding may be balanced and sensed by same methods as does in a rotating loop direction finders. The two types of direction finder, (rotating loop and goniometer) are operated similarly when taking bearings.

A typical dial of Goniometer is shown in Fig. 5.25. This Goniometer dial fitted with three pointers, one is direction finder pointer, which indicates the true bearings. Second pointer is 'sense' pointer which is 90° to direction finder pointer, it will only sense a bearing and dial reading is not necessary for this. The third one is, the Magnetic Reciprocal (MR) pointer, which is set according to the local magnetic variation so as to given direct reading of magnetic reciprocal bearings. The search coil is connected through the slip rings and spring contacts so that it can rotate in any direction.

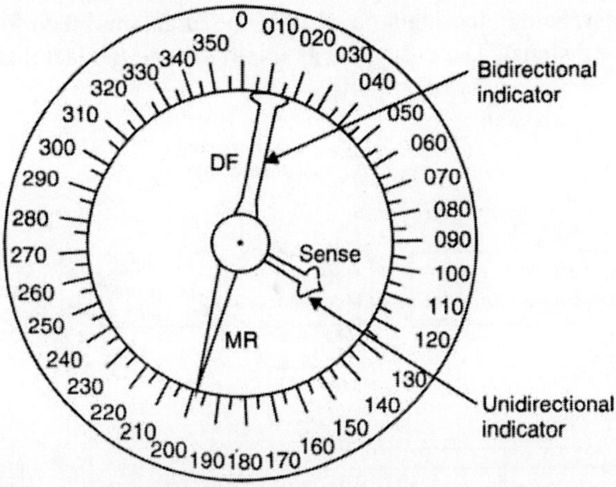

Fig. 5.25. The DF Pointer, Shows a Goniometer Dial Calibrated 360°.

5.8. AUTOMATIC DIRECTION FINDERS (ADF)

Introduction. In distinction to the direction finders, there are numerous types in which the proper bearing is determined by means of a visual device. In this device, the operator is not required to listen for a minimum signal as in an aural-null type, reception with phones being needed only for station identification or for monitoring.

Visual direction finders may be grouped in two main classes. They are as follows :

1. Automatic direction finders (Radio compass) airborne equipment
2. Right-left type (Homing) NDB (Nondirectional beacons) ground equipment.

In Automatic direction finders indicate directly on a 360° scale the station bearing. In Right-left types utilize a zero center meter or similar indicator and indicates deviation in bearing from the manually established loop position.

Basically the air borne equipments such as ADF consists of the following :

1. Antenna system
2. Receiver
3. Display

The antenna system consists of an omnidirectional sense antenna and rotatable loop, but now a days fixed loop and electronic goniometer are mostly used, which perform the same function. The receiver consists of RF and detector sections. The frequency range is between 195 to 1500 KHz. The output of the receiver is directional signal and indent, in the some cases voice. Speaker and headphone are used for voice and indent, where else the directional signals are fed to the display part of the receiver.

The display section can be Relative Bearing Indicator (RBI) or Radio Magnetic Indicator (RMI).

5.8.1. Elements of Automatic Direction Finders (ADF)

Fig. 5.26 shows the typical circuit elements of automaic direction finder. The first successful commercial automatic direction finder in general use in the United States was developed in 1937 jointly by Radio Corporation of America and Sperry Cysoslope Company. Later designs have followed for same principles.

The A.C. Supply serves to modulate the RF center in the loop channel by means of balanced modulator. In this the input consists of the large amplitude local a.c. signal and a much smaller RF signal. The difference in amplitude and the fact that two different frequencies are fed in are points of importance.

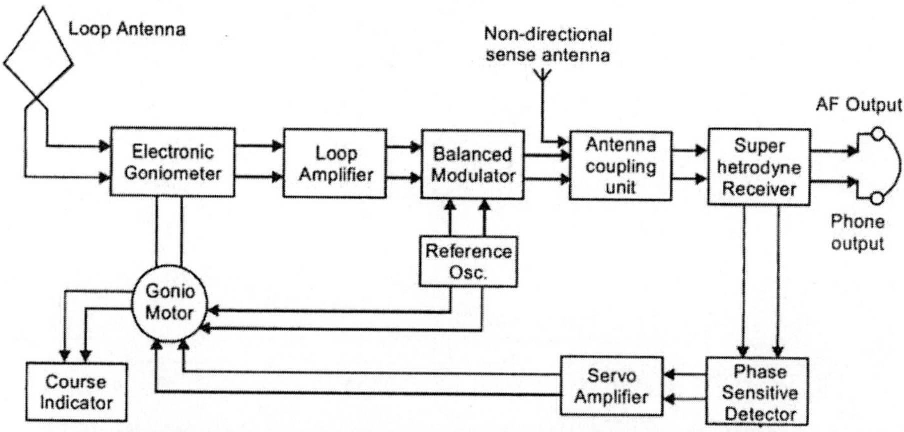

Fig. 5.26. Block Diagram of Radio Compass Receiver System.

The incoming RF signals are amplified by loop amplifier and fed to the balanced modulator which produces the product of goniometer output with an oscillator output at frequency supplying a reference phase. The output of the balanced modulator is given by the following expressions.

$$E_{(mod)} = E_g \, . \, E_{ref}$$
$$= A_0 \sin \Delta\phi \; e^{j\omega t} \, . \, E_{ref} \cos \omega_{ref} \, . \, t$$
$$= A_0 \, E_{ref} \sin \Delta\phi \cos \omega_{ref} \, . \, t \, . \, e^{j\omega t}$$

The output of the balanced modulator $E_{(mod)}$ is further added with signal pickup of non-directional sense antenna which is independent of the azimuth direction of arrival of the radiowave. The total voltage at the input of RF amplifier in the superhetrodyne receiver, is may be given by

$$E_{in} = E_{(mod)} + h_f \, . \, E_0 \, . \, e^{j\omega t}$$
$$= (E_{ref} \, . \, A_0 \sin \Delta\phi \cos \omega_{ref} \, . \, t + h_f \, . \, E_0) \; e^{j\omega t}$$

$$E_{in} = \left[1 + \frac{E_{ref} \, . \, A_0 \, . \sin \Delta\phi}{h_f E_0} \, . \cos \omega_{ref} \, t \right] h_f E_0 \, . \, e^{j\omega t}$$

E_{in} (Input voltage of RF amplifier) represents an amplitude modulated signal which is modulated at frequency of Reference oscillator with the amplitude proportional to the sin $\Delta\phi$. If the $\Delta\phi$ lying between $\pi < \Delta\phi < 2\pi$, sin $\Delta\phi$ is negative and the demodulated signal will have negative amplitude which shows that the phase is reversed. Other then this case if $\Delta\phi$ lying between 0 and π, the sin $\Delta\phi$ is positive and the demodulated signal at ω_{ref} will be in phase within the reference signal. Now to control the reversible motor, it is coupled mechanically to the rotor of goniometer. The demodulated and reference signals are fed to a pair of coils which are right angle with a 90° phase shift between them to produce a rotating magnetic field at ω_{ref} for a synchronous motor. The rotations of the motor is being controlled by the sign of $\Delta\phi$. As the rotor is coupled to the goniometer rotor so as to turn it in a direction to reduce $\Delta\phi$, tending to orient the plane of the goniometer coil along the direction of the net magnetic field. The bearing or course will be indicated by the position of the rotor of Goniometer.

Errors : In the following condition the ADF will not be able to give accurate bearing :

(*i*) In case of heavy lighting it fails to indicate correct direction due to interference by the lighting.

(*ii*) In case of heavy wind storm which makes aircraft to drift perpendicular to the direction of motion causes misleading of homing direction because the radio compass will always pointed towards the runway field.

5.9. RADIO BEACON

Radio beam works on principle of the wireless light noise. It is being used in various forms such as

1. Flashing radio beacon
2. Revolving radio beacons
3. Radio-range beacons.

5.9.1. Flashing Radio Beacons

Flashing radio beacons are same as omnidirectional flashing lighthouse. The morse signal radiated at predefined intervals takes the place of the flashing of light houses and makes recognition easier. These beacons are operated at 300 KHz, frequency and it may operate continuously in fog. These beacons are generally better to use the Direction finder in the ship or aircraft having equiped with direction finding instruments.

5.9.2. Revolving Beacons

Revolving beacons are used to determined the direction or bearing from a fixed point without the use of direction finding apparatus in the craft. A very simple receving equipments is required to observe either the instant at which a sharply focussed beam sweeps the sector or to detect a rise and fall in signal strength according to the beam rotates. This may be called as directional transmission.

The production of revolving beacons depends upon the frequency to be operate. If the operating frequency in the radar of 30,000 KHz, it is possible to design narrow unidirectional beam transmitters, in which the directional effect is obtained by using some reflector system which will rotates around the radiating aerial. The first revolving beacon was installed by the Marconi company at Inchkeith. It had single wire vertical aerial transmitter, with a parabolic reflector which could be rotated about this aerial. The reflector consisted of a number of vertical wires in the form of a parabola, with the transmitting aerial at the focus. Although the aperture of the parabola was only about two wavelengths. The beam was sufficiently sharp to give an angular accuracy of 2° to 3° at a range of 10 miles.

All forms of revolving loop transmitter are liable to night effect, it is not possible to prevent the radiations of an electric field from the horizontal parts of the loop and which is horizontally polarised.

5.9.3. Radio Range Beacons

Radio range beacons are operating at medium frequencies, are fixed in direction, and are used to assist the aircraft to land safely *i.e.* homing. Broadly, there is the type of beacon that gives an aural indication of the direction, and another that gives a visual indication such as frequency meter. This types of beacons are discussed in details in the following chapter in this book.

5.10. RADAR BEACONS

A radar beacon system is similar in operation to conventional radar except that the return signal is not reflected from the target but it is transmitted by a transmitter on board, sometime it is known as secondary radar that distinguish, it from a primary radar (the usual radar). A radar system cooperative passive target, while the beacons system operates with an active target. A radar beacon systems consists of transmitter (interrogator) sub-system and a receiver (transponder) sub-system. Generally, this system is used in conjunction with the primary radar. So in this case the interrogating transmitter might be the radar itself. If this system is used independently than a separate transmitter specifically designed as part of the system. The transponder is located on boards or aircraft, upon receiving a signal from the interogating sub-system (transmitter), the aircraft transponder automatically transmits a reply in the form of code. The reply from transponder is in coded form for the purpose of identification, is received back at the interrogating sub-system, where it is decoded and display on a radar screen. This type of beacon is known as **IFF (Identification of Friendly or fol) system.**

Radar beacon system provides better target signals, confirmed means of target identification and a means whereby information may be conveyed from the target back to the interogator. The reply signal from the beacon is better than the signal received by conventional radar system. The magnitude of reply signal does not fluctuate as in the convntional radar system. An important information may be encoded in the reply is the target altitude. The reply code might also indicate which air-traffic controller has responsibility for the control of the aircraft.

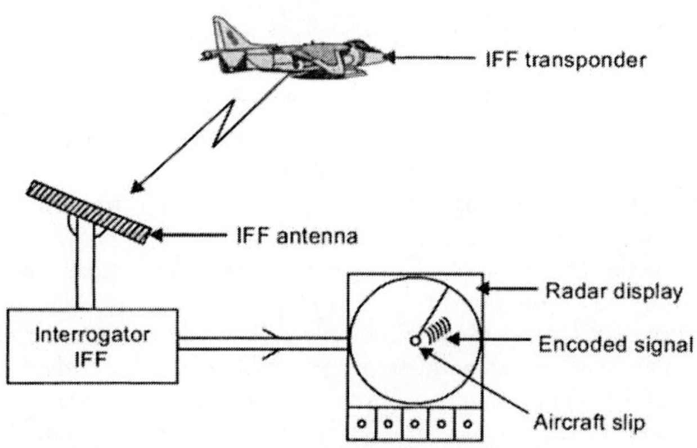

Fig. 5.27.

5.11. COMMUTATED AERIAL DIRECTION FINDER (CADF)

Principle : Commutated Aerial Direction Finder (CADF) works on doppler principle. It can be appreciated by considering a single vertical dipole rotating along a circular path in a horizontal plane. The motion of the aerial adds a sinusoidal FM component to a signal transmitted from an aircraft. This is because the signal path length varies from transmitter to aerial with time. The phase of this sinusoidal FM modulated signal is directly related to the azimuth angle of arrival of signal.

The phase deviation is function of antenna rotation rate, and diameter of the circular path. The physical rotation of an antenna at the required angular velocity over a large circular path to produce effective doppler shift is not practical because of mechanical limitations. Hence, in the CADF system, a set of 16 dipole elements is located at the circumference of a circle with a diameter of 1.17 m, at a minimum frequency, and at angular intervals of 22.5°. They are cyclically switched in sequence, by a commutator. Thus this system is called **Commutated Aerial Direction Finder**. The effect produced is the same as to that of a single vertical aerial revolving along a circular path. Fig. 5.28 shows the arragements of 16 dipoles.

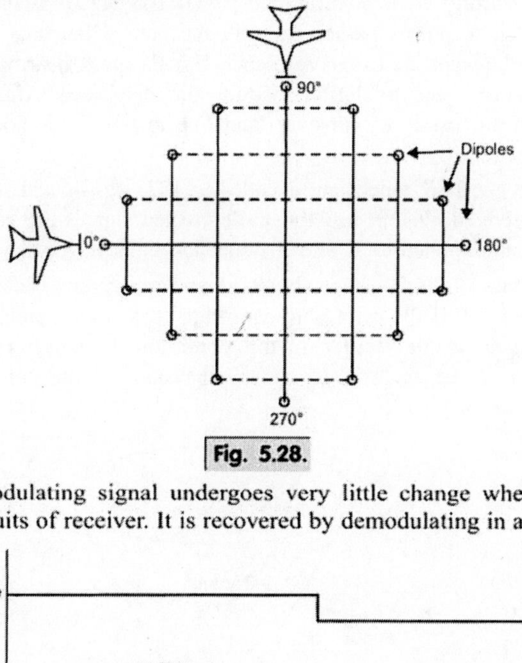

Fig. 5.28.

The phase modulating signal undergoes very little change when passing through the RF and IF circuits of receiver. It is recovered by demodulating in a FM discriminator.

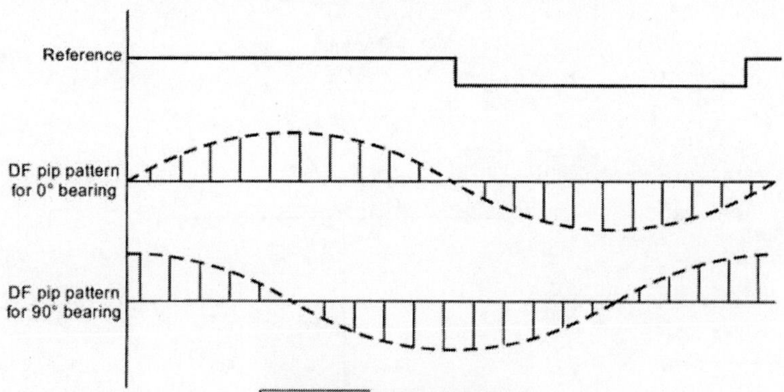

Fig. 5.29. Pip Pattern.

The demodulated signal is in the form of phase steps is called **Pip pattern** or DF signal as shown in Fig. 5.29 and it is not suitable for bearing computation at this stage. It is becuase the phase change occurs only during the time when activity of one dipole ceases and other dipole starts; for remaining period (during the time the dipole is ON) there is no phase change taking place and DF signal contains only noise. The S/N ratio is improved by passing the signals through gating circuit in Data Extractor PCB in DF porocessor.

The gated DF signal is passed through other circuits like filter, amplifier, limiter for determination of invalid and overhead pass condition. The limiter output is sinusodial and it contains phase jitters and it will make bearing computation erraneous. In order to stabilize the signal is multiplied with two reference square wave signals (216 Hz 0° and 216 Hz 90°) and the resultant output is integrated over a time period in DF signal processor card. The output of the integrator is a d.c. voltage proportional to sine and cosine functions of bearing angle. This d.c. voltage is passed through two FET switches controlled by two reference square waves (108 Hz 0° and 108 Hz 90°) and summed up. This output after passing through an filter/amplifier is given to a zero crossing detector to produce a square wave called SCL. The starting edge of SCL opens a gate through which high frequency clock, one clock pulse corresponding to 0.1°, is allowed to pass through it. The starting edge of reference signal 108 Hz 0° is used to close the gate. The number of clock pulses passed through the gate when it is open (determined by phase difference between SCL and reference signal) gives bearing angle directly up to 0.1° accuracy. This is fed to digital readout and displayed. After suitable parallel to serial conversion process it is sent over land line to ATC (Air-Traffic Control) sides in the form of FSK tones.

At ATC side, the FSK tonnes are decoded to TTL signals and suitable converted back to parallel form. For displaying in CRT, the corrected signals are converted into sine and cosine functions and applied to X and Y deflection coils for displaying the vector form.

Overhead Pass : The 50° cone above the antenna system is called "**cone of silence**" or "**OH pass region**". If the pilot is transmitting from this region, it means his aircraft is very much nearer to the facility. In this condition the bearing is to be blanked and OH pass alarm is to be actuated to convey the same to the controller. The Fig. 5.30 shows the cone of silence.

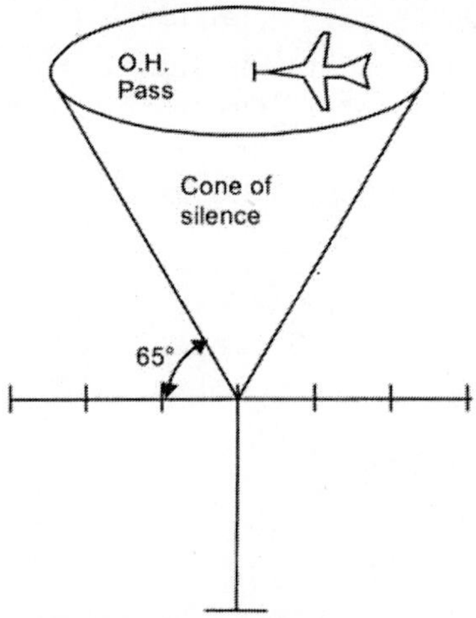

Fig. 5.30. Cone of Silence.

5.11.1. General Description of CADF system

A simple block diagram of CADF system is shown in the Fig. 5.31 and Fig. 5.32.

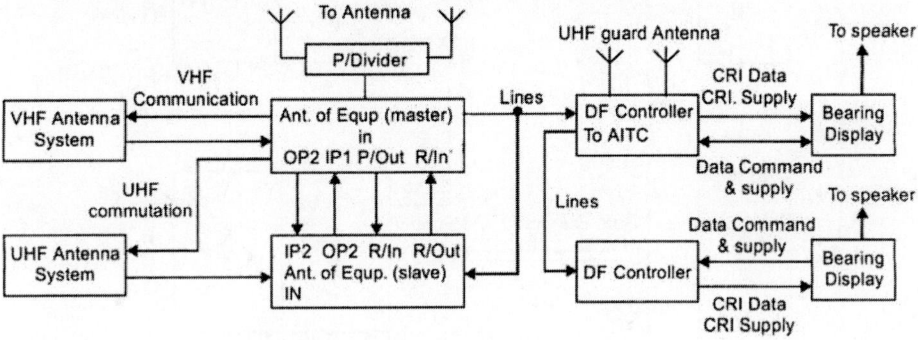

Fig. 5.31. Block Diagram of CADF.

The complete system is consisting of DF equipment which computes the bearing from RF received through Antenna system VHF and Antenna system UHF depending upon the selected. The bearing data thus computed is sent to bearing display and control unit and data is displayed in digital and analog form in display unit in ATC room. The system is dual channel and can simultaneously handle two aircrafts of different RF channel. Antenna simulators are also provided for maintenance of the systems one each for VHF and UHF.

Some Technical specification of CADF :

(*i*) Frequency range :

VHF 100 to 156 MHz

UHF 225 to 399.975 MHz

(*ii*) Commutation pulse rate—216 Hz

(*iii*) The Antenna unit consists—16 dipole elements

(*iv*) System error—±2° in Normal sites.

(*v*) Vertical coverage—65°

(*vi*) Facilities :

(*a*) Bearing store and recall facility

(*b*) Overhead pass indication

(*c*) Speech monitoring and recording facility

(*d*) System failure alarm.

(*e*) Invalid data indication

(*vii*) **Guard Channel :** Automatic changeover of the system to VHF and/or UHF guard channel in case of guard-channel transmission with audio and visual alarm.

(*viii*) Remote operation—up to 10 km.

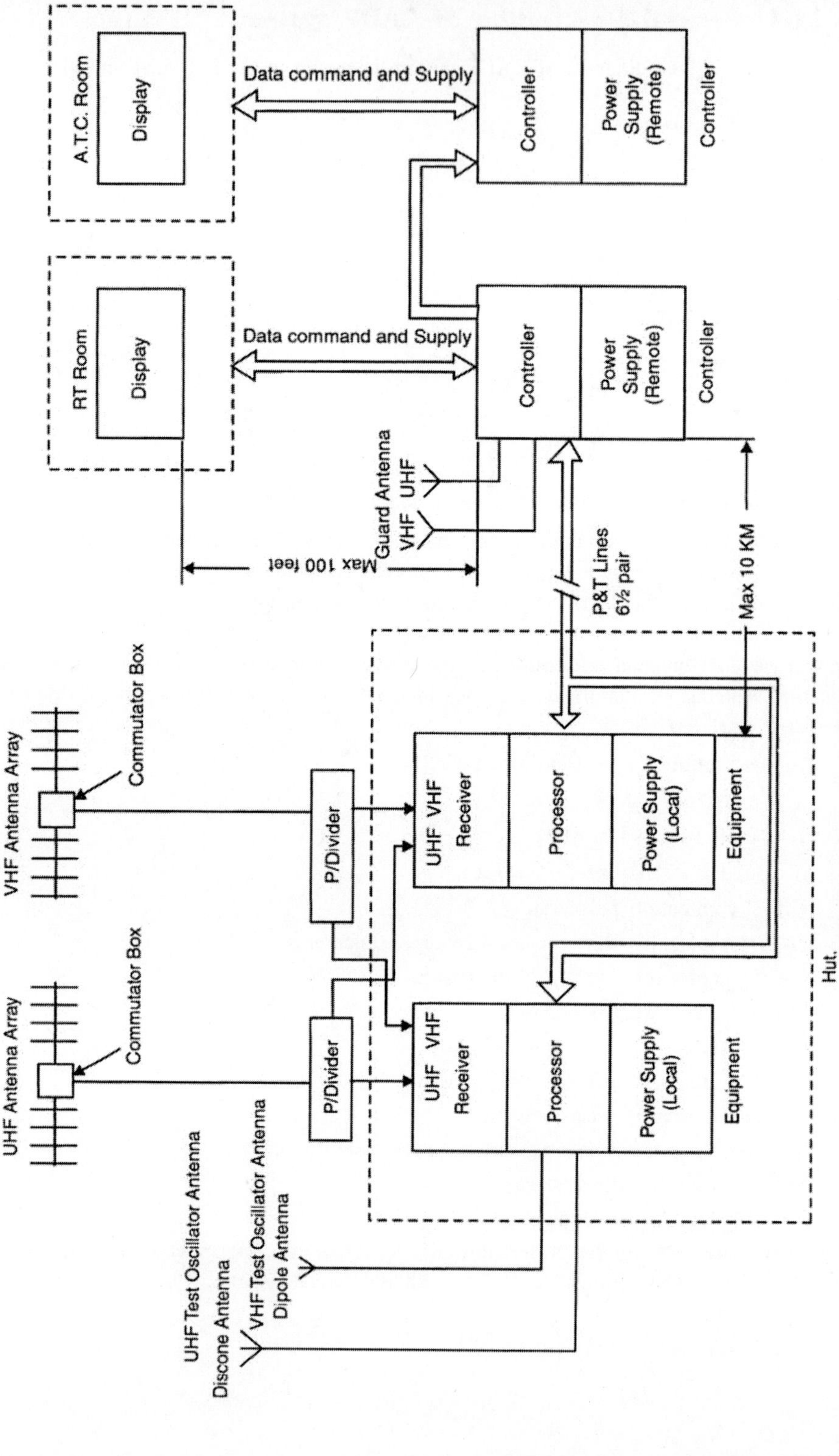

Fig. 5.32. Deployment of Typical CADF System.

SOLVED EXAMPLES

Example 1. A circular loop 24 inch in diameter with eight turns operates at 1500 KHz. Calculate the *he* is required.

Solution. $\lambda = \frac{c}{f} = \frac{3 \times 10^8}{1500} = 200 \text{ m}$

The loop diameter $d = 0.61 \text{ m}$

$$A = \pi r^2 = 3.14 \times (0.31)^2 = 0.292 \text{ m}^2$$

$$h_e = \frac{2\pi NA}{\lambda} = \frac{2\pi \times 8 \times 0.292}{200}$$

$$h_e = 0.0734 \text{ m} = 7.34 \text{ cm}$$

Example 2. A circular loop with 10 turns at the height of 1 m. Find the operating frequency if the diameter of loop is 2 m.

Solution. $f = ?$

$$d = 2 \text{ m}$$

Area of circular loop $= \pi r^2 = 3.14 \times 1^2$

$$A = 3.14 \text{ m}^2$$

$$h_e = \frac{2\pi \times N \times A \times C}{f}$$

$$f = \frac{2\pi NAC}{h_e} = \frac{2 \times 3.14 \times 10 \times 3 \times 10^8}{1} = 18.84 \times 10^9 \text{ Hz}$$

$$f = 18.84 \text{ GHz}$$

Example 3. Calculate the maximum voltage induced in the loop antenna having

Number of turns $N = 100$

Quality factor $Q = 10$

Frequency of operations = 300 KHz

Area of the loop $A = 1 \text{ m}^2$

and Field strength $E_0 = 3.146\ \mu\text{V/m} = \pi\ \mu\text{V/m}$

Solution. Given, $N = 100$

$$Q = 10$$

Frequency = 300 KHz $\left(\lambda = \frac{3 \times 10^8}{300 \times 10^3} = 1000 \text{ m}\right)$

Area of the loop $A = 1 \text{ m}^2$

$$E_0 = \pi\ \mu\text{V/m}$$

Now put the values in the following equations

$$EV = \frac{2\pi E_0 ANQ}{\lambda}$$

$$EV = \frac{2\pi E_0 \times 1 \times 100 \times 10}{1000} = 2\pi E_0 = 2\pi \times \pi \ \mu V$$

$$\simeq 2 \times 10 \ \mu V \qquad (\pi^2 \simeq 10)$$

$$EV = 20 \ \mu V$$

SUMMARY

- A Radio Direction Finder (RDF) is a device for determining the directions of the energy flux of an EM wave.
- A direction finder when used at a receiving station consists of a directive array, a receiver and a means for coupling the array to the latter.
- The allotted frequencies and in the Low frequency/Medium frequency, High frequency and VHF bands.
- The loop antenna is a radiating coil of any convenient cross-section of one or more turns carrying radio frequency current. It may assume any shape such as rectangular, triangular, square, circle and hexagonal.
- A loop of more than one turn is often called as Frame.
- $e = \frac{2\pi . N . A . E_{max}}{\lambda} \cos\theta \sin wt$
- The voltage e is directly proportional to the frequency. If the effective height is defined as the voltage amplitude for unit field when $Q = 0$,

$$h_e = \frac{2\pi NA}{\lambda} = \frac{2\pi NAf}{C}$$

- The process of finding the phase of signal with respect to the reference field is known as sense-finding.
- Errors in loop direction finder are as follows :
  (*i*) Antenna or vertical effect
  (*ii*) Night effect or polarisation error
  (*iii*) Error due to abnormal propagation.
- Disadvantages of Loop direction finders :
  (*i*) Suitable for LF and MF
  (*ii*) Spurious induced voltages are produced if nearly loop, wires and conduction are present.
  (*iii*) Its transmission efficiency is very poor.
  (*iv*) It is subjected to antenna effect and night effect which are minimised by using Adcock antenna.
- An aural null direction finder comprises essentially a directive array confined with a sensitive and highly stable receiver.
- If the horizontal member of a loop is removed and shielding or balancing the effect of the lower horizontal wire than such a directive array is known as an Adcock Antenna.
- **Goniometer :** When it is desirable to avoid the necessity of physically rotating a loop or Adcock antenna, as when their dimensions are very large, it is possible

to obtain the effect of rotations by employing two such antenna, at sight angles to each and combining their output in a goniometer, used to read the bearing.

- In Automatic Direction Finders (ADF) the operator is not required to listen for a minimum signal as in an aural-null type, reception with phones being needed only for station identifications or for monitoring. ADF indicate directly on a 360 scale the station bearing.
- The air borne equipments such as ADF consists of the following :

  (*i*) Antenna system (*ii*) Receiver, (*iii*) Display
- The operating frequency of ADF is 195 to 1500 KHz.
- Commutated Aerial Direction Finder (CADF) works on doppler principle. It can be appreciated by considering a single vertical dipole rotating along a circular path in a horizontal plane.
- CADF system has a set of 16 dipole elements is located at the circumference of a circle with a diameter of 1.17 m at angular intervals of 22.5°.
- In CADF system dipoles are cyclically switched in sequence by a commutator that why it is called Commutated Aerial Direction Finder (CADF).
- The Operating frequencies of CADF are

  VHF — 100 to 156 MHz

  UHF — 225 to 399.975 MHz.
- Commutation pulse rate is 216 Hz.
- Accuracy of CADF is ±2° in Normal sites.

UNSOLVED PROBLEMS

1. A circular loop 2 feet in diameter with ten turns operates at 3.0 KHz. Find the effective height (h_e). **[Ans.** 0.19 m]
2. Find the voltage induced by a plane wave of field strength 0.01 volt/m and operating frequency is 1 MHz.

   (*a*) A frame aerial 1 meter square of 12 turns, the plane of the frame being in the plane of propagations of the wave.

   (*b*) A vertical aerial 8 meters high. **[Ans.** (*a*) 2.51 mV; (*b*) 80 mV]
3. A frame aerial has 20 turns of diameter 6 ft., the resistance of the coil being 2.5 ohms at 100 KC/s. If a signal of this frequency has a field strength of 30 μ volts/meter, what current will it produce in the coil when it is tuned to the signal ? **[Ans.** 1.32 μA]
4. A loop of area 1 m^2 and having 10 turns has its plane at 45° off the line of bearing of a distant station which produces an incident magnetic field of strength 10^{-9} lines per square cm at a frequency of 0.3 MHz. Find the EMF induced in the loop. **[Ans.** 1.33 μV]

REVIEW QUESTIONS

1. Describe loop antenna as a directions finder. **[UPTU, 2003]**
2. Describe an Aural-Null direction finder.
3. Explain the working of Goniometer.

4. Explain the working of loop antenna as a direction finder and enumerate the errors occurs in direction finding.
5. Describe the ADF with the help of simple block diagram.
6. Derive the equation to find out the resultant output voltage of loop antenna.
7. Define the relation between the height of loop antenna and the frequency used for operation.
8. Describe the comparison between Adcock and Loop direction finder.
9. Write short notes on :
   (*i*) Adcock direction finders
   (*ii*) Polarisation error
   (*iii*) Night effect
   (*iv*) Accuracy of loop antenna
   (*v*) Pip pattern in CADF
   (*vi*) Overhead Pass in CADF
   (*vii*) Goniometer **[UPTU, 2006]**
   (*viii*) Sense finder
   (*ix*) Radar Beacons **[UPTU, 2006]**
10. Describe the radio directions finding method using loop antenna. **[UPTU, 2004]**
11. Explain the principle of CADF in brief.
12. Describe the working of CADF with the help of Block Diagram.
13. Why CADF system is called commutated Aerial Direction Finder ?
14. Explain the errors encountered in direction finding. How are these removed. **[UPTU 2006]**
15. What is radar beacons ? What are the various applications of this system ? Explain. **[UPTU 2006]**

6

CHAPTER

RANGE SYSTEMS

INSIDE THIS CHAPTER

6.1. INTRODUCTION

Radio ranges are navigational aids and it is specifically used by aircraft (Air Navigation). Three types of radio ranges are in use, which are as follows :

1. Low frequency four course radio range
2. VHF omni directional radio range (VOR) and
3. Dopper VOR (Latest development)

LF/MF four course radio range operator in LF/MF range of 200-400 KHz. It is obsolescent and now-a-days VOR (VHF omni range) is in use in the world. The latest introduction in radio navigation is DVOR (Doppler VHF omni Range).

6.2. THE LOW FREQUENCY/MEDIUM FREQUENCY RADIO RANGE

The low frequency range is one of the earliest electronic aids to air navigation. The LF radio station sends out "on course" signals which interlock with those of surrounding stations to form easily followed airways. Radio range station operates on frequencies between 200 KHz to 400 KHz.

The low frequency radio range depends on the super positions of two figure-8 directional antenna patterns at right angles to provide on-course signals. When two vertical antenna towers are properly fed with radio-frequency energy they produce a figure-8 radiation pattern as shown in Fig. 6.1. This pattern clearly indicates the directional nature of the radiation.

In the Fig. 6.1, the relative strength of the signal radiated in any direction is represented by the length of the line drawn in that direction from the center (O) to the edge of the pattern. Thus maximum signals are radiated in direction OA and OB, while minimum signals are radiated in direction OC and OD. In intermediate directions, the signal radiated is of intermediate strength, as shown by lines OE and OF. The signal radiated in the direction OE, for example, is much stronger than the signal in the direction OC but weaker than that radiated in direction OB.

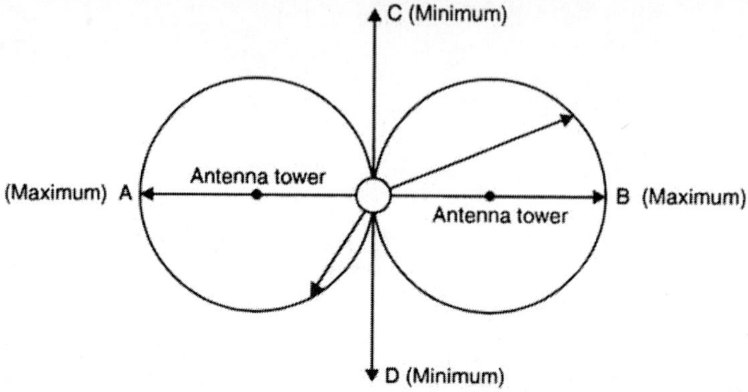

Fig. 6.1. Figure-8 Radiation Pattern from Two Vertical Antenna Towers.

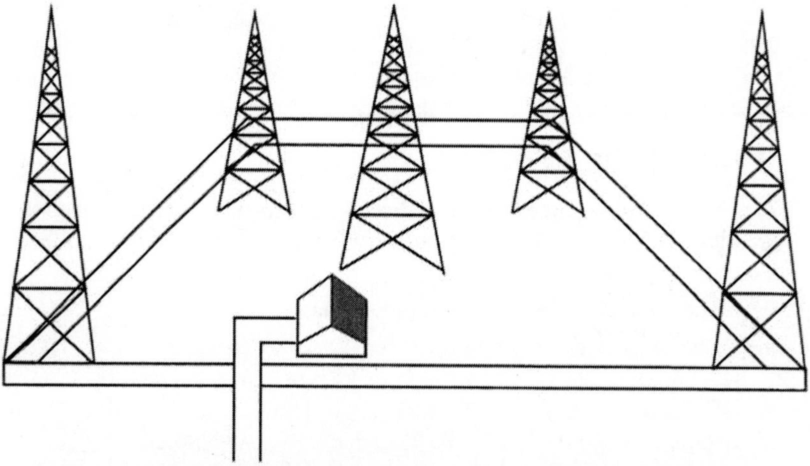

Fig. 6.2. Typical Radio Range Station. The four outside towers are used for transmission of "on-course" signals; the center tower permits separate voice communications.

If two pairs of vertical towers are placed in a square pattern, as shown in Fig. 6.2, the result will be two figure 8-pattern available at right angles to each other. The center tower shown in the Fig. 6.2 is used for transmitting weather reports. With two figure-8 pattern available the production of on course signals becomes a simple matter. The code signal A (which looks like this •—) is transmitted over one pair of diagonally opposite towers, and the code signal *N* (which looks like this —•) is transmitted over the other pairs of towers. In this way four signal zones, called sectors are produced, two of which carry the 'A' signal, while the other two carry the '*N*' signal. Between the four sectors there are four regions of overlapping, where both signals are present as shown in Fig. 6.3.

The '*A*' and '*N*' code signals consist of a medium pitched tone (about 1020 cycles) are, therefore, audible in the airborne radio range receiver. From Fig. 6.3 it is apparent that the following signals are heard in various zones :

1. In the '*N*' section a clear '*N*' is heard;
2. in the '*A*' sectors a clear '*A*' is heard;
3. in the overlapping *AN* regions these are narrow (3° wide) lanes where the '*A*' and '*N*' sound equally strong. However, the '*A*' and '*N*' signals are not heard

individually in those overlapping zones, but the keying is so arranged that they blend together to give a continuous tone, the "on-course" signal.

As long as the aircraft flies in the direction of the on course signal lanes, the pilot will hear the continuous tone, but as soon as the craft deviates from the course either the '*A*' or the '*N*' will predominate, depending on the direction of the course deviation. As shown in Fig. 3.4, the double figure-8 defines "four on-course signals", where a continuous tone is heard. The *A* and *N* transmissions are interrupted frequently by a three-letter call signal, identifying the station from which the transmission originates.

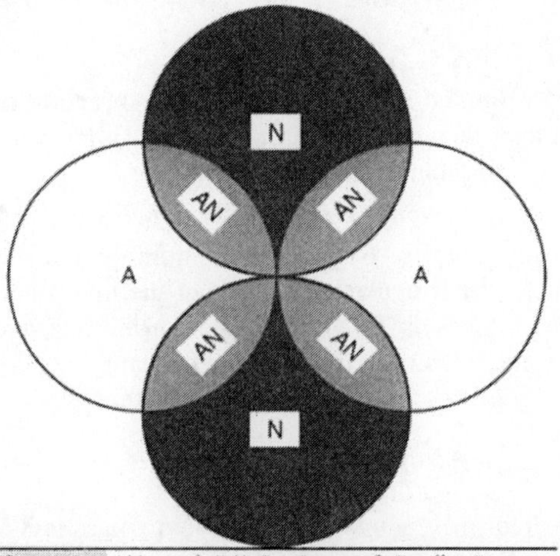

Fig. 6.3. 'A' and 'N' Sectors of Radio Range.

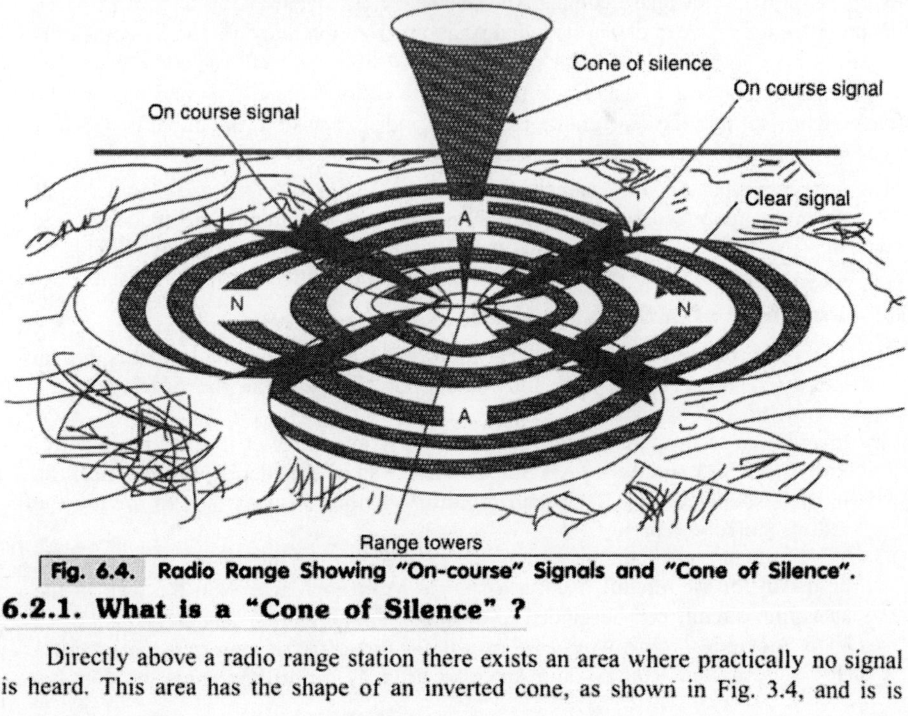

Fig. 6.4. Radio Range Showing "On-course" Signals and "Cone of Silence".

6.2.1. What is a "Cone of Silence" ?

Directly above a radio range station there exists an area where practically no signal is heard. This area has the shape of an inverted cone, as shown in Fig. 3.4, and is is heard. This area has the shape of an inverted cone, as shown in Fig. 6.4, and is known as the **cone of silence**. A pilot passing through a cone of silence would know that he was directly over the range station whose signal he had been receiving.

6.2.2. How Radio Range Signals are Received ?

There is no separate equipment needed to receive radio range signals. Any receiver that can be tuned to the frequency of radio range station (200 to 400 KHz) can receive the aural radio ranges signals. To provide visual indication, however, special indicators are required. These generally consist of a "radio compass" that indicates the course and a "left-right indicator" which shows whether the station is to left or right of the aircraft.

Disadvantages of LF Range

1. It provides only a limited number of courses (4-direction only) in preselected arbitrary directions; likewise, the aircraft has no simple means of determining which of the four possible course it is following.
2. It is tidious to listen the tones continuously which caused fatigue to the pilot.
3. Some major defects are the "Night effect", multiple courses and bent courses. Night effect, caused by irregular reflections of the ionosphere, results in fading of the on course signals is an erratic pattern. Multiple and bent courses are caused by multiple reflection from mountains and bending of radio waves, when they pass from one type of terrain to another.

6.3. VHF OMNI RANGE (VOR)

It means a very high frequency omni directional radio ranges and operates in the VHF band of 112-118 Mhz, thus being used as medium to short range radio Navigational aid. It works on the principle of phase comparision of two 30 Hz signal *i.e.* an aircraft provided with appropriate receiver can obtain its radial positions from the range station by comparing the phases of the two 30 Hz sinusoidal signals obtained from the VOR radiation. Any fixed phase difference defines a Radial/Track (an outward vector from the ground station into space). VOR in fact provides an infinite number of radials/Tracks to the aircrafts, the four provided by a LF/MF radio range.

There are two types of VOR, namely conventional (VOR) (C-VOR) and Doppler VOR (DVOR). Even though both serve the same purpose as far as the aircrafts are concerned, the selection of C-VOR and D-VOR depends upon various site conditions in an airfield. D-VOR is more accurate than C-VOR, but it is complex and costilier than C-VOR.

6.3.1. Purpose of VOR and Uses

1. The main purpose of the VOR is to provide the navigational signals for an aircraft receiver which will allow the pilot to determine the bearing of the aircraft to a VOR facility.
2. In additions to this, VOR enables the Air-Traffic controller in the Area Control Radar (ARSR) and ASR (Air Surveillance Radar) for identifying the aircrafts in their scopes easily. They can monitor whether aircrafts are following the radials correctly or not.
3. VOR located outside the airfield on the extended centre line of the runway would be useful for the aircraft making a straight VOR approach. With the help of the auto pilot aircraft can be guided to approach the airport for landing.
4. VOR located enroute would be useful for airtraffic to maintain their PDR's (Pre determined Routes) and are also used as reporting points.

5. VOR located at radial distance of about 40 miles in different directions around an International Airport can be used as holding VOR for regulating the aircraft for their landing in quickest time. They would be of immense help to the aircraft for holding overhead and also to the ATCO for handling the traffic conveniently.

6.3.2. Principle of Operation

A/c determines its bearing by comparing phase of reference 30 Hz and variable 30 Hz signals, References 30 Hz signal has the same phase at all the 360° points around VOR whereas the phase of variable 30 Hz changes at the rate of 1° for one degree deviation of azimuth angle. The reference 30 Hz signal and variable 30 Hz signal are in the same phase in the direction of magnetic north. In fact this direction is taken as zero degree for VOR and other azimuth angles are measured in clockwise direction as shown in Fig. 6.5.

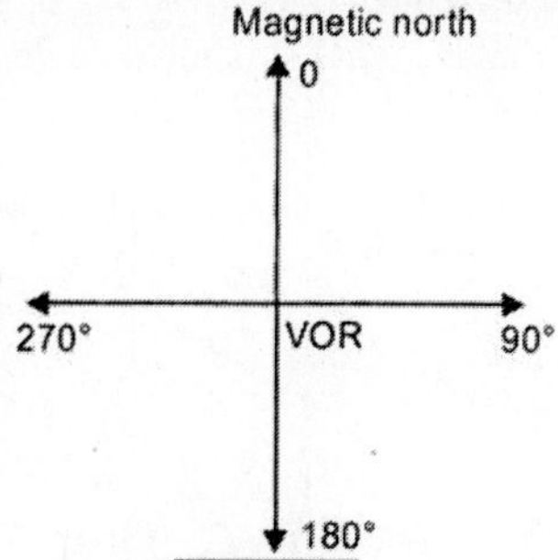

Fig. 6.5.

The VHF Radio Transmitter radiates two patterns, distinguishable by different modulations, one of which is omnidirectional and carries the modulations of 30 Hz reference signal, while the second pattern is a figure of eight rotating at 30 revolutions per second. The RF phases of the two are locked. The omni radiation is a much stronger field than the figure of eight radiation as shown here in Fig. 6.6.

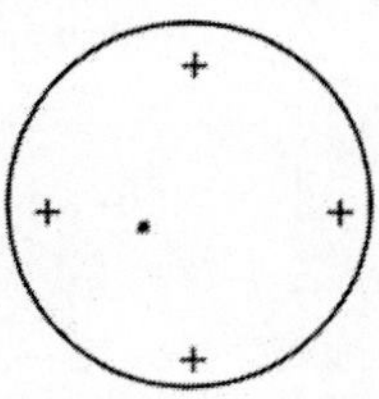

Fig. 6.6(*a*). Omni Radiation Pattern.

Fig. 6.6(*b*). Figure of Eight Pattern Rotating at 30 rev./sec.

The 30 Hz signals cannot be directly radiated into space. These signals modulates carrier frequencies in a specific manner and aircraft receiver processes the signal and derive the desired information. Besides, station identification and voice signals are also radiated.

The signals radiated from the VOR are as follows :

1. Reference 30 Hz } Navigation
2. Variable 30 Hz }
3. Station identification
4. Voice signal

The above mentioned signals modulates the carrier and following RF signals are radiated from the antenna system :

(*a*) Reference carrier

(*b*) Variable carrier

6.3.2.1. Reference Carrier

(*i*) Reference 30 Hz sinusoidal signal frequency modulates 9960 Hz. 9960 Hz will be referred as subcarrier in VOR system. The sub carrier modulation consists of a 9960 Hz plus and minus 480 Hz at 30 Hz rate as shown in Fig. 6.7.

This 9960 Hz subcarrier amplitude modulates on RF frequency.

(*ii*) **Station identification and Voice Signal.** Every VOR station radiates its own identification. This signal indicates to pilot the VOR station to which he has tuned. The station code is a combination of English alphabets. For example Calcutta VOR has identification of —•—• • • —, These alphabets are morse coded and in this, the code audio frequency of 1020 Hz is generated. The morse code corresponding to letter C and its audio signal is shown in Fig. 6.8.

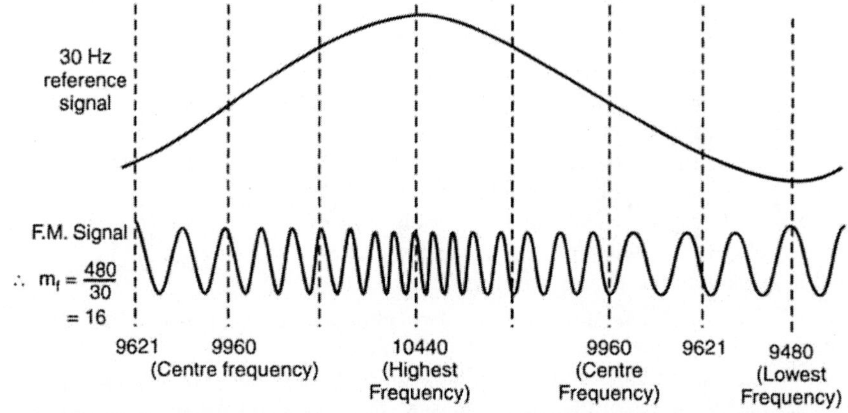

Fig. 6.7. 9960 Hz Sub-Carrier Amplitude Modulates on RF Frequency.

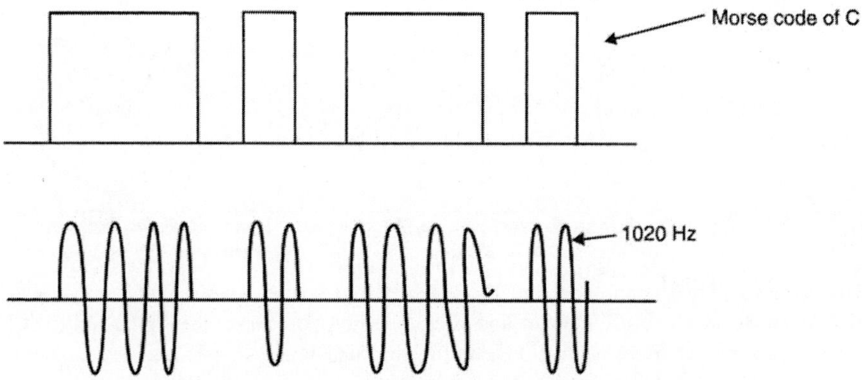

Fig. 6.8. Representation of Morse Code using ASK Modulation.

This coded audio signal amplitude modulates the carrier. The depth of modulation by this signal is adjusted 10%. Occasionally audio signal corresponding to voice meant for communicating with aircraft also amplitude modulates the carrier. Now a day voice communication in VOR radiation is seldom done. Whenever voice signal is radiated, the depth of modulation by it is adjusted to 30% and at that time identification transmission is stopped.

The reference carrier modulated by all the above signals are fed to aural system, which has omni radiative pattern. It may be noted that phase of 30 Hz reference signal, if demodulated by airborne receiver, will remain the same in all directions.

Summary of signal radiated on reference carrier :

(*a*) 30 Hz reference sinusoidal frequency modulates sub carrier 9960 Hz. This sub carrier amplitude modulates reference carrier.

(*b*) Morse coded 1020 Hz station identification tone amplitude modulates reference carrier.

(*c*) Voice signal amplitude modulates the references carrier.

6.3.2.2. Variable Carrier

Variable 30 Hz signal is obtained by the techniques known as **space modulation** which means the modulation is not generated inside transmitter but it is achieved by combination of R.F. signals in space. Earlier versions of VOR (LORENTZ VOR) does it by rotating dipole fed with pure carrier at the rate of 30 revolutions/sec. The radiation pattern of dipole is as shown in Fig. 6.9.

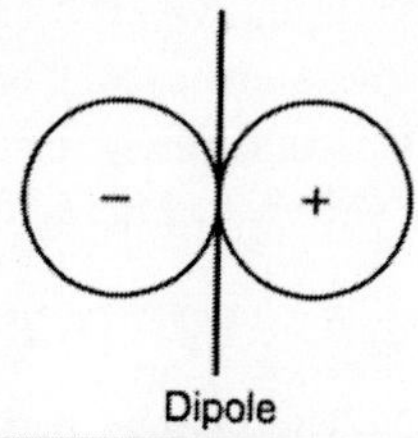

Fig. 6.9. Radiation Pattern of Dipole.

As the dipole is rotated, so this pattern also rotates. The plus radiation adds to omni refernece carrier and minus radiation substracts. The resultant radiation is a rotating Limacon as shown in Fig. 6.10.

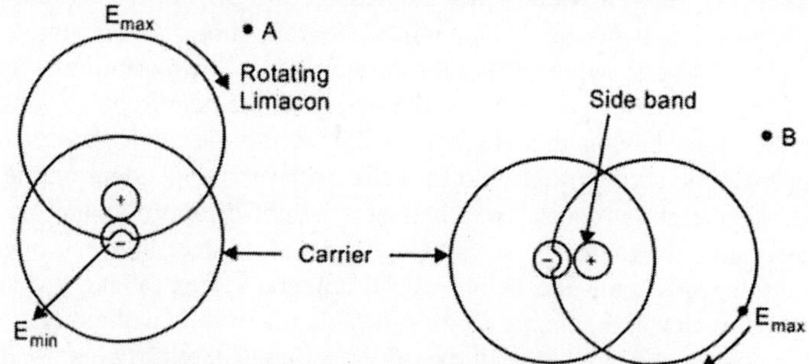

Fig. 6.10. The Physical Orientation of Arrival.

An observer placed at point '*A*' in space will be illuminated by R.F. of rotating Limacon and, therefore, it will see RF of varying amplitude. Since the rate of rotation is 30 rev/sec hence RF variation as observed is also 30 Hz. The variation of RF is nothing but amplitude modulation by 30 Hz and demodulation of this RF produces 30 Hz variable signal. The occurance of peak and trough amplitude of RF in different directions will differ in time, hence the demodulated 30 Hz phase will be varying.

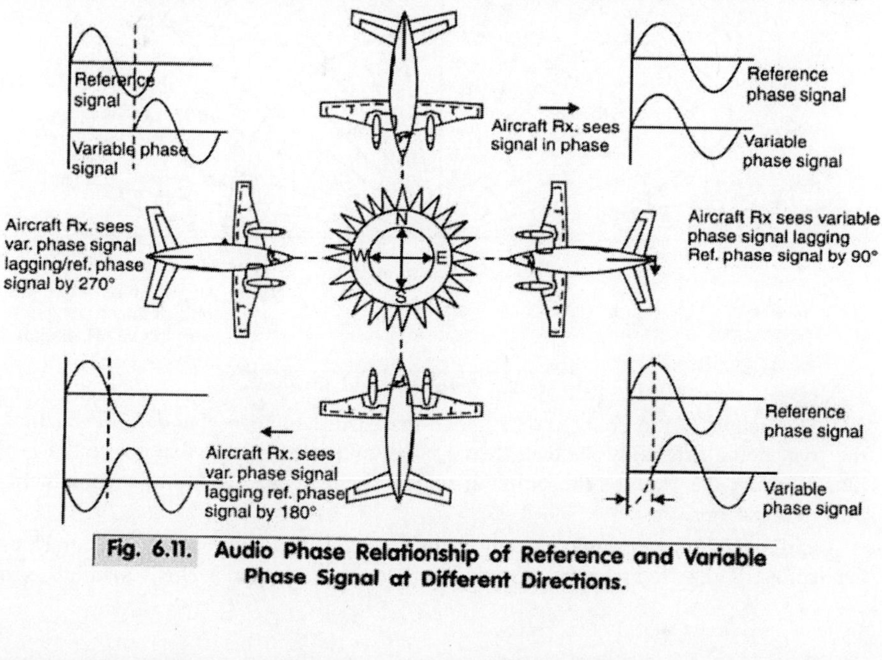

Fig. 6.11. Audio Phase Relationship of Reference and Variable Phase Signal at Different Directions.

In modern C-VOR, rotating arrivals are not used but stationary two pairs of slotted arrivals are used. The physical orientation of arrivals are shown in Fig. 6.10.

Audio phase relationship of reference 30 Hz and variable 30 Hz at different directions is shown in Fig. 6.11.

6.4. VERTICAL RADIATIONS PATTERN

The vertical radiation pattern of the VOR array is determined in its shape by two factors :

(*i*) The free space vertical pattern of the individual antennas in the array.

(*ii*) The vertical dimensions of the array w.r.t. Earth and the counter poise.

From the Fig. 6.12, it is clear that the actual earth affects the vertical pattern at low elevation angles *i.e.* between 0–17°, while the reflections at higher elevation angles between 17°–19°, occurs wholly from the counterpoise. The discontinuiy angle (Minima) between 12°–15° depends primarily on the height of the counterpoise above the ground, and the major lobe having the Maxima at 25° in this particular case, is the result of reflections from the counterpoise and thus this pattern is dependent on the height of the antenna above the counterpoise, which is at a height of approximately 5 ft. above the counterpoise and the counterpoise is at a height of 15 feet above ground. The minor lobe pattern depends upon the height of the antenna system above ground, because at various points in elevation, energy from VOR antenna in reflected either from the ground or from the counterpoise, whereas the earth is not considered to be a good reflector like the metallic counterpoise, thus appreciable energy gets lost upon reflection from the earth's surface. Therefore, we neither get all lobes of equal size nor nulls in between.

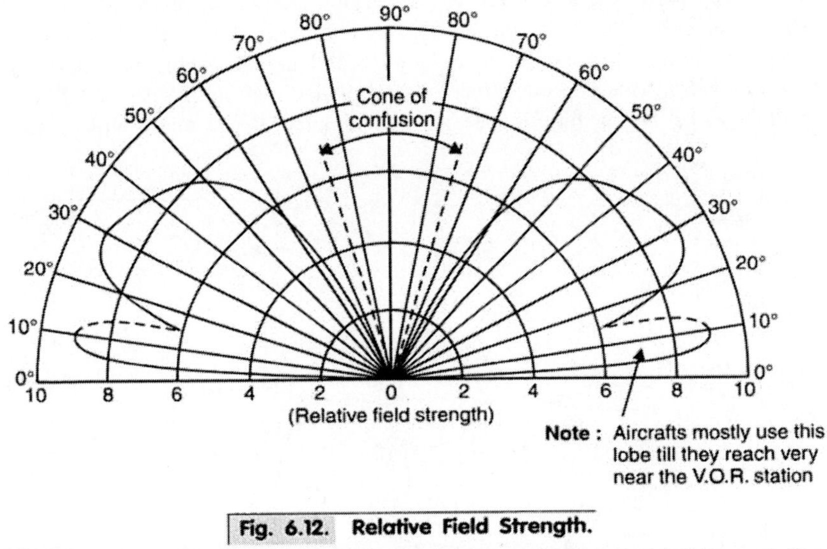

Fig. 6.12. **Relative Field Strength.**

The height of the counterpoise above the earth has been changed several times in the past, being originally 30 feet, then 15 feet and at present 8 feet above the ground, thus forming the roof of the building which shelters the transmitting equipment and other accessories.

Earlier VOR's were constructed with "Octagonal counterpoise", but after careful evaluation of the flight check, it was found responsible for a slight azimuth error at eight points of the compass, hence circular shape was standardised. It can be formed of a solid disc of copper or tin alloy of 35 feet diameter but going into the economical factors, they can be designed into a steel-ribbon mesh.

6.5. CONE OF CONFUSION

In the vicinity of the VOR transmitter, the vertical radiations may be rather weak and the airborne equipment is insensitive to compare phase. This region is cone shaped and is called a **cone of confusion/cone of silence**. When flying in this region the sense indicator will flip up and down and the deviation indicator will swing to left and right till this region is passed, thus indicating the time of passing the station.

This cone of confusion was quite wide in the earlier systems, but in recent systems using cylindrical antenna, considerable improvement has been achieved and this zone of silence (no signal area) or cone of confusion has been narrowed down to angles above 80 degree of the horizon. Thus the transition time through this inverted cone is reduced to few seconds or maximum two minutes, depending upon the flight level of the aircrafts and "To-From Flag" movement is very smooth from "TO-FROM" when aircraft passes over the VOR site.

So the design of the antenna system plays an important role in governing the width of this cone of confusion, because it has been observed that more the amount of vertically polarised components in the VOR radiation, the cone of confusion, which normally indicates over the station position to the pilot, will become extremely erratic and cannot be clearly marked. Since vertical polarised radiation is undesirable, thus old alford loop antenna array was replaced by 4-slot cylindrical antennas where it was kept to a minimum level by the slot dimensions of the antenna. Another factor, in the size of the counterpoise under the antenna system and the distance of the counterpoise from the antenna. So necessary steps have also been taken in this regard. The equipment shelter is designed and positioned to reflect as small a vertical radiation components as is practicable.

6.6. THE VOR GROUND EQUIPMENT

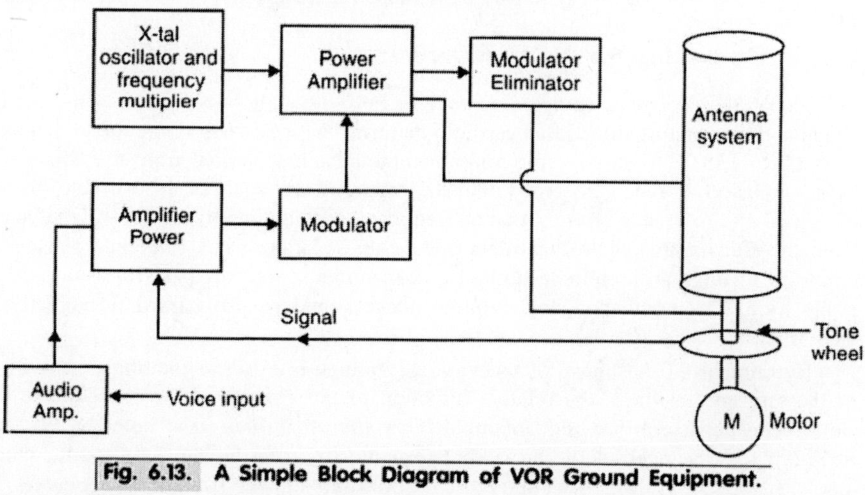

Fig. 6.13. A Simple Block Diagram of VOR Ground Equipment.

A simple block diagram of VOR ground equipment is shown in Fig. 6.13. The above diagram consists of crystal controlled oscillator, frequency multiplier, power amplifiers and the antenna system along with the tone wheel signal (*i.e.* 9960 Hz sub carrier) which is coupled to the rotating part of the antenna system. The power amplifiers which gets an input consisting of the tone wheel signal and a voice signal, is amplitude modulated by the modulator. Further the output of power amplifier is given to the antenna system (omni directional antenna) and some part (approximately 10%) of the output goes to the modulation eliminator and energize the rotating antenna. Modulation eliminator consists of a bridge of coaxial lines of 50 Ω characterstics impedence and a load consisting of two special VHF diodes which act as non linear resistors. if the bridge is balanced there will be no output. The modulation eliminator has efficiency of about 23% and deliver about 15 W to the rotating antenna.

6.7. VOR AIRBORNE RECEIVER

Airborne receiver is typically a high quality double superhetrodyne type along with a broadband omnidirectional antenna (u-shape or horn type). It is omnidirectional VHF AM/FM Rx, which can be used for both VOR and ILS localizer facilities. It is provided with the means for recovering, separating and interprating the navigational information received from the ground station. The main units of this receiver are shown in Fig. 6.14.

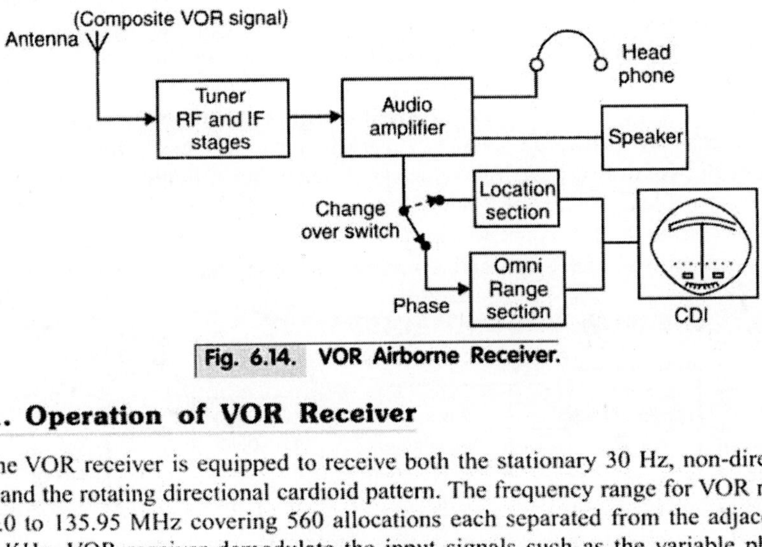

Fig. 6.14. VOR Airborne Receiver.

6.7.1. Operation of VOR Receiver

The VOR receiver is equipped to receive both the stationary 30 Hz, non-directional signal and the rotating directional cardioid pattern. The frequency range for VOR receiver is 108.0 to 135.95 MHz covering 560 allocations each separated from the adjacent one by 50 KHz. VOR receiver demodulate the input signals such as the variable phase 30 Hz signal and reference phase signal as frequency modulation on the 9960 Hz subcarrier. Both the signals are fed to the filters and being separated. The reference phase signal is passed through an amplitude limiter, a discriminator and low pass filter to get 30 Hz modulation. The another signal, variable phase signal is also passed through the low pass filter.

By comparing the phase of the variable voltage due to the rotating field with that of the reference voltage, the relative direction of the aircraft with respect to the range station can be determined. For example if an aircraft is located at point 'X; as shown in Fig. 6.15. The peak of the heart shaped pattern will be facing the aircraft, which is due east of the station. At this point maximum signal strength is received. This corresponds to a certain phase (time) difference between the break of the rotating pattern (variable voltage) and the reference voltage. For any other location of the aircraft with respect to the station the phase difference between the variable signal and the reference voltage will be different because of the rotation of the pattern. Thus every direction of the aircraft with respect to the stations correspond to a definite phase relationship between the two signals. This is translated by the indicators in the receiver in terms of compass heading of the aircraft with respect to the range station.

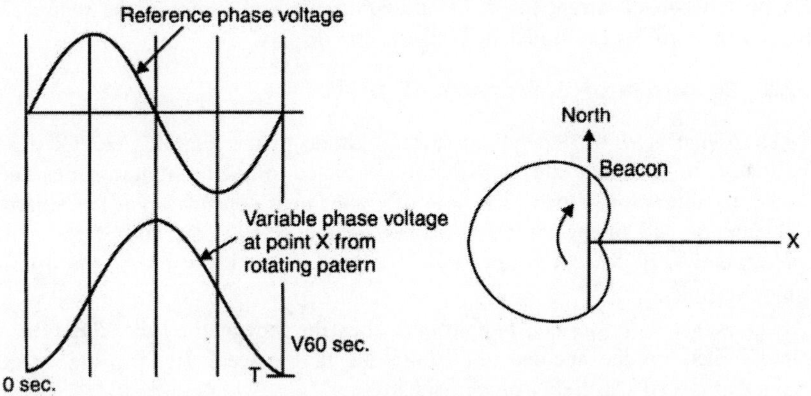

Fig. 6.15. VOR Phase Relations and Rotating Cardioid Pattern (Plane at X).

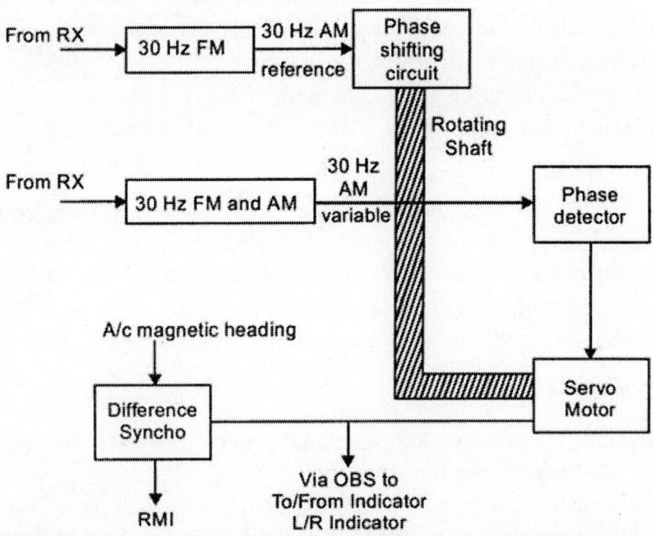

Fig. 6.16. Phase Comparision by VOR Airborne Equipment.

A brief description of phase comparision by VOR is explained with the help of block diagram shown in the Fig. 6.16. The essential elements of the receiver for phase comparision are shown in Fig. 6.16. They are as follows :

(*i*) Phase shifting cet

(*ii*) Phase detector

(*iii*) Servomotor

(*iv*) Difference synchro.

The demodulated output of the receiver fed to the instrumentation unit contains the variable phase 30 Hz signal and the reference phase signal as FM on the subcarrier of 9960 Hz. These signals are divided by filters in two channels. The reference signal is passed through the amplitude limiter, a discriminator and a LP filter to obtain the 30 Hz modulation. Sameway the variable phase signal also amplified by a LP amplifier. To have correct course reading the phases of both the branches are must be equal. By utilising a feedback arrangement, a phase-detector and a motor, the phase difference between the two 30 Hz signal is available for display.

6.7.2. Range and Accuracy of VOR

The range of VOR depends upon the altitude of the aircraft. As VOR operates in VHF range, in this range the VOR facility restricted to line of sight range extended upto 10-15% by refraction effects. The line of sight range depends upon the altitude of the VOR antenna and of the aircraft. At 5000 feet it is 100 NM and at 15000 feet it is approximately 170 NM. A better range is also possible by installing the transmitter at an elevated place.

Generally $\pm 2^\circ$ errors is available in the bearing indicated by the VOR. But with the overall effects on the accuracy of VOR, due to the weak signals, site irregularities, terrain features, it can be as much as $\pm 20^\circ$.

Advantages of VOR

There are some advantages of the VOR, which are summed up below :

1. Any number of aircraft can use facility at the same time.
2. Visual indications of bearings.
3. No special charts or Tables required.
4. It is free from night effect, static and coastal refraction.

Limitations of VOR

There are some disadvantages of the VOR system, which are summed up in following points :

1. Its range depends upon the altitude of the aircraft.
2. It is subjected to high ground error.
3. Frequent air calibration is necessary.

6.8. DVOR

Introduction. Doppler VORs are second generation VORs, the main aim being to improve the accuracy of the signals.

Conventional transmitter suffer from reflection from objects in the vicinity of the site. It was found that the errors due to this could be reduced if the horizontal dimensions of the axial system were increased. However, this could not be achieved with the conventional method of transmission and a new approach was necessary.

In the DVOR system the reference (or constant phase) signal is transmitted from a central aerial and it is amplitude modulated. The variable signal is transmitted from a system of 50 aerials encircling the central and it is frequency modulated. Thus the modulations are employed in reverse rules. The resultant propagation is much less sensitive to obstructions in the vicinity. Otherwise, the transmission frequencies are the same, the airborne equipment can receive and process the signal and as far as the airborne operation is concerned, there is no difference.

6.8.1. Development of DVOR

As pointed at the serious course errors caused by reflection affecting the phase of the variable signal in a conventional VOR. Investigations to eliminate the effects of siting conditions and signal reflections, on omnicourse errors led to the development of the Doppler VOR which minimize variable phase shift as a small of course error.

In early 1950, a project was undertaken at the Technical Development center in Indianapolis to determine the characteristics of a compatible DVOR. In April 1958, a DVOR utilising 20 loops with an aperture of 17 feet was flight tested and in June 1958, a 50 loop DVOR having an aperture of 44 feet was tested as possible solution to VOR siting problems at different locations.

Then, in August 1958, a prototype of the experimental DVOR was installed and tested at Charleston, S.C., where a conventional 4 loop VOR was already installed. Flight test showed definite improvement in course scalloping. Under the same conditions where the conventional VOR had a maximum scallopping error of ±2.8°, the maximum course scalloping from the DVOR was ±0.4°.

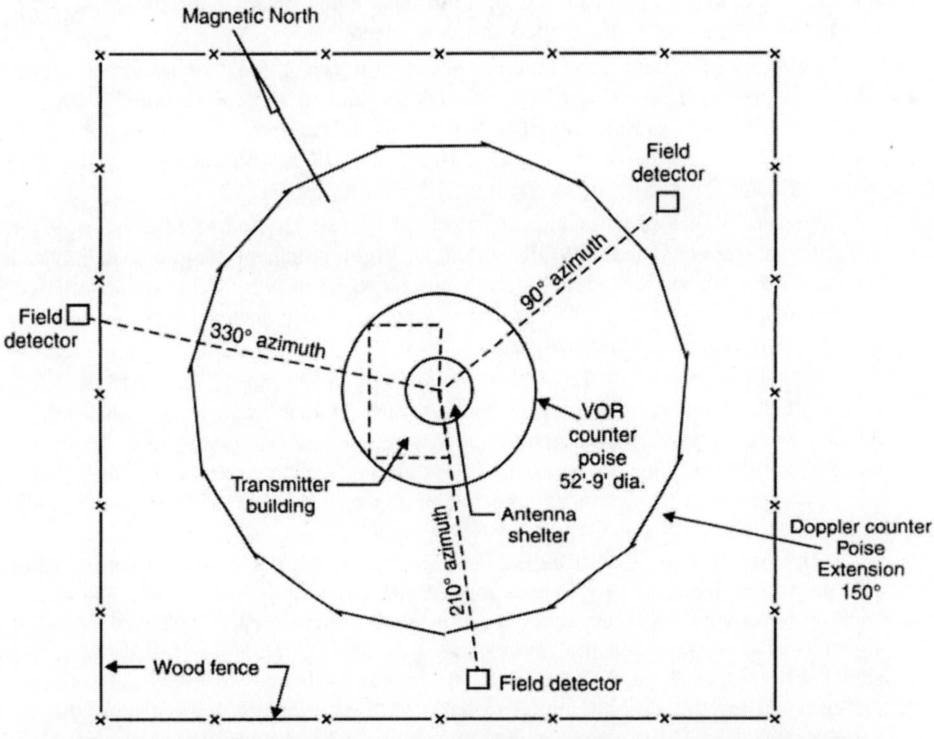

Fig. 6.17. Typical plot layout of a doppler VOR facility (Location of the field detectors are shown at Magnetic Azimuths 90°, 210° and 330°. Alternate field detector locations are Magnetic Azimuths of 30°, 150°, and 270°.

The results of the tests at Charleston, SC and at Indianpolis provided the background information for the development of the present 50 loop DVOR having an antenna system diameter of 44 feet and a counterpoise diameter of 150 feet.

In Fig. 6.17, a typical plot layout of DVOR is shown.

Theory of Operation

The course forming signals received by an aircraft from a Doppler VOR are compatible with the signals received by an aircraft from the conventional VOR. The aircraft receiver should respond in the same manner to signals received from either system. Considering the same siting conditions for both systems, reflections will cause considerably less error in the resultant variable signal from the Doppler VOR. This is because the variable signal from the DVOR is produced by frequency modulation.

The FM variable signal from the DVOR is received with a minimum phase shift due to reflections, because of the capture effect in an FM system. This means in the DVOR system that the stronger direct path FM signal will predominate in the aircraft receiver when it is received along with the weaker reflected path FM signal. The output of FM detector in the receiver will contain principally the 30 Hz information carried by the stronger direct signal, not the 30 Hz information carried by the reflected signal. The 30 Hz variable signal contained in the direct path FM signal, will give a true omni cause indication when compared in phase with the omnidirectional AM reference signal.

The DVOR 30 Hz reference signal is produced by amplitude modulation. However, due to the fact that it is transmitted with the same audio phase in all directions, reflected signals will have small effect upon the desired course.

The most important consideration when comparing the two systems is that the variable signal is a result of FM in the DVOR and in the conventional VOR, it is a result of AM. FM variable signal is much less susceptible reflection errors than the variable signal in the conventional VOR. This is the primary reason for installing DVOR, and to replace some existing conventional VORs as well.

Fig. 6.18 shows phasors representation of the 30 Hz course forming signal from both the conventional and DVOR system at eight points of magnetic azimuth. It is assumed that both VOR's are to be free from course erros. The definition for omnicourse in the conventional VOR is the number of degrees that detected 30 Hz variable signal at some particular magnetic azimuth.

Fig. 6.18 shows the 30 Hz signal produced by space modulation (AM) lagging the FM produced reference signal as a receiver moved clockwise around the VOR. The degree of lag increases with a greater clockwise movement around the VOR. It is to be noted that the FM detected 30 Hz signal is always shown in the same phase (0°) and is, therefore, called the reference signal. The phase of the AM 30 Hz signal is different at every azimuth and is called the variable signal.

Referring the Fig. 6.18 (labelled doppler) notice the FM produced signal leads the AM produced signal as a receiver is moved clockwise around the VOR. The degree of lead increases with a greater clockwise movement around the VOR. Here the AM 30 Hz signal is always shown in the same phase (0°) and is, therefore, called the reference signal. The phase of the FM detected 30 Hz signal changes at every azimuth and is, therefore, called the variable signal in the DVOR system. By comparisons, the 30 Hz variable signal is AM produced in the conventional VOR and 30 Hz variable signal is FM produced in the DVOR.

This condition of the FM variable signal leading is in a clockwise receiver movement, is necessary in order to have the DVOR system compatible with the conventonal VOR system. This means a VOR receiver NE of a conventional VOR will receive an AM 30 Hz signal that lags the received FM 30 Hz signal by 45°. The same receiver NE of a DVOR will receive an FM detected 30 Hz signal that leads the received AM detected 30 Hz signal by 45°. This is as the AM signal lagging the FM signal as in the conventional

VOR. The omnicourse indicated by the receiver will be 45° in either case. The two system are compatible because the AM detected 30 Hz signal whether the original source of signal is from a conventional or from a DVOR. The FM detector in the VOR receiver will produce a 30 Hz signal whether the signal was originally generated by the electromechanical action of a tone wheel in a conventional VOR or by the special method employed in the DVOR system.

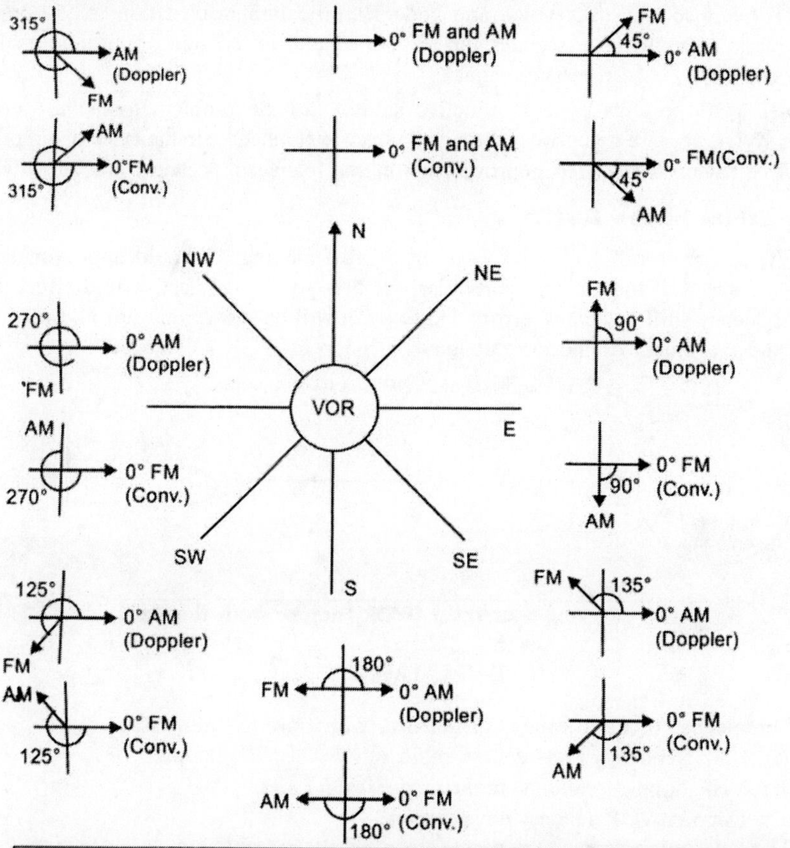

Fig. 6.18. Phasor Diagrams of VOR Systems (Doppler and Conventional).

The special method of generating the signal in the DVOR system utilizes the Doppler effect. The doppler effect is a change in wavelength of a sound, as heard by a listener, during the period of time a change in distance occurs between the source of sound and the listener. A change of wavelength constitutes a change in pitch of a sound as heard by a listener. This same condition holds true for either audio or radio frequencies.

During the period of time that a change of distance occurs between an RF source and RF receiver, there will be a change in wavelength of the energy as it is received. A change of wavelength will result in a change in the frequency being received because

$$f = \frac{C}{\lambda}$$

where 'C' is the velocity of light.

The distance between the aircraft receiver and the DVOR RF source is made to change 30 times per second. The doppler RF source is changed in position by feeding, in sequence. 50 antennas which are arranged in a circular manner. The system is designed so that the change in distance between the RF source and receiver causes the received frequency to increase and decrease by approximately 480 cycles. This doppler effect produced frequency modulation is detected by the aircraft receiver to produce a 30 Hz course forming signal. The phase of the 30 Hz signal will be different at each azimuth around the Doppler VOR.

The DVOR also transmits voice and code identification information to the craft. This information amplitude modulates the RF corner and is radiated omnidirectionally alongwith the 30 Hz AM reference signal.

Doppler VOR presents a very attractive solution to the problem associated with conventional VOR because it generates the signal which is compatible to the existing airborne receiver while having the desired improvements in site tolerance, accuracy and stability.

Doppler Effect in the DVOR

In DVOR, the aircraft receiver (observer) is also moving. It would approximately show that the speed of the aircraft, particularly if it is going very fast, would affect the Doppler frequency shift and cause errors. However, it will be shown that, for all practical purposes, the movement of the aircraft may be ignored.

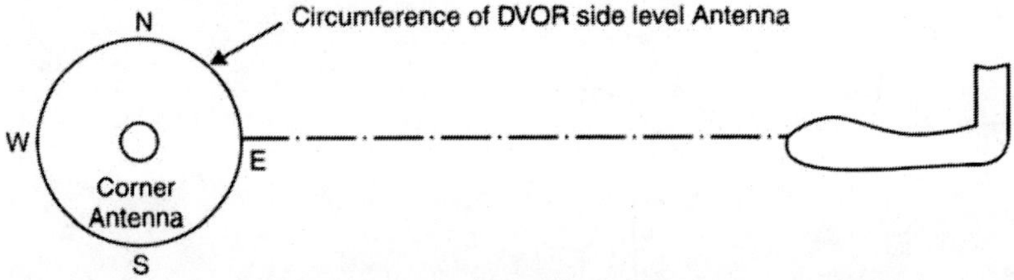

Fig. 6.19. A/c Approaching a DVOR Feature from the East.

SUMMARY

- These types of radio ranges are in use, which are as follows :
  (*i*) Low frequency four course radio range
  (*ii*) VHF omni directional radio range (VOR) and
  (*iii*) Doppler VOR (Latest development)
- The low frequency range is one of the earliest electronic aids to air navigations. The LF radio station sends out "on course" signals which interlock with those of surrounding station to form easily followed airways. Radio range station operates on frequencies between 200 KHz to 400 KHz.
- Directly above a radio range station there exists an area where practically no signal is heard. This area has the shape of an inverted cone and is known as the cone of silence.
- VHF Omni Range (VOR) operates in VHF band of 112-118 MHz.
- VOR works on the principle of phase comparision of two 30 Hz signal.
- There are two types of VOR :
  (*i*) Conventional VOR (C-VOR)
  (*ii*) Doppler VOR (D-VOR)
- The main purpose of the VOR is to provide the navigational signals for an aircraft receiver which will allow the pilot to determine the bearing of the aircraft to a VOR facility.

- A space modulation technique is used to obtained the variable 30 Hz signal. This type of modulation is not generated inside the Tx but it is achieved by combination of RF signals in space.
- **Range and Accuracy of VOR :** The range of VOR is 100 NM at 5000 feet and it is 170 NM at 15,000 feet. A better range is also possible by installing the transmitter at an elevated place. Generally ±2° erors is available in the bearing but with overall effects on the accuracy of VOR, it can be as much as ±20°.
- **Advantage of VOR :**
  (*i*) Any number of aircraft can use facility at the same time.
  (*ii*) Visual indication of bearings.
  (*iii*) No special charts and tables required.
  (*iv*) It is free from night effect, static and coastal refraction.
- **Limitations of VOR :**
  (*i*) Its range depends upon the altitude of the aircrafts.
  (*ii*) It is subjected to high ground error.
  (*iii*) Frequent air calibration is necessary.
- DVOR is second generation of VORs, the main aim being to improve the accuracy of the signals. Conventional transmitter suffer from reflection from objects in the vicinity of the site. The errors due to this could be reduced if the horizontal dimension of the axial system were increased. However, this could not be achieved with the conventional method of transmissions. A new approach was necessary at it came as DVOR.

REVIEW QUESTIONS

1. Explain the LF/MF four course radio range.
2. Explain the VOR ground equipment with the help of block diagram.
3. Write the errors in VOR system.
4. Describe the DVOR with the help of block diagram. **[UPTU, 2003]**
5. Explain "Cone of Silence".
6. Define the purpose of VOR in navigation.
7. Explain the VOR Air borne Receiver with the help of block diagram.
8. Describe the frequency range of VOR receiver with the total number of allocation of channels.
9. Explain the advantages and the limitations of VOR.
10. What is radio range ? Explain long wave radio range. What are its shortcomings and how they are removed ? **[UPTU, 2006]**
11. Write short notes on :
    (*a*) Ground equipment of VOR (*b*) Advantages of DVOR
    (*c*) Range and accuracy of VOR (*d*) Doppler effect in DVOR
    (*e*) Cone of confusion.
12. Explain how, in a long-wave radio range system, it would be possible for an aeroplane pilot, following a particular equisignal course, to determine whether he was approaching or receiving from radio range transmitter. **[UPTU, 2006]**
13. Is there anything inherent in the very high frequency omnidirectional range that limits the use of this principle to very high frequencies ? Explain. **[UPTU, 2006]**

7

CHAPTER

HYPERBOLIC ELECTRONIC NAVIGATIONAL SYSTEMS

INSIDE THIS CHAPTER

7.1. INTRODUCTION

Hyperbolic systems are based on the measurement of the difference in the time of reaching of EM waves from the two or more transmitters to the receiver in the craft.

Before we proceed to the hyperbolic navigation systems, let us first understand the principle of hyperbolic systems. In this chapter you will be learning about the various hyperbolic electronic navigational systems such as LORAN, DECCA and Omega.

7.2. PRINCIPLE OF HYPERBOLIC ELECTRONIC SYSTEMS

When you know the difference in your distance from two objects, you know you're on a curved line defined by that difference. By definition, a hyperbolic curve is formed by joining all points that are the same difference in distance between two points. In hyperbolic navigation systems, your receiver measures the difference in time of receipt of radio waves, and you or the set translate the time difference into a distance difference using the known speed of the waves.

It's easy to grasp the concept naturally by placing yourself on an imaginary hockey field. If the sound of bells rung simultaneously under each goalpost reached you at the same time, you'd know that you were equidistant from each goalpost, or somewhere on the 50-yard line. If the sound from the bell at the north goalpost reached you sooner than the other one, you'd know that you were closer to that bell. The amount of time difference would determine how much closer you were. Just as the receipt of the sounds at the same time places you on a line, the straight 50-yard line, so, too, the receipt at a specific difference in time places you on the field somewhere on a different line - **a hyperbolic curve.**

If we observe the pattern shown in the Fig. 7.1 below. In it, a pair of transmitters takes the place of the goalposts, and the line between them is called **the baseline**, which is bisected by a line perpendicular to it (centerline) - the 50-yard line in our analogy. All the other lines are hyperbolic curves.

Unless you received the signals from each transmitter simultaneously, you'd be on one of these curves. If you received the signal from transmitter X, say, 50 microseconds before the one from M, then you'd be on line CD; if 150 seconds before, then on line EF.

Note that the distance between the lines in this pattern increases with increasing distance from the baseline. Because it does, the potential accuracy of the system is greater when you're nearer the baseline - where the lines converge. In fact, when you're located in the vicinity of the baseline extension (shown as dashed lines) where the divergence of the lines is most pronounced, accuracy may be reduced to unacceptably low levels.

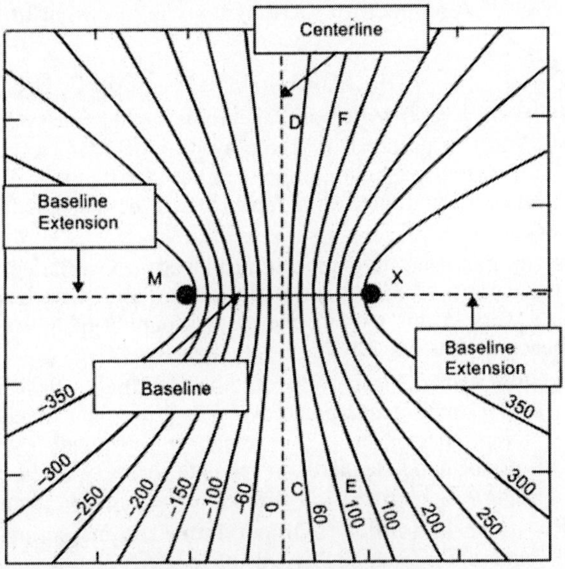

Fig. 7.1. Hyperbolic Basics.

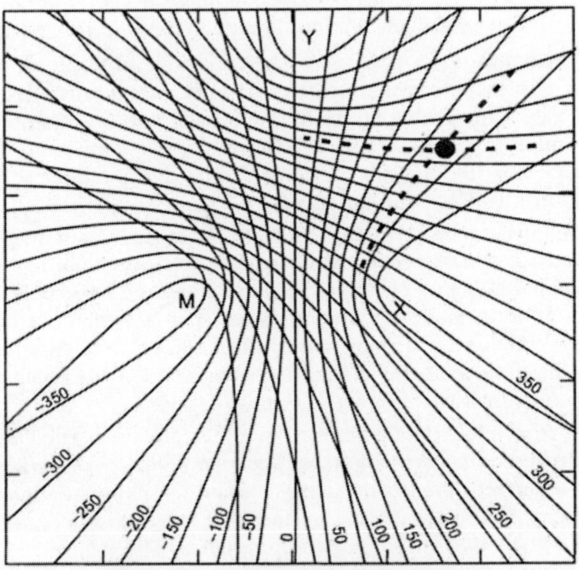

Fig. 7.2. A Sample Hyperbolic Fix.

The reading you get from one pair of transmitters gives you a single LOP - the curve you're on. To get a fix, you need to measure the time difference from a second pair that places you on another curve. Your fix is at the intersection of the two curves, as shown in Figs. 7.1 and 7.2.

Hyperbolic basics and a sample hyperbolic fix. (See Fig 7.1 and Fig.7.2.)

7.3. HYPERBOLIC ELECTRONIC NAVIGATIONAL SYSTEMS

There are some hyperbolic navigational systems, which are being developed in the long span of time. A brief description of these systems is discussed in this chapter.

1. Decca
2. Consol
3. Omega
4. LORAN (LORAN-A, B, C, D and F)

Decca was the first of several hyperbolic electronic navigation systems developed during and after World War II. Decca was first employed in the 1944 Normandy invasion. Its coverage was later extended beyond the English Channel to the waters off Newfoundland, the North Sea, and the Sea of Japan. It will be discussed in brief in this chapter in the later sections.

The Consol system, based on an approach developed in Germany during the war, was established in the eastern North Atlantic. Because of the short baseline distance between transmitting towers in this system, the close grouping of towers serves much like a single radio beacon - but one having a very long range.

The Omega system was devised to overcome the major drawback of other hyperbolic systems - their limited geographic coverage. With eight transmitters located some 5,000 to 6,000 miles apart, Omega's coverage is worldwide - a spread made possible by the use of atomic time standards to regulate the very low-frequency, long-range signals. This system will be discussed further section in this chapter.

Hyperbolic LORAN lines of position (LOP) are formed by measuring the difference in reception times of synchronized signals. Groups of LORAN stations are used to form intersecting LOP to provide cross fixing. A LORAN net, or chain, requires a master station, initiating the pulse, and a series of slave stations. In very simplified terms, the master transmits and the slave responds. Various delays are built in for technical purposes.

In principle the signals of a pair arrive at the same time along the base line connecting the pair. It is at the base line where the apex of each hyperbolic LOP lies. Beyond the stations they flare so that no useful hyperbolic rate information is provided. This is the unusable area of the **base line extension**. At any point other than the base line and base line extension the difference in arrival time of the master and slave pulses form the LOP in the form of hyperbolic curves.

It is the calculated difference in arrival times of master-slave pairs that form charted rates and the observed arrivals that form the navigational information. If the observation is exactly that of the charted value one is somewhere along the hyperbola of that value. That information is of little value without a second pair's hyperbola intersecting to provide a definitive fix along the first hyperbola. The station configurations thus must provide a geometry that allows such intersection. Where LORAN chains overlap one may obtain cross fixes from two different master/slave pairs.

LORAN suffers from electronic effects of weather and in particular atmospheric effects related to sunrise and sunset. The most accurate signal is the ground wave that following the Earth's surface, preferably along a seawater path. At night the indirect sky wave, taking paths bent back to the surface by the ionosphere, is a particular problem as multiple signals may arrive via different paths. The ionosphere's reaction to sunrise and sunset accounts for the particular disturbance during those periods. Magnetic storms have serious effects as with any radio-based system.

7.4. LORAN-A

LORAN-A was a less accurate system operating in the 1,750-1,950 KHz frequency band prior to deployment of the more accurate LORAN-C system.

7.4.1. Principles of Operation

Loran provided facilities whereby ships and aircraft derived their position at long distances. The system required at least three transmitting stations for each 'chain', and the observer used a special Loran receiver. A chain consisted of one master and two slave stations. Differences in the arrival time of pulses from a pair of stations was measured and displayed on the face of a cathode ray tube. Each fix required two observations and the operation normally took about five minutes. The readings were then transposed to a Loran lattice chart and position could be plotted. In some cases readings were referenced to special Loran tables. Because Loran-A signals were pulsed and not continuous transmissions, a relatively small transmitter could achieve tremendous peak power levels. The maximum reliable range for Loran-A was 700 miles by day and 1,400 miles at night.

7.4.2. Signal Characteristics

Each transmission pulse lasted about 40 microseconds and reoccurred at regular, accurately controlled intervals. This interval, called the **Pulse Repetition Interval (P.R.I.)** varied for each station and lasted between 29,000 and 40,000 μs. These pulses provided precise index marks for use in time measurements. The transmissions of corresponding master and slave pulses were separated by a fixed time interval, which consisted of the time for a signal to travel from the master to the slave, plus one-half the P.R.I., plus an additional small time, called the **'coding delay'**. It should be noted that the observer is interested only in measuring the difference between the time of arrival of the two pulses, and not the actual time taken for each pulse to reach the receiver. There was no need, therefore, for an absolute synchronization of the receiver time base with the transmitter.

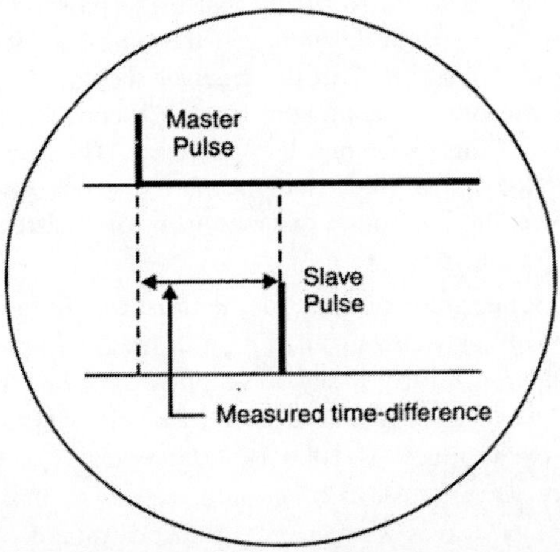

Fig. 7.3. Display of Time Difference in LORAN-A.

At all points in the coverage area, the time interval between a master pulse and the next slave pulse was greater than the interval between a slave pulse and the next master pulse. That methodology provided a positive method of identifying the signals arriving from each station, even though their actual appearance was similar. In the measuring process, the time difference was always measured from the master pulse to the slave pulse, and the time delay of one half of the pulse recurrence interval was automatically removed. The lines of constant time difference for each pair of stations were pre-computed, taking into consideration the curvature and eccentricity of the Earth, the time for the master pulse to reach the slave station, and the coding delay. These "hyperbolic" lines were made available in the form of overprinted charts and tables.

7.4.3. Arrangement of Station Pairs

When a common master controlled two slaves, the master was called a 'double pulsed' station because it transmitted two entirely separate sets of pulses, one set paired with the pulses from each adjacent station. Pairs of Loran stations were situated up to 600 miles and more apart.

Loran transmitters emitted 40 microsecond pulses. For H-rate pairs, the basic recurrence interval was every 30,000 microseconds; for L-rate pairs, 40,000 microseconds; and for S-rate pairs, 50,000 microseconds Each pulse had a peak power in excess of 200,000 watts, but since the duty cycle (the ratio of time the transmitter is on duty) for an L-rate pair, for example, is only 40/40,000, or 0.001, of the time, the transmitter has an average power output of only 200 watts or so. (200,000 × 0.001).

If any trouble occurred at either the master or the slave station that might impair the accuracy of the pulse timing, the transmitters operated on a 2 sec ON then 2 second OFF mode. This appeared to the operator as a blinking signal. Blinking signals were not used for navigation.

7.4.4. Reception of Signals

In order to properly display the pulses to be measured as shown in Fig. 7.3, the receiver's time base had to synchronize so the length of the trace on the C.R.T. matched the P.R.I. of the station. Failing to do so would cause the pulses to appear as if they were drifting to the left or to the right depending if the time base was too short or too long respectively. The face of the C.R.T. in the receiver displayed two time base lines because a pair of stations was always being compared. For convenience, the upper trace was called the "A" trace and the lower one the "B" trace. By convention, the master station was displayed on the upper trace and the slave on the lower one. The time difference measurement was the horizontal distance from the master pulse to the slave pulse.

In an attempt to gain longer-range navigation, a variant of Loran-A was developed. It was known as SS (**sky-wave-synchronized**) Loran In the SS Loran system, the slave station of a pair was synchronized by a sky-wave pulse reflected from the 'E' layer, rather than by the ground wave as in standard Loran. This allowed the master and slave stations to be separated by as much as 1000 to 1200 miles. The Loran charts were calibrated in terms of sky waves, instead of ground waves, so that correction factors were unnecessary when sky waves were used. A disadvantage of the system was encountered when the indicator was located close to either or both stations, since erratic reception resulted when the angle of reflection of the sky wave from the *E* layer approached the critical angle. As the critical angle was approached, the radio waves exhibited increasing penetrating power and would go entirely or part way through the '*E*' layer.

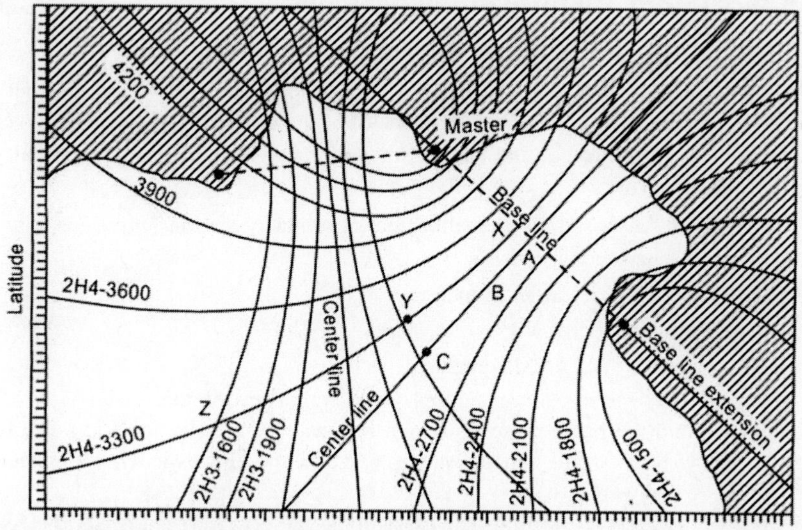

Fig. 7.4. A sample Loran chart showing the location of a master and two slave stations. Also shown are station identifiers and time differences on the curves.

7.4.5. Identification of LORAN-A Pairs

Loran-A stations did not transmit call signs. Instead, identification was made entirely by two distinguishing characteristics: (*a*) radio frequency channel; (*b*) pulse repetition rate.

(A) By Channels. Different groups of Loran stations operated on different frequencies Four fixed frequencies were available between 1,750 and 1,950 kc/s. The receiver was fitted with a channel selector switch for tuning to the desired frequency. They were assigned the following designations:

Channel 1 - 1,950 kc/s

Channel 2 - 1,850 kc/s

Channel 3 - 1,900 kc/s

Channel 4 - 1,750 kc/s

(B) By Pulse Repetition Rate. In order to economize on frequency channels, a number of pairs of Loran stations were operated on the same frequency, but each pair operated at a different pulse repetition rate. That meant that signals from all stations on the same frequency within range appeared on the indicator, but they drifted across the scan at varying speeds. The operator selected a particular pair of stations by means of switches on the receiver which make the sweep repetition rate of the indicator the same as the pulse repetition rate of the desired pair. The desired signals would now be stationary, while the remainder still drifted across the scan and could be ignored.

Two switches were provided. The first one adjusted for the basic pulse repetition rate, of which there were three in advanced Loran sets: High, Low and Slow. The second switch adjusted for a specific pulse repetition rate differing from the basic by a small amount. There were eight of these specific rates, numbered 0 to 7, for each basic pulse repetition rate. This system thus provided 96 separate station pairs using the four frequency channels available.

7.4.6. Station Identification Symbols

Each pair of Loran-A stations was given a three character identification symbol, of which the first character was the channel; the second was the *basic* pulse repetition rate, and the third for the *specific* pulse repetition rate. These symbols were given in the Loran Tables, on printed on the charts. All stations were listed in the Admiralty List of Radio Signals - Vol 5.

In addition, each station was allocated an arbitrary station letter. For example, the N.E. Atlantic. chain consisted of:

Master station : U, at Skuvanaes

Slave station : K, at Vik

Slave station : A, at Mangersta.

These letters were not transmitted. The frequency of this chain was 1,950 kc/s (Channel 1) and the basic pulse repetition rate was LOW. The specific pulse repetition rate of pair U-K is 5, and that of U-A is 6. Thus, the U-K pair was designated 1 L 5, and the U-A pair was 1 L 6.

In order to receive the first pair of stations, the operator had to set the receiver as follows:

1. Set the Channel switch to 1.
2. Set the basic P.R.R. switch to L.
3. Set the specific pulse repetition rate (5)

7.4.7. Pulse Repetition Rates

The following *base* pulse repetition rates were used:

H = **33 1/3 pps**

L = **25 pps**

S = **20 pps (for later equipment)**

The eight variants of PRR and three ranges (H, L, S) produce the following PRR variants:

H0 = 33 3/9 pps	L0 = 25 pps	S0 = 20 pps
H1 = 33 4/9 pps	L1 = 25 1/16 pps	S1 = 20 1/25 pps
H2 = 33 5/9 pps	L2 = 25 2/16 pps	S2 = 20 2/25 pps
H3 = 33 6/9 pps	L3 = 25 3/16 pps	S3 = 20 3/25 pps
H4 = 33 7/9 pps	L4 = 25 4/16 pps	S4 = 20 4/25 pps
H5 = 33 8/9 pps	L5 = 25 5/16 pps	S5 = 20 5/25 pps
H6 = 34 pps	L6 = 25 6/16 pps	S6 = 20 6/25 pps
H7 = 34 1/9 pps	L7 = 25 7/16 pps	S7 = 20 7/25 pps

Therefore 4 basic frequencies × 3 PRR rates × 8 PRR variations = Maximum of 96 pairs of stations. On an ordinary radio receiver, a Loran station sounded like a continuously firing machine gun that changed tone slowly.

7.4.8. Equipment Types

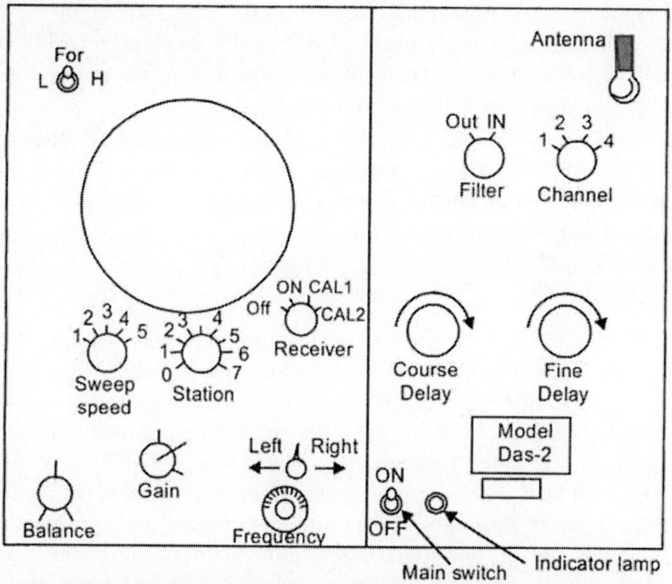

Fig. 7.5. The Model DAS-2 was a Popular Loran-A Receiver. ***(Graphic Courtesy of Admiralty Manual of Navigation).***

Fig. 7.6. DAS-3 Receiver.

7.4.9. The End of LORAN-A

The accuracy of Loran-A varied according to location, time of day, weather and relative geometry of transmitting stations. Aside from some testing by the USCG, the follow-on system Loran-B, never made it as a commercial system of navigation due to technical problems. It was eventually surpassed by Loran-C which provided longer range, greater accuracy when it first came into operation in 1957. Loran-A was phased out in December 1980 in North America and world-wide by 1985. The last holdout was Japan where the plug was pulled on May 9, 1997.

Loran-A closures came much to the delight of amateur radio operators who had to share their 160 meter band with Loran on a secondary basis for so many years. During that era, amateurs were required to reduce power substantially in the bands 1,800 to 2,000 kHz. In Canada, power levels of 375 watts were permissible during daylight hours and 150 watts at night. Depending on the region, daytime power was 500/200 watts in the United States and 200/50 watts by night.

At the end of the system's life cycle, the development of receivers has spanned right up to the APN-35. The set had been reduced to the size of a shoebox and had automated the process of performing the fix while including ground speed, distance traveled, distance remaining, and ETA.

7.5. LORAN-B

LORAN-B was basically a phase comparison variation of Loran-A and an attempt to re-design the system properly plus add some new features. Wartime development of the Loran-A system had been rushed. Even in ground wave range, it was never a very accurate system. One shortcoming was it's pulse length which was approximately 40 microseconds. It was noted however, that Loran was providing usable, steady ground wave signals to at least 400 miles out to sea. In the original design, no attempt had been made to achieve phase coherence in the transmitted pulses, which made phase comparison within the pulse (as in Loran-C) impossible. With Loran-B the transmitters were redesigned to achieve this function and a new receiver was developed as well. Phase comparison increased accuracy to about half a cycle, (about 0.25 microsecond) a considerable improvement. Another new feature of the system had the transmitters operating in chains instead of pairs, thus allowing simultaneous measurement of two time differences.

The unique feature of Loran-B was that, although it operated on the same frequencies as Loran-A, the timing control and the timing measurements were made on the pulse carrier, after cycle identification from the pulse envelope. The process made for very precise timing measurements. However, at the land-sea interface, there was considerable phase distortion. This resulted in major differences in the phase to envelope timing depending on the azimuth from the transmitter. There were probably instrumentation problems resulting from the temperature and vibration sensitivity of components, which were available at the time. In any case, the work was abandoned in favor of the Loran-C system in the 1950.

7.6. LORAN-C

LORAN-C, operating at 100-110 kHz, was initially developed under U.S. Navy sponsorship and deployed to cover large portions of the northern hemisphere for both marine and air navigation. Originally just a marine navigation system, LORAN-C determines present position by the intersection of Lines of Position (LOPs) that are hyperbolic curves. At least three stations, (a Master and two Secondaries) are needed. Accuracy is plus or minus 2.5 miles. The LORAN-C uses triangulation to measure the location of an aircraft or boat.

LORAN-C provides the following informations :

1. Range (RGE)
2. Track (TRK)
3. Ground speed (GS)

4. Estimated time en route (ETE)
5. Cross track error (XTD)
6. Track angle error (TKE)
7. Desired track (DTK)
8. Winds and drift angle (DA)

7.6.1. Basic Theory of Operation

The Loran C system consists of a chain of transmitting stations, each separated by several hundred miles. Within the Loran chain, one station is designated as the **master station** and the others as **secondary stations**. There must be at least two secondary stations for one master station; therefore, every Loran transmitting chain will contain at least three transmitting stations. The master and secondary stations transmit radio pulses at precise time intervals. A Loran receiver measures the **time difference** (TD) in reception at the vessel between these pulses; it then displays either this difference or a computed latitude and longitude to the operator.

The signal arrival time difference between a given master-secondary pair corresponds to the difference in distance between the receiving vessel and the two stations. The locus of points having the same time difference from a specific master-secondary pair forms a hyperbolic line of position (LOP). The intersection of two or more of these LOP's produces a fix of the vessel's position. There are two methods by which the navigator can convert these time differences to geographic positions. The first involves the use of a chart overprinted with a Loran **time delay lattice** consisting of time delay lines spaced at convenient intervals. The navigator plots the displayed time difference by interpolating between the lattice lines printed on the chart. In the second method computer algorithms in the receiver's software convert the time delay signals to latitude and longitude for display.

Early receiver conversion algorithms were imprecise; however, modern receivers employ more precise algorithms. Their position output is usually well within the 0.25 NM accuracy specifications for Loran C.

7.6.2. Components of The Loran System

The components of the Loran system consist of

1. The land based **transmitting stations**,
2. The Loran **receiver**
3. **Antenna**, and
4. The **Loran charts.**

Land-based facilities include master transmitting stations, at least two secondary transmitters for each master transmitter, control stations, monitor sites, and a time reference. The transmitters transmit the Loran signals at precise intervals in time. The control station and associated monitor sites continously measure the characteristics of the Loran signals received to detect any anomalies or any out-of-specification condition. Some transmitters serve only one function within a chain (i.e., either master or secondary); however, in several instances, one transmitter can serve as the master of one chain and secondary in another. This dual function lowers the overall costs and operating expense for the system. Loran receivers exhibit varying degrees of sophistication; however, their signal processing is similar. The first processing stage consists of **search and acquisition**, during which the receiver searches for the signal from a particular Loran chain, establishing the approximate location in time of the master and secondaries with sufficient accuracy to permit subsequent settling and tracking.

After search and acquisition, the receiver enters the **settling phase**. In this phase, the receiver searches for and detects the front edge of the Loran pulse. After detecting the front edge of the pulse, it selects the correct cycle of the pulse to track. Having selected the correct tracking cycle, the receiver begins the **tracking and lock** phase, in which the receiver maintains synchronization with the selected received signals. Once this phase is reached, the receiver displays either the time difference of the signals or the computed latitude and longitude as discussed above.

7.6.3. Description Of Operation

The Loran signal consists of a series of 100 kHz pulses sent first by the master station and then, in turn, by the secondary stations. For the master signal, a series of nine pulses is transmitted, the first eight spaced 1000 msec apart followed by a ninth transmitted 2000 msec after the eighth. Pulsed transmission results in lower power output requirements, better signal identification properties, and more precise timing of the signals. After the time delays discussed below, secondary stations transmit a series of eight pulses, each spaced 1000 msec apart. The master and secondary stations in a chain transmit at precisely determined intervals. First, the master station transmits; then, after a specified interval, the first secondary station transmits. Then the second secondary transmits, and so on. Secondary stations are given letter designations of W, X, Y, and Z; this letter designation indicates the order in which they transmit following the master. When the master signal reaches the next secondary in sequence, this secondary station waits an interval, defined as the **secondary coding delay, (SCD)** or simply **coding delay (CD)**, and then transmits. The total elapsed time from the master transmission until the secondary emission is termed the **emissions delay (ED)**.

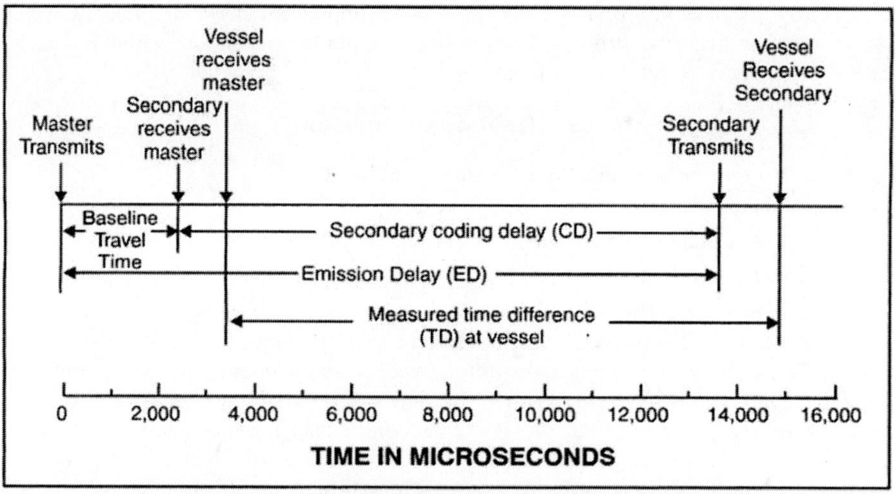

Fig. 7.7. The Time Axis for LORAN –C TD for Point "A".

The ED is the sum of the time for the master signal to travel to the secondary and the CD. The time required for the master to travel to the secondary is defined as the **baseline travel time (BTT)** or **baseline length (BLL)**. After the first secondary transmits, the remaining secondaries transmit in order. Each of these secondaries has its own CD/ED value. Once the last secondary has transmitted, the master transmits again, and the cycle is repeated. The time to complete this cycle of transmission defines an important characteristic for the chain: the **group repetition interval (GRI)**. The group repetition interval divided by ten yields the chain's designator. For example, the interval between successive transmissions of the master pulse group for the northeast US chain is 99,600 msec. From the definition above, the GRI designator for this chain is defined as 9960. The GRI must be sufficiently large to allow the signals from the master and secondary stations in the chain to propagate fully throughout the region covered by the chain before the next cycle of pulses begins.

Other concepts important to the understanding of the operation of Loran are the baseline and baseline extension. The geographic line connecting a master to a particular secondary station is defined as the station pair **baseline**. The baseline is, in other words, that part of a great circle on which lie all the points connecting the two stations. The extension of this line beyond the stations to encompass the points along this great circle not lying between the two stations defines the **baseline extension**. The importance of these two concepts will become apparent during the discussion of Loran accuracy considerations below.

As discussed above, Loran C relies on time differences between two or more received signals to develop LOP's used to fix the ship's position. This section will examine in greater detail the process by which the signals are developed, transmitted, and ultimately interpreted by the navigator. The basic theory behind the operation of a hyperbolic system is straightforward. First, the locus of points defining a constant difference in distance between a vessel and two separate stations is described by a mathematical function that, when plotted in two dimensional space, yields a hyperbola.Second, assuming a constant speed of propagation of electromagnetic radiation in the atmosphere, the time difference in the arrival of electromagnetic radiation from the two transmitter sites to the vessel is proportional to the distance between the transmitting sites and the vessel. The following equations demonstrating this proportionality be ween distance and time apply :

$$\text{Distance} = \text{Velocity} \times \text{Time}$$

or, using algebraic symbols

$$d = c \times t$$

Therefore, if the velocity (c) is constant, the distance between a vessel and two transmitting stations will be directly proportional to the time delay detected at the vessel between pulses of electromagnetic radiation transmitted from the two stations.

7.6.4. Pulse Pattern of LORAN-C

As mentioned above, Loran uses a pulsed signal rather than a continuous wave signal. Fig. 7.8 represents the Loran signal. The master station transmits nine of these signals and the secondary stations every transmission cycle transmit eight. The pulse exhibits a steep rise to its maximum amplitude within 65 msec of emission and an exponential decay to zero within 200 to 300 msec. The signal frequency is nominally defined as 100 kHz; in actuality, the signal is designed such that 99% of the radiated power is contained in a 20 kHz band centered on 100 kHz.

Since all Loran-C transmitters operate on the same frequency, the GRI is the key by which a receiver can identify and isolate signal groups from a specific chain. In naming the chains, the GRI is included. As an example the Great Lakes chain has a GRI of 8970. This means the time interval is 89700 microseconds. The rightmost zero is always implied and the GRI is always in multiples of 10 microseconds. In old Loran-C receivers, the operator had to actually set this number to receive the chain. In cases where the Loran signals were observed on an oscilloscope, pulses from the desired chain would be stationary while those from other chains would be drifting down the time base at varying speeds. It is in fact, the only way of positively identifying a chain. In modern receivers, this is now done automatically.

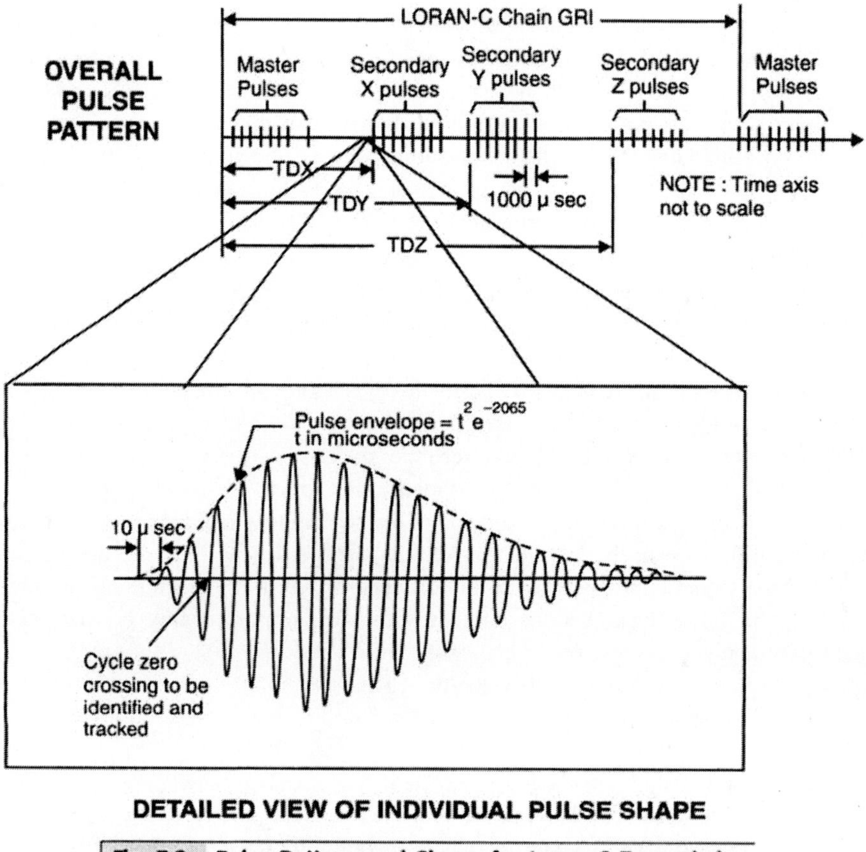

Fig. 7.8. Pulse Pattern and Shape for Loran C Transmission.

GRI's are chosen on the basis of :

(*a*) Baseline lengths between master and secondaries. If the distance between the master and first secondary is say 1000 kms, the radio signal will take 33,000 microseconds to get to the slave so the GRI cannot possibly be less than that.

(*b*) Number of slaves that have to be accommodated - they all have to have delays so that there is no possibility of them crossing over anywhere in coverage area.

(*c*) Geography.

(*d*) Other nearby chains with consideration given to interference.

(*e*) Sky wave cross-rate interference.

(*f*) Duty cycle of the transmitters - a faster GRI means the average power of the transmitted signal is higher so the final stage in the transmitter requires more cooling. With average baseline lengths and three slaves, the minimum GRI cannot be much less than 50,000 microseconds.

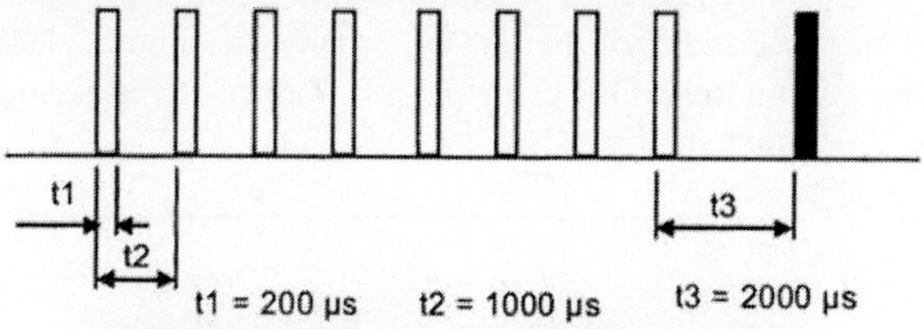

Fig. 7.9. (Graphic Courtesy of NELS).

Each Loran-C pulse has an approximate duration of 200 microseconds (μs). The interval between pulses within a pulse group is 1000 μs, except for the last two pulses at the Master, which have a 2000 μs, interval. The graphic below illustrates one pulse.

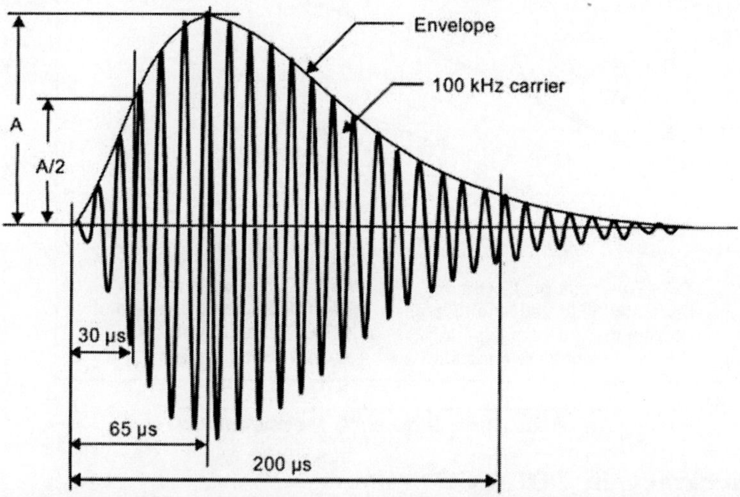

Fig. 7.10. This graphic illustrates the points on the Loran-C pulse envelope that define the start time, the time of maximum envelope power and the stop time of the pulse. *(Graphic courtesy of NELS).*

7.6.5. Phase Coding

To reduce the effects of interference and noise on time difference measurements, and to assist in distinguishing between master and secondary stations, the carrier phase of selected transmitted pulses is reversed in a predetermined pattern. The pattern is shown in Fig. 7.11, where a minus sign indicates an inverted pulse (180° phase shift), and a plus sign means no phase shift. This pattern is repeated every two GRI's. Simply stated, phase coding determines whether the first peak in the pulse is upwards or downwards.

	Station	
	Master	Secondary
First GRI Period	+ + − − + − + − +	+ + + + + + − − +
Second GRI Period	+ − − + + + + + −	+ − + − + + − −

Fig. 7.11. Phase Coding (Graphic by Orblet Technologies AB).

7.6.6. Time Difference Measurements

The basic measurements made by Loran-C receivers are to determine the difference in the time-of-arrival (TD) between the master signal and the signals from each of the secondary stations of a chain. Each TD value is measured to a precision of about 0.1 microseconds (100 nanoseconds) or better. As a rule of thumb, 100 nanoseconds corresponds to about 30 metres. The principle of time difference measurements in hyperbolic mode is illustrated in Fig. 7.12.

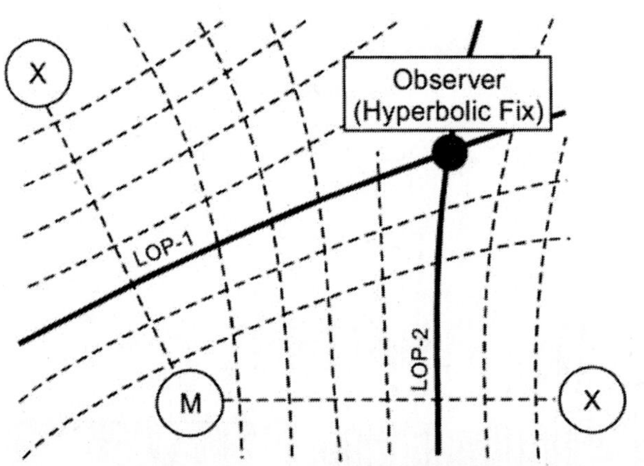

LOP - The locus of all positions where the observed time difference between the arrival of signals from two stations are constant.

The LOP forms a hyperbola which gives rise to the designation of LORAN-C as a hyperbolic radionavigation system

Fig. 7.12. Time Difference Measurements.

7.6.7. Accuracy in LORAN-C

The Loran-C service will support an absolute accuracy varying from 185 meters to 463 meters (0.1 to 0.25 nautical miles), depending on where the observer is within the coverage area. Absolute accuracy defines a user's true geographic position (latitude and longitude). Repeatable accuracy is a measure of an observer's ability - by using a navigation system such as Loran-C - to return to a position visited previously using the same navigation system. Loran-C repeatable accuracy is sometimes as good as 18 meters and is usually better than 100 meters within the coverage area.

7.7. LORAN-D

LORAN-D was a short range, high accuracy, low power, tactical system designed for use as a bombing aid by the United States Air Force. It used "portable" transmitters and much shorter baselines than Loran- C, plus a sixteen pulse transmission instead of eight. Loran-D modified the Loran-C pulse pattern by interleaving an additional 8 pulses after each pulse of the current Loran-C pulse pattern. To interleave the additional pulses within the standard group interval, Loran-D used a modulation scheme known as **Supernumerary Interpulse Modulation (SIM)**. The additional interleaved additional pulses were coded in a pattern to which the normal Loran-C receiver was non-responsive. Thus conventional Loran-C receivers designed to receive only the 8-pulse Loran-C pattern operated on the imbedded Loran-C portion of the Loran-D signal.

In standard Loran-C, the pulses of a Loran group are spaced 1000 μs apart for all Secondaries. The Master's pulse spacing is the same for the first 8 pulses. Having an additional 9th pulse located 2000 μs after the 8th pulse identifies the Master station. The most obvious spacing to interleave additional pulses between the standard Loran-C pulses is to use the previous Loran-D format. In Loran-D the spacing was 500 μs apart. The original Loran-D pulses peaked at 80 μs, with the standard sampling point located at 60 μs. Later, Loran-D pulses were transmitted identically to Loran-C.

The system was used for several war exercises in Europe in the 1960's and later. One station was temporarily erected in the UK.

7.8. LORAN-F

LORAN-F. At the same time that the U.S. Air Force was developing Loran-D, Motorola experimented with the Multi-User Tactical Navigation Systems (MUTNS), a continuously pulsed pseudorandom coded low frequency navigation system used for drone control. It was sometimes referred to as Loran-F in proposals but this was not an official designator. Loran-D came out ahead during system evaluations; therefore no further work was done on MUTNS

7.9. DECCA NAVIGATOR

7.9.1. Introduction To DECCA Navigator

The Decca Navigator System, herein abbreviated as **DNS** for convenience, is a hyperbolic radio navigation system which was established in the United Kingdom after World War 2 and later used in many areas around the world. It operates by measuring the phase differences between continuous signals from a master and slave stations. The differences are then related to a hyperbolic lines printed on a chart. By plotting the readings from two pairs of hyperbolas at any particular instant, users could plot their position instantly.

The system uses groups of at least three shore based transmitter stations called chains operating in the 70-130 KHz band. Each chain comprises one Master and two or three Slave stations, usually located 80 to 110 km from the Master station. The accuracy of DNS ranges from 50 meters during daytime to 200 meters at night. It can decrease up to 800 meters as the distance from the baseline increases. Accuracy is also affected by seasonal effects which generally reduces the accuracy by a factor of 6 to 8. Typically the range for DNS is 400 miles with a reproducibility of 200 meters.

7.9.2. System Overview : DECCA

This system, which was primarily developed as a long distance marine navigational aid, is basically a 'range difference' system, but as c.w. is used the range differences have to be interpreted in the form of phase differences. Lines of constant phase difference will once again take the form of a family of hyperbolae with their foci at the transmitters. A fix is obtained by making use of two or more families of constant phase hyperbolae established by a master and two or more slave transmitters. A difficulty inherent in the use of c.w. is at once apparent; if all transmitters were on the same frequency it would be impossible to sort out their phase relationships. To overcome this difficulty all the transmitters work on different, but related, frequencies. The frequencies used lie in the band 70 - 130 kc/s.

7.9.3. The Chain

A Decca chain normally consisted of a master station controlling the phase of three slaves, which were situated about 120 degrees apart, at a radius of 60 to 100 miles from the master. That provided all-round coverage, and, because ground waves of this frequency band are not seriously attenuated by passing over land, the stations could advantageously be situated well inland.

Each Station in the chain would normally transmit a particular unmodulated phase stable carrier wave. These carriers were all harmonically related to an internal station reference, which was about 14.2kHz, referred to as "f".

- The **Master** Station normally transmitting a "**6f**" unmodulated carrier wave signal in the 85 kHz band.
- **Red** Slave Station transmitting a "**8f**" signal in the 112 kHz Band.
- **Green** Slave transmitting a "**9f**" signal in the 127 kHz band.
- **Purple** Slave transmitting a "**5f**" signal in the 71 kHz band.

The Slave stations received and phase locked their station reference oscillators to the Master "6f" Transmission.

The frequency ranges for the master and slave stations were:

Master:	84 - 86 kHz
Red slave:	112 - 115 kHz
Green slave:	126 - 129 kHz
Purple slave:	70 - 72 kHz

Since the signals were continuous wave (CW), 150 Hz spacing was sufficient to ensure there would be no interference.

These transmissions were received by a special receiver and frequency multiplying circuits therein produced phase comparisons of:

24f for the Master and Red

18f for Master and Green

30f for the Master and Purple

The Master and Slave stations are usually spaced many wavelengths apart so that the family of hyperbolae will contain many that correspond to zero (or 2np) phase difference and there will therefore be a multiple ambiguity. The area between two adjacent lines of zero phase difference is known as a lane and phase comparisons carried out at the receiving point will merely show the position within a lane. To get complete information it is necessary that the phase comparator be of such a kind that it integrates the total phase change. Provided that the lane in which the point of departure lies is known, and continuous reception of the signals is maintained, the phase comparator will indicate not only position within a lane but also the number of lane boundaries that have been crossed. The latest type of indicator, having had the 'Decca co-ordinates' of the point of departure set in, shows both the lane and the position within the lane. There are of course three such indicators to show the positions in the Red, Green and Purple lanes.

For very long range navigation such as is involved in crossing the North Atlantic the Decca Navigator Company have proposed a system to which they have given the name.

7.9.4. The Block Diagram of the Basic Decca Receiver

The block diagram of the basic Decca receiver for survey and marine use (not including Multipulse) as shown in Fig. 7.13. The master and slaves were processed by superheterodyne stages in the left column. The resultant outputs were applied to frequency multiplying circuits in the middle column. The outputs of the frequency multipliers were applied to the discriminators in the rightmost columns. The difference in phase would be amplified and read on a decometer. It is very easy to see how all the slave signals are compared to the master in this diagram. *f* is some frequency around 14.2 KHz. *(Graphic courtesy Decca Navigator Co. Modified by Jerry Proc).*

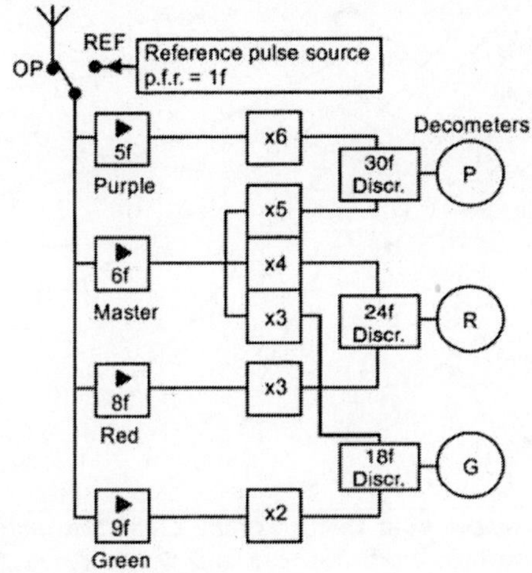

Fig. 7.13. The Block Diagram of the Basic Decca receiver.

Although for most of the time the stations only transmitted their single carrier, during part of the transmission cycle, each station would transmit what was termed a Multi-pulse also called "Mark 10" transmission. In the Mark 10 context, it meant that Multipluse could only be received on Mk X receiver equipment and higher.

The Multi-pulse was transmitted by each of the stations in turn during the 20 second transmission cycle to provide a coarse reading, or a Zone reading, and was generated by all 5 transmitters at the given station briefly transmitting simultaneously. During transmission of the Multi-pulse by, say the Red Slave, all transmissions from the other stations in the Chain would be suppressed.

7.9.5. Decca Lattice

An example of a Decca lattice chart showing the lines of hyperbola from the red and green slaves.

The transmissions from the chain are received by a special ship borne receiver, which measures the difference in phase of signals arriving from master and slaves. All stations in a Decca chain must 'phase locked', and this has to be done over an appreciable distance separating the stations, sometimes up to 100 nautical miles, the phase difference being determined by this distance. Each slave station is fitted with equipment which receives the master signal, converts it to the slave frequency, and uses it to control the drive oscillator of the slave transmitter. Thus a constant phase relationship is maintained. To ensure that this relationship is maintained accurately, a monitoring station checks the transmissions.

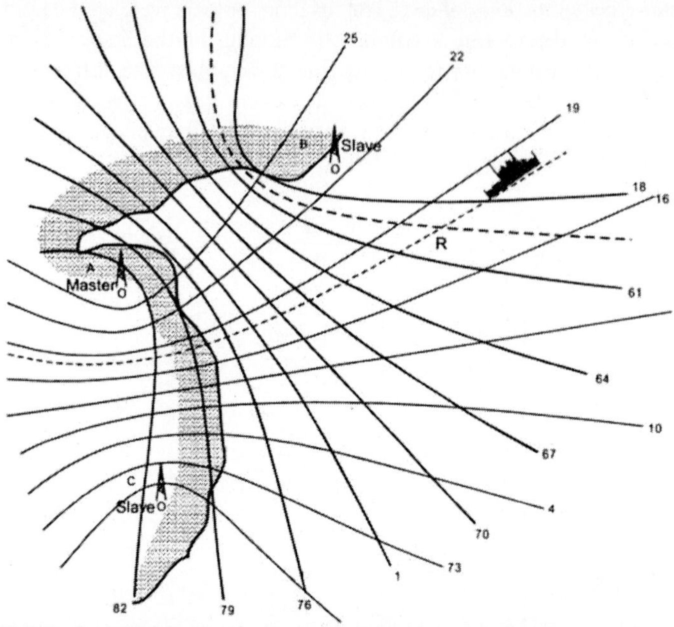

Fig. 7.14. An Example of a Decca Lattice Chart Showing the Lines of Hyperbola from the Red and Green Slaves.

7.9.6. Decometers

The detected phase differences are displayed on phase meters called **'decometers'**, and the readings may be plotted onto Decca lattice charts, on which the lines of position are numbered in the same units as those shown on the decometers.

The decometer indications are continuous, and depending on the position in the coverage, readings of the two appropriate decometers can be taken simultaneously whenever a fix is required. The third decometer can give some additional information, but usually its readings are disregarded in the wide sector around the base-line extension. The lattice patterns are formed by hyperbolic position lines similar to those previously described. They are overprinted on ordinary Mercator charts. The slave stations are known as Red, Green and Purple slaves, according to the printed colour of the lattice lines derived from their transmissions.

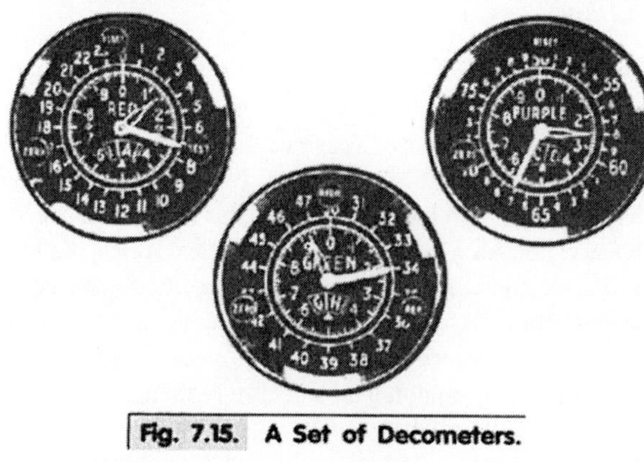

Fig. 7.15. A Set of Decometers.

A very high degree of instrumental accuracy is obtained by the use of continuous wave transmissions, the phase of which, on arrival, can be measured to within 4 degrees. Near the base-line, between a pair of stations, this may represent a distance as small as 10 yards, though it must be borne in mind that constant and variable errors due to operational causes exist in the system, which, in practice, does not normally give an accuracy as good as +/–10 yards. It is, however, considerably more accurate than any system employing pulse transmissions (*i.e.* Loran).

7.9.7. LANES

The areas between the lines of zero phase difference in a Decca pattern are known as "**lanes**". It is essential that the signals from the two stations should be received separately in order to preserve their individual phase properties. Since the C.W. transmissions are simultaneous, this can only be achieved by transmitting on different frequencies: but these frequencies must have an exact common multiple. The transmissions are received by what are virtually four separate receivers within the Decca Navigator receiving equipment. The frequencies of these signals are then multiplied up to their lowest common multiple, the so-called "comparison frequency" on which the phase comparison is made. As the pattern is traversed by the Decca receiving equipment so the reading will be observed to alter steadily from 0 degrees to 360 degrees between the limits of each lane; the decometer, from which this reading is obtained, is therefore graduated in fractions of a lane instead of in degrees.

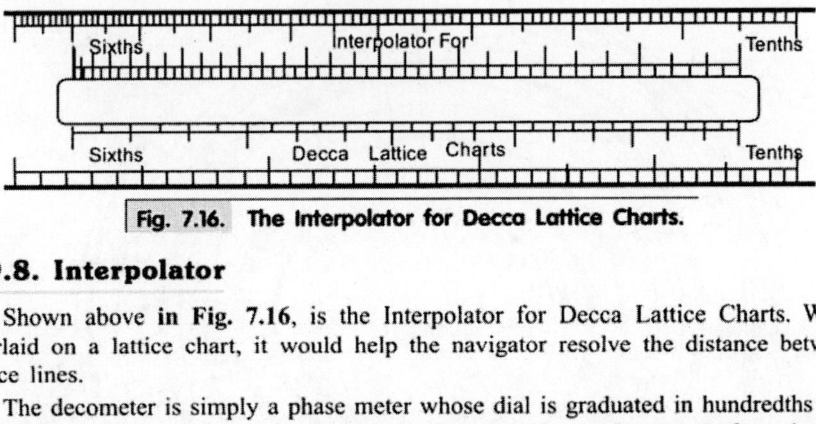

Fig. 7.16. The Interpolator for Decca Lattice Charts.

7.9.8. Interpolator

Shown above **in Fig. 7.16**, is the Interpolator for Decca Lattice Charts. When overlaid on a lattice chart, it would help the navigator resolve the distance between lattice lines.

The decometer is simply a phase meter whose dial is graduated in hundredths of a lane width; one revolution of the fractional pointer represents the extent of one lane. It will, therefore, indicate very accurately a receiver's position between two lattice lines, but it is unable to identify the particular lane in which it is situated. Since lane width varies from less than a mile near the base line to 3 miles or so at 300 miles range, this would cause a high degree of ambiguity, which a ship, entering the coverage area after an ocean passage, might not be able to resolve. Once the initial position has been established, however, the decometer, which is capable of continuous rotation, can integrate its movements in the lattice pattern by a set of counters geared to the fractional pointer.

7.9.9. Errors and Accuracy in DECCA System

The presence of a sky wave component in the received signal may cause variable errors at the receiver. In the case of Decca, there is no means of distinguishing between the two waves. This leads to an inaccuracy in the decometer readings, which will vary, with the range from the transmitters and with the time of day. At ranges greater than 75 miles, the accuracy at night is noticeably lower than it is by day, since sky wave effect is normally only experienced during the night. Beyond a range of between 150 and 220 miles there is a serious danger of "lane-slip".

Disregarding systematic errors and transmission failures, the accuracy of a fix from the Decca system can be considered to depend upon :

(*a*) Instrumental errors

(*b*) Propagation errors caused by either Sky Wave or Coastal effects.

(*c*) Lane width.

(*d*) Angle of cut of the hyperbolae.

Thus Decca errors are subject to many variables and cannot be summarized precisely. The following figures give a guide to the accuracy that should generally be expected:

(*a*) By DAY 0-100 miles or by NIGHT 0-75 miles: +/- 10 feet near base line; 1 mile at limits

(*b*) By NIGHT 75-240 miles: Up to a maximum of about 5 miles, depending on sighting of slave stations.

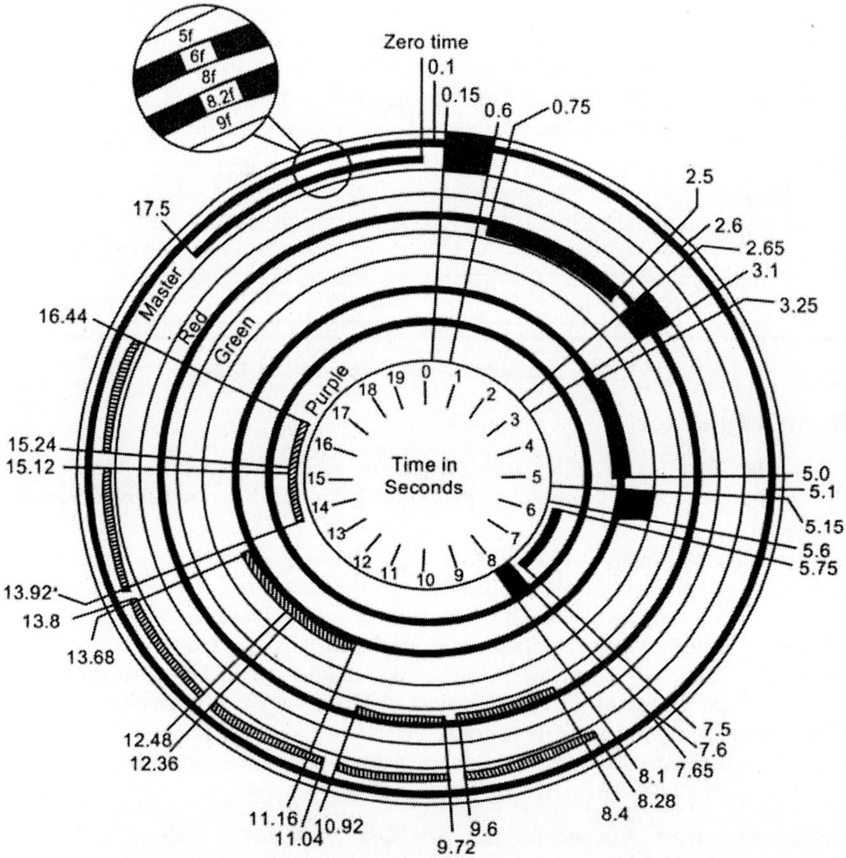

Fig. 7.17. This figure illustrates the 20 second transmission sequence for MP Decca Navigation chain. Hatched periods denote 8.2f transmissions for chain control and surveillance. (Graphic courtesy of The Decca Navigator Co).

Fig. 7.18. DECCA Slave Site.

7.9.10. DECCA Slave Site

This photo shown in Fig. 7.18, Illustrates a typical slave site. At the left is the transmitter building. The antenna in the foreground received the phase control signal sent by the master station thus providing input for the Phase Control Unit. Due to the distance and the contrast in the photo, the centenary of support wires is not visible on the main antenna tower in the background.

7.9.11. Dectra

The Dectra system is designed to give navigational assistance over a particular route. To do this a master and slave pair of stations is located at each end of the route in such a way that the route perpendicularly bisects the line joining the slave to the master of each pair. The sets of hyperbolae due to each pair of stations make a pattern such as is shown in Fig. 7.19.

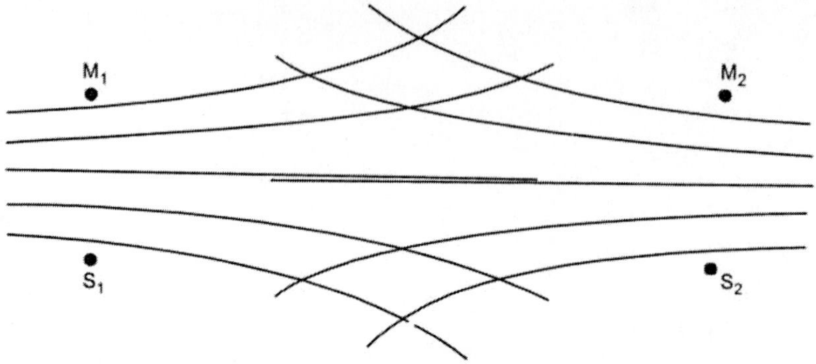

Fig. 7.19. System of Hyperbolae Established in a Dectra System.

The method of establishing the hyperbolae, and of measuring phase differences in the air, differs from the normal Decca system in that the master and slave stations do not transmit simultaneously but alternately. Thus the master transmits for a period and then, by momentarily changing frequency, signals to the slave to start transmitting. The slave, which has been using the master station's transmissions to control the phase of a stable oscillator, radiates in phase coherence with the master by using this stable oscillator to drive its transmitter. After a period the slave shuts down and the cycle repeats.

In the air a somewhat similar process is followed. The transmissions from the master station are used to phase lock an oscillator with very good short-term stability; then when the slave station radiates, the phase of its transmission, at the aircraft, is compared with that of the oscillator.

Different frequencies are used for the master-slave pair at either end of the route. in order to provide range measuring facilities the two frequencies *F1* and *F2* are made to differ by a frequency *f* which is also an integral sub-multiple of both *F1* and *F2*. A typical set of frequencies might be *F1* = 85.100 kc/s, *F2* = 84.915 kc/s and *f* = 185 c/s.

Now if the stable oscillator is tuned to, and is being phase locked at *F1* its frequency is subdivided until the frequency *f* is produced. A second r.f. amplifier is tuned to *F2* and its output mixed with that of the other r.f. amplifier tuned to *F1*. This also gives rise to signals at frequency *f*. We now have two signals at frequency *f*. One is derived directly from the *F1* signals and its phase is directly related to that of the *F1* signals. The other is derived from both the *F1* and *F2* signals and its phase will be related to the phases of both these signals. If we compare the phases of the two signals at frequency *f*, the phase determining effects of the *F1* signal cancel out and we are left with a phase dependent only on the *F2* signal and a phase meter will plot the intersections of the course with a series of hyperbolae with the two master stations as foci and with a lane spacing determined by *F2*. By tuning the stable oscillator to *F2* (when approaching that end of the route) a second set of hyperbolae, with lane spacing *F1* can be established.

A block diagram of the DECTRA Range Measuring system :

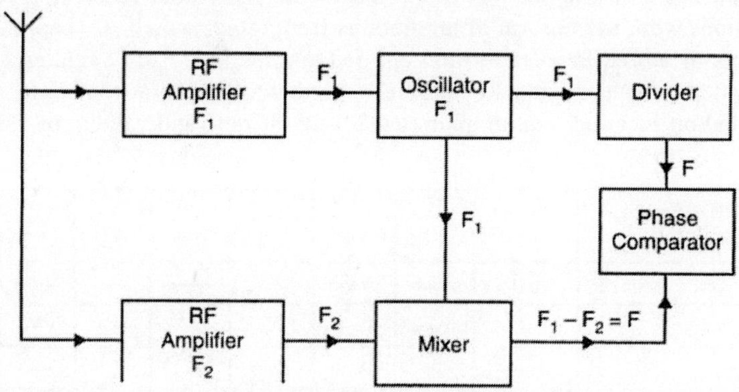

Fig. 7.20. A Block Diagram of the DECTRA Range Measuring System.

A block diagram of the system is given in **Fig. 7.20.**

In case, at either end of the route, the signals from the distant master stations arc not audible, an alternative ranging system is provided. This consists of a high stability, crystal-controlled, oscillator which is set up, before departure, to be in phase with the transmissions of the master station at that end of the route. Thereafter it is free running and by keeping it in an accurately controlled over its frequency stability is assured to about one part in 10^8 for a period of an hour. After take-off comparison of the phase of the master station's transmission with that of the crystal-controlled oscillator gives a direct measure of the distance traveled.

In practice the crystal controlled oscillator works at a frequency *NF*1 or *NF*2, and is divided down to *F*1 or *F*2 for phase comparison.

7.10. OMEGA

Omega is a worldwide, internationally operated radio navigation system. It operates in the Very Low Frequency (VLF) band between 10 and 14 kHz. It provides an all weather, medium-accuracy navigation service to marine navigators. The system consists of eight widely-spaced transmitters.

There is no master-secondary relationship between the Omega stations as there is between Loran C stations. The navigator is free to use any station pair that provides the most accurate line of position. Additionally, Omega measures phase differences between the two signals whereas Loran C measures time delays between signal receptions.

Each Omega station transmits on the following frequencies: 10.2 kHz, 11.05 kHz, 11.3 kHz, and 13.6 kHz.

7.10.1. Basic Operation : OMEGA

An Omega receiver determines position in either the **direct ranging mode** or the **hyperbolic mode**. Some call the direct ranging mode the rho-rho mode. In the direct ranging mode, the receiver measures ranges from stations by measuring phase shifts between transmitted signals and an internal reference signal. In the hyperbolic mode, the receiver measures position relative to transmitter pairs by making phase comparisons between signals coming from these pairs.

The **inherent accuracy** of the OMEGA system was limited by the accuracy of the propagation corrections that were applied to the individual receiver readings. These corrections were in the form of predictions from tables which were applied to manual receivers or stored in memory and applied automatically in computerized receivers. The system was designed to provide a predictable accuracy of 2 to 4 nm which depended on location, station pairs used, time of day, and validity of the propagation corrections.

Transmission Inverval	0.9	1.0	1.1	1.2	1.1	0.9	1.2	1.0	0.9
A	10.2	13.6	11.33	f^1	----	----	----	----	----
B	f^2	10.2	13.6	11.33	f^2	----	----	----	----
C	----	f^3	10.2	13.6	11.33	f^3	----	----	----
D	----	----	f^4	10.2	13.6	11.33	f^4	----	----
E	----	----	----	f^5	10.2	13.6	11.33	f^5	----
F	----	----	----	----	f^6	10.2	13.6	11.33	f^6
G	11.33	----	----	----	----	f^7	10.2	13.6	11.33
H	13.66	11.33	----	----	----	----	f^8	10.2	13.6

Fig. 7.21. Transmission Format of OMEGA System.

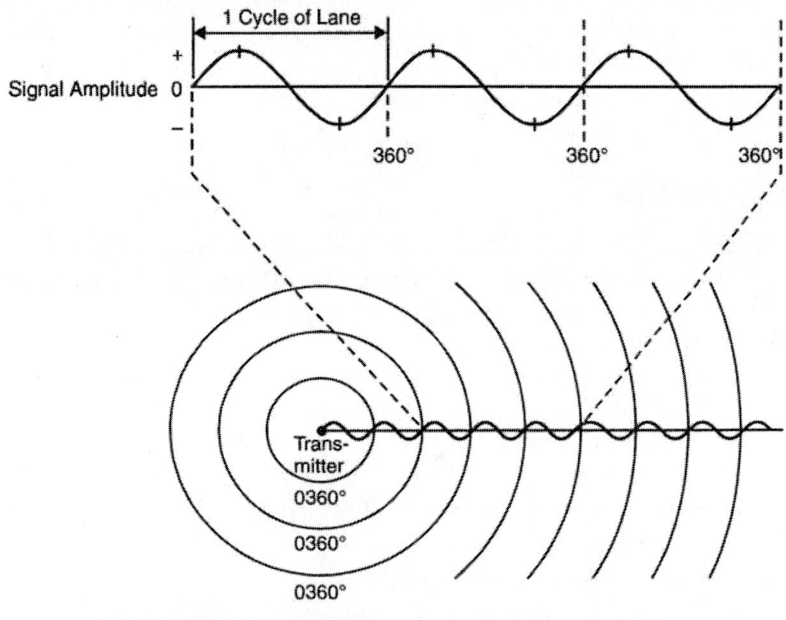

Fig 7.22. Omega Lanes Formed by Radio Waves.

7.10.2. Direct Ranging Mode

The Omega wavelength, at 10.2 kHz, is approximately 16 miles long. The wavelength defines the width of each Omega "lane." See Fig. 7.22. This figure shows the lanes as concentric circles formed around the transmitting station. An Omega receiver measures the phase of the received signal within a known lane. This phase shift allows the receiver to determine its position's fraction distance between lanes. Knowing which lane it is in and the fractional distance between lane boundaries, the receiver can calculate an LOP. The LOP is the line of points corresponding to the fractional distance between lanes calculated by the receiver.

The schematic of **Fig. 7.22** does not take into account that the transmitted navigation signal forming the Omega lane, is not stationary. Rather, it propagates at the speed of light. To account for this moving wave, the receiver generates a reference signal at the same frequency of the Omega navigation signal. This reference signal "freezes" the Omega signal from the receiver's perspective in a manner analogous to the way a strobe light flashing at the same frequency of a rotating disk freezes the disk from an observer's perspective. Comparing these "frozen" reference and navigation signals allows the receiver to measure the phase difference between the navigation signal and the reference signal. This phase difference, in turn, is proportional to the receiver's fractional distance between two Omega lanes.

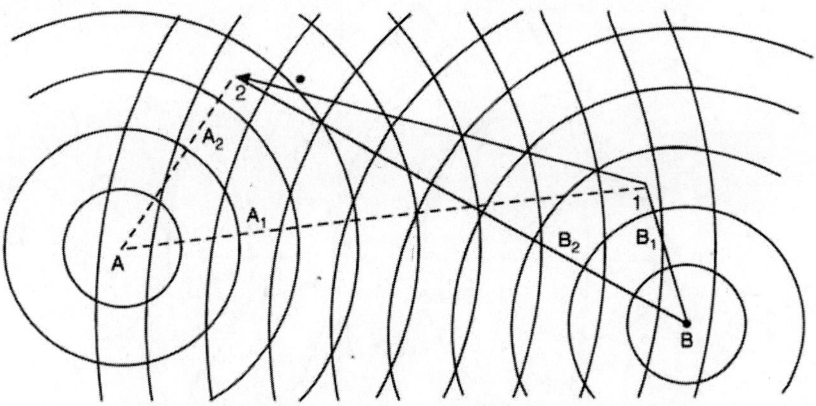

Fig. 7.23. Position Fixing in the Direct Ranging Mode.

During the trip to point 2, the receiver keeps track of how many lanes it crosses. When it stops, it determines the fractional distance between lane boundaries at point two. From the lane counting and the phase comparison at point 2, the receiver calculates *A*2 and *B*2, the distances between the receiver and stations *A* and *B*, respectively. Figure 7.23 shows the position fixing in the direct ranging mode.

7.10.3. Hyperbolic Mode

In the direct range mode discussed above, the receiver measured the distance between it and two or more transmitting stations to determine lines of position. In the hyperbolic mode, the receiver measures the difference in phase between two transmitters. See Fig. 7.24. This figure shows two transmitting stations, labeled *A* and *B*. Both of these stations transmit on the same frequency. Additionally, the stations transmit such that their waves' phase is zero at precisely the same time. Because each signal's phase is zero at each wave front, the phase difference where the wave fronts intersect is zero.

Connecting the intersecting wave fronts yields a line along which the phase difference between the two signals is zero. This line forms a hyperbola called an **isophase contour**. At any point along this contour, the phase difference between the stations is zero. At any point between the isophase contours, there is a phase difference in the signals proportional to the fractional distance between the contours. The set of isophase contours between station pairs forms a series of lanes, each corresponding to one complete cycle of phase difference. The hyperbolic mode lane width on the stations' baseline equals one-half the signal wavelength. For a 10.2 kHz signal, the baseline lane width is approximately 8 miles. Each of these 8 mile wide lanes is divided into 100 **centilanes (cels)**. The receiver measures the phase difference between stations in hundredths of a cycle. These units are termed **centicycles (cec)**. In Fig. 7.24 omega lanes formed by hyperbolic isophase contours are shown.

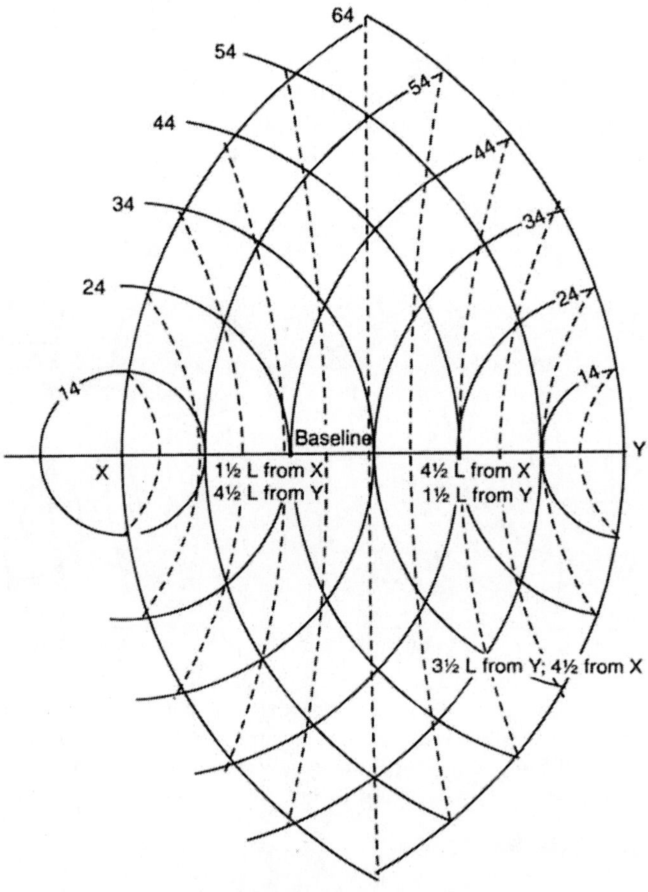

Fig. 7.24. Omega Lanes Formed by Hyperbolic Isophase Contours.

7.10.4. Direct Ranging And Hyperbolic Operation

Originally, the hyperbolic mode was more accurate because the direct ranging mode required a precise receiver internal oscillator to remain synchronized with the atomic oscillators used by the transmitting station. Since these oscillators would have made the receiver prohibitively expensive, the receiver carried an oscillator that was subject to clock error. In the direct ranging mode, this clock error would have been critical because this mode relies on a direct comparison between one transmitter's signal and the clock internal oscillator. In the hyperbolic mode, the receiver measures the phase difference between two transmitted signals and the receiver's internal oscillator. When the receiver subtracts one phase difference from another to calculate the difference, the clock error is mathematically eliminated. In other words, as long as the clock error remained constant between the two measurements, subtracting the two phase differences canceled out the error.

The microprocessing of modern receivers, however, allows the direct ranging mode to be used. The methodology used is similar to that used by the Global Positioning System to account for inaccuracies in GPS receiver clocks.

The Omega receiver makes three ranging measurements and looks at the intersection of the three resulting LOP's. If there were no clock error present, the LOP's would intersect at a pinpoint. Therefore, the receiver subtracts a constant clock error from each LOP until the fix is reduced to a pinpoint. This technique allows a receiver to use the direct ranging method without a precise atomic oscillator. This technique works only if the clock error is constant for each phase difference measurement.

7.10.5. Ambiguity

In this CW system, ambiguous LOP's occur since there is no means to identify particular points of constant phase difference which reoccur throughout the coverage area. The area between lines of zero phase difference were called lanes. Single frequency receivers use the 10.2 kHz signals whose lane width is about eight nautical miles on the baseline between stations.

Multiple frequency receivers extended the lane width, for the purpose of resolving lane ambiguity. Lane widths of approximately 288 nm along the baseline could be generated with a four-frequency receiver. Because of the lane ambiguity, a receiver had to be preset to a known location at the start of a voyage. The accuracy of that position had be known with sufficient accuracy to be within the lane that the receiver was capable of generating (*i.e.*, 4 nm for a single frequency receiver or approximately 144 nm for a four-frequency receiver).

Once set to a known location, the Omega receiver counted the number of lanes it crossed in the course of a voyage. This lane count was subject to errors which could be introduced by an interruption of power to the receiver, changes in propagation conditions near local sunset and sunrise, and other factors. To use the single frequency Omega receiver effectively for navigation, it was essential that a dead reckoning plot or similar means be carefully maintained and the Omega positions compared to it periodically so that any lane ambiguities could be detected and corrected.

The accuracy of an Omega phase difference measurement was independent of the elapsed time or distance since the last update. Unless the Omega position was verified occasionally by comparison to a fix obtained with another navigation system or by periodic comparison to a carefully maintained plot, the chance of an error in the Omega lane count increased with time and distance. These errors were reduced in multiple frequency receivers since they were capable of developing larger lane widths to resolve ambiguity problems.

SUMMARY

- Hyperbolic systems are based on the measurement of the difference in the time of reaching of EM waves from the two or more transmitters to the receiver in the craft.
- A hyperbolic curve is formed by joining all points that are the same difference in distance between two points. In hyperbolic navigation systems, your receiver measures the difference in time of receipt of radio waves, and you or the set translate the time difference into a distance difference using the known speed of the waves
- There are some hyperbolic navigational systems, which are being developed in the long span of time. A brief description of these systems is discussed in this chapter.
  1. Decca
  2. Consol
  3. Omega
  4. LORAN (LORAN-A, B, C, D and F)
- The Omega system was devised to overcome the major drawback of other hyperbolic systems - their limited geographic coverage. With eight transmitters located some 5,000 to 6,000 miles apart, Omega's coverage is worldwide - a spread made possible by the use of atomic time standards to regulate the very low-frequency, long-range signals.
- Loran provided facilities whereby ships and aircraft derived their position at long distances. The system required at least three transmitting stations for each 'chain', and the observer used a special Loran receiver
- LORAN-B **was basically a phase comparison variation of Loran-A and an attempt to re-design the system properly plus add some new features**
- LORAN-C, operating at 100-110 kHz, was initially developed under U.S. Navy sponsorship and deployed to cover large portions of the northern hemisphere for both marine and air navigation. Originally just a marine navigation system, LORAN-C determines present position by the intersection of Lines of Position (LOPs) that are hyperbolic curves. At least three stations, (a Master and two Secondaries) are needed. Accuracy is plus or minus 2.5 miles. The LORAN-C uses triangulation to measure the location of an aircraft or boat.
- The components of the Loran system consist of
  1. **The land based** transmitting stations,
  2. **The Loran** receiver
  3. Antenna**, and**
  4. **The** Loran charts
- The Loran-C service will support an absolute accuracy varying from 185 meters to 463 meters (0.1 to 0.25 nautical miles), depending on where the observer is within the coverage area.
- Repeatable accuracy is a measure of an observer's ability - by using a navigation system such as Loran-C - to return to a position visited previously using the same

navigation system. Loran-C repeatable accuracy is sometimes as good as 18 meters and is usually better than 100 meters within the coverage area.

- LORAN-D was a short range, high accuracy, low power, tactical system designed for use as a bombing aid by the United States Air Force. It used "portable" transmitters and much shorter baselines than Loran- C, plus a sixteen pulse transmission instead of eight. Loran-D modified the Loran-C pulse pattern by interleaving an additional 8 pulses after each pulse of the current Loran-C pulse pattern.
- **LORAN-F.** At the same time that the U.S. Air Force was developing Loran-D, Motorola experimented with the Multi-User Tactical Navigation Systems (MUTNS), a continuously pulsed pseudorandom coded low frequency navigation system used for drone control. It was sometimes referred to as Loran-F in proposals but this was not an official designator.
- The Decca Navigator System, herein abbreviated as DNS for convenience, is a hyperbolic radio navigation system which was established in the United Kingdom after World War 2 and later used in many areas around the world. It operates by measuring the phase differences between continuous signals from a master and slave stations. The differences are then related to a hyperbolic lines printed on a chart. By plotting the readings from two pairs of hyperbolas at any particular instant, users could plot their position instantly.
- The frequency ranges for the master and slave stations were :

Master:	84 - 86 kHz
Red slave:	112 - 115 kHz
Green slave:	126 - 129 kHz
Purple slave:	70 - 72 kHz

Since the signals were continuous wave (CW), 150 Hz spacing was sufficient to ensure there would be no interference

- The **Dectra** system is designed to give navigational assistance over a particular route. To do this a master and slave pair of stations is located at each end of the route in such a way that the route perpendicularly bisects the line joining the slave to the master of each pair.
- The detected phase differences are displayed on phase meters called 'decometers', and the readings may be plotted onto Decca lattice charts, on which the lines of position are numbered in the same units as those shown on the decometers.
- The areas between the lines of zero phase difference in a Decca pattern are known as **"lanes".**
- The **Interpolator** for Decca Lattice Charts when overlaid on a lattice chart, it would help the navigator resolve the distance between lattice lines.
- Thus Decca errors are subject to many variables and cannot be summarized precisely. The following figures give a guide to the accuracy that should generally be expected:

  (*a*) By DAY 0-100 miles or by NIGHT 0-75 miles: +/- 10 feet near base line; 1 mile at limits

  (*b*) By NIGHT 75-240 miles: Up to a maximum of about 5 miles, depending on sighting of slave stations.

- **Omega** is a worldwide, internationally operated radio navigation system. It operates in the Very Low Frequency (VLF) band between 10 and 14 kHz. It provides all weather, medium-accuracy navigation service to marine navigators. The system consists of eight widely spaced transmitters.
- Omega measures phase differences between the two signals whereas Loran C measures time delays between signal receptions.

  Multiple frequency receivers extended the lane width, for the purpose of resolving lane ambiguity. Lane widths of approximately 288 nm along the baseline could be generated with a four-frequency receiver. Because of the lane ambiguity, a receiver had to be preset to a known location at the start of a voyage. The accuracy of that position had been known with sufficient accuracy to be within the lane that the receiver was capable of generating (*i.e.*, 4 nm for a single frequency receiver or approximately 144 nm for a four-frequency receiver).

REVIEW QUESTIONS

1. Explain briefly the principle of the hyperbolic electronic system.
2. Name the different navigational system used the principle of hyperbolic electronic navigational system.
3. Define the principle of LORAN-A briefly.
4. How the identification of LORAN-A pairs carried out.
5. What are the information provide by the LORAN-C.
6. Define the working principle of LORAN-C and write the components of the LORAN system.
7. Write the accuracy of LORAN-C.
8. Explain the working block diagram of DECCA receiver in brief.
9. Explain the principle of OMEGA system and define the Direct ranging mode.
10. Describe LORAN as hyperbolic navigator. **[UPTU, 2003]**
11. Write short notes on :
    (*i*) DECTRA
    (*ii*) DECCA Lattice
    (*iii*) Frequency ranges for the master and slave stations in Decca.
    (*iv*) Decometers
    (*v*) Errors and accuracy in Decca system
    (*vi*) Modes of operation in OMEGA system.
    (*vii*) LORAN **[UPTU, 2004, 2006]**
12. Explain the operation and typical application of LORAN. **[PTU, 2004]**
13. Describe the working of a hyperbolic navigation system. **[UPTU, 2004]**

8

CHAPTER

DME AND TACAN

INSIDE THIS CHAPTER

8.1. INTRODUCTION

DME, as the name indicates is an electronic equipment used to measure slant range distance between aircraft and a ground station. The DME operates in the UHF range between 960 MHz and 1215 MHz.

DME **stands for** Distance Measuring Equipment**. It is a transponder-based radio navigation technology generally colocated at VORs.**

Aircraft use DME to determine their distance from a land-based transponder by sending two pulses with a random time interval between them. The transponder echoes the pulses. The DME receiver then searches for two pulses with the correct time interval between them and measures the delay between the pulse transmission and the pulse reception.

Once the receiver is locked on, it has a narrower window in which to look for the echoes and can retain lock. A typical DME transponder can provide concurrent distance information to about 100 aircraft.

Airborne DME measures elapsed time required for exchange of signals and converts into distance and ground speed

Fig. 8.1.

DME frequencies are paired to VOR frequencies. So generally a DME interrogator is designed to automatically tune to the corresponding frequency when the colocated VOR is selected. Airplane's DME instrument uses frequencies from 1025 to 1150 MHz. DME transponders transmit on 962 to 1150 MHz and receive on 962 to 1213 MHz. The channel width is 100 kHz.

One important thing to understand is that DME provides the physical distance from the aircraft to the DME transponder. This distance is often referred to as 'slant range' and depends trigonometrically upon both the altitude above the transponder and the ground distance from it.

DME as its name states is an electronic device that measures "slant range" from the DME station. Slant range is a measure of an aircraft's position relative to the DME station that incorporates the height of the aircraft, its angle from the ground station and its unknown ground range based upon a 90° angle. The farther the aircraft is from the station and the lower the aircraft's altitude, the more accurate the distance reading. An aircraft could be directly over the DME station at an altitude of 10,500 feet above ground level (AGL) and the DME would correctly indicate the aircraft is two miles from the station.

8.1.1. History of DME

As early as 1946 many organisations in the West took an active part in the development of DME system. The first experimental L band DME was designed in 1946 by the Combined Research Group (CRG) at the Naval Research Laboratory (NRL).

The L band, between 960 MHz and 1215 MHz was chosen for DME operation mainly because :

Lower side of L band of Radar frequency. (UHF Range)

(*i*) Nearly all other lower frequency bands were occupied.

(*ii*) Better frequency stability compared to the next higher frequencies in the microwave band.

(*iii*) Less reflection and attenuation than that experienced in the higher frequencies in the microwave band; and

(*iv*) More uniform omnidirectional radiation pattern for a given antenna height than that possible at higher frequencies in the microwave band.

8.1.2. Application of DME

The DME is used as a :

(*i*) Navigation aid in conjunction with the Very High Frequency Omni Range (VOR).

(*ii*) Component of the Instrument Landing System (ILS), where DME is also an acceptable alternative to marker beacons where provisions of marker beacons are impracticable.

(*iii*) Component of the Microwave Landing System (MLS).

The important applications of DME are :

(*i*) Provide continuous navigation fix (in conjunction with VOR).

(*ii*) Permit the use of multiple routes on common system of airways to resolve traffic.

(*iii*) Permit distance separation instead of time separation between aircraft occupying the same altitude facilitating reduced separation thereby increasing the aircraft handling capacity.

(*iv*) Expedite the radar identification of aircraft; and

(*v*) Provide DME distance in lieu of fan marker beacons and radio range intersections in connection with instrument approaches and holding operations respectively.

8.1.3. Rho-Theta Navigation System

DME's navigation aid is based on the principles of Rho-Theta Navigation System.

The Rho-Theta Navigation System is based on the Polar coordinate system of azimuth and distance as shown in Fig. 8.2.

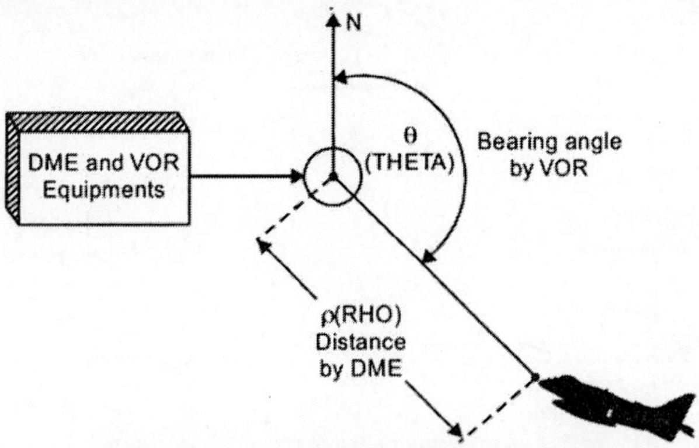

Fig. 8.2. Rho-Theta Navigation System.

The Very High Frequency Omni Range (VOR) and DME constitute the basic components of the Rho-Theta Navigation System. While the VOR provides azimuth information (Theta) to the pilot, the DME provides the distance information (Rho) so that the pilot receives a continuous navigational fix relative to a known ground location.

The DME equipment on the aircraft is very simple to use, the pilot having only to tune the equipment to the appropriate frequency and read the display once the DME has **Locked on** to the ground transponder. The frequency control for the DME receiver is normally derived from the VOR/Localizer receiver, the DME channels being paired with the VHF frequencies used by the VOR/Localiser equipment.

Because the distance measurement taken by the aircraft DME receiver is from Air-to-Ground, DME records Slant Ranges which are greater than the actual distance between the ground facility and the ground position of the aircraft. The difference between the slant and actual range is dependent on the relationship of the aircraft height and distance from the ground equipment. Unless the aircraft is 5000 feet or more, close to the ground transponder, then the discrepancy is small and can, for all practical purposes, be ignored.

8.2. DME PRINCIPLES

The operating principle of DME system is based on the Radar principle i.e. the time required for a radio pulse signal to travel to a given point and return. In fact it is a Secondary Radar.

8.2.1. Principles of Secondary Radar

In **Secondary Radar** system the target's **active** participation is necessary for its detection as against **Primary Radars** where the **target's role is passive.**

Secondary Radar system basically consists of two principle components namely the 'Interrogator' which is ground based and the 'Transponder' which is carried on the targets. Each of these components consists of a set of one pulse transmitter and one receiver. The Interrogator radiates pulses which when received by a corresponding transponder on a target will initiate a reply from that transponder. These replies are then collected by the interrogator to extract information about the targets.

8.2.2. Basic Block Diagram of DME System

DME is a Secondary Radar with the location of the Transponder and Interrogator reversed. Fig. 8.3 shows the elements of a DME system.

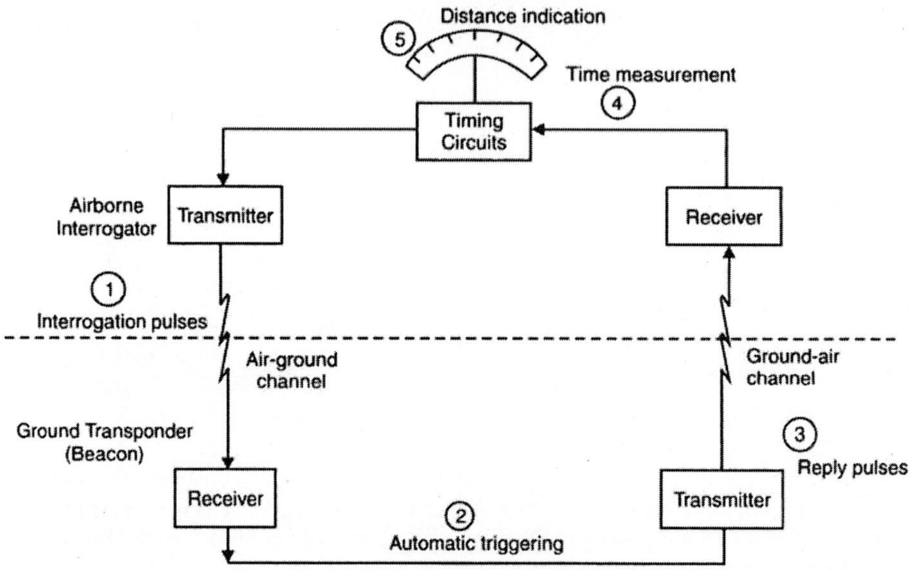

Fig. 8.3. Elements of DME System.

Referring to Fig. 8.3 the airborne transmitter repeatedly initiates a process of sending out very short, very widely spaced **interrogation** pulses. These are picked up by the ground transponder receiver whose output triggers the associated transmitter into sending.

8.2.3. Modes of Operation

There are two modes of aircraft interrogations. They are called **Search** and **Track**.

The **Search** mode is automatically established whenever the airborne equipment is tuned to a new DME ground Transponder, or if for some reason, a major interruption in the replies occur.

When the aircraft's transmitter is in **Search** mode, it transmits interrogations at a higher rate (**about 150 interrogations per second**). When the aircraft receives **at least 65%** replies to its interrogations **Lock-on** will be established and the transmitter changes to the **Track mode of operation**. This process may take **up to 30 seconds**. Only when this is achieved, the cockpit readout of the DME range is turned on.

In the **Track** mode the aircraft's interrogation rate reduces considerably (about 30 interrogations per second). The reduced interrogation rate of transmission in the track mode will allow more aircraft to use the DME station.

While in Track mode, if the signal is lost momentarily, the equipment enters Memory State. There are two types of memory states. They are called **Velocity Memory and Static or Position Memory states**. During the **Velocity Memory** state, the distance display continues to get updated at the same rate as it was prior to the time of signal loss. In the **Static Memory** state, the position display is held stationary at the same reading as it was prior to the time of signal loss. If the signal is not restored within about 10 seconds, the equipment goes into Search mode of operation again.

8.3. SYSTEM TIMING

The airborne equipment measures the elapsed time between the transmission of the interrogation pulse and the receipt of the reply pulse and converts time into a distance indication. This measurement can be easily explained by the system timing diagram shown in Fig. 8.4.

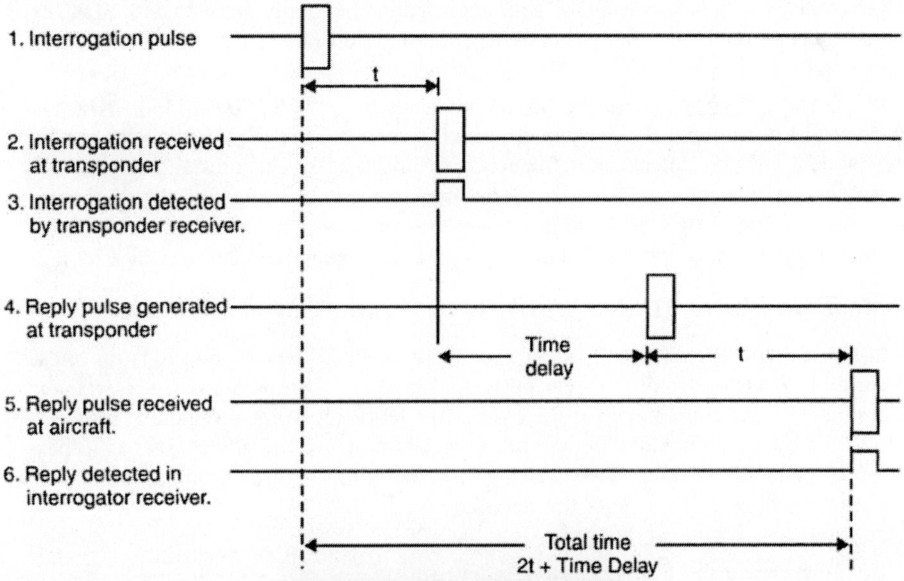

Fig. 8.4. DME System Timing.

Timing starts at the range circuits of the airborne equipment with the starting of the interrogation pulse. After a time delay *t*, depending on the distance between the aircraft and the ground equipment, the interrogation pulse is received at the antenna of the ground Transponder. The interrogation pulse is detected and after a fixed time delay, the reply pulse is generated. After time delay *t*, the **reply pulse** reach the airborne equipment. The distance between the aircraft and the ground Transponder is thus determined by measuring the total time elapsed between initial transmission of the interrogation pulse and the reception of the corresponding reply pulse minus the time delay.

The range in nautical miles, between the airccraft and the transponder is obtained by the simple formula

$$\text{Range} = \frac{\text{Total time } (\mu\,\text{sec}) - \text{Time delay } (\mu\,\text{sec})}{12.36}$$

The denominator 12.36 μ sec is the time taken by the pulse to travel **1 nautical mile to and fro**. This time is also called **Radar Mile**.

8.4. CHARACTERISTIC FEATURES OF THE DME SYSTEM

Following are the characteristic features of the DME systems.

8.4.1. DME Channels

Although the frequency band allocated for DME is 960 MHz – 1215 MHz, the lowest DME operating frequency is 962 MHz and the highest operating frequency is

1213 MHz leaving 2 MHz on either side of the band. This resultant band of 962 MHz – 1213 MHz is divided into 126 one-MHz channels for interrogation, and 126 one-MHz channels for transponder replies with the interrogation frequency and reply frequency always differing by 63 MHz.

The number of channels available is doubled by the use of *X* and *Y* channels which define specific pulse separation for the interrogator and reply pulses. The pulse separation for *X* channels is 12 microseconds for both interrogation and reply pulses. For *Y* channels, the pulse separation is 36 microseconds for interrogation pulses and 30 microseconds for reply pulses.

The airborne interrogator operates in the frequency band **1025 MHz – 1150 MHz**. The frequency of operation of the *X* channel ground transponder is 63 MHz lower than the corresponding airborne interrogator frequency for the first 63 channels and for the next 63 channels, it is higher by 63 MHz. The frequency of operation of the *Y* channel ground transponder is 63 MHz higher than the corresponding airborne interrogator frequency for the first 63 channels and for the next 63 channels, it is lower by 63 MHz.

8.4.2. Twin Pulse (Pulse Pair) Technique

In the present international system specification for DME, the interrogation pulse and the reply pulse is actually a twin pulse (pulse pair). There is a fixed, pre-arranged spacing between the components of the twin. Most DME receivers, ground and airborne, are followed by discriminators or twin pulse decoders, which are adjusted so as to pass only pulse pairs of the prescribed spacing. Isolated single pulses or pulse pairs with some other spacing will not pass the decoder.

The purposes of the twin pulse technique are :

(*i*) To increase the average power radiated.

(*ii*) To make the system less susceptible to errors or interference caused by false signals.

By using twin pulse technique, the DME operating channels can be increased. This is done by establishing new operating channels by using differently spaced twins to define a number of channels on each radio frequency.

8.4.3. Pseudo-Gaussian Pulse Shape

A wave form similar to Gaussian in shape is used for the DME pulses. Because of its shape there is **less side-band energy radiated, therefore, less interference with other DMEs operating on adjacent channels.** The true Gaussian waveform is a symmetrical bell-shaped statistical curve. However, it is possible to create a facsimile of the Gaussian curve by squaring the amplitude at each point in time of a sine wave. The resultant will be a co-sine squared waveform which will be in close approximation of a Gaussian waveform as shown in Fig. 8.5.

The other reason for selecting Gaussian pulse shape for the DME pulses is due to that the fact the Noise also has a Gaussian wave shape. **Noise has been an essential ingredient in most of the DMEs to generate Squitters.**

The Gaussian Pulse is characterised by the following parameters :

(*i*) **Partial Rise Time.** The time as measured between the 5 and 30 percent amplitude points on the leading edge of the pulse envelop, *i.e.* between points *h* and *i*.

(*ii*) **Pulse Rise Time.** The time as measured between the 10 and 90 percent amplitude points on the leading edge of the pulse envelope *i.e.*, between points *a* and *c*.

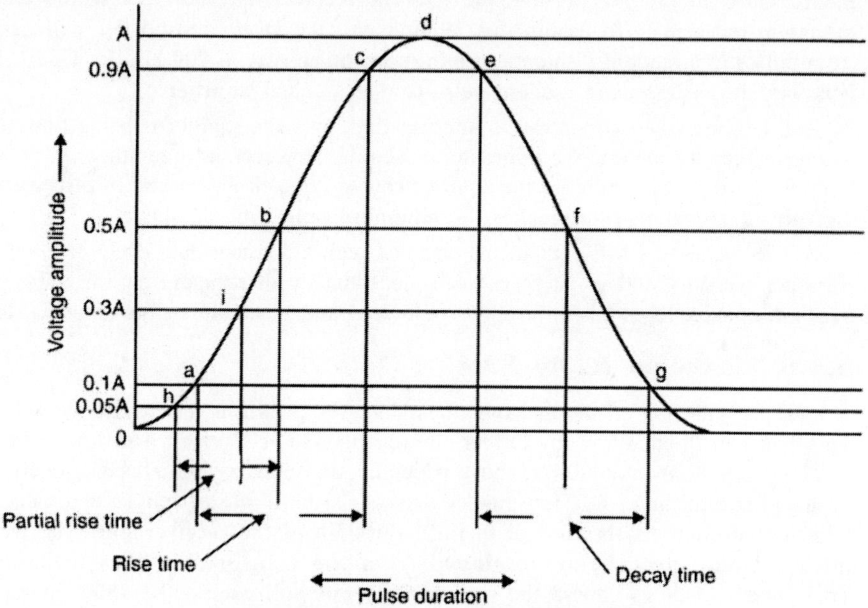

Fig. 8.5. Pseudo-Gaussian Waveform.

(*iii*) **Pulse Duration.** The time interval between the 50 percent amplitude point on leading and trailing edge of the pulse envelope, *i.e.* between points *b* and *f*.

(*iv*) **Pulse Decay Time.** The time as measured between the 90 and 10 percent amplitude points on the trailing edge of the pulse envelope, *i.e.* between points *e* and *g*.

8.4.4. Pulse Jittering

Since all aircraft using the ground DME Transponder operate on the same pair of receiving and transmitting frequencies, all replies of the Transponder to all aircraft interrogations are received by all these aircraft. It is, therefore, necessary for each aircraft to select those replies which result from its own interrogations. This is done as follows :

Transmission of aircraft interrogation pulse pairs is continuous and, in turn, aircraft reception of the ground Transponder replies is also continuous. Transmission of the aircraft interrogation pulse pairs is **semi-random**; that is, **the number of pulse pairs per second (PRF) transmitted** by a particular aircraft remains fairly constant, but the **intervals between pulse pairs vary**. The variation in time spacing of the pulse pairs is **unique** to each aircraft and permits the aircraft to select the replies to its particular interrogations. The aircraft receiver continuously compares the time spacing pattern of the reply pulse pairs with the time-spacing pattern of interrogation pulse pairs. Only those pulse pairs which lie in matching patterns are processed to extract the range information.

The variation in time spacing of the pulse pairs of the aircraft interrogation is termed as **Pulse Jittering.**

8.4.5. Duty Cycle of the Transponder

Airborne DME receivers require a minimum number of random pulses to ensure its correct operation by providing appropriate AGC signal. However, unless sufficient interrogating aircraft are present, the airborne receiver may not receive the required minimum pulse rate. To ensure this requirement, the DME transponder will generate extra pulses in a random fashion at a **minimum pulse rate of 700 Hz** (for GCEL DME it is 945 Hz). These extra random pulse-pairs are called **Squitter**.

At the time when no aircraft is interrogating, only the squitter is being transmitted, at an average rate equal to the minimum pulse rate. However, as the number of authentic aircraft interrogations increase **the squitter rate is reduced and becomes zero when the aircraft interrogation rate reaches the minimum pulse rate or above.**

(Older versions DME equipments operated with a constant duty cycle of 2700 pulse pairs per second (2800 in later versions), done mainly to regularly activate high power transmitters using Klystrons for example to increase the mean time before failures (MTBF).

8.4.6. Maximum Reply Rate

As the pulse rate of the interrogations increases, a limit is reached above which the Transponder will not allow any further interrogations to be serviced. This limit is reached at a reply rate of about 2800 Hz, above which the transponder would become overloaded. To avoid overloading, the Transponder detects the high rate of replies and causes the **receiver automatic gain control to limit the gain of the receiver until the weaker, more distant, aircraft, are excluded from the transponder,** thus lowering the Transponder loading. Should the system reply rate still exceed the 2800 limit, **video output pulses** are randomly suppressed to limit the maximum **reply rate to 2800 Hz.**

Because of this limitation in heavy traffic the aircraft may not receive 100 percent replies to their interrogations. In DME system operation it is assumed that on an average 95 percent of aircraft interrogating a ground transponder at any given time will be in the track mode and 5 percent will be in search mode. So for 100 aircraft the total interrogation rate of the ground transponder will be

(30 pps × 95 aircraft) + (150 pps × 5 aircraft) = 3600 interrogations per second.

As the maximum reply rate of the transponder is limited to 2800 pulse pairs per second, the ratio of replies to interrogations is 78 percent. However, the airborne DME receiver is designed to operate safely with a reply ratio as low as 70 percent. Hence providing a safe margin of operation.

In the extreme case, when 100 aircraft are using the DME and the ground equipment, fails all aircraft will be in the search mode and will produce a total of 15000 (100 × 150) interrogations per second. When the ground transponder returns to normal service, its receiver gain will drop until only the 2800 **strongest** interrogations are serviced. This means that replies will be provided to the seventeen or so aircraft which are likely to be nearest to the ground transponder. Once these aircraft enter the track mode their interrogations will drop from 150 to 30 pulse pairs per second, thus gradually relieving the ground transponder of about 2100 interrogations and allowing it to increase its sensitivity and reply to more distant aircraft.

8.4.7. Transponder Identification

The need for transmitting Identification Message on a DME is to associate it with its paired facility. This is essential because the pilot should know that he has selected the right DME associated with a particular VOR or ILS that he is using.

Each operational DME is identified by a 2 or 3-character long Morse Code message which is transmitted at regular intervals. During this time, the squitter and reply pulses are replaced by regularly placed pulses at 1350 pulse pairs per second. These pulses activate a 1350 Hz tuned circuit in the aircraft receiver producing an audio signal. Each identification code (ident) is unique and identifies a specific DME site.

Frequently, DME is co-located with VOR or ILS equipment and for this reason the DME may derive the identification code from the co-located facility. In such cases **any failure of the external ident generator** should cause the DME to internally generate and transmit the ident in place of the **failed unit.**

8.4.8. Fixed Time Delay

In the ground Transponder, the time delay of a received aircraft interrogation pulse pair to the corresponding reply pulse pair is adjusted to **exactly 50 microseconds.** This is done to facilitate the aircraft receiver to deduct a constant time delay from the total time elapsed between interrogation and reply while calculating the **slant range distance.**

8.4.9. Echo Suppression

The normal slant range for a normal DME operating at optimum efficiency is about 200 NM. However, this maximum range may be seriously degraded by the terrain surrounding the installations. A major contributing factor to distance accuracy degradation is the effect of echoes on the interrogation pulses arriving at the Transponder. Fig 8.6 shows the generation of echoes.

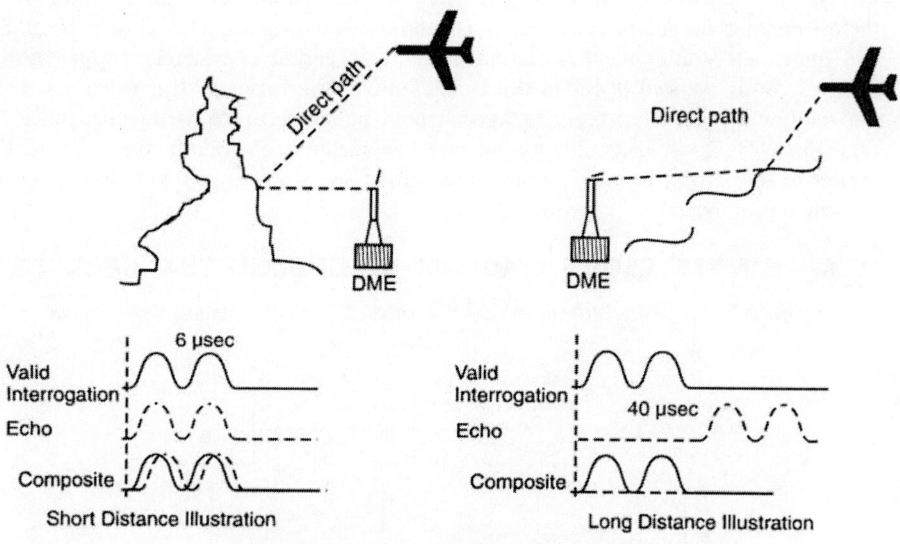

Fig. 8.6. Echo Generation Diagram.

Because of siting problems, the interrogations from aircraft may be reflected by obstructions and arrive at the DME station delayed in time from the direct interrogations. Therefore, it is possible for the aircraft to lock on to the replies from these echoes and provide the aircraft with erroneous distance information.

There are two types of echoes that may cause interference. They are short distance and long distance echoes.

(*a*) **Short Distance Echoes.** This type of echo is defined as an echo that occurs within 12 μ sec of the interrogations. This is caused by an obstruction behind and within close proximity to the DME station difference is < 2 NM.

(*b*) **Long Distance Echoes.** This echo is defined as any echo occuring 60 μ sec after the valid interrogations. This is caused by an obstruction normally appearing between the DME station and the aircraft difference is > 10 NM.

Echo suppression is employed to eliminate the effect of echoes. The **Short Distance Echo Suppression (SDES)** is used to eliminate echoes with short delays and **Long Distance Echo Suppression (LDES)** is used to eliminate echoes with long delays.

8.4.10. Priority of Reply Pulse

There are three modulating inputs to the transponder transmitter. The encoder section of the transponder attaches priority to these signals. The order of priority is as given below :

(*i*) Identification Pulses
(*ii*) Reply Pulses
(*iii*) Squitter

8.4.11. Transponder Receiver Dead Time

In order to avoid oscillations of the transponder and also to some extent eliminate the effects of echo pulses, the receiver is suppressed during the process of transmission and immediately after the transmitted pulse. This period of **receiver suppression is typically 60 μ sec and is called the Dead Time of the receiver**. It is to be noted that the transponder does not reply to interrogation pulses received during this time. The total time lost is typically 20 percent and this means that aircraft may only receive replies to 80 percent of their interrogation pulses, even when well within range of the ground transponder.

8.5. SIMPLE BLOCK DIAGRAM OF GROUND TRANSPONDER

Figure 8.7 shows the simple block diagram of a DME ground transponder.

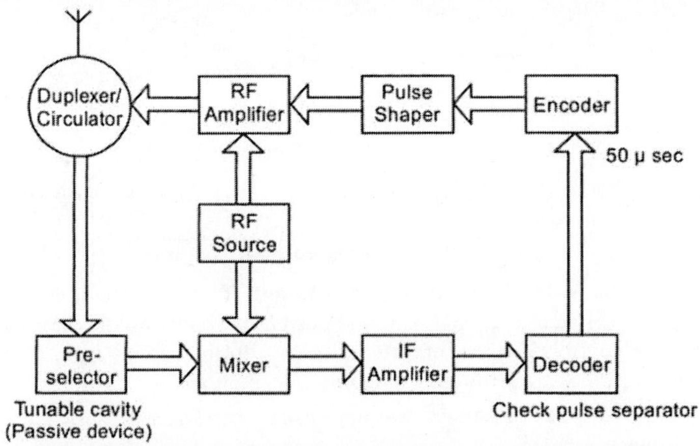

Fig. 8.7. Simple Block Diagram of Ground Transponder.

The interrogations from the aircraft are routed through the duplexer (cirulator) to the receiver section. The pre-selector filter accepts only the operating channel fequency interrogations and routes them to the mixer stage. Here the interrogations are converted into IF signals amplified and the video pulse pairs are detected. The detected video pulses are passed on to the decoder stage. The decoder validates the interrogations and triggers the encoder stage to produce the reply pulse pairs. These reply pulse pairs are shaped into gaussian pulse pairs before modulating the UHF carrier. The modulated carrier is then routed through the duplexer/circulator to the antenna for radiation.

8.6. SIMPLE BLOCK DIAGRAM OF AIRCRAFT INTERROGATOR

Figure 8.8 shows the simple block diagram of a DME aircraft interrogator.

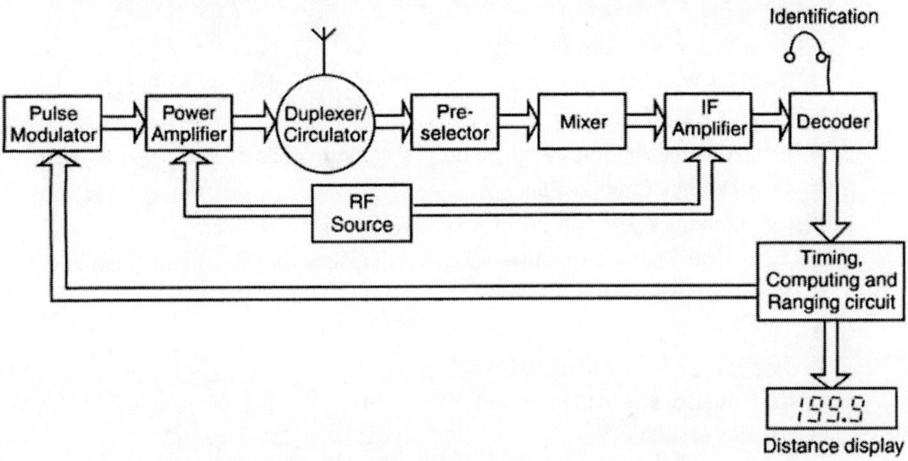

Fig. 8.8. Simple Block Diagram of Aircraft Interrogator.

The RF Source feeds the power amplifier stage with the RF carrier. In the power amplifier stage, the RF carrier is pulse modulated as well as amplified before being transmitted through the antenna system as aircraft interrogations. The timing and control circuits provide the required signals to produce the coded pulses to modulate the transmitted RF. They also control the pulse interrogation rate.

The replies from the ground transponder enter the receiver section of the interrogator. After heterodyning and amplifying the IF is applied to the decoder stage. The decoder separates the distance replies from the noise and other information transmitted from the ground transponder. The separated distance pulses are passed on to the computing circuit. The computing circuit measures the time lapse between interrogation and reply pulses and converts this into distance to drive the distance indicator.

8.7. FUNCTIONAL DESCRIPTION OF AWA/GCEL DISTANCE MEASURING EQUIPMENT

Introduction. The DME equipment selected for discussion is the series GCEL-752 type A69870, manufactured in India by the Gujarat Communications and Electronics Limited under the licence from AWA, Australia.

The GCEL 752 DME equipment is designed and manufactured to meet the requirements laid down by the ICAO for this type of equipment. It is fully solid state and uses digital techniques to minimise the number of adjustable controls. The equipment is available in several standard configurations depending on RF monitoring as well as primary power requirements. The basic transponder provides modulation and drive to a power amplifier assembly which raises the power output either to 100 Watt or 1 kWatt. Duplication of transponders, monitors, mains power supplies and battery banks are all possible in the GCEL 752 series equipment. A common controller is used for *all* equipment configurations.

8.8. EQUIPMENT STRUCTURE

The equipment structure of the GCEL 752 dual transponder dual monitor DME is given below :

The equipment consists of

- Two 100 Watt transponders;
- Two 1 kWatt RF Amplifiers;
- One Control Panel;
- One RF Distribution Panel behind the Control Panel; and
- One DC Distribution Panel.

All these modules are housed in a standard 19 inch cabinet.

The basic 100 Watt transponder assembly consists of five modules namely :

- Receiver Video Module;
- 100 Watt RF Amplifier Module;
- 100 Watt Power Supply Module;
- Test Interrogator Module; and
- Monitor Module

The AC Power Supply Module is housed within a cabinet external to the main DME rack.

8.9. FUNCTIONAL DESCRIPTION

A block diagram of the Dual 1 kWatt transponder with dual monitor is shown in Fig. 8.9 of the two transponders, one is active and the other is kept as standby.

Incoming interrogations from aircraft are intercepted by the antenna and are passed on to the preselector filter of the operating transponder through a directional coupler, a coaxial relay, and a circulator. Additionally, the two Test Interrogator Modules each generate RF interrogation pulses which are fed to the receiver path via the directional coupler. Both Test Interrogators operate at a **PRF of 100 Hz** each and are interconnected to prevent the **interrogation of one falling in the Transponder deadtime** produced by the other.

The preselector filter consists of three coupled resonant cavities and can be tuned for each channel in the frequency band of **1025 MHz to 1150 MHz**. The pre-selector filter rejects **IF, image frequency** and other **unwanted input signals**. The interrogation signals are then routed to the **Receiver/Video Module.**

The Receiver/Video Module detects and decodes on-channel interrogations and encodes the synchronous reply trigger pulse pair. Random reply pulse pairs are added, if necessary, to the synchronous replies to maintain a minimum reply rate of **945 Hz**. The maximum reply rate is limited to **2800 Hz by reducing receiver sensitivity**. The keyed ident signal occurs **every 30 seconds** and mark of the ident replaces normal pulses with a **1350 Hz pulse train.**

Receiver output reply pulse pairs trigger the modulation generator in the 100 Watt RF Amplifier Module producing RF pulses which are connected to the 1 KW Power Amplifier Module. The excitation frequency for the 100 W RF Amplifier Module is provided by the local oscillator in the Receiver/Video module.

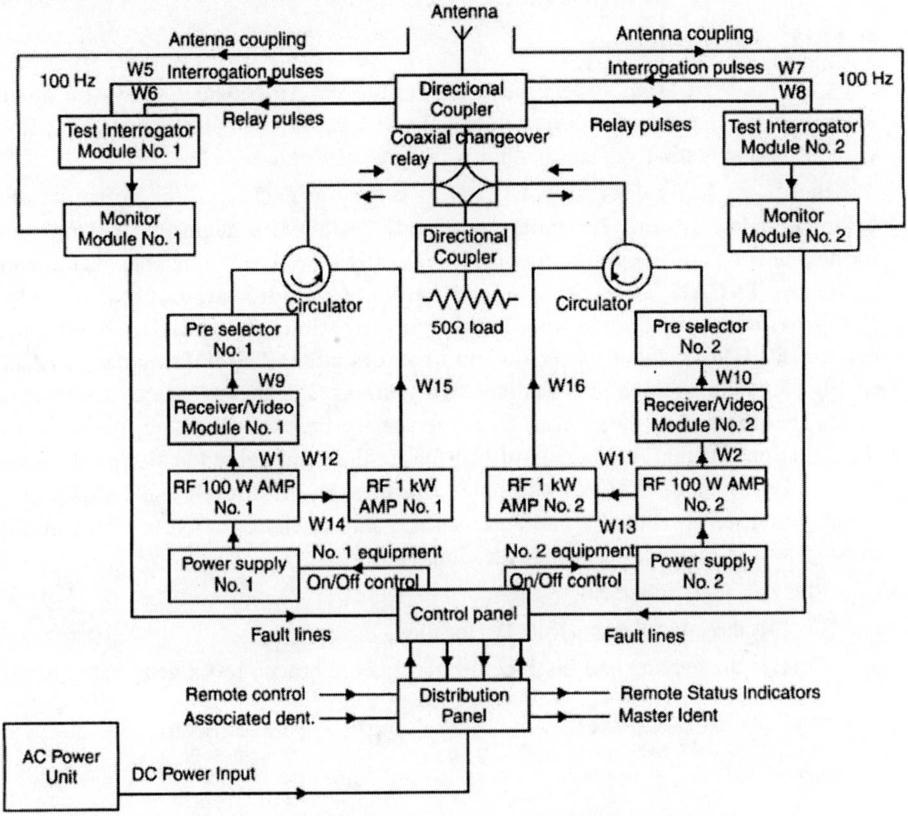

Fig. 8.9. Block Diagram of the Dual 1 KW Transponder.

The output from the 1 KW Power Amplifier passes through the circulator, changeover relay and the directional coupler to the antenna connector.

Transponder **output** pulses are sampled in the directional coupler and detected in the reply detector of the Test Interrogator Module. Processing within the Test Interrogator Module separates **synchronous** replies from all other **non-synchronous** replies and **squitter**.

The Monitor Module processes **output** signals from the Test Interrogator Module and sends a **pass** or **fail** signal to the Control Panel for each monitored parameter.

The other transponder is on **standby** and if **manually** turned on for test purposes, operates through the other circulator, changeover relay, directional coupler and is terminated in a dummy load. A coaxial patch panel allows any one of the **Test Interrogator to be connected to coupled ports on the directional coupler for testing off the air transponder. A monitor disable switch** on the Monitor Module alerts the Control Unit that only one Monitor is in use on the online equipment and the selection of SINGLE auto sequence in the Control Panel instructs the Control Panel not to transfer to the equipment under test and take shut down action if **online equipment develops a fault-condition.**

8.10. TACTICAL AIR NAVIGATION (TACAN)

8.10.1. Introduction

Like the VOR-DME system the TACAN (Tactical Air Navigation) system provides an aircraft with continuous distance (range) and bearing (Azimuth) information from a ground beacon station, located within a line-of-sight range.

Before we begin discussing TACAN, it is need to understand the definition of the polar-coordinate system. The **polar-coordinate system** is a geometric system used to locate points on a plane. In electronics, it is usually used for plotting antenna directional patterns. **TACAN** is a polar-coordinate type radio air-navigation system that provides an aircrew with distance information, from distance measuring equipment **(DME),** and bearing (azimuth) information. This information, as shown in **Fig. 8.10**, is usually provided by two meters. One meter indicates, in nautical miles, the distance of the aircraft from the surface beacon. The other meter indicates the direction of flight, in degrees-of-bearing, to the geographic location of the surface beacon. By using the TACAN equipment installed in the aircraft and TACAN ground equipment installed aboard a particular surface ship or shore station, a pilot can obtain bearing to and distance from that location.

Pilot can then either :

(1) Fly directly to that particular location, or.

(2) Use the bearing and distance from a specific beacon to fix geographic location.

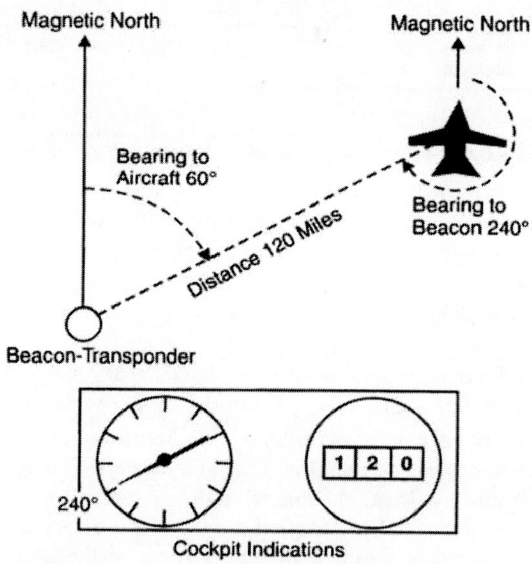

Fig. 8.10. TACAN Aircraft Position Indication.

TACAN operates completely in the UHF band from 962 to 1213 Mhz and thus overcomes the limitations of VOR caused by relatively low frequency operation. The system has a total of 126 channels in the UHF band ,each being capable of providing full service for 100 aircrafts. TACAN is remarkably accurate. Its provides compass direction with less than one degree error and is capable of indicating true distance from the ground station within about 600 feet because of its UHF operation TACAN can use a very small ground station antenna which permits its installations on AIR FORCE mobile units and Navy carriers.

8.10.2. TACAN Principles

The distance measuring concept used in TACAN equipment is an outgrowth of radar-ranging techniques. Radar-ranging determines distance by measuring the round-trip travel time of pulsed RF energy. The return signal (echo) of the radiated energy depends on the natural reflection of the radio waves. However, TACAN beacon-transponders generate artificial replies instead of depending on natural reflection. Now look at **Fig. 8.11.** The airborne equipment generates timed interrogation pulse pairs that the surface TACAN system receives and decodes. After a 50-μsec delay, the transponder responds with a reply. The airborne DME then converts the round- trip time to distance from the TACAN facility. The frequency and identification code provide the geographic location of the transmitting beacon.

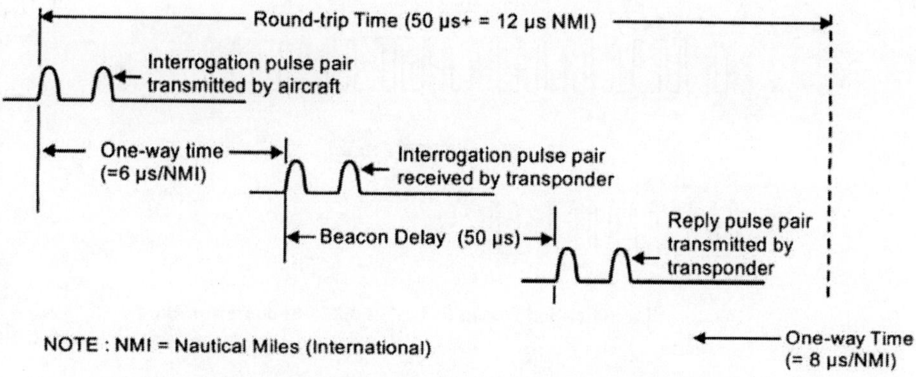

Fig. 8.11. Distance Measuring Round-Trip Travel Time.

8.10.3. TACAN Pulse Pairs

TACAN transponders use twin-pulse decoders to pass only those pulse pairs with the proper spacing. The purpose of this twin-pulse technique is to increase the average power radiated and to reduce the possibility of false signal interference. After the receiver decodes an interrogation, the encoder generates the necessary pulse pair required for the transponder's reply. A TACAN pulse pair generated by airborne or ground equipment is shown in **Fig. 8.12**.

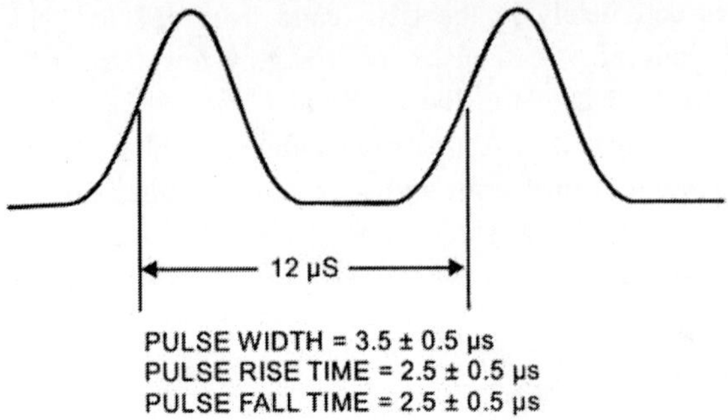

Fig. 8.12. TACAN Pulse Train.

8.10.4. Constant Transponder Duty-Cycle

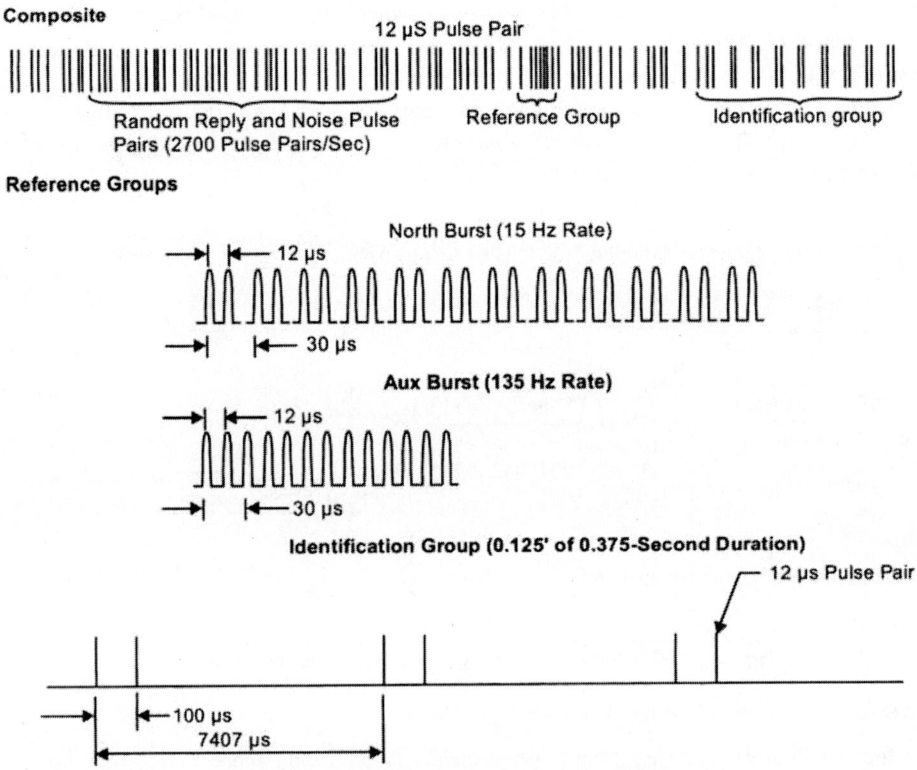

Fig. 8.13. Transponder Output Pulse Train.

In principle, reply to aircraft the TACAN transponder need only interrogations at 30 pulse pairs-per-second, per airborne equipment, to supply the necessary distance data. However, the total pulse out put of the transmitter constantly varies, according to the number of interrogating aircraft. In addition, random noise may trigger the transmitter. For the transponder to provide azimuth information, the average power supplied to the antenna must be relatively uniform over time. To accomplish this, the transponder is operated on the constant-duty-cycle principle. In this method of operation, the receiver uses automatic gain and squitter (noise generated output) controls to maintain a constant pulse output to the transmitter, as shown in Fig. 8.13. If few interrogations are being received, the gain and squitter of the receiver increase and add noise-generated pulses to the pulse train. If more interrogating aircraft come into range, the gain and squitter decrease and reduce the number of noise-generated pulses. The relationship between the gain and the number of pulses is such that only a 2-dBm change in sensitivity occurs between reception from 1 aircraft and those from 100 aircraft. An added advantage of using a constant duty cycle is that overall transmitter power drain remains constant.

8.10.5. Beacon-Transponder Identification Code

Before an aircrew can use TACAN information that its equipment receives, it must positively identify the transmitting TACAN station. To meet this need, the ground station transmits an identification code at approximately one-half minute intervals. It does this by momentarily interrupting the transponder distance data and squitter-generated output with pulse groups spaced at a 1350-pps rate. Each pulse group contains two sets of 12-μsec pulse pairs spaced 100 μsec apart. The duration of the identification pulse groups varies, to represent Morse-coded characters. The duration for a dot (.) is 100 to 125 ms, and for a dash (-) 300 to 375 ms. An identification group is shown in Fig. 8.13.

8.10.6. 15-Hz-Bearing Information

The timing of the transmitted pulses supplies the actual distance information to the aircraft. This leaves amplitude modulation as another medium for the transponder to convey other information to the aircraft. The TACAN beacon-transponder modulates the strength of the pulse to convey bearing information by producing a specific directional-radiating pattern rotated around a vertical axis. This signal, when properly referenced, indicates the aircraft's direction from the TACAN facility. This signal and the distance data give a two-piece fix (distance and direction) for determining specific aircraft location. The RF energy from the TACAN transmitter is fed to the antenna central element, which has no directivity in the horizontal. Parasitic elements positioned around the central element are electronically rotated (switched on and off) at 15 revolutions per minute. The distance between the central element and the parasitic elements selected to obtain a cardioid radiation pattern. To an aircraft at a specific location, the distance pulses appear to contain a 15-Hz amplitude-modulated signal because of the rotation of the cardioid radiation pattern. This pattern is shown in **Fig. 8.14** view A and view B.

The aircraft TACAN equipment obtains bearing information by comparing the 15-Hz modulated signal with a 15-Hz reference burst signal it receives from the ground facility. The phase relationship between the 15-Hz modulated signal and the 15-Hz reference burst signal depends on the location of the aircraft in the cardioid pattern. The 15-Hz reference burst signals are transmitted when the maximum signal of the cardioid pattern aims due East. This group of 12 pulse pairs is commonly referred to as the North or main reference burst. You can see the relationship between the reference pulses and the cardioid pattern by comparing view *A* and view *B* of **Fig. 8.14**.

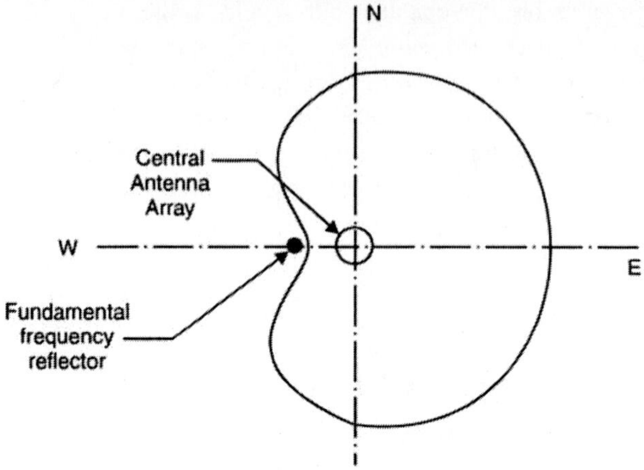

(*a*) Cardioid pattern

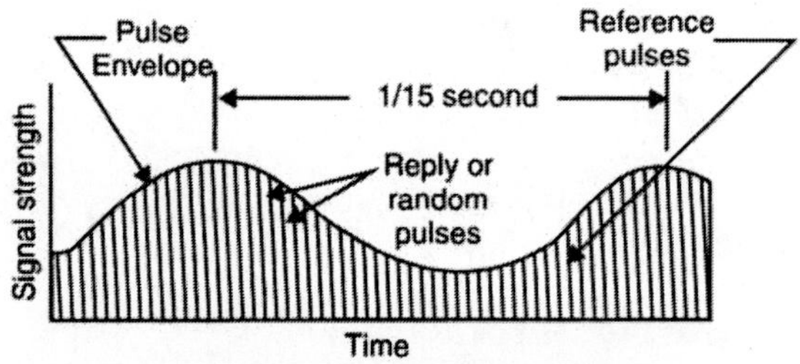

(*b*) Amplitude-modulated pulse pairs

Fig. 8.14. TACAN Radiation Pattern: A. Cardioid Pattern; B. Amplitude-Modulated Pulse Pairs.

8.10.7. 135-Hz Bearing Information

Errors arising from imperfections in the phase measuring circuits and radio propagation effects are known as site error. These errors are significantly reduced by the addition of 32 outer parasitic elements added to the electronically scanned antenna. Electronically switching these elements modifies the antenna cardioid pattern. Though the cardioid pattern is still predominant, it is altered by superimposed ripples. The aircraft now receives the 15-Hz signal with a 135-Hz ripple amplitude modulated on the distance data pulses (Fig. 8.15). To furnish a suitable reference for measuring the phase of the 135-Hz component of the envelope wave, the transponder is designed to transmit a coded 135-Hz reference burst similar to that explained for the 15-Hz reference. The 135-Hz reference group is commonly referred to as the auxiliary or aux reference burst. The composite TACAN signal is composed of 2700 interrogation replies and noise pulse pairs-per-second, plus 180 North burst pulse pairs-per-second, 720 auxiliary burst pulse pairs-per-second, for a total of 3600 pulse pairs-per-second, or 7200 pulses-per-second spacing are a characteristic of that particular TACAN signal element. However, it is important to understand that proper spacing between pulses and pulse pairs is what actually provides the aircraft with the means to distinguish between the TACAN pulses and any other pulses that might be present on the received radio frequency. Check the reference data in the appropriate technical manual for specific pulse characteristics and spacing.

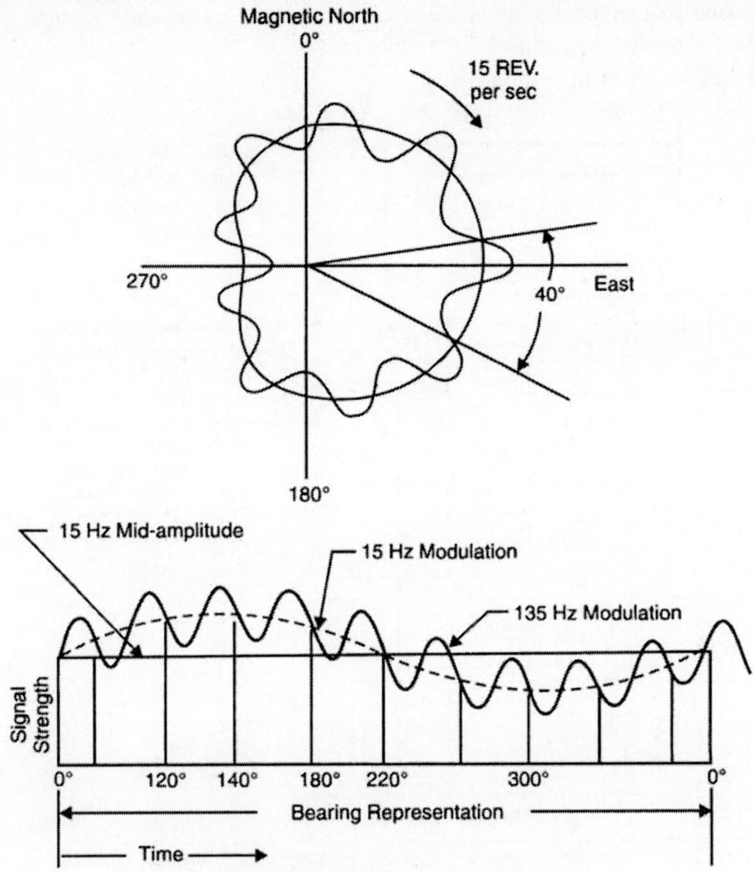

Fig. 8.15. TACAN Modulation Envelope.

8.10.8. TACAN Signals Priorities

Priorities have been established for transmission of the various types of TACAN signals. These priorities are as follows:

1. Reference bursts (North and auxiliary)
2. Identification group
3. Replies to interrogations
4. Squitter

Therefore, the identification group, replies, or squitter will be momentarily interrupted for the transmission of either the main or auxiliary reference group. The transmission of replies or squitter will be interrupted every 37.5 seconds during the transmission of an identification code dot or dash.

8.10.9. Simple Block Diagram of TACAN

A simple block diagram of TACAN system describes the functional operation of an electronics system in the same way they do in electrical systems. Fig. 8.16 shows a simplified block diagram of a shipboard tactical air navigation (TACAN) system.

The TACAN system is an electronic air navigation system that provides a properly equipped aircraft with bearing and distance from a shipboard or ground radio beacon selected by the pilot. The system is made up of

1. A radio beacon (consisting of the receiver-transmitter group, the antenna group, and the power supply assembly) and
2. The radio set in the aircraft.

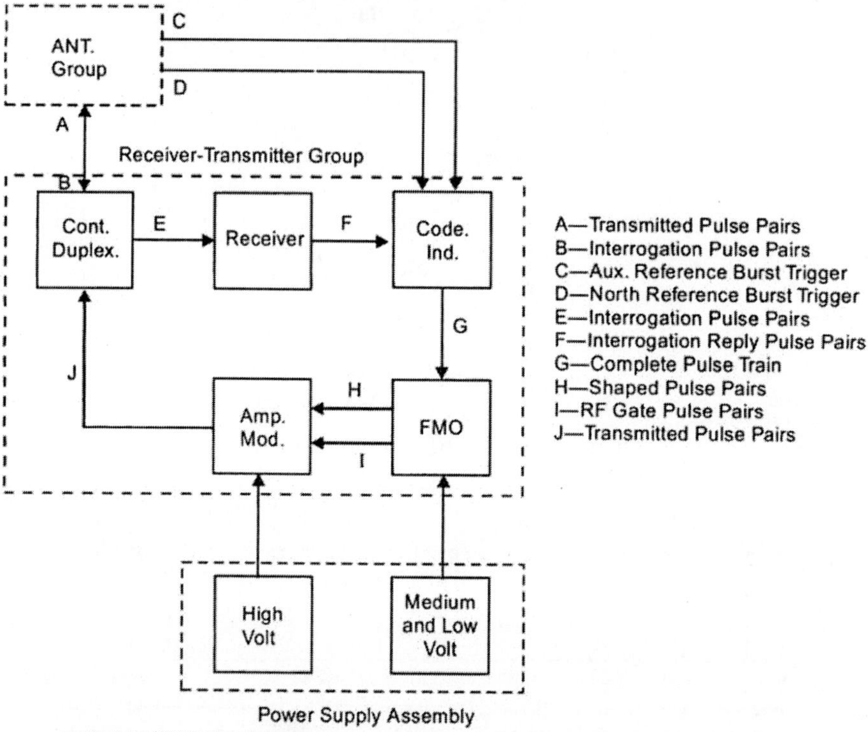

Fig. 8.16. A Simplified Block Diagram of Shipboard TACAN System.

Duplexer is used to isolate the Rx and Tx whole using same antenna for both receptors and transmission. The duplexer has tuned cavities, which makes the rejection of undersigned signals. Antenna system used in TACAN is quite different from DME as it has to provide both range and bearing information.

The antenna group develops the coarse and fine bearing modulation electronically. Rather than forming the TACAN radiations pattern by the old mechanical rotation method. This can be achieved by digital switching of parasitic elements arranged in concentric arrange around the central radiation. Twelve inner elements provide the 15Hz modulation and 32 outer elements provide the 135Hz modulation (replacing the nine outer elements of the rotated antenna.) the 15Hz and 135Hz modulation pattern is provided by electronically switching the diodes in each of the parasitic elements in prescribed time sequence, which is repeated once in each 15Hz interval.

The groups of timing pulses, which are being generated, by the antenna group, feed to the TACAN Rx, which generated the reference sinusoidal. The both 15Hz modulation and 135Hz ripple modulated are being transmitted as per the pattern describe in the earlier sections. Code identifier will identify the station, which is in range.

8.10.10. TACAN Equipment (AN/URN-25)

Many different types of TACAN equipment have been used for air navigation. Today, the AN/URN-25 is taking over the task of tactical air navigation from the

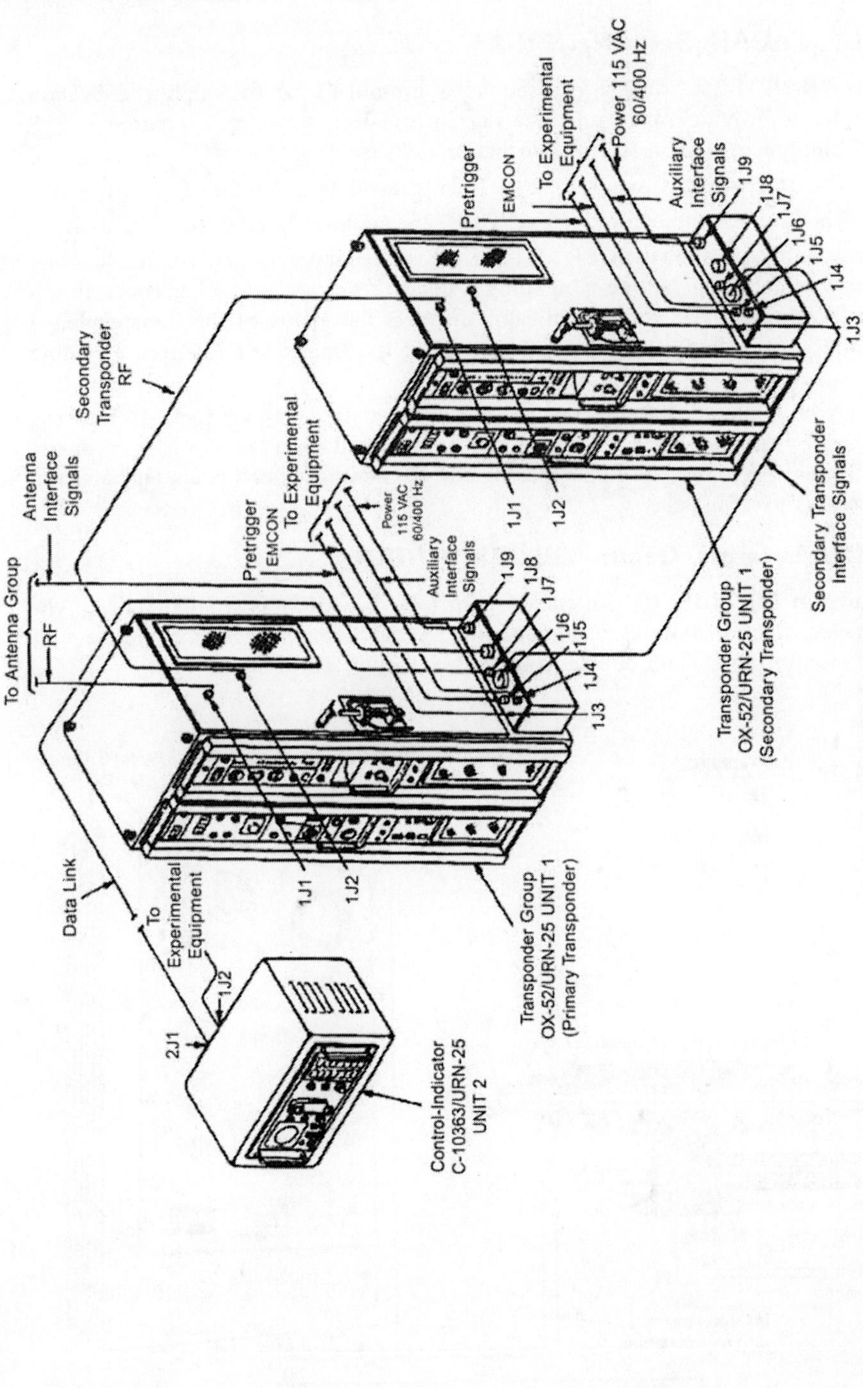

Fig. 8.17. TACAN SET AN/URN-25.

older AN/URN-20 on new construction ships and as ships complete overhaul. Brief explanation of a practical TACAN system AN/URN-25is described in the following sections.

8.10.11. TACAN Set AN/URN-25

The AN/URN-25 TACAN is used as a ground-based or shipborne beacon transponder to provide range and bearing information to aircraft equipped with TACAN equipment. It consists of two major units :

1. The Transponder Group OX-52/URN-25, commonly referred to as unit 1, and
2. The Control-Indicator C-10363/URN-25, commonly referred to as unit 2.

These units are shown in **Fig. 8.17.** Each transponder is housed in a cabinet with two vertical drawers, one containing a coder keyer and the other containing a receiver-transmitter. The control-indicator displays the status of the transponder(s) and failure alarms, and allows limited control of the transponder(s) from a remote location.

It may be mounted in its own cabinet or in a standard 19-inch rack. To increase the channels available, the TACAN set can be operated in either the X or Y mode. The Y mode changes the pulse pair spacing and the auxiliary burst count and spacing, and increases system delay.

8.10.12. Antenna Group OE-273(V)/URN

Shown in **Fig. 8.18,** the Antenna Group OE- 273/URN is a solid-state, high-performance, electronically-scanned, all-band TACAN antenna system, complete with integral monitoring system and built-in fault isolation capability.

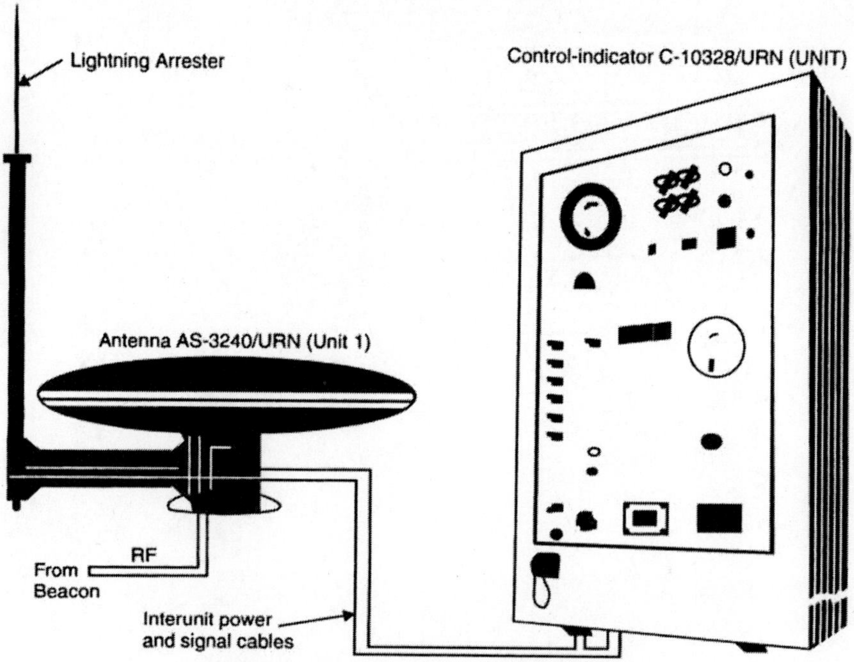

Fig. 8.18. Antenna Group OE-273(V)/URN.

The antenna group develops the coarse and fine bearing modulations electronically. Rather than forming the TACAN radiation pattern by the old mechanical rotation method, the AS-3240 achieves the same effect by digital switching of parasitic elements arranged in concentric arrays around the central radiator. Twelve inner elements provide the 15-Hz modulation (replacing the single-phase rotating parasitic element in the mechanically rotated antenna), and 32 outer elements provide the 135-Hz modulation (replacing the nine outer elements of the rotated antenna). The 15- and 135-Hz modulation pattern is provided by electronically switching the diodes in each of the parasitic elements in prescribed time sequence, which is repeated once in each 15-Hz interval.

In effect, the elements are rotated electrically, rather than mechanically. An advantage this provides is the elimination of the bandwidth limitations inherent in the old mechanically-rotated antennas. In the electronically-scanned antenna, the appropriate ring for a given frequency segment is activated by a fast electronic switch, based on information from the TACAN frequency synthesizer. This allows instantaneous band switching and all-band operation.

8.10.13. Capabilities and Limitations

In the X mode of operation, the TACAN set transmits on one of 126 discrete channel frequencies (which are 1-MHz apart) from 962 to 1024 MHz and from 1151 to 1213 MHz. In the Y mode of operation, the set transmits on one of 126 discrete channel frequencies (which are 1-MHz apart) within the range of 1025 to 1150 MHz. The navigation set receiver, operating in the 1025- to 1150-MHz range for both the X and Y modes, is always displaced 63 MHz from the transmitter frequency.

The TACAN set can simultaneously provide individual distance measuring service for up to 100 interrogating aircraft. Of the 3,600 pulse pairs-per-second transmitted by the TACAN, 900 pulse pairs (MAIN and AUXILIARY bursts) contain the bearing information; the remaining 2,700 pulse pairs are either random noise pulses, identity pulses, or replies to interrogating aircraft. Once every 30 seconds, the interrogation replies and random noise pulses are interrupted for the transmission of identity pulses.

The navigation set has a receiver sensitivity of –92 dBm or better and a nominal peak power output of 3 kilowatts at the transponder cabinet output. (Power output may limited to less than peak by directives). Since the bearing and identification signals are delivered spontaneously and not in response to interrogations, an unlimited number of properly equipped aircraft can derive this information from the TACAN set over a line-of-sight (los) range up to 200 nautical miles.

SUMMARY

- DME (Distance Measuring Equipment), as the name indicates is an electroic equipment used to measure slant range distance between aircraft and a ground station.
- The DME operates in the UHF range between 960 MHz to 1215 MHz. Its transmit on 962 to 1150 MHz and receive on 962 to 1213 MHz. The channel width is 100 KHz.
- **Application of DME :**
  - (*i*) Navigation aid in conjunction with the VOR.
  - (*ii*) Components of ILS (Instrument Landing System) where DME is also alternate to marker beacom.

(*iii*) Components of MLS (Microwave Landing System)

(*iv*) Expedite the radar identifications of aircrafts.

- **DME Principles :** The operating principle of DME system is based on the Radar principle, the time required for a radio pulse signal to travel to a given point and return. In fact, it is a secondary radar.
- **Modes of Operation :** There are two modes of aircraft interrogations.
  (*i*) Search mode and (*ii*) Track mode
- The variations in time spacing of the pulse pairs of the aircraft interrogation is termed as pulse jittering.
- Like the VOR-DME system the **TACAN (Tactical Air Navigation)** system provides an aircraft with continuous distance (range) and bearing (Azimuth) information from a ground beacon station, located within a line-of-sight range.
- TACAN operates completely in the UHF band from 962 to 1213 Mhz and thus overcomes the limitations of VOR caused by relatively low frequency operation. The system has a total of 126 channels in the UHF band ,each being capable of providing full service for 100 aircrafts.
- TACAN is remarkably accurate. Its provides compass direction with less than one degree error and is capable of indicating true distance from the ground station within about 600feet.because of its UHF operation.
- TACAN can use a very small ground station antenna which permits its installations on AIR FORCE mobile units and Navy carriers.
- The distance-measuring concept used in TACAN equipment is an outgrowth of radar-ranging techniques. Radar ranging determines distance by measuring the round-trip travel time of pulsed RF energy.
- **TACAN Signals Priorities.** Priorities have been established for transmission of the various types of TACAN signals. These priorities are as follows :
  1. Reference bursts (North and auxiliary)
  2. Identification group
  3. Replies to interrogations
  4. Squitter

REVIEW QUESTIONS

1. Write the applications of DME.
2. Write the principle of DME.
3. Explain the working of DME system with the help of block diagram.
4. Write the modes of operation in DME and explain them briefly.
5. Explain briefly the characteristic features of the DME system.
6. Draw the neat block diagram of ground transponder of DME and briefly explain the working of this.
7. Draw the neat block diagram of Aircraft Interrogaor of DME and briefly explain the working of this.
8. Briefly explain the principle of TACAN.
9. Explain the working of TACAN with the help of simple block diagram.
10. Write short notes on :
    (*i*) Modes of operation in DME
    (*ii*) Working frequency of DME
    (*iii*) Accuracy DME
    (*iv*) Limitation and capabilities of TACAN system.
    (*v*) DME **[UPTU, 2004]**

9

CHAPTER

ELECTRONICS AIDS TO APPROACH AND LANDING SYSTEM

INSIDE THIS CHAPTER

9.1. INTRODUCTION

In this chapter we will be discussing about the various aids used to guide the aircraft to approach and land safely. If the visibility is good, *i.e.* approximately 5 Km or more horizontally and 300 m vertically, than the landing operation is carried out by visual observations of ground features and landing lights, either in day or night, is known as "Visual Fligh Rules" (VFR) conditions.

When the visibility is less and it is not possible to carry out landing using VFR, than this operation is carried out by using Instrument Landing Systems (ILS).

The ability to land on aircraft under conditions of low or zero visibility is one of the most important factors determining the realibility of air travel. There are two electronics systems now in use to provide atleast a partial solution to this problem. They are Ground Controlled Approach (GCA) and the Instrument Landing System (ILS). Although they can be used in emergencies to accomplish a true blind landing, both of these arrangements are fundamentally blind-approach systems, in which the final landing is normally carried out visualy after the electronic system has brought the aircraft out of the overcast in the current position to complete a landing.

Electronics systems GCA's, ILS and Microwave Landing System (MLS) are altogether responsible for the approaching and landing safely at the runway. These system will be discussed in brief in the further section of this chapter. GCA's and ILS are being used in India for the approach and landing by Microwave Landing System (MLS) is still not implemented in India but it is likely to be introduced in course of time.

9.2. GROUND CONTROLLED APPROACH (GCA)

The GCA system consists of two microwave radar sets, which are usually installed adjacent to the runway. The first is for general survillance or search and for control of the traffic pattern of aircraft in the vicinity of the landing field. It is known as

Surveillance Radar Element (SRE) and it has a range of approximately 70 NM. The second is high resolution short-range set (normally 20 to 30 Km) that is used to conduct the actual landing. This radar is known as **Precisions Approach Radar (PAR)**. It has got two display system, one is arranged to present elevtions as a vertical displacement, and range as a horizontal displacement, while the other presents azimuth displacement from the glide path. Glide path is a path which is to be strickly followed by the aircraft upto the touchpoint of the runway. Glide path is predefined for each type of aircraft.

9.3. SRE (Survelllance Radar Element)

SRE (Surveillance Radar Element) which is usually designated the Airport Surveillance Radar (ASR), is the type of radar with a similar specification to that of surveillance radar is that used to control the approach of aircraft to an airfield. The requirement here is two fold. First, all aircraft flying within a range of thirty or forty miles must be detected; secondly, they must be accurately 'talked down' on to the correct runway. The detection of the aircraft is made easier by the fact that they will be flying in well-defined height zones. The radar will not have to deal with very low flying targets and will not need to take an interest in very high targets. Consequently the detection beam is usually set with its maximum at an angle of about 3° to reduce the likelihood of ground clutter. MTI techniques are widely used for GCA Radars in order that the controllers shall have the clearest possible picture of aerial activity in the vicinity of the airfield. The shape of the beam is 'cosecant squared', designed to give good coverage up to a height of about 5000 feet.

With the ever-increasing density of air traffic it is essential that the information on all targets be frequently renewed and scanning rates of as much as 30 RPM are used. Apart from this high scanning rate, which reduces the number of pulses per paint, the other features of the specification make it easier to meet than that of surveillance radar. The targets are large, and co-operative, the range is not great and no extreme accuracy is required.

To get the required results a peak power of about 100 kW at a wavelength in the 10-cm band is normally used.

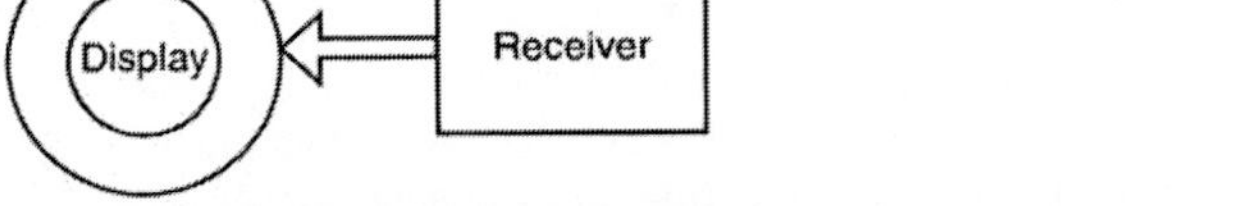

Fig. 9.1. Simple Block Diagram of SRE.

A simple block diagram of SRE is shown in Fig. 9.1. As usually transmitter, producing the RF pulses and are being radiated in the air by using antenna (normally parabolic reflector using cosecant squarred pattern) through the duplexer,which is acting as switch, closes the way to the receiver and the rf energy is connected to the antenna. During the silence period the antenna receives the reflected energy from the target and it is connected to the receiver through the Analog and the Digital signal processing un:'s. The processed videos are finally displayed on the PPI for the radar controller to guide the aircraft efficiently.

9.4. PAR (PRECISION APPROACH RADAR)

To obtain the more accurate data for taking the aircraft down on to the runway a second radar is used known as PAR, which scans a limited angle in azimuth about the required direction of approach, and a limited angle in elevation about the recommended glide path. It is common practice to use one radar which is continuously switched between the azimuth and elevation aerial systems, the output being switched in synchronism between the bearing and elevation displays as shown in diagrammatic form in Fig. 9.2.

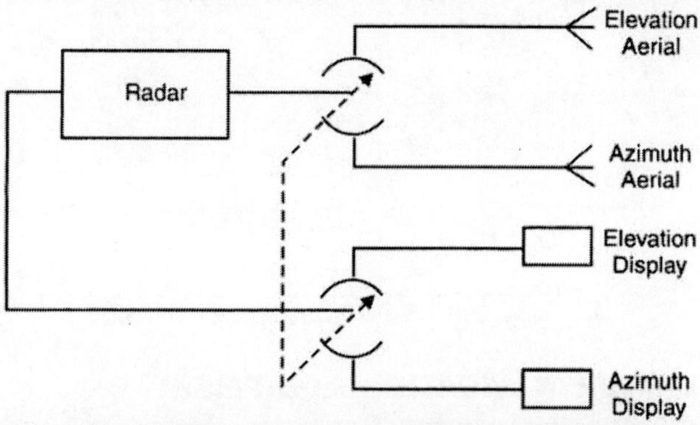

Fig. 9.2. Diagrammatic Representation of Switching used to Enable One Radar to Provide Both Azimuth and Elevation Information.

The azimuth beam has a width of something less than 1° and commonly scans ±10° about the required direction of approach. The elevation beam has a width (in the vertical plane) of the same order as the azimuth beam and normally scans ±3.5° about the required glide path. The displays used are both of the expanded, partial, PPI type with the required direction and angle of approach, and range markers, superimposed either electronically or by means of a transparent mask suitably engraved. The range scales on the two displays should be identical to facilitate the work of the final approach controller.

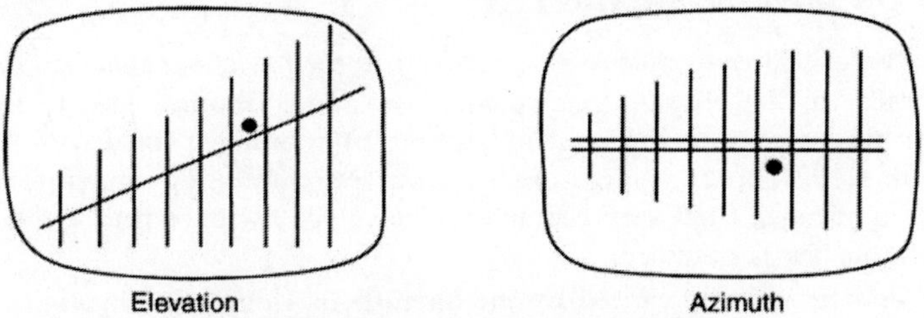

Fig. 9.3. The Display used in PAR.

Fig. 9.3 shows the appearance of the two displays with an aircraft approaching rather high and to the left of the correct approach.

The Precision Approach Radar information's are shown in a Beta Scan Scope in **Fig. 9.4** with two "pictures". The upper picture is a side view of the landing approach up to the touch down point on the runway with an imaginary line as a reference. This upper picture shows the data determined by the antenna to the height scanning. It is also called **Elevation Picture** therefore.

The top view of the imaginary prolongation of the centerline of the runway, the touch down and the range marks, is corresponding in the lower picture. This lower picture shows the data recorded by the antenna to the horizontal sector scanning. It also is named **Azimuth- Picture** therefore.

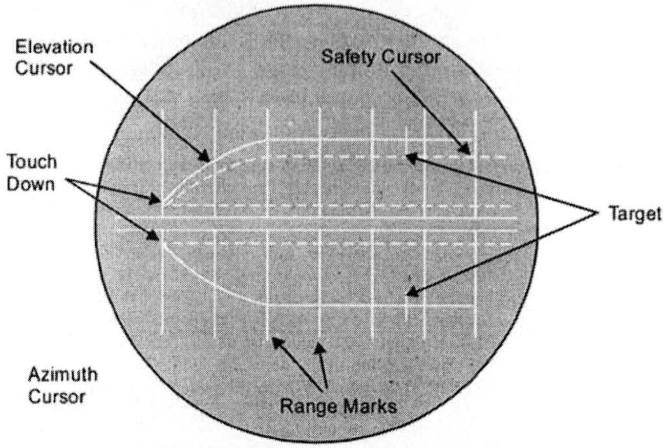

Fig. 9.4. **Beta Scan Scope for PAR.**

9.5. DOPPLER NAVIGATION EQUIPMENT

Doppler navigation is based on a radar wave transmissions beamed toward the earth behind the aircraft. This radar does not sense range and bearing (direction) as ordinary search radar does. Instead it uses a CW (continuous wave) transmission to measure the ground speed and drift angle of the aircraft. The doppler navigation system operates anywhere. It is relatively unaffected by weather conditions, and is independent of ground based navigation aids. This permits an aircraft crew to compute an aircraft's track. The track is projected on the ground from any known position (usually the position of take-off) to any posiion desired. Therefore, long distance navigation is possible. For more detail on doppler radar, student may refer chapter 3 and 4 in Part-I of this book.

9.5.1. Doppler Navigation

As we have already discussed in the radar section about the Doppler effect and doppler radar in the chapter four of this book. The **Doppler effect** is employed to determine the velocity of the craft in a frame of coordinate fixed with respect to the aircraft in the Doppler Navigation system. By combining this information with information obtained from a gyrocompass. The velocity with respect to the conventional earth co-ordinates is obtained.

The doppler radar is carried by the aircraft for Doppler Navigation. The Doppler radar directs several beams of EM waves towards the earth. The energy is received back by the aircraft after scattered from the earth. The signal received by the aircraft is compared with the transmitted signal to determine the Doppler shift of frequency, which is proportional to the velocity component of the aircraft in the direction of beam.

The Fig. 9.5 illustrated the principle of Doppler Radar. Let us assume that if the velocity component of the aircraft in the direction of the beam is V_r, than for transmitted frequency f_t, the frequency of the wave is measured at the ground is will be $f_t\left(1+\frac{Vr}{c}\right)$.

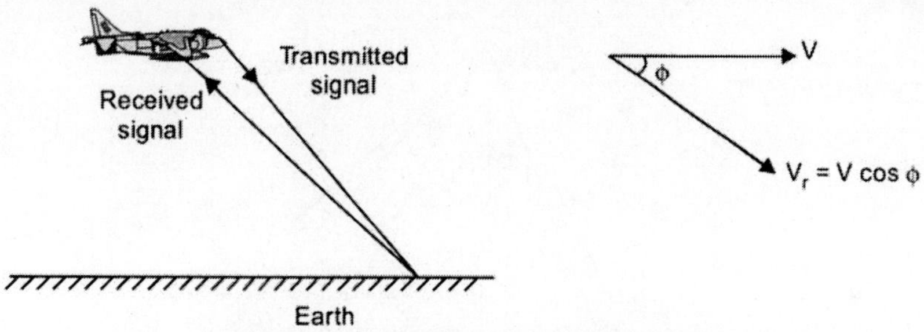

Fig. 9.5. Principle of Doppler Radar.

where c is the velocity of EM wave. The reflected wave have this frequency but when they are received by the moving radar in the aircraft, the received frequency is increased in the ratio $1+\frac{Vr}{c}$. The received frequency, f_r is therefore given as

$$f_r = f_t\left(1+\frac{Vr}{c}\right).\left(1+\frac{Vr}{c}\right)=f_t\left(1+\frac{Vr}{c}\right)^2$$

or

$$f_r = f_t\left(1+\frac{2Vr}{c}+\frac{V_r^2}{c^2}\right) \quad ...(9.1)$$

Practically, $V_2 << c$ so that $\frac{V_r^2}{c^2}$ is almost negligible, so the Doppler shift f_D is given by

$$f_D = f_r - f_t \cong f_t\frac{2V_r}{c}=\frac{2V_r}{\lambda} \quad ...(9.2)$$

where λ is the wavelength of the transmitted signal. The received signal at f_r is mixed with a part of the transmitted signal at f_t to produce a beat frquency signal,

$$f_D = f_r - f_t$$

Now if the beam is directed backwards then the aircraft, radar is moving away from a fixed point on the ground, and V_r is negative, than the Doppler shift is also negative. Positive sign indicates movement of radar in same direction of beam.

9.5.2. Beam Configuration

The Fig. 9.6 shows that the aircraft is flying over the earth and transmitting EM waves in a narrow beam making an angle ϕ with the horizontal. Now if the aircraft is in level flight and the beam is directed in the vertical plane, having the forward velocity V of the aircraft, the velocity component in the direction of beam is $V\cos\phi$ and the Doppler shift, f_D is $\frac{2V\cos\phi}{\lambda}$. In this case only one beam is considered to obtain V and ϕ angle is known. But when the aircraft has a velocity component perpendicular to the longitudinal axis of the aircraft, than the component of the velocity along the beam is zero (cos 90° = 0) so it would not be possible to obtained the doppler shift. Therefore we need more than one beam to obtain the velocity in the general case.

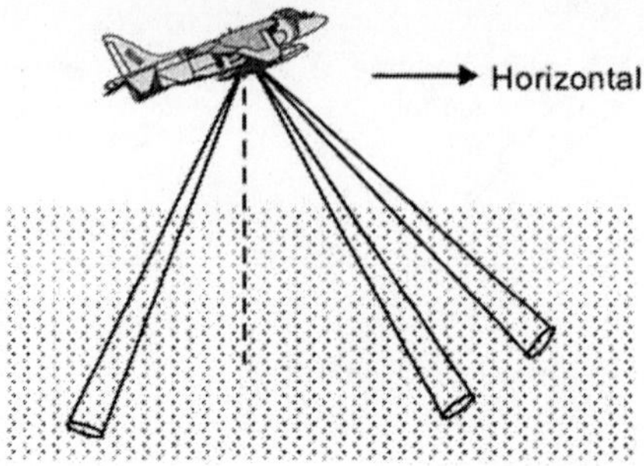

Fig. 9.6. Antenna System Fixed to Aircraft.

Figure 9.6 shows that a number radar antenna beams directed simultaneously in different directions towards the ground, allows the determination of the velocity component of the aircraft from the Doppler shift of frequency measured at the aircraft. Figure 9.7 (*a*), (*b*), (*c*) and (*d*) shows the plan view of the radar beams for the various multi beam configurations.

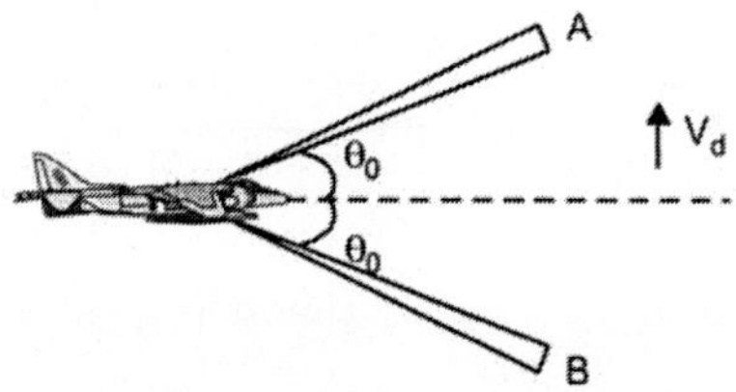

(*a*) Two-beam non-Janus system.

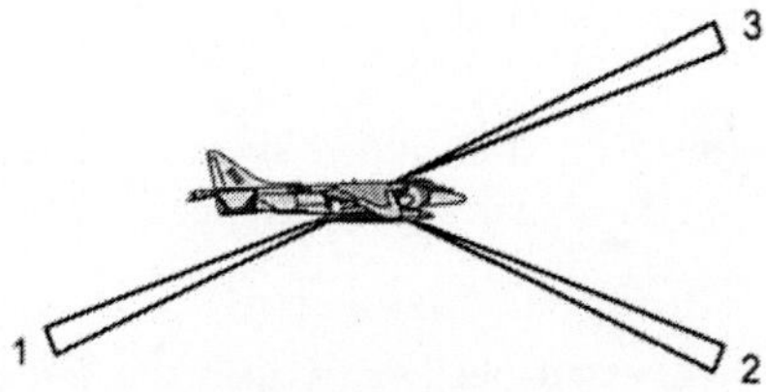

(*b*) Three-beam Janus-λ.

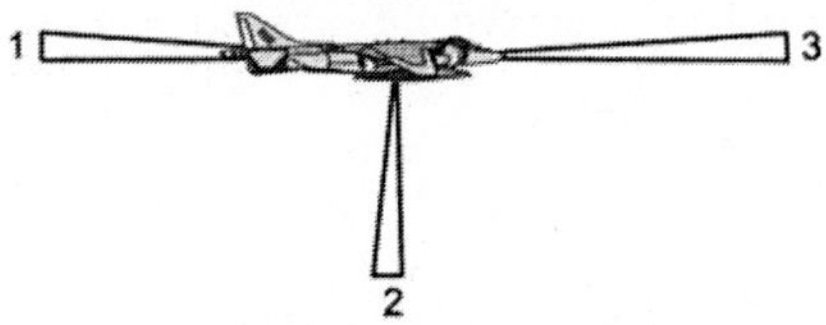

(*c*) Three-beam Janus-T.

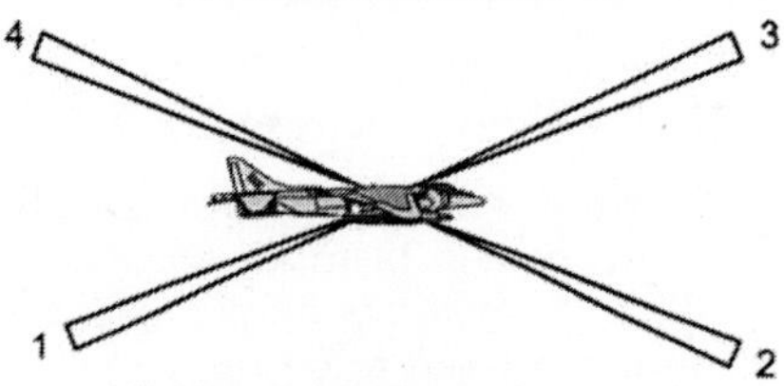

(*d*) Four beam Janus (Janus X)

Fig. 9.7.

9.5.3. Errors in Systems

The two beam system is subjected to errors in the presence of pitching, which causes changes in the altitude or orientation of a beam with the horizontal plane, as shown in Fig. 9.7(*a*)

Also a lateral drift velocity of the aircraft as shown in the Fig. 9.7, causes the Doppler shift of '*A*' to increase while that of '*B*' decrease by the same amount in proportion to the drift velocity. The heading velocity of the aircraft may be obtained from the average of the doppler shifts, whereas the lateral drift velocity can be obtained from the difference of doppler shifts. It is to be noted here that a vertical velocity component changes both the doppler frequencies, which may wrongly appear to be as due to a forward or backward velocity depending on the sign (either +ve or –ve) of the doppler shift, unless the information of vertical velocity is also obtained by other navigational instrument such as the radio altimeter.

The Janus system is named after a Greek God who could look both ways, minimises the errors due to charge in altitude of the aircraft arising from pitch and roll. In the Janus system, the sign of the doppler shift of the front beam is, if fact opposite that of the rear beam.

In the case of level flight the both shifts are equal and their magnitudes are added during the computations of the velocity components.

In the case of a pitch angle making the front of the aircraft lower than its rear, the front beam depression angle increases, while the rear beam depression angle decreases. This results to decrease the magnitude of the Doppler shift of the front beam and increase that of the rear beam, tending to keep the sum of the magnitudes unaffected by the pitch angle. Besides this with a stabilized antenna, the Doppler shift due to vertical velocity cancels out in the Janus system. For the vertical components of the shifts in the forward and backward beam are equal cancel out if the difference is calculated.

9.5.4. Equation for Doppler Shift

The doppler shift is obtained with any beam by summing the components due to each of three orthogonal velocity components of the aircraft, resolved along the direction of beam. To have simplication, let us assume that the antenna is stablized for altitude i.e. pitch and role. In the case of fixed antenna, the similar equations may be obtained but the velocities obtained will have to be converted to those in earth co-ordinates by taking consideration of attitude stablization. The three components of aircraft velocity are the vertical components V_v and two horizontal components V_H and V_D, V_H being in the vertical plane containing the longitudinal axis of the aircraft and V_D being perpendicular to this. Figure 9.8(*a*) shows the velocity components for a single beam and Fig. 9.8(*b*) shows that for the four beam of a Janus-χ system.

From the Fig. 9.8(*b*), shows that for the four beams of a Janus-χ system and we can have the velocity components along beams 1, 2, 3 and 4 are given as

$$v_1 = -V_H \cos\gamma_H + V_D \cos\gamma_D + V_v \cos\gamma_v \qquad ...(9.3)$$

$$v_2 = V_H \cos\gamma_H + V_D \cos\gamma_D + V_v \cos\gamma_v \qquad ...(9.4)$$

$$v_3 = V_H \cos\gamma_H - V_D \cos\gamma_D + V_v \cos\gamma_v \qquad ...(9.5)$$

$$v_4 = -V_H \cos\gamma_H - V_D \cos\gamma_D + V_v \cos\gamma_v \qquad ...(9.6)$$

where γ_H, γ_D and γ_v the angles of the velocity components V_H, V_D and V_v respectively with the beams. The direction consines of the beams, for a perfect symmetry of the four beams, are given as

$$\cos\gamma_H = \cos\alpha_0 \cos\theta_0;\ \cos\gamma_D = \cos\alpha_0 \sin\theta_0;\ \cos\gamma_v = \sin\alpha_0,$$

where, α_0 = Depression angle of the beam (the angle between the beam and the plane of the antenna)

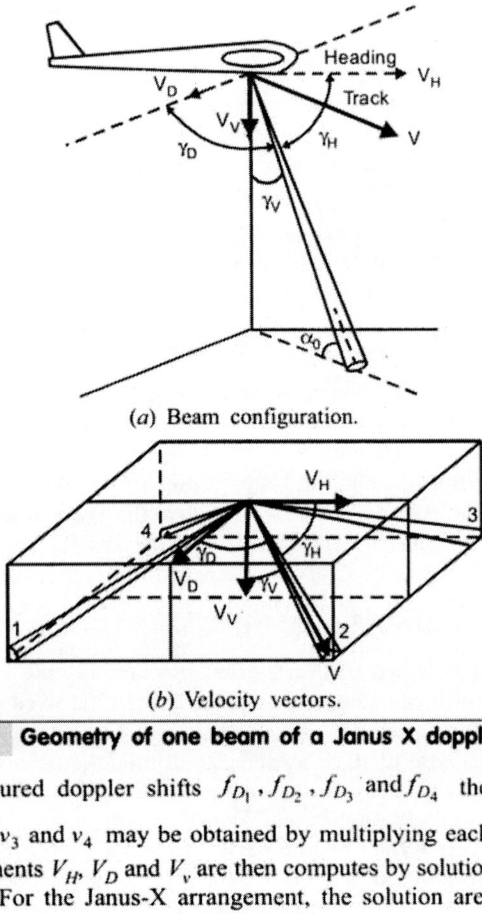

(*a*) Beam configuration.

(*b*) Velocity vectors.

Fig. 9.8. Geometry of one beam of a Janus X doppler radar.

From the measured doppler shifts $f_{D_1}, f_{D_2}, f_{D_3}$ and f_{D_4} the respective velocity components v_1, v_2, v_3 and v_4 may be obtained by multiplying each by the factor $\lambda/2$. The velocity components V_H, V_D and V_v are then computes by solution of the simultaneous eqn. (9.3) to (9.6). For the Janus-X arrangement, the solution are given by,

$$V_H = \frac{(v_2 + v_3) - (v_1 + v_4)}{4\cos\gamma_H} = \frac{\lambda}{8}\frac{(f_{D_2} + f_{D_3}) - (f_{D_1} - f_{D_3})}{\cos\alpha_0 \sin\theta} \quad ...(9.7)$$

$$V_D = \frac{(v_1 + v_2) - (v_3 + v_4)}{4\cos\gamma_H} = \frac{\lambda}{8}\frac{(f_{D_1} + f_{D_2}) - (f_{D_1} + f_{D_2})}{\cos\alpha_0 \sin\theta_0} \quad ...(9.8)$$

$$V_v = \frac{(v_1 + v_2 + v_3 + v_4)}{4\cos\gamma_v} = \frac{\lambda}{8}\frac{(f_{D_1} + f_{D_2} + f_{D_3} + f_{D_4})}{\sin\alpha_0} \quad ...(9.9)$$

It is to be mentioned that any three equations from Eqn. (9.4) to (9.6) are sufficient to solve for V_H, V_D and V_v. Whenever we use all the four equation are considered the solution is simple and given in Eqn. (9.7) to (9.9). Similar solutions may be obtained for the 3 beam Janus-λ antenna system, using only the first three equation from Eqn. (9.2) to (9.4).

9.5.5. Effect of Beam Angle

An important parameter of a Doppler radar is the angle between the longitudinal axis of the aircraft axis of the antenna and the beam centroid i.e. the angle γ_H in the Janus systems. The Doppler shift for a given velocity will be larger for a smaller value of γ_H. However, a smaller value of γ_H makes the back scatter signal reflected from the water surface smaller. A compromise value of the beam angle α_H is between 65° and 85°. The value of the θ_0 affects the sensitivity to lateral drift velocity, with larger values of θ_0 giving greater sensitivity to drift.

9.5.6. Heading And Track Stabilization

An antenna system having its principal axis in the same vertical plane as the longitudinal axis *i.e.* heading of the aircraft is called **heading stabilization.** In this case horizontal components V_H and V_D are computed separately and the actual direction of motion or track of the aircraft can be computed. The angle between the heading and track is called **drift-angle (δ)** and it is given by

$$\tan\delta = \frac{V_D}{V_H}$$

or

$$\delta = arc\tan\frac{V_D}{V_H} \qquad ...(9.10)$$

Track speed is given by

$$V_g = (V_H^2 + V_D^2)^{1/2} = V_H \sec\delta \qquad ...(9.11)$$

If the antenna system is rotated about a vertical axis by a servo mechanism obtain by the drift component V_D, the rotation will continue until the drift components becomes nil. The axis of the antenna is then oriented along the direction of the track and antenna is referred as **track stabilized** or drift-angle stabilized antenna. Fig. 9.9 (*a*) and (*b*) shows the orientation of the beam cluster, when there is a drift, for the two types of antennas.

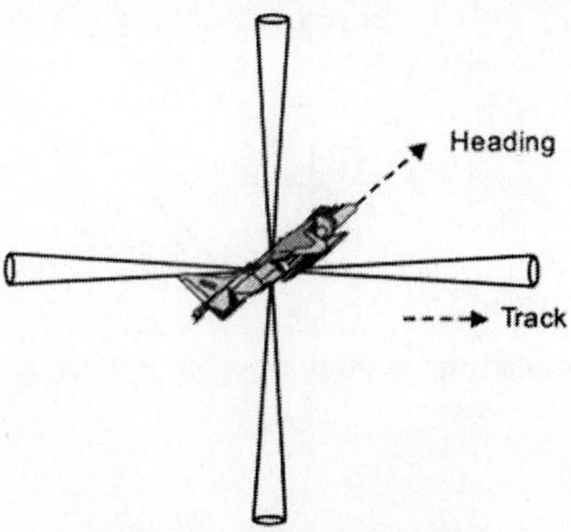

(*a*) Antenna heading-stablized.

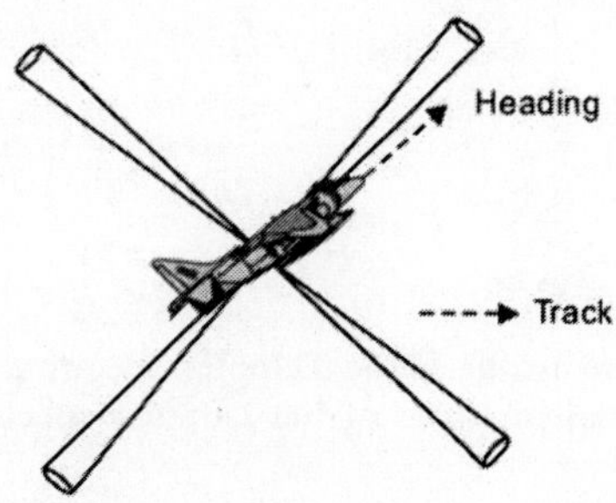

(*b*) Antenna track-stabilized.

Fig. 9.9. Beam Configuration.

9.5.7. Doppler Spectrum

The antenna system, in practice, have a finite beamwidth (generally 4°) correspond to a number of values of depression angle and causes. The doppler shift also be different. The various scattering elements in the area illuminated by the beam produce doppler shifted returns in random phase relationship to each other. The resultant shift has the nature of a narrow band frequency spectrum. The spectrum of the resulting return signal therefore appear like that due to a narrow band noise superimposed on the Gaussian noise background which is always present. Figure 9.10 shows the type of spectrum obtained with Doppler radar. The shape of the main peak of spectrum is nearly Gaussian in nature. The amplitude of the peak is a function of radar parameter and the back scattering property of the terrain. The mean frequency of the spectral distribution is fd and the angle corresponding to the Doppler shift f_d defines the beam centroid.

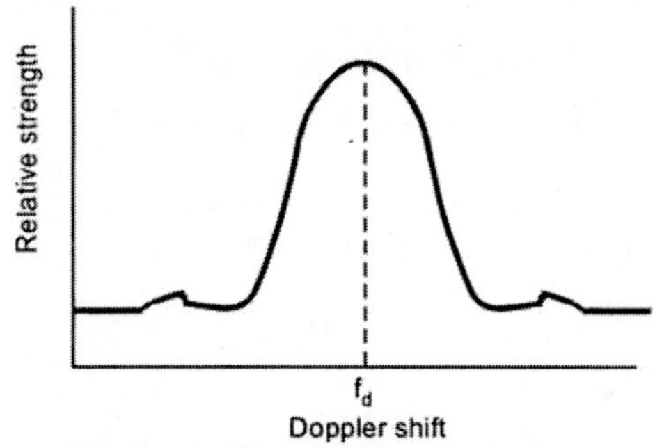

Fig. 9.10. The Doppler Spectrum.

If it is assume that the scattering property of the terrain back along the beam to be uniform over the area illuminated by beam, and that the returns from the nearest and farthest illuminated points are nearly the same, than the half power of the Doppler spectrum is may be given as

$$f_d = \frac{2V\cos\gamma}{\lambda} \qquad ...(9.12)$$

If the HPBW (Half power bandwidth) is $\Delta\gamma$ differences between the upper and lower half power points of the Doppler spectrum, f_{D_1} and f_{D_2} for respectively is given by

$$\Delta f_D = f_{D_1} - f_{D_2}$$

$$= \frac{2V}{\lambda}\left[\cos\left(\gamma \frac{\Delta\gamma}{2}\right) - \cos\left(\gamma + \frac{\Delta\gamma}{2}\right)\right]$$

$$= \frac{2V}{\lambda}\sin\Delta\gamma\,.\,\Delta\gamma$$

The width of the Doppler spectrum is proportional the central Doppler frequency. The relative width of the Doppler spectrum is given as :

$$\frac{\Delta f_D}{f_D} = \frac{\frac{2V}{\lambda}\sin\gamma\,.\,\Delta\gamma}{\frac{2V\cos\gamma}{\lambda}}$$

or $$\frac{\Delta f_d}{f_d} = \frac{\sin\gamma\,.\,\Delta\gamma}{\cos\gamma} = \tan\gamma\,.\,\Delta\gamma$$

The relative width $\frac{\Delta f_d}{f_d}$, typically lies between 15% to 25%.

In case of water, the back scatter cross-section cannot be assumed to be uniform over the area illuminated by the beam, but will increase the increasing values of γ, tending to distort the peak of the Doppler spectrum of the lower side of the peak. There are some more sources of error in Doppler navigation over water surface include, the errors due to water currents, winds blowing over the surfaces and wind down particle, all these leading to change of the Doppler spectrum. However, by knowing the surface wind velocity automatic correction of the error involved may be incorporated.

9.5.8. Frequency Tracker

Frequency Tracking is process that make use of doppler shift information. Coherent radar can also used to track the doppler frequency shift of the moving target echo. Frequency tracking is especially useful for airborne pulse doppler radar and missile guidance radars. The tracking doppler filter not only provides a measure of relative velocity, but it also improves the ability of the radar to isolate the moving target echo from the much larger clutter echoes. The frequency tracker may also be used to locates the centre of the noise like doppler spectrum and provide as output of the pure signal of this frequency. There are many configurations of the frequency tracker, we will discuss the mostly used configuration, that is using a tracking oscillator. The frequency of the tracking oscillator is compared with the spectrum in a discriminator type devices and error signal is generated whenever the oscillator deviates from the center of the spectrum, is fed back to correct the oscillator frequency. In Figure 9.11 shown a two-filter frequency tracker. The main components of this type of configuration are as follows :

(*i*) Mixer
(*ii*) Filter and envelope detector
(*iii*) Square wave generator
(*iv*) Phase detector
(*v*) Tracking oscillator
(*vi*) Integrator

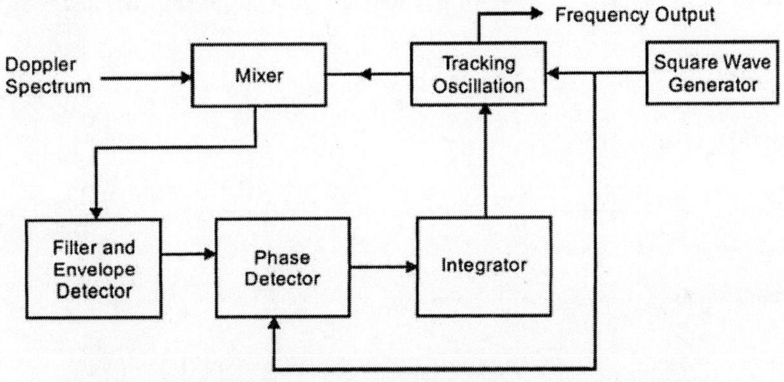

Fig. 9.11. Block Diagram of a Two Filter Frequency Tracker.

In this arrangement the frequency of tracking oscillator is switched by a square wave, between the two values which are separated by the spectrum width.

The output of tracking oscillator and incoming doppler spectrum is mixed in the mixer and passed through a low pass filter and envelope detector. The square wave output of filter and the output of square wave generator are fed to the phase detector. The phase of the first input with relative to the second input determines the sign of the phase detector output, which applied as the error signal. This error signal is integrated and fed to the voltage control oscillator (VCO) of the tracking oscillator, to change its mean frequency. When two frequencies of the oscillator straddle the center frequency of the Doppler spectrum, the zero-error condition is obtained.

9.5.9. The Doppler Radar Range Equation

The magnitude of the energy received from the ground by the doppler radar is an important parameter as it compute the maximum range at which the radar operates.

The range of a doppler radar depends on the scattering area of the ground illuminated by a radar beam. As the beams are inclined to the vertical by an angle γ, the illuminated area, is given by

$$S = \frac{r^2\omega}{\cos\gamma}$$

where $r^2\omega$ is the area normal to the beam control for a range 'r' and angular beamwidth 'ω'. As in the radar range equations, the effective scattering area, Se will be given by

$$Se = \sigma_0 s = \frac{\sigma_0 r^2\omega}{\cos\gamma}$$

where σ_0 = Scattering cross-section per unit area illuminated.

The power received is therefore may be given by

$$Pr = \frac{P_T G_r}{(4\pi r^2)^2} \cdot \frac{\sigma_0 \omega_r^2}{\cos\gamma} \cdot A_e$$

We know, $$G = \frac{4\pi A_e}{\lambda^2}$$ and antenna gain 'G' is also defined by

$$G = \frac{4\pi}{\omega}$$

By both the relation of gains, we get

$$\frac{4\pi A_e}{\lambda^2} = \frac{4\pi}{\omega} \quad \text{or} \quad A_e = \frac{\lambda^2}{\omega}$$

Replacing A_e in eqn. by $A_e = \frac{\lambda^2}{\omega}$, we get

$$P_r = \frac{P_t G_T \lambda^2 \sigma_0}{16\pi^2 r^2 \cos\gamma}$$

The SNR at the IF is obtained by dividing the P_r by the inherent available noise power $N = KT_0BF_n$, we get

$$\frac{S}{N} = \frac{P_t G_T \lambda^2 \sigma_0}{16\pi^2 r^2 KT_0 BF_n \cos\gamma} \qquad ...(9.13)$$

where, P_t = Transmitted peak power

G_T = Antenna gain of transmitter

λ = Wavelength

σ_0 = Scattering cross-section

T_0 = Standard room temperature 290°

K = Boltzman constant

B = Effective : IF bandwidth

F_n = Noise figure of the receiver

It is to be noted that the range equation (9.13), indicates an inverse square law dependance of (S/N) on range instead of the inverse fourth power law expected for a pulse radar detecting a target much smaller from the beamwidth.

The typical parameters of a doppler navigation equipment are shown as

Tx Power (peak) = 1100 W

Average power = 50 W

Operating frequency = 8.7 to 8.9 GHz

PRF = 50 KHz

PW = 0.9 μ sec

Altitude range = 20 km

9.6. INSTRUMENT LANDING SYSTEM (ILS)

9.6.1. General Description

Instrument Landing System, ILS is an air navigation system for landing. Instrument landing system (ILS) facilities are a highly accurate and dependable means of navigating to the runway in IFR conditions. When using the ILS, the pilot determines aircraft position primarily by reference to instruments.

ILS is classified by category in accordance with the capabilities of the ground equipment. *Category* I ILS provides guidance information down to a decision height (DH) of not less than 200 ft. Improved equipment (airborne and ground) provide for *Category* II ILS approaches. A DH of not less than 100 ft. on the radar altimeter is authorized for Category II ILS approaches. The ILS consists of following components :

1. Localiser. Informs the pilot with the airplane's horizontal position relative to runway centreline. Localiser broadcasts from beyond the far end of the runway with a horizontal array of antenna. It operates on frequences 108.0 to 111.95 MHz with the channel width of 50kHz.

2. Glideslope. Informs the pilot with the airplane's vertical positition relative to the ideal approach. Glideslope is broadcasted from the side of the runway. The operating frequences reach from 329.15 to 335.0 MHz with the channel width being 150 kHz.

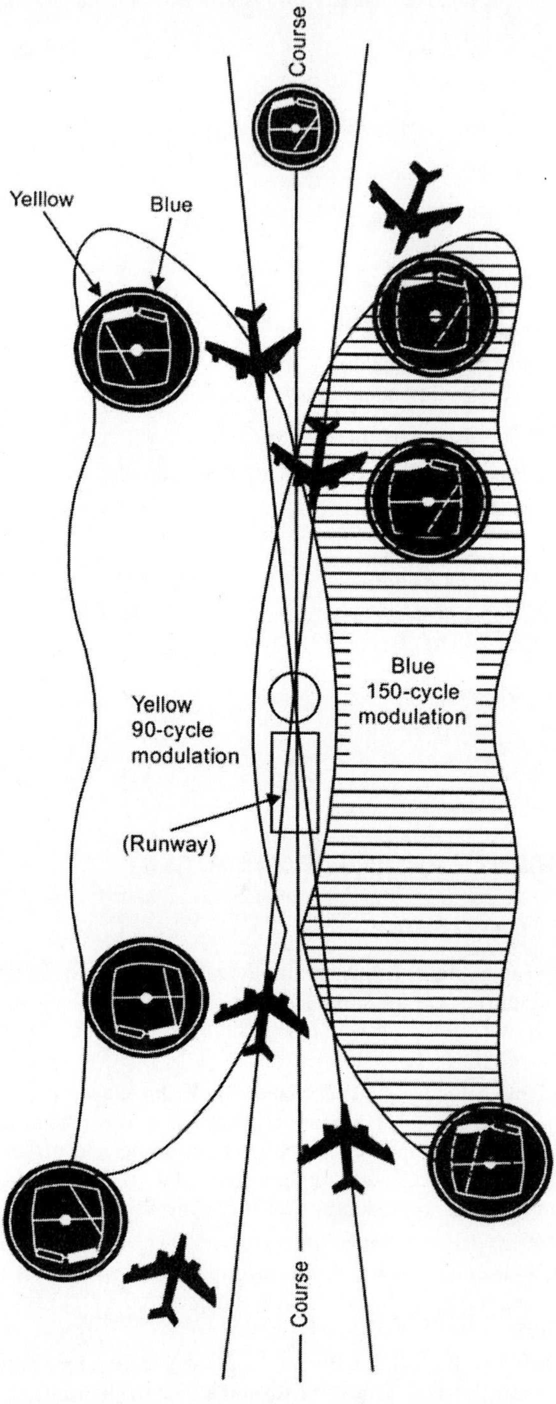

Fig. 9.12. Localizer Field Pattern and Indications.

3. Outer Marker. Situated on the same line with the localiser and the runway centerline, four to seven miles before the runway. Transmit low powered 400 Hz tone signal on the 75 MHz frequency. When the airplane approaches the runway low enough from the right direction, marker beacon receiver hears the signal. This is shown for the pilot by blinking the outer marker light of the receiver.

4. Middle Marker. Functions as the outer marker, positioned 0.5 to 0.8 miles before the runway. When the airplane is above the middle marker, the receiver's middle marker light starts blinking. In this situation the pilot has the last chance to decide if the landing will be made or not.

5. Inner marker

6. Approach lights. Most installations include medium- or high intensity approach light systems.

The ILS provides the lateral and vertical guidance necessary to fly a precision approach, where glide slope information is provided. A precision approach is an approved descent procedure using a navigation facility aligned with a runway where glide slope information is given. When all components of the ILS system are available, including the approved approach procedure, the pilot may execute a precision approach.

Localizer

1. Ground Equipment. The primary component of the ILS is the localizer, which provides lateral guidance. The localizer is a VHF radio transmitter and antenna system using the same general range as VOR transmitters (between 108.10 MHz and 111.95 MHz). Localizer frequencies, however, are only on odd-tenths, with 50 kHz spacing between each frequency. The transmitter and antenna are on the centerline at the opposite end of the runway from the approach threshold.A horizontal section of the three dimensional localizer field pattern is shown **in Fig. 9.12.**

The localizer *back course* is used on some, but not all ILS systems. Where the back course is approved for landing purposes, it is generally provided with a 75 MHz back marker facility or NDB located 3 to 5 NM from touchdown. The course is checked periodically to ensure that it is positioned within specified tolerances.

2. Signal Transmission. The signal transmitted by the localizer consists of two vertical fan-shaped patterns that overlap, at the center (see ILS Localizer Signal Pattern below in Fig. 9.13(*a*)). They are aligned with the extended centreline of the runway. The right side of this pattern, as seen by an approaching aircraft, is modulated at 150 Hz and is called the "blue" area. The left side of the pattern is modulated at 90 Hz and is called the "yellow" area. The overlap between the two areas provides the on-track signal.

The width of the navigational beam may be varied from approximately 3° to 6°, with 5° being normal. It is adjusted to provide a track signal approximately 700 ft wide at the runway threshold. The width of the beam increases so that at 10 NM from the transmitter, the beam is approximately one mile wide.

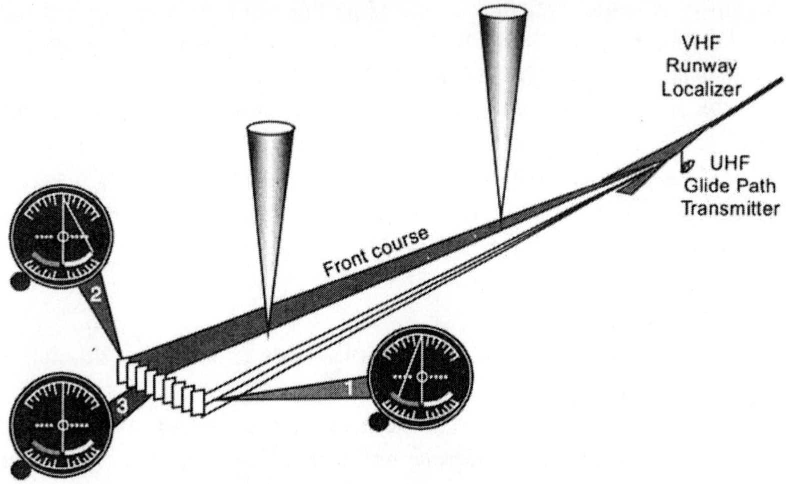

Fig. 9.13(a). ILS Localizer Signal Pattern.

The *localizer is identified* by an audio signal superimposed on the navigational signal. The audio signal is a two-letter identification preceded by the letter "I", e.g., "I-OW".

The *reception range* of the localizer is at least 18 NM within 10° degrees of the on-track signal. In the area from 10° to 35° of the on-track signal, the reception range is at least 10 NM. This is because the primary strength of the signal is aligned with the runway centerline.

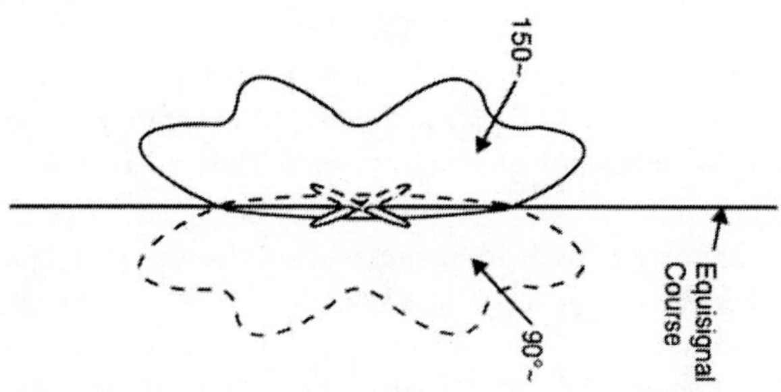

Fig. 9.13(b). Directional Pattern used in Runway Localizer of ILS.

3. Localizer Receiver. The localizer signal is received in the aircraft by a localizer receiver. The localizer receiver is combined with the VOR receiver in a single unit. The two receivers share some electronic circuits and also the same frequency selector, volume control, and ON-OFF control.

The localizer signal activates the vertical needle called the *track bar* (TB). Assuming a final approach track aligned north and south (see **ILS Localizer Signal Pattern** Fig. 9.13(*b*) above), an aircraft east of the extended centerline of the runway (*position 1*) is in the area modulated at 150 Hz. The TB is deflected to the left. Conversely, if the aircraft is in the area west of the runway centerline, the 90 Hz signal causes the TB to deflect to the right (*position 2*). In the overlap area, both signals apply a force to the needle, causing a partial deflection in the direction of the strongest signal. Thus, if an aircraft is approximately on the approach track but slightly to the right, the TB is deflected slightly to the left. This indicates that a correction to the left is necessary to place the aircraft in precise alignment.

At the point where the 90 Hz and 150 Hz signals are of equal intensity, the TB is centered, indicating that the aircraft is located precisely on the approach track *(position* 3).

When the TB is used in conjunction with the VOR, full-scale needle deflection occurs 10° either side of the track shown on the track selector. When this same needle is used as an ILS localizer indicator, full-scale needle deflection occurs at approximately 2.5° from the center of the localizer beam.

Thus the sensitivity of the TB is approximately *four times greater* when used as a localizer indicator as opposed to VOR navigation.

In the localizer function, the TB does not depend on a correct track selector setting in most cases; however, the pilot should set the track selector for the approach track as a reminder of the final approach.

When an OFF flag appears in front of the vertical needle, it indicates that the signal is too weak, and, therefore, the needle indications arc unreliable. A momentary OFF flag, or brief TB needle deflections, or both, may occur when obstructions or other aircraft pass between the transmitting antenna and the receiving aircraft.

9.7. GLIDE SLOPE EQUIPMENT

1. Transmitter. The glide slope provides vertical guidance to the pilot during the approach. The ILS glide slope is produced by a ground-based UHF radio transmitter and antenna system, operating at a range of 329.30 MHz to 335.00 MHz, with a 50 kHz spacing between each channel. The transmitter is located 750 to 1,250 feet (ft) down the runway from the threshold, offset 400 to 600 ft from the runway centerline. Monitored to a tolerance of ± 1/2 degree, the UHF glide path is "paired" with (and usually automatically tuned by selecting) a corresponding VHF localizer frequency.

Like the localizer, the glide slope signal consists of two overlapping beams modulated at 90 Hz and 150 Hz (see **Glide Slope Signal Pattern** Fig. 9.14 below). Unlike the localizer, however, these signals are aligned above each other and are radiated primarily along the approach track. The thickness of the overlap area is 1.4° or .7° above and .7° below the optimum glide slope.

This glide slope signal may be adjusted between 2° and 4.5° above a horizontal plane. A typical adjustment is 2.5° to 3°, depending upon such factors as obstructions along the approach path and the runway slope.

False signals may be generated along the glide slope in multiples of the glide path angle, the first being approximately 6° degrees above horizontal. This false signal will be a reciprocal signal (*i.e.* the fly up and fly down commands will be reversed). The false signal at 9° will be oriented in the same manner as the true glide slope. There are no false signals below the actual slope. An aircraft flying according to the published approach procedure on a front course ILS should not encounter these false signals.

2. Signal Receiver. The glide slope signal is received by a UHF receiver in the aircraft. In modern avionics installations, the controls for this radio are integrated with the VOR controls so that the proper glide slope frequency is tuned automatically when the localizer frequency is selected.

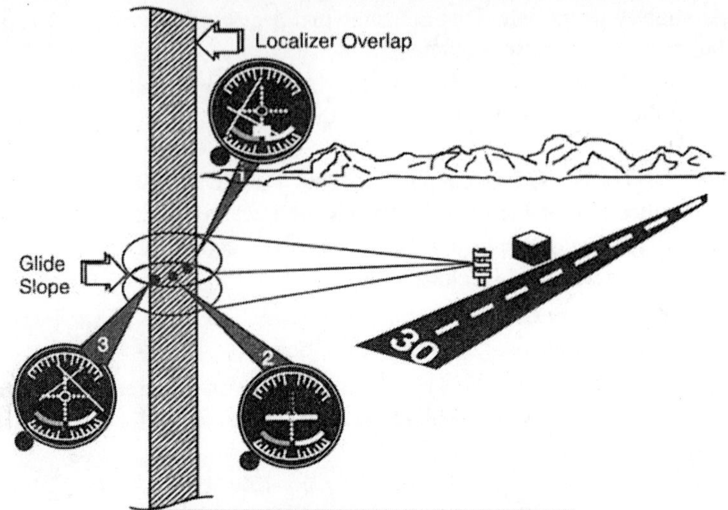

Fig. 9.14. Glide Slope Signal Pattern.

The glide slope signal activates the glide slope needle, located in conjunction with the TB (see **Glide Slope Signal Pattern** Fig. 9.14 above). There is a separate OFF flag in the navigation indicator for the glide slope needle. This flag appears when the glide slope signal is too weak. As happens with the localizer, the glide slope needle shows full deflection until the aircraft reaches the point of signal overlap. At this time, the needle shows a partial deflection in the direction of the strongest signal. When both signals are equal, the needle centers horizontally, indicating that the aircraft is precisely on the glide path.

The pilot may determine precise location with respect to the approach path by referring to a single instrument because the navigation indicator provides both vertical and lateral guidance. In the **Glide Slope Signal Pattern** Fig. 9.14 above, *position* 1, shows both needles centered, indicating that the aircraft is located in the center of the approach path. The indication at *position* 2 tells the pilot to fly down and left to correct the approach path. *Position* 3 shows the requirements to fly up and right to reach the proper path. With 1.4° of beam overlap, the area is approximately 1,500 ft thick at 10 nautical miles (NM), 150 ft at 1 NM, and less than one foot at touchdown.

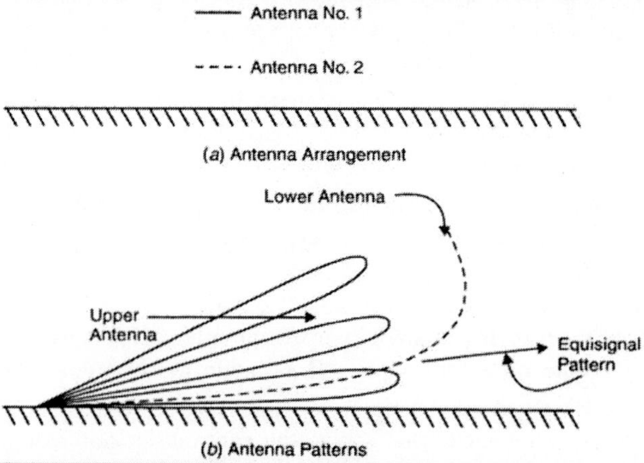

Fig. 9.15. Antenna Patterns Producing Equisignal Glide Path for ILS.

The apparent sensitivity of the instrument increases as the aircraft nears the runway. The pilot must monitor it carefully to keep the needle centered. As said before, a full deflection of the needle indicates that the aircraft is either high or low but there is no indication of how much it is high or low.

3. A simple working B/D of Instrumentation of ILS receiver. A simple working B/D of Instrumentation of ILS receiver is shown in Fig. 9.16. The modulated output signal received from the receiver (90 hz +150 hz), applied to the filters which separate both the signals 90 hz and 150 hz. A variable resistance is put at the input of received signals, which will compassionate for different losses in the rectifiers and filters. Each of the signals is being rectified by using bridge rectifier. The rectified voltages further connected to the circuits, which provides the differential voltage for the coil of indicator. The resistance in series with the thermistor, which is used to compassionate due to temperature in circuit, carries total rectified current. When the signal is low or due to equipment failure, the indicator coil is not energized than the voltage across resistances R_1 given for flag alarm. This voltage makes to mask the pointer.

The similar circuits are being used in the instrumentation part of the Localizer and glide slope.

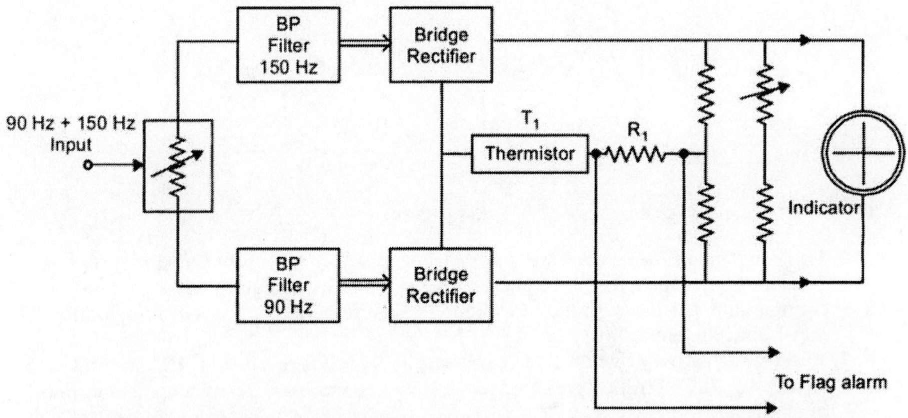

Fig. 9.16. A Simple Working Block Diagram of ILS Receiver Instrumentation.

9.8. ILS MARKER BEACONS

1. General. Instrument landing system marker beacons provide information on distance from the runway by identifying predetermined points along the approach track. These beacons are low-power transmitters; that operate at a frequency of 75 MHz with 3 W or less rated power output. They radiate an elliptical beam upward from the ground. At an altitude of 1,000 ft, the beam dimensions are 2,400 ft long and 4,200 ft wide. At higher altitudes, the dimensions increase significantly.

2. Outer Marker (OM). The outer marker (if installed) is located 3 1/2 to 6 NM from the threshold within 250 ft of the extended runway centerline. It intersects the glide slope vertically at approximately 1,400 ft above runway elevation. It also marks the approximate point at which aircraft normally intercept the glide slope, and designates the beginning of the final approach segment. The signal is modulated at 400 Hz, which is an audible low tone with continuous Morse code dashes at a rate of two dashes per second. A 75 MHz marker beacon receiver receives the signal in the aircraft. The pilot bears a tone over the speaker or headset and sees a blue light that flashes in synchronization with the aural tone (see the **Marker Beacon Lights** Fig. 9.17. Where geographic conditions prevent the positioning of an outer marker, a DME unit may be included as part of the ILS system to provide the pilot with the ability to make a positive position fix on the localizer. In most ILS installations, the OM is replaced by an NDB.

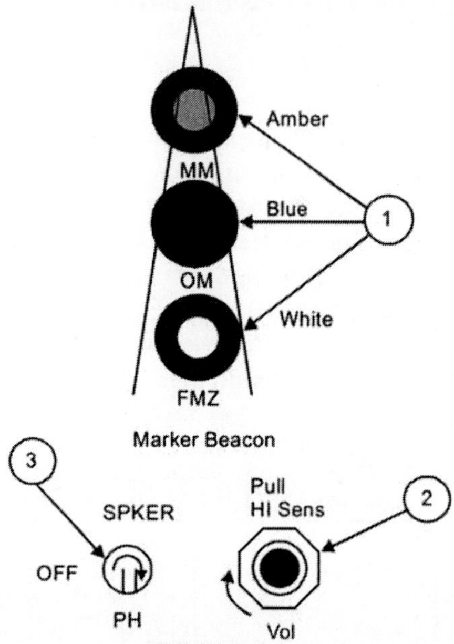

Fig. 9.17.

1. **Marker Beacon Indicator Lights :** Indicate passage of OM, MM and airway bacons. The OM light is blue, the MM light is amber and the airway light is white. Pictured configuration has Press-To-Test and dimming features. Other types and configuration may be encountered.
2. **Receiver sensitivity Switch :** Increases sensitivity of the receiver in "HI" position for airway use. "LO" position preferred for ILS work to more accurately pinpoint location of OM and MM.
3. **Speaker/Phone Switch :** Selects speaker or phones for reception of aural tones. In some aircraft centre position turns receiver off. May be located remotely from indicator lights.

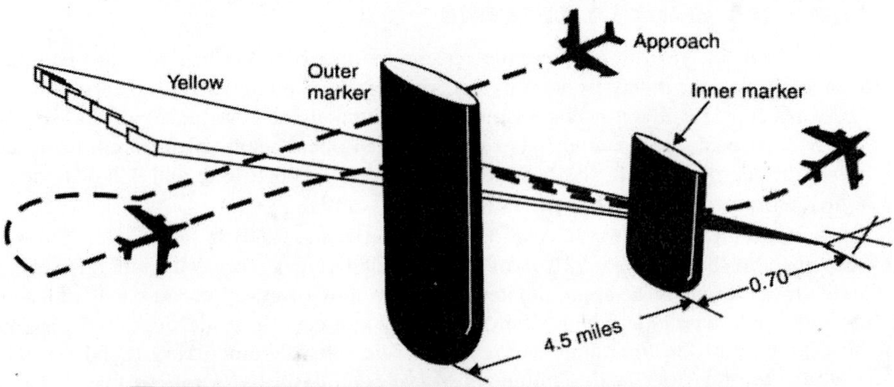

Fig. 9.18. Glide Path Showing Outer and Inner Markers.

3. Middle Marker (MM). Middle markers have been removed from all ILS facilities in Canada bur are still used in the United States. The middle marker is located approximately 0.5 to 0.8 NM from the threshold on the extended runway centerline. The middle marker crosses the glide slope at approximately 200 to 250 ft above the runway elevation and is near the missed approach point for the ILS Category I approach.

4. Back Marker (BM). The back course marker (BM), if installed, is normally located on the localizer back course approximately four to six miles from the runway threshold. The BM low-pitched tone (400 Hz) is heard as a series of dots. It illuminates the aircraft's white marker beacon light. An NDB or DME fix can also be used and in most locations replace the BM.

9.9. LIGHTING SYSTEMS

1. General. Various runway environment lighting systems serve as integral parts of the ILS system to aid the pilot in landing. Any or all of the following lighting systems may be provided at a given facility: approach light system (ALS), sequenced flashing light (SFL), touchdown zone lights (TDZ) and centerline lights (CLL-required for Category II [Cat II] operations.)

2. Runaway Visibility Measurement. In order to land, the pilot must be able to see appropriate visual aids not later than the arrival at the decision height (DH) or the missed approach point (MAP).

Until fairly recently, the weather observer simply "peered into the murk", trying to identify landmarks at known distances from the observation point. This method is rather inaccurate; therefore, instrumentation was developed to improve the observer's capability.

The instrument designed to provide visibility information is called a **transmissometer**. It is normally located adjacent to a runway. The light source is separated from the photoelectric cell receiver by 500 to 700 ft. The receiver, connected to the instrument readout in the airport tower, senses the reduction in the light level between it and the light source caused by increasing amounts of particulate matter in the air. In this way the receiver measures the relative transparency or opacity of the air. The readout is calibrated in feet of visibility and is called runway visual range (RVR).

3. Runaway Visual Range (RVR). The RVR is the maximum distance in the direction of take-off or landing at which the runway or the specified light or markers delineating it can be seen from a height corresponding to the average eye-level of pilots at touchdown.

Runway visual range readings usually are expressed in hundreds of feet. For example, "RVR 24" means that the visual range along the runway is 2,400 ft. In weather reports, RVR is reported in a code: R36/4000 FT/D; meaning RVR for Runway 36 is 4000 ft and decreasing. Because visibility may differ from one runway to another, the RVR value is always given for the runway where the equipment is located. At times, visibility may even vary at different points along the same runway due to a local condition such as a fog bank, smoke, or a line of precipitation. For this reason, additional equipment may be installed for the departure end and mid-point of a runway.

Runway visual range reports are intended to indicate how far the pilot can see along the runway in the touchdown zone; however, *the actual visibility at other points along the runway may differ* due to the siting of the *transmissometer*. The pilot should take this into, account when making decisions based on reported RVR.

Runway visual range is not reported unless the prevailing visibility is less than two miles or the RVR is 6,000 ft or less. This is so because the equipment cannot measure RVR above 6,000 ft. When it is reported, RVR can be used as an aid to pilots in determining what to expect during the final stages of an instrument approach. Instrument approach charts state the advisory values of visibility and RVR.

Runway visual range information is provided to the ATC, arrival control sector, the PAR position, and the control tower or FSS. It is passed routinely to the pilot when conditions warrant. RVR information may be included in aviation weather reports.

Ground visibility will continue to be reported and used in the application of take-off and landing minima. At runways with a transmissometer and digital readout equipment or other suitable means, RVR is used in lieu of prevailing visibility in determining the visibility minima unless affected by a local weather phenomenon of short duration.

The normal RVR reading is based on a runway light setting of strength 3. If the light settings are increased to strength 4 or 5, it causes a relative increase in the RVR reading. No decrease in the RVR reading is evident for light settings of less than setting 3. Pilots shall be advised when the runway light setting is adjusted to 4 or 5. If the RVR for a runway is measured at two locations, the controller identifies the touchdown location as "ALFA and the mid runway location as "Bravo".

In all cases, the pilot can request a light setting suitable for his or her requirements. When more than one aircraft is conducting an approach, the pilot of the second aircraft may request a change in the light setting after the first aircraft has completed its landing.

Because of the complex equipment requirements, RVR usually is only available at more active airports and not necessarily for all runways. If RVR equipment is not available or temporarily out of service for a given runway, the pilot uses the observer method to provide visibility information. In this case, the visibility is expressed as miles or fractions of a mile.

9.10. NDBS AT MARKER BEACON SITES

Additional aids may be available to assist the pilot in reaching the final approach fix. One of these aids is the NDB, which can be co-located with or replace the outer marker (OM) or back marker (BM). It is a low-frequency non-directional beacon with a transmitting power of less than 25 watts (W) and a frequency range of 200 kilohertz (kHz) to 415 kHz. The reception range of the radio beacon is at least 15 nautical miles (NM). In a number of cases an en route NDB is purposely located at the outer marker so that it may serve as a terminal as well as an en route facility.

Distance Measuring Equipment (DME) can be used for distance information instead of Marker beacons.

A standard ILS is termed "Category I", allowing landings for suitably equipped aircraft in weather as low as a half-mile visibilty and 200-ft ceiling (cloud base or vertical visibility). More advanced Category II and III systems allow operations in near-zero visibility. Category III is flown by an autoland system on board the landing aircraft. Category II/III installations include in-runway centreline and touchdown zone lighting, as well as other aids and enhancements.

9.11. LIMITATION OF ILS

A limitation of ILS is its sensitivity to obstructions in the signal broadcast area. In the 1970's there was a major US & European effort to establish Microwave Landing

Systems, which are not similarly limited, and which allow curved approaches. Procedures based on the extreme three-dimensional accuracy of the Global Positioning System are now being developed and implemented. It is widely expected that eventually these procedures, with related ground-based enhancements, will replace ILS.

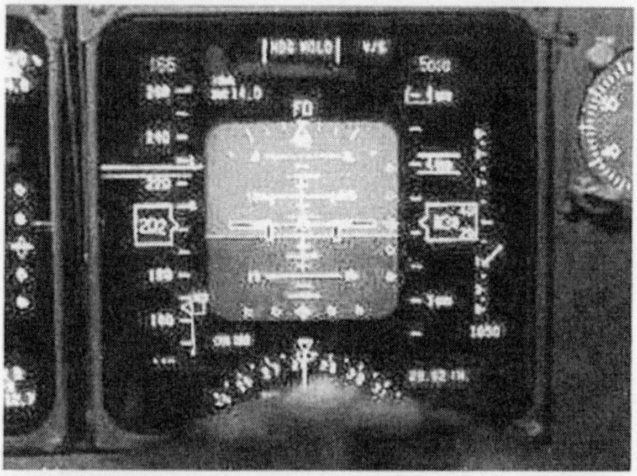

Fig. 9.19. A Cockpit with Computerized Instrument Landing Equipment.

An aircraft on an instrument landing approach has a cockpit with computerized instrument landing equipment that receives and interprets signals being from strategically placed stations on the ground near the runway.

9.12. MICROWAVE LANDING SYSTEM (MLS)

System Description. The time-referenced scanning beam Microwave Landing System (MLS) has been adopted by ICAO as the standard precision approach system to replace ILS. MLS provides precision navigation guidance for alignment and descent of aircraft on approach to a landing by providing azimuth, elevation and distance. The system may be divided into five functions :

1. Approach azimuth;
2. Back azimuth;
3. Approach elevation;
4. Range; and
5. Data communications.

With the exception of DME, all MLS signals are transmitted on a single frequency through time sharing. Two hundred channels are available between 5031 and 5090.6 Megahertz (MHz). By transmitting a narrow beam which sweeps across the coverage area at a fixed scan rate, both azimuth and elevation may be calculated by an airborne receiver which measures the time interval between sweeps. For the pilot, the MLS presentation will be similar to ILS with the use of a standard CDI or multi-function display,

The standard configuration of MLS ground system includes:

(*a*) An azimuth station to perform the functions of Approach azimuth and Data communications. In addition to providing azimuth navigation guidance, the station transmits basic data which consists of information associated directly with the operation of the landing system, as well as advisory data on the performance and state of the ground equipment. The standard version of MLS is capable to provide a maximum coverage region of +/- 40 degrees.

(*b*) An elevation station which performs the function of the Approach Elevation. The elevation signal can provide coverage up to +30 degrees above the ground level.

(*c*) DME to perform the range guidance, both standard DME/N and precision DME/P information.

The standard configuration can be upgraded to give enhanced capabilities by adding one or more of the following functions or characteristics :

1. Back Azimuth—provides lateral guidance for missed approach and departure navigation.
2. Auxiliary data transmission—provides additional data, including refined airborne positioning, meteorological data, runway status and other supplementary information.
3. Expanded Service Volume (ESV) proportional guidance to 62 degrees, that is, the azimuth coverage region is extended from +/- 40 degrees to +/- 62 degrees.

Similar to other Navigational Aids the MLS also use a Morse Code Identification which consist of a four-letter designation starting with the letter M. The code is transmitted at least six times a minute by the approach azimuth sector.

A fully MLS equipped runway has four signal transmitters. Two transmitters provide MLS Azimuth functions; they are located at each end of the runway, and are positioned facing the runway. With both runway directions equipped, the azimuth antenna facing the approaching aircraft is configured as the Approach Azimuth Transmitter (AZ) and the opposite antenna becomes the Back Azimuth Transmitter (BAZ). The AZ transmitter is used to guide the aircraft during an instrument (non-visual) approach to the active runway. It may be used in precision area navigation (RNAV) when coupled with the elevation and distance measuring equipment that are the remaining transmitted signals.

The Azimuth station transmits MLS angle and data on one of 200 channels within the frequency range of 5031 and 5091MHz. The equipment is normally located about 1,000 feet beyond the stop end of the runway, but there is considerable flexibility in selecting the site. The coverage in range is 20 Nmiles.

The AZ and BAZ signal beam shapes are both fan-shaped in a vertical plane formed along any of the antenna's radials.

As viewed by the pilot, the azimuth beam swept from the right-most coverage angle to the left-most in the "TO" scan and is then returned from the left to the right coverage angle in the "FRO" scan after a specified delay at the left-most limit. The time difference between the two scans is determined. Information derived from the data words transmitted to the aircraft by the MLS gives the Aircraft MLS Receiver specific information with regard to site geometry. This information is used along with beam timing to determine accurately the azimuth angle of an aircraft.

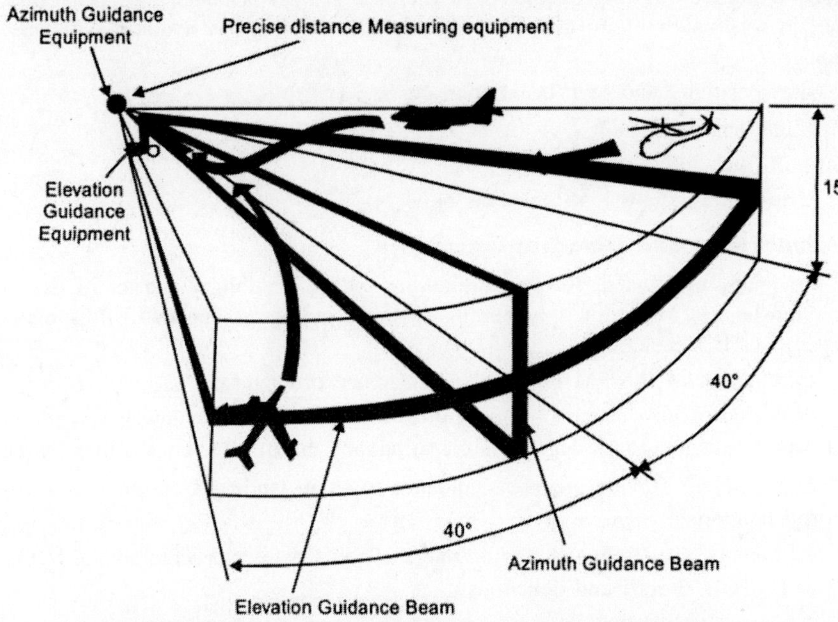

Fig. 9.20. Concept of Microwave Landing System (MLS).

The data transmissions can include both the basic and auxiliary data words. All MLS facilities transmit basic data. Where needed, auxiliary data can be transmitted.

The basic data consists of :

1. Station Identification
2. Exact location of azimuth, elevation and DME/P
3. Stations 9used for MLS receiver processing)
4. Ground equipment performance level, and
5. DME/P channel and status.

The auxiliary data consist of :

1. 3D locations of MLS equipment
2. Waypoint co-ordinates
3. Runway conditions, and
4. Weather information (e.g. RVR, ceiling, altimeter setting, wind, wake vortex, wind shear)

The MLS has the capability to fulfill a variety of needs in the approach, landing, missed approach and departure phases of the flight. For example:

(*a*) Curved and segmented approaches

(*b*) Selectable glide path angles, *i.e.* can be adaptable to type of aircraft, such as its speed and size.

(*c*) The establishment of boundaries to ensure clearance from obstructions in the terminal area. (such as high buildings, nearby mountains, etc.).

ILS Limitations

The Instrument Landing System (ILS) has served as the standard precision approach and landing aid for the last 40 years. During this time it has served well and has undergone a number of improvements to increase its performance and reliability. However, in relation to future aviation requirements, the ILS has a number of basic limitations :

1. site sensitivity and high installation costs;
2. single approach path;
3. multi path interference; and
4. channel limitations - 40 channels only.

MLS Advantages and Disadvantages

As previously mentioned, ILS has limitations which prohibit or restrict its use in many circumstances. MLS not only eliminates these problems; but also offers many advantages over ILS including:

1. Elimination of ILS/FM broadcast interference problems;
2. Provision of ail-weather coverage up to ±60 degrees from runway centerline, from 0.9 degree to 15 degrees in elevation, and out of 20 nautical miles (NM);
3. Capability to provide precision guidance to small landing areas such as roof-top heliports;
4. Continuous availability of a wide range of glide paths to accommodate STOL and VTOL aircraft and helicopters;
5. Accommodation of both segments and curved approaches;
6. Availability of 200 channels - five times more than ILS;
7. Potential reduction of Category I (CAT I) minimums;
8. Improved guidance quality with fewer flight path corrections required;
9. Provision of back-azimuth for missed approaches and departure guidance;
10. Elimination of service interruptions caused by snow accumulation; and
11. Lower site preparation, repair, and maintenance costs.

The main disadvantages of MLS are:

1. Its cost
2. It is not available worldwide
3. Only few airlines are equipped for MLS
4. It has limitations of channels for the DME/P component of the system

9.12.1. Accuracy and Coverage of MLS

Approach Azimuth Guidance

The approach azimuth antenna normally provides a lateral coverage of 40° either side of the center of scan . Coverage is reliable to a minimum of 20 NM from the runway threshold and to a height of 20,000 feet (ft). The antenna is normally located about 1000 feet beyond the end of the runway.

Back Azimuth Guidance

The back azimuth antenna provides lateral guidance for missed approach and departure navigation. The back azimuth transmitter is essentially the same as the approach azimuth transmitter. However, the equipment transmits at a somewhat lower data rate because the guidance accuracy requirements are not as stringent as for the landing approach. The equipment operates on the same frequency as the approach azimuth bur at a different time in the transmission sequence. On runways that have MLS approaches on both ends, the azimuth equipment can be switched in their operation from the approach azimuth to the back azimuth and vice versa.

Elevation Guidance

The elevation station transmits signals on the same frequency as the azimuth station. The elevation transmitter is normally located about 400 ft from the side of the runway between the threshold and the touchdown zone.

Range Guidance

Range guidance, consistent with the accuracy provided by the azimuth and elevation stations, is provided by the MLS precision DME (DME/P). DME/P has an accuracy of $\pm$100 ft as compared, with $\pm$ 1200 ft accuracy of the standard. DME system. In the future it may be necessary to deploy DME/P with modes which could be incompatible with standard airborne DME receivers.

While many of these capabilities are available to any MLS-equipped aircraft, the more sophisticated capabilities such as curved and segmented approaches are dependent upon the particular capabilities of the airborne equipment. Unfortunately due to the problems explained in our last contribution in the way the MLS was incepted, the airlines were quite rightly very sceptic to invest in expensive elaborated airborne MLS systems to harvest the best benefits. Thus airborne MLS systems are still not widely used mainly because the MLS failed to have a world-wide operational use.

SUMMARY

- Electronic aids to approach and Landing system are of an aircraft being used when the landing operations in difficult to carry out by visual observations of ground features and landing lights. The main purpose of these aids to guide the aircraft to approach and land safely specially in bad weather conditions.
- Electronics systems GCA, ILS and MLS are altogether responsible for the approaching and landing safely at the runway.
- GCA system consists of two microwave radar sets as SRE and PAR.
- The purpose of SRE (Surveillance Radar Element) to search and control of air traffic in the vicinity of the runway. Its range is approximately 70 NM.
- The purpose of PAR (Precision Approach Radar) to guide the aircraft for actual langing up to their touchdown point of the runway. This radar provides the horizontal displacement and vertical displacement in range and azimuth from the glide path of an aircraft.
- **SRE (Surveillance Radar Element)** which is usually designated the Airport Surveillance Radar (ASR),is the type of radar with a similar specification to that of surveillance radar is that used to control the approach of aircraft to an airfield.
- To obtain the more accurate data for talking the aircraft down on to the runway a second radar is used known as **PAR (Precision Approach Radar)**, which scans a limited angle in azimuth about the required direction of approach, and a limited angle in elevation about the recommended glide path.
- **Doppler navigation** is based on a radar wave transmission beamed toward the earth behind the aircraft. This radar does not sense range and bearing (direction)

as ordinary search radar does. Instead it uses a continuous wave (CW) transmission to measure the ground speed and drift angle of the aircraft

- **Instrument Landing System,** ILS is an air navigation system for landing. Instrument landing system (ILS) facilities are a highly accurate and dependable means of navigating to the runway in IFR conditions. When using the ILS, the pilot determines aircraft position primarily by reference to instruments.
- The ILS consists of following components :
  1. **Localiser.** Informs the pilot with the airplane's horizontal position relative to runway centreline. Localiser broadcasts from beyond the far end of the runway with a horizontal array of antenna. It operates on frequences 108.0 to 111.95 MHz with the channel width of 50kHz.
  2. **Glideslope.** Informs the pilot with the airplane's vertical positition relative to the ideal approach. Glideslope is broadcasted from the side of the runway. The operating frequences reach from 329.15 to 335.0 MHz with the channel width being 150 kHz.
  3. **Outer marker.** Situated on the same line with the localiser and the runway centerline, four to seven miles before the runway. Transmit low powered 400 Hz tone signal on the 75 MHz frequency. When the airplane approaches the runway low enough from the right direction, marker beacon receiver hears the signal. This is shown for the pilot by blinking the outer marker light of the receiver.
  4. **Middle marker.** Functions as the outer marker, positioned 0.5 to 0.8 miles before the runway. When the airplane is above the middle marker, the receiver's middle marker light starts blinking. In this situation the pilot has the last chance to decide if the landing will be made or not.
  5. Inner marker
  6. Approach lights
- Instrument landing system **marker beacons** provide information on distance from the runway by identifying predetermined points along the approach track. These beacons are low-power transmitters; that operate at a frequency of 75 MHz with 3 W or less rated power output.
- Various runway environment lighting systems serve as integral parts of the ILS system to aid the pilot in landing. Any or all of the following lighting systems may be provided at a given facility: approach light system (ALS), sequenced flashing light (SFL), touchdown zone lights (TDZ) and centerline lights (CLL-required for Category II [Cat II] operations.)
- **NDBs at Marker Beacon Sites.** Additional aids may be available to assist the pilot in reaching the final approach fix. One of these aids is the NDB, which can be co-located with or replace the outer marker (OM) or back marker (BM). It is a low-frequency non-directional beacon with a transmitting power of less than 25 watts (W) and a frequency range of 200 kilohertz (kHz) to 415 kHz. The reception range of the radio beacon is at least 15 nautical miles (NM).
- **The ILS has number of basic** limitations :
  1. Site sensitivity and high installation costs;
  2. Single approach path;
  3. Multi path interference; and
  4. Channel limitations - 40 channels only.
- The time-referenced scanning beam **Microwave Landing System (MLS)** has been adopted by ICAO as the standard precision approach system to replace

ILS. MLS provides precision navigation guidance for alignment and descent of aircraft on approach to a landing by providing azimuth, elevation and distance. The system may be divided into five functions:

1. Approach azimuth;
2. Back azimuth;
3. Approach elevation;
4. Range; and
5. Data communications.

- The data transmissions can include both the basic and auxiliary data words. All MLS facilities transmit basic data. Where needed, auxiliary data can be transmitted.

  The basic data consists of:

  1. Station Identification
  2. Exact location of azimuth, elevation and DME/P stations 9used for MLS receiver processing)
  3. Ground equipment performance level, and
  4. **DME/P channel and status.**

  The auxiliary data consist of:

  1. 3D locations of MLS equipment
  2. Waypoint co-ordinates
  3. Runway conditions, and
  4. Weather information (e.g. RVR, ceiling, altimeter? setting, wind, wake vortex, wind shear)

- **MLS Advantages :**
  1. Elimination of ILS/FM broadcast interference problems;
  2. Provision of ail-weather coverage up to ±60 degrees from runway centerline, from 0.9 degree to 15 degrees in elevation, and out of 20 nautical miles (NM);
  3. Capability to provide precision guidance to small landing areas such as roof-top heliports;
  4. Continuous availability of a wide range of glide paths to accommodate STOL and VTOL aircraft and helicopters;
  5. Accommodation of both segments and curved approaches;
  6. Availability of 200 channels - five times more than ILS;
  7. Potential reduction of Category I (CAT I) minimums;
  8. Improved guidance quality with fewer flight path corrections required;
  9. Provision of back-azimuth for missed approaches and departure guidance;
  10. Elimination of service interruptions caused by snow accumulation; and
  11. Lower site preparation, repair, and maintenance costs.

- The main disadvantages are :
  1. Its cost
  2. It is not available worldwide
  3. Only few airlines are equipped for MLS
  4. It has limitations of channels for the DME/P component of the system.

- **Accuracy and Coverage of MLS.** The approach azimuth antenna normally provides a lateral coverage of 40° either side of the center of scan. Coverage is reliable to a minimum of 20 NM from the runway threshold and to a height of 20,000 feet (ft). The antenna is normally located about 1000 feet beyond the end of the runway.

Range guidance, consistent with the accuracy provided by the azimuth and elevation stations, is provided by the MLS precision DME (DME/P). DME/P has an accuracy of ±100 ft as compared, with ±1200 ft accuracy of the standard. DME system.

REVIEW QUESTIONS

1. Describe the electronics aids to approach and landing system and briefly explain the GCA system.
2. Briefly explain the ILS (Instrument Landing System).
3. Write and explain the components of ILS equipment.
4. Write short notes on ILS and MLS. **[UPTU, 2003]**
5. Briefly explain the working of ILS receiver instrumentation with the help of simple block diagram.
6. Explain the working of MLS and write the improvements acheived over ILS.
7. Compare the ILS and MLS and which one is better option for proper landing of an aircraft during adverse weather condition.
8. Explain ILS and compare it with GCA. **[UPTU, 2006]**
9. What do you understand by doppler navigation ?
10. Define the beam configuration in doppler radars.
11. Describe the doppler frequency equation.
12. How we can achieve heading and track stabilization in doppler radars.
13. Explain the block diagram of frequency tracker.
14. Give the doppler range equations.
15. Short notes on :
    (*i*) GCA system **[UPTU, 2004]**
    (*ii*) Doppler Navigation equipment
    (*iii*) Localiser
    (*iv*) Glide slope
    (*v*) ILS Marker Beacon
    (*vi*) Lighting system
    (*vii*) Limitation of ILS system
    (*viii*) Advantages of MLS
    (*ix*) MLS
    (*x*) Operating frequencies of ILS and MLS system
    (*xi*) PAR **[UPTU, 2004]**
    (*xii*) ILS **[UPTU, 2004]**
    (*xiii*) Accuracy and coverage of MLS
    (*xiv*) Disadvantages of MLS

10
CHAPTER

TRACKING RADAR

INSIDE THIS CHAPTER

10.1. INTRODUCTION

In this chapter students will be studying about the Tracking of the targets with the help of Tracking Radar. Tracking means tracking the path of the targets by means of the radar. The following characteristics of the targets are important:

1. Range of the target
2. Velocity of the target
3. Azimuth angle of the target

Different types of techniques of tracking the targets (such as lobe switching, conical switching, monopulse tracking) are being explained in this chapter.

10.2. TRACKING WITH RADAR

A tracking radar system measures the different co-ordinate of a target such as range, elevation angle, azimuth angle, and doppler shift frequency to determine the target path and to predict its future position. All or any of the coordinates may be used to predict the future position. In general it is the method by which angle tracking is accomplished may be called the tracking radar. One of the most useful features of radar is the ability of a radar set to continuously predict the next location of its target from the information being received from the target and to align itself to continuously point at that predicted location. When this is occurring, the radar set is said to be **tracking the target.**

10.2.1. Radar Servo Tracking System

One of the most basic tracking system designs is the servo tracking system, as shown in the Fig. 10.1. Here, the radar antenna is initially trained on a target after which it automatically remains pointed at the target as it follows its motion. Furthermore, the system provides continuous position information to the operator and possibly to a fire control system. The antenna is rotated by a motor which provides a negative position feedback signal to a controller. This subsystem is known as a **servo-mechanism.**

The range and doppler frequency shift can also be continuously tracked, if desired, by a servo–control–loop actuated by an error signal generated in the radar receives. In the continuous radar a servomechanism actuated by an error signal make the antenna beam in angle. There are various methods for generating the error signal may be classified as given below :

1. A sequential lobbing
2. A conical scan
3. A mono pulse.

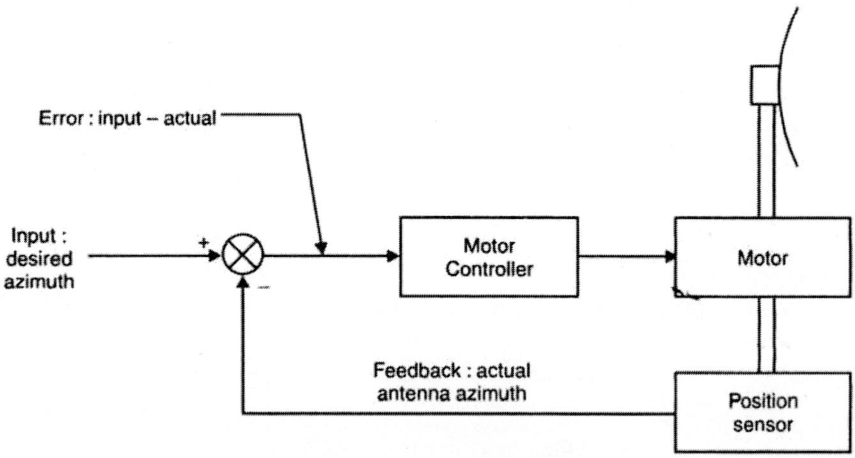

Fig. 10.1. Servo Mechanism.

Tracking radars information may be given to the display unit or may be supplied to an automatic computer, which determines the target path and predicts its probable future course. Tracking radar must find its target before it can track. There are some radar, which works as search radar to find the target before switching to a- tracking mode. Due to some operational limitation, single radar for both the search and the tracking mode is not favorable. If the antenna pattern is a narrow pencil beam and if the reach area is large, a relatively long time might be required to find the target. Therefore, separate search radar is employed to provide the information related to the position of the target to the tracking system. When separate search radar employed for giving the information of the target to the tracking radar system, is called acquisition radar.

10.2.2. Types of Tracking Radar

There are many types of tracking radar, which are being used to track the targets. They are as follows :

1. Single Target Tracker (STT)
2. Track While Scan (TWS)
3. Automatic Detection And Track (ADT)
4. Phased Array Tracking

10.2.2.1. Single Target Tracker (STT)

A single target tracker is used where continuous tracking of a single target with the higher data rate is required. Generally this type of tracking is used in controlling the missile movements as it is carried out by a Missile Guidance Radar (MGR).

In this type of tracker a closed loop servomechanism is used to keep the angle-coordinates error very small. If the errors are less than there will be more accuracy.

10.2.2.2. Track While Scan (TWS)

TWS radar scans their beam over relatively large areas. It rapidly scans in the angular sector to keep track of the targets, may be move one target at a time. The radar computer still measures returned power as a function of beam location to provide tracking but the large scanning area enables the radar to still see the target even if track has beam broken or lost. However, this large scan makes the TWS highly vulnerable to ECM jamming. An illustration of TWS radar is shown in Fig. 10.2.

10.2.2.3. Automatic Detection and Track

This type of technique used in the most of the air traffic control radar. This technique is employed with the air surveillance radar, which has 7.5 to 15 RPM (Revolution per minute). It does provide updating data according to the rotation of the antenna. It can track enormous number of targets simultaneously may be few hundreds. In this antenna position is not controlled by closed loop, it works in the open loop system. It is being further discussed in chapter 10(detection of signal in noise).

10.2.2.4. Phased Array Radar Tracking

In this technique multiple targets are tracked in the time-sharing basis by the computer. Large numbers of tracks can be scanned rapidly in this technique because of the electronically steered phased array method is used here. The antenna beam is scanned electronically make to switched on from one angular direction to another in a very small time such as few microseconds.

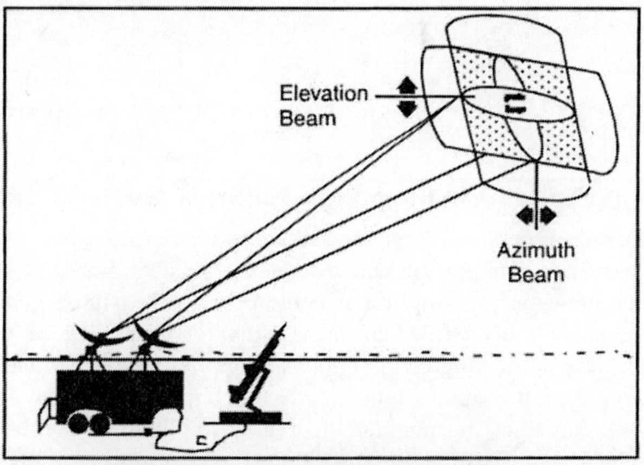

Fig. 10.2. Track-While-Scan Radar.

10.3. BLOCK DIAGRAM OF TRACKING RADAR

Figure 10.3 shows the typical block diagram of Tracking Radar. Most tracking radar use angular information as the basis for tracking operation. For better accuracy it is important that radar concentrates on one target at time. Range gating/Doppler filtering can be used for that purpose. Time and frequency control for range and doppler gating, is done in range and doppler trackers respectively. The angular error signal for the desired target to be tracked, are developed in the error demodulator block which is also controlled by range/doppler gate generation block and than fed back to the steerable antenna in a closed loop for tracking.

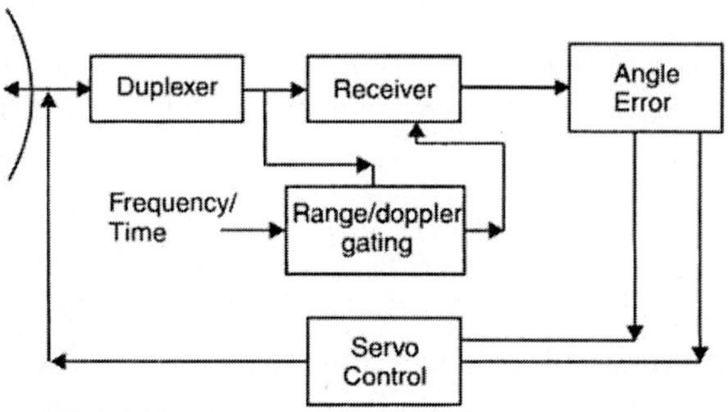

Fig. 10.3. Typical Block Diagram of Tracking Radar.

10.4. SEQUENTIAL LOBBING

A single beam is switched between two sequential angular positions to obtain on angle measurement. This is called sequential lobbing, lobe switching or sequential switching. Here the direction of the antenna beam is rapidly switched between two position *A* & *B* as shown in Fig. 10.4. The echo signal from the target will fluctuate at

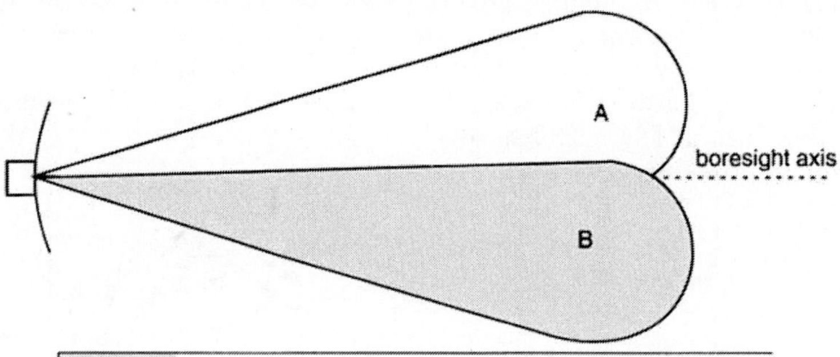

Fig. 10.4. Dual Beam Polar Pattern in Sequential Lobbing.

the switching rate, unless the target reaches exactly between two directions. The error signal obtained from a target not located on the switching axis is shown in Fig. 10.5. The difference between the amplitude of these two echo signal given the error signal. The error signal may be defined as the angular displacement of the target from the switching axis. These tracking error signals are applied to a servomechanism unit, which attempts to position the antenna beam on the target. When the target is located at references direction than the angular error is zero. Thus sequential lobbing with two position switched beam is used for tracking a target in only one place. The sign of the difference determine the direction of the antenna to move in order to align the switching axis with direction of the target.

Four switching position are required to obtain the angle measurement in the orthogonal coordinate. Thus a two–dimensional sequentially lobbing radar consists of a cluster of four feed horns illuminating a single reflector antenna, arranged so that the right-left, up-down sectors are covered by successive antenna position. A cluster of five feed horns might also be used, with a control feed used for transmission and four outer feeds used for reception on a sequential basis.

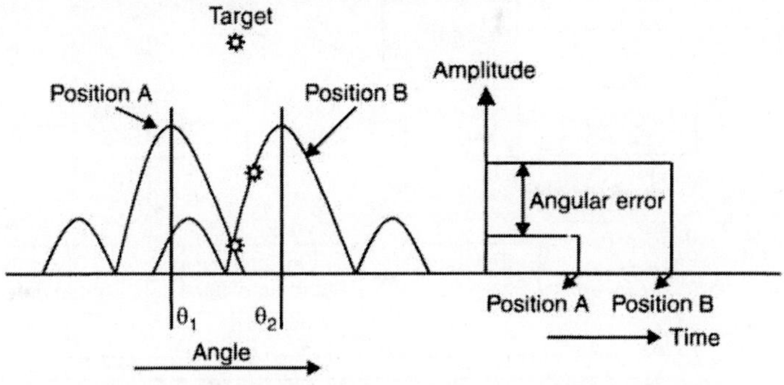

Fig. 10.5. Rectangular Representation and Error Signal.

10.5. CONICAL SCANNING

In the conical scanning, the squinted beam is scanned rapidly and continuously in a circular path around the axis. The angle between the axis of rotation and the axis of the antenna beam is called the squint angle. If a target is present within squint angle, then the echo signal from the target will be amplitude modulated at a frequency equal to the rotation frequency of the beam also called conical scan frequency. The amplitude of the modulation depends on the angular distance between the target direction and the rotation axis. The location of the target in two angle co-ordinate determines the phase of the conical –scan modulation is extracted from the echo signal and applied to a servo control system which positions the antenna on the target both in azimuth and elevation Two servo motors are required one for azimuth and the other for elevation. The conical scan modulation becomes zero and thus the target is tracked accurately by conical scanning, both in azimuth and elevation.

For true tracking, the scan frequency must be as per the Nyquist criteria for the sampling rate. In pulsed radar it must be at least are four pulses per scan, two for generating azimuth error signal and two for generating elevation error signal. The actual seen frequency also depends upon the scan mechanism. The antenna beam is squinted and scanned either mechanically by offsetting the feed and rotating it or electronically with the help of phase shifters. Electronic scan is very fast than the mechanically scan. Typical scan rates 30-to 40-scan/sec.

Block diagram. A B/D of the conical scan tracking radar system is as shown in Fig. 10.6 The antenna is mounted such that it can be positioned both in elevation with the help of respective motors. The scan antenna beam generates two signals, the for azimuth detector and the other for elevation detector, which are 90 degree out of phase.

The receiver is a conventional superhetrodyne except for features related to the conical scan tracking radar. The error signal is extract in the video after the second detector. The error signal is compared with the elevation and azimuth references signal, in the angle detectors, which are phase sensitive detection. Phase sensitive detector is a non linear device in which the input signal is mixed with a reference signal. This angle detection produces a D.C. voltage, which is proportional to the error, and the sign is an indication of the direction of the error. The angle error detector output is amplified and drive the antenna elevation and azimuth servomotors. The angular position of the target may be determined from the elevation and azimuth of the antenna axis.

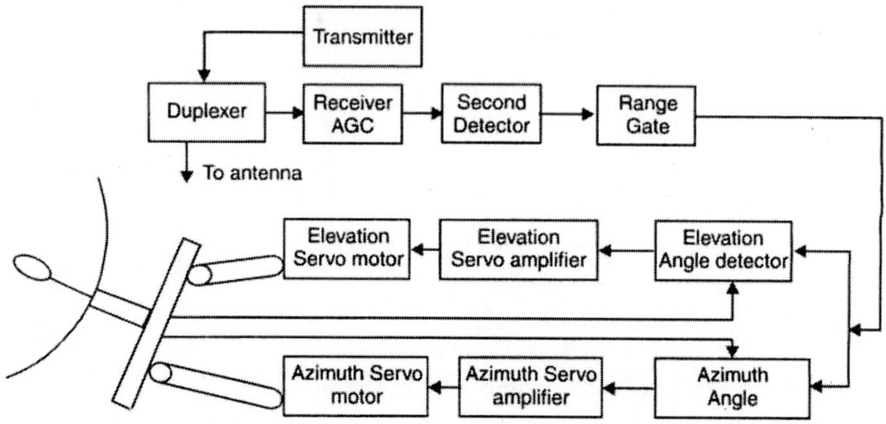

Fig. 10.6. Block Diagram of Conical Scan System.

Conical Scan systems require a minimum amount of hardware and therefore are commonly used on inexpensive, mobile systems such as AAA or mobile SAM sites. They suffer the serious disadvantage of not being able to see a target outside their narrow scan patterns. This means that not only is a second radar required to help it find the target (to "acquire the target") but also the tracked aircraft can easily "escape" if it is successful in breaking track since the Conical Scan radar cannot see the target, except in the track mode.

10.6. MONO PULSE TRACKING

There are two disadvantages connected with the use of lobe scanning and conical scanning which are as follows :

1. The motion of the antenna is more complex in lobe switching and conical scanning.
2. In conical scan, a minimum of four pulses is required. The difficulty here is that if the targets cross-section, the tracking accuracy may be degraded.

The effect of the fluctuating echo can be sufficiently serious in some application severely limits the accuracy of those tracking radar which requires many pulses to be processed in extracting the error signal.

The above problems can be overcome by using the single pulse only. There are several methods by which angle error information might be obtained with a single pulse. The tracking techniques which derive angle error information on the basis of single pulse, is known as mono pulse tracking or simultaneous lobbing more than one antenna beam is used simultaneously in there method where as in conical scan or lob switching tracker, which uses one antenna beam on a time–shared basis. There are several methods by which a mono pulse angle measurement can be made. The most popular method is the amplitude–comparison mono pulse, which compares the amplitude of the signals simultaneously received in multiple squinted beams to determine the angle.

10.6.1. Amplitude Comparison Mono Pulse

In an amplitude comparison mono pulse system, four feeds are used with one paraboloid reflector. There are four horn antennas displaced about the central focus of the reflector. The Transmitter feeds the horns simultaneously so that a sum signal is transmitted. The echo signal is received by a receiver duplexer using a hybrid ring (or rat-race) to provide the following three signals :

1. Sum signals $(A + B + C + D)$
2. Azimuth error signal $= (A + C) - (B + D)$
3. Elevation error signal $= (A + B) - (C + D)$

A diagram of Monopulse Radar Beam Patterns is shown in Figure 10.7.

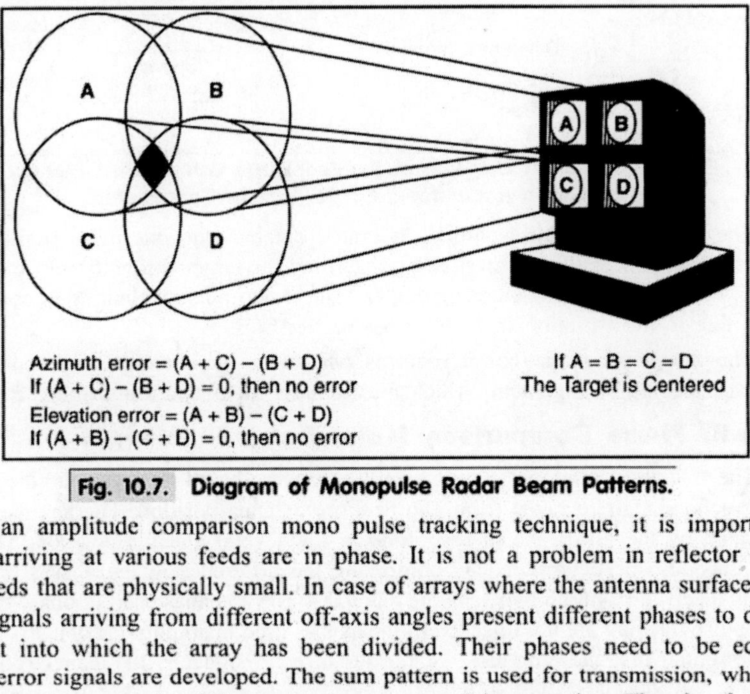

Fig. 10.7. Diagram of Monopulse Radar Beam Patterns.

In an amplitude comparison mono pulse tracking technique, it is important the signal arriving at various feeds are in phase. It is not a problem in reflector antenna with feeds that are physically small. In case of arrays where the antenna surface is very large signals arriving from different off-axis angles present different phases to different segment into which the array has been divided. Their phases need to be equalized before error signals are developed. The sum pattern is used for transmission, while both the sum pattern and the difference pattern are used in reception. The signal received with the difference pattern provides the range measurement and is also used as references to extract the sign of the error signal. Signals received from the sum and difference patterns are amplified separately and combined in a phase sensitive detector to produce the error signal characteristics.

A block diagram of the amplitude comparisons monopulse tracking radar for a single angular coordinates shown in Fig. 10.8.

The receiver has three separate input channels consisting of three mixers, common local oscillator, three IF amplifiers and three detector. The elevation and azimuth error signals are used to drive a servo amplifier and a motor in order to position the antenna in the direction of target. The o/p of sum channel is used to provide the data generally obtained from a radar receiver so that it can be used for application like automatic control of the firing weapon.

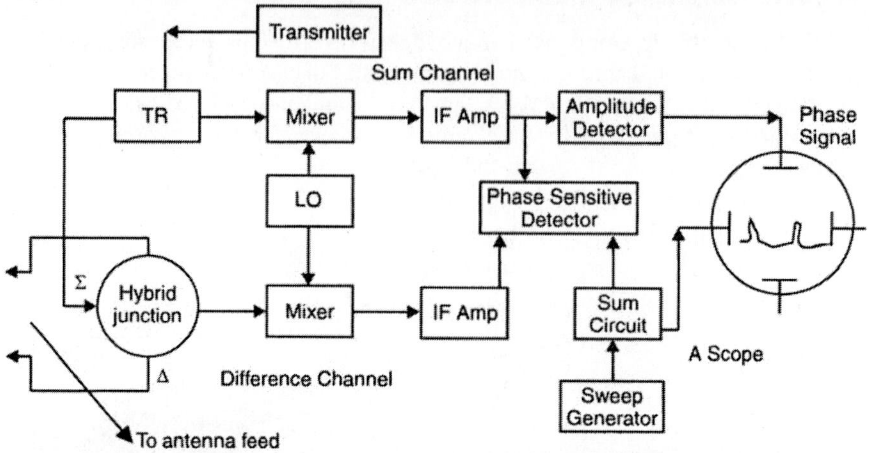

Fig. 10.8. A Block Diagram of the Amplitude Comparisons Monopulse Tracking Radar for a Single Angular Coordinates.

The advantage of this technique is that it obtains with one pulse (unlike 4 pulses in conical scanning) all the information regarding the target and able to locate the target in less time comparing to other methods. Also, mono pulse technique is not subject to errors due to the variation in target cross section.

The disadvantage is that it requires two extra Rx channels and more complex duplexer feeding arrangement, which makes the system bulky and more expensive.

10.6.2. Phase Comparison Mono Pulse Tracking

The tracking techniques discussed till now were based on a amplitude comparison of the echo signals received from two or more antenna position. The angular error was generated by the difference in amplitude in the several antenna position. The angular error may also be determined by comparing the difference in the phases between the signals received from the two separate antennas. The antennas used in phase comparison monopulse tracking are not offset from the axis as used in amplitude comparison trackers. It is essential that the amplitude of the target echo signal is the same from the each antenna, but the phases are different.

Tracking radar, which operates with phase information, might be called **simultaneous phase comparison radar**, or phase comparison mono pulse. The measurement of angle of arrival by comparison of the phase relationship in the signals from the separated antennas of a radio interfero meter has been widely used by the radio astronomers for precise measurement of the position of radio stars. The interferometer as used by the radio astronomer is a passive instrument the source of energy being radiated by the target itself.

Tracking radar, which operates with phase information, is similar to an active interferometer.

An additional antenna and receiving channel is necessary in order to track in two orthogonal co-ordinates. In the phase comparison radar, four antennas are required, one of these antenna for transmitter only, while the other three for receiver.

This technique has not been as widely used as the other techniques discussed earlier. There are two main drawback why this might be so.

1. The side lobes levels, which result higher than the single antenna.
2. The phase comparison radar does not usually make efficiently use of the total available antenna aperture.

10.7. TRACKING IN RANGE

In almost all the tracking radar , the target is continuously tracked in as well as in azimuth. An operator who keeps watching the desired target in the "A" –scope and maintain a marker over the target by moving the hand wheel accomplished range tracking in a simple manner. The position of hand wheel is the range of the target and it may be converted to a voltage which is being supposed to the data processor to predicts the future position of the target increases, it is difficult to sustained the level of efficiency by the operator. So the automatics tracking become necessary to track the targets accurately. Due to such limitations, it was replaced by a closed loop automatic tracking, which is known as split-gate tracker.

The split-gate tracker technique used in automatic range tracking. This technique uses two gates, one is early gate and the other is late gate as shown in Fig. 10.9. The video echo pulse is shown in Fig. 10.9(a), the relative position of the two gates at a particular instant is in Fig. 10.9(b), and the difference signal in Fig. 10.9(c). The amplitude of two samples collected on the leading and trailing edge of the echo, which after filtering is approximately a triangle, are compared to generate the corrections signal.

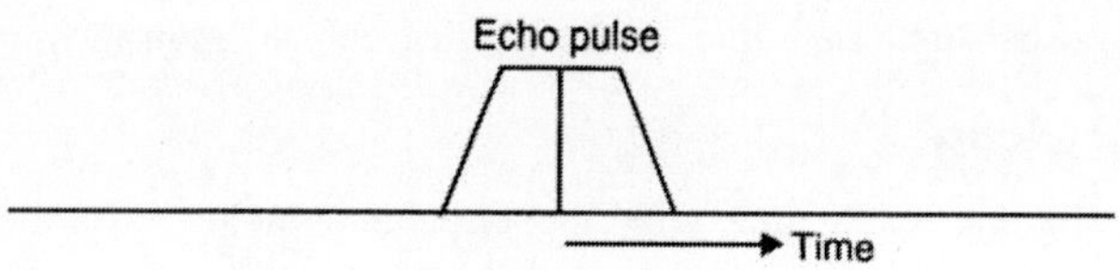

(a) Echo Pulse.

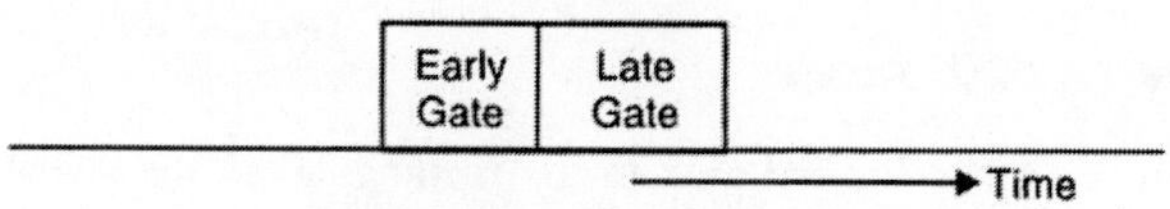

(b) Early-late-range gates.

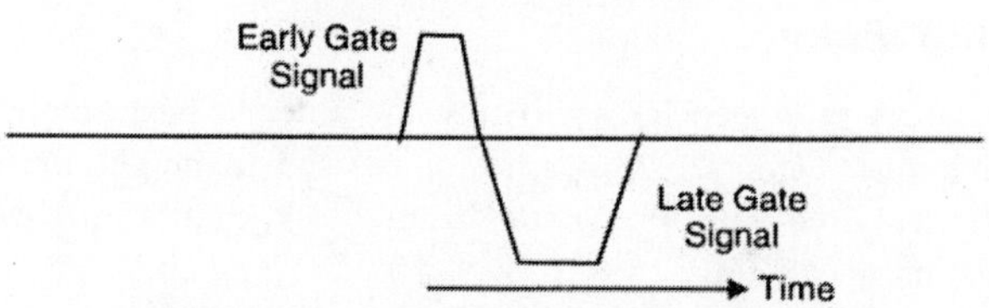

(c) Difference between early and late range gate.

Fig. 10.9. Split-Range–Gate-Tracking.

10.8. ACQUISITION

The acquisition radars are those radar which are being empyaled to search the target to be tracked and then designate the target position to the tracker. There radar have large search large an surveillance radar. Normally a tracker radar uses a narrow beam (pencil) to track the target in angle accurately. So it is difficult for the tracker to search the target in large area due narrow beam width. The acquisition of the target for the purpose of tracking is performed by the acquisition radar, and in known as the acquisition of the target.

There are various method of scanning pattern used with narrow beam (pencil) antennas to find a target.

1. Helical scanning beam
2. Palmer scan
3. Spiral scan
4. Raster or TV scan
5. Nodding scan

10.8.1. Helical Scan Beam

In this method antenna is scanning in azimuth continuously and at the same time it is raising or lowering in elevation scanning. It makes a helix track in the space. If fire control radar used in the search made than the antenna will rotate at certain RPM and covered some elevation angle.

10.8.2. Palmer Scan

This type of scanning is used where a search area is larger in one dimension than another. Palmer scan consists of a conical scan around the axis of rotation.

The palmer scan reduces to the conical scan where the axis of rotation is stationary. To generate palmer scan pattern, same mechanism of conical scan may be employed.

10.8.3. Spiral Scan

It covers an angular reach area with the circular symmetry. In the spiral scan volume do not receive the same signal unless the speed of the scanning is changed during the complete scan cycle.

10.8.4. Raster or TV Scan

The raster scan reach the limited area in the rectangular in the shape. In this scan the reach area is painted in uniform manner unlike the spiral and palmer scan.

10.8.5. Nodding Scan

The nodding scan is produced by oscillating the antenna beam slowly in azimuth and rapidly in elevations. Generally this method is used in height finding radars. It covers the hemispherical area with the narrow beam width, that is, in elevation it is up to 90° and in azimuths up to 360°.

SUMMARY

- A tracking radar system measures the different co-ordinate of a target such as range, elevation angle, azimuth angle, and Doppler shift frequency to determine the target path and to predict its future position.
- One of the most basic tracking system designs is the servo tracking system.

- There are various methods for generating the error signal may be classified as given below:
  1. **A sequential lobbing :** A single beam is switched between two sequential angular positions to obtain on angle measurement. This is called sequential lobbing, lobe switching or sequential switching.
  2. **A conical scan:** In the conical scanning, the squinted beam is scanned rapidly and continuously in a circular path around the axis. The angle between the axis of rotation and the axis of the antenna beam is called the **squint angle**. If a target is present within squint angle, then the echo signal from the target will be amplitude modulated at a frequency equal to the rotation frequency of the beam also called conical scan frequency.
  3. **A mono pulse.** The tracking techniques that derive angle error information on the basis of single pulse, is known as **mono pulse tracking or simultaneous lobbing**. There are several methods by which a mono pulse angle measurement can be made.

     (*a*) Amplitude Comparison Mono Pulse

     (*b*) Phase Comparison Mono Pulse Tracking
- **Acquisition radar.** When separate search radar employed for giving the information of the target to the tracking radar system, is called acquisition radar.
- **Types of Tracking Radar**
  1. Single target tracker (STT)
  2. Automatic detection and track (ADT)
  3. Phased array radar tracking (multiple target tracking)
  4. Track while scan. (TWS)
- **Tracking in Range :**
  1. Split-gate-tracker
  2. Range Glint
- **Acquisition.** A tracking radar must first find and acquire its target before it can operates as tracker.
- There are different types of scanning pattern employed with pencil beam antennas
  1. Helical scanning beam
  2. Palmer scan
  3. Spiral scan
  4. Raster or TV scan
  5. Nodding scan

REVIEW QUESTIONS

1. What do you understand by term Tracking Radar ?
2. Explain the radar servo tracking system with the help of simple block diagram.
3. Explain various types of tracking radar ? **[M.D. University, 2006]**
4. Describe the sequential lobbing method of tracking of an acquired target.

5. Describe the conical scanning method of tracking of an acquired target. How is this an improvement over lobe switching ? [M.D. University, 2004]
6. Explain the working of mono pulse tracking radar for one angle coordinates. [M.D. University, 2004]
7. Explain the working of mono pulse radar with the help of a block diagram. [UPTU, 2003]
8. Describe the various methods of mono pulse tracking technique.
9. Define tracking in range and explain split gate tracker method.
10. What do you understand by acquisition of target and explain the various method of antenna scanning ?
11. What are the various Antenna scanning ? Explain each in detail. [M.D. University, 2001]
12. What are the various techniques of tracking ? [PTU, 2004]
13. Explain the basic difference search radar and tracking radar. [PTU, 2004]
14. Compare different types of trackers system. [PTU, 2004]
15. With the aid of a sketch, describe the equipment and technique used in monopulse method of target tracking ? [M.D. University, 2005]
16. How is conical scanning an improvement over lobe switching ? [M.D. University, 2005]
17. Explain the block diagram of amplitude comparison in monopulse in one angle coordinate. [M.D. University, 2006]
18. Explain the block diagram of conical scan tracking radar. [M.D. University, 2006]
19. Explain split gate tracker and range glint. [M.D. University, 2006]
20. Write shorts on :
    (*a*) Sequential lobbing [M.D. University, 2003]
    (*b*) Conical scanning
    (*c*) Amplitude comparison mono pulse tracking
    (*d*) Phase comparison mono pulse tracking
    (*e*) Range Glint
    (*f*) Split-gate tracker
    (*g*) Helical scanning
    (*h*) Palmer scanning
    (*i*) Nodding scanning
21. Explain the following in detail :
    (*a*) Sequential Lobbing
    (*b*) Monopulse Tracking Radar. [M.D. University, 2007]

QUESTION BANK

TELEVISION ENGINEERING

1. Draw the block diagram of TV transmitter and explain the function of each block.
2. Why is scanning necessary in TV transmission? Why is it carried out at a fast rate?
3. What is the basic principle of operation of TV camera tube?
4. Why are synchronizing pulses transmitted along with the picture signal?
5. What is a raster and how it is produced on the picture tube screen?
6. Why is FM preferred to AM for sound signal transmission?
7. Describe briefly the functions of various controls provided on the front panel of a TV receiver.
8. Describe the basic principle of colour TV transmission and reception.
9. Describe the functions of saturation and hue controls in a NTSC colour TV receiver.
10. What is the compatibility in TV transmission?
11. Why TV transmission is carried out in the UHF and VHF bands?
12. Justify the choice of a rectangular frame with width to height ratio for TV transmission and reception.
13. How the illusion of continuity is created in 625 lines TV system?
14. What is frequency interleaving?
15. Determine the number of lines get traced during each vertical retrace in 625 lines TV system.
16. What is a composite video system?
17. What do you understand by compatibility?
18. Explain how sync pulses are generated in a horizontal deflection circuit.
19. What is flicker? How it is reduced in 625 lines TV system.
20. What is high frequency compensation? Discuss various techniques for high frequency compensation.
21. What are synchronization pulses and why they are needed?
22. What do you understand by interlaced scanning?
23. Why has the frame repetition rate been chosen to be 25 and not 24 as in motion pictures?

24. Draw composite video signal waveforms for the picture information.
25. How is the sync pulses transmitted along the picture signal?
26. Justify the choice of 625 lines for the TV transmission. Why the total number of lines is kept odd in all TV systems?
27. What is the significance of choosing the number of lines 625 not 623 or 627 in a TV system.
28. What is the need of interlaced scanning?
29. Establish the relationship between the picture cell and the bandwidth.
30. What is the purpose of the synchronizing circuits?
31. Describe the various steps in the evolution and construction of a composite signal.
32. How the interlaced scanning reduces flicker and conserve bandwidth.
33. What do you understand by active and blanking periods in horizontal and vertical scanning?
34. What is meant by vertical and horizontal resolution? Derive an expression for the highest modulating frequency in a TV system and show that it is nearly 5MHz in the 625-B monochrome system.
35. What do you understand by interlace error and how it affects the quality of picture?
36. What are front porch, horizontal sync pulses, back porch and active line periods in a video signal?
37. Explain how the vertical and horizontal sync pulses are separated and shaped at the receiver.
38. Sketch the complete pulse trains that follow at the end of both even and odd fields. Fully label them and explain how the half line discrepancy is removed by insertion of pre-equalizing pulses.
39. Justify the need of pre and post equalizing pulses. Why is it necessary to keep their duration equal to half-line period?
40. Justify the need of a blanking period corresponding to 20 complete lines after each active field of scanning.
41. Why does the vertical retrace not begin with the incoming of the first serrated vertical sync pulse.
42. Justify the name vestigial sideband.
43. Why is it necessary to effect economy in channel bandwidth? Why SSB is not used for picture signal transmission.
44. Enumerate various advantages and disadvantages of VSB transmission.
45. Why is a portion of lower side band of AM picture signal transmitted along with the carrier and full VSB? Does it need any correction somewhere in the TV link? If so, where is it carried out?
46. Why is it necessary to modulate the picture and sound signal before transmission?
47. Show that in the 625-B system, a total channel bandwidth of 11.25MHz would be necessary if both the sidebands of the amplitude-modulated picture are fully radiated along with the frequency modulated picture signal.

48. Show that a total channel bandwidth of 7MHz is necessary for successful transmission of both picture and sound signal in 625 lines TV system.
49. Why is the sound carrier located at 5.5MHz away from the picture carrier? Sketch frequency distribution of the channel and mark the location of picture and sound signal carrier frequencies.
50. Justify the allocation of 8MHz in British TV system and 6MHz in the American system for each TV channel. What is the separation between picture and sound carriers in each of these systems?
51. What is modulation index in FM transmission and how does it affect the bandwidth required for each FM channel.
52. Explain how you would proceed to determine the channel bandwidth for transmission of sound signals by frequency modulation.
53. What do you understand by an electron multiplier?
54. What is the operating principle of a black and white camera tube?
55. Differentiate negative modulation from positive modulation.
56. Draw and explain cross-section view of Image Orthicon camera tube.
57. What is photoelectric effect in video cameras?
58. What do you understand by dark current in Vidicon?
59. Explain how the Image Orthicon develops video signal when light from any scene is focused on its faceplate.
60. What precautions must be observed while handling the picture tube? Why is it necessary to provide spark gap protection between the various electrodes?
61. What do you understand by degaussing?
62. Discuss the merits of negative modulation.
63. Which material is preferred to be used as camera lens?
64. Compare the salient features of Orthicon, Vidicon and Plumbicon camera.
65. Sketch the sectional view of a picture tube that employs electrostatic focusing and electromagnetic deflection and label all the electrodes.
66. Explain briefly, how the electron beam is focused on the tube screen. What do crossover points in the electron gun mean?
67. Which material is preferred to be used as a camera lens?
68. What is the function of Aquadag coating on the inner side of the tube well? Why is a grounded coating provided on the outer surface of the picture tube?
69. Why is aluminized coating provided on the phosphor screen? How are any stray ions prevented from hitting the screen?
70. Discuss the merits of electromagnetic deflection over electrostatic deflection in TV picture tubes. Why is cosine winding used for deflection coils?
71. Show with a circuit diagram how DC potentials are supplied to the various electrodes of the picture tube.
72. What are the functions of brightness and contrast controls?
73. What do you understand by image multiplication and signal multiplication in an Image Orthicon camera tube?
74. In an Image Orthicon, what is the function of wire mesh screen and why is it located very close to the target plate?

75. Explain with the help of sketches, how video signal is developed in a Vidicon camera tube.
76. How is Vidicon different from the Image Orthicon and what are its special applications.
77. What are the existing TV standards?
78. What is the need of trap circuits?
79. How does low frequency compensation take place in TV receiver?
80. Explain the functional significance of the blocks of a TV receiver.
81. Discuss the constructional features of the various types of tuners.
82. What are the various components of a TV studio?
83. Explain the vertical deflection circuits and the process of alignment of TV receivers.
84. Write a note on the various coupling methods used in the receivers.
85. Draw a block diagram of an RF tuner and explain how incoming signals from different stations are translated to common picture IF and sound IF frequencies.
86. What are the advantages of AGC in TV receivers and what do you understand by delayed AGC.
87. Why is the design of horizontal output stage very much different than that of vertical stage, while their functions are very similar to each other? What are the special features which make the operation of line output stage efficient?
88. Write short notes on: EHT and TV studio.
89. What is the function of RF tuner in TV receiver?
90. Discuss the various techniques for high frequency compensation.
91. Differentiate between high frequency and low frequency compensation.
92. Draw the block diagram of monochrome TV receiver and explain each section.
93. What do you understand by direct coupled video amplifier? Also discuss its main features.
94. Discuss the basic principle of AGC and show how it is applied to the tube and transistor amplifiers.
95. Explain with circuit diagram how the high voltage pulse induced in the output transformer windings, are used to generate EHT supply.
96. Write short note on VHF tuner.
97. What do you understand by image rejection ratio? Explain how by providing an RF amplifier, image signal rejection is greatly minimized.
98. What are the merits of using an RF amplifier before the frequency converter?
99. What are the essential functions which are assigned to IF section of the receiver.
100. What do you understand by inter carrier sound system. Where and how is the inter carrier sound signal separated from the video signal?
101. Draw a block diagram of sound channel in a TV receiver. Explain briefly how the inter carrier sound signal as obtained at the video detector, is processed to produce sound output.
102. Describe briefly how EHT and boosted supply voltages are developed from the horizontal output circuit of the sweep amplifier.
103. Discuss the significance of a chromaticity diagram

104. How the mixing of colours takes place.

105. Describe the constructional features of the colour TV transmission and reception.

106. Write a note on colour fundamentals in TV systems.

107. Explain the difference between primary colours and complimentary colours.

108. Explain how the 'Y' and colour difference signals are developed from the camera outputs. Why is the 'Y' signal set =0.3R +0.59G -0.11B.

109. Write a short note on PAL and NTSC systems.

110. Explain the terms chrominance and luminance.

111. What are the characteristics by which coloured visual information can be represented?

112. What are the limitations of NTSC system?

113. Discuss the merits and demerits of SECAM system.

114. Explain how by frequency interleaving the colour information is accommodated within the same channel bandwidth of 7MHz.

115. Draw block diagram of colour TV receiver.

116. Explain with the block diagram how both (B-Y) and (R-Y) signals are combined around the same sub carrier frequency by Quadrature modulation. Why is the colour signal bandwidth requirement much less than those of Y signal?

117. Discuss the factors which influence the choice of sub carrier frequency in a colour TV system.

118. Justify the choice of 3.579545MHz as the sub carrier frequency in the NTSC system. How does it affect the line and field frequencies?

119. Explain with a suitable diagram the coding process in PAL colour system. Why the colour is burst signal transmitted after each scanning line.

120. Explain the delay line method of separating U and V signals in a PAL receiver. What is the function of a colour killer circuit in the path of chrominance signal in the receiver?

121. Describe the manner in which the APC circuit functions to keep the reference sub carrier frequency constant in the PAL system.

122. What is the basic difference between the SECAM and other colour TV systems? Describe briefly the encoding and decoding processes of the SECAM system.

123. Discuss the relative merits and demerits of the three TV systems. Explain the factors which influence the choice of any one of three systems.

INDEX

A

B

C

D

E

N

O

P

Q

R

S

T

U

V

W

Y

Win Prizes

ATTENTION READERS

Win Prizes

We request for your frank assessment regarding various aspects of the book as given below :

Fundamentals of
TELEVISION AND RADAR ENGINEERING
By : K.K. Sharma

Please Fill up the Sheet in Neat and Clean Writing

(i) What topics of your syllabus, important for your examination point of view, are not covered in this book ?

....

....

....

....

(ii) What are the chapters wherein the treatment of the subject matter is not systematic or organised or updated ?

....

....

....

....

(iii) Have you come across misprints/mistakes/factual inaccuracies in the book ? Please specify the chapters and the page numbers.

....

....

....

....

(iv) What other books on the same subject have you found/heard better than the present book ? Please specify in terms of price and quality.

....

....

....

....

(v) Further suggestions and comments for the improvement of this book.

....

....

....

....

(PS : If need be please attach a separate sheet to write your views)

OTHER DETAILS

(i) Who recommended you this book ? (Please tick (✓) in the box near the option relevant to you.

☐ Teacher ☐ Friends ☐ Bookseller

....

....

....

....

(ii) Name of the recommending teacher, his designation and address:

....

....

....

....

(iii) Name and address of the bookseller you purchased the book from :

....

....

....

....

(iv) Name and address of your institution (Please mention the University or Board as the case may be)

....

....

....

....

(v) Your name and complete address :

....

....

....

....

(vi) Course you are enrolled in :

....

....

....

....

The best assessment will be awarded each month : The award will be in the form of our publications as decided by the Editorial Board.

Please mail the filled up coupon at your earliest to :

S.K. KATARIA & SONS®
4885/109, Prakash Mahal, Dr. Subhash Bhargav Lane,
Daryaganj, New Delhi-110002
Tel.: +91-11-43551243
Mobile : +91-9871775858
Email : katariabook@gmail.com